Nucleic Acid Biochemistry and Molecular Biology

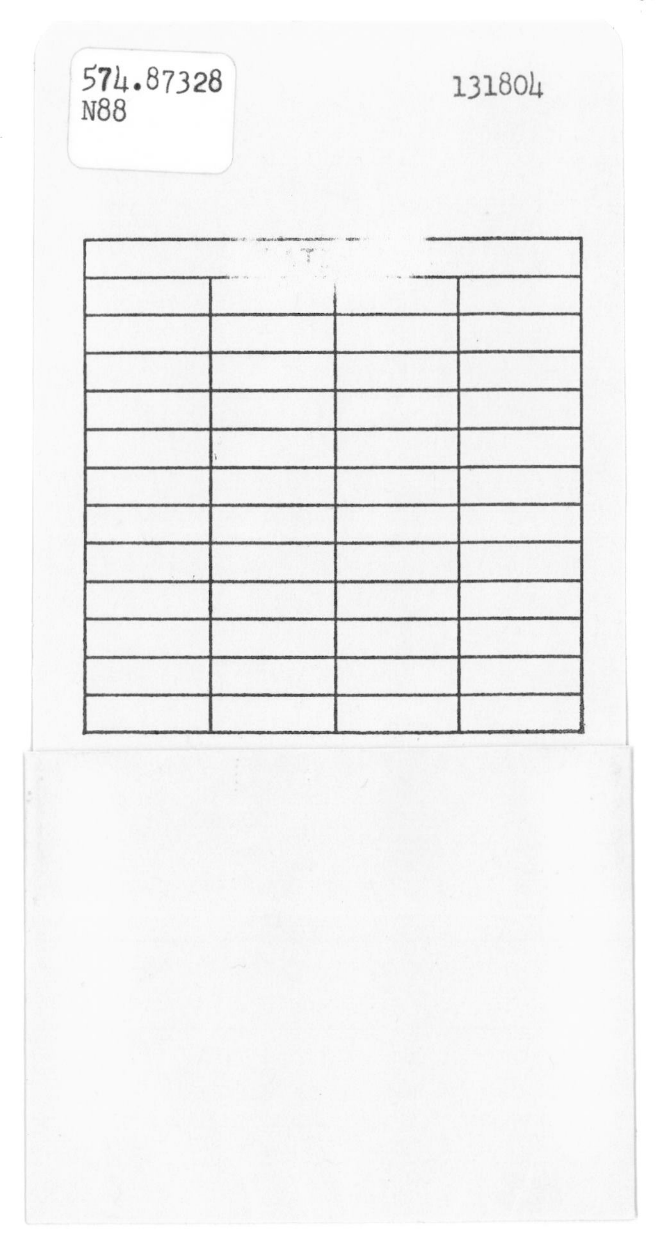

Nucleic Acid Biochemistry and Molecular Biology

W.I.P. MAINWARING, J.H. PARISH &
J.D. PICKERING Department of Biochemistry
University of Leeds

N.H. MANN Department of Biological Sciences
University of Warwick

BLACKWELL SCIENTIFIC PUBLICATIONS
OXFORD LONDON
EDINBURGH BOSTON MELBOURNE

Editorial offices:
Osney Mead, Oxford, OX2 0EL
8 John Street, London, WC1N 2ES
9 Forrest Road, Edinburgh, EH1 2QH
52 Beacon Street, Boston, Massachusetts 02108, USA
99 Barry Street, Carlton, Victoria 3053, Australia

First published 1982

Photoset by Enset Ltd, Midsomer Norton, Bath
and printed in Great Britain by
Butler & Tanner Ltd,
Frome and London

DISTRIBUTORS

USA
Blackwell Mosby Book Distributors
11830 Westline Industrial Drive
St Louis, Missouri 63141

Canada
Blackwell Mosby Book Distributors
120 Melford Drive, Scarborough
Ontario, M1B 2X4

Australia
Blackwell Scientific Book Distributors
214 Berkeley Street, Carlton
Victoria 3053

British Library
Cataloguing in Publication Data

Nucleic acid biochemistry and molecular biology.
1. Nucleic acids
I. Mainwaring, W.I.P.
547.7′90442 QD435

ISBN 0-632-00632-3

Contents

Contents

Preface

This book is intended primarily as a text for university students of biochemistry and other courses with a major component of molecular biology. We believe that it may also be useful to those embarking on research projects involving nucleic acids. The book draws on our experience in teaching the methodology, biochemistry and molecular biology of nucleic acids to undergraduates in two universities. We recognize that readers will differ in their scientific backgrounds and for this reason we have introduced, in appropriate places, elementary chemical and genetic detail. However, we are bound to assume a core of biochemical knowledge and understanding such as that acquired from a comprehensive introductory course in biochemistry at a university or from a study of a comprehensive introductory biochemistry textbook.

The aim of the text is to present a working knowledge of those branches of nucleic acid science that are most important in current research. Of course this statement involves a value judgement on our part. We have to try and look into the future and to provide background material for a person reading the scientific literature over the next few years. Our only comment on our choice of emphasis is a suggestion for the reader: see if we have got it right by keeping up with the literature. The subject is expanding rapidly and any textbook can only complement a study of current papers and reviews.

Of several ways that could have been used to divide up the material, we have chosen a treatment suggested by our teaching experience. After an introductory overview and a brief summary of nucleotide metabolism, we describe the properties of nucleic acids and the exploitation of these properties in nucleic acid methodology and manipulation. Three chapters then cover the molecular biology of bacteria as a unified topic and these are followed by four chapters on eukaryotes. The final chapter concerns recombinant DNA methods and applications ('genetic engineering').

We are grateful to everyone who has helped with the production of the book and particularly to our colleagues who have so generously supplied figures and photographs. Our special thanks go to Dr A. Kingsman for a critical reading of a draft and for contributions to sections 3.1.2, 3.2, 3.4 and 3.5 and to Drs S. Hardy and S. Baumberg for helpful advice and comment.

John Howard Parish
W. Ian P. Mainwaring
Janet D. Pickering
Nicholas H. Mann

Chapter 1 Introduction

1.1 NUCLEIC ACID STRUCTURE

1.1.1 Primary structure

In all forms of life, there are only two forms of nucleic acid, ribonucleic acid (RNA) and deoxyribonucleic acid (DNA). Both contain three types of structural units: (a) pentose (5C sugar), (b) nitrogenous bases, pyrimidines (6-membered rings) and purines (5-membered rings), and (c) phosphate groups. The structures of these molecules are given in Fig. 1.1. The pentoses and nitrogenous bases have distinctive numbering systems; by convention, the C atoms of the *pentoses* are given a dash or *prime* (e.g. 5′) to distinguish them from the atoms making up the rings of the nitrogenous bases.

RNA and DNA have certain constituents in common, namely phosphate groups, the purines, adenine (A) and guanine (G), and the pyrimidine, cytosine (C). RNA and DNA differ in containing a unique pentose and a unique pyrimidine base: RNA contains ribose and uracil (U), whereas DNA contains 2′-deoxyribose and thymine (T).

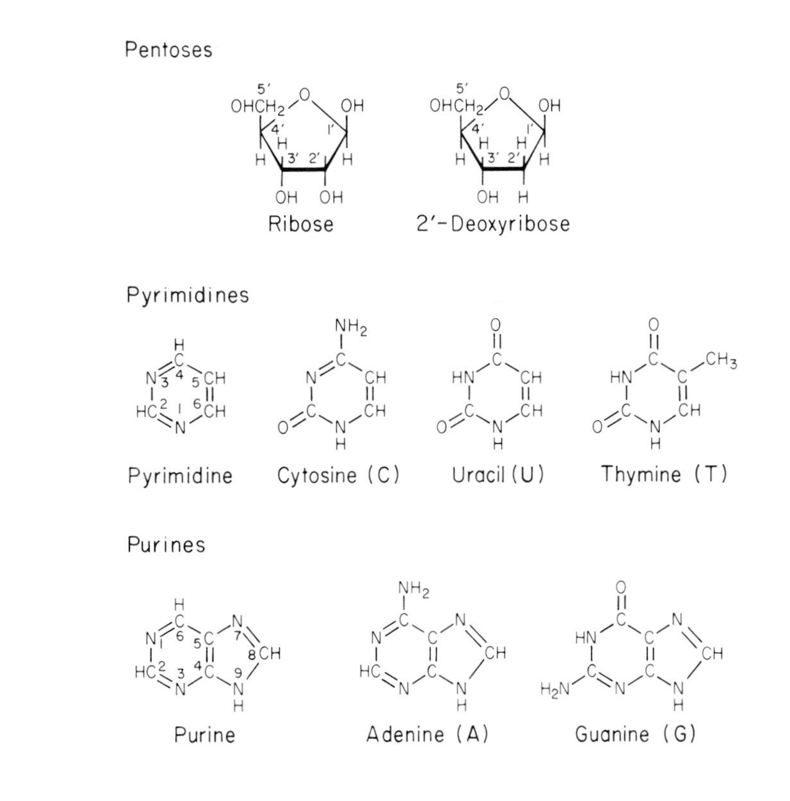

When nitrogenous bases are covalently linked to a pentose, they form a *nucleoside*. The bond is between C-1 (i.e. 1′) of the pentose and either N-1 of the pyrimidine or N-9 of the purine base. Depending on the nature of the nitrogenous base and the pentose, a series of either ribonucleosides or 2′-deoxyribonucleosides are formed. The common ribonucleosides are adenosine (A), guanosine (G), cytidine (C) and uridine (U). The common 2′-deoxyribonucleosides are similarly named, but prefixed by deoxy- (abbreviated to d), e.g. 2′-deoxyadenosine (dA). The 2′-deoxynucleoside derived from thymine (T) is 2′-deoxythymidine (dT). When a phosphate group is attached by an ester linkage to a nucleoside, a *nucleotide* is formed. The most common site of esterification is the hydroxyl group at the 5′ position of ribose and 2′-deoxyribose. Up to a maximum of three phosphate groups can be added at the 5′ position, thus forming 5′-mono-, -di- and tri-nucleotides. The abbreviated forms of the nucleotides derived from 2′-deoxyadenosine, for example, are dAMP, dADP and dATP. The first (or α) phosphate group at the 5′ position is attached via an ester linkage, but the next two phosphate groups (β and γ) are attached via phosphoanhydride linkages. The β and γ bonds are *high-energy* bonds, each releasing a large amount of free energy on their hydrolysis; by contrast, the phosphate ester (α) bond is not a high-

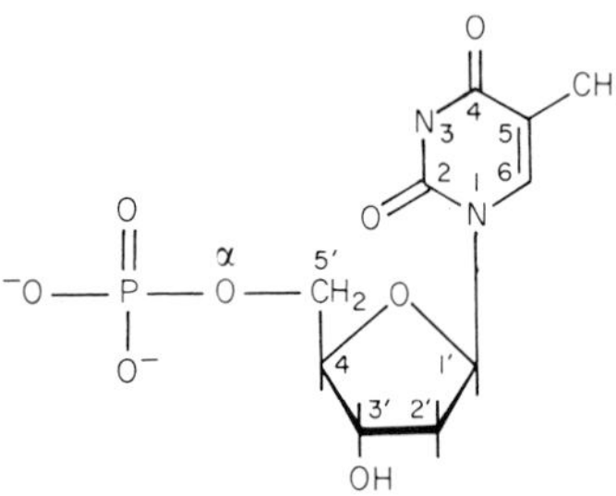

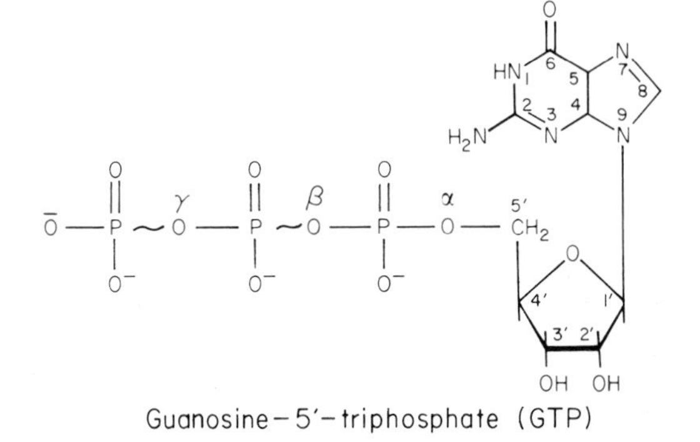

Fig. 1.2. Representative structures of nucleotides. In these more complex structures it is not usual to include all the carbon and hydrogen atoms given in nucleotide constituents (compare Fig. 1.1). High-energy bonds are frequently denoted by a distinctive symbol, ~, as here.

energy bond. The high-energy bonds in ribo- and 2′-deoxyribonucleoside-5′-triphosphates are essential for nucleic acid synthesis, described later. Representative structures of nucleotides are presented in Fig. 1.2.

Nucleic acids are polynucleotides. They are long polymerized chains of nucleotides, joined by phosphodiester bonds spanning from the 5′ position of one nucleotide to the 3′ position of the adjacent nucleotide. The structure of nucleic acids is, therefore, maintained by repeated 5′ → 3′ phosphodiester bonds. The fundamental, repeating structure of a polynucleotide is given in Fig. 1.3(a), and the detailed structures of a polyribonucleotide and a polydeoxyribonucleotide are given in Fig. 1.3(b).

It is important to emphasize that nucleic acids have polarity, meaning that their ends (or termini) are not the same. One end of a polynucleotide chain bears a 5′ phosphate group, whereas the other

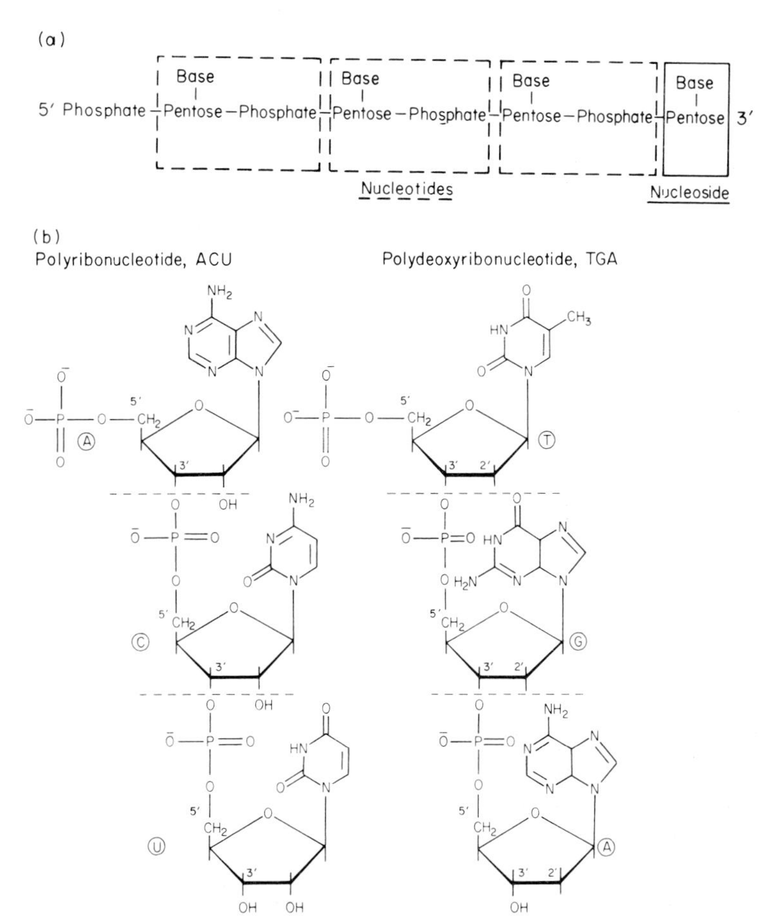

Fig. 1.3. The structure of polynucleotides. (a) The basic structure of a polynucleotide. Note that the 5′-end bears a phosphate group, whereas the 3′-end does not. In fact the 3′-end is a nucleoside. (b) The detailed structure of a polyribonucleotide, 5′ ACU 3′ and a polydeoxyribonucleotide, 5′ TGA 3′.

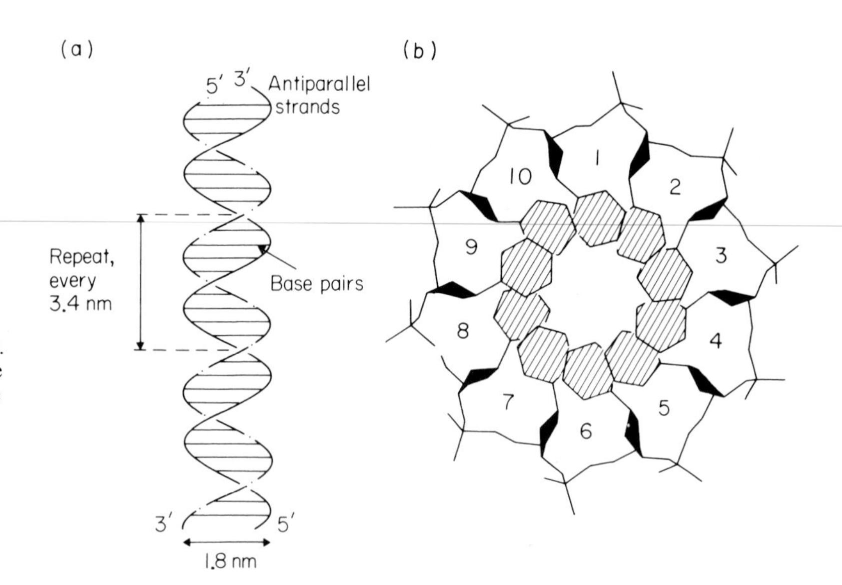

Fig. 1.4. Simplified structure of a DNA double helix. (a) Geometry of the helix. (b) Diagram of one of the strands of the helix, viewed down the central axis. The sugar-phosphate backbone is outside, the bases (all pyrimidines here; in hatching) are inside. The tenfold symmetry is evident, thus enforcing a repeat in the helix after ten nucleotide residues. For details of the geometry see Fig. 3.2.

bears a 3′ hydroxyl group. By convention, the sequence of nucleic acids is written in the 5′ → 3′ direction, the 5′ terminus always being to the *left*. Thus a pentaribonucleotide containing two residues of guanosine (G) and one each of adenosine (A), uridine (U) and cytidine (C) could have the formal sequence 5′ pApUpGpCpG 3′, each p representing a 5′→3′ phosphodiester bond. However, for simplicity the sequence is written, AUGCG. As a corollary, it should be added that AUGCG and GCGUA are *different* polynucleotides. Sequences of polydeoxyribonucleotides are written similarly, 5′→3′, but for simplicity, the small d prefix of 2′-deoxyribonucleotides is omitted. A short sequence of DNA, for example, is written as TCAGT, the presence of thymidine (T) rather than uridine (U) making it clear that this is a short sequence of DNA.

1.1.2 Secondary structure

DNA has a pronounced secondary structure, consisting of two independent, covalently linked chains, coiled around a common axis and associated together by hydrogen bonds. This pronounced secondary structure is termed the double helix or the Watson-Crick structure. The purine and pyrimidine bases are on the inside of the helix, whereas the phosphate and 2′-deoxyribose moieties are on the outside (Fig. 1.4). The planes of the bases are perpendicular to the axis of the helix and the planes of the 2′-deoxyribose molecules are nearly at right angles to those of the bases. The diameter of the helix is 1.8 nm. Adjacent bases are separated by 0.34 nm along the helix

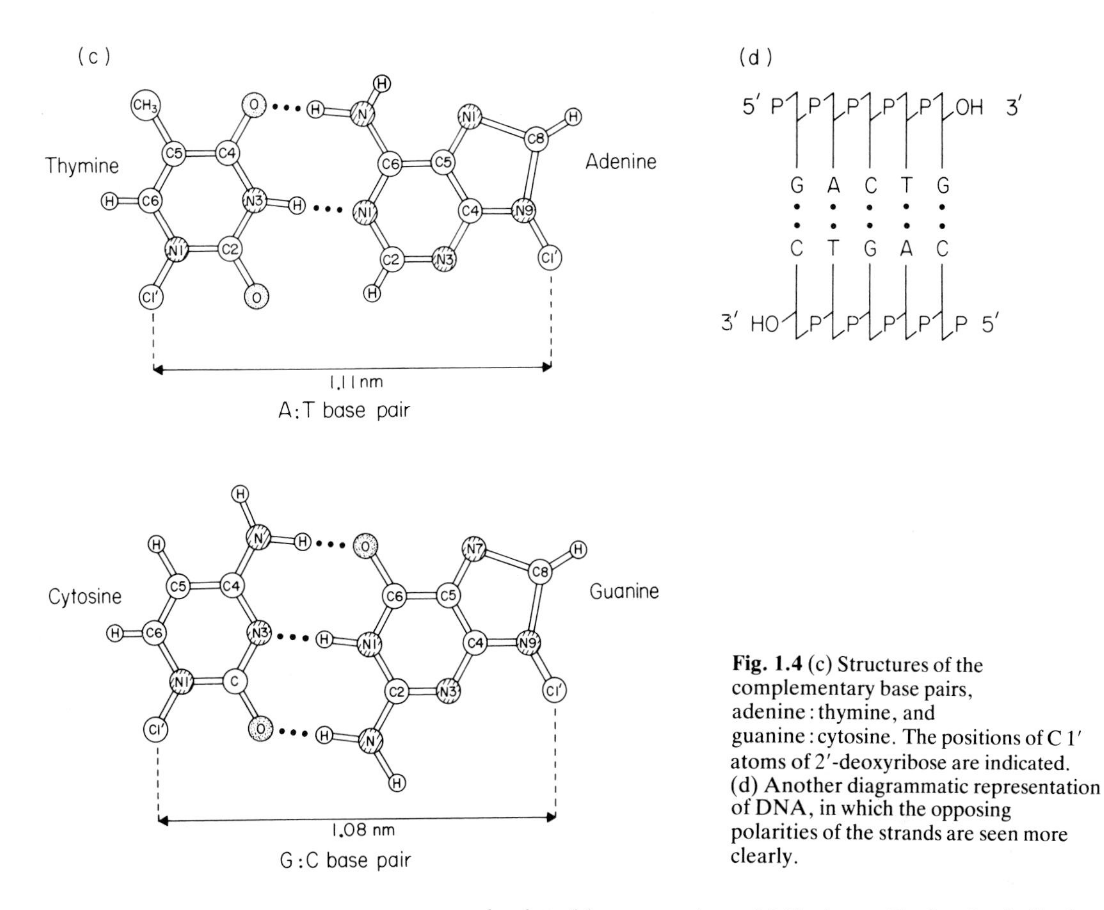

Fig. 1.4 (c) Structures of the complementary base pairs, adenine : thymine, and guanine : cytosine. The positions of C 1′ atoms of 2′-deoxyribose are indicated. (d) Another diagrammatic representation of DNA, in which the opposing polarities of the strands are seen more clearly.

and related by a rotation of 36°. Accordingly, the helical structure is repeated after ten nucleotide residues on each chain, or every 3.4 nm. Two important features of the double helix should be stressed. First, the two chains run in opposite directions, that is they are *antiparallel.* Second, the two strands have different sequences, but the sequence of one specifies precisely the sequence of the other because of the rigid rule of *complementary base pairing. Adenine* always pairs with *thymine* (A:T) and *guanine* always pairs with *cytosine* (G:C). An example of two complementary, antiparallel DNA sequences that could form a stable double helix is as follows:

5′ ATACAGGGTAT 3′
3′ TATGTCCCATA 5′

The DNA in prokaryotes, and chloroplasts and mitochondria of eukaryotes, is in the form of a closed, *circular* superhelix. By contrast, the nuclear DNA of eukaryotes is in long, *linear* threads, with the ends completely free.

With the exception of certain viral RNA molecules, RNA does not exist in the form of two complementary strands. As a generalization, therefore, native DNA is usually a double-stranded molecule, whereas RNA is usually a single-stranded molecule. RNA in prokaryotes and the cytoplasm of eukaryotic cells is of three types, messenger (mRNA), transfer (tRNA) and ribosomal (rRNA). There is some secondary structure in rRNA, but complementary base pairing, adenine:uracil, and guanine:cytosine, is so extensive in tRNA that this adopts a unique, clover-leaf structure (see Fig. 1.9). Some less conventional base pairing is also found in tRNA. The relative amount of these three classes of RNA is approximately 90–95% rRNA, 5–10% tRNA and 1–3% mRNA.

1.1.3 Organization of nucleic acids in cells

With the notable exception of tRNA, the vast majority of the nucleic acids in cells are present as complexes with proteins or nucleoprotein complexes. In prokaryotes, the DNA is organized in a condensed tertiary structure, the nucleoid, by its association with proteins. The majority of prokaryotic RNA is present as rRNA in the form of polyribosomes; these are strings of ribosome monomers (ribonucleoprotein particles) associated with a delicate strand of mRNA. Each ribosome monomer is composed of large and small ribosome subunits, such that under the electron microscope the ribosomes are found free in the cytoplasm but some are attached to the inner surface of the bacterial cell wall. In eukaryotic cells, the majority of the DNA, associated with large amounts of protein, is found within the nucleus, the distinctive, membrane-enclosed organelle of all higher organisms. Indeed, the name, eukaryote, is taken from the Greek (eu with, karyon nucleus). DNA also occurs in mitochondria, and in plants, in chloroplasts. Polyribosomes again represent the majority of eukaryotic RNA and many of these are attached to membranes of the endoplasmic reticulum. Proteins for secretion, or export, are generally made on membrane-bound polyribosomes.

1.2 TRANSFER OF GENETIC INFORMATION IN NORMAL CELLS

1.2.1 The flow of genetic information

The store of genetic information in cells resides in the sequence of the bases in DNA. Thus any change in the base sequence of DNA has profound, even lethal, effects on the regulation of macromolecular synthesis in cells. Crick first proposed the essential stages of informational transfer, in what has now become known as the central dogma (Fig. 1.5). The original ideas of Crick have had to be changed to accommodate new and unexpected concepts of contemporary molecular biology, including the Nobel prize-winning researches of Temin and Baltimore on the replication of the genes of certain RNA viruses.

The fundamental concept of informational transfer is that the base sequence of DNA provides an initial *template* for specifying absolutely the synthesis of more DNA and other templates, which in turn direct and control the synthesis of other macromolecules in the overall chain of informational transfer. The ultimate objective of the process is to synthesize proteins of defined amino acid sequence. The various types of RNA are the 'middlemen' of informational transfer; tRNA and rRNA constitute vital components of the machinery for protein synthesis and mRNA provides the template for faithful and accurate translation. Each of the six processes given in Fig. 1.5 will now be described in broad terms to provide the necessary foundations for detailed discussion of these topics in the chapters which follow.

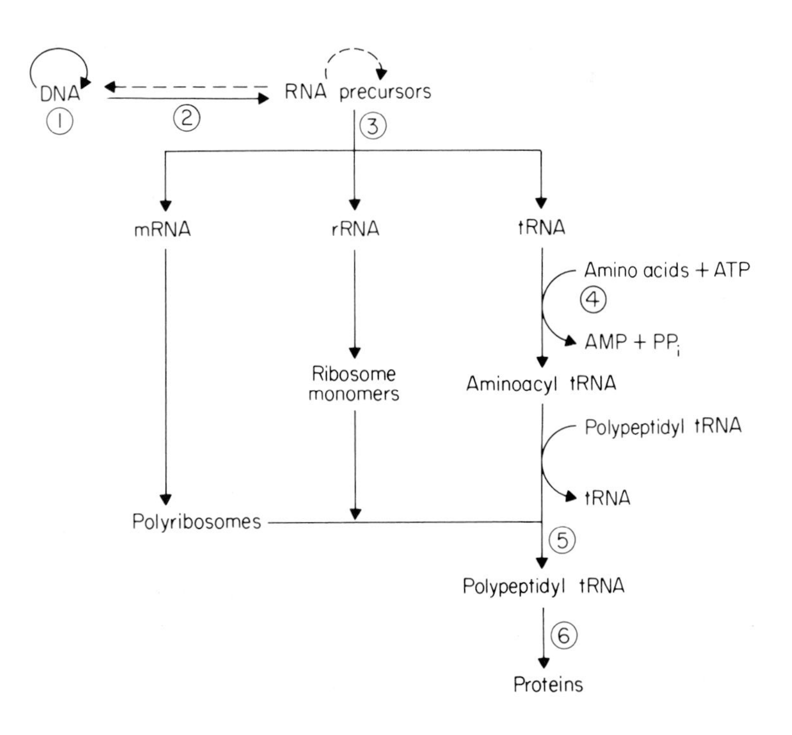

Fig. 1.5. The flow of genetic information in normal cells. (1) DNA replication (DNA synthesis); (2) transcription (RNA synthesis); (3) post-transcriptional modification; (4) activation of amino acids; (5) translation (protein synthesis) and (6) post-translational modification. The processes indicated by the arrows were first identified in certain RNA-containing viruses; small amounts of the enzymes engaged in these unusual processes have since been detected in normal, virus-free cells.

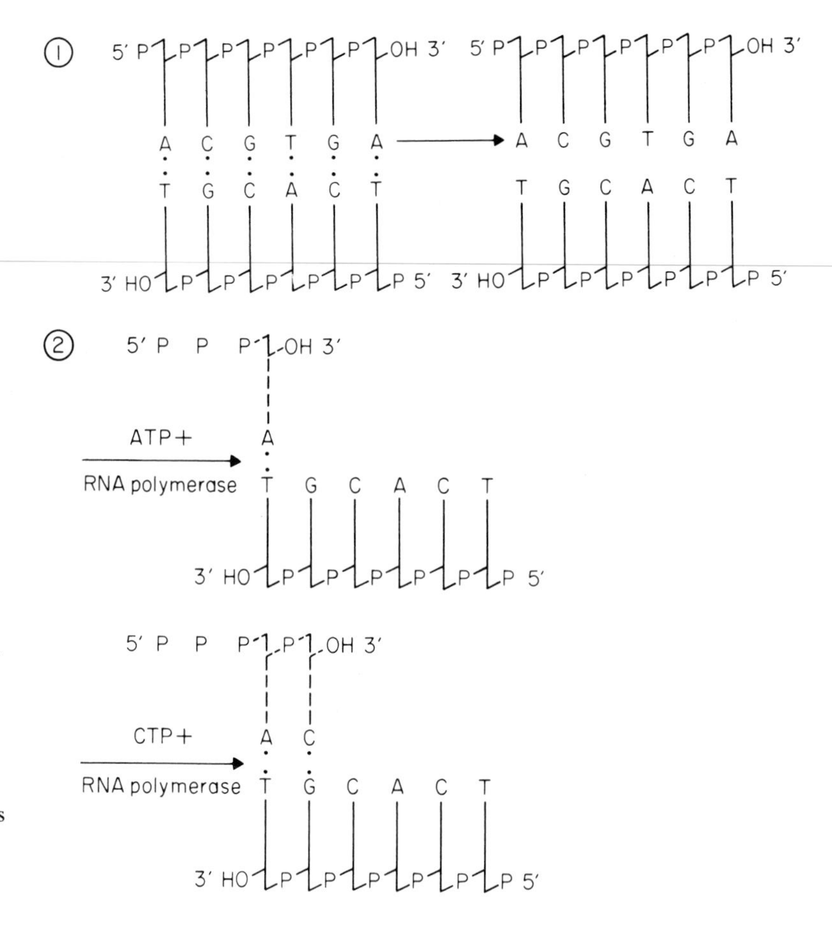

Fig. 1.6. The essential features of DNA replication. This is best envisaged as proceeding in a series of stages. (1) Destabilization (unwinding) of the helical DNA template; for simplicity, only the replication of one strand is shown. (2) A short, complementary sequence of RNA primer is formed; it is illustrated here as a dinucleotide, but the primers are actually about 100 nucleotides long.

1.2.2 DNA replication

During growth and cell division, each daughter cell must receive its complement of DNA, identical in amount and structure, especially in terms of base sequence, to that of the parent cell. This demand for DNA is met by the process of DNA replication. The essential features of the process are as follows.

(a) DNA replication must be preceded by unwinding or destabilization of the DNA.

(b) The DNA, with *short primer sequences of RNA,* serves as a template for DNA polymerase enzymes, which condense 2′-deoxyribonucleotide-5′-triphosphates (dATP, dCTP, dGTP and dTTP) together, starting from the 3′ OH terminal of the RNA primer (Fig. 1.6). As each base is inserted into the growing chain, in the *5′→3′ direction,* two high-energy bonds are consumed and inorganic pyrophosphate is eliminated. The base-pairing rules (A:T

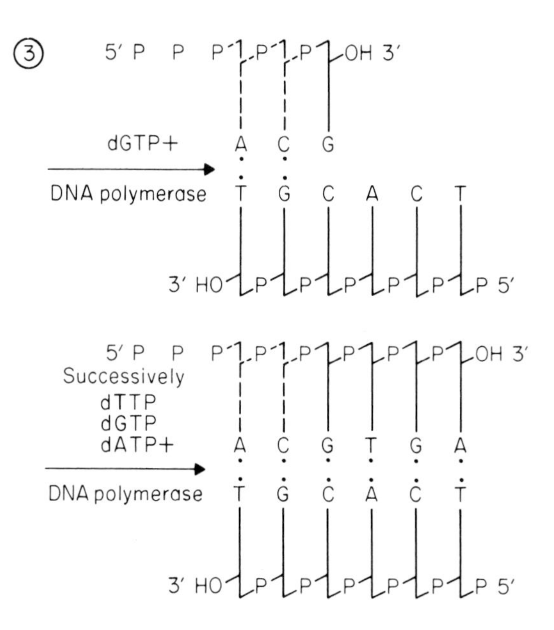

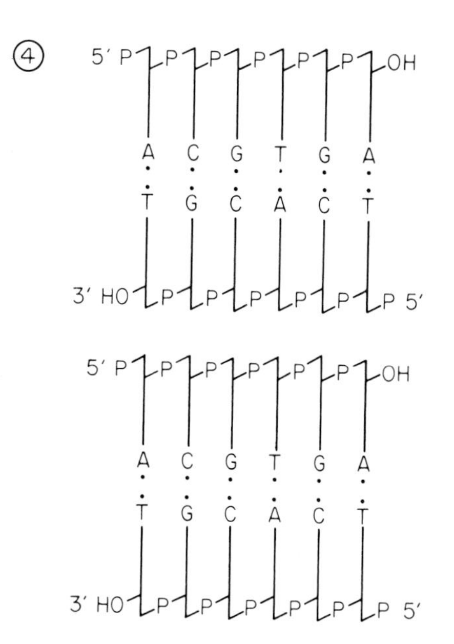

Fig. 1.6 (cont.) (3) As dictated by base-pairing rules, DNA is synthesized one nucleotide at a time, by interaction of the 5′-triphosphate group of the incoming 2′-deoxyribonucleotide and the 3′ OH group of the RNA primer, or later in synthesis, of the newly synthesized DNA strand itself. (4) As DNA replication proceeds, the template DNA becomes progressively unwound, but only at completion of the replication of both parent strands do they finally separate (see also Fig. 1.7).

and G:C) ensure that the newly synthesized DNA is *complementary* in base sequence to the template DNA. The RNA primer sequences are then excised and DNA polymerases fill in the gaps in the daughter DNA strands.

(c) The overall process is *bidirectional* and *symmetrical,* in that both DNA template strands are replicated, and *semi-conservative*, in that the new double helix consists of one template, parental strand and one, new daughter strand. The semi-conservative nature of DNA replication was first demonstrated in the classical work of Meselson and Stahl (Fig. 1.7). Their experiment was based on the fact that DNA molecules of differing density can be completely resolved in an analytical ultracentrifuge. The bacterium, *Escherichia coli,* was grown for several generations in a medium in which the only source of nitrogen was in the form of the heavy isotope, as $^{15}NH_4Cl$. The DNA was then 'heavy', as ^{15}N-DNA. The bacteria were then transferred to a normal medium containing the usual and lighter nitrogen atom as $^{14}NH_4Cl$. At various times, small samples of bacteria were harvested and, after extraction, their DNA was

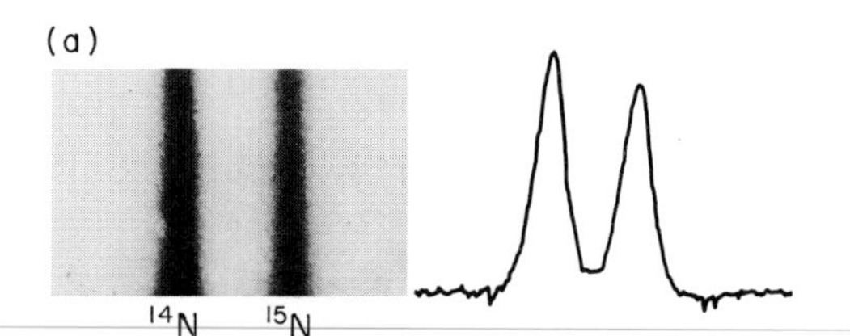

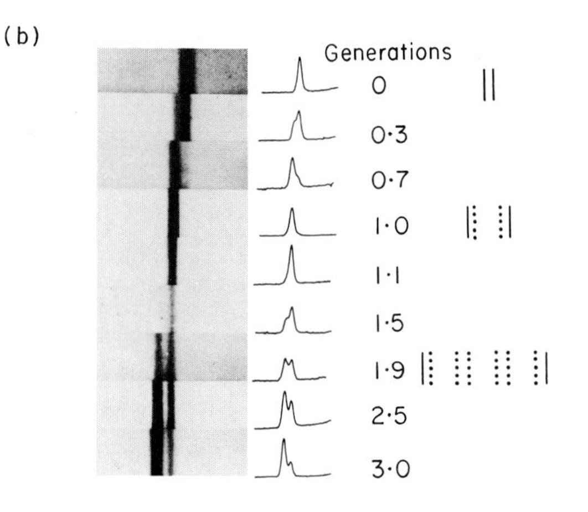

Fig. 1.7. The semi-conservative nature of DNA replication. (a) The separation of heavy ^{15}N- and light ^{14}N-labelled DNA in an analytical ultracentrifuge; left, an ultraviolet absorption photograph and right, a tracing of such a photograph in a densitometer
(b) Analysis of *E. coli* DNA at various times after transferring the bacterium from a medium containing $^{15}NH_4Cl$ to $^{14}NH_4Cl$. (Reproduced, with permission, from Meselson & Stahl, 1958.) To the right a schematic interpretation of double-stranded DNA molecules after 0, 1 and 2 generations (parental ^{15}N-DNA as solid lines, and replicated ^{14}N-DNA as dotted lines).

analysed in the ultracentrifuge. After one generation, the DNA sedimented as a new species, midway between pure ^{15}N-DNA and ^{14}N-DNA, containing equal amounts of heavy and light isotopic nitrogen, as a ^{15}N-DNA:^{14}N-DNA hybrid. This result could only be obtained if the parental ^{15}N-DNA was not maintained as an intact unit during replication. After two generations, there were equal amounts of pure ^{14}N-DNA and the hybrid ^{15}N- and ^{14}-DNA. Only a semi-conservative mode of DNA replication can explain this distribution of the isotopes of nitrogen with time.

(d) The template DNA is not replicated as an extremely long, unbroken strand of DNA. Rather, DNA replication is a *discontinuous* process, meaning that replicated DNA is initially a series of short fragments, perhaps only a few hundred nucleotides long, named Okazaki fragments, after their discoverer. These fragments are then joined together, 5′ P to 3′ OH, by the enzyme DNA ligase. In bacteria, the energy for this reaction is provided by the concomitant breakdown of NAD^+; in eukaryotes and viruses, ATP fulfils this role:

$$\text{DNA—3}'\text{ OH} + 5'\text{Ⓟ—O—DNA} + \begin{Bmatrix}\text{ATP}\\ \text{NAD}^+\end{Bmatrix} \rightleftharpoons$$

$$\text{DNA—O—Ⓟ—O—DNA} + \text{AMP} + \begin{Bmatrix}\text{PP}_i\\ \text{NMN}\end{Bmatrix}$$

The reaction proceeds through three steps and irrespective of the energy source, ATP or NAD^+, it involves the formation of an activated or high-energy form of DNA ligase, as the enzyme-adenylate complex:

$$E + ATP\ (\text{or}\ NAD^+) \rightleftharpoons E—AMP + PP_i\ (\text{or NMN})$$
$$E.AMP + P—O—5'\ DNA \rightleftharpoons E + AMP—P—O—5'\ DNA$$
$$DNA\ 3'—OH + AMP + P—O—5'\ DNA \rightleftharpoons DNA\ 3'—O—P—O—5'\ DNA + AMP.$$

The process of DNA replication requires the concerted activities of many enzymes and controlling factors.

1.2.3 Transcription

RNA synthesis is carried out by RNA polymerases. Transcription has certain similarities to the mechanism of DNA replication, but there are some important differences. The essential features of transcription are as follows.

(a) RNA polymerases catalyse the step-wise condensation of ribonucleoside-5′-triphosphates (ATP, CTP, GTP and UTP) through nucleophilic attack by the free 3′ OH group of the growing chain by the innermost nucleotidyl phosphate of the incoming nucleoside-5′-triphosphate. The sequence of the RNA is dictated by the base sequence of the template DNA. Accordingly, there is *complementary base pairing* between the template DNA and newly transcribed RNA, dG:C, dC:G, dT:A and dA:U (see Fig. 1.8).

(b) Transcription is a continuous process as the RNA is transcribed in the *5′ → 3′ direction* without a break and the RNA products (or *transcripts*), especially in eukaryotes, are extremely long, containing thousands of polymerized ribonucleotides.

(c) The initial ribonucleotide is usually GTP or ATP and their associated phosphate groups are not removed during the subsequent completion and post-transcriptional processing of the RNA. Consequently, RNA molecules have a distinctive 5′ terminus, ppp$\left\{\begin{matrix}A\\G\end{matrix}\right.$ and always a 3′ terminus with a free hydroxyl group.

(d) Transcription is a *conservative* process because the DNA template is fully conserved and neither DNA strand appears in the RNA product.

(e) RNA polymerases, unlike DNA polymerases, do not require an RNA or other primer sequence on the DNA template.

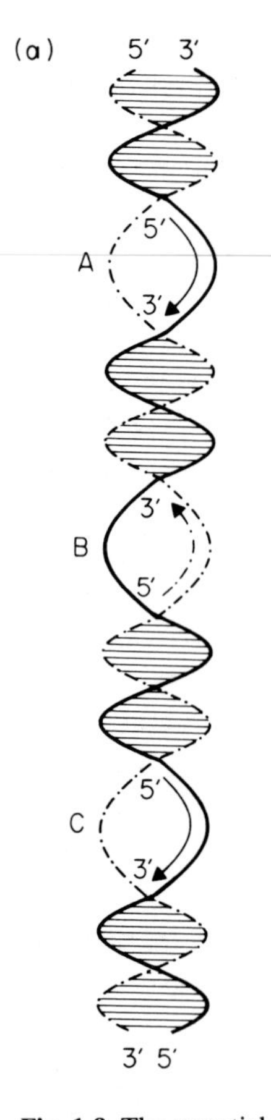

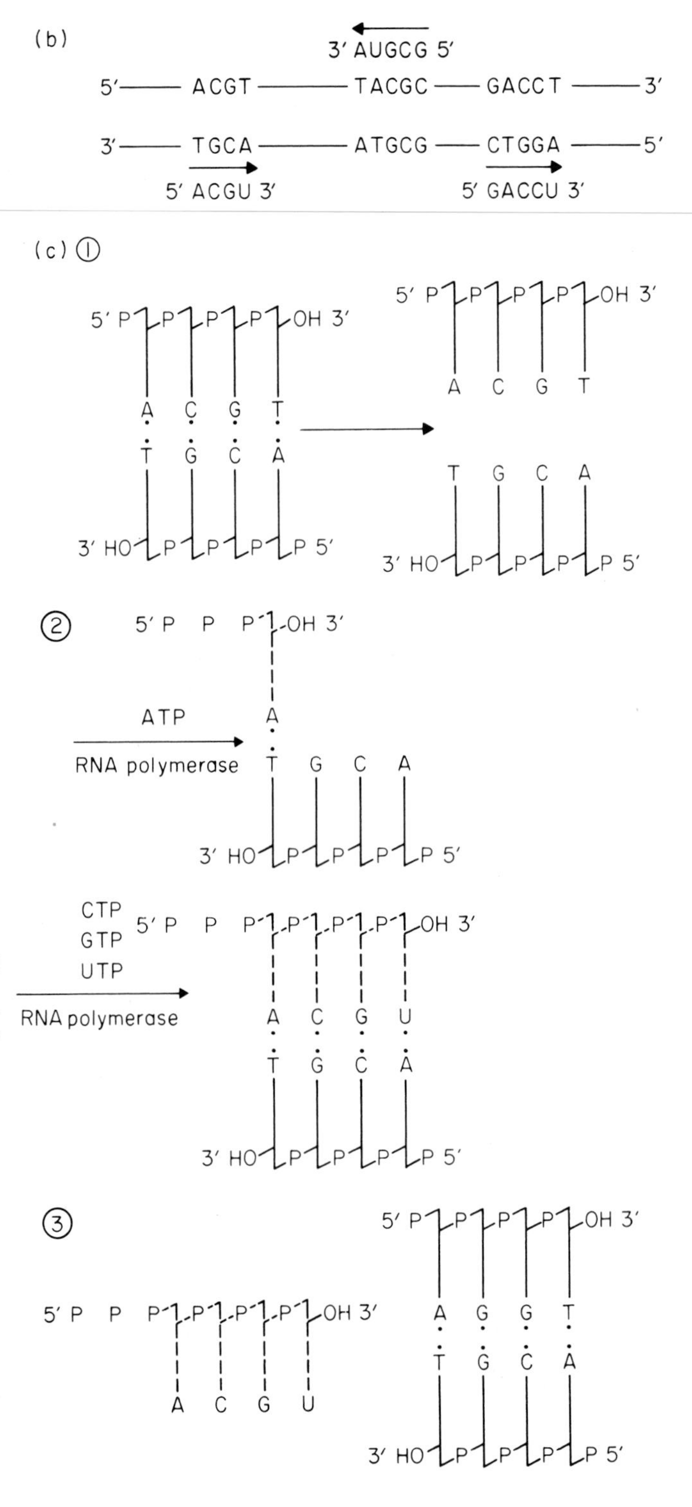

Fig. 1.8. The essential features of transcription. (a) Strand selection. During the expression (transcription) of genes, the template DNA must be destabilized whereas the DNA in non-expressed regions retain their helical, H-bonded secondary structure. The DNA strand actually copied during transcripton, the transcribing strand, need not be the same in different genes. To illustrate this, the two antiparallel strands of template DNA are presented as solid or dashed lines; three genes (A, B and C) are being expressed from different DNA template strands. The transcripts from genes A, B and C are also presented as solid or dashed lines, with arrows, to indicate which template DNA strand was transcribed and the direction of transcription. (b) The transcription of the three genes in (a) to illustrate both the 5′ → 3′ direction of

(cont. on facing page)

(f) Transcription is an *asymmetric* process because *only one* strand of template DNA is transcribed in a particular region of the genome. In addition, active or *expressed* genes need not necessarily all be on the same DNA strand (Fig. 1.8) so that the RNA polymerases must be directed to the appropriate transcribing strand with remarkable accuracy by a combination of their own intrinsic properties, regulatory proteins and other factors. How this rigorous strand selection is carried out in even the simplest organisms remains one of the most challenging, yet unsolved, questions of molecular biology. Certainly *both* DNA strands in the template must be present for accurate transcription; the part played by the 'inactive', non-transcribing strand remains unclear.

The synthesis of RNA proceeds in a series of steps; binding of RNA polymerase to the DNA template; initiation; elongation and termination. The enzyme binds to specific regions of the DNA template called *promoter* sites and initiates transcription close to these important areas of a gene. The single RNA polymerase *holoenzyme* of prokaryotes has a complex subunit or quaternary structure, $\alpha_2\beta\beta'\sigma$, functionally divisible into the single *σ subunit* and the *core enzyme, $\alpha_2\beta\beta'$*. The dissimilar β and β' subunits are involved in binding to the DNA template, whereas the σ subunit selects the initiation sites for transcription. Another distinct protein, ρ, composed of four identical subunits, participates in chain termination. These characteristics of bacterial transcription are not applicable to eukaryotes. Higher organisms have multiple RNA polymerases of specific function and none of them contain a σ subunit; furthermore, ρ factor is not present in eukaryotes. Nonetheless, the essential pattern of transcription is similar in all forms of life, but the *control of gene expression* and even *gene structure* are fundamentally different. Discussion of these contrasting aspects of gene expression in prokaryotes and eukaryotes is one of the principal features of this book. With the advent of genetic engineering (Chapter 11) and the technology to produce purified eukaryotic genes in reasonable quantity, the transcriptional units of higher organisms, particularly their promoter sites, are now being investigated in remarkable detail.

Fig. 1.8. (cont.) transcription and the operation of base-pairing rules between DNA template and RNA transcript. (c) Detail of the transcription of gene A. This is best envisaged as proceeding in stages. (1) Destabilization of the DNA template, and the separation of transcribing and non-transcribing strands. (2) Formation of the transcript, one nucleotide at a time, from the transcribing strand in the $5' \rightarrow 3'$ direction. (3) Release of the completed RNA transcript and the return of the template DNA to the helical, H-bonded form.

1.2.4 Post-transcriptional modification

One of the most exciting yet unexpected findings of contemporary biology is that the genes of eukaryotes often have a unique complexity. Many genes of higher organisms contain *intervening sequences* or *introns* interspersed between the remaining gene sequences or *exons*. Since the exons and introns are transcribed as an *intact transcriptional unit*, the initial RNA transcripts of active genes in eukaryotes are extremely long. However, the RNA equivalent to the *introns* is not present in the mature species of functional RNA. Accordingly, the long transcripts must be *cleaved* precisely at the positions corresponding to the intron–exon boundaries, and the RNA equivalent to the exons *spliced* or ligated back together. This topic is considered thoroughly in section 8.3, but post-transcriptional modification is a most important process in eukaryotes and must be carried out with remarkable, even faultless, accuracy. Any defect in the splicing enzymes would result in a complete breakdown in the transfer of genetic information. Post-transcriptional modification is less extensive in prokaryotes, but occurs nonetheless. In all forms of life, the final products are shorter, modified and mature forms of rRNA, tRNA and mRNA.

1.2.5 Activation of amino acids

Unlike the immediate precursors of nucleic acids, the nucleoside 5′-triphosphates, amino acids are not high-energy molecules in their own right. Accordingly, an important prerequisite for protein synthesis is the activation of the 20 amino acids, often referred to as the 'magic 20', which are the universal building units of all proteins. The magic 20 amino acids, plus their abbreviations, are given in Table 1.1. The activated intermediates for protein synthesis are amino acid esters, in which the carboxyl group of an amino acid is linked to either the 2′ or 3′ OH group at the 3′ terminus of tRNA. The activated intermediates are *aminoacyl-tRNA* molecules and their synthesis is catalysed by highly specific enzymes, *aminoacyl-tRNA synthetases*. For each of the 20 amino acids, there is at least 1

Table 1.1. The amino acids present in proteins, and their abbreviations.

Alanine	Ala	Leucine	Leu
Arginine	Arg	Lysine	Lys
Asparagine	Asn	Methionine	Met
Aspartic acid	Asp	Phenylalanine	Phe
Cysteine	Cys	Proline	Pro
Glutamine	Gln	Serine	Ser
Glutamic acid	Glu	Threonine	Thr
Glycine	Gly	Tryptophan	Trp
Histidine	His	Tyrosine	Tyr
Isoleucine	Ile	Valine	Val

specific aminoacyl-tRNA synthetase and also at least 1 specific or *cognate* tRNA. The overall reaction may be written as follows:

$$\text{amino acid} + \text{ATP} + \text{tRNA} + H_2O \rightleftharpoons \text{aminoacyl-tRNA} + \text{AMP} + 1P_i$$

The overall reaction is thus favoured in the direction of synthesis. The free energy of hydrolysis of the ester bond of aminoacyl-tRNA is similar to that of the terminal phosphyl group of ATP, so that at first sight, the reaction could be seen to be delicately balanced. The hydrolysis of inorganic pyrophosphate provides the additional energy to drive the reaction in favour of synthesis. Individual steps in these reactions have now been established.

$$\text{Amino acid} + \text{ATP} \rightleftharpoons \text{Aminoacyl-AMP} + PP_i$$
$$\text{Aminoacyl-AMP} + \text{tRNA} \rightleftharpoons \text{Aminoacyl-tRNA} + \text{AMP}$$
$$PP_i + H_2O \rightleftharpoons 2P_i.$$

The extreme specificity of the aminoacyl-tRNA synthetases in only attaching an amino acid to its correct (cognate) tRNA makes a vital contribution to the overall accuracy of protein synthesis.

tRNA molecules have several distinctive features: an overall clover-leaf structure; extensive secondary structure resulting from intramolecular base pairing (H-bonding); many modified bases and the base sequence, CCA, at their 3′ OH terminus (Fig. 1.9). A vitally important feature of tRNA molecules is a triplet of bases forming the *anticodon* loop. Each specific tRNA molecule has a *unique* sequence in its anticodon. Indeed, it is the anticodon loops, rather than the associated amino acids in aminoacyl-tRNA complexes, which dictate that the amino acid is introduced into proteins in the correct sequence.

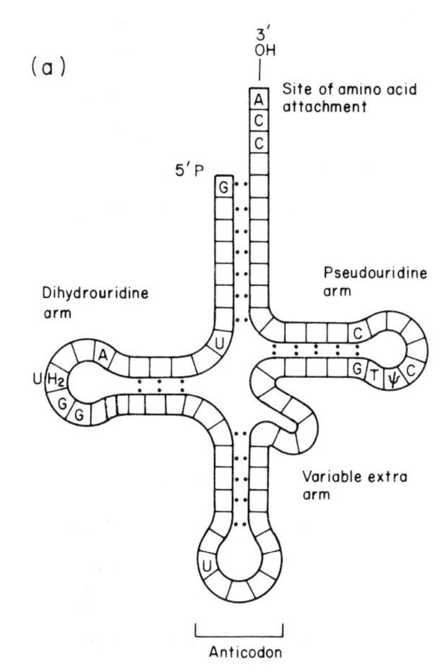

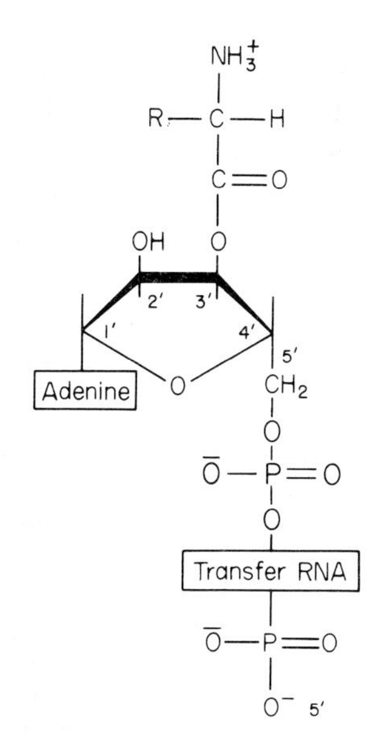

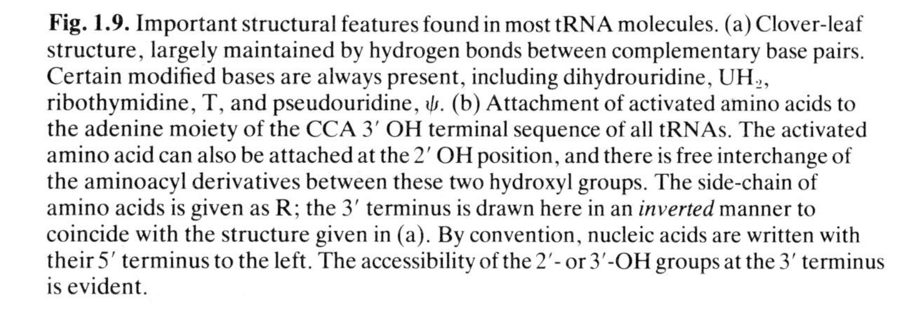
Fig. 1.9. Important structural features found in most tRNA molecules. (a) Clover-leaf structure, largely maintained by hydrogen bonds between complementary base pairs. Certain modified bases are always present, including dihydrouridine, UH_2, ribothymidine, T, and pseudouridine, ψ. (b) Attachment of activated amino acids to the adenine moiety of the CCA 3′ OH terminal sequence of all tRNAs. The activated amino acid can also be attached at the 2′ OH position, and there is free interchange of the aminoacyl derivatives between these two hydroxyl groups. The side-chain of amino acids is given as R; the 3′ terminus is drawn here in an *inverted* manner to coincide with the structure given in (a). By convention, nucleic acids are written with their 5′ terminus to the left. The accessibility of the 2′- or 3′-OH groups at the 3′ terminus is evident.

1.2.6 Translation

Apart from the essential function of tRNA, the remaining two forms of RNA also play crucial roles in the translational process. rRNA provides the basic machinery for protein synthesis in the form of the small and large ribosome subunits. These are somewhat larger in eukaryotes, so that the combined subunits, constituting the ribosome monomer, have a sedimentation coefficient of 80*s* in higher organisms but 70*s* in bacteria. mRNA acts as a *template* for the synthesis of polypeptides, thus providing the link between the genetic information residing in the base sequence of mRNA and the amino acid sequence of proteins. This informational link between nucleic acids and proteins was made through the brilliant work of Nirenberg and Khorana which led to the cracking of the *genetic code* (Table 1.2). The important features of the code are as follows.

(a) The code is *essentially universal*; the only departures so far

Table 1.2. The genetic code. This is in triplets of bases or codons. The codon for a given amino acid is read from the first (5′) base by reading across→, then down, ↓, and finally the last (3′) base by reading backwards, ←. Where the last two lines intersect, the amino acid encoded in the codon sequence is found. There is thus a single codon, AUG, for methionine and a single codon, UGG, for tryptophan. By contrast, there are no less than six alternative codons (UCU, UCC, UCA, UCG, AGU and AGC) for serine. Three codons (UAA, UGA and UAG) are stop signals.

First position (5′-end)	Second position				Third position (3′-end)
	U	C	A	G	
U	Phe	Ser	Tyr	Cys	U
	Phe	Ser	Tyr	Cys	C
	Leu	Ser	Stop	Stop	A
	Leu	Ser	Stop	Trp	G
C	Leu	Pro	His	Arg	U
	Leu	Pro	His	Arg	C
	Leu	Pro	Gln	Arg	A
	Leu	Pro	Gln	Arg	G
A	Ile	Thr	Asn	Ser	U
	Ile	Thr	Asn	Ser	C
	Ile	Thr	Lys	Arg	A
	Met	Thr	Lys	Arg	G
G	Val	Ala	Asp	Gly	U
	Val	Ala	Asp	Gly	C
	Val	Ala	Glu	Gly	A
	Val	Ala	Glu	Gly	G

reported are in the autonomous and self-contained process for protein synthesis in mitochondria (see section 10.4).

(b) The code is in the form of triplets of bases or *codons*, which are read without a break; the code does not contain pauses or 'punctuation marks'.

(c) The code is *unambiguous*, because a given codon designates only one amino acid.

(d) Certain codons (UAG, and UGA) do not specify an amino acid, but constitute *stop signals*.

(e) From 4 bases in mRNA (A, C, G and U) a maximum of 64 codons is possible, yet only 61 are used to encode the 20 amino acids incorporated directly into proteins.

Accordingly, the code is highly *degenerate,* with certain amino acids (e.g. Leu, Arg and Ser) being designated six codons each. Codons specifying the same amino acid are called *synonyms.* Most synonyms differ only in the last base of the triplet and the reason for this will be discussed later. The extensive degeneracy of the code is significant in two ways. First, it minimizes the deleterious effects of mutations, or changes in the base sequence of DNA, introduced either naturally, by evolutionary selection, or artificially, by X-rays or chemicals (mutagens). If the code were not degenerate, only 20 codons would direct the insertion of amino acids into proteins, leaving a theoretical maximum of 44 to serve potentially as stop signals. The chance of mutations leading to the termination of protein synthesis is clearly higher with a non-degenerate than a degenerate code. Second, degeneracy of the code allows DNA of varying base composition (for example, between species) to code nevertheless for proteins of similar function and thus similar amino acid sequence. Genomes containing DNA of different base composition and sequence can encode similar proteins if synonyms are widely used during the translational process.

(f) The code is read in one direction only, with translation of the mRNA in the *5′ → 3′ direction*. This important finding was established by the translation of synthetic polyribonucleotides of known base sequence. Take, for example, the sequence:

$$5' \text{ AAA } (\text{AAA})_x \text{ } \underline{\text{AAC}} \text{ } 3',$$

a polymer of adenylic acid, with a single cytidine residue at the 3′ terminus. This means there is a distinctive codon at the 3′-end (as underlined). By inspection of the genetic code (Table 1.2), translation 5′→3′ would form N-terminal:Lys.$(\text{Lys})_x$.Asn:C-terminal whereas translation 3′ → 5′ would form N-terminal:Gln.$(\text{Lys})_x$.Lys:

C-terminal. Analysis of the translation products established unequivocally that translation proceeded $5' \rightarrow 3'$. The fact that proteins are made in the *N-terminal → C-terminal direction* was established prior to the availability of synthetic mRNA templates by Dintzis. The complete amino acid sequence of the α- and β-globin chains of rabbit haemoglobin had been fully determined and reticulocytes synthesize these chains exclusively, even under conditions *in vitro*. He exposed reticulocytes to exogenous [^{3}H]leucine for periods of time shorter than necessary to complete the synthesis of the globin chains. Chains being synthesized prior to exposure to the [^{3}H] label would contain endogenous, non-radioactive leucine. The α- and β-chains were then separated and digested with trypsin. The amount of [^{3}H]leucine in the peptides, relative to their total content of leucine, was determined. Peptides known to be near the C-terminal ends of the globin chains were more radioactive, clearly indicating that these were synthesized last and, therefore, protein synthesis proceeds from the N-terminal → C-terminal residues (Fig. 1.10).

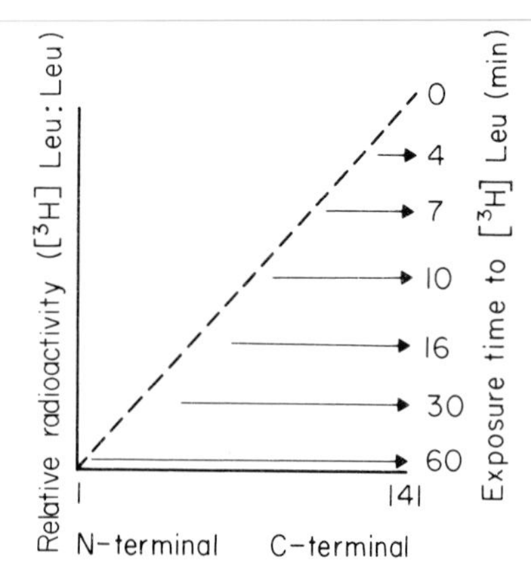

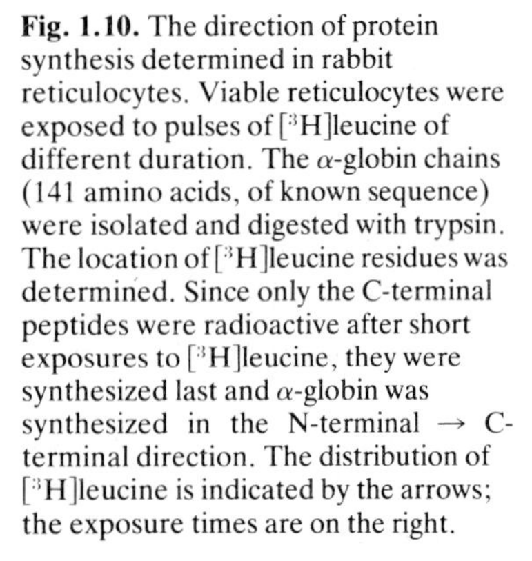

Fig. 1.10. The direction of protein synthesis determined in rabbit reticulocytes. Viable reticulocytes were exposed to pulses of [^{3}H]leucine of different duration. The α-globin chains (141 amino acids, of known sequence) were isolated and digested with trypsin. The location of [^{3}H]leucine residues was determined. Since only the C-terminal peptides were radioactive after short exposures to [^{3}H]leucine, they were synthesized last and α-globin was synthesized in the N-terminal → C-terminal direction. The distribution of [^{3}H]leucine is indicated by the arrows; the exposure times are on the right.

Although the DNA comprising the *genome*, or genetic store, of an organism contains vast numbers of discrete genes, each of unique base sequence, not all of these are active, or *expressed*, in the transcription of specific template mRNAs at any one time. Indeed, discussion of the molecular mechanisms for the control of gene expression from complex genomes is one of the most important features of this book (notably in Chapters 6 and 8).

Translation proceeds in three distinct phases, initiation, elongation and termination, each of which will now be briefly summarized. Each of these phases requires specific factors, many of which are integral components of ribosome subunits. Whereas ATP provides the essential energy for the activation of amino acids, the three phases of protein synthesis need the concomitant hydrolysis of the high-energy (γ) bond of GTP.

Initiation

In prokaryotes and eukaryotes, the initiating amino acid at the N-terminal end of all proteins during their biosynthesis is *N-formylmethionine* or *methionine*. Special *initiator tRNAs*, $tRNA^{fMet}$ (bacteria) or $tRNA_i^{Met}$ (eukaryotes) introduce the first amino acid; internal methionine residues are introduced by another tRNA, $tRNA^{Met}$. Since there is only a single codon for methionine, AUG (Table 1.1), it has long been speculated that initiator tRNAs must have a unique structure for their distinctive role in translation.

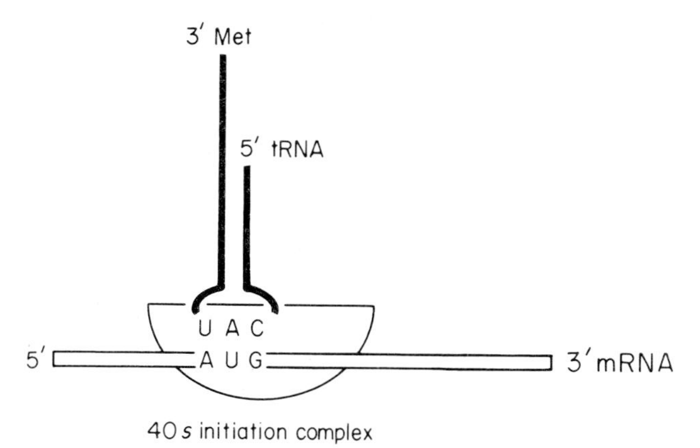

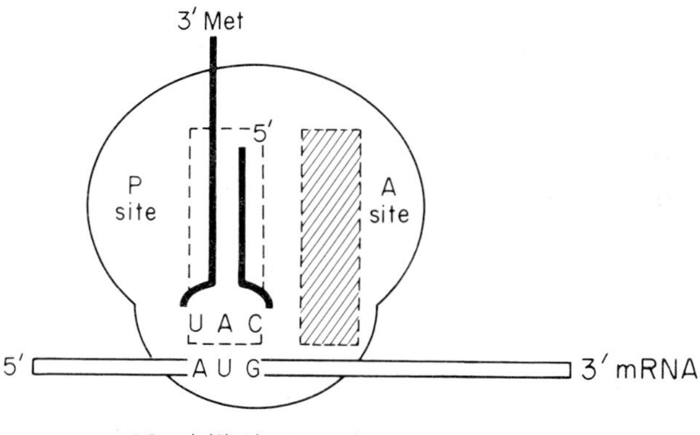

Fig. 1.11. The formation of the initiation complex of translation in eukaryotes. The mRNA is shown, by convention, with the 5′-end to the left. The initiator aminoacyl complex, Met–tRNA^{fMet}, interacts by codon–anticodon base pairing in an antiparallel manner. Once the large ribosome subunit is attached, the overall alignment is such that the initiator tRNA occupies the P site in the 80*s* ribosome monomer; the A site is empty. In prokaryotes, the initiator tRNA bears N-formylmethionine and the initiation complex is of lower sedimentation coefficient (70*s*).

Indeed, this has recently been established (see section 9.2) in the form of a G + C-rich region in the anticodon stem. The formyl group in bacteria is introduced into methionine, *in situ,* in the initiating aminoacyl-tRNA, Met-tRNA^{fMet}.

In eukaryotes, protein synthesis begins by the interaction between a small 40*s* ribosome subunit, an mRNA and Met-tRNA^{fMet}, forming a 40*s* initiation complex. A large 60*s* subunit is then joined, forming an 80*s* initiation complex (Fig. 1.11). In bacteria, the ultimate initiation complex is slightly smaller at 70*s*. It should be noted that the formation of a stable initiation complex requires the mRNA and initiator tRNA to be associated together in an *antiparallel* manner. This functional association is dependent on a precise interaction between the initiating *codon,* AUG, in the mRNA, with the *anticodon,* CAU, in the initiator tRNA, as follows:

3′ UAC 5′ tRNA anticodon
5′ AUG 3′ mRNA codon.

Such codon–anticodon interactions provide a further means (after amino acid activation) of maintaining the fidelity of the translational process.

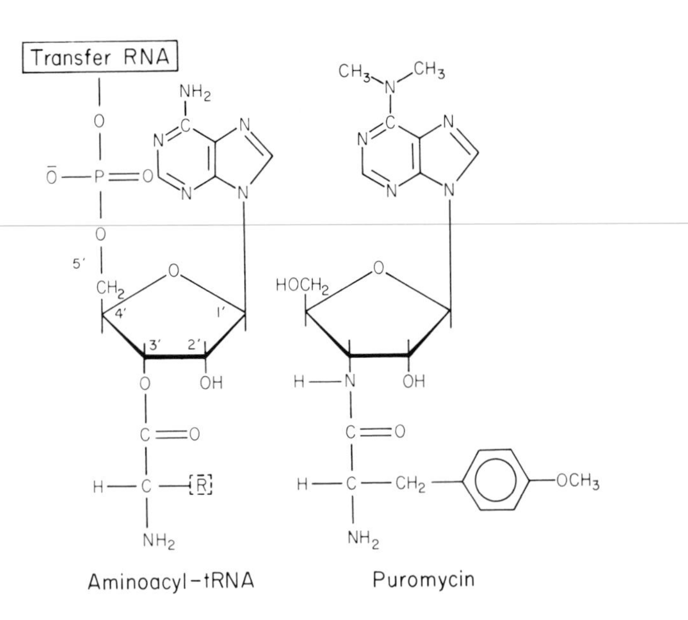

Fig. 1.12. The structural similarity between puromycin and the 3′-terminus of an aminoacyl-tRNA. R indicates the side-chain of an amino acid.

The antibiotic, puromycin, is a structural analogue of the 3′ aminoacyl-adenosine terminus of aminoacyl-tRNA. In addition, puromycin contains an α-amino group (Fig. 1.12) and therefore can mimic an amino acid and be incorporated into peptide linkage. Consequently, puromycin is an extremely powerful inhibitor of protein synthesis, releasing nascent polypeptide chains before their synthesis is complete. The released polypeptide has puromycin linked at its C-terminal end. Studies with puromycin led to the functional identification of two distinct sites in the ribosome monomers engaged in translation, namely an *A (aminoacyl) site* and a *P (peptidyl) site.* The structural features of the A and P sites still remain unclear. Puromycin can prevent the entry of aminoacyl-tRNAs, but not peptidyl-tRNAs, into the A site. The initiator aminoacyl complex is *unique* in being able to occupy the P site *directly,* emphasizing still further its crucial function in translation. In the initiation complex, the A site is empty.

Elongation

During each round of elongation, an amino acid is introduced into polypeptide linkage, in the N → C-terminal direction, or 5′ → 3′ along the strand of mRNA. Each elongation cycle has three steps (Fig. 1.13):

(a) binding of an incoming aminoacyl-tRNA in an antiparallel

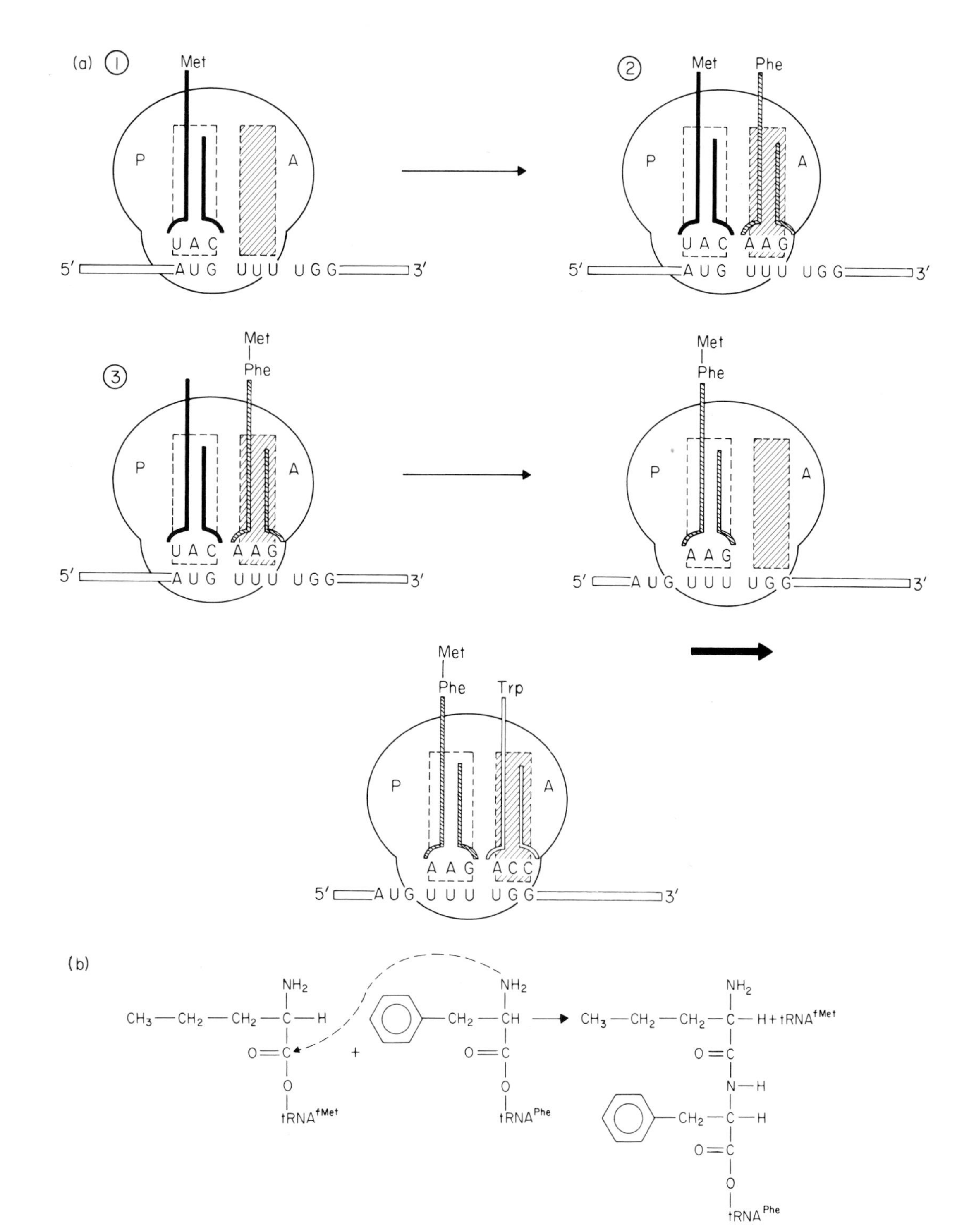

Fig. 1.13. The elongation of protein synthesis in eukaryotes. (a) The elongation phase of translation on an mRNA molecule containing three codons, the initiator codon, AUG, followed by UUU(Phe) and UGG(Trp). Elongation proceeds through three steps, starting from the 80*s* initiation complex (Fig. 1.12). (1) Binding of the aminoacyl–tRNA, Phe-tRNA^{Phe}, to the A site. (2) Peptide bond formation, catalysed by peptidyl transferase, and the formation of peptidyl-tRNA in the A site. (3) Translocation of the peptidyl-tRNA to the P site, catalysed by translocase, leaving the A site empty and displacing the initiator tRNA. The ribosome monomer moves along the mRNA in the $5' \rightarrow 3'$ direction, ending one cycle of elongation. Binding of the next aminoacyl-tRNA, Trp-tRNA^{Trp}, would begin the next cycle of elongation. (b) Detail of the formation of the peptide bond in (a) between methionine and phenylalanine.

manner, as dictated by codon (mRNA)–anticodon (tRNA) interactions, into the vacant A site;

(b) peptide bond formation by transfer of the amino acid in the P site on to the aminoacyl-tRNA in the A site;

(c) translocation of the peptidyl-tRNA into the P site leaving the A site free for the next incoming aminoacyl tRNA.

Steps (a) and (c) involve the hydrolysis of GTP. During the translocation reaction, the ribosome monomer moves along the mRNA in the $5' \rightarrow 3'$ direction, a distance of three nucleotides, equivalent to a codon. By this process, a new mRNA codon occupies the A site and the next round of elongation can begin.

Many ribosome monomers can be recruited on to a single strand of mRNA forming *polyribosomes*, or, more simply, polysomes. Each ribosome functions independently. Ribosomes near the 5′-end have synthesized only the N-terminal sequence of the protein, whereas those closer to the 3′-end contain protein nearing completion (Fig. 1.14).

Since the genetic code is degenerate, some tRNA molecules must have to recognize and combine with *several different codons* in mRNA sequences. Codon–anticodon interactions generally obey strict base-pairing rules, but the interactions between the third codon base and the first anticodon base are less stringent. As Crick first pointed out, steric criteria permit some *wobble* or flexibility in the formation of hydrogen bonds at the third codon base. This is achieved by unusual base pairing (e.g. G:*U* rather than G:*C*) or by the presence of inosine (I) as the first or *wobble base* of an anticodon.

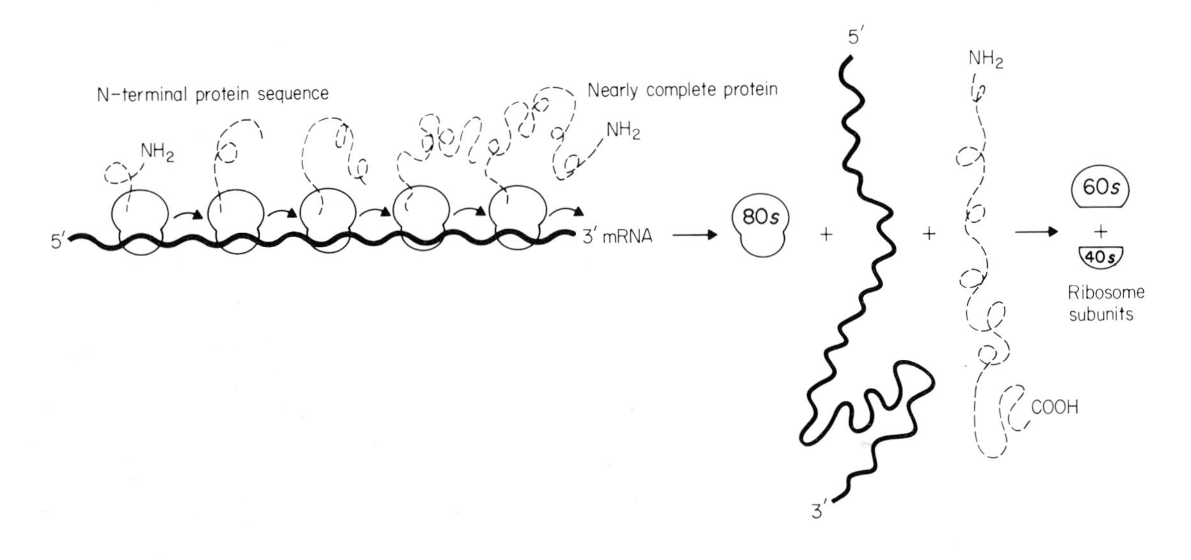

Fig. 1.14. The structure of eukaryotic polyribosomes and the ribosome cycle. The same picture would apply to prokaryotes, except that the ribosome subunits are 30*s* and 50*s*, and the ribosome monomer 70*s*. Progression of the individual ribosomes is in the $5' \rightarrow 3'$ direction along the mRNA.

Inosine is formed by the post-transcriptional deamination of adenosine residues in tRNA.

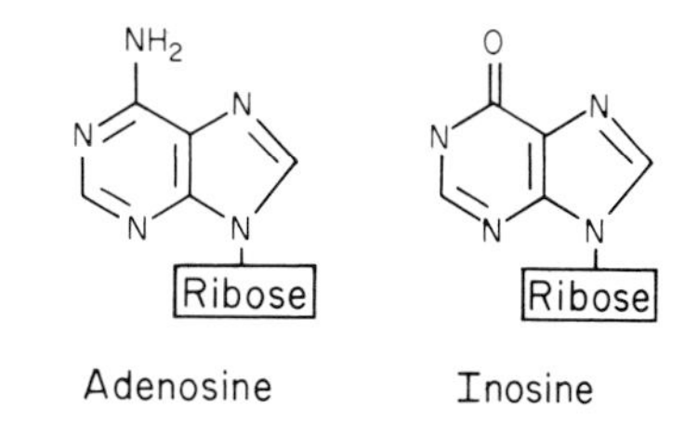

In his 'wobble hypothesis', Crick proposed that a restricted number of unusual base pairings was permitted.

First anticodon base		Third codon base
C	G	
A	U	
U	A or G	
G	C or U	
I	C, U or A	

Some examples of codon–anticodon interactions are presented below. The mRNA and tRNA must interact in an antiparallel manner and unusual base pairings are indicated by the boxes.

(1) $tRNA^{Ala}$, anticodon IGC

C G [I]	C G [I]	C G [I]	3′ Anticodon 5′
G C U	G C C	G C A	5′ Codons 3′

(2) $tRNA^{Phe}$, anticodon GAA

A A G	A A G	3′ Anticodon 5′
U U C	U U [U]	5′ Codons 3′

Three final points should be made about codon–anticodon interactions. First, the first two bases of a codon pair precisely and conventionally, A:U and G:C. From this it follows that codons different in either of their first two bases must be recognized by *different* tRNAs; for example, UUA and CUA are codons for leucine, but they are recognized by different species of $tRNA^{Leu}$. Second, a particular species of tRNA recognizes one, two or three codons, depending whether the first base of its anticodon is C or A (one codon), U or G (two codons) or I (three codons). Hence, the imprecise pairing or wobble at the third codon base partially explains the degenerate nature of the genetic code. The presence of inosine maximizes the number of codons that may be recognized by a particular species of tRNA. Third, amino acids complexed with their cognate tRNA play no part in controlling the accuracy of translation and in codon–anticodon interactions in particular. It is possible to modify specifically the amino acid moiety of aminoacyl–tRNA complexes, leaving the recognition function of the tRNA unimpaired. In these circumstances, the surrogate amino acid is incorporated into polypeptide linkage, because the mRNA–tRNA interaction remains completely accurate; for example:

Cys-$tRNA^{Cys}$: Cysteine incorporated

Cys-$tRNA^{Cys} \xrightarrow[\text{reduction}]{\text{Gentle}}$ Ala-$tRNA^{Cys}$: alanine incorporated.

Termination

When one of the stop codons (UAA, UGA or UAG) is at the ribosome A site, the binding of further aminoacyl-tRNAs is prevented. Rather, the stop codons are recognized by termination factors and translation is halted. The ribosome monomer dissociates from the mRNA, releasing both the synthesized polypeptide chain and its two ribosome subunits (Fig. 1.14). These then reassociate randomly with another mRNA strand, if available. The components of ribosomes, including the ribosome subunits, are not of major importance in mRNA selection.

1.2.7 Post-translational modification

Nascent, newly synthesized proteins are then subjected to further processing reactions, producing the final, functional form of the proteins. Post-translational modifications can range from relatively simple processes, such as removal of the initiator residue of methionine or the formation of disulphide bonds, to complex processes, such as the modification of amino acids or the cross-linking of individual polypeptide chains.

1.2.8 The interrelationship between transcription and translation, and their control

Since mRNA is translated in the $5' \rightarrow 3'$ direction, the *same direction* in which it is synthesized, this enables transcription and translation to be intimately coupled in prokaryotes (Fig. 1.15). Incisive studies by Yanofsky emphasized the additional importance of coupling in the transfer of genetic information in bacteria. The transfer of genetic information in *E. coli* was found to be precisely

(a)

(b) Regulatory gene Operator gene Structural genes

DNA 5′ 3′ r p 1 o a 2 b c 3 3′ 5′

mRNA 5′

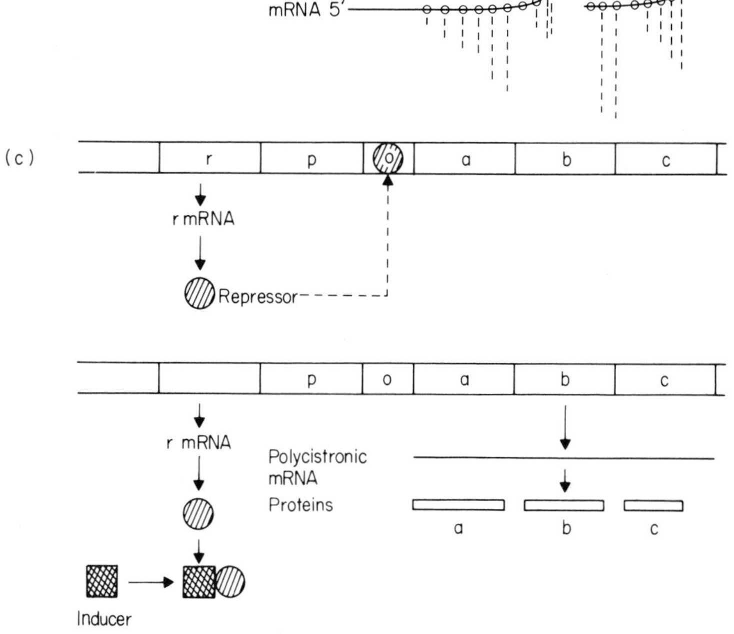

Fig. 1.15. The essential features of information transfer in bacteria and its regulation. (a) Coupled transcripton and translation. Transcription of just one DNA strand is shown, and even before the transcription of mRNA is completed, the nascent mRNA associates with ribosomes and begins to be translated. (Reproduced, with permission, from Miller *et al.* 1970.) (b) Features of a bacterial operon, consisting of regulatory, operator and structural genes, with p (promoter) and o (operator) sites within the operator gene; RNA polymerase molecules (hatched boxes), numbered 1, 2 and 3; CAP protein (solid box); ribosomes (circles) and proteins undergoing synthesis (dashed lines). Of the RNA polymerase molecules, 1 is bound to the promoter but has not begun transcription, 2 has transcribed much nascent RNA and coupled translation is in progress and 3 has transcribed nearly all of the operon but the mRNA has been progressively degraded from its 5′-end by ribosome-associated ribonucleases. (c) An operon in the repressed and induced states. The single strand of polycistronic mRNA has separate initiation and termination points for each of the three proteins it encodes. Thus translation alternatively stops and starts on one long mRNA sequence, as in (b).

colinear, the fundamental basis of the *colinearity concept.* In prokaryotes, there is a perfect match of base sequence in DNA (gene), of base sequence in mRNA (transcriptional product) and of amino acid sequence in the encoded protein (translational product):

```
DNA (gene)  5′ A|T|G   A|A|A ———————— 3′
            3′ T|A|C   T|T|A ———————— 5′
               | |     | | ← absolutely precise
                              alignment
mRNA        5′ A|U|C   U|U|U - - - - - 3′
               | |     | |
protein  N-T   M|e|t   P|h|e -·-·-·-·- C-T
               | |     | |
```

It is essentially the logical outcome of one code being transferred to another, without a break. The complexity of gene structure in eukaryotes, notably in the presence of transcribed but not translated introns, precludes the direct application of the colinearity concept to higher organisms.

The classical studies of Jacob and Monod established the important concept of mRNA in informational transfer and led to the identification of the characteristic transcriptional unit of bacteria, the *operon* (Fig. 1.15). An operon is a tightly coordinated unit of transcription in both structural and functional terms. Each operon is composed of a *regulator* gene, an *operator* gene and a sequence of *structural* genes or *cistrons*. The mRNA is transcribed from the set of cistrons as a long continuous strand which encodes for several polypeptide chains; such a bacterial mRNA is termed *polycistronic*. The regulator gene encodes for a regulatory protein, the *repressor*. By contrast, the operator gene does not encode any protein but contains two distinct intragene sequences, the p- and o-sites. The p- (promoter) site is where RNA polymerase binds and initiates transcription; regulatory proteins, including the cyclic AMP-binding protein (CAP protein), also bind within the p-site, destabilizing the operator and enhancing transcription. The repressor binds to the o-site and impairs transcription of the polycistronic mRNA encoded in the set of cistrons. These revolutionary ideas of bacterial regulation stemmed from studies on the lactose (*lac*) operon of *E. coli,* which regulates the synthesis of three enzymes involved in carbohydrate metabolism, namely β-galactosidase, permease and transacetylase. Remarkably, all of the components of the *lac* operon have now been completely sequenced. Other widely investigated operons are *gal* and *ara,* containing the structural genes for the enzymes of galactose and arabinose metabolism, respectively.

These operons enable bacteria to respond quickly and efficiently to changes in the supply of nutrient sugars in their environment. Under conditions of culture in artificial media, bacteria utilize glucose as the prime source of both carbon and metabolic energy, but in its absence, they can survive on galactose or arabinose. This metabolic flexibility is brought about by selective changes in genetic transcription, favouring the synthesis *de novo* of the enzymes most suited to the metabolism of the predominant environmental sugar. These changes in gene expression are alternatively known as the *catabolite effect* or *catabolite repression.* Thus, certain enzymes are *inducible,* their synthesis being regulated by the presence or absence of *inducers.* Fluctuations in enzyme biosynthesis may be triggered in two ways, either by changes in the activities of repressors or by changes in the synthesis of cyclic AMP. The activity of repressor proteins can be modulated in one of three ways, by *inducers, co-repressors* or *attenuators. Inducers* are generally *intermediates* of metabolic pathways whereas *co-repressors* are *end-products* of metabolic pathways. Known *attenuators* include molecules necessary for the translation of the polycistronic mRNA of a given operon, including specific species of tRNA. Changes in the intracellular synthesis of cyclic AMP will regulate the attachment of CAP protein to the p-site and thus control the initiation of transcription.

If a bacterium is grown on a mixture of glucose and a less favourable carbon and energy source, such as galactose, it will show diauxic growth (Fig. 1.16). When the most favourable sugar, glucose, is present, the bacterium does not need high levels of the enzymes for lactose and arabinose metabolism. Consequently, intracellular concentrations of cyclic AMP are low, the repressors of the *lac* and *gal* operons are fully active and transcription of these operons is extremely limited or *repressed.* Galactose is a selective inducer of the *gal* operon. Accordingly, growth on galactose leads to the specific combination of the inducer, galactose, with the *gal* repressor, relieving the constraint on transcription of the *gal* operon and inducing more enzymes for galactose metabolism. In addition galactose somehow stimulates an increase in the intracellular synthesis of cyclic AMP and, aided by CAP protein, this leads to further destabilization and activation of the p-site of the *gal* operon. By such mechanisms, diauxic patterns of bacterial growth are plausibly explained.

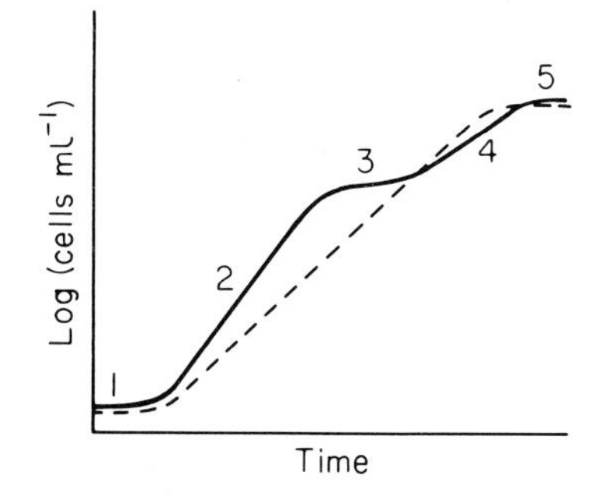

Fig. 1.16. Diauxic growth in bacterial cultures. This is a plot of cell density (ml^{-1}) against time of bacterial growth on a medium containing limited quantities of glucose, the favoured energy source, and galactose, a less favourable substrate. Phase 1, lag phase; 2, growth at the expense of glucose with no expression of galactose genes (*gal* operon); 3, lag phase, with the glucose consumed but no induction yet of the *gal* operon; 4, induction of the *gal* operon but growth is slower than on glucose; 5, stationary phase, since both sugars are now consumed. The broken line shows the pattern of growth in the absence of catabolite repression. This can be achieved either by adding exogenous cyclic AMP to the culture or by using a strain with a mutation (for example in the structural gene for CAP protein) that overrides catabolite repression.

Co-repressors and *attenuators* are needed to enable certain repressors to work at maximal efficiency in inhibiting the transcription of their particular operon. Excellent examples include the *trp*, *his* and *phe* operons, containing the structural genes for the biosynthesis of these amino acids in bacteria. The elegant studies of Yanofsky have provided a penetrating insight into the regulation of the *trp* operon. The polycistronic mRNA of the *trp* operon encodes five enzymes for tryptophan biosynthesis (Fig. 1.17). This *trp* mRNA has a short half-life of only some three minutes, thus enabling bacteria to respond rapidly to fluctuations in their demand for tryptophan. The *trp* operon is fundamentally different from the *lac* operon in two ways. First, the *trp R* gene, encoding the specific *trp* repressor is not part of the operon but far removed elsewhere in the bacterial genome. Second, the *trp* operator (containing p- and o-sites) is separated from the five structural genes by a *leader* (L) gene and this contains a novel *attenuator* (a-site) for precise control of transcription of the *trp* operon. The operon is regulated by independent mechanisms operating at the o- and a-sites, but both reflecting changes in the availability of intracellular tryptophan. The *trp* repressor alone cannot bind to the o-site but essentially needs the presence of tryptophan as well. Tryptophan is thus the *co-repressor* of the *trp* operon. With high concentrations of tryptophan in the cell, the *trp* repressor is active and transcription of the operon is prevented. At low concentrations of tryptophan, the repressor is inactive and transcription of the operon is initiated.

The attenuator site was revealed by studies on mutants of the L gene region which could synthesize a short leader peptide but not the enzymes encoded in the structural *trp* genes. When the mRNA corresponding to the leader peptide was sequenced, it was surprisingly found to contain two consecutive codons for tryptophan, UGCUGC (Fig. 1.17). These tandem codons are recognized with extreme accuracy by the aminoacyl derivative Trp-tRNA^{Trp}; in fact this specific species of tRNA is the *attenuator* of the *trp* operon. A critical feature of transcriptional control by the attenuator site is the close coupling of transcription and translation in bacteria, with ribosomes translating the leader mRNA following closely behind the RNA polymerase transcribing the L gene and the a-site. Indeed, the secondary structure of the mRNA dictates whether RNA polymerase can transcribe the a-site or not. At low concentrations of tryptophan, there is little Trp-tRNA^{Trp} available and consequently ribosomes are halted at the tandem tryptophan codons. Translation is reduced but the 'stalled ribosomes' change the secondary structure

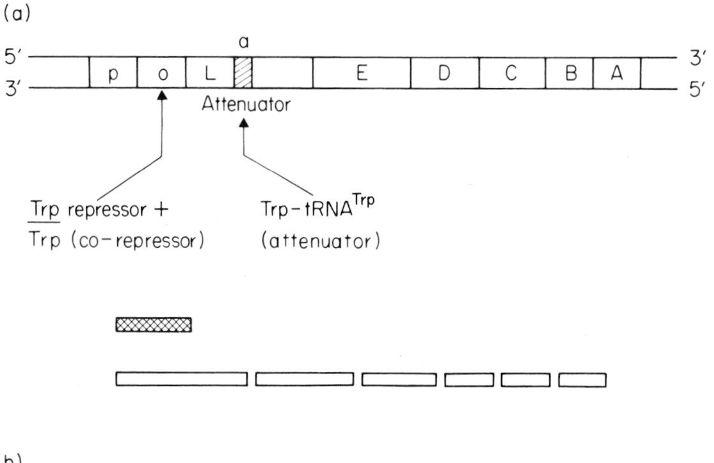

(b)

5′ AUG AAA GCA AUU UUC GUA CUG AAA GGU UGG UGG CGC ACU UCC UGA 3′

Met Lys Ala Ile Phe Val Leu Lys Gly Trp Trp Arg Thr Ser Stop

Fig. 1.17. Features of the *trp* operon in *E. coli.* (a) the arrangement of the promoter (p), operator (o) and attenuator (a) control sites, together with the structural genes for the leader (L) sequence and five enzymes necessary for the synthesis of tryptophan. The regulator gene, *trp* R, encoding the *trp* repressor is not part of the operon but found elsewhere in the bacterial genome. In the presence of the *trp* repressor plus tryptophan (co-repressor) at the o-site, only a short leader peptide can be produced. However, the presence of Trp-tRNA^{Trp} (attenuator) at the a-site leads to the production of a normal leader peptide and the five enzymes encoded in the *trp* operon. (b) The base sequence of part of the mRNA of the leader sequence, containing the tandem codons for tryptophan (underlined) and the corresponding amino acid sequence of the *trp* leader peptide.

of the mRNA in such a way that the coupled process of transcription is increased. RNA polymerase is now able to override the inhibition of transcription normally maintained at the a-site and the *trp* genes are expressed to meet the demand for enzyme synthesis *de novo.* At high concentrations of Trp-tRNA^{Trp}, translation proceeds efficiently, but transcription is halted. By these subtle mechanisms, intracellular concentrations of tryptophan regulate the coupled transcription and translation of the *trp* operon (for additional details, see section 6.5).

Certain operons are not inducible and are always expressed. Such operons are said to be *constitutive.* Certain types of mutation can result in otherwise inducible operons becoming constitutive. The regulatory genes encoding repressor proteins are constitutive, being permanently transcribed, if only at a very low rate. RNA polymerase cannot initiate transcription of regulatory genes very effectively.

Despite the sophistication of these investigations on bacterial transcription, they have not provided detailed insights into the control of gene expression in higher organisms. There are several reasons for this. First, the presence of the nucleus prevents the direct coupling of transcription and translation in eukaryotes. Second, while eukaryotic genes contain promoter sites for RNA polymerase these are not regulated by induction and repressor mechanisms. Indeed, higher organisms contain no strict equivalents to either repressors or CAP proteins. Third, the transcriptional units in eukaryotes are *monocistronic* and thus encode only one polypeptide chain. In addition, structural genes are not in clusters like a bacterial operon. Differences such as these prompted investigators to seek alternative and novel explanations for the regulation of gene expression in eukaryotes.

1.3 VIRUSES AND THEIR GENES

Viruses are small infectious particles or *virions* surrounded by a protective coat or *capsid.* In certain more complex virions, the capsid of protein is surrounded by an additional *envelope,* containing lipid and glycoprotein. Viruses can only be replicated within a *host cell* because they cannot synthesize proteins or generate energy to meet their metabolic needs. The replication of viruses essentially requires the help of enzyme systems in their host cell, and constituents of the host, including histones and glycoproteins, are sequestered into virions during their replication and assembly. Viruses are effective, self-reproducing parasites which differ from obligately parasitic bacteria and fungi in two striking ways. First, viruses contain *only one* type of nucleic acid, either DNA or RNA. Consequently, the store of genetic information in certain viruses resides in the base sequence of their RNA. The complexity of viral genomes ranges considerably, from R17 with only 3 genes, to complex pox viruses with approximately 240 genes. Second, virions can be assembled within host cells starting from viral nucleic acid *alone.* The viral genome is expressed to produce viral proteins and then the genome is replicated. The viral proteins of the capsid and the replicated genome then *spontaneously assemble* into the structurally complete and infective virion.

This remarkable *self-assembly* process in viruses was first demonstrated by Fraenkel-Conrat and Williams with tobacco mosaic virus (TMV). This virion is a rod, 300 nm long and 18 nm in diameter. The genome is a single-stranded RNA, containing only 6400 ribonucleotides. In sharp contrast, the protein contains about 340 000 amino acid residues. This small genome is insufficient to code for one large protein or indeed many different, small proteins. TMV, like all viruses, overcome their paucity of genetic information by having capsids composed of large numbers of a few protein subunits. In TMV, the capsid is made up of 2130 *identical* protein subunits, each containing 158 amino acid residues. These subunits are arranged round the RNA gene in a geometrically perfect, helical array (Fig. 1.18). Indeed, as Klug first postulated, it is the beautifully ordered, symmetrical assembly of the capsid subunits together which provides the molecular basis for self-assembly of the virion. The subunits of the capsid protein assemble together to form a critical, *two-layered*

Fig. 1.18. The self-assembly process of tobacco mosaic virus. (a) A model of part of the TMV virion, with a helical array of capsid subunits around the single-stranded RNA genome. (b) The assembly of capsid subunits firstly into small, geometric polymers and later, into a two-layered disc containing 34 protein subunits. In the presence of TMV RNA, the two-layered disc is converted by a slipping motion, as arrowed, into a helical, lockwasher disc. (c) The further assembly of the TMV virion begins by the association of an initiation region of TMV RNA with a 34-subunit disc, forming a lockwasher disc. More discs are added to the looped end of the RNA, forming a series of lockwasher discs, which together constitute the virion itself. One end of the RNA is reeled through the partially assembled virion to meet a continuous flow of oncoming discs. The path taken by the RNA during virion assembly is shown on the extreme right. (d) The base sequence of the initiation region of TMV RNA, containing hairpin, stem and loop structures.

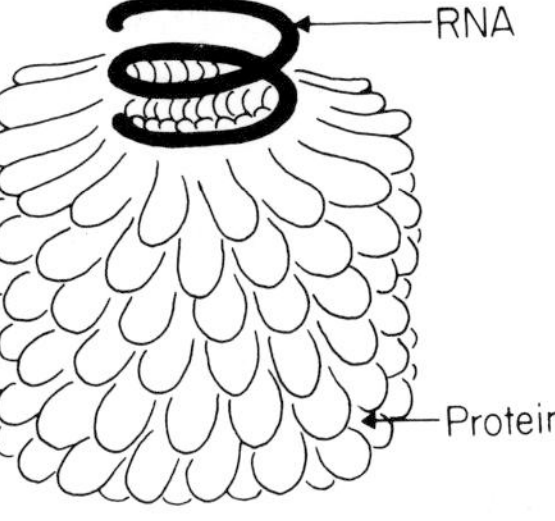

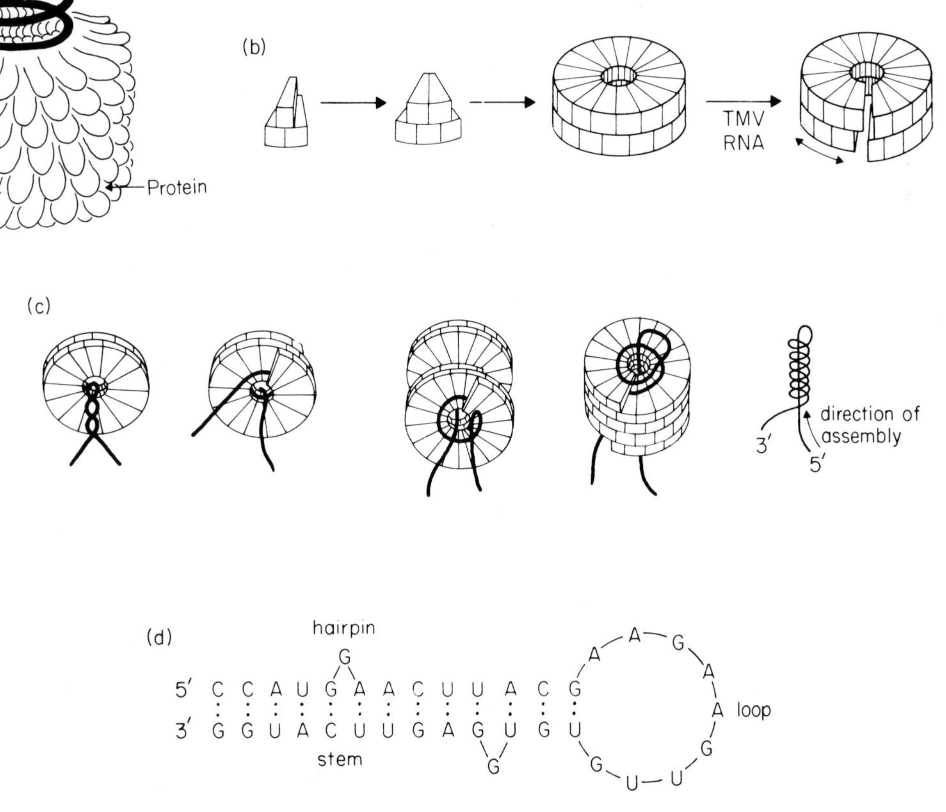

disc, containing 34 subunits. Each layer of the disc is a ring of 17 subunits, about the same number as the number of subunits, actually 16⅓, per turn of the TMV helix. The two discs then slide over each other, forming a *two-turned helical structure,* known as a *lockwasher.* RNA associates with both of these early two-layered structures and more discs are added sequentially, in the lockwasher form. The discs interact rapidly and specifically with TMV RNA but not other molecules of RNA. TMV RNA contains a unique *initiation region* that binds specifically to the first disc and sets the self-assembly process in motion. The initiation region has an unusual hairpin structure with a base-paired stem and loop; this loop is probably the initiation point itself and, interestingly, contains guanine as every third base (Fig. 1.18). The principles of self-assembly established with TMV are almost certainly applicable to other virions, including those with a circular rather than a rod-like structure.

There are many ways of classifying viruses. They may be classified according to the hosts they parasitize, e.g. animal or plant viruses, etc. Bacterial viruses are referred to as bacteriophages, or simply as *phages,* whereas fungal viruses are termed *mycophages.* The nature of the host influences the mechanism of entry of virions. Phages are unique in that after attachment to the bacterial cell wall, *only* the phage genome is introduced into the host. During their infection of eukaryotes, all other viruses enter their hosts by pinocytosis and thus enter *intact.* The capsid and envelope are then digested away, releasing the viral genome within the host cell. Viruses may also be classified on the basis of morphological criteria. However, the most widely used classification is that proposed by Baltimore and based on the nature of the viral genome. On this genetic criterion, six classes of

Table 1.3. The classification of viruses proposed by Baltimore. The complementarity of viral genomes and their replicative intermediates is established from the convention that mRNA is (+)RNA.

Type	Nature and complementarity of the viral genome	Nature and complementarity of the replicative intermediate	Examples
I	Double-stranded (±) DNA	Double-stranded (±) DNA	Most bacteriophages Many animal viruses (e.g. cowpox virus)
II	Single-stranded (+) DNA	Double-stranded (±) DNA	Some bacteriophages (e.g. ϕX174 and M13)
III	Double-stranded (±) RNA	Double-stranded (±) RNA	Animal reoviruses All mycophages
IV	Single-stranded (+) RNA	Single-stranded (−) RNA	Some bacteriophages (e.g. phage Qβ) Animal picornaviruses (e.g. polio virus) Most plant viruses (e.g. TMV)
V	Single-stranded (−) RNA	Single-stranded (+) RNA	Some animal viruses (e.g. influenza and rabies viruses)
VI	Single-stranded (+) RNA	Double-stranded (±) DNA	Animal retroviruses (e.g. Rous sarcoma virus)

viruses may be identified (Table 1.3). Certain type I animal viruses (e.g. herpes viruses) have attracted considerable attention because they may cause certain types of human cancer; some type VI animal viruses (e.g. Rous sarcoma virus) cause cancers in birds and mammals. By convention, mRNA is defined as (+) RNA and its complement in base sequence is (−) RNA; furthermore, during normal transcription, the transcribing strand is (−) DNA and the non-transcribing strand is (+) DNA. Using these conventional criteria, the complementarity of viral genomes can be unequivocally described. During viral infection and multiplication, viruses must synthesize essential templates for the replication of their genomes. The nature and complementarity of the replicative intermediates for each of the six types of virus are also included in Table 1.3.

The single- and double-stranded DNA of viral genomes is readily replicated by the DNA polymerases of the host cell. The strategy underlying the replication of RNA-containing viruses is not only more complex (Fig. 1.19), but raises some interesting and unexpected aspects of the transfer of genetic information. Studies on the replication of many RNA-containing viruses, types III to V, led to the identification of an RNA-dependent RNA polymerase, *RNA replicase*. Replication of type VI retroviruses provided an even more complex sequence of reactions, involving the novel RNA-dependent

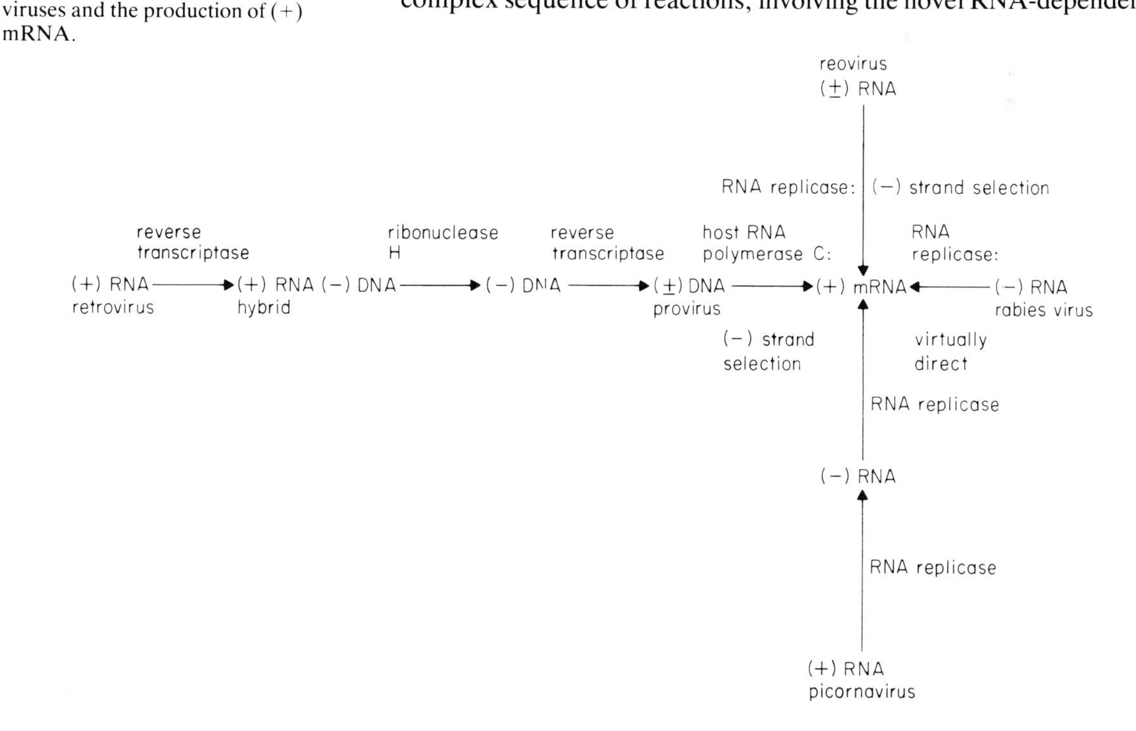

Fig. 1.19. The sequence of events necessary for the replication of RNA viruses and the production of (+) mRNA.

DNA polymerase, *reverse transcriptase*. Some of the unusual, double-stranded (±) DNA intermediate or *provirus* is permanently incorporated into the genome of the host cells. This genetic change results in malignant transformation, e.g. the experimental production of sarcomas in chickens first reported by Rous. RNA replicase and reverse transcriptase are not found in appreciable amounts in host cells and both are, in fact, encoded in the viral genomes. Expression of these viral genes within the host cell is a prerequisite for replication of RNA viruses. These unusual enzymes provide two points of attack in virus-specific chemotherapy.

Most viruses are *virulent*, since infection consequently leads to propagation of the virus within the host cell. The new generation of virus particles gradually accumulates until the strained host cell bursts, releasing the new virions. Some viruses are said to be *temperate*, however, the best known example being phage λ. Once within its host bacterium, two courses of development are available to phage λ; it can destroy its host (*lytic* pathway) or it can integrate its genome of DNA into the genetic material of the host (*lysogenic* pathway). In the lytic pathway, the viral genes are fully expressed and the rapid production of about 50 new virions leads to lysis of the bacterium. In the lysogenic pathway, the viral genes are not expressed significantly and the viral genome is integrated as a *prophage* into the host chromosome; the resultant genetic union of host and prophage produces a lysogenic bacterium of *lysogen*. The presence of the prophages renders the host cell immune to further infection by the same phage. In lysogenic bacteria, the prophage is replicated as part of the host chromosome, but the lytic functions of temperate phages are only dormant in the lysogen. Chemicals and physical treatments which interfere with DNA replication in the host induce the prophage to undergo lytic development. This 'induction' process changes the pattern of transcription so that the phage DNA is replicated out of phase with the host's chromosomal DNA; ultimately, many new phage particles are formed and the host cell is then destroyed by lysis.

1.4 GENETIC TECHNIQUES AND NOMENCLATURE

1.4.1 Introduction

The disciplines of genetics, molecular biology and nucleic acid biochemistry are inextricably entwined and this blend of research endeavour has resulted in the emergence of a new branch of biological science, often referred to as 'molecular genetics'. In fact, it is only by the free interchange between such classical branches of science that new insights have been gained into the organization and expression of genetic information. This section defines terms and explains phenomena referred to later in various chapters of this book.

1.4.2 Basic genetics of eukaryotes

Individual sequences of DNA encoding information for a specific sequence of RNA or protein are *genes,* and copies of genes are arranged together in extremely long tracts of DNA, visible under the microscope, called *chromosomes.* In higher eukaryotes, there are *two sets* of equivalent genes for controlling the inheritance of a given characteristic, e.g. eye colour, synthesis of an enzyme, height, and so on. These two sets of equivalent genes are carried on two sets of chromosomes, joined together as a *homologous chromosome* pair; each chromosome strand of DNA is called a *chromatid* and the two chromatids are held together by a structure called the *centromere.* With the exception of germ cells, involved in sexual reproduction, all eukaryotic species contain a constant and distinctive number of paired (2n) chromosomes, or *diploid* chromosome number. The two copies of a given gene in each chromosome pair may not be identical, meaning that, in molecular terms, the base sequences of their DNA are not identical. The alternative copies of a given gene are called *alleles.* During normal cell division or *mitosis*, the genes of the original parent cell are distributed exactly, both in terms of amount and base sequence, between the two daughter cells to preserve the constancy of their store of genetic information. Individual members of a species with two similar alleles of a given gene are termed *homozygous*; those with different alleles are *heterozygous.*

In his historic experiments on the cross-breeding of the common pea, *Pisum sativum,* Mendel first noted that pairs of factors (later termed genes) segregate independently after fertilization; examples of these allelic pairs included the shape of the seeds (round or wrinkled) and height of the plant (tall or short). During artificial fertilization on cross-breeding, Mendel noted that one allele was always expressed over the other; in modern terminology, one allele is *dominant* (expressed) whereas the other allele is *recessive* (not expressed). Indeed, a recessive gene is only expressed if the individual is homozygous with respect to that gene, containing two similar recessive alleles. If we assume, for example, that eye colour in humans is attributable to the genetic interaction between two alleles, *brown* (dominant) and *blue* (recessive), we can define the genetic constitution or *genotype* of an individual with respect to eye colour by citing the paired alleles, e.g. brown-eye/blue-eye. As a consequence of dominance, the eye colour of progeny produced by the mating of parents of known genotype can be predicted (Table 1.4). The observable outcome of the expression of allelic genotypes, here eye colour, is called phenotypic expression, or simply, the *phenotype*. These principles of Mendelian inheritance, genetic segregation and dominance are applicable to lower eukaryotes, such as the pea, without reservation. However, phenotypic expression in more advanced eukaryotes is very complex; even in the example cited, eye colour in humans, many pairs of gene alleles are involved, rather than just one. Mendel's contribution to genetics nonetheless remains enormous.

Table 1.4. The phenotypic expression of eye colour in humans. Two alleles, brown and blue, are assumed to control the inheritance of eye colour; brown is the dominant allele.

Genotype	Allelic relationship	Phenotype (eye colour)
Brown-eye/brown-eye	Homozygous	Brown
Brown-eye/blue-eye	Heterozygous	Brown
Blue-eye/brown-eye	Heterozygous	Brown
Blue-eye/blue-eye	Homozygous	Blue

The continuance of most eukaryotes is by way of sexual reproduction. Certain organs produce two types of sex cell or *gamete*, these being termed female-type *ova* and male-type *spermatozoa* in animals. Reproduction in plants can be more complex, involving both sexual and asexual processes. In animals, a distinctive pair of *sex chromosomes* determines the sexes, male and female (for further details, see section 7.2). Each gamete contains only *one* set of unpaired (1n) chromosomes and is, therefore, said to contain *haploid* chromosomes. During sexual reproduction, there is fusion of male- and female-type gametes, forming a *diploid zygote,* from which the diploid adult is formed by cell division (or mitosis). Gametes are formed by *reductive divisions* or *meiosis* in specialized, reproductive cells. In meiosis the diploid chromosomes divide once, but the reproductive cells twice, so that only one half of each homologous chromosome pair or *chromatid* appears in each gamete (Fig. 1.20(a)). Equally important, however, meiosis permits the segregation of alleles by the *cross-over* of segments of chromatids (Fig. 1.20(b)). The

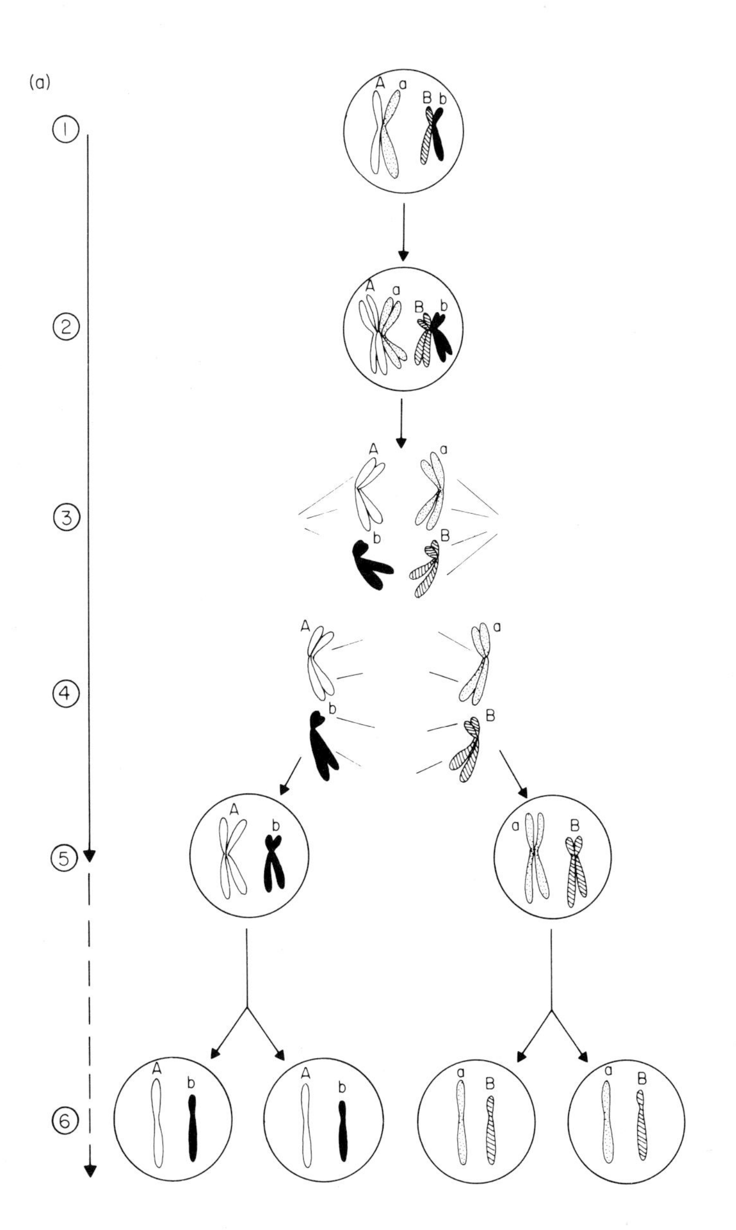

Fig. 1.20. The process of meiosis in eukaryotes forming the sexual gametes. (a) Meiosis results in the formation of four haploid gametes of differing genetic constitution (or genotype) by reductive division of one diploid reproductive cell. For clarity, the meiotic division of only two homologous chromosome pairs is shown, each pair being distinguished by the composite chromatid, Aa and Bb. The first round of meiotic division, solid arrow (———→), is accompanied by DNA replication and proceeds via four stages. (1) Interphase, the diploid reproductive cell at rest, containing homologous pairs. (2) Prophase, where each chromosome pair is replicated but the chromatids remain together. (3) Metaphase, in which the nucleus disintegrates and the replicated chromosomes separate randomly. (4) Anaphase, involving complete separation of the replicated chromosomes and cell division, bringing the first round of meiotic division to an end. The second round of meiotic divisions then starts, dashed arrow (----→), and is not accompanied by DNA replication. (5) This stage produces two daughter nuclei, each with one diploid chromosome pair (two chromatids). (6) The last stage produces two more daughter nuclei which are haploid since they receive only one chromatid each.

cross-over of chromatids can be seen under the microscope as distinctive chromosome configurations called *chiasmata.* This process introduces considerable *genetic diversity* into sexual reproduction in eukaryotes. This diversity is accentuated by additional processes. In mammals, for example, only one haploid ova

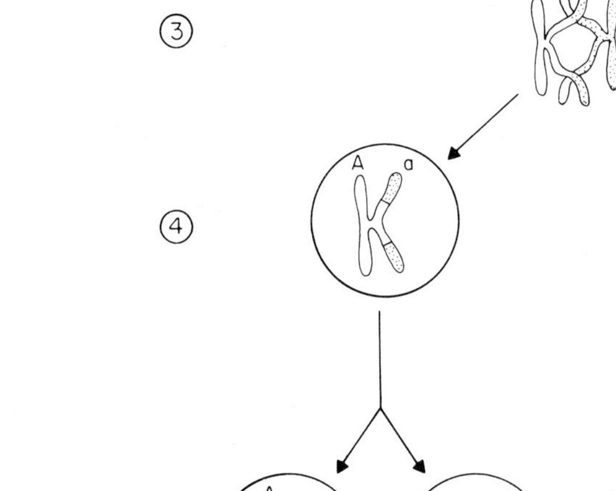

Fig. 1.20. (b) An illustration of how meiosis can produce gametes of markedly different genotype by crossing-over of chromatids during the first meiotic division. (1) Diploid reproductive cell containing, for simplicity, only the chromosome pair, Aa. (2) Replication of the chromatids. (3) Cross-over of chromatid DNA by means of contact points or chiasmata. (4) Cell division produces two daughter nuclei with diploid chromosomes yet partial exchange between chromatids. (5) Further division produces four haploid gametes of distinctive genotype.

of the four produced by meiosis persists, the other three disintegrate and become polar bodies. In fungi, one diploid reproductive cell produces eight haploid gametes or spores.

Sexual reproduction has two advantages. First, by mixing of alleles it enables a defective gene to be cancelled out by a correct gene. Second, it creates considerable genetic variety in the offspring and thus encourages the chance of improving a species by the selective pressures of evolution, a concept embodied in the work of Darwin. The sexual cycle can also be exploited for two types of experimental analysis, *recombination* and *complementation*.

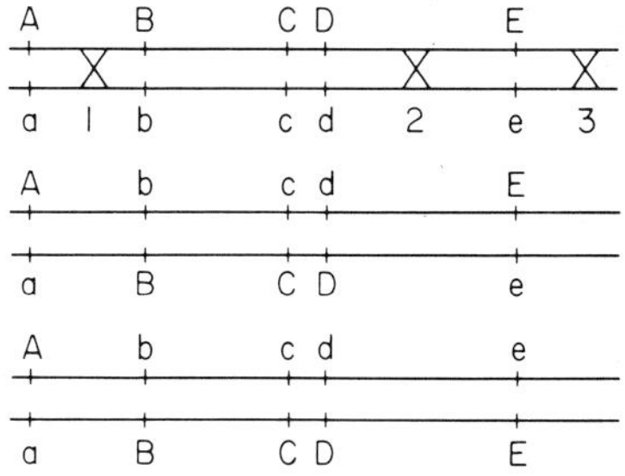

A B C D E
a 1 b c d 2 e 3

A b c d E
a B C D e

A b c d e
a B C D E

Fig. 1.21. The principal of genetic recombinants in eukaryotes. The large and small letters represent alternative alleles. The top part of the figure illustrates the heterozygote ABCDE/abcde, with three hypothetical cross-over points. Cross-overs at points 1 and 2 produce the recombinant AbcdE/aBCDe (middle pattern) and at points 1 and 3 produce the recombinant Abcde/aBCDE (lower pattern). Recombinants with the allelic arrangements cD or Cd would be rare because of the low probability of cross-over occurring between such closely linked genes as C and D.

During the segregation of gene alleles in meiosis, genetic recombination occurs and sections of chromosomes are interchanged with the result that the alleles appear in new patterns. From the frequency with which such recombinants arise, it is possible to determine the actual order of the genes (Fig. 1.21).

Complementation testing is possible in lower eukaryotes, such as yeast, a unicellular fungus. Yeasts can reproduce by the fusion of opposing types of gametes, but under the appropriate conditions of culture it is possible to grow haploid forms of yeast indefinitely. However, if the stable, haploid cells of the opposing mating type are mixed they can fuse and give rise to a stable diploid generation. Haploid yeasts permit detailed genetic analysis of experimental mutants to be undertaken. Two strains of homozygous yeast could have an identical phenotypic defect, for example, in the absence of a metabolic pathway, yet this could be attributable to different genotypic defects, because the defective genes could be encoding different enzymes in the same pathway. If haploid cells of opposing mating types from the two defective strains are mixed, it is easy to test for the restoration of normal phenotype, in this case by measuring the different enzymes in the pathway. Provided that the possibility of genetic cross-over (recombination) can be eliminated, the restoration of the normal phenotype can only be explained by a non-defective allele in one chromosome *complementing* the defective allele in the other chromosome and vice versa. Complementation testing of various yeast mutants has enabled the genetic control of metabolic pathways to be determined precisely. In fact complementation testing can reveal very subtle inter-relationships between genes, because complementation may occur between different subunits of a multimeric protein.

1.4.3 Genetic analysis of bacteria and their phages

Bacterial genetics

Bacteria are haploid organisms, and so the problem of genetic analysis and complementation testing is how to achieve a diploid state. Full diploidy is rarely achieved but all the processes of genetic transfer in bacteria allow the construction of either permanent or transient strains which are diploid over a restricted region of their genome. Such strains are called partial diploids or *merodiploids*. There are three means of genetic transfer; *transformation, conjugation* and *transduction*.

Transformation consists of incubating a culture of recipient bacteria with purified DNA from a donor culture. If the recipient cells are capable of taking up the DNA, and are, therefore, said to be *competent,* genetic transformation of the recipient cells occurs. These transformants arise by the purified DNA lining up with the corresponding homologous region of the recipient haploid chromosome, and producing transient, partial diploidy. Genetic exchange can occur and if the donor and recipient have different alleles for certain genes, then genetic recombinants can be recognized. This approach is not suitable for complementation analysis because the transformants are transient. Nonetheless, genes can be ordered on the basis of co-transformation frequencies. If the recipient is defective in three genes *a, b* and *c* and the phenotypes of these mutations can be easily determined, for example, as growth requirements for certain nutrients, analysis of the transformants gives a clear indication of the linkage or arrangement of the three genes. If 100% a^+ transformants are also 30% b^+ but only 0.1% c^+, it is likely that genes *a* and *b* are closely linked, whereas *a* and *c* are not. Only if *a* and *b* are closely linked can there be a chance that the transformation of one gene is accompanied by transformation of the other.

Conjugation only occurs in certain bacteria, notably *Escherichia coli* and *Streptomyces coelicolor.* In this process, donor and recipient bacteria contact each other closely, or conjugate, and DNA passes from donor to recipient. Often, the DNA passing into the recipient is not the donor chromosome's, but, rather, small, independent molecules of extrachromosomal DNA called *plasmids.* The DNA of certain plasmids contains *F (fertility) genes* for the formation of sex pili and other components for conjugation. If a plasmid with F factor (containing the F genes) is incorporated into the bacterial chromosome itself, such bacteria, harbouring F factor in their haploid chromosome, are called *Hfr* strains because they have a high frequency (*Hfr*) of recombination and are particularly good donors in bacterial conjugation. The crucial advantage of *Hfr* strains is that the *entire chromosomal DNA,* with the integrated F factor, is transferred via the sex pili into the recipient bacterium. An unstable merodiploid cell is formed and recombinant analysis performed. This is possible because the transfer of donor genes is a lengthy but ordered process in which some genes invariably enter the recipient cell early and others late. It is, therefore, possible to interrupt the conjugation process and to order genes by timing their entry into a suitably chosen recipient that has many defective alleles. As just described, bacterial conjugation is not suitable for complementation testing. However, with the advent of recombinant DNA technology, it is possible to incorporate purified genes, including those from bacteria, into the DNA of plasmids. The plasmid–gene complexes may be transferred from one bacterial cell to another by both conjugation and transformation. The establishment in a bacterial cell of a plasmid containing chromosomal genes results in a *stable* merodiploid and thus such systems may be used in complementation experiments. This is a new and exciting aspect of contemporary studies on bacterial genetics.

Transduction is the process by which a phage mediates the transfer of bacterial chromosomal genes from donor to recipient bacterium. There are two mechanisms of transduction. Generalized transduction can be effected by only a few types of phage, including phage P1. The assembly process of such phages in their bacterial host cell can sometimes be erroneous and the final phage particles contain short, random pieces of the bacterial chromosome rather than the normal phage genes. When this surrogate phage attaches to a suitable bacterial host cell, a fragment of the bacterial DNA, rather than phage DNA, is inserted. The bacterial host cell will, therefore, not propagate the phage but rather becomes an unstable merodiploid from which recombinant genes can, nevertheless, be recovered. Generalized transduction is not always successful in the sense that genetic recombination between the incoming and recipient bacterial chromosomes does not occur; this is abortive transduction. If the recipient bacterium is carefully chosen because it is unable to grow on a medium defective in some particular nutrient, abortive transductant colonies can be recognized. These colonies are small because only one cell, the original transductant, must supply all of the limiting nutrient required for growth of the defective recipient bacteria. Despite their small size, abortive transductants can be used for complementation testing. *Specialized* transduction may be effected by phages containing bacterial genes incorporated into their genome by a highly specific type of DNA recombination. The unique feature of specialized transducing phages is that their genome contains both viral and bacterial genes. This combined bacterial plus viral genome is termed a *chimaera* and many examples of their usefulness in studies on the regulation of gene expression in bacteria will be presented in Chapter 5. These chimaeras can only be formed between specialized transducing phages and a restricted number of bacterial genes; in a given chimaera, the complement of bacterial genes is absolutely constant. Such phage chimaeras can be used in complementation testing.

Phage genetics

Recombination between the genomes of phages can be readily detected. If a bacterial culture is infected with a large number of mixed phage particles, from two different strains, the average host cell will contain one or more phage chromosomes of each viral strain. This is the equivalent of diploidy. Provided that the infection can be conducted in such a way to distinguish complementation and recombination, both types of analysis can be separately performed with great ease.

In summary, although bacterial chromosomes are haploid, the processes of genetic transfer applicable to bacteria, namely transformation, conjugation and transduction, provide the means for the experimental analysis of genetic recombination and gene complementation—phenomena requiring merodiploid or diploid chromosome formation.

1.4.4 Mutations in prokaryotes and eukaryotes

Prokaryotic mutations

Changes in the base sequence of DNA, the genetic store of information, clearly have a profound and even lethal effect on all organized forms of life; changes in the base sequence of the normal or *wild-type* DNA of a given species are called *mutations.* Mutations occur naturally at a very low rate of probability, say 1 chance in 10^8, but nonetheless mutation provides the natural basis for evolution. The rate of mutation can be accelerated experimentally, by chemicals (or *mutagens*) or by physical means (such as exposure to ultraviolet light). Classical studies by Benzer first established the important principle that the unit of genetic inheritance, the gene, is not the same as the unit of *mutation.* Indeed, only one base need be changed in the base sequence of a gene to alter its biological and genetic properties. Equally important research by Beadle and Tatum established that one gene encodes for only one functional polypeptide chain, and from this it follows that changing just one base in a gene results in the synthesis of an aberrant, even non-functional polypeptide chain. All available evidence supports this view of mutation. The change of one base is called a *point mutation* and four alternative types of single-base change are known (Fig. 1.22). Of these, the insertion and deletion mutations ultimately promote changes in the reading of the genetic code and are called *frame-shift* mutations, because they

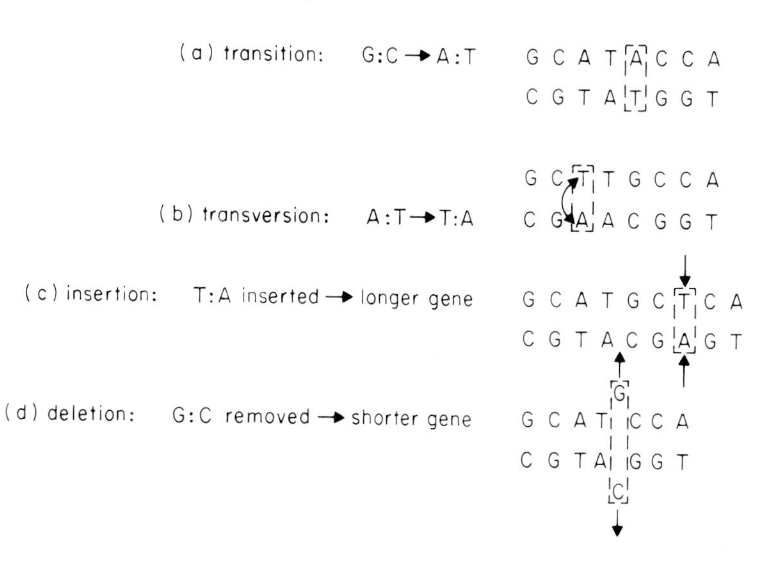

Fig. 1.22. The four types of single-base or point mutations. A short sequence of a hypothetical, normal wild-type gene is presented and below are listed the four types of mutation; in each case, the base changes are enclosed in boxes. The four changes are as follows: (a) transition, one complementary base pair replaces another; (b) transversion, the complementary base pairing in each DNA strand is changed over; (c) insertion, one more complementary base pair appears in the mutant gene; (d) deletion, one complementary base pair is lost from the mutant gene.

change the 'reading-frame' of translation. Other mutations can be more extensive and involve the deletion, addition or modification of several bases in a gene. Such mutations, with extensive changes in base sequence, are collectively termed *deletions,* the same nomenclature used for one type of point mutation. Among the characteristic features of a deletion is the fact that recombination is *impossible* within the deleted region.

The impact of studies on bacterial mutants has been remarkable. By such means, investigators have determined the basis of metabolic pathways, the nature of vital biosynthetic intermediates and the mechanisms for controlling gene expression. These three examples merely underline the value of mutants in the advance of genetics and molecular biology.

Bacterial mutants can be identified in a variety of ways. Mutation may be accompanied by a morphological change observable under the microscope or by a change in the staining properties of a bacterium. It is more usual, however, to identify mutations by changes in biochemical and genetic properties, meaning by changes in phenotypic expression. If the mutation generates a requirement for a vitamin or other nutrient, it is an *auxotrophic* mutation. Conversely, if the change in DNA sequence spares the bacterium from its usual dependence on a nutrient, this is a *prototrophic* mutation. Another readily detected type of mutation is one where the mutant is resistant to a drug normally lethal to the wild-type

Fig. 1.23. The effects of various types of mutation on the amino acid sequence of a polypeptide. The hypothetical gene illustrated here encodes for a polypeptide of seven amino acids only; very short peptides, even tripeptides, can be biologically active. The transcribed mRNA contains a stop signal (underlined) at its 3′-end. The effects of mutations in three base pairs (enclosed in dashed boxes) are indicated. The genetic code was given earlier in Table 1.2. Changed amino acids in the aberrant, mutant polypeptide are enclosed in solid boxes. (a) Transition mutant, base pair 1, A:T → G:C; second codon UUU(Phe)→UCU(Ser). (b) Transversion mutant, base pair 1, A:T→T:A; second codon UUU(Phe)→UAU(Tyr). (c) Transversion mutant, base pair 3, T:A→G:C, fifth codon UUG(Leu)→UAG(stop signal) and causes premature termination of translation as an aberrant tetrapeptide. (d) Deletion mutant, base pair 2, creates a frame-shift mutant, in which all residues beyond the point of mutation in the third codon are changed, e.g. CAA(Gln) → AAA(Lys) and GUA(Val) → UAA(stop signal); the product is a pentapeptide of anomalous sequence.

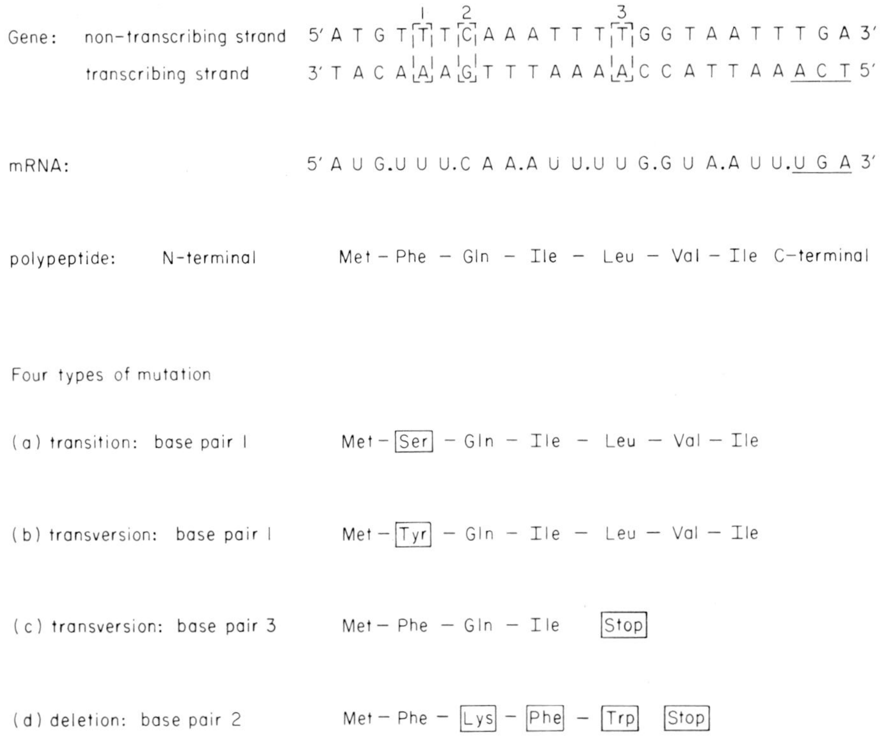

bacterium. Accordingly, many mutants can be most easily detected by changes in bacterial growth in chemically defined media.

Examples of the effects of various types of bacterial mutation on the normal flow of genetic information are presented in Fig. 1.23. These examples are restricted to the 'local' effects of mutation within a single bacterial gene. However, because of the prevalence of bacterial genes in operons, the effect of a mutation in one structural gene (or cistron) has serious implications for the correct expression of other cistrons in the operon, which may themselves be genetically unaltered. An example of such a *polar* mutation is presented in Fig. 1.24. In essence, a polar mutation is one in which the 'roll on' effect of a single, crucial mutation is seen throughout an operon. Polar mutations usually follow from the modification of the base sequence

Fig. 1.24. This figure is an extension of Fig. 1.23. Here the gene encoding the septapeptide is now termed gene A, but it is now one structural gene in an operon, followed by gene B, a structural gene encoding for a second, tetrapeptide. The polycistronic mRNA of the hypothetical operon contains two initiator codons (AUG, doubly underlined) and two stop signals (singly underlined). The effects of mutations in two base pairs are indicated, previously given in the simpler case in Fig. 1.22. (a) Non-polar mutant, with transition A:T → G:C at base pair 1; the second codon, gene A UUU(Phe) → UCU(Ser) is the only change. Thus, one amino acid in the septapeptide is altered but translation of the tetrapeptide is normal. (b) Polar mutant, with deletion at base pair 2 setting a profound change in the reading frame of the entire operon. From mutant gene A, an anomalous peptapeptide is produced. In mutant gene B, the frame-shift destroys the initiation of translation, because the first codon in the second initiator region of the polycistronic mRNA is altered, AUG(Met) → UAU(Tyr). No tetrapeptide is made, because the ribosomes cannot utilize Tyr $tRNA^{Tyr}$ to initiate protein synthesis.

Gene A
Gene B

1 2

Genes: 5' A T G T T T C A A A T T T T G G T A A T T T G A T A A A A A T A T A T G G C A C A G T T T T A G 3'

3' T A C A A A G T T T A A A A C C A T T A A A C T A T T T T T A T A T A C C G T G T C A A A A T C 5'

mRNA: 5' A U G.U U U.C A A.A U U.U U G.G U A.A U U.U G A.U A A.A A A.U A U.A U G.G C A.C A G.U U U.U A G 3'

polypeptides: Met – Phe – Gln – Ile – Leu – Val – Ile   Met – Ala – Gln – Phe

Two types of mutation

(a) transition base pair 1
(non-polar mutant)

Met – Ser – Gln – Ile – Leu – Val – Ile   Met – Ala – Gln – Phe

(b) deletion, base pair 2
(polar mutant)

Met – Phe – Lys – Phe – Trp – Stop   Tyr

No tetrapeptide
Initiator, AUG, is lost

of promoter and initiator regions of genes. As shown in Fig. 1.24, mutations usually result in the substitution of one amino acid in a polypeptide chain by another; these are *mis-sense mutations*. Clearly, the change of a codon into a stop signal prematurely terminates translation; these are *nonsense mutations*.

There are two extremes of mutations. Since the genetic code is degenerate, a change in base sequence may not alter the meaning of a codon, for example, CGU and *A*GU, or AGA and AG*G*, would still code for arginine. These mutations are *silent*, or are simply undetectable. In sharp contrast, other mutations are *lethal*. Bacteria contain only one type of RNA polymerase and the synthesis of a totally non-functional enzyme would be incompatible with life. Such a key enzyme may be synthesized with modified amino acid sequence after mutation in such a way that the mutant enzyme is partially active but with markedly changed physicochemical properties. Mutations of this type in key enzymes are often *conditionally lethal*; for example, bacteria can normally be grown at temperatures between 30 and 40 °C and wild-type RNA polymerase is fully active over this temperature range. However, a mutant RNA polymerase may be thermally denatured at temperatures above 36 °C and thus such mutants are *temperature-sensitive*. This mutation is conditionally lethal in that the bacteria can survive the effects of the genetic change at temperatures of 36 °C and below.

Eukaryotic mutations

Since eukaryotic cells cannot be grown so easily in defined media *in vitro*, mutations are not so easy to create or detect in higher organisms as in bacteria. Mutations can be more readily produced, however, in more primitive fungi. Inherited clinical syndromes in humans provide useful mutants for experimental investigation. Inherited disorders in human haemoglobin have featured prominently in the progress of molecular genetics over the last two decades. Normal adult haemoglobin (HbA) is a tetramer, $\alpha_2\beta_2$. That mutant proteins contain a modified amino acid sequence was first established by Ingram and Hunt from studies on sickle-cell haemoglobin (HbS). In this point mutation, the change is in the gene coding for the β-globin chain; only one amino acid at position 6 is changed, GAA(Glu) $\rightarrow$ GUA(Val). The deoxygenated form of the HbS is 'sticky' and spontaneously forms long, insoluble fibres which cannot conduct the normal function of gaseous exchange in capillaries. The abnormal gene has not been corrected during the course of evolution because individuals with the sickle-cell trait are resistant to malaria. The precipitates of deoxy HbS distort the erythrocyte into the characteristic sickle shape from which the syndrome takes its name. During the life cycle of the malarial mosquito, *Anopheles,* the sporozoite cannot gain access into the interior of sickled erythrocytes. This interesting form of resistance indicates why mutations are not invariably 'bred out'—they can have surprising advantages. Studies on the wide range of abnormal human haemoglobins have shown that the types of mutation detected in bacteria are also found in higher organisms (Table 1.5).

Table 1.5. The molecular basis of many inheritable disorders in the synthesis of human haemoglobin. Adult human haemoglobin, HbA, has the oligomeric structure, $\alpha_2\beta_2$. Each identical α chain has 141 amino acids, from N-terminal valine as 1; each identical β chain has 146 amino acids, from N-terminal valine as 1.

Hb	Type of mutation	Globin chain affected	Number of residues changed	Molecular defect	Effect on haemoglobin structure and function
	Transition	β6	1	GAA(Glu) → G*G*A(Gly)	Small microcythaemic erythrocytes
M	Transition	α or β	1–4	{CAU(His) / CAC} → {*U*AA(Tyr) / *U*AC}	Permanent ferric (oxidized) state as methaemoglobin
Riverdale-Bronx	Transition	α or β	1 or 2 but Gly → Arg e.g.	Many possibilities {GGU(Gly) / GGG} → {*C*GU(Arg) / *A*GG}	Distorted tertiary structure and poor haem attachment; extremely unstable Hb
Constant Spring	Transition	α141	1	UAA(Stop) → *C*AA(Gln)	Defective Hb with abnormally long α chain (171 amino acids)
Guntlid	Deletion	β94–98	5	Deletion of pentapeptide sequence Ala – – Lys	No haem attachment to β chain; extremely unstable Hb

Apart from mutations affecting protein structures, other genetic abnormalities in eukaryotes can change the course of differentiation and even the lifespan of higher organisms. A mutation on the X- (sex) chromosome (the Tfm, testicular feminization mutant) can impair the development of all male characteristics in mammals (see section 8.5 for details). In Amphibia, inheritable defects in several genes can prevent the nucleolar organizers from arranging together properly; this is a lethal type of mutation, because the animals are unable to synthesize ribosomes. Offspring manage to survive briefly because of the remarkable numbers of preformed ribosomes, no less than 10^{12}, present in amphibian oocytes (see section 7.2).

Chapter 2 Nucleotide Metabolism

2.1 INTRODUCTION

The biosynthesis, interconversion and degradation of pyrimidine and purine compounds are important for nucleic acid biochemistry for three reasons: nucleic acid biosynthesis is regulated in part by the availability of nucleoside triphosphate substrates for the polymerases; the inhibition of reactions in nucleotide metabolism by drugs is important for certain types of chemotherapy and also for a variety of experimental approaches; and many of the pathways are exploited in experiments that involve radioactive labelling of nucleic acids *in vivo*. These experiments require precursors (frequently nucleosides) because cells are not permeable to nucleoside triphosphates. This chapter outlines nucleotide metabolism insofar as it applies to the rest of the book. For a systematic account see Henderson and Paterson (1973).

2.1.1 Nomenclature

Pyrimidines and purines ('the bases') and their ribonucleosides all have three-letter abbreviations, for example Ura and Urd for uracil and uridine respectively. A non-ribose nucleoside is indicated by a single-letter prefix. In practice the main use of this is to indicate deoxyribonucleosides (dUrd is deoxyuridine) but in principle it can be extended to other pentoses; for example, aUrd and lUrd are those nucleosides in which uracil is attached via an *N*-glycoside bond to arabinose and lyxose respectively. Nucleotides are abbreviated so that a single capital letter (with a prefix for cases other than ribo-) represents the nucleoside component and MP, DP and TP represent mono-, di- and tri-phosphates. A number is used to indicate the position of the phosphate substitution (3′-AMP, 5′-dATP, etc.). If the number is omitted, assume 5′-substitution. The abbreviations are summarized in Table 2.1.

2.2 BIOSYNTHESIS

2.2.1 5-Phosphoribose 1-pyrophosphate

Biosynthesis of nucleotides involves the formation of a β-nitrogen glycoside bond between a ribose moiety and a nitrogen atom. The most usual way of achieving this is by using the extremely reactive compound, 5-phosphoribose 1-pyrophosphate (prpp). The product of the reaction (Fig. 2.1) is a 5′-nucleotide. The biosynthesis

Table 2.1. Names and abbreviations of bases, nucleosides and nucleotides.

Structure of base	Name and abbreviation	Ribonucleoside	Ribonucleotides	Deoxyribonucleoside	Deoxyribonucleotides
	Uracil Ura	Uridine Urd	Uridylic acid UMP	Deoxyuridine dUrd	Deoxyuridylic acid dUMP
	Cytosine Cyt	Cytidine Cyd	Cytidylic acid CMP	Deoxycytidine dCyd	Deoxycytidylic acid dCMP
	Thymine Thy	Thymine riboside Thd	Thymine ribotide TMP	Thymidine dThd	Thymidylic acid dTMP
	Orotic acid Oro	Orotidine Ord	Orotidine monophosphate OMP		
	Adenine Ade	Adenosine Ado	Adenylic acid AMP	Deoxyadenosine dAdo	Deoxyadenylic acid dAMP
	Guanine Gua	Guanosine Guo	Guanylic acid GMP	Deoxyguanosine dGuo	Deoxyguanylic acid dGMP
	Hypoxanthine Hyp	Inosine Ino	Inosinic acid IMP		
	Xanthine Xan	Xanthosine Xao	Xanthosine monophosphate XMP		
	Uric acid Uri				

of prpp is an extension of the oxidative branch of the pentose pathway (Fig. 2.2). The key enzyme, ribose phosphate pyrophosphokinase, is a regulatory enzyme that is inhibited by purine ribonucleoside diphosphates and by pyrimidine ribonucleotides. It is also the

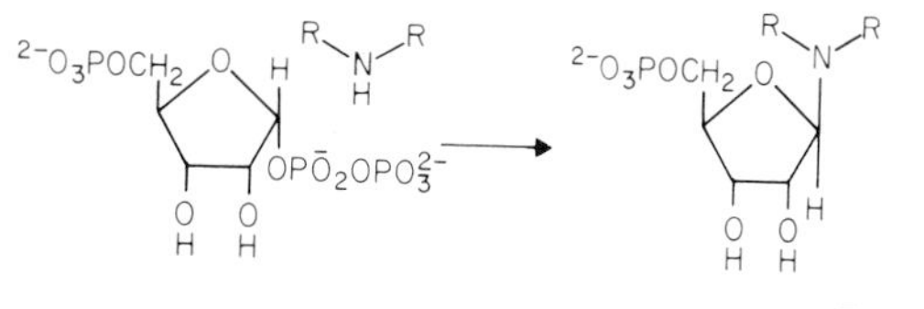

Fig. 2.1. Enzymic formation of an *N*-glycoside bond from a base (HNR_2) and prpp.

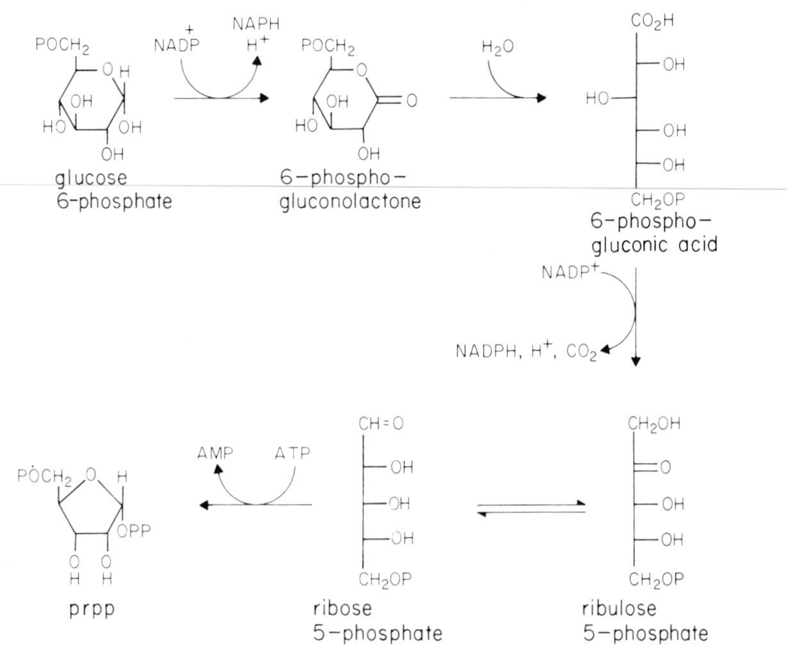

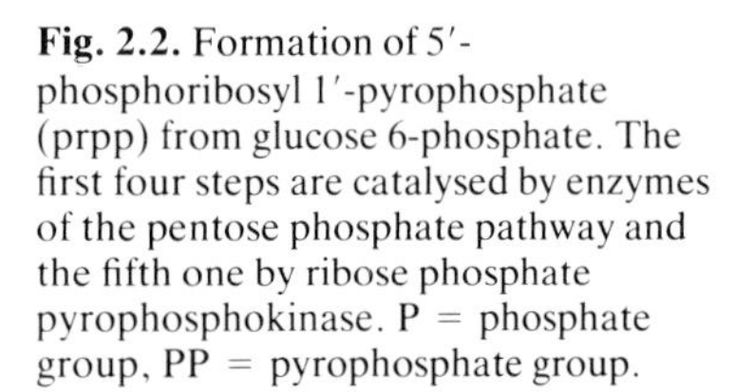

Fig. 2.2. Formation of 5′-phosphoribosyl 1′-pyrophosphate (prpp) from glucose 6-phosphate. The first four steps are catalysed by enzymes of the pentose phosphate pathway and the fifth one by ribose phosphate pyrophosphokinase. P = phosphate group, PP = pyrophosphate group.

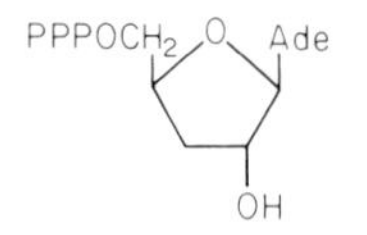

Fig. 2.3. Cordycepin 5′-triphosphate. PPP = triphosphate group.

site of action of certain purine nucleoside antibiotics, such as cordycepin, 3′deoxyadenosine. Such nucleosides are metabolized in cells to form the corresponding 5′-triphosphates (Fig. 2.3); these are inhibitors that compete with ATP in the reaction catalysed by ribose phosphate pyrophosphokinase.

2.2.2 Biosynthesis of pyrimidine ribonucleotides

The pyrimidine skeleton is assembled from carbamyl phosphate and aspartate in a reaction catalysed by aspartate transcarbamylase. A simple cyclization and oxidation of the product of this reaction, carbamyl aspartate, yields orotic acid (Oro) which reacts with prpp to form 5′-OMP. Further simple conversions yield the two pyrimidine ribonucleoside triphosphates (Fig. 2.4). The central regulatory enzyme is aspartate transcarbamylase. In *E. coli* the enzyme is activated by ATP (which acts by reducing the apparent K_m of the enzyme for carbamyl phosphate) and is inhibited by CTP (which increases the apparent K_m for carbamyl phosphate). In some other organisms different pyrimidine nucleotides act as allosteric inhibitors.

6-Azauracil is an antimetabolite that specifically inhibits pyrimidine biosynthesis. It is metabolized by a ‘scavenging pathway’

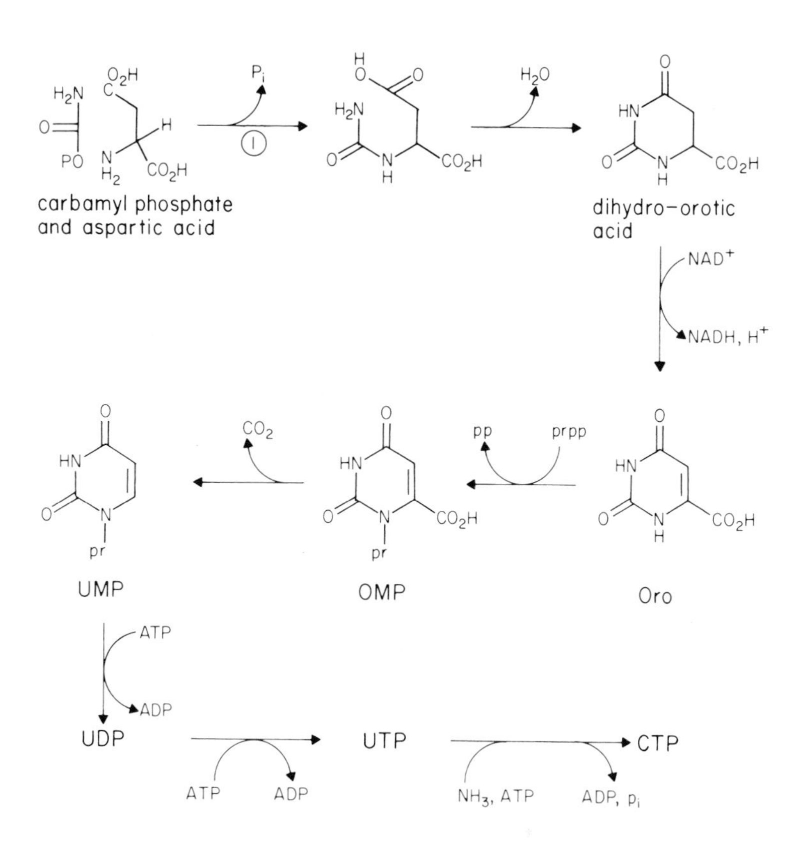

Fig. 2.4. Biosynthesis of pyrimidines. The symbol pr attached to an N-atom means a 5′-phosphoribose attached via a β-riboside bond. Enzyme 1 is aspartate transcarbamylase.

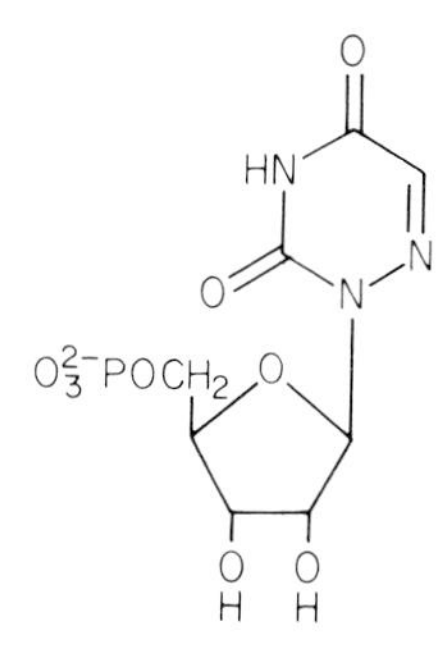

Fig. 2.5. 6-Azauridylic acid.

(see section 2.3.2) to the corresponding 5′-ribonucleotide (Fig. 2.5), which is an inhibitor of OMP decarboxylase (Fig. 2.4).

2.2.3 Biosynthesis of purine ribonucleotides

Purine biosynthesis is radically different from pyrimidine biosynthesis. The most obvious difference is that it is a great deal more complex (Fig. 2.6). This increased complexity derives largely from the fact that whereas the pyrimidine skeleton is constructed from two precursors, the nine atoms of the purine skeleton are derived from seven different precursor molecules and only nos. 2, 1 and 7 have a common precursor (glycine). However, the greatest difference between the two pathways is that the pyrimidine skeleton is assembled completely (in the Oro molecule) before the reaction with prpp, whereas the first step in purine biosynthesis is the formation of an *N*-glycoside bond between ribose 5-phosphate and NH_3. The glycine skeleton, a carbon from the C1-FH_4 pool and a glutamine nitrogen complete the aminoimidazole part of the purine

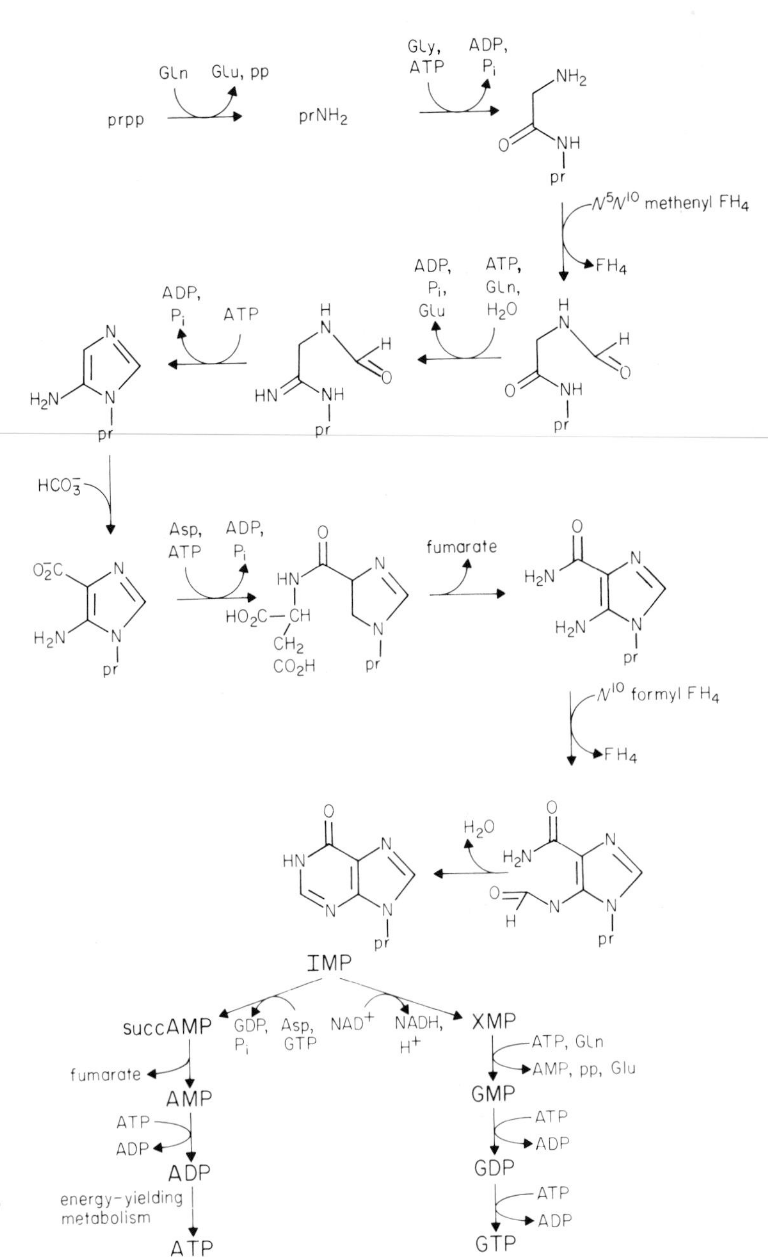

Fig. 2.6. Biosynthesis of purines; pr is 5′-phosphoribose (see Fig. 2.4); succAMP is AMP with one H on N-6 substituted by HO_2C-CH-CH_2CO_2H and the subsequent reaction is similar to the one, shown in full earlier in the pathway, that gives rise to N-1 in IMP.

ring system. The pyrimidine side of the molecule is completed with a C-atom from CO_2, an N from aspartate and a C from the C1-FH_4 pool, to produce IMP. The pathway then branches to form the adenine and guanine ribonucleotides (Fig. 2.6).

The pathway is regulated partly by the allosteric properties of the first enzyme, ammonia phosphoribotransferase, which is separately inhibited by adenine and guanine nucleotides. The balance between ATP and GTP is maintained partly by substrate availability (IMP to AMP requires GTP and conversely IMP to GMP requires ATP) and partly by the allosteric properties of the enzymes in the two branches.

Purine biosynthesis is inhibited by purine analogues such as 8-azaguanine and 6-mercaptopurine (Fig. 2.7); both of these are

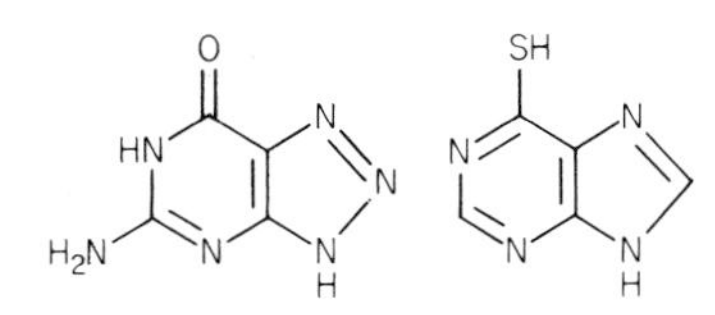

Fig. 2.7. 8-Azaguanine and 6-mercaptopurine.

metabolized to the corresponding 5′-ribonucleotides, which inhibit conversion of XMP to GMP and IMP to succAMP respectively. Purine biosynthesis is also inhibited by the antibiotic azaserine, $N_2CHCO_2CH(NH_2)CO_2H$; this inhibits all glutamine-dependent amination reactions.

2.2.4 Biosynthetic origin of deoxyribose

In the majority of organisms the substrates for the reduction of a ribose moiety to a deoxyribose moiety are the ribonucleoside diphosphates. The enzyme responsible, nucleoside diphosphate reductase (or 'nucleotide reductase') contains non-haem iron and benzyl-free radicals. The mechanism involves direct displacement of the 2′-OH by an H-atom derived from an SH-group in the enzyme. There are two enzyme subunits, B1 (catalytic) and B2 (regulatory). The reducing power is derived ultimately from NADPH. It is commonly imagined that the only electron transport chain that links oxidation of NADPH to reduction of ribose involves the oxidized and reduced forms of thioredoxin (Holmgren 1981). However, in *E. coli,* and presumably some other organisms as well, there is an alternative pathway involving a different peptide, glutaredoxin. The evidence is derived from the finding that mutants of *E. coli* that lack thioredoxin are able to synthesize DNA in unsupplemented medium.

Nucleotide reductase is regulated in a complicated fashion that ensures maintenance of the proper balance between the four deoxyribonucleoside triphosphates (Fig. 2.8). This is achieved by a cascade

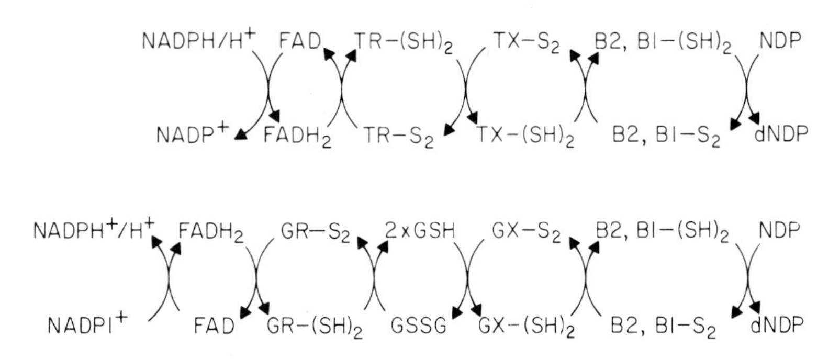

Fig. 2.8. Electron transport for the ribonucleotide reductase system linked via thioredoxin (upper scheme) or glutaredoxin (lower scheme). The symbols $-S_2$ and $-(SH)_2$ refer to oxidized and reduced forms of a polypeptide with two cysteine residues. B1 and B2 are the subunits of ribonucleotide reductase; TX, thioredoxin; TR, thioredoxin reductase; GX, glutaredoxin; GSH and GSSG are the reduced and oxidized forms of glutathione; GR, glutathione reductase.

involving nucleoside triphosphates as allosteric effectors. Thelander and Reichard (1979) have written an authoritative review of the enzyme (Fig. 2.9). Nucleotide reductase can be inhibited *in vivo* by hydroxyurea.

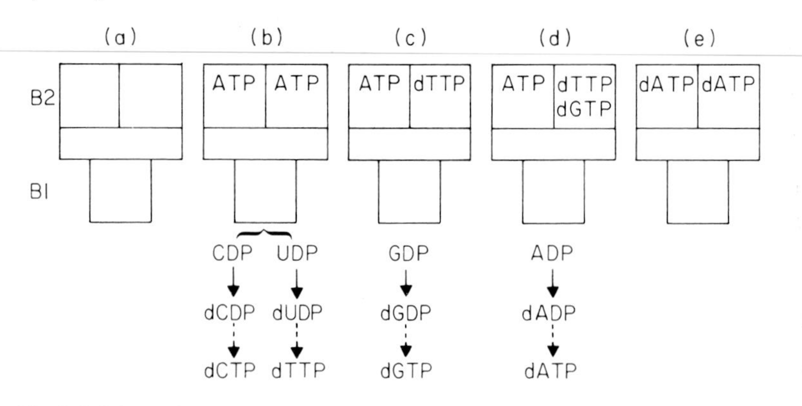

Fig. 2.9. Scheme for the regulation of ribonucleotide reductase. The subunit B2 contains the binding sites both for the effectors and the nucleotide substrates. There are two classes of effector binding sites. ATP and dATP can bind to either; both the other effectors can only bind to one. ATP, dCTP, dTTP and dGTP are positive effectors; dATP is an inhibitor. The reactions catalysed are shown below the figure. Dashed arrows refer to reactions not catalysed by the enyme. The sequence shows a cycle of reactions. (a) No effectors are bound and there is little or no reaction. (b) ATP is bound and allows reduction of the pyrimidine nucleotides. (c) dTTP accumulates and allows reduction of GDP. (d) dGTP has accumulated and dTTP and dGTP together allow reduction of ADP. (e) This leads to accumulation of the inhibitor (dATP) and all reduction ceases.

In *Lactobacillus leichmanii* and certain other bacteria, the substrates for ribonucleotide reductase are the ribonucleoside triphosphates and also the enzyme is vitamin B12 dependent.

2.2.5 Thymidylic acid biosynthesis

Thymidylate synthase catalyses the conversion of dUMP to dTMP. The reaction is the replacement of H-5 in a uracil moiety by a CH_3 group. Other biological methylations, including most nucleic acid modifications (see section 5.3.2), employ *S*-adenosyl methionine as methyl donor. However, thymidylate synthase employs N^5N^{10} methylene FH_4. Hence thymidylic acid biosynthesis is linked to FH_2 reduction (Fig. 2.10). This fact is extremely important both experimentally and clinically. Inhibitors of FH_2 reductase inhibit DNA synthesis. Two such inhibitors are the folic acid analogues aminopterin (the same structure as folic acid except that the OH at C-4 is

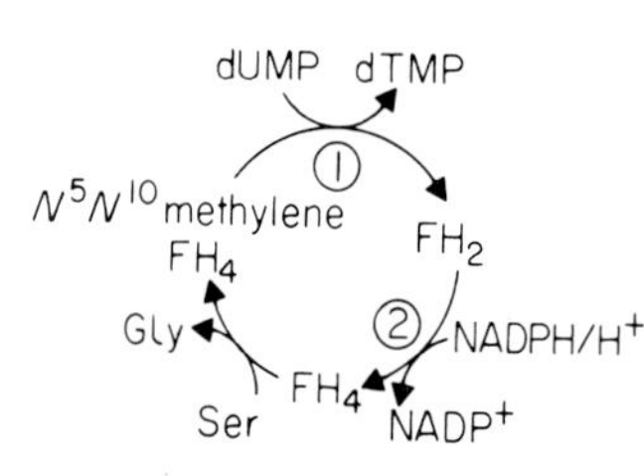

Fig. 2.10. Biosynthesis of thymidylic acid. Enzyme (1) is thymidylate synthase and enzyme (2) is dihydrofolate reductase.

replaced by NH_2) and amethopterin (N^{10}-methylaminopterin). These substances inhibit rapidly dividing cells and amethopterin (often referred to by its pharmaceutical trade name 'methotrexate') is employed as an anti-cancer drug. In bacteria, DNA synthesis can be inhibited by sulphonamides. These drugs inhibit the biosynthesis of *p*-aminobenzoate and consequently of folic acid as well.

2.2.6 Pathways leading to deoxyribonucleoside triphosphates

Because of the different levels of phosphorylation of the substrates for nucleotide interconversions, the synthesis *de novo* of the four nucleotide substrates for DNA polymerase, dCTP, dTTP, dATP and dGTP, involve phosphorylations and dephosphorylations. The pathways are summarized in Fig. 2.11. In addition to steps already covered in this chapter, we have introduced an additional step, the deamination of dCMP. In those organisms and tissues that have been examined so far, dCMP deaminase is found and there are consequently two routes to dUMP. Presumably this is exploited in the regulation of the ratio of dCTP/dTTP.

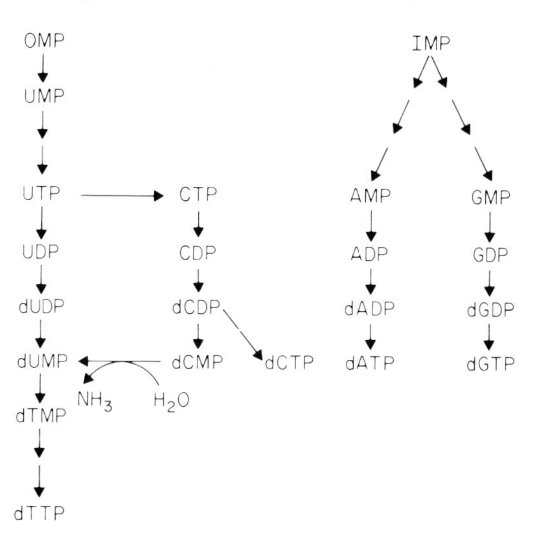

Fig. 2.11. Summary of pathways leading to the four nucleotide substrates for DNA polymerase.

2.3 CATABOLISM AND SCAVENGING

2.3.1 Nucleotide catabolism

Nucleotides are hydrolysed to form nucleosides in reactions catalysed by phosphatases, and nucleosides are hydrolysed to form bases and sugars in reactions catalysed by nucleosidases.

Further catabolism of pyrimidine bases is reductive and is quantitatively relatively unimportant. Cyt is converted to Ura by deamination and the 5,6-double bond of Ura is reduced (NADPH) and then the lactam bonds are hydrolysed (starting with the 1,6-bond) to yield NH_3, CO_2 and $H_2NCH_2CH_2CO_2H$ (β-alanine). Thymine catabolism follows a parallel course; the final products are NH_3, CO_2 and $H_2NCH_2CHMeCO_2H$ (β-amino *iso*butyrate).

In contrast, purine catabolism is oxidative and is quantitatively important. Not only do purine nucleotides and their degradation products occur in large amounts because of the roles of ATP and GTP in intermediary metabolism, but also the biosynthesis and degradation of purines forms the pathway of nitrogen excretion in birds and other 'uricotelic' animals.

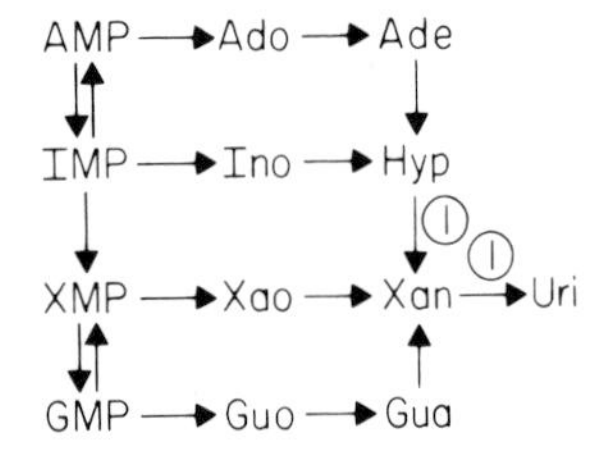

Fig. 2.12. Interconversion of purines during their degradation. The interconversions of nucleotides (left-hand side of figure) occur during biosynthesis (see Fig. 2.6). Both reactions marked (1) are catalysed by the same enzyme, xanthine oxidase.

The pathways of purine interconversions and the formation of uric acid (Uri) are summarized in Fig. 2.12. In many animals the pathway proceeds beyond Uri by one or more subsequent steps of oxidation and hydrolysis.

The breakdown of nucleotides generates sugars or sugar phosphates. Ribose is metabolized by the transketolase and transaldolase branch of the pentose phosphate pathway. Deoxyribose 5-phosphate is catabolized by deoxyriboaldolase (Fig. 2.13). Note that this enzyme, which we shall consider again in section 2.3.4, catalyses a reaction that, on the face of it, could be used for deoxyribose biosynthesis. However, the equilibrium position lies so far in the direction of acetaldehyde and glyceraldehyde 3-phosphate formation that the reaction is of no biosynthetic consequence whatever under physiological conditions. All the deoxyribose moieties in cells are derived from reactions catalysed by nucleotide reductase (see section 2.2.4).

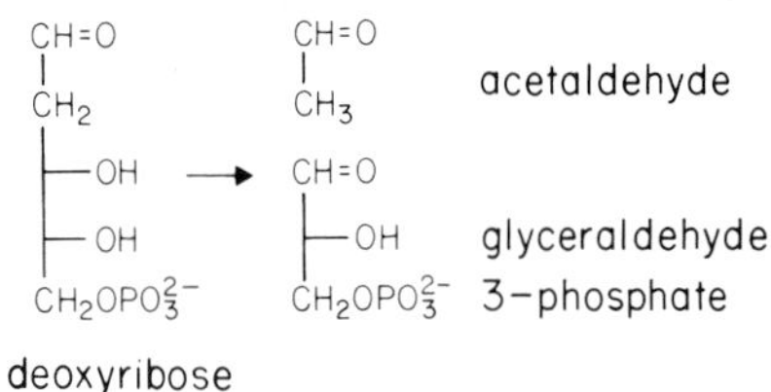

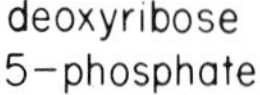

Fig. 2.13. Catabolism of deoxyribose 5-phosphate in the reaction catalysed by deoxyriboaldolase.

2.3.2 Scavenging reactions

In addition to the enzymes for biosynthesis *de novo* of nucleotides and for their degradation, there are enzymes that catalyse a variety of nucleotide and nucleoside interconversions that serve the purpose of scavenging bases and nucleosides for the purpose of nucleotide synthesis. There are four types of scavenging enzyme.

Phosphoribosyl transferase catalyses the conversion of a base and prpp to form the 5′-nucleotide and pyrophosphate. Orotic acid phosphoribosyl transferase is such an enzyme of the pyrimidine biosynthetic pathway (Fig. 2.4). However, there are two enzymes that catalyse corresponding reactions for purines. One enzyme is specific for adenine, the other for either hypoxanthine or guanine. The latter enzyme (hypoxanthine/guanine phosphoribosyl transferase, HGPRT) is of considerable importance in mammals. The serious inherited disorder in humans, the Lesch-Nyhan syndrome, is due to a deficiency of HGPRT. The gene for HGPRT is of importance in mammalian somatic cell genetics because $HGPRT^-$ cells can be selected positively. These cells are resistant to 6-mercaptopurine because they are unable to convert it into the active metabolite, 6-mercaptopurine 5′-ribonucleotide (see section 2.2.3). There is no enzyme corresponding to phosphoribosyl transferase for deoxyribose compounds.

Nucleoside kinase catalyses the conversion of nucleosides to 5′-nucleotides; the phosphate comes from ATP. Every one of the eight common ribo- and deoxyribo-nucleosides can be converted to its corresponding 5′-nucleotide by a kinase.

Nucleoside transferase only catalyses reactions in the deoxyribose series. It equilibrates a mixture of a base (e.g. Thy) and a nucleoside (e.g. dAdo) with the converse mixture (dThd and Ade).

Nucleoside phosphotransferase occurs in both the ribose and deoxyribose series. It equilibrates a mixture of nucleoside and inorganic phosphate with the free base and the sugar 1-phosphate. Different nucleoside phosphotransferases are specific for one of the two sugars and for the type of base (either purine or pyrimidine).

2.3.3 Radioactive labelling of nucleic acids *in vivo*

Nucleic acid of high specific activity can be obtained by growing cells at the expense of carrier-free [^{32}P]inorganic phosphate as sole phosphorus source. To obtain nucleic acid labelled with either ^{14}C or ^{3}H the cells are grown in a medium containing radioactive bases or nucleosides. Alternatively an animal may be injected with these substances. The label does not necessarily end up in the residue it started off in; for example, [^{14}C]Oro, [^{14}C]Ura and [^{14}C]Urd will all label C- and U-residues in RNA and C- and T-residues in DNA. The efficacy of a label depends upon a number of variables including permeability, pool size and the amounts of scavenging enzymes. For most systems there is a shortage of information and one proceeds by trial and error or by following laboratory lore handed down by previous experimenters. An example is that rat liver RNA can be labelled with radioactive Oro; in the mouse this does not work and one uses Urd. In many eukaryotic microorganisms the only successful label is radioactive Ade.

It is frequently convenient to label either DNA or RNA specifically. DNA is specifically labelled with [Me-^{3}H]dThd. RNA can be specifically labelled with [5-^{3}H]Urd; that fraction of the nucleoside that ends up as [5-^{3}H]dUMP is converted, by thymidylate synthase, to unlabelled dTMP, because H-5 is lost during the reaction (see section 2.2.5).

2.3.4 Thymine auxotrophy in bacteria

Thymidylate synthase mutants (Thy$^-$) of *E. coli* and certain other bacteria are relatively easy to isolate. The mutants are at a selective advantage over the wild type when grown in a rich medium, supplemented with Thy or dThd, in the presence of an inhibitor of FH_2 reductase such as trimethoprim (see section 2.2.5). Thy$^-$ mutants will grow in a medium supplemented with either low concentrations of dThd or relatively high concentrations of Thy. These findings correlate with the fact that whereas radioactive dThd is a good label for bacterial DNA, radioactive Thy is usually a poor one. This is because the pool size of the essential metabolite, deoxyribose 1-phosphate is low. In *E. coli* this is kept very low by the operation of deoxyribomutase and deoxyriboaldolase (Fig. 2.14). Thy$^-$ mutants survive in high concentrations (50 μg ml^{-1}) of Thy because the deoxyribose 1-phosphate is forced into the formation of dThd by the law of mass action. However, it is possible to select mutants of Thy$^-$

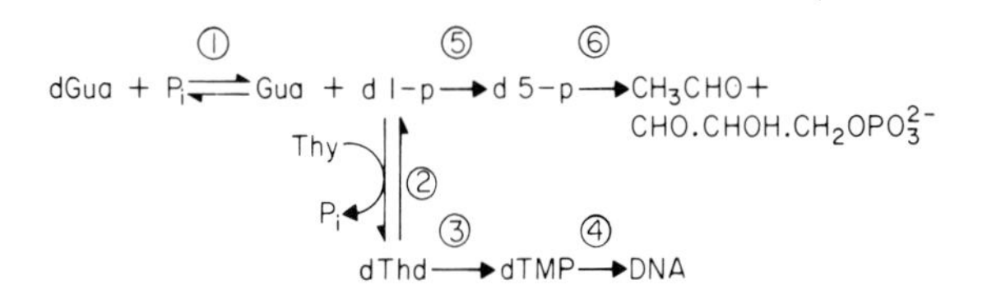

Fig. 2.14. Pathway for the incorporation of thymine into DNA in a bacterium. Enzymes are as follows: (1) purine deoxyribonucleoside phosphotransferase; (2) thymidine phosphotransferase; (3) thymidine kinase; (4) nucleoside mono- and di-phosphate kinases and DNA polymerase; (5) deoxyribomutase; (6) deoxyriboaldolase; d 1-p and d 5-p, deoxyribose 1- and 5-phosphates.

strains of *E. coli* that still require thymine but will grow in very low concentrations (2 $\mu g\ ml^{-1}$) of the base. These 'very low thymine requirers' are double mutants; the second mutation occurs in one of the genes for either deoxyribomutase or deoxyriboaldolase. Such organisms are of experimental importance in studies on the bacterial cell cycle and for obtaining very high specific activity [^3H]DNA.

Methods for labelling nucleic acids *in vitro* are discussed in section 3.4.2.

Chapter 3 Nucleic Acid Chemistry

3.1 PRIMARY, SECONDARY AND TERTIARY STRUCTURE

3.1.1 Introduction

The primary structure of nucleic acid consists of pyrimidine or purine bases linked via *N*-glycoside bonds to a sugar residue, the sugars being linked by phosphodiester residues (see section 1.1.1). There is one exception to the *N*-glycoside bonding in the residue pseudouridine (see section 3.5.2) that occurs in tRNA.

Secondary structure refers to the two characteristic polynucleotide conformations, the double helix and the base-stacked structure (see section 3.1.2). Any further folding of the molecules is described as tertiary structure.

3.1.2 Secondary structure

The double helix

Most cellular DNA consists of two covalently independent and complementary molecules arranged in an antiparallel helical double molecule. It is an example of an intermolecular hydrogen-bonded structure. Cellular RNA has considerable double helical content, also consisting of stretches of antiparallel complementary sequence, but in this case the arrangement is achieved by the molecule bending back on itself and is an intramolecular hydrogen-bonded structure (Fig. 3.1).

The geometry of the DNA double helix was deduced from X-ray diffraction patterns generated by DNA fibres. Analysis of such patterns reveals the three different conformations of the DNA double helix, A-DNA, B-DNA and C-DNA. The A and B forms of

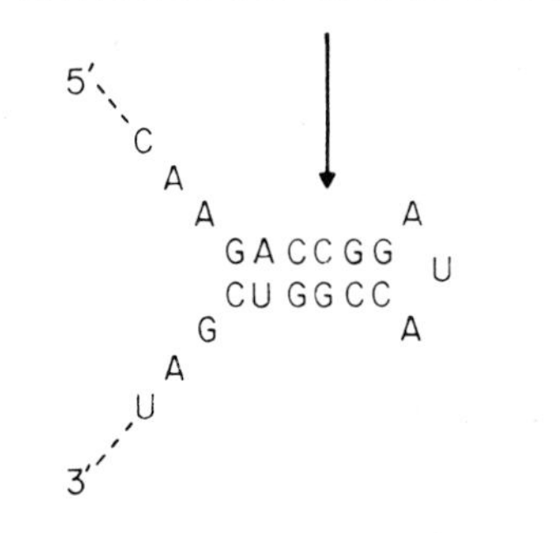

Fig. 3.1. Intramolecular complementarity and hydrogen bonding in a part of an RNA molecule.

DNA are illustrated in Fig. 3.2. B-DNA was the form deduced by Watson and Crick. The base pairs are inclined at 6° to planes perpendicular to the helix axis; there are 10 base pairs per full turn; the 2 strands are antiparallel. The structure forms a right-handed helix. A-DNA differs from the B structure in the following respects. The base pairs, although parallel to one another and spaced by the same amount (0.36 nm) as in the B form, are inclined by about 20° to planes perpendicular to the helix axis. The number of base pairs per full turn is 12 and the structure is fatter but longitudinally more compact (more base pairs per unit length along the helix axis). The two grooves that run round the outside of any double helix are of approximately the same size in A-DNA but very different from one another in B-DNA. Perhaps the most striking difference between these two structures is the fact that base pairs overlap in B-DNA so

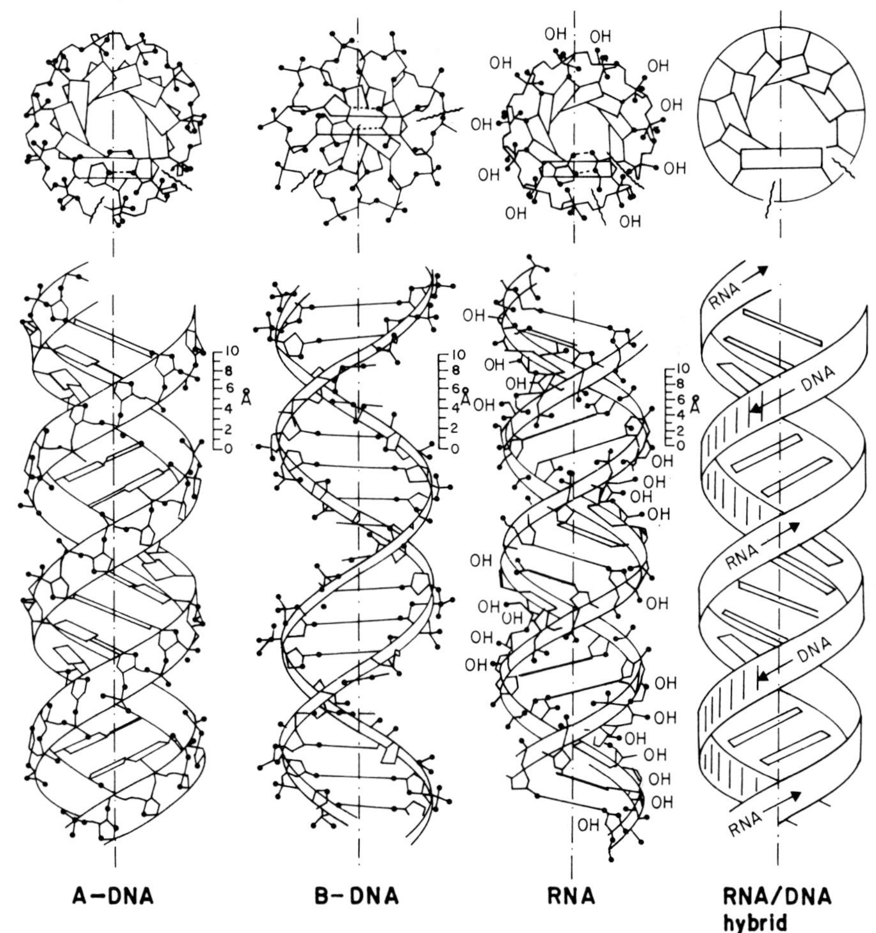

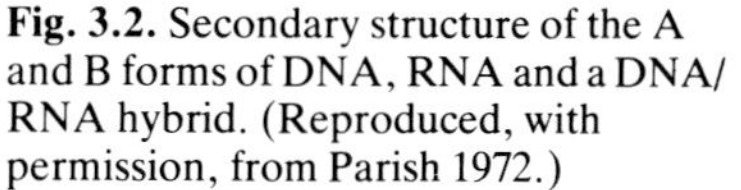
Fig. 3.2. Secondary structure of the A and B forms of DNA, RNA and a DNA/RNA hybrid. (Reproduced, with permission, from Parish 1972.)

that a top view of the helix appears to be 'full of base pairs' but A-DNA has a hole down the middle.

C-DNA is a distorted B structure with a non-integral number of bases per full turn and base pairs are somewhat more inclined to planes parallel to the axis.

The conformations of A-, B- and C-DNA can be described in terms of torsional angles between bonds. These parameters and full details of the helix geometry are described by Arnott (1970).

There is a great deal of interest in other possible double helical structures for DNA. Until relatively recently, proponents of models that differ radically from A-, B- or C-DNA have run the risk of being regarded by other molecular biologists in the way that astronomers regard members of the Flat Earth Society. However, the question must now be regarded as open following a very high-resolution X-ray crystallographic analysis of the synthetic DNA fragment d(C—G—C—G—C—G) by Wang *et al.* (1979). In the presence of Na^+, Mg^{2+} and spermine salts, this substance crystallizes in a form in which the oligonucleotides (that are self-complementary) form H-bonded double helical fragments stacked on top of one another in an approximation of a continuous helix. The helix is unusual in three important respects: it is left-handed; the *N*-glycoside linkage of the dGuo residues is in the unusual *syn* conformation; there are two nucleotides in the asymmetric unit. Another way of expressing this last point is to say that the phosphorus atom in the sequence G—C is a different distance from the helix axis than that in the sequence C—G. Otherwise the structure is not unlike A- and B-DNA: the strands are antiparallel; there are 12 base pairs per full turn; the base pairs are inclined by 7° to planes perpendicular to the axis; the sugar residues have the same conformations as in B-DNA. This new structure, as built into a theoretical model for a DNA molecule, has been christened Z-DNA.

The biological significance of Z-DNA is still uncertain. However, there is theoretically no reason why a Z-sequence should not exist in the middle of a B-sequence provided there is an even number of base pairs in the Z-region and it is possible that the binding of specific proteins to GC-rich sites in DNA might effect a change of conformation from B to Z.

The RNA double helix can only assume the A form. Model building reveals that it is impossible to accommodate the 2′-OH groups of the ribose moieties in a B or C structure and the A structure has been confirmed experimentally by X-ray diffraction studies of synthetic polymers, such as poly I: poly C, or naturally occurring, intermolecular double-stranded RNA from reovirus and crystals of tRNA (see section 3.1.4). Similarly a DNA:RNA hybrid helix assumes the A conformation (Fig. 3.2).

Base stacking

An unpaired sequence of nucleotide residues, such as that in a loop of an RNA molecule (Fig. 3.1) does not assume a random conformation but rather the bases stack parallel to one another spaced by 0.36 nm. The bases in the double helix are stacked in this way. Indeed the double helix can be regarded as a special case of a stack in which the bases are paired with complementary bases in a different stack.

Stability of secondary structure

What holds the double helix together? One factor is the hydrogen bonding. There are many ways of drawing H-bonds between nucleic acid bases. The two Watson–Crick pairs (Fig. 3.3) assume that the lactam and amine tautomers apply (Fig. 3.3(a,b)). This is thermodynamically correct. However, why choose the atoms of the Watson–Crick schemes as H-bond donors and acceptors? These do not necessarily form the most stable arrangements. Indeed the A:T structure shown in Fig. 3.3(c) is more stable than the corresponding Watson–Crick structure (Fig. 3.3(a)). The answer lies in the

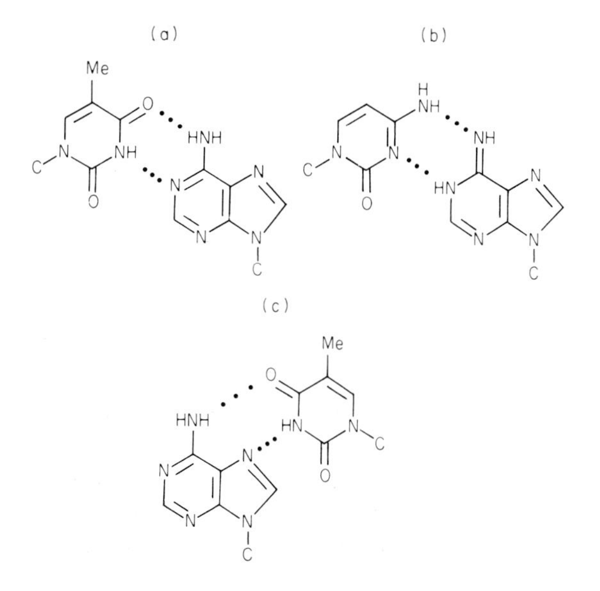

Fig. 3.3. Base pairs. (a) Watson–Crick base pair for A:T. (b) The effect of using the wrong tautomeric form for an N atom: the N^6 of the A-residue is in the imine (= NH) form and A now pairs with C. The wrong tautomer for an O atom also results in incorrect pairing (see Fig. 3.15(d)). (c) Structure of an alternative A:T hydrogen bonding scheme (the Hoogsteen structure).

C-1′–C-1′ distance. This is the same (1.11 nm) in both G:C and A:T in the Watson–Crick structures so the overall helix geometry is independent of sequence provided the two strands are complementary.

Hydrogen bonding is not the only stabilizing factor in the double helix. The other factors are sometimes called 'hydrophobic interactions' which refers to two fundamentally different types of association. The first is that purines and pyrimidines are actually attracted to one another in stacks. There is an optimum spacing (0.36 nm) for two such rings. The critical molecular process is the interaction of the clouds of π electrons that lie on either side of the rings. If the rings are more closely spaced, the benefits of the π–π interactions are cancelled by the repulsion between the molecules. These are sometimes referred to as the out-of-plane interactions between the pyrimidine or purine rings (in contrast to the H-bonds or 'in-plane' interactions).

The second 'hydrophobic' interaction follows from the structure of water. Water is an extensively H-bonded structure. In liquid water there are large ice-like domains. Within each of these domains, each oxygen atom is surrounded by four more oxygen atoms in tetrahedral array. There is a hydrogen atom between each pair of oxygen atoms. In terms of the formal valency of the H_2O molecule of the four H's surrounding every O, two 'belong' to the O itself and the other two 'belong to other O's' and are associated by H-bonding. (This last sentence has no real meaning in terms of the electron orbitals forming the O—H bonds.) Substances dissolved in water disrupt the ice-like domains. In the cases of certain molecules, including ribose and phosphate, the solution process is achieved by the formation of new H-bonds between the OH or O^- groups of the dissolving molecule and surrounding H_2O. However, in other cases (including the bases of nucleic acids) the extent of this type of H-bonding is limited and so, if molecules of these types are distributed in water, discontinuities or 'holes' are formed in the H-bonding of the ice-like domains without any compensatory H-bonding for the H_2O molecules. It is more efficient to generate one large discontinuity of this type than to form many little ones. Thus those parts of the structure of a macromolecule that cannot H-bond to water readily will tend to associate with one another. The term for this associative force is 'cavitation energy'. It is the acquisition of minimum cavitation energy that contributes to the stability of regions of protein structure in which amino acid residues such as leucine, isoleucine and phenylalanine are concentrated. Likewise, cavitation energy contributes significantly to the stability of the DNA double helix.

What are the relative contributions to double helix stability of in-plane (H-bonding), out-of-plane (base stacking) and cavitation energy interactions? It is difficult to be certain because it is hard to quantify the contribution from cavitation energy from known thermodynamic parameters, but as a rough approximation, the contribution of each is approximately one-third of the total. It is certain that the in-plane and out-of-plane interactions have about the same contribution as one another. For a more detailed review with references to the original literature see Parish (1972). For the purposes of this book, two important things follow from this discussion. Watson–Crick base-pairing interactions are of critical and unique importance in the recognition processes involved in the generation of a new double helix, either through renaturation of denatured nucleic acids or during the biosynthesis of nucleic acids. However, during processes in which secondary structure is lost (denaturation), the relaxation of each of three constraints contributes to the denaturation process.

3.1.3 Denaturation and renaturation

Denaturation

Nucleic acids can be denatured by heating, by exposing the solution to extremes of pH, or by certain proteins. Proteins that affect nucleic acid conformation are discussed in section 4.2.2. Experimentally, nucleic acid solutions are denatured by heating (the process is frequently referred to as melting) or, in the case of DNA only, by raising the pH to approximately 12. RNA is hydrolysed by alkali and both DNA and RNA are hydrolysed by acid (see section 3.3.1).

The progress of denaturation can be followed from changes in the absorption of the solution at the absorption maximum for the purines and pyrimidines (approximately 260 nm for molecules of 'average' base composition), by measuring the sensitivity of samples to single-strand specific nucleases (see section 3.4.3) and by chromatography (see section 3.2.2). The last two methods have the advantage of being applicable to trace amounts of radioactively labelled preparations.

The reason for the use of absorption spectroscopy for measuring the extent of denaturation of a nucleic acid sample is that the absorbance of purines and pyrimidines is reduced if the bases are stacked. This is referred to as the hypochromic effect. When a nucleic acid is denatured the stacking is lost and the absorbance increases.

This increase is described as a hyperchromic shift. For the denaturation of a totally double-stranded sample of DNA the hyperchromic shift is of the order of 30%. If a solution of DNA is slowly heated and if the absorbance (which can be converted to a measure of the extent of denaturation) is plotted as a function of temperature, the DNA is observed to denature in a cooperative fashion; that is to say that over a very narrow temperature range the total secondary structure is lost. The temperature at which the solution contains 50% denatured and 50% double-stranded DNA is called the melting temperature (T_m). Such a melting profile is illustrated in Fig. 3.4(a). The value of T_m is a function of the nature of the ions in solution and the ionic strength and also the nature of the DNA. If the first two variables are fixed, the T_m can be used as an indirect method of determining the nucleotide composition (see section 3.5.3).

RNA shows a less straightforward denaturation profile (Fig. 3.4(b)). There is a cooperative melting with an apparent T_m that corresponds to the denaturation of the double-stranded part of the molecule. Superimposed on this is a non-cooperative loss of base stacking from those parts of the molecule that are not base paired.

The denaturation of nucleic acid solutions can be followed by monitoring other optical properties of the solution (optical rotatory dispersion and circular dichroism; Parish 1972). However, the contribution of these sophisticated spectroscopic variants on the measurement of optical activity to nucleic acid biochemistry has been minimal and they are not discussed further in this book.

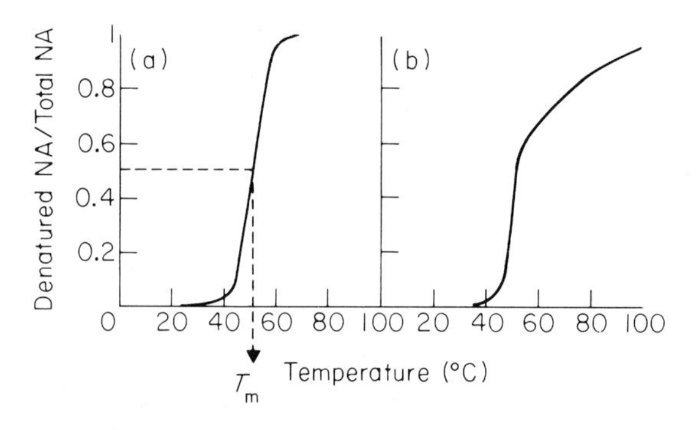

Fig. 3.4. Denaturation profiles ('melting curves') of (a) DNA and (b) RNA.

Renaturation

Renaturation is not simply the reversal of denaturation. If denatured nucleic acids are maintained at about 5–10 °C below their melting temperature, the complementary strands will slowly reassociate and the double helical structures will re-form. The process for intermolecular double helices (DNA) is clearly concentration dependent. The process of two complementary sequences 'finding one another' is dependent upon the frequency with which matched molecules collide. For high molecular weight DNA, the process is more complicated than that because partial complementarity may occur between regions of the molecules and form an incomplete double-stranded molecule. This accounts for the finding that if a denatured solution of high molecular weight DNA, from a vertebrate for example, is cooled to below the melting temperature, the absorbance at 260 nm falls but not to the value of the original native solution. However, if the DNA is fragmented into short lengths (this is achieved experimentally by subjecting the solution to high shear gradients by squirting the solution in and out of a syringe several times through a narrow needle) the solution will renature completely. What is more, the rate of renaturation can be used as a measure of the complexity of the sequence. The principle can be seen by considering two solutions of denatured DNA of the same concentration (i.e. nucleotide residues per unit volume) and in identical buffers but from a very simple source (for example a small virus) and a very complex source (for example a vertebrate). If the two solutions are maintained at the same temperature, the simple DNA will renature more rapidly because the chances of two colliding sequences being complementary are clearly much higher if the DNA comes from the simpler source. The procedure is known either as the Britten and Kohne method (after its inventors) or the C_0t method (pronounced 'cot'). The symbol, C, represents the concentration of single-stranded DNA and C_0 is the total DNA concentration. The corresponding rate equation is:

$$C/C_0 = 1/(1 + kC_0t),$$

where k is the reassociation rate constant. A plot of C/C_0 (i.e. the proportion of DNA that is single-stranded) versus C_0t constitutes a C_0t curve (Fig. 3.5). The value of C_0t for which C/C_0 is 0.5 is referred to as $(C_0t)_{1/2}$, 'cot-a-half'; this value is related to the genome molecular weight. C_0t curves therefore provide a method for measuring genome molecular weight without a requirement for intact DNA molecules. They also reveal fractions of DNA (from a single source) with different sequence complexity (Fig. 3.5(b); Britten *et al.* 1974); this is of particular importance in animal DNA (see section 7.2).

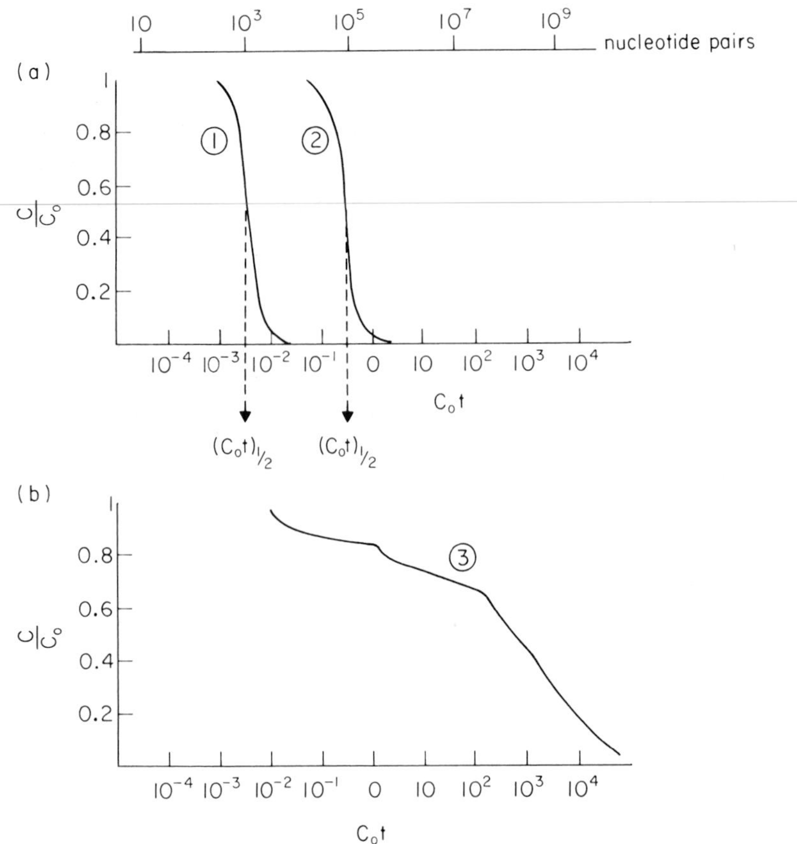

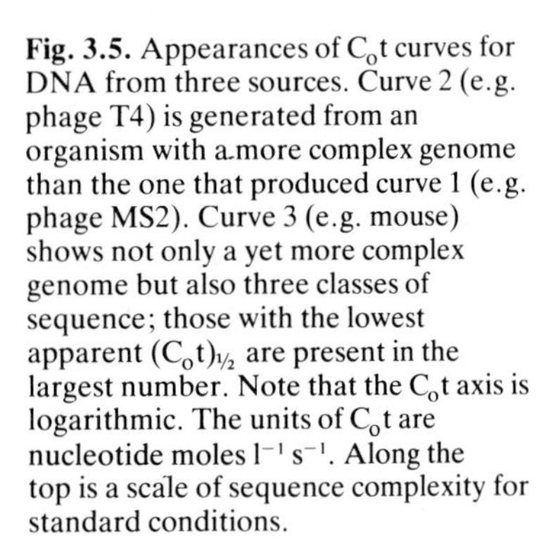

Fig. 3.5. Appearances of C_0t curves for DNA from three sources. Curve 2 (e.g. phage T4) is generated from an organism with a more complex genome than the one that produced curve 1 (e.g. phage MS2). Curve 3 (e.g. mouse) shows not only a yet more complex genome but also three classes of sequence; those with the lowest apparent $(C_0t)_{1/2}$ are present in the largest number. Note that the C_0t axis is logarithmic. The units of C_0t are nucleotide moles l^{-1} s^{-1}. Along the top is a scale of sequence complexity for standard conditions.

The complexity of a genome is defined as the sum of the lengths of its unique sequences plus the unit lengths of each of its repetitive sequence families. Thus a genome comprising sequences *a, b, c* and *d* would have a complexity of $a+b+c+d$. This genome is composed entirely of unique sequences and so the complexity is the same as the genome molecular weight. This would be true for our example of a suitable virus. Higher organism genomes frequently contain repetitive and unique sequences such that there may be 10^6 copies of *a*, 10^3 copies of *b*, 10 copies of *c* and 1 copy of *d, e* and *f*. The complexity of this genome would be $a+b+c+d+e+f$ and may, therefore, be significantly less than the genome molecular weight. The units of complexity are usually nucleotides for RNA and nucleotide pairs for DNA. In Fig. 3.5 we see that genome 1 has a lower complexity than genome 2 and indeed we can assign complexity values to $(C_0t)_{1/2}$ values as long as the reassociation reactions are performed under standard conditions (0.18M Na^+, 60 °C) or are standardized by calculation to these conditions. When a genome is composed of unique sequences the reassociation curve in a C_0t analysis has a single kinetic component. However, a similar analysis of the genome of a higher organism may reveal several components each representing portions of the genome with similar reiteration frequencies. Short, highly repetitive sequences present in the genome in thousands of copies will reassociate faster than longer, middle repetitive sequences present in tens of copies, which will, in turn, reassociate faster than the unique sequences. In this example then we would see three kinetic components as for genome 3 in Fig. 3.5. It should be emphasized that a single kinetic component is an average value (usually identified with a non-linear least squares programme) and may be comprised of many different sequence families related by similar repetition frequency. They are seen as a separate component simply because they form a significant proportion of the total DNA. The repetition frequency of sequences within a kinetic component can be obtained from the ratio of the repetitive sequence reassociation rate constant to the single copy rate constant. So C_0t analysis can yield a number of useful pieces of information, it can give the genome complexity, the number of repetitive classes and the proportion of the total genome represented by those classes.

The other important technical method that derives from renaturation studies is the process of nucleic acid hybridization (see Grossman & Moldave 1971 for technical details). This is the formation of a double helical molecule from two separate populations of denatured preparations. Several examples of hybridization methods appear in later parts of the book. Here we may merely note the more important uses of the procedure and comment on the theory behind the more common types of hybridization experiment. Hybridization is a method for recognizing complementary sequences. DNA:DNA hybridization can be used to evaluate the evolutionary relatedness of different organisms, for recognizing the presence of a sequence of a part of one replicon in another (for example bacterial DNA in a phage genome; see section 4.1) and in the mapping of restriction fragments (see section 3.4.2).

DNA:RNA hybridization can be used for probing for the presence of a transcribed strand of DNA with its RNA transcript. One application is taxonomic: all organisms contain rRNA and the sequences of these molecules seem to evolve slowly so that the degree of homology between rRNA and the ribosomal cistrons of a different organism can reveal phylogenetic relatedness of widely separated species.

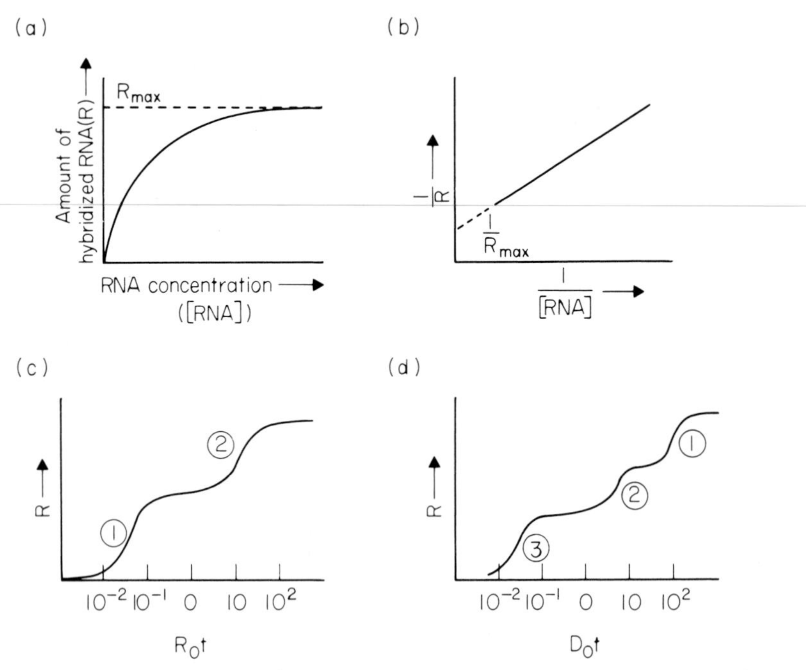

Fig. 3.6. (a and b) RNA:DNA hybridization experiment in which the amount of RNA is increased to attempt to obtain the saturation value. (a) Plot of hybrid yield as a function of RNA concentration. (b) Double reciprocal plot of data such as those employed in (a). (c) A R_0t curve: in this case there are two abundance classes of transcripts—there are more copies of RNAs that form hybrids at 1 than there of those that form hybrids at 2. (d) A D_0t curve showing three abundance classes of DNA sequence. If, for example, hybrids that form at 3 are between RNA and DNA from single-copy genes, hybrids that form at 2 are between RNA and DNA from genes present in several copies and those at 3 from genes present in large numbers of copies.

DNA:RNA hybridization is frequently used to measure the number of genes for a particular RNA transcript. There are two approaches designed to ensure that all the hybrids are accounted for. One obvious problem is that RNA:DNA hybrid formation may be competed for by the formation of DNA:DNA double helical structures. This can be overcome because RNA:DNA hybrids are more stable and by using as an incubation temperature one just below the T_m, the sought after RNA:DNA hybrids form preferentially. The other part of the problem is most simply solved by using varying RNA concentrations and extrapolating the extent of hybridization to infinite RNA concentration (in practice this is most easily achieved with a reciprocal plot; Fig. 3.6(a,b)).

Usually radioactive RNA is employed. Unhybridized RNA is removed by digestion with RNAase A and the RNA:DNA hybrids are either precipitated and counted or isolated on nitrocellulose (see section 3.2.2).

For many purposes, rather than measuring the amount of hybrid formed at equilibrium, the rate of hybrid formation is measured under conditions in which one nucleic acid is in vast excess. The methods are analogous to the C_0t procedure and are referred to as the R_0t and D_0t ('rot' and 'dot') methods. In the R_0t method (Fig. 3.6(c)) RNA is in excess and the rate of hybrid formation (measured as insensitivity of the DNA radioactivity to S1 nuclease; see section

3.4.3) is plotted as a function of R_0t (R_0 = total RNA concentration). In the D_0t method (Fig. 3.6(d)) DNA is in excess and the rate of hybrid formation (insensitivity of the RNA radioactivity to RNAase A; see section 3.4.1) is measured as a function of D_0t (D_0 = total DNA concentration). Note from the figures that R_0t and D_0t curves are usually plotted 'upside down' with respect to the C_0t curves, that is the vertical axis shows increasing hybridization whereas in a C_0t curve it shows increasing single-strandedness. Values of $(R_0t)_{1/2}$ and $(D_0t)_{1/2}$ are calculated, usually by computer fitting of the curves. The number of components in a R_0t curve gives the number of hybridization classes and $(R_0t)_{1/2}$ gives the concentration of RNA for each class in a mixture. $(D_0t)_{1/2}$ gives the number of DNA sequences homologous to different RNA species. This last number is the number of gene copies and is a value that can, in principle, be arrived at by RNA saturation equilibrium hybridization.

3.1.4 Tertiary structure of DNA

The majority (probably all) of naturally occurring DNA molecules have a tertiary structure that involves helical turns other than those present in the Watson–Crick double helix. The nature of the tertiary structure is most readily seen by considering covalently closed, relatively small continuous DNA molecules (plasmids). Four possible tertiary structures for such a molecule are sketched in Fig. 3.7. In considering these pictures, remember that the *single* line in each structure represents a *double*-stranded molecule. Let us assume that each of these molecules is topologically equivalent. In experimental terms, that means that we could take a molecular model of a plasmid and twist it into any one of the shapes in Fig. 3.7 without breaking any covalent bonds. If we were to do that, we would observe that in order to interconvert (a), (b) and (c) the manipulations of the model would involve altering the total number of turns in the Watson–Crick helix. The nature of the changes would depend upon whether the superhelix was in the same or the opposite sense to the Watson–Crick helix. If the model represented a real plasmid, the number of Watson–Crick turns would increase as we went from (a), through (b) to (c). In other words, the supercoiled forms (a) and (b) are 'underwound'. (For the purposes of building the model we could have produced an 'overwound' supercoil so the effect of doing the former experiment would be to decrease the number of Watson–Crick turns.) Before we return to the more quantitative implications of these facts, consider the manipulation that results in the structure

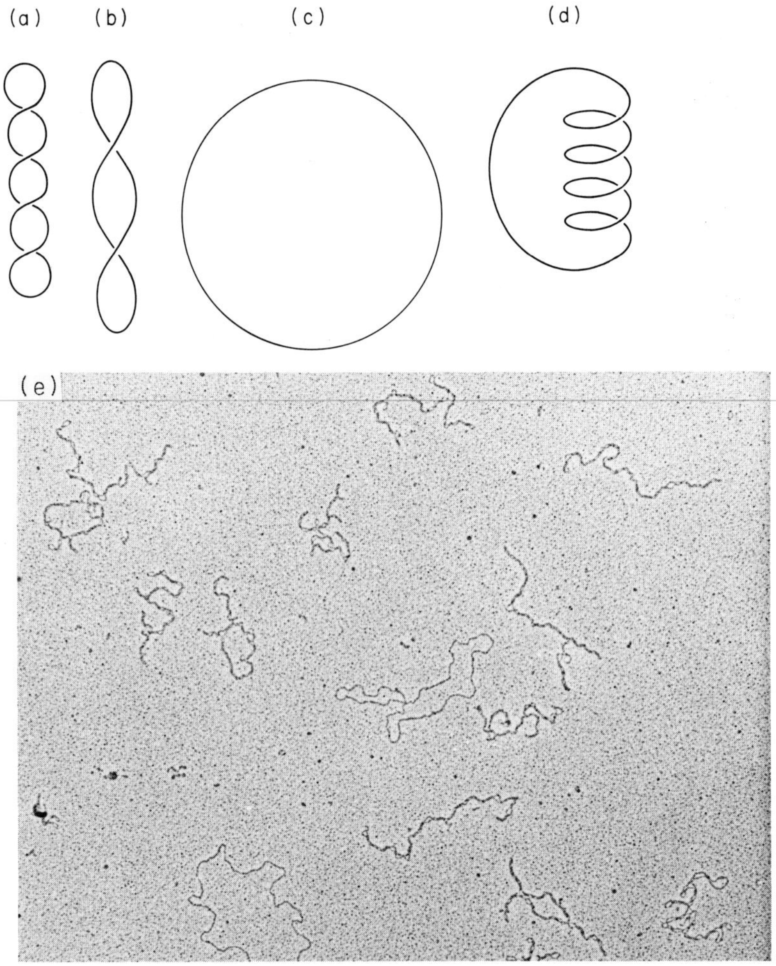

Fig. 3.7. (a–d) Tertiary structures for a covalently closed DNA molecule. The single line represents a Watson–Crick double helix—see text for discussion. (e) An actual example of supercoiled molecules of the plasmid ColE1 visualized by electron microscopy. Two of the molecules have been 'nicked' in one strand and have the structure shown in (c); the others demonstrate supercoiling as in (a). (We are grateful to Dr V. Virrankoski-Castrodeza for providing the photograph.)

in Fig. 3.7(a) converting into that in Fig. 3.7(d). Two differences would be observed:

(1) the transition would be easier to accomplish; less work is needed;

(2) the number of Watson–Crick turns would be unchanged.

Thus although all the structures are topologically equivalent, only two of them, (a) and (d), are of comparable stability. The structure shown in (a) is found in plasmids (Fig. 3.7(e); see section 4.1.2), mitochondrial and chloroplast DNA molecules (see section 8.2) and the bacterial chromosome (see section 4.1.1); the structure shown in (d) represents the conformation of DNA in eukaryotic nucleosomes (see section 7.2). In this last case, the whole DNA molecule is not a continuous loop. However, it is constrained by DNA-binding proteins in such a way that DNA domains behave as though they were parts of closed loops.

For quantitative purposes, three parameters are used to describe DNA topology. The topological winding number (α) is the number of times one single strand crosses the other one; the helix winding number is the number of helical turns in an unconstrained molecule

(β; this would be the number of Watson–Crick turns in a linear double-stranded molecule of the same number of base pairs); the difference (α–β) is the number of superhelical turns (τ). Note that by twisting a model, without breaking any bonds, α remains invariant but it is possible to alter β and τ. Fig. 3.7(c) represents a conformation in which τ is zero; equally by introducing a very high density of superhelical turns, it is possible to make β zero and construct a conformation in which there is no Watson–Crick helix (just base pairing) so that all the helices are supercoils.

We can now re-state the generalization about the 'underwinding' of natural DNA molecules—τ is invariably negative. The supercoil is therefore 'unstable'. This has two experimental consequences: if a supercoiled molecule has its topological constraint removed by introducing a single-strand discontinuity with a nuclease (see section 3.4.2) it will relax (a structure such as Fig. 3.7(a) will open up to form Fig. 3.7(c) as visualized in Fig. 3.7(e)); secondly the T_m of a natural plasmid (Fig. 3.7(a)) is invariably lower than the T_m for its relaxed form. This last point follows from the fact that by removing one Watson–Crick helix by denaturation β is reduced and the magnitude of τ is reduced.

The tertiary structure of DNA and the properties of supercoiled molecules with a negative τ are exploited by organisms in two ways:

(1) supercoiled DNA is more compact and takes up less space in the cell or nucleus;

(2) the facility of denaturation represents a device for making the activity of proteins that unwind the double helix easier; these proteins include the nucleic acid polymerases (see sections 4.2 and 4.3). However, there is an energetic deficit to be made up; the introduction of supercoils involves expenditure of metabolic energy in the form of ATP hydrolysis (see section 4.2.2).

3.1.5 Tertiary structure of RNA

There are two methods for trying to establish the further folding of RNA molecules. One is to study the optical properties and chemical reactivities of the molecules in solution; the other is to obtain a high-resolution X-ray diffraction diagram of the crystals and work out the structure systematically. From the history of the attempts to come up with a tertiary structure for tRNA, one is bound to conclude that the former approach is hardly satisfactory. None of the many models for the molecule that were deduced from spectroscopic and chemical studies corresponded with the true conformation

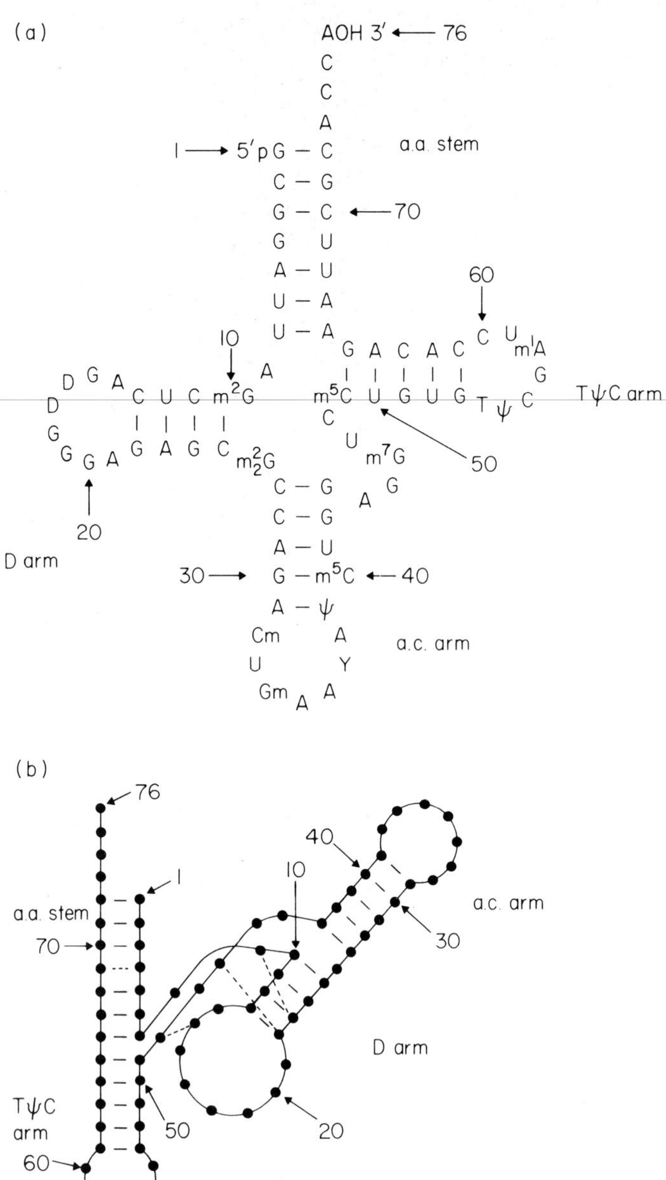

Fig. 3.8. Yeast tRNA^{Phe}. (a) The sequence written in its clover-leaf secondary structure. See Fig. 3.29 for the structures of the modified residues (m_2^2G, D, ψ, etc.). The molecule shares with all tRNA molecule three 'arms' (each composed of a double helical 'stem' and a 'loop') and one additional stem. The arms and stem are named after characteristic residues or functions: a.a. (amino acid acceptor), D (dihydrouracil), a.c. (anticodon) and TψC (from that sequence). The D-arm is invariably so-called even though at least one tRNA lacks the D-residues. The region between residues 44 and 48 in the tRNA^{Phe} sequence varies greatly between different tRNA sequences. In some molecules it is very much longer and is then referred to as the 'extra arm'. Hyphens (-) represent G:C or A:U base pairs. (b) The clover-leaf distorted so as to stack the a.a. and TψC stems and also the a.c. and D stems. Dotted lines represent additional H-bonds in the tertiary structure.

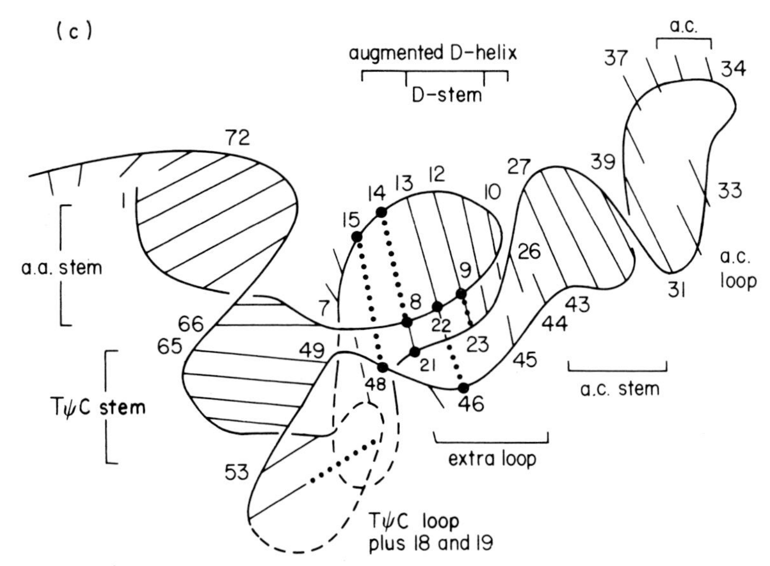

Fig. 3.8. (cont.) (c) The tertiary structure of yeast tRNA^{Phe} as determined by X-ray crystallography. (Reproduced, with permission, from Roberts *et al.* 1974.)

that was determined from tRNA crystals. For this reason, tRNA is the only class of cellular RNA whose conformation is worth considering. The structure is constructed from the clover-leaf secondary structure by base stacking of the stems of the molecule and is held together by many unusual H-bonded interactions, some involving the 2′-OH of ribose groups (Fig. 3.8).

3.2 ISOLATION, FRACTIONATION AND CHARACTERIZATION OF NUCLEIC ACIDS

3.2.1 Isolation

Nucleic acids are water-soluble polyanions; their sodium and potassium salts are insoluble in most organic solvents, including mixtures of ethanol and water but they are not denatured by organic solvents. They are thus different from other commonly encountered cellular components. Indeed the only other group of substances that resemble nucleic acids at all in their solubility properties are the teichoic acids. These latter are polyalditol phosphates found in the cell walls of Gram-positive bacteria. Detailed recipes for nucleic acid extraction can be found in Parish (1972) or in the several volumes of *Methods in Enzymology* devoted to nucleic acids. Only the principles are discussed here.

The isolation of relatively large quantities of nucleic acid from several grams of tissue or cells is usually achieved by extraction of the material with an organic solvent (either phenol or chloroform) which will denature proteins and leave the nucleic acids intact and in solution in the aqueous phase. The details of the different recipes are designed to overcome the two main obstacles to successful isolation of nucleic acids, namely damage by hydrodynamic shear and hydrolysis by nucleases. In general, shearing is more likely to damage DNA than RNA and, conversely, enzymic hydrolysis is likely to cause more problems with RNA than DNA. The tissue or cells are thus disrupted as gently as possible. With many bacteria this involves preliminary treatment with lysozyme, to remove, or at least to damage, the cell walls. Nucleases are inhibited by including in the solution detergents or chelating agents (or both). Repeated extractions are normally required and the nucleic acids can be precipitated from the aqueous phase with alcohol. The only serious contaminants are likely to be glycogen and teichoic acids. Special procedures are available for removing these substances. Alternatively the cells can be disrupted and subcellular fractions containing, for example, nuclei, polysomes or microsomes, can be obtained by differential or rate-zonal centrifugation and these fractions can then be used as a starting material for the extraction.

On a small scale, the solvent extraction procedures can also be used. However, there are several alternative methods of removing protein from small samples of cells or subcellular fractions; for example the preparation can be treated with a proteolytic enzyme or, in the case of polysomes, with a detergent. The detergent-treated sample is simply layered over a sucrose gradient or an electrophoresis gel and the RNA species are resolved.

Any RNA molecule or any viral nucleic acid should be obtained in an undegraded state. Similarly plasmid DNA should be recovered intact; the special procedures for obtaining plasmids are described in section 3.3.5. Chromosomal DNA is usually broken into fragments. Special methods are available for obtaining the bacterial chromosome in one piece but the preparations contain other macromolecular species (see section 4.1.1).

3.2.2 Fractionation

In this section and the following one (3.2.3) only those methods in more-or-less everyday use in laboratories will be described. Birnie (1972) and *Methods in Enzymology* are good sources for general technical detail.

Separation of the major classes of nucleic acids

DNA, tRNA, rRNA and mRNA are normally separated by suitable modification of the isolation procedure; for example, ribosomes, isolated by subcellular fractionation are the obvious starting point for rRNA—polysomes are the starting point for a mixture of this with mRNA. Moreover, the major classes of nucleic acid have different solubilities in salts and certain solvents. RNA of high molecular weight is insoluble in 4M NaCl whereas tRNA and DNA are soluble. Most purification procedures for DNA include an incubation with RNAase to destroy contaminating RNA; the ribonucleotide fragments can then be removed by dialysis. Transfer RNA can be selectively leached out of many cells by mild treatment with phenol.

Rate-zonal centrifugation

Zonal centrifugation through sucrose gradients is the standard procedure for fractionating amounts of RNA of the order of milligrams or less. Such gradients can be designed to resolve molecules differing in sedimentation coefficient by about 4 S-units. Sucrose gradient centrifugation of DNA is beset by difficulties owing to the extreme dependence of sedimentation coefficient on DNA concentration; the consequence is that any quantity of DNA greater than radioactively labelled trace amounts forms distorted zones with a steep tailing edge containing an unresolved mixture slowly sedimenting as a meshwork of molecules. However, sucrose gradients are particularly useful for the fractionation of radioactively labelled denatured DNA molecules. The gradients are made up in alkali (pH 12.0). One use of this is to detect the presence of small single-stranded DNA fragments, which may occur in unrepaired DNA from cells that have suffered radiation or chemical damage (see sections 3.3.2 and 3.3.3), or the fragments that are produced during DNA replication (see section 4.2.2). The other application of alkaline sucrose gradients is in the fractionation of DNA-containing plasmids. When the plasmid denatures, the two strands cannot come apart and the plasmid assumes a tangled but rather tightly constrained ball that sediments very rapidly and leaves the linear, denatured fragments of chromosomal DNA in a relatively slowly sedimenting zone. Fig. 3.9 (see next page) summarizes the sedimentation properties of various forms of a small DNA molecule.

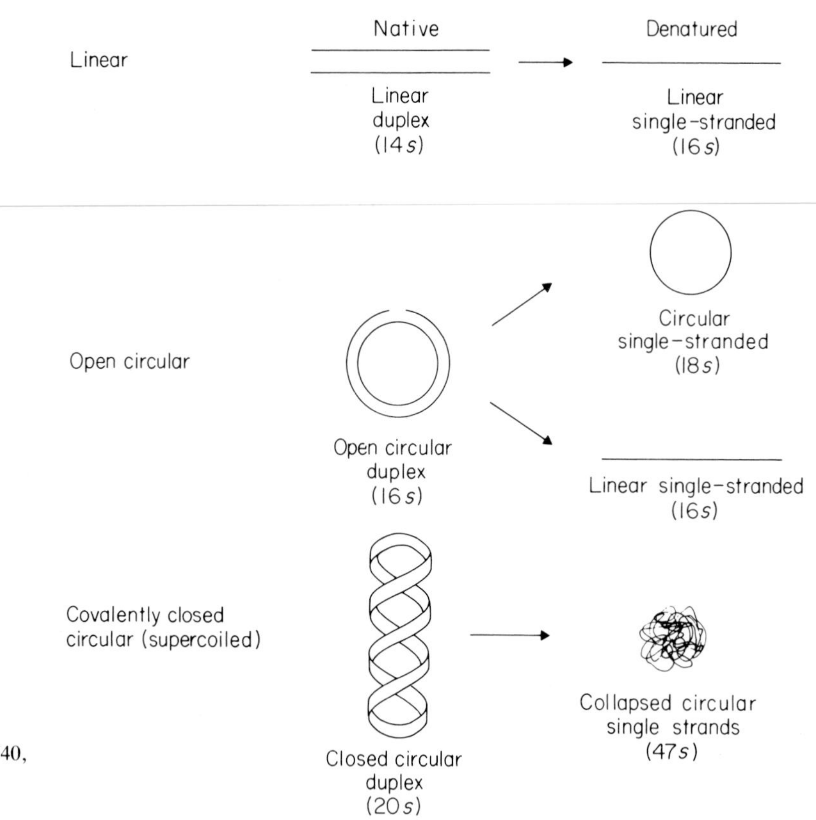

Fig. 3.9. Forms of the DNA from SV40, a small animal virus, with their sedimentation coefficients.

Buoyant density centrifugation

Buoyant density centrifugation is a technique for resolving nucleic acids on the basis of density. The material most commonly used for establishing the gradient is CsCl. As buoyant density centrifugation is an important analytical method as well as an important method for fractionation (see section 3.2.4), the theory has been considered in great detail (Vinograd & Hearst 1962). It is an equilibrium method and relies upon the facts that strong solutions (5M) of CsCl form a density gradient at high centrifugal fields and that the Cs salt of DNA has a density intermediate between the extremes of concentration (and hence of density) at the top and bottom of the centrifuge. CsCl gradients will resolve DNA species whose densities differ because of differences in base composition (see section 3.5.3), differences in tertiary structure in the presence of a dye (see section 3.3.5) or because of differences in density generated by differences in the components of the growth medium of cells. There are two ways of achieving this 'density labelling'. In the classical procedure of

Meselson and Stahl (see section 1.2.2), the difference is produced by growing bacteria at the expense of $^{15}NH_4Cl$ as sole nitrogen source and then transferring the cells to $^{14}NH_4Cl$. A CsCl gradient can resolve the dense ^{15}N-labelled DNA from ^{14}N-DNA and the hybrid $^{14}N/^{15}N$ molecules. An alternative to using isotopes of nitrogen for density labelling is to grow cells in the presence of either 5-bromouracil or 5-bromodeoxyuridine. These compounds are metabolized by scavenging reactions (see section 2.3.2) to 5-bromodeoxyuridine 5′-triphosphate; this is a substrate for DNA polymerase and results in the synthesis of DNA containing 5-bromouracil residues in place of thymine. Such DNA is denser than normal DNA.

Column chromatography of RNA

In contemporary nucleic acid methodology, ion-exchange chromatography of RNA is limited to a number of procedures for the resolution of tRNA species. Sephadex and Sepharose gel exclusion chromatography have relatively few applications although Sephadex is the only convenient method for resolving tRNA and 5*s* rRNA.

The formation of hybrids between nucleic acids in solution and immobilized, complementary oligonucleotide sequences is exploited in a special variety of affinity chromatography. The most important application of this technique is in the purification of eukaryotic mRNA. The resin most commonly used is oligo (dT) cellulose, that is oligothymidylic acid covalently attached to cellulose. The resin is gently stirred with a solution of unfractionated or partially fractionated RNA. Hybrids are formed between the oligo dT and the poly (A) sequences at the 3′-end of mRNA (see section 9.3.6). The resin is placed in a little chromatography column and eluted with buffer until all the unbound RNA (tRNA and rRNA) is removed. The ionic strength is then reduced and the column warmed to melt the hybrids and mRNA is eluted.

Separation of denatured and native DNA

There are two materials in common use that have different affinities for single- and double-stranded DNA: hydroxyapatite (one of the crystalline states of calcium phosphate) and nitrocellulose. In neither case is the physicochemical basis of the fractionation fully understood. Hydroxyapatite is most commonly used as a stationary phase in column chromatography. At low (0.12M) phosphate concentrations double-stranded DNA has a much higher affinity for hydroxyapatite than does single-stranded DNA and so builds to the column. The method is valuable for removing residual double-stranded DNA from a denatured sample or for separating the reactants and products of a hybridization reaction. By increasing the phosphate concentration to 0.5M or more, double-stranded DNA can be eluted from the column and recovered. Alternatively, the temperature of the column is raised to the above-melting temperature of the bound DNA so that it is eluted as it denatures. Nitrocellulose binds single-stranded DNA but not double-stranded molecules. A common procedure is to bind denatured, single-stranded DNA to a nitrocellulose membrane filter, then to 'fix' the DNA to the filter by baking at 80 °C for 2–4 hours. DNA prepared in such a way is available to participate in hybridization reactions with other single-stranded nucleic acids (RNA or DNA) making the separation of hybrids from the reaction mix a simple matter of removing the filter. If a radioactive RNA is used in a hybridization reaction with the immobilized DNA, the hybrids formed can be quantified simply by counting the radioactivity bound to the filter in a scintillation counter.

Fig. 3.10. Southern hybridization. (a) Pattern of restriction fragments obtained from a recombinant λ phage carrying yeast DNA, fractionated by electrophoresis in 1% agarose gel and visualized by staining with ethidium bromide. The numbers refer to the size of the fragments in kilobase (kb) pairs. The DNA in these bands was transferred to a nitrocellulose sheet by first denaturing the DNA (by immersion in NaOH solution) and then blotting the DNA from the gel to the filter using apparatus shown in (c). This apparatus is designed to draw buffer through the wick, through the gel and through the nitrocellulose into the paper towels. The buffer carries the DNA with it in a vertical direction and the DNA binds to the nitrocellulose. The nitrocellulose sheet was then incubated with denatured ^{32}P-labelled yeast 'probe' DNA and hybrids were allowed to form. Hybrids were identified by autoradiography (b). The 3.7 and 3.4 kb pairs sequences can be seen to be homologous with the probe while the remaining bands are composed of λ sequences and, therefore, do not hybridize with the labelled DNA probe. (Data were kindly supplied by Dr A. Kingsman.)

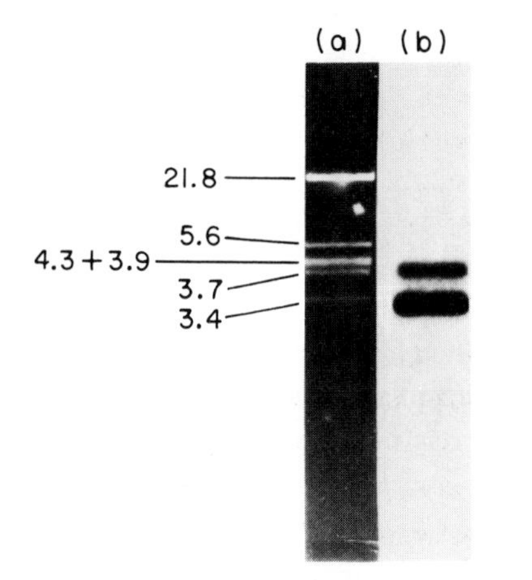

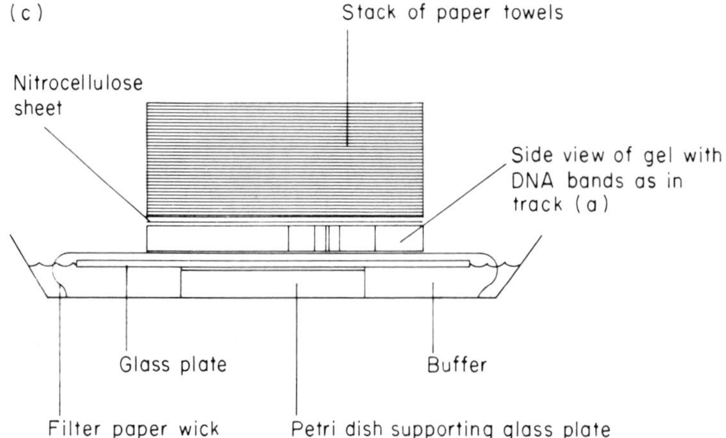

Non-specific binding of the single-stranded component in solution to the nitrocellulose filter used in the hybridization reaction is prevented by blocking binding sites with a mixture of ficoll, serum albumin and polyvinylpyrrolidone (Denhardt's solution). The affinity of nitrocellulose for DNA is also exploited in the 'Southern transfer' method for blotting DNA fragments, separated by gel electrophoresis, out of the gel and on to a nitrocellulose sheet. This procedure allows the hybridization of fragments fixed in the same array as in the gel and by the use of autoradiography hybrids formed by specific fragments can be identified (Southern 1980; Fig. 3.10).

This powerful technique for transfer hybridization is limited only by the difficulty of immobilizing low molecular weight DNA and RNA molecules in nitrocellulose. One alternative is to bind the nucleic acid covalently to chemically modified paper (see section 3.3.2). Alternatively the Southern transfer can be modified by using glyoxal and dimethyl sulphoxide to denature the DNA under conditions that promote transfer and retention of even small DNA molecules and (with modifications) RNA molecules to cellulose nitrate (Thomas 1980).

Gel electrophoresis

Gel electrophoresis is a powerful method for the fractionation of nucleic acids. It is limited by scale but for most purposes only small quantities (μg) are needed as the technique is largely analytical although there are some important preparative applications (see section 3.5). Usually one of two materials is used as the gel matrix: polyacrylamide is generally used for the fractionation of relatively low molecular weight molecules (i.e. less than 1 kb (kilobase)) so that 3% polyacrylamide gels might be used to fractionate molecules in the 1 kb range while 15% polyacrylamide might be used to separate molecules only a few nucleotides long. The resolving power of high percentage polyacrylamide gels is such that molecules differing in length by a single nucleotide may be separated; this property is exploited to great effect in the rapid sequencing methods described in section 3.5. Larger molecules (i.e. greater than 1 kb) are usually fractionated in agarose gels so that 0.1% agarose is used to fractionate fragments of between 40–79 kb while 1.5% agarose would be used to fractionate molecules around 1 kb. Gels may be run under native or denaturing conditions. A variety of denaturants may be incorporated into gels for the fractionation of single-stranded nucleic acids so that secondary structures do not interfere with mobility through the gel. Among the most commonly used denaturants are formamide, urea and methyl mercury hydroxide (Maniatis & Efstratiadis 1980). An alternative strategy involves the irreversible denaturation of nucleic acids by glyoxalation followed by electrophoresis on a non-denaturing gel (Carmichael & McMaster 1980). This has the advantage that denatured and native nucleic acids can be resolved on the same gel.

Nucleic acids migrate in gels largely on the basis of differences in molecular weight. All nucleic acids have an approximately equal charge density (-1 per nucleotide residue if we neglect the partial charges on the basis of residues). Moreover, in most cases, in any one preparation the conformations of the molecules will be the same. Thus all molecules will experience the same force pulling them through the electric field but the large molecules will experience more difficulty in penetrating the matrix of the gel. They are, therefore, retarded with respect to the smaller ones. The mobility of a molecule is approximately inversely proportional to the log of its molecular weight (see section 3.2.4), this relationship generally holds only for a restricted range of molecular weights.

Gel electrophoresis may also be used for the separation of conformations of the same molecule; for example the covalently closed and open circular forms of SV40 may be separated from each other and from linear molecules. The precise relationship of the mobilities depends on the voltage gradient (Fig. 3.11).

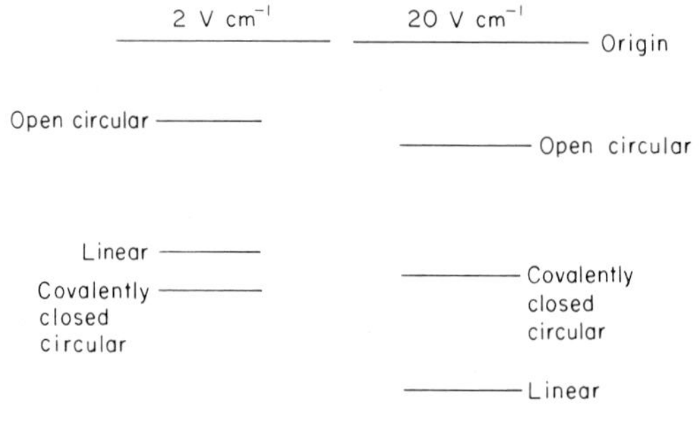

Fig. 3.11. Variation in the mobility of three conformations of SV40 DNA (see Fig. 3.9 for details) with the voltage gradient. The genome length is 5.2 kb pairs.

3.2.3 Characterization

Detection and measurement

Unlabelled nucleic acids are most commonly detected and measured by two techniques. The first depends on their absorbance at 260 nm. The relationship between concentration and absorbance depends on the secondary structure of the molecule because of the hyperchromic effect (see section 3.1.3). Double-stranded nucleic acid at a concentration of 1 mg ml^{-1} has an absorbance of about 20, whereas a solution of a single-stranded nucleic acid at the same concentration has an absorbance of about 26. Fractions of sucrose or caesium chloride gradients can be monitored or gels may be scanned for material absorbing at 260 nm. The second detection method involves staining nucleic acid with dyes, which in association with nucleic acids fluoresce under UV-irradiation. These procedures are most commonly used for the detection of nucleic acids in gels. Ethidium bromide is frequently used although acridine orange is useful for some applications (Carmichael & McMaster 1980).

Radioactive nucleic acids are detected by scintillation counting or by autoradiography. We shall see later (see section 3.5) that the bulk of structural and chemical studies on nucleic acids employ ^{32}P-labelled material. The isotope is convenient to use because it can be detected very readily by exposing gels to pieces of X-ray film.

Of the colour reactions for nucleic acids, two are convenient for measuring the nucleic acid content of nucleoproteins or cells. These are the hydrolysis of RNA and recognition of the pentose by the green colour produced with orcinol and the reaction between deoxyribose and diphenylamine to produce a blue colour.

Biological and biochemical assay and detection

There are a variety of biological or biochemical methods available for the detection of nucleic acids that depend on their specific structures or functions; for example tRNA can be assayed by the aminoacyl tRNA synthetase reaction (see section 5.4), which might measure the incorporation of radioactivity of labelled amino acids into an acid-insoluble fraction, and messenger RNA is frequently assayed by its ability to direct the synthesis of proteins in a crude cell-free system. The only commonly used biological assay is transformation or, in the case of viral DNAs, transfection. In transformation procedures it is necessary to quantify the number of recipient cells that inherit a selectable marker and relate this figure to quantities of DNA, for example if 1 μg of plasmid DNA carrying a drug resistance gene is used to transform *E. coli* under standardized conditions, about 10^5 cells receive the plasmid and become drug resistant. In transfection experiments infection centres (plaques) are quantified and then related to quantities of DNA (see section 11.5).

3.2.4 Determination of molecular weight

There are two approaches to determining the molecular weight of a macromolecule. One can do the job systematically and thoroughly by making measurements of physical properties of solutions, such as sedimentation coefficients, light-scattering properties, osmotic pressure, and calculating the molecular weights from appropriate equations. Alternatively, one can use reliable standards to calibrate some appropriate fractionation protocol. The disadvantages of the former approach are that the methods are frequently time consuming, there may be a requirement for specialized apparatus that is not available and the calculations may be complicated, especially by the need to get round the contributions of conformational parameters. In fact the last of these is a strength rather than a disadvantage, especially of hydrodynamic methods, because if the molecular weights are certain from independent measurements, the techniques can be exploited to measure conformational parameters in solution.

The routine practice in most nucleic acid laboratories is to measure the molecular weights of unknown molecules by using gel electrophoresis or direct contour length measurement in the electron microscope. In both cases molecules of known molecular weight are used as internal standards for the calibration of the technique.

As mentioned previously native gels are used to fractionate double-stranded nucleic acids and denaturing gels are used for single-stranded molecules so that mobility measurements, and, therefore, molecular weight estimates, are not influenced by conformational variations. This subject is treated thoroughly by McDonell *et al.* (1977) and Johnson and Grossman (1977).

In the electron microscopic techniques the nucleic acid is spread out on a film of a basic protein and then stained and shadowed and examined in the microscope. As with the electrophoretic techniques a denaturant (usually formamide) is used in the analysis of single-stranded molecules to prevent 'bunching-up' of the molecules due to base-pairing interactions. An internal standard molecule of known molecular weight is incorporated in the sample and the lengths of the unknown and stranded molecules are measured in microns. Contour lengths can be converted to molecular weights or numbers of bases or base pairs; for double-stranded DNA the relationship is the following:

$$1 \text{ micron } (\mu\text{m}) = 2 \times 10^6 = 3000 \text{ base pairs (3 kb pairs).}$$

The use of electron microscopy in nucleic acid research will be dealt with in more detail in section 3.2.5.

The molecular weight of a nucleic acid may also be calculated from its $S_{20,w}$ value measured either in an analytical centrifuge or by comparison with marker molecules of known $S_{20,w}$ value in a sucrose gradient in a preparative centrifuge. In this second case the distance travelled down the centrifuge tube is directly proportional to $S_{20,w}$. While this procedure is much less accurate than making the measurement in an analytical centrifuge it is quick and easy and is sufficient for most purposes. The most reliable relationships are for linear nucleic acids: for native DNA M_r between 0.3×10^6 and 3×10^6

$$S_{20,w} = 0.116 M_r^{0.325}$$

and for native DNA M_r 3×10^6 to 130×10^6

$$S_{20,w} = 0.034 M_r^{0.405}.$$

The need for two equations arises from the fact that duplex DNAs of different sizes behave in different ways. Short molecules behave as stiff rods whereas the longer molecules adopt a loosely coiled structure (Eigner 1968). Reliable equations also exist for single-stranded nucleic acids but only in denaturing conditions, e.g. 0.1M NaOH for DNA and 1.1M formamide for RNA (Boedtker 1968, Eigner 1968). The relationship is

$$S_{20,w} = 0.052 M_r^{0.4}.$$

In this case one equation may be used across a wide molecular weight range as single-stranded molecules of different sizes exhibit similar flexibilities.

3.2.5 Electron microscopy

The visualization of nucleic acids in the electron microscope, pioneered by Kleinschmidt (1968) has provided techniques of major importance in molecular biology. The methods allow direct measurement of molecular weight, the assessment of identity of two molecules, the mapping of transcriptional units and the localization of origins of replication. For a review of techniques see Davis *et al.* (1974).

Nucleic acids can be visualized in the electron microscope in a field of protein particles. The basic protein, cytochrome c, is most commonly used and the image is composed of a column of protein particles surrounding the DNA or RNA. The nucleic acid is mixed with cytochrome c to form the spreading solution or hyperphase. This is spread over a buffer solution (the hypophase) usually by running the spreading solution down a glass slide. The basic protein carries the nucleic acid with it and spreads and extends the molecules. With suitable modifications, the spreading can release nucleic acid from nucleoprotein particles, notably phages (Fig. 3.12). The extension of the nucleic acids is critical for contour length measurements. The film of protein and nucleic acid is picked up on to a parlodion covered grid, stained with uranyl acetate and (in most protocols) is subsequently shadowed with a heavy metal.

If the spreading solution and hypophase are both aqueous, double-stranded DNA appears as a gently curved molecule but single-stranded DNA forms condensed structures owing to random base interactions. If, however, single-strand contour lengths are to be measured these random interactions need to be eliminated and this is most commonly achieved by incorporating 40% (v/v) formamide in the 10% (v/v) formamide in the hypophase. Single-stranded nucleic acids appear slightly thinner than double-stranded molecules but have a rather kinked appearance.

Pretreatment of the nucleic acid molecule can be used to identify unique sites and sequences in the electron microscope image. As the base composition of DNA molecules is not uniform along their

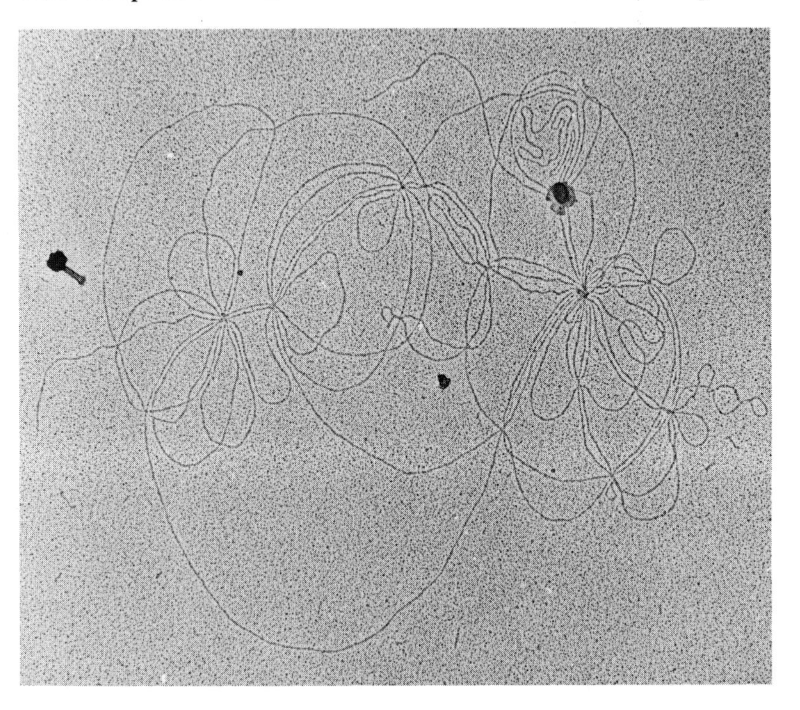

Fig. 3.12. Visualization of DNA released from *Myxococcus* phage Mx1 by the method of Kleinschmidt. The figure contains one complete Mx1 DNA molecule. An intact particle of the same phage can be seen on the left. (We are grateful to Dr V. Virrankoski-Castrodeza for providing the photograph.)

lengths, partial denaturation will result in denatured loops that locate high AT regions of the molecule. This technique is called denaturation mapping (Inman 1974) and can be used, for example, to locate origins and termination regions of chromosomes by examining partially denatured replicating molecules. A different approach involves predigestion of the molecule with a restriction enzyme. This is particularly valuable if the molecule (such as a viral or plasmid DNA) has a single site for a restriction enzyme. In conjunction with denaturation mapping it is possible to orient replication forks and their direction of movement (Fareed & Kasamatsu 1980). A rather different use of a restriction enzyme to visualize a recombinational intermediate is illustrated in Fig. 4.19(b).

Electron microscopy of hybridized nucleic acid molecules provides a method for examining the extent and nature of the complementarity between two molecules. A non-complementary region appears as a single-stranded loop. These are referred to as R- (for RNA) and D- (for DNA) loops. Analysis of loops is important in the characterization of recombinant DNA molecules (see section 11.6) and in recognition of introns and exons in eukaryotic coding regions (see section 8.3). The technical details, as applied to the construction of a transcriptional map of adenovirus, are described by Sharp *et al.* (1980).

3.3 THE ORGANIC CHEMISTRY OF NUCLEIC ACIDS

This subject is important for two reasons. Many of the reactions are exploited in sequencing techniques and many are also the basis of action for many cytotoxic, mutagenic and carcinogenic substances and of certain chemotherapeutic drugs. In this chapter, we have concentrated on reactions that are genuinely important for these reasons. For more systematic accounts of nucleic acid chemistry see Parish (1972) and Zorbach and Tipson (1968, 1973).

3.3.1 Hydrolysis and hydrazinolysis

Nucleic acids are phosphodiesters. Most phosphodiesters, for example diethyl phosphate, $(EtO)_2PO_2H$, are remarkably resistant to acid and base catalysed hydrolysis. DNA is a typical phosphodiester in this respect; in contrast, RNA is quite readily hydrolysed by alkali. The reason is shown in Fig. 3.13: the 2′-OH group participates in the reaction to form a 2′, 3′-cyclic phosphate intermediate; this is

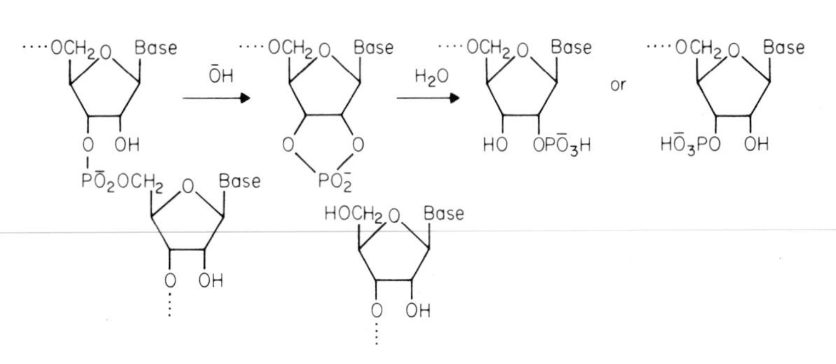

Fig. 3.13. Hydrolysis of a phosphodiester bond in RNA in a reaction catalysed by alkali.

ring-opened to yield a mixture of 2′- and 3′-mononucleotides. Note that if the RNA molecule contains 2′-methyl ribose residues (Fig. 3.29), this hydrolysis cannot occur and the presence of these residues can be detected by the presence of dinucleotides in an alkaline digest of the RNA.

The *N*-glycoside bond is acid labile. Nucleic acids might thus be expected to lose their bases in acid solution and leave polyribose phosphate (or polydeoxyribose phosphate). DNA is more acid labile than RNA; purines are more readily released than pyrimidines; and the polypentose phosphates are not stable in aqueous solution. The rationale for the different rates of hydrolysis of different *N*-glycosides is outside the scope of this book. It is sufficient to observe that dilute acid selectively removes purines from DNA to leave untouched the stretches of pyrimidines. This used to form the basis of a sequencing method. The gentlest and most selective conditions involve using formic acid and diphenylamine (see section 3.2.3). Pyrimidine residues are not removed from RNA except by extreme conditions above the boiling point of water and this is the reason why when RNA is hydrolysed with acid prior to doing an orcinol test only the ribose residues originally attached to purines are detected (see section 3.2.3). The depurinated sequences are unstable because the sugar residues in these molecules are now reducing sugars at C-1′ and the phosphate ester bonds can be lost by a β-elimination reaction. The consequence of this process for a partial hydrolysate of DNA, for example, is that the oligopyrimidine nucleotides that are produced have terminal phosphomonoester groups at both 5′- and 3′-ends.

We have seen that by carefully manipulating the conditions of acid hydrolysis we can selectively remove purines from a nucleic acid. There is a comparable reaction for the removal of pyrimidines and it is used in one of the most versatile methods for DNA sequencing (see section 3.5.4). Hydrazine ($NH_2.NH_2$) reacts with pyrimidine residues as shown in Fig. 3.14. The conditions for this reaction can be modified to make the reagent C-specific.

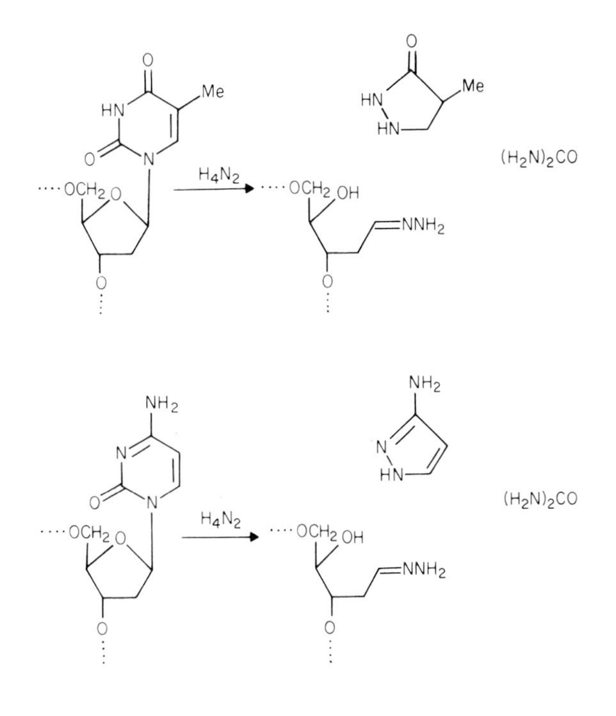

Fig. 3.14. Reaction between hydrazine and pyrimidine residues in DNA.

Large DNA molecules can also be fragmented by other methods, notably hydrodynamic shear. Additionally the most important consequence of X-irradiation of DNA is the production of double-strand breaks and consequent fragmentation of the molecule.

3.3.2 Reactions with electrophilic reagents

An electrophilic reagent is one that reacts with a complementary ('nucleophilic') reagent in such a way that an 'electrophile', or 'electron lover', is generated as an intermediate. As this type of reaction is of great importance in the biological chemistry of DNA, let us consider the general reaction in some detail. A good example of an electrophilic reagent is an ester of a sulphonic acid, for example methyl methanesulphonate, $MeOSO_2Me$. This reacts with acetate ions, $CH_3CO_2^-$, to form methyl acetate and methanesulphonate as follows:

$$MeOSO_2Me + CH_3CO_2^- = CH_3CO_2Me + {}^-OSO_2Me.$$

The reaction involves prior dissociation of the $MeOSO_2Me$ to form Me^+ and $^-OSO_2Me$. The Me^+ (methyl carbonium ion) is the electrophile that will react with any molecule with a high density of electrons that can neutralize the charge on the unstable Me^+ ion. Such a molecule is the nucleophile (the acetate ion in the present example). The reason that $MeOSO_2Me$ is a powerful electrophilic reagent is that the counter-ion produced ($^-OSO_2Me$) is very stable; another way of expressing this is to say that methanesulphonate is a 'good leaving group'. Let us now consider the general case and use ElX to represent the electrophile reagent (X^- being the leaving group), and let Nu^- represent the nucleophile. The reaction will occur in two stages:

$$ElX = El^+ + X^-$$
$$El^+ + Nu^- = ElNu,$$
overall $$ElX + Nu^- = ElNu + X^-.$$

In the cases we shall be considering, the nucleophile is not negatively charged overall (unlike the acetate ion) but rather is a molecule with a high density of electrons at one point. In this case, the general reaction is exactly the same except that we can represent the nucleophile by Nu and the product must necessarily become positively charged:

$$ElX = El^+ + X^-,$$
$$El^+ + Nu = ElNu^+$$
overall $$ElX + Nu = ElNu^+ + X^-.$$

These reactions imply that ElX is a powerful electrophilic reagent, that is that X^- is an extremely good leaving group, El^+ forms readily and the dissociation occurs readily. Note that the rate of reaction is determined solely by ElX concentration. In the case where El^+ does not form so readily, Nu is required to assist the dissociation as follows.

$$Nu + ElX = Nu...El...X = {}^+NuEl + X^-.$$

In this case El^+ is not formed as an intermediate, Nu...El...X is a transition state and the reaction rate is dependent upon the concentrations of both Nu and El. The truth about most electrophilic reagents is that they lie somewhere between these two extremes. They can, just, dissociate to form El^+ but the participation of Nu considerably accelerates the reaction rate. Moreover, it is possible to grade electrophilic reagents according to the extent to which the first type of mechanism predominates over the second. Now consider the case where the reagent has a mixture of nucleophiles to choose from. In other words we mix ElX with Nu^1, Nu^2, Nu^3, etc. If ElX can dissociate on its own, all the products, $^+Nu^1El$, $^+Nu^2El$, $^+Nu^3El$..., should be formed in equal amounts. However, the greater the requirement for nucleophile involvement in the dissociation, the greater will be the predominance of one product, say $^{+}Nu^{1}El$, where Nu^{1} is the strongest nucleophile. Such an alkylating agent is thus more *selective* for the nucleophile that is the target for its substitution. Let us consider now the reaction between some electrophilic reagents and DNA.

The nucleophilic centres in the DNA molecule are N and O atoms in bases. By far the most electron-dense and hence nucleophilic centre is N-7 in guanine; other sites are O^6 and N-3 in guanine, N-7, N-1 and N-3 in adenine and O^4 in thymine. The 'biological alkylating agents' are electrophilic reagents. They can be ordered according to their selectivity. The following four substances are arranged in order of decreasing selectivity: $MeOSO_2Me$, $EtOSO_2Me$, $Me(NO)NCONH_2$ and $Et(NO)NCONH_2$. The first two compounds are methyl and ethyl methanesulphonates; the last two are methyl and ethyl *N*-nitrosoureas. In these latter cases the electrophiles are the methyl and ethyl diazonium ions (MeN_2^+ and EtN_2^+) and the leaving group is the carboxamate ion ($H_2NCO_2^-$). If DNA viruses or cells are treated with these compounds and the DNA is recovered, hydrolysed and analysed, the amounts of minor reaction products relative to the predominant product, *N*7-alkylguanine, increase in the same order as the decreasing selectivity ($MeOSO_2Me$ to $Et(NO)NCONH_2$). Three of these reactions are illustrated in Fig. 3.15. With the increase of diversity of alkylation products goes an increase in biological activity. $Et(NO)NCONH_2$ is the most effective mutagen and the most hazardous carcinogen of this set. The reason is that whereas N7-substitution of guanine seems to be a relatively innocuous molecular lesion in DNA, certain of the products, notably O^6-alkylguanine, cause mutation owing to the mis-pairing of this with a T-residue because the molecule is constrained in the 'wrong' tautomeric form (Fig. 3.15(d)). See Lawley (1976a,b) for reviews.

A different sort of alkylating agent is one that contains the group $-S(CH_2)_2Cl$ or alternatively $-NMe(CH_2)_2Cl$. In these cases the leaving group is Cl^- and the electrophile is a cyclic sulphonium or ammonium ion. These reagents are selective, that is to say they alkylate DNA predominantly at N7 of guanine. The importance of this type of linkage is that it occurs in certain bifunctional alkylating agents of which the simplest examples are $Cl(CH_2)_2S(CH_2)_2Cl$ (sulphur mustard or mustard gas) and $Cl(CH_2)_2NMe(CH_2)_2Cl$ (nitrogen mustard). The reaction between sulphur mustard and a guanine residue is shown in Fig. 3.15(a). Once a substitution reaction has occurred, the product, choloroethyl thioethyl guanine, is itself an

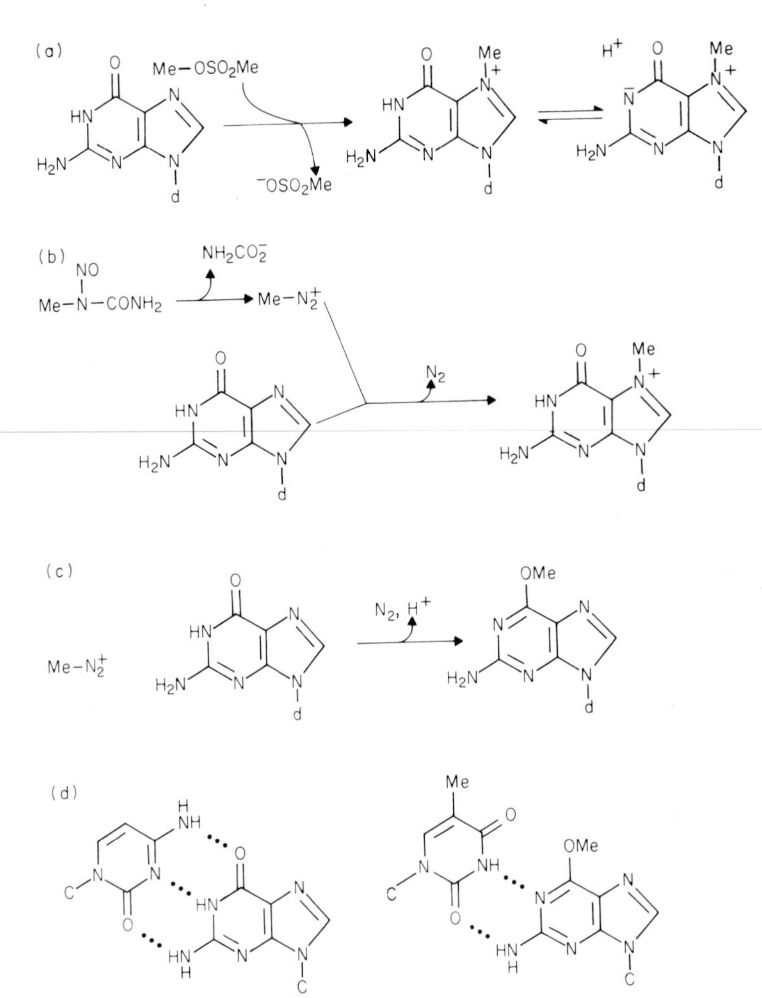

Fig. 3.15. Reaction between G-residues in DNA and alkylating agents. (a) Reaction with methyl methane sulphonate (d = deoxyribose). There is only one product (7 = methyl G). The H^1 of 7-Me G is weakly acidic (pK approximately 7) and the product exists partly as a zwitterion. (b) Reaction between methyl nitrosourea and a G-residue to form 7-Me G. The methane diazonium ion (MeN_2^+) is an intermediate (the other intermediate product, $NH_2CO_2^-$—the hydroxamate ion—decomposes in a reaction not shown here). (c) Formation of O^6-Me G from methyl nitrosourea and a G-residue. (d) Comparison of a G:C base pair and the pairing of O^6-Me G with T; the molecule locked into the 'wrong' tautomeric form (a lactim).

alkylating agent because it contains the grouping $-S(CH_2)_2Cl$ and can react with another nucleophile. This can either be water or another guanine residue. There are two situations in a DNA double helix in which a pair of guanine residues are such a distance apart that the gap between the two N-7 atoms can be bridged by a mustard molecule: the sequence G—G in one strand and the sequence 5′—G—C— in one strand and the complementary 3′—C—G— in the other one. In this case the effect of bridging the residues with a mustard is to tie the two strands of the double helix together by a covalent cross-link (Fig. 3.15(b)). DNA cross-linked in this way can be recognized by its very rapid renaturation kinetics. Also, provided the damage is not repaired (see section 4.3.4), one cross-link of this sort is lethal because the strands cannot separate during DNA replication. Certain cytotoxic drugs used in the chemotherapy of cancer cross-link DNA in this way. Sulphur mustard is far too toxic. (The acute toxicity of sulphur mustard, which was exploited when it was used as a war gas, is not due to the cross-linking of DNA.) Nitrogen mustard, although extremely toxic, has been used in attempted cancer therapy in the past under the designation HN2. However, it has been superseded by drugs such as melphalan, in which the methyl group of nitrogen mustard is replaced by a phenylalanine residue. Other anticancer drugs that are based on the cross-linking of DNA include myleran, $MeSO_2O(CH_2)_4OSO_2Me$. This substance is a methanesulphonate electrophilic reagent; the guanines are bridged not by a mustard linker but rather by $—(CH_2)_4—$. The principle behind all of these drugs is the same; rapidly dividing cells, including those in tumours, should be exceptionally sensitive to an agent that inhibits cell division (Lawley & Brookes 1965).

An alternative cross-linking agent, with promising antineoplastic properties is BCNU (bis chloroethyl nitrosourea), $Cl(CH_2)_2N(NO)C(=O)NH(CH_2)Cl$. The compound has potentially three alkylating parts. This and other chloroethyl nitrosoureas are effective cross-linking agents although the chemistry of their reactions with bases remains to be elucidated (Ewig & Kohn 1977). A quite different kind of cross-linking agent is represented by the antineoplastic drug, *cis* diaminechloroplatinum (II):

```
       Cl
       |
  Cl—Pt—NH3
       |
      NH3
```

The two Cl, two N and Pt atoms in the molecule all lie in the same plane. The reaction is complicated and involves several types of association with DNA including a type of cross-linking (Mong *et al.* 1980).

So far we have only considered compounds that effect electrophilic substitution spontaneously in solution. There is another group of electrophilic reagents that are generated *in situ* following metabolic activation of the drug, mutagen or carcinogen. The most familiar mutagen of this type is *N*-methyl, *N*-nitroso, *N*′-nitro guanidine, $Me(NO)N—C(=NH)—NHNO_2$, variously abbreviated to NTG, NMNG and NMG. Something of a mystery surrounds the true mechanism of action of NTG. The activation event is a hydroxylation and the proximal electrophile is MeN_2^+. However, in bacteria the mutations occur in the region of the DNA replication fork; multiple mutations are often produced and are of varied type. Precautions must be taken in working with NTG; like all electrophilic mutagens it is a potent carcinogen. Many carcinogenic substances are metabolically activated to produce substances that substitute into DNA bases. Examples of alkylating carcinogens are the so-called alkyl nitrosamines such as 'dimethyl nitrosamine', Me_2NNO, properly called *N*-nitrosodimethylamine. Again the alkylating species is generated from a metabolically activated intermediate generated by hydroxylation. Different types of carcinogen, for example aromatic amines and polycyclic aromatic hydrocarbons, are also metabolized to form reactive species that will substitute bases (usually guanine) in DNA. In these cases the reactions are arylations rather than alkylations. The correlation between carcinogenicity, reaction with DNA and mutagenicity is the basis for the somatic mutation theory of cancer, that is the view that a clone of cancer cells is derived from a mutant somatic cell, and also for the mutagenicity screening tests for carcinogens (see section 4.3.2).

Finally mention should be made of the naturally occurring antibiotic mitomycin c. This substance requires metabolic activation (in this case a reduction) to form a bifunctional alkylating agent that cross-links DNA in a similar manner to the mustards.

In contrast to alkane diazonium ions that alkylate DNA (Fig. 3.15(b)), the aromatic diazonium ions condense with the amino groups (mainly N^2 in Gua) to form azo compounds. This reaction is the basis of a method for attaching nucleic acids to paper for hybridization as an alternative to Southern transfer (see section 3.2.2). The compound NBPC, m-nitrobenzyloxymethylpyridinium chloride, condenses with alcohols, including the OH groups in paper to form compounds of the form $mO_2N{-}C_6H_4CH_2OCH_2OR$ (Fig. 3.16). In

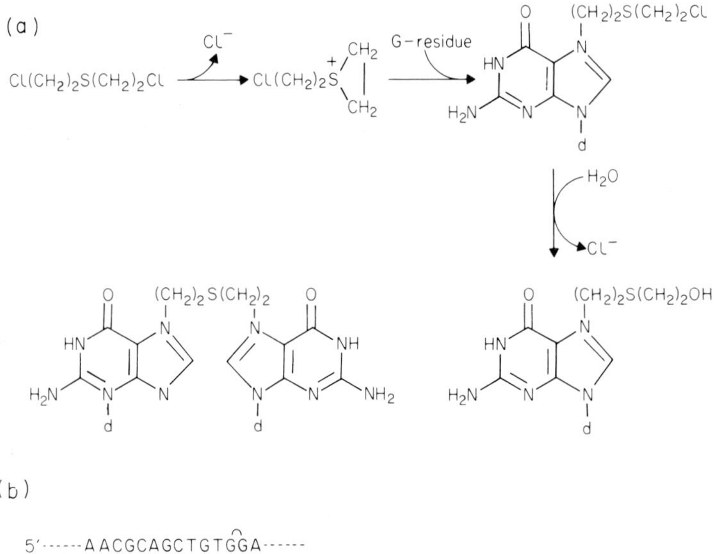

Fig. 3.16. (a) Reactions between G-residues and sulphur mustard. (b) Positions of G-residues in a DNA sequence that, for stereochemical reasons, may be cross-linked by a mustard. The G-residues not cross-linked in the diagram cannot be bridged in this way.

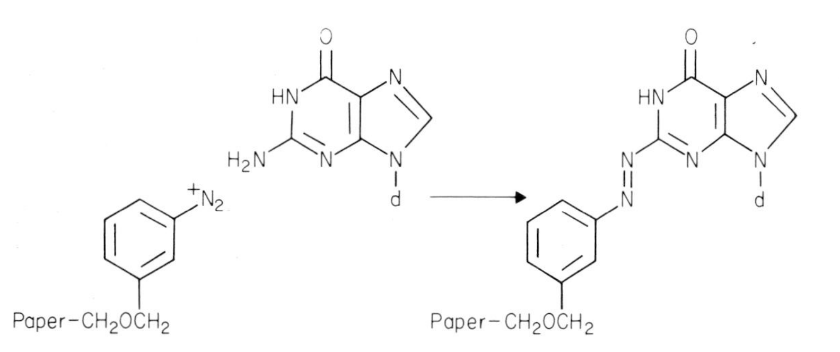

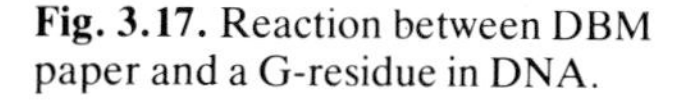
Fig. 3.17. Reaction between DBM paper and a G-residue in DNA.

the case where the alcohol is paper, the product is called NBM paper. The nitro group is reduced to an amine and this (ABM paper) is diazotized to form DBM paper which will react with G-residues in nucleic acids (Fig. 3.17) and fix the nucleic acid to the paper (Alwine *et al.* 1980).

3.3.3 Addition reactions

The 5,6 double bond in pyrimidines

The 5,6 double bond of thymine, uracil and, to a lesser extent, cystosine has considerable olefinic character, that is to say that it is involved in addition reactions. One can, for example, hydrogenate a nucleic acid with hydrogen in the presence of a palladium catalyst. Of greater biological importance is the self-addition of two pyrimidine residues to form from the two double bonds a cyclobutane structure. This reaction requires ultraviolet light. If the sequence T—T occurs in a double-stranded DNA molecule, the two thymine bases are in approximately the orientation and appropriately spaced for the reaction to occur. When the DNA is UV-irradiated, therefore, such cyclobutane structures are formed and when the DNA is hydrolysed with acid the compound 'thymine dimer' is found in the hydrolysate (Fig. 3.18). The formation of thymine dimers is by far the most important reaction that occurs during the UV-irradiation of cells with the consequent repair (see section 4.3.4).

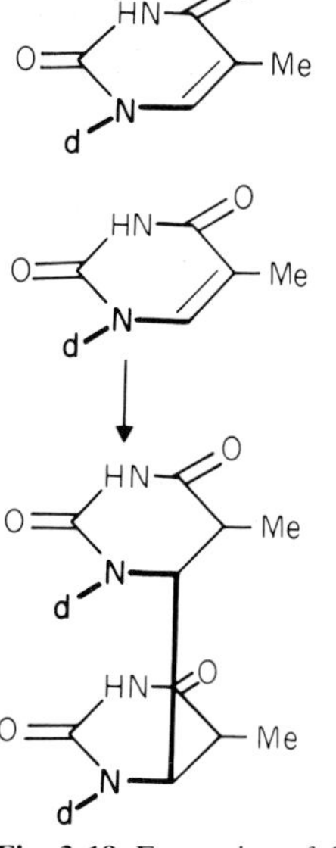

Fig. 3.18. Formation of thymine dimer from two adjacent T-residues in DNA.

Amides

Amides have a characteristic reversible addition reaction with carbodiimides (substances of the type R—N=C=N—R′). The addition occurs across the C=O double bond of the amide. The character of the C=O bond at C-4 of guanine and C-4 of uracil and thymine is of such a character that this reaction can occur. The reaction is useful for increasing the specificities of nucleases during RNA sequencing (see section 3.5.2).

3.3.4 Amino groups

Amino groups in nucleic acids react with sodium nitrite at low pH (about 4.5). The effect is to convert A to I, C to U and G to X (Fig. 3.19). The first two events are mutagenic in DNA because I and U mimic G and T respectively in base pairing; production of X (a xanthine residue) from G is an inactivating event.

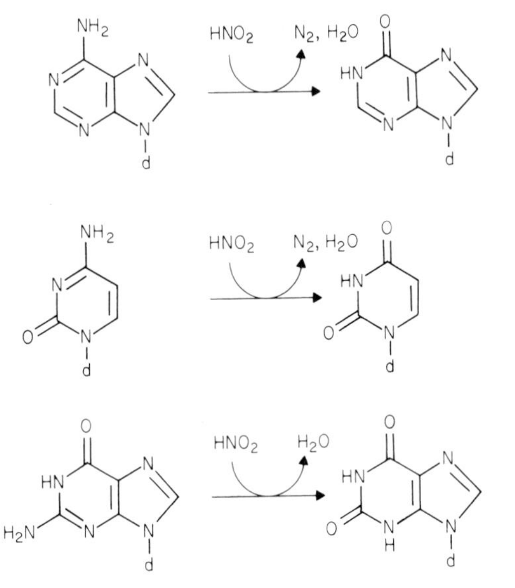

Fig. 3.19. Deamination reactions; d is deoxyribose.

3.3.5 Intercalation

Intercalation is a type of non-covalent interaction between double-stranded DNA and the chromophores of certain dyes and drugs. The process is important for a variety of technical reasons and enables us to extend the discussion of aspects of DNA conformation beyond the point at which it was left in section 3.1.4.

Intercalation is the sandwiching of a flat molecule between two adjacent pairs of bases in the DNA double helix. The effect is a lengthening and a partial unwinding of the helix. Successful intercalating agents have a flat aromatic chromophore about the same size as a base pair and substituted with at least one polar group. Synthetic intercalating agents usually contain either the acridine or phenanthridine ring system; examples of these and also the antibiotic actinomycin D are shown in Fig. 3.20(a). Actinomycin contains a phenoxazine ring; it also differs from the synthetic molecules in that it is sequence specific—it binds to G-residues in the sequence:

$$\begin{array}{l} ....5'—G—C—.... \\ ....3'—C—G—.... \end{array}$$

This is a very simple example of a 'palindromic' sequence (it reads the same either way) and actinomycin D is the first example of a molecule that recognizes such symmetry features (see section 4.3.5).

Actinomycin D is an inhibitor of RNA polymerase. All RNA polymerases are inhibited by actinomycin D; certain organisms are insensitive to the antibiotic but this is invariably because the cells are impermeable to it.

Many other intercalating agents are mutagens; they specifically induce frame-shift mutations. Drugs such as those shown in Fig. 3.20(a) do not usually produce many mutations in bacteria, although they are effective with phage. However, very efficient frame-shift mutagens are the 'acridine mustards'. These substances are both intercalating dyes and also alkylating agents (Fig. 3.20(b)).

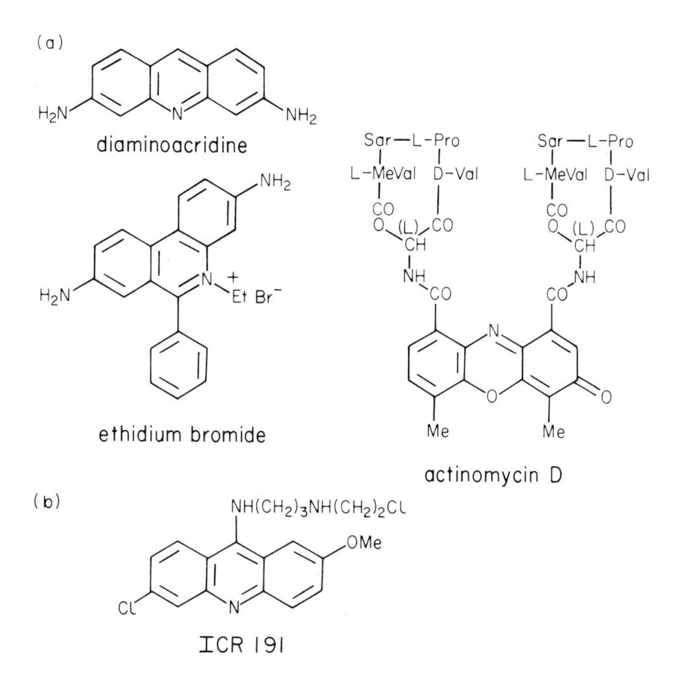

Fig. 3.20. (a) Three intercalating agents. (b) An acridine mustard.

Intercalation and plasmids

For a supercoiled plasmid, the degree of supercoiling, τ, is equal to $(\alpha - \beta)$ (see section 3.1.4). Intercalation reduces β (this is the unwinding process). If we reduce β by an amount, δ, the new value of τ will be $(\alpha - \beta + \delta)$. As τ is negative, the magnitude of $|\tau|$ will be reduced. This is shown schematically in Fig. 3.21(a).

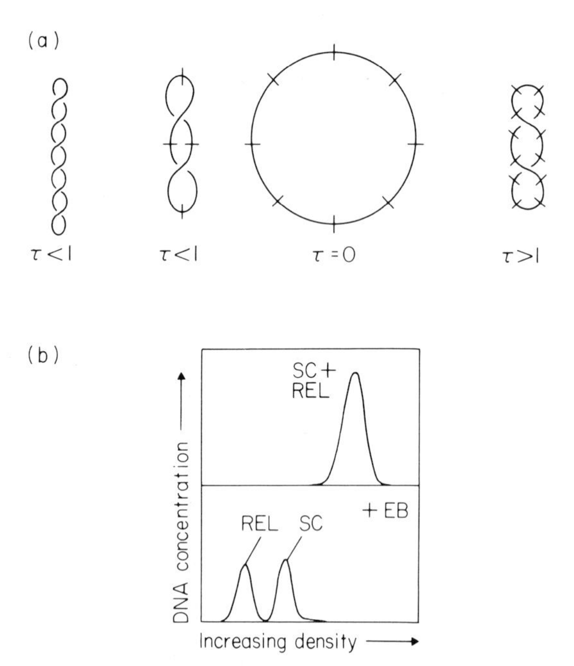

Fig. 3.21. Intercalation and plasmids. (a) Effect of progressive intercalation in a supercoiled plasmid. The continuous loop represents double-stranded DNA and the short lines represent molecules of an intercalating dye, such as ethidium bromide. (b) CsCl centrifugation of a mixture of equal amounts of supercoiled and non-supercoiled (either linear or relaxed) molecules of the same base composition in the absence and presence of ethidium bromide (EB). SC, supercoiled; REL, relaxed. Note that the DNA has a lower density in the presence of the dye.

The structures shown in Fig. 3.21(a) will have different sedimentation coefficients. The supercoiled molecule without any dye bound will have the highest sedimentation coefficient because the conformation is the most compact. The sedimentation coefficient will proceed through a minimum (corresponding to the open circular conformation) and will then increase again as the plasmid starts supercoiling in the opposite direction. This is the basis for a simple method of measuring the value of τ; the sedimentation coefficient is titrated against increasing concentration of a dye (such as ethidium bromide); a correction needs to be made for the effect of intercalation of dye molecules upon sedimentation coefficient irrespective of conformational changes. The method is described by Waring (1970) and extended to diagnose intercalation by a variety of dyes and drugs.

The intercalation of DNA by ethidium bromide is exploited in the most universally applicable method for the preparation and purification of plasmids. It is based on the fact that it is harder to intercalate a supercoiled molecule than a linear piece of double helix. The reason is simply that more work is involved in the unwinding of the superhelix than the unwinding of the Watson–Crick helix alone. Thus if a preparation contains a mixture of linear fragments of bacterial chromosomal DNA and supercoiled plasmids and the two species of molecules are competing for ethidium ions, the chromosomal fragments will bind more ethidium than the supercoiled plasmids. As intercalation by ethidium reduces the density of DNA, plasmids of the same base composition as the chromosomal DNA will become relatively more dense in ethidium bromide and can be separated in a CsCl gradient (Fig. 3.21(b)).

Mechanism of intercalation

Sobell and his colleagues have undertaken a detailed X-ray crystallographic analysis of model substances consisting of actinomycin D or other intercalating dyes co-crystallized with nucleotides. Their data not only confirm the structure of the intercalated complexes and provide accurate molecular parameters but also suggest an answer to the question of how the dye actually finds its way into the double helical structure. The suggestion is that the helix does not unwind to separate two base pairs in such a way as to form a convenient parallel-sided slot for the dye, but rather the helix bends to produce a hinged 'mouth' into which the dye can fit. When the mouth shuts round the dye the parallel-sided intercalation complex is generated (Fig. 3.22; Sobell *et al.* 1977).

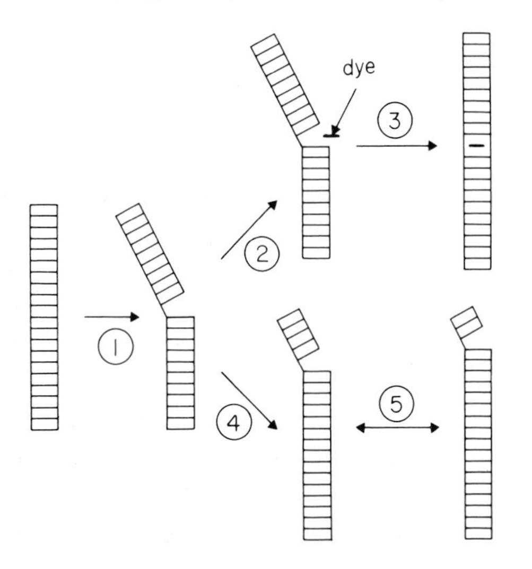

Fig. 3.22. Creation of a hinge in a double helix. The pictures represent a projection of a side view of the helix and the 'rungs of the ladder' are base pairs. Process (1) involves expenditure of energy; the others do not. Processes (2) and (3) show the insertion of an intercalating dye. Processes (4) and (5) show the migration of a hinge.

The implications of the presence of hinges in the DNA double helix extend beyond understanding how intercalating dyes can enter the helix. Hinged DNA is an important concept for understanding the flexibility of the molecule and also for understanding how DNA goes round corners in tightly packed structures such as phage heads. With regard to the former question, thermodynamic considerations suggest that although work is involved in creating a hinge (by producing a discontinuity in the base stack; see section 3.1.2), a hinge can be moved along the helix without further expenditure of free energy (Fig. 3.22).

3.3.6 Chemical synthesis of nucleic acids

The chemical synthesis of oligonucleotides and nucleic acids is a specialized topic not covered in this book. Owing largely to the work of Khorana and his colleagues, there have been dramatic achievements and the technology of DNA synthesis is important in recombinant DNA research (see section 11.5). For an introduction to the chemical literature on the subject see Davies (1976) and Reese (1978). The technical methods are introduced by Narang *et al.* (1980) and the strategy for the synthesis of a whole gene (tyrosine suppressor tRNA) is described by Brown *et al.* (1980). The most significant current technical development is the invention of semi-automated devices for the synthesis of defined DNA sequences. Such devices have been called 'gene machines' and are already in use for applied recombinant DNA research.

3.4 ENZYMIC MANIPULATION OF NUCLEIC ACIDS

3.4.1 Enzymes

This section summarizes the enzymes that are used for the experimental manipulation of nucleic acids. The list of enzymes is not exhaustive but contains all the enzymes commonly used in molecular biology laboratories.

The majority of the enzymes that we shall list can be divided into two groups. Those that hydrolyse phosphoester linkages in nucleic acids (the nucleases) and those that create phosphoester linkages (the polymerases, transferases and kinases). Nucleases are classified as either endo- or exo-nucleases depending on whether they require a free nucleic acid terminus for activity. Thus, endonucleases would be

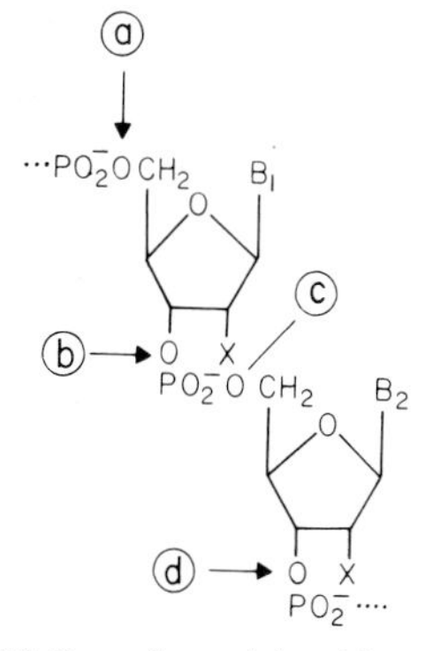

Fig. 3.23. Part of a nucleic acid sequence (X = OH in RNA and H in DNA) showing four phosphate ester bonds, (a)–(d), discussed in the text as sites of action of nucleases.

active on both circular and linear DNA but exonucleases would only be active on linear molecules. Endonucleases are further classified on the basis of whether they cleave only at specific sites (e.g. endonuclease *Eco* RI only cleaves at the sequence GAATTC) or at any phosphodiester linkage. Exonucleases are subclassified on the basis of their terminus specificity, i.e. whether they require free 5′ or 3′ termini. Other important considerations when one is discussing nucleases are whether they are specific for DNA or RNA, single- or double-stranded molecules and whether their cleavage products have 5′- or 3′-terminal phosphates (i.e. whether cleavage occurs at (b) or (c), respectively, in Fig. 3.23). A third group of enzymes, the phosphatases, are also used to hydrolyse phosphoester bonds.

Phosphatases and nucleases

The enzymes are dealt with in alphabetical order except for restriction endonucleases, which appear in a subsection at the end.

Bacterial alkaline phosphatase (BAP) catalyses the removal of 5′-phosphate groups from a variety of substrates including DNA, RNA, NTPs and dNTPs. The product has a 5′ group. This enzyme is frequently used to remove a terminal phosphoryl group prior to replacing it with a ^{32}P-labelled group.

Exonuclease III (from *E. coli*) has three major activities. It is a 3′–5′ exonuclease releasing 5′-mononucleotides from the 3′-termini of duplex DNA. It has a 3′-phosphatase activity removing phosphate groups from 3′-termini and it has an endonucleolytic activity specific for apurinic and apyridiminic sites which results in new 5′-termini with base-free deoxyribose 5-phosphate residues. The enzyme also has an RNAase activity degrading RNA in RNA:DNA hybrids. The 3′ exonucleolytic activity is the most useful as under controlled conditions it turns duplex DNA into a template: primer molecule suitable for DNA polymerase (Rogers & Weiss 1980):

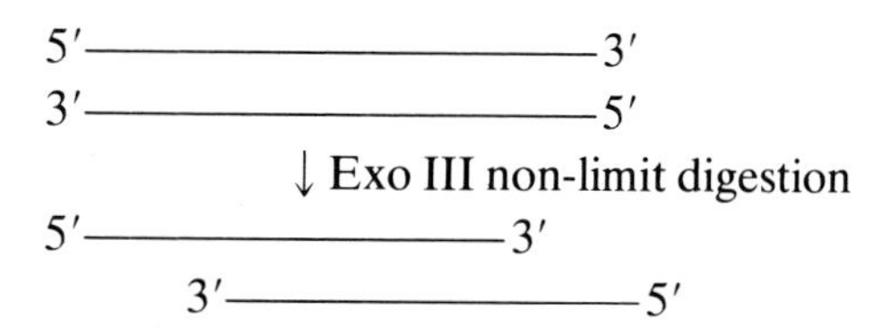

Exonuclease VII (from *E. coli*) hydrolyses single-stranded DNA from 5′- and 3′-termini. It can, therefore, be used to trim a single-stranded molecule to the size of a shorter complementary strand. Berk and Sharp (1978) used the enzyme to map transcripts of SV40

by hybridizing mRNA to SV40 DNA, digesting the DNA with Exo VII and then identifying the sequences protected by the mRNA.

Lambda 5′-exonuclease catalyses the stepwise release of 5′-NMPs from the 5′-ends of duplex DNA. The enzyme is frequently used to produce DNA molecules with protruding 3′-termini prior to their use in terminal transferase reactions (Lobban & Kaiser 1973).

Micrococcal (Staphylococcal) nuclease cleaves single- or double-stranded DNA or RNA at random sites to yield 3′-phosphate and 5′-OH termini. Greatest activity is achieved with single-stranded nucleic acid and A-T rich regions. This enzyme is commonly used in the analysis of chromatin (Bellard *et al.* 1977).

Pancreatic DNAase I produces random single-strand nicks in double-stranded DNA in the presence of Mg^{2+} ions and double-strand cleavage in the presence of Mn^{2+} ions. Products have 5′-phosphate and 3′-OH groups. The enzyme is also used for analysis of chromatin and random cleavage of purified DNA molecules.

Pancreatic ribonuclease cleaves RNA at position (c) on Fig. 3.23 when nucleoside N_1 is a pyrimidine. Products have 5′-OH and 3′-phosphate groups. This enzyme and ribonucleases T1, T2 and U2 are used in RNA sequencing (Brown 1979).

Ribonuclease T1 (from *Aspergillus oryzae*) cleaves RNA at position (c) on Fig. 3.23 when nucleoside N_1 is Guo. Products have 5′-OH and 3′-phosphate groups. The specificity for guanosine is not absolute.

Ribonuclease T2 (from *Aspergillus oryzae*) cleaves RNA at position (c) on Fig. 3.23 when nucleoside N_1 is Ado. Products have 5′-OH and 3′-phosphate groups. The specificity for adenosine is not absolute.

Ribonuclease U2 (from *Ustilago sphaerogena*) cleaves RNA at position (c) on Fig. 3.23 when nucleoside N_1 is a pyrimidine nucleoside. Products have 5′-OH and 3′-phosphate groups.

S1 nuclease (from *Aspergillus oryzae*) hydrolyses single-stranded nucleic acids; DNA is the preferred substrate. The enzyme that occurs, a minor component of amylase (Vogt 1980), has a low pH optimum (4.0–4.3) and requires Zn^{2+} ions. The products of S1 nuclease digestion are 5′ mononucleotides. This enzyme is used in a variety of techniques including the mapping of mismatched regions in heteroduplex DNA and the elimination of unreacted components of a hybridization experiment (see section 3.1.3).

Snake venom phosphodiesterase (SVP) hydrolyses single-stranded DNA and RNA exonucleolytically from a 3′-OH end to produce 5′-nucleoside monophosphates.

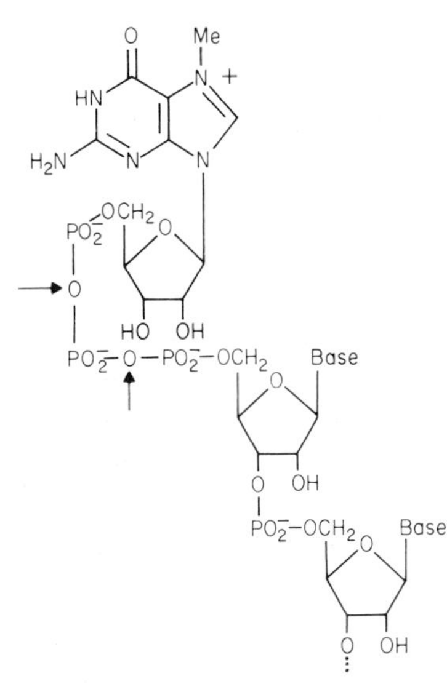

Fig. 3.24. A typical eukaryotic mRNA 5′ 'cap' showing linkages (arrows) hydrolysed by TAP.

Spleen phosphodiesterase hydrolyses single-stranded DNA and RNA exonucleolytically from a 5′-OH terminus to yield 3′-nucleoside monophosphates. Both SVP and spleen phosphodiesterase are used in RNA sequencing studies (Brown 1979).

Tobacco acid pyrophosphatase (TAP) hydrolyses the pyrophosphate linkages of the 'cap' structure on the 5′-ends of eukaryotic mRNAs (see section 7.2.2) to yield p^7Me G, pp^7Me G, ppN-pN mRNA and pN-pN mRNA. The structure of a 5′ 'cap' and the linkages hydrolysed by TAP are shown in Fig. 3.24. TAP is used in the 'TAP, BAP and kinase strategy' for labelling 5′-termini of eukaryotic mRNAs (see section 3.4.2; Efstradiatis *et al.* 1977).

Restriction endonucleases

Restriction endonucleases (or 'restriction enzymes') are sequence specific nucleases that are named from their role in the microbiological processes of restriction and modification. Restriction is a limitation to the efficiency of propagation of foreign DNA (for example that of a phage) in a bacterium (Fig. 3.25). It is technically

Fig. 3.25. Restriction and modification. A and B are two strains of a bacterium that is host to a phage, ϕ. (a) Schematic diagram of the appearance of ϕ plaques on plates of strains A and B. In (1), few plaques are formed on strain B; if, however, phage from one of these are plated, it forms plaques with equal efficiency on both strains. The change is only transitory because phage last propagated on strain A, plate on B with low efficiency again. The following terms are used in describing such results. Strain B is *restrictive*; A is *unrestrictive*. Phage ϕ used in an experiment such as (2) is called ϕ.B; that used in experiment (3) is ϕ.A. Phage ϕ.B is *modified*; ϕ.A is unmodified and is consequently *restricted* on strain B. The ratio of (no. of plaques on A)/(no. of plaques on B) in an experiment such as (3) is the *restriction coefficient*. (b) Representation of the chromosomes of host (a circle) and phage (a line). The restriction sites are represented by rectangles; in this simple case there are just three such sites in the bacterial chromosome and one in the phage chromosome. The modified chromosomes (those of strain B and ϕ.B) have methylated residues in the restriction sequence.

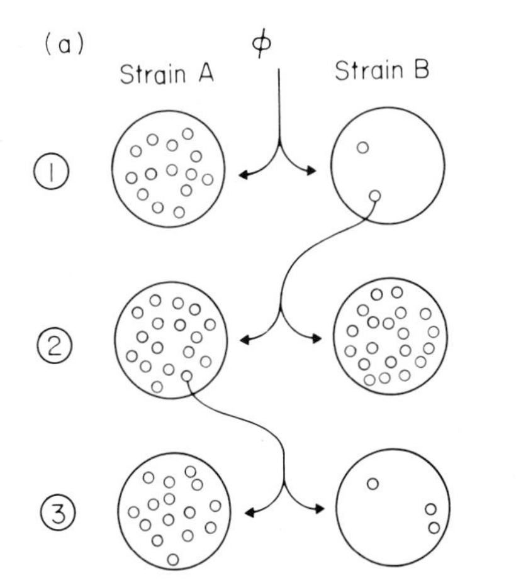

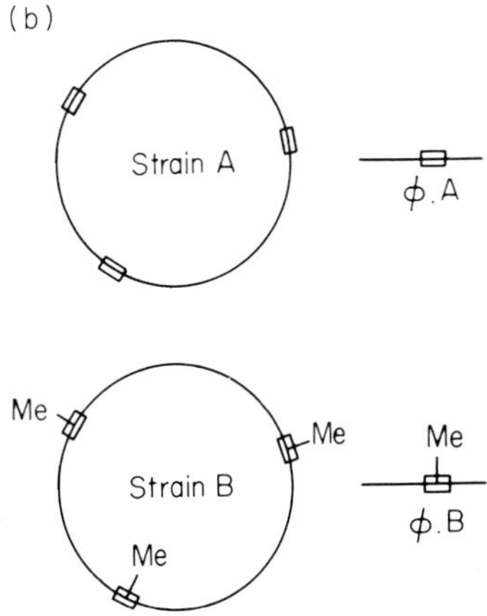

important in recombinant DNA experiments (see section 9.2.1); moreover, restrictive bacteria (and that seems to include the majority of species) are a source of restriction enzymes. The enzymes are reviewed by Roberts (1980a).

Restriction endonucleases cleave both strands of duplex DNA. Type I enzymes recognize a specific target site and then cleave unpredictably in the vicinity of that site. In the case of the *E. coli K* enzyme there are three subunits, one each for the endonuclease activity, site recognition and site methylation. The enzyme requires S-adenosyl methionine and ATP. Type II enzymes recognize a specific target site, usually a 4, 5 or 6 nucleotide palindromic sequence, and cleave the duplex DNA within that site. The cleavage points on each strand are within the palindromic site and frequently at different, but symmetrical phosphodiester linkages. This results in the production of DNA fragments with mutually cohesive ends, e.g. the restriction endonuclease *Hin*d III cleaves double-stranded DNA as follows:

```
5′———AAGCTT—————3′
3′———TTCGAA—————5′
        ↓ Hind III
 5′ AGCTT———————3′
        A———————5′
        +
 3′—————A
 5′—————TCGA5′
```

Most commonly used Type II enzymes produce protruding 5′-ends but some produce protruding 3′-ends (e.g. *Pst* I) or flush ends (e.g. *Hae* III). Type II enzymes are usually simple dimeric proteins and their corresponding modification enzymes are distinct proteins. A short list of some commonly used restriction endonucleases is shown in Fig. 3.26. Type II enzymes are used to map physically and restructure DNA molecules (see Chapter 9).

Enzymes that form phosphate ester bonds

With the exception of the kinase and ligases the enzymes described here that create phosphoester linkages either require a template for the synthesis of polynucleotides (generally the polymerases) or do not require such a template (the transferases). All the transferases require a 3′-OH primer terminus to which they add nucleotides in a stepwise fashion. The polymerases, on the other

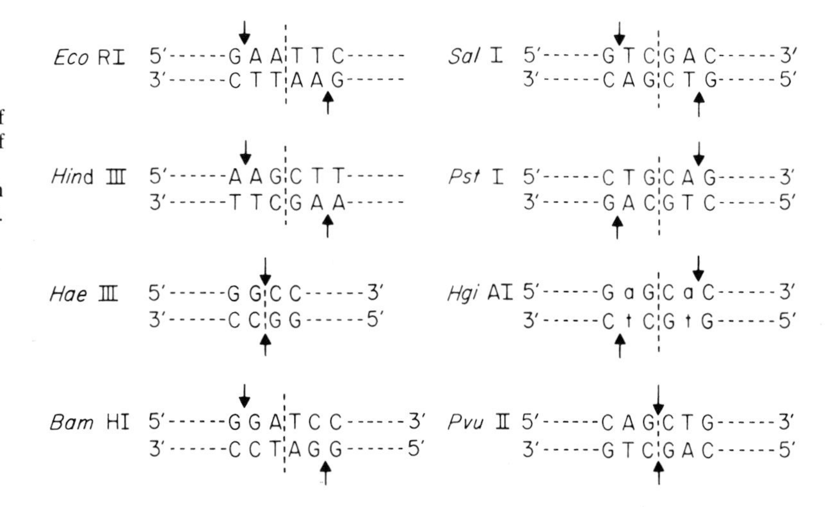

Fig. 3.26. Examples of restriction sequences. Arrows indicate positions of cuts and vertical dotted lines are axes of twofold rotational symmetry. The symbol $^{a}_{t}$ represents a base pair in which either residue may be A and the other T. Abbreviations for organisms: *Eco, E. coli; Hin, Hemophilus influenzae; Hae, H. aegyptius; Bam, Bacillus amyloliquefaciens; Sal, Streptomyces albus; Pst, Providencia stuartii; Hgi, Herpetosiphon giganteus,* and *Pvu, Proteus vulgaris.* In every case the enzyme cuts a bond such as (b) in Fig. 3.23. The *Hgi* enzyme is unusual in that the palindrome does not need to be perfect.

hand, may or may not require a primer terminus. We describe now a few enzymes commonly used as experimental tools, which create phosphoester linkages, including kinase and ligases.

ATP: adenyltransferase, otherwise terminal riboadenylate transferase, catalyses the stepwise addition of poly-A to the 3′-OH terminus of single-stranded RNA. This enzyme has been used to polyadenylate RNA molecules so that they may be used as templates in a polymerase reaction using oligo-dT as a primer molecule.

DNA polymerase I (pol I) catalyses the polymerization of deoxyribonucleotides from triphosphate precursors in a 5′→3′ direction. It is dependent on a 3′-OH terminus primer and a template. The enzyme also has 3′- and 5′-exonuclease activities (see section 4.2.2). Partial digestion of the enzyme with subtilisin yields a large fragment of the protein, 'the Klenow fragment', which retains the polymerase and 3′-exonuclease activities but lacks the 5′-exonuclease. The holoenzyme is commonly used for replacing 'cold' nucleotides with radioactive ones by the process of nick translation (see section 3.4.2) while the Klenow fragment is frequently used for polymerization reactions *in vitro* where it is important not to degrade newly synthesized polynucleotides (see section 3.5.4).

E. coli DNA ligase catalyses the formation of a phosphodiester bond between juxtaposed 5′-phosphate and 3′-OH termini in duplex DNA containing single-strand nicks. The enzyme requires NAD^+ as co-factor (Fig. 3.27).

T4 DNA ligase catalyses the formation of a phosphodiester bond between juxtaposed 5′-phosphate and 3′-OH termini in duplex DNA. The enzyme requires ATP.

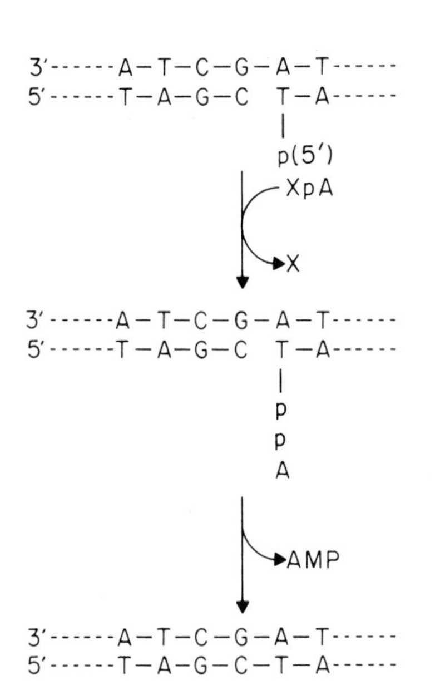

Fig. 3.27. Reaction catalysed by DNA ligase. In the case of the *E. coli* enzyme, XpA and X are Nad^+ and nicotinamide mononucleotide respectively; in the case of the T4 enzyme they are ATP and AMP.

T4 RNA ligase catalyses the ATP-dependent ligation of a 5′-phosphate terminus of a nucleic acid or oligonucleotide to the 3′-OH of another such molecule. Originally described as an 'RNA ligase' the enzyme can also use denatured DNA or single-stranded oligodeoxyribonucleotides as substrates (Hinten *et al.* 1978). (The use of ligases in recombinant DNA technology is described in section 11.4.)

Polynucleotide kinase (T4) catalyses the transfer of the γ-phosphate group of ATP to the 5′-OH terminus of DNA, RNA, and synthetic polynucleotides. The enzyme is frequently used to label nucleic acids by transferring [γ-^{32}P]ATP phosphate to 5′-termini (Efstradiatis *et al.* 1977).

Reverse transcriptase or *RNA-dependent DNA polymerase* of avian myoblastosis virus (AMV) catalyses the polymerization of deoxyribonucleotides from a DNA or RNA single-stranded template. The enzyme requires a primer with a free 3′-OH terminus. The enzyme is frequently used to make DNA copies of polyadenylated mRNA molecules using oligo-dT as a primer (Ghosh *et al.* 1980).

RNA polymerase (E. coli) catalyses the polymerization of ribonucleotides from triphosphate precursors. The enzyme does not require a primer but does require a template. The enzyme can use duplex DNA, single-stranded DNA and RNA as templates but transcript lengths are greatest with duplex DNA containing specific RNA polymerase binding sites (promoters).

Terminal deoxynucleotidyl transferase (TdT) catalyses the polymerization of deoxynucleotides at the 3′-OH termini of single-stranded DNA. In the presence of Co^{2+} ions instead of Mg^{2+} ions the enzymes will catalyse the stepwise addition of homopolymer tails to the 3′-termini of duplex DNA. The enzyme is frequently used to label DNA molecules with radioactive nucleotides and to restructure DNA molecules (see Chapter 11; Boseley *et al.* 1980).

One additional enzyme that should be mentioned is *Eco* RI methylase. This enzyme catalyses the transfer of methyl groups from S-adenosyl methionine to the 3′proximal adenine (marked by *) in the *Eco* RI endonuclease recognition sequence:

$$5'\ GA\overset{*}{A}TTC\ 3'$$

so protecting the DNA from cleavage by the *Eco* RI restriction endonuclease. The enzyme is used in elegant procedures for restructuring DNA molecules (Maniatis *et al.* 1978).

3.4.2 Labelling nucleic acids *in vitro*

Many of the uses of the enzymes described in the previous section will become apparent in subsequent chapters. Here we deal with an important area of nucleic acid manipulation, that of radioactive labelling *in vitro*. The advantage of these techniques is that much higher specific activities can be achieved than by labelling *in vivo*. Four procedures are commonly used.

(1) Labelling by nick translation makes use of the 5′-exonuclease and polymerase activities of DNA polymerase I. Duplex DNA to be labelled is first nicked (single-strand cleavage) with a very low concentration of pancreatic DNAase I. The DNA polymerase I exonucleolytically degrades one chain of the duplex in a 5′–3′ direction starting at the 5′-terminus of a nick and then 'fills in' behind the progressive degradation using the other strand as template. If α-labelled nucleoside triphosphates are included in the reaction mix, they will be incorporated by the polymerase activity and so label the DNA (Rigby *et al.* 1977; Fig. 3.28).

(2) 5′-End labelling with polynucleotide kinase is a simple procedure commonly used for labelling nucleic acids prior to sequencing by the chemical cleavage methods (see section 3.5). Usually 5′-termini carry phosphoryl groups and these have to be removed, generally

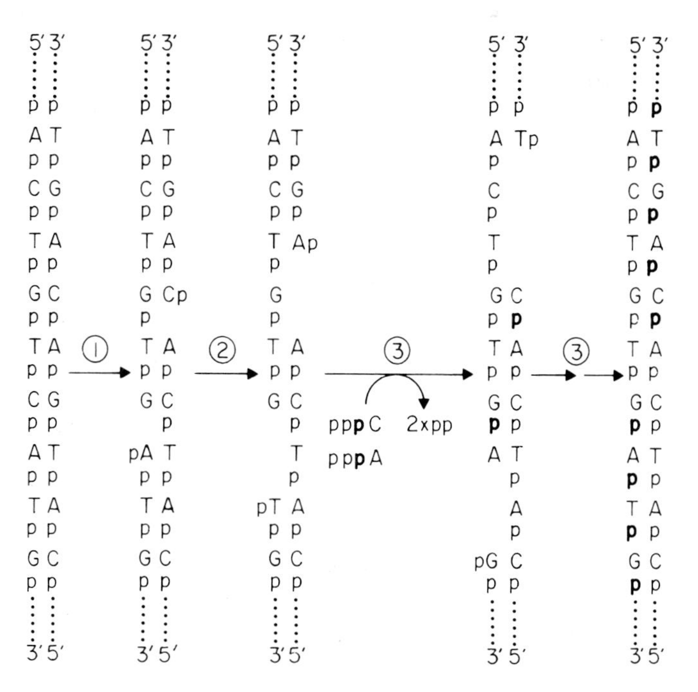

Fig. 3.28. Nick translation. (1) DNA is nicked with a small amount of bovine pancreatic DNAase. (2) The preparation is incubated with DNA polymerase—the 5′-exonuclease activity erodes the chains. (3) The polymerase activity uses the other strand as template. Bold type represents a ^{32}P-labelled residue; pppA is 5′-dATP, etc.

using BAP (see section 3.4.1); [^{32}P]phosphate is then transferred from γ-labelled ATP using kinase. If eukaryotic mRNAs are to be labelled in this way, their 5′-cap structure has to be removed with TAP (see section 3.4.1) prior to BAP treatment.

(3) 3′-End labelling can be achieved using the large fragment of DNA polymerase I to add labelled nucleotides to the 3′-ends of a duplex molecule with protruding 5′-termini (e.g. the reaction products of restriction endonuclease *Hind* III cleavage) in which the protruding 5′-ends act as template:

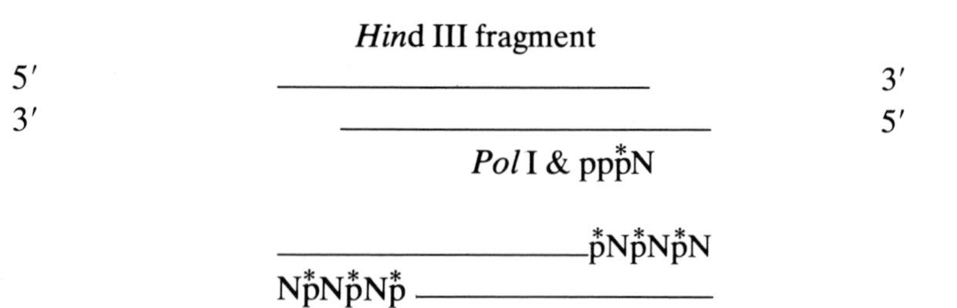

This procedure effectively labels both ends of a restriction fragment. The fragment may then be identified by autoradiography. If the strands are separated, the individual molecules will only be labelled at one end.

(4) Terminal deoxynucleotidyl transferase is also used for labelling 3′-termini of single- or double-stranded molecules although the enzyme is considerably less active on recessed 3′ termini of duplex DNA. Again, labelled nucleotides are incorporated.

3.5 NUCLEIC ACID SEQUENCING

3.5.1 Introduction

Techniques for determining the nucleotide sequences of nucleic acids were developed in the early 1960s. At that time the effort was concentrated on tRNA and 5*s* rRNA because these are the smallest naturally occurring nucleic acids and were natural starting points for working out new methods. Initially the sequencing of RNA was a time-consuming and tedious process but the rapidity of the methods soon caught up with protein sequencing. From the mid-1970s long-range sequencing of DNA became possible and at present the sequencing of DNA is the most rapid technique for determining the sequence of any informational macromolecule. The greatest single reason for this change is that alternative approaches to the traditional method for sequencing have been devised. The traditional approach is to fragment the macromolecule specifically (with base-specific nucleases in the case of nucleic acids), to sequence the fragments and then to piece the sequence together from the fragments. The new approaches involve reading the sequence off from electrophoresis gels of fragments differing from one another by a single residue. The impact of these sequencing methods on molecular biology may be greater than any other technique or theoretical idea since Watson and Crick proposed the double helical structure for DNA. In this section a 'traditional' method for sequencing RNA will be described followed by examples of rapid sequencing techniques.

3.5.2 RNA sequencing following oligonucleotide fractionation

References to experimental details dealt with in this section can be found in Brown (1979). Initial cleavage of the RNA molecule is usually with RNAase T1 (see section 3.4.1). The T1 oligonucleotides must then be separated. In contrast to protein sequencing the difficult part of RNA sequencing by traditional methods is the separation of the oligonucleotides. The identification of the monomeric constituents (mononucleotides) is simple because there are only four main ones; in contrast, the time-consuming and technically difficult part of protein sequencing is the recognition of the monomers as there are 20 amino acids commonly found in proteins. The most effective method of separating the oligonucleotides from RNA is by two-dimensional electrophoresis. The first dimension consists of electrophoresis on cellulose acetate paper at pH 3.5; at this pH the differences in charge between the different bases are maximized. Thus although all phosphate charges will be the same and all nucleotides will migrate towards the positive electrode, guanine residues will contribute the greatest positive charge and thus retard the migration to the greatest extent, cytosine and adenine will be intermediate and uracil will be uncharged. The cellulose acetate papers have a very low capacity for nucleotides and it is for this reason that the method can only be applied to minute quantities of very high specific activity ^{32}P-labelled RNA. The partially resolved oligonucleotides are transferred by a blotting process to DEAE ion-exchange paper or DEAE thin-layer chromatography plates. The second dimension of the fractionation consists either of electrophoresis on the DEAE-paper or of an unusual modification of paper chromatography called 'homochromatography' in which the paper is eluted with a strong solution of a partial digest of unlabelled RNA. The principle behind the technique is that each class of oligonucleotide in the mixture will move up the DEAE saturating its binding sites. At any time, each oligonucleotide will produce the equivalent of a solvent front. The different oligonucleotides in the radioactive sample will travel at these fronts. The name, homochromatography, was coined to refer to the fact that each component is eluted by itself (or at least by an oligonucleotide with the same chromatographic properties).

The second dimension in DEAE (either electrophoresis or homochromatography) resolves the fractions partially separated in the cellulose acetate dimension largely on the basis of molecular weight. The electrophoretic method is useful for relatively small oligonucleotides (up to around ten residues) and homochromatography resolves larger oligonucleotides. The particular advantage of the two-dimensional electrophoresis system is that the various oligonucleotides from a complete T1 digest, all of which contain only one G residue and that at the 3′-end, lie in a set of 'graticules' in the map of spots and from the position of the spots the nucleotide composition (but not the sequence) can be deduced.

Radioactive oligonucleotides are identified by autoradiography, eluted from the paper and sequenced with an exonuclease (see section 3.4.1). Peptides are usually sequenced by the removal of terminal amino acid residues sequentially and characterization of the residue removed or a derivative of it, for example by the Edman procedure. In contrast, the sequence of residues in an oligonucleotide is usually determined by analysing the residual oligomers. The exonuclease used most commonly is the spleen enzyme. Conditions are chosen so that when the enzymic reaction is stopped a mixture of oligomers is obtained. If we call our oligonucleotide 5′—(1)—(2)—(3)—..., where the numbers in parentheses represent nucleosides and hyphens are phosphates, the mixture of oligomers will be: (1)—(2)—(3)—..., (2)—(3)—..., (3)—..., etc. By electrophoresis in a suitable system it is possible to resolve this mixture in such a way that the mobility of any one residue relative to the oligomer running immediately behind it, for example (the mobility of (3)—(4)...)/(the mobility of (2)—(3)...) is diagnostic of the residue missing from the more rapidly migrating oligomer (i.e. the nature of (2) in the above hypothetical reaction). The complete sequence of the RNA molecule is pieced together by looking for overlaps between oligomers obtained with different base-specific nucleases and also by fractionating partial T1 digests and grouping T1 oligonucleotides in these larger fragments. Several modifications of the procedures have been employed for difficult or ambiguous regions of RNA molecules. As a particular example, a carbodiimide reagent (see section 3.3.3) called

CMCT is useful for increasing the specificity of RNAase A, or RNAase U2. RNA (or an oligonucleotide) modified with this reagent and used as a substrate for RNAase A will be hydrolysed only at C; such a substrate will only be hydrolysed at A by RNAase U2. As the CMCT group can be removed by very dilute alkali, oligomers from such a digest can be converted to unmodified compounds for analysis and further enzymic digestion.

Transfer RNA and, to a lesser extent, high molecular weight rRNA contains several nucleoside residues other than C,U, A and G. Minor nucleosides are a mixed blessing for sequencers. As the modified residues only occur rarely in a sequence, they are useful 'handles' for piecing the sequence together. On the other hand, as the sequencing is performed with minute quantities of RNA, detectable only by ^{32}P-labelled phosphate residues, the chemical nature of the modified

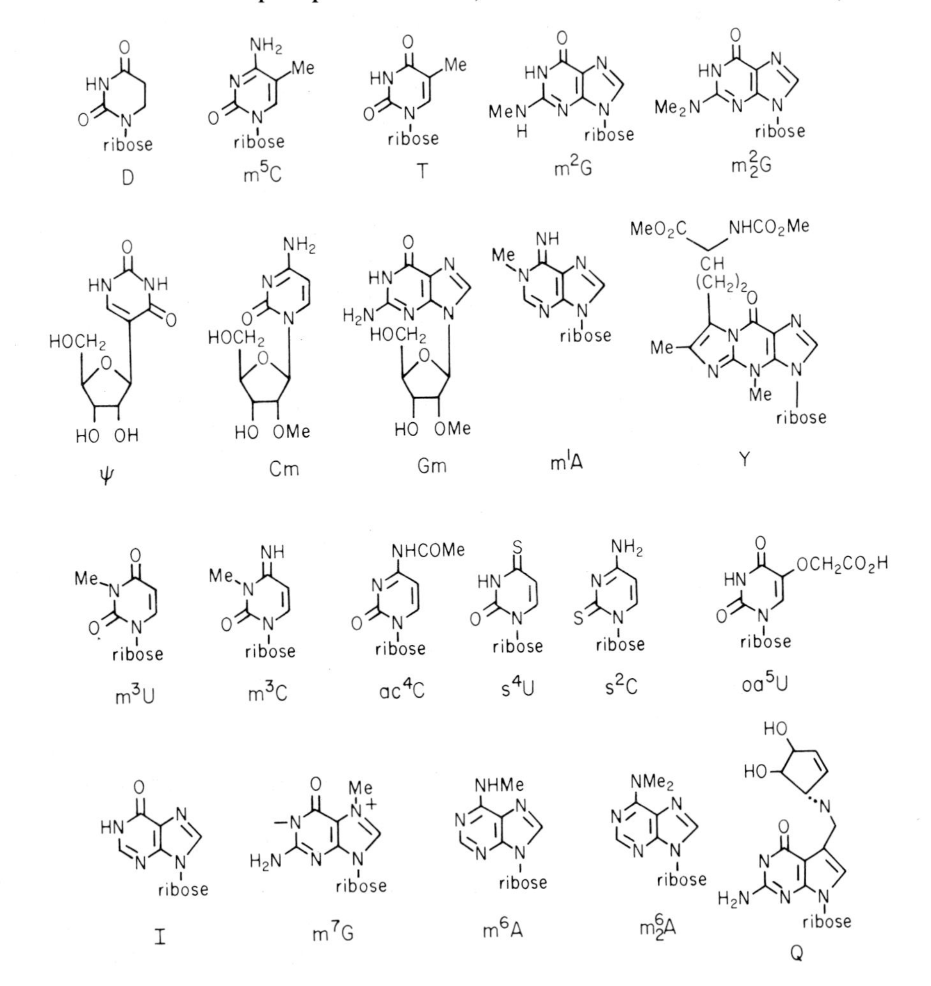

Fig. 3.29. Examples of modified nucleoside residues in RNA molecules. The first ten structures are the modified residues of yeast tRNA^{Phe} (Fig. 3.8(a)). The remaining residues are all fairly common in different tRNAs and rRNAs. The symbols are used to represent the residue in a sequence. A symbol before a capital letter (e.g. m^2 N^2-methyl, $m^2_2N^2$-dimethyl) represents a modification to the base and one following the capital letter represents 2′-substitution of ribose. The exceptions are D (dihydrouridine), T (thymine ribonucleoside), ψ (pseudouridine) and very highly modified residues (e.g. Y) for which any rational abbreviation would be clumsy.

nucleosides cannot be determined from the data obtained from the sequencing. The nucleosides have been previously isolated in chemical quantities and their structures established. Their presence in the labelled digests is recognized by the electrophoretic mobilities of their corresponding nucleotides and, in appropriate cases, by the change in the mobilities of these nucleotides following chemical reactions specific to the modified residues. The modified nucleosides commonly encountered in tRNA and rRNA are shown in Fig. 3.29.

3.5.3 DNA base composition

From the complementary nature of the two chains of double-stranded DNA, it follows that A/T = G/C = 1, where the letters refer to the molar proportions of the four residues. It thus follows that given the value for any one of these, the other three can be calculated; for example, if T = 0.2, it follows that A = 0.2 also and, as T + A + G + C = 1, then it also follows that C = G = 0.3. A convention has arisen that the data are expressed as 'GC-content'. This is normally expressed as a percentage. Thus a DNA with a GC-content of 40% (approximately true for all mammalian DNA) has molar proportions of 0.2 G, 0.2 C, 0.3 A and 0.3 T.

The base composition can be determined by either direct or indirect methods. The two direct methods are as follows. The DNA can be hydrolysed in acid, the bases separated by paper chromatography or ion exchange chromatography and their relative amounts can be determined spectrophotometrically. Alternatively, the DNA can be hydrolysed to a mixture of nucleotides by digestion with a mixture of nucleases and the nucleotides can be fractionated by electrophoresis and the relative amounts can be determined. This has the advantage that it can be applied to very small amounts of ^{32}P-labelled DNA.

The indirect methods consist of measuring a physical parameter that can be related to GC-content by an empirical equation. One possibility is to measure T_m (see section 3.1.3); another is to measure the density of the Cs salt (see section 3.2.2). This latter measurement can be made either in the analytical centrifuge or (in the case of radioactively labelled trace amounts of DNA) in a preparative centrifuge. The gradient in this case is fractionated; the position of the DNA peak is determined and the density of CsCl in the peak fraction can be measured from the refractive index of the solution.

Indirect methods of measuring base composition do not detect minor residues. Moreover, there are certain phage DNA molecules that have one of the normal residues (A,T,C or G) totally replaced by a different one. In these cases indirect methods give meaningless values for apparent GC-content. All the modified residues found in DNA so far are modified in the base. Some minor and replacement bases found in DNA are shown in Fig. 3.30. The roles of minor bases in DNA are not all fully understood. The two methylated bases in the figure are produced by the 'modification enzymes', which protect DNA from restriction (see section 3.4.1). Also the presence of 5-methyl C residues favours the transition from B to Z DNA (see section 3.1.2) and it has been suggested this might be significant in the relatively highly methylated DNA of eukaryotes (Behe & Felsenfeld 1981; see section 8.8).

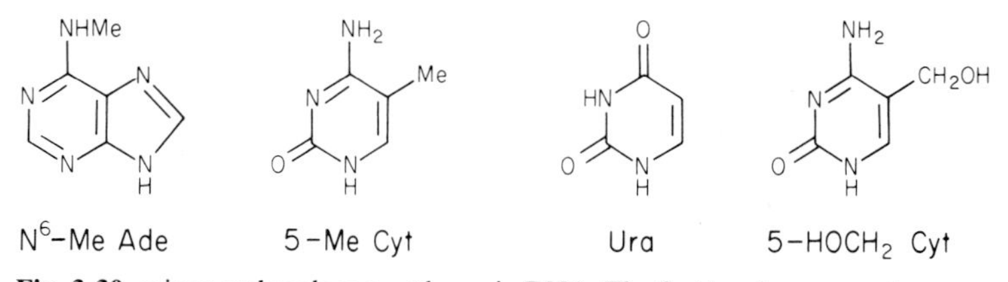

Fig. 3.30. minor and replacement bases in DNA. The first two bases are minor components, formed by post-replicative modification of A- and C-residues, in most DNA molecules. The second two are examples of 'replacement bases'. Ura replaces Thy in *Bacillus subtilis* phage PBS1 and 5-$HOCH_2$Cyt replaces Cyt in the T-even phages of *E. coli*. These bases are incorporated as such during DNA replication; in the latter case the hydroxyl group is glycosylated in a post-replicative modification.

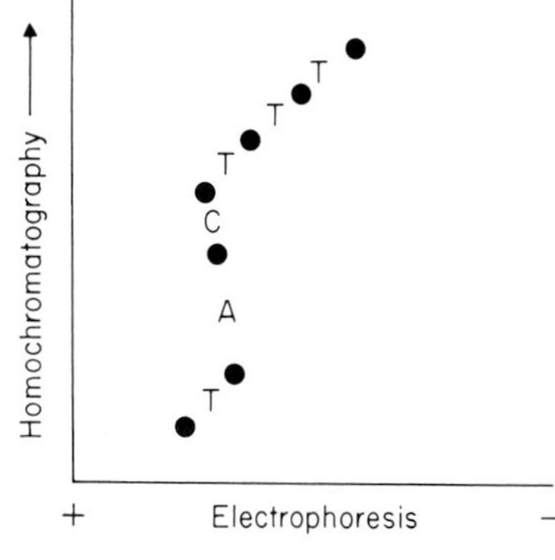

Fig. 3.31. Separation of products from partial digestion of an oligonucleotide with the composition ACT_4 by snake venom phosphodiesterase showing characteristic mobility shifts after loss of specific residues. The difference between the products is shown between the schematic autoradiograph spots. The sequence of the oligonucleotide can be read off as TTTCAT.

3.5.4 DNA sequencing

The methods used for 'traditional' sequencing of RNA can also be applied to short DNA fragments (Fig. 3.31). However, nucleic acid sequencing involving the isolation and analysis of small oligonucleotides has now been superseded by two rapid techniques by which 200–300 nucleotides can be 'read' in one experiment taking about a day. The two techniques are known as the Maxam and Gilbert chemical cleavage technique (Maxam & Gilbert 1980) and the dideoxy-chain termination method (Sanger *et al.* 1977b). A third technique, also due to Sanger, is of historical interest in the evolution of rapid sequencing methods. The technique is known as the plus–minus method and is reviewed by Brown (1979).

Both the Maxam and Gilbert and the Sanger techniques rely on partial reactions whose products are polynucleotides that have one constant terminus and one variable but base terminus.

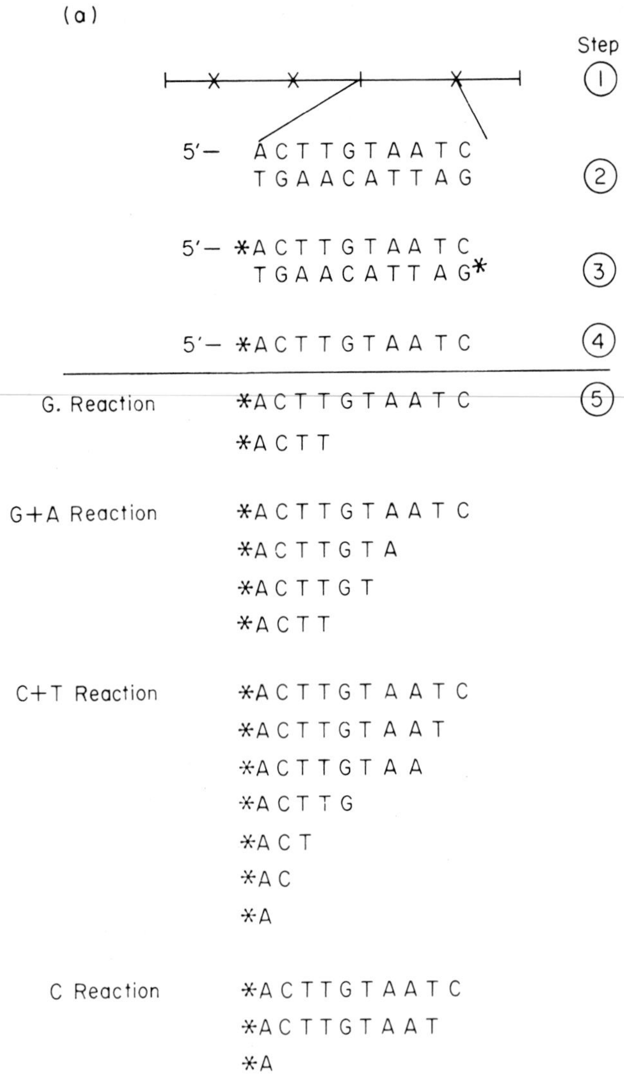

(a)

Fig. 3.32. (a) The steps in a Maxam and Gilbert sequencing experiment. Step 1 shows a length of DNA of about 1 kb pair with the positions of two kinds of restriction site marked as bars and crosses. Step 2 is the purification of a sub-fragment generated by restriction. This fragment would routinely be 100–500 bp; we have drawn just 10 for convenience. Step 3 shows both 5′-termini labelled (*) with ^{32}P by a BAP-and-kinase reaction. Step 4 shows the strands separated and one of them purified to be used in the cleavage reactions. Step 5 shows all the products of each of the four partial cleavage reactions.

Maxam and Gilbert chemical cleavage

Our aim is to sequence a length of DNA about 1000 nucleotides long shown at the top of Fig. 3.32(a). The first step is to construct a map of restriction endonuclease cleavage sites as shown in Step 1. The bars and crosses represent such sites. Each fragment that is generated by such cleavage is sequenced independently and then the entire sequence constructed from the positions of the fragments

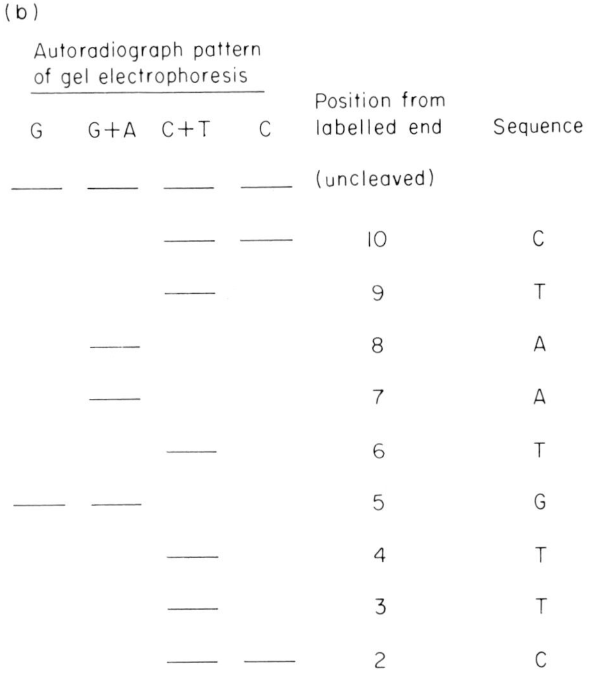

(b)

Autoradiograph pattern of gel electrophoresis

G	G+A	C+T	C	Position from labelled end	Sequence
—	—	—	—	(uncleaved)	
		—	—	10	C
		—		9	T
	—			8	A
	—			7	A
		—		6	T
—	—			5	G
		—		4	T
		—		3	T
		—	—	2	C

Fig. 3.32. (b) The pattern obtained when the products of step 5 in (a) are run on a polyacrylamide gel and identified by autoradiography.

on the map. Each fragment is purified by gel electrophoresis and each strand is then end-labelled with ^{32}P. Either the 5′- or 3′-termini may be labelled depending on which end of the strand one is interested in. For small fragments whose entire sequence can be read in one experiment the choice of which end to label is equivocal but for larger fragments (> 400 nucleotides) both ends might be labelled and the sequence from each end determined in separate experiments. In the example shown in Fig. 3.32(a) we have shown the 5′-termini labelled with ^{32}P. In the next step the strands are separated by electrophoresis in a denaturing polyacrylamide gel and each strand purified by elution from the gel. A single strand is then cleaved in each of the chemical cleavage reactions (Step 5). Purines are selective targets for electrophiles (see section 3.3.2) and pyrimidines are selective targets for hydrazine (see section 3.3.1). The versions of these reactions used in the sequencing method are shown in Fig. 3.33.

In the G-specific reaction (Fig. 3.33(a)) dimethyl sulphate methylates guanine at the N-7 position (arrowed). A positive charge is distributed in the imidazole portion of the purine ring. Base attack at C-8 breaks the C-8—N-9 bond and then piperidine displaces the opened 7-methylguanine and catalyses the β-elimination of both phosphates from the sugar. Adenine also becomes methylated at N-3 but not N-7 with piperidine to cause strand breakage. In the G+A reaction the bases are protonated and the glycosidic bond broken by treatment with acid. Piperidine is finally used to eliminate phosphates and so cleave the DNA.

Hydrazine is used to attack pyrimidines in the C+T reactions (Fig. 3.33(b)) at positions C-4 and C-6. The pyrimidine ring opens and then joins with hydrazine to include C-4, C-5, C-6 in a new five-membered ring. Further reaction with hydrazine releases this new pyrazolone ring and N-2, C-2 and N-3 are released as urea (in Fig. 3.33(b)) leaving sugars as hydrazones in the DNA backbone. Piperidine reacts with these glycosides and so catalyses the β-elimination

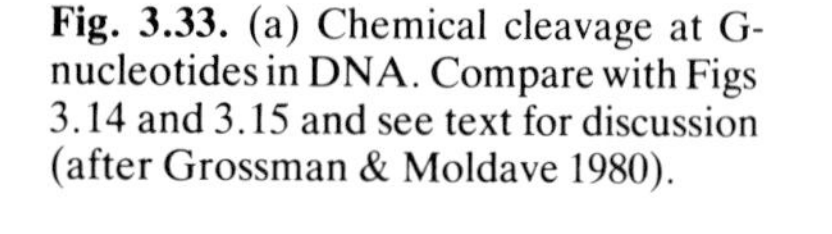

Fig. 3.33. (a) Chemical cleavage at G-nucleotides in DNA. Compare with Figs 3.14 and 3.15 and see text for discussion (after Grossman & Moldave 1980).

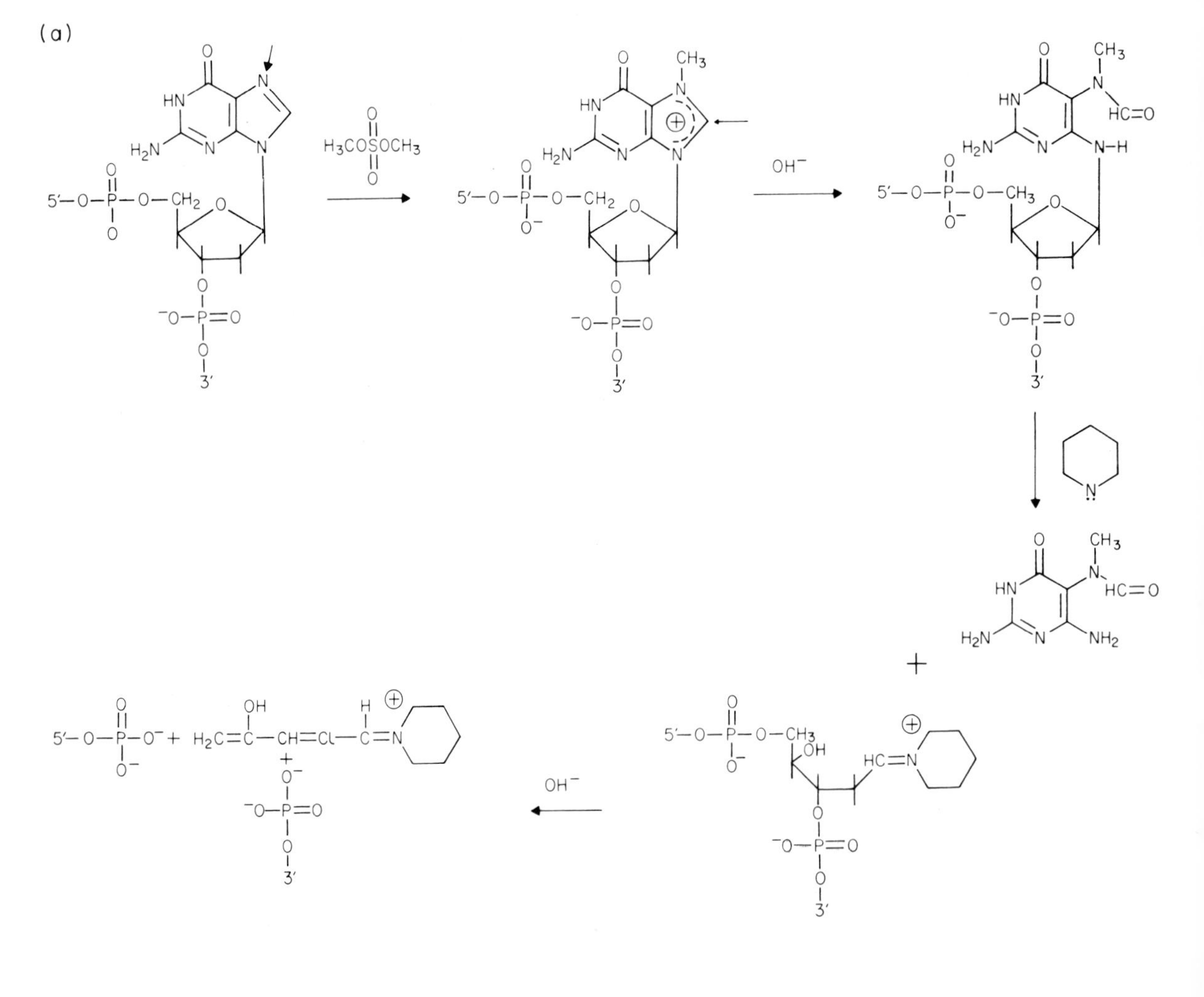

of both phosphates from the sugar. The reaction can be made C specific by totally inhibiting the reaction with thymine with 2M NaCl.

The four reactions are not taken to their limits so that all possible partial reaction products are formed (Fig. 3.32(a) step 5). Only those containing the ^{32}P-labelled 5′ termini are shown as these are the only products that will be detected by autoradiography. The partial reaction products are then fractionated on a polyacrylamide gel (6–20%) and the gel is then autoradiographed. A schematic diagram of the pattern generated by the fractionation is shown in Fig. 3.32(b) remembering that the gel is capable of resolving fragments that differ by a single nucleotide. The sequence can be read-off from the gel pattern from the nested set of fragments from each reaction.

The dideoxy chain termination method

This procedure also relies on the electrophoretic separation of nested sets of products from partial base-specific reactions, but in this

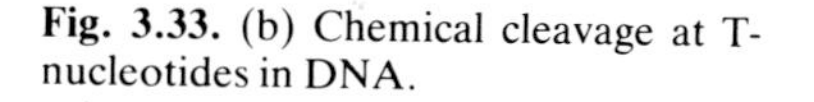

Fig. 3.33. (b) Chemical cleavage at T-nucleotides in DNA.

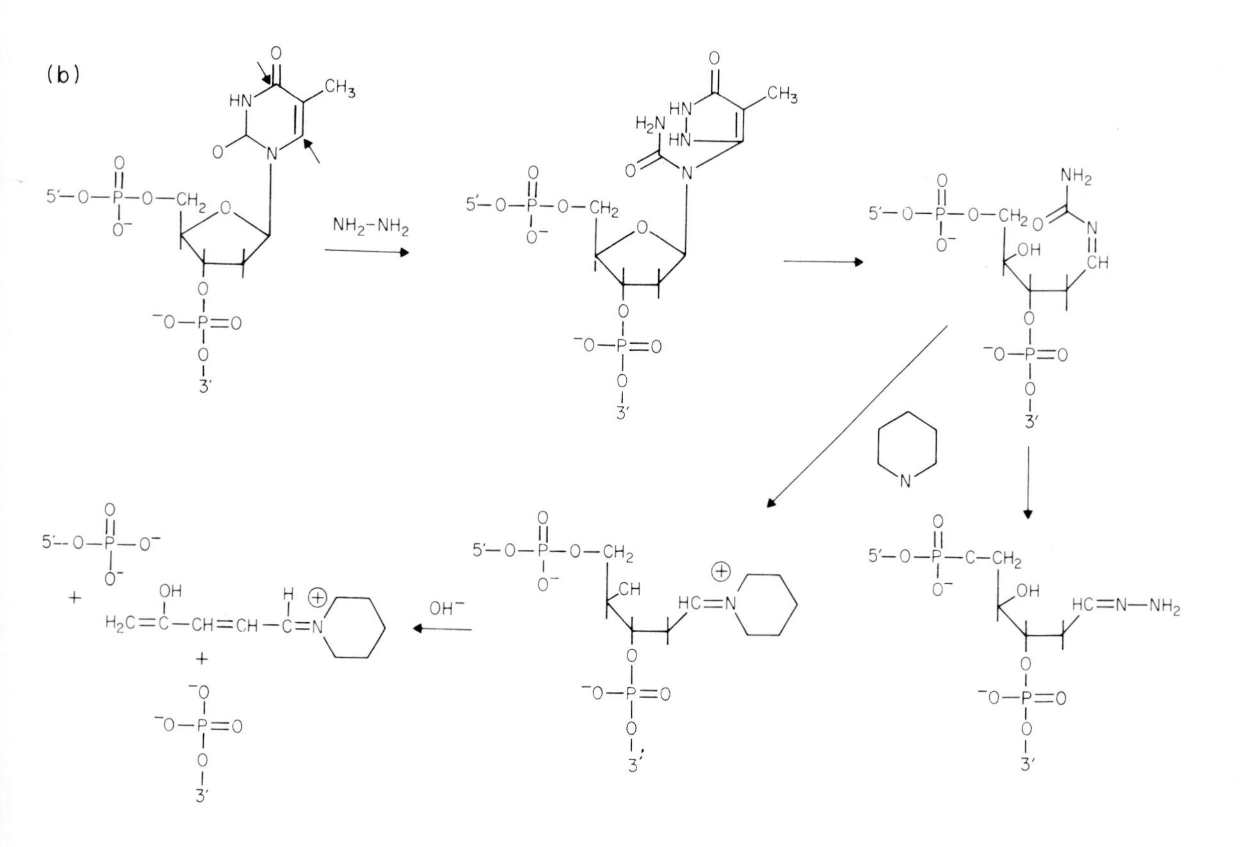

case the reactions are catalysed by a DNA polymerase. The nested set of products in this case is caused by the competition for incorporation into the nascent chain by dideoxy derivatives of the four nucleoside triphosphates. These dideoxy (dd) derivatives have the general structure:

so that they are deoxy at both the 2′ and 3′ positions. This means that if a dd-nucleotide is incorporated by DNA polymerase the chain terminates at that point as there is no available 3′-OH group to participate in the next phosphodiester linkage.

A scheme for sequencing a length of DNA is shown in Fig. 3.34. As in Fig. 3.32 our aim is to sequence a stretch of about 1000 nucleotides. The first step is to construct a detailed restriction endonuclease cleavage map of the sequence and then, as in the Maxam–Gilbert procedure, to determine the nucleotide sequence of each fragment. To achieve this the fragment to be sequenced must be in a single-strand form. This can be achieved in a variety of ways (Smith 1980) but perhaps the most elegant and useful procedure is to insert the

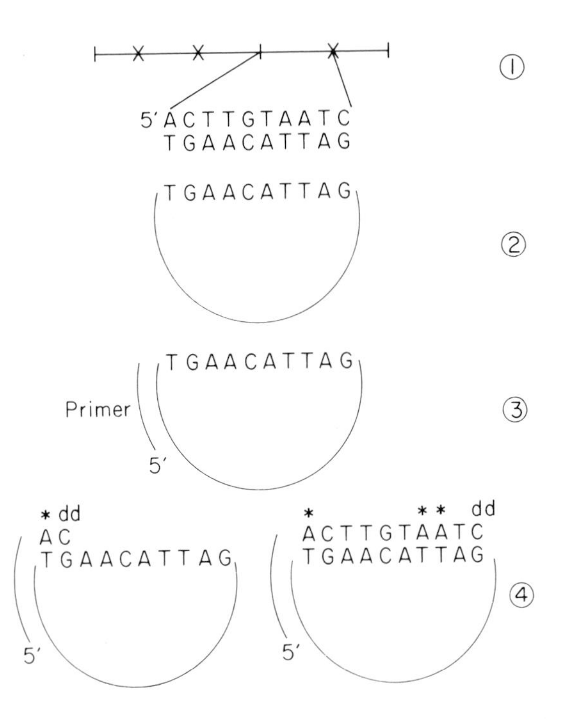

Fig. 3.34. A scheme for sequencing by the dideoxy chain termination procedure. Step 1 shows a length of DNA similar to that shown in Fig. 3.32(a) with two classes of restriction sites. Step 2 shows a fragment from this sequence inserted in the single-stranded phage, M13. The fragment shown is only 10 nucleotides long but in practice it could be up to 500. Step 3 shows the primer fragment annealed to the single-stranded recombinant phage DNA. Step 4 shows the products of a DNA polymerase reaction in which dideoxyCTP is included in the reaction mixture at a concentration that yields a nested set of products with a dideoxyC residue (C^{dd}) incorporated at every possible position. The products are labelled because [α-^{32}P]ATP was included to produce radioactive adenine nucleotide residues ($\overset{*}{A}$).

fragment into the genome of the single-stranded DNA phage M13 (see section 11.3.2; Messing *et al.* 1981). The fragment is inserted into a specific site in the double-stranded replicative form of the phage; when the so-formed hybrid phage genomes are packaged into phage particles only one specific strand is packaged. This means that an M13 lysate contains phage particles each of which contains a single-stranded DNA molecule composed of the M13 genome plus restriction fragment (step 2 in Fig. 3.34). These molecules will act as template in our sequencing reactions. In the next step we anneal to the hybrid single strands a previously purified DNA fragment that comes from a site on the genome close to the site insertion of our fragment to be sequenced (step 3). This fragment is our primer and is so arranged that it will prime synthesis across our sequence. This primer/template system is then used in each of four reactions in which the Klenow fragment of DNA polymerase I (see section 3.4) synthesizes the complementary sequence to our 10 nucleotide (Fig. 3.26) fragment using the short M13 fragment as primer. Each of these four reactions contains an α^{32}P-labelled nucleoside triphosphate in addition to the other three non-radioactive nucleoside triphosphate and a low concentration of one dd-nucleoside triphosphate. This concentration is such that a dideoxy derivative is incorporated relatively rarely so that a nested set of dideoxy terminated chains are produced from each reaction. The reaction products from the ddC reaction are shown in step 4 of Fig. 3.34. The products of each of the four reactions are denatured and fractionated on a polyacrylamide gel and the sequence read off in a similar way to that in Fig. 3.32. An example of an actual gel is shown in Fig. 3.35.

3.5.5 Rapid sequencing of RNA

The device of reading the sequence from a gel of fragments resolved on the basis of chain length can be applied equally to RNA sequencing. In this case the specific cuts are made with the base specific RNAases (see section 3.4.1). An alternative to using enzymic hydrolysis is very limited digestion with alkali. This is the basis of the very ingenious 'readout sequencing of RNA' (Gupta *et al.* 1979). The principle of the readout method can be seen by considering a theoretical molecule of 10 residues, 5′ (1)(2) . . . (10) 3′. The molecule is treated with very dilute alkali, to generate a mixture of oligonucleotides with free 5′-OH groups. It is important that the conditions are such that the chances of two breaks occurring in one chain are remote. The 5′-ends are labelled with ^{32}P (polynucleotide

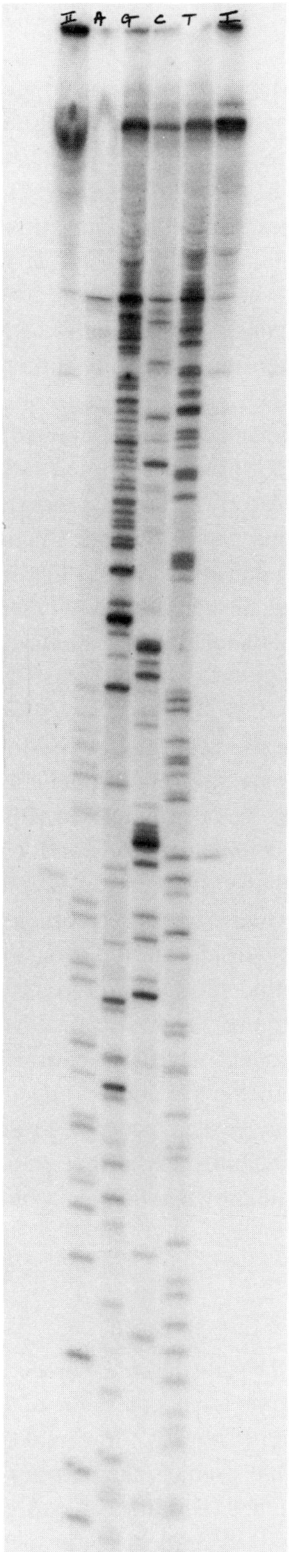

Fig. 3.35. Actual example of a dideoxy sequencing gel. The left-hand tracks of the gel were used to determine a part of the ϕX174 sequence (Sanger *et al.* 1978) and the right-hand ones to deduce the site specificity of the Hgi A1 restriction enzyme (Fig. 3.26) by the method of Brown and Smith (1977). The figure was generously supplied by Dr N.L. Brown.

kinase and radioactive ATP) and the mixture is run on a gel to produce a 'ladder' of ten rungs corresponding to the undigested molecule, (2) . . . (10), (3) . . . (1), etc. Note that oligomers such as (1) . . . (5) will not be detected because they are not radioactive. The whole ladder is now printed on to an ion-exchange thin layer chromatography plate (polyethylene imine cellulose) and digested *in situ* with a non-specific RNAase. The plate is now developed and a two-dimensional autoradiogram can be obtained. The mobility of the mononucleotide spots in the second dimension defines the nature of the residue and the sequence can be read off.

Other good examples of rapid sequencing methods applied to RNA molecules are by Gross *et al.* (1978) and Krupp and Gross (1979). An alternative is to make a cDNA copy and to sequence this. For many purposes it is convenient to clone the cDNA (see Chapter 11). However, in the case of the RNA of dangerous viruses this may be inconvenient or even prohibited and the direct cDNA reverse transcripts must then be sequenced. Application of this technique led to the determination of the complete sequence of polio virus RNA (Kitamura *et al.* 1981).

The paper by Gross *et al.* describes the sequence of a viroid. Viroids are small RNA molecules that are the smallest infectious agents known. They are single-strand covalently closed RNA molecules with much intramolecular complementarity and consequent secondary structure. Unlike other viruses they contain no capsid protein. Also the size of the molecules is too small for them to be informational. The potato spindle viroid referred to above contains only 180 nucleotides. The viroid must, therefore, be a transcription product of the host cell's genome which is expressed (or alternatively a normal transcription product which is abnormally processed) by a process that is induced by the presence of infecting viroid RNA in the diseased cell. The only viroids recognized so far are plant pathogens. However, the possibility exists that similar agents may be the cause of diseases of unknown aetiology in other organisms, including perhaps, ourselves.

NOTE ADDED IN PROOF

DNA sequencing has advanced beyond the state described in section 3.5.4. The most popular rapid technique is an extension of the method of p. 119 but employing random ('shotgun') fragments. A computer program finds the overlaps and predicts the patterns of restriction digests for the complete sequence (R. Larsen & J. Messing, Apple II software for M13 shotgun DNA sequencing, available from Bethesda Research Laboratories Inc., Bethesda, Maryland).

Chapter 4

Prokaryotic DNA, Replication, Repair and Recombination

4.1 ORGANIZATION OF THE PROKARYOTIC GENOME

4.1.1 The bacterial chromosome

The bacterial chromosome is a single covalently closed double-stranded DNA molecule. Given its cellular environment and a full complement of enzymes, the chromosome is capable of replication as a single unit. DNA molecules with this property are termed replicons. Rapidly growing *E. coli* cells contain several copies of the chromosome (see section 4.2.4). However, the pattern of segregation into daughter cells does not allow the maintenance of stable sets of two or more chromosomes (and hence of alternative sets of alleles in any one line of cells) and consequently like many (perhaps all) prokaryotes, *E. coli* is haploid.

The *E. coli* chromosome has a DNA molecular weight of approximately $M_r\,2.4 \times 10^9$ which corresponds to approximately 4×10^6 base pairs or a contour length of about 1.3 mm. As the *E. coli* cell is a little cylinder about 2 μm long and 0.5 μm wide, it is clear that the chromosome must have an extensively folded tertiary structure. The topology of the molecule and the factors that maintain its tertiary structure can be examined by making preparations of bacterial nucleoids. These are obtained by gentle lysis of the cells and are purified by rate-zonal centrifugation. They have a sedimentation constant of approximately 3200*s* and consist of DNA with some RNA and phospholipid. In the electron microscope, the nucleoids have a characteristic and remarkable appearance (Fig. 4.1). By studying the effects of degradative enzymes on these structures, the following conclusions can be reached. The DNA is present in approximately 50 superhelical domains. Each 'petal' in Fig. 4.1 is such a domain. The blob in the middle contains material derived from the plasma membrane and also RNA. If the nucleoids are treated with RNAase, the supercoiling remains intact but the central part of the structure is relaxed and the sedimentation coefficient decreases.

Is the nucleoid in Fig. 4.1 an accurate picture of the organization of the DNA in the intact cell? Is the RNA involved in maintaining the structure a special 'structural' molecule or is it mRNA in the process of transcription? It is not possible to answer these questions unequivocally. Provided that the DNA is not nicked during the isolation of the nucleoid, the topological winding number must be unchanged (see section 3.1.4). However, the 'centre' of the nucleoid could be an artefact. Strong salt is required for the preparation of

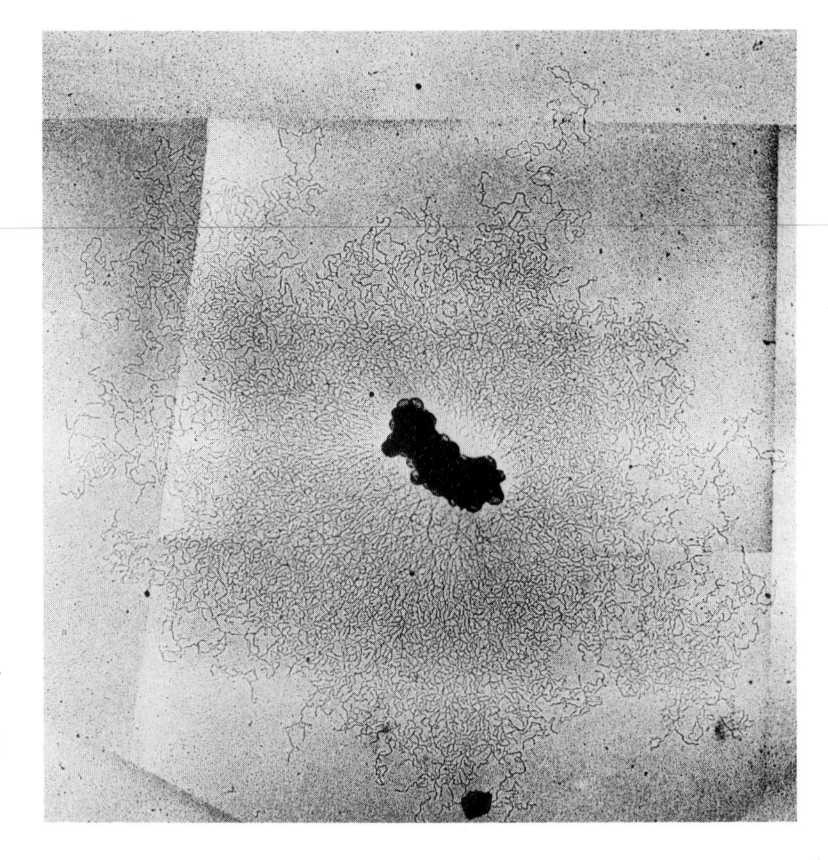

Fig. 4.1. The *E. coli* nucleoid. This electron micrograph is of an 'envelope-bound folded chromosome', the sedimenting form of the nucleoid. Magnification ×12 000. (We are grateful to Dr N. Nanninga for this print, reproduced, with permission, from Meyer *et al.* 1976.)

nucleoids and RNA aggregates at high ionic strength (see section 3.2.2). It is, therefore, possible that the chromosome is attached via newly transcribed mRNA and ribosomes involved in the biosynthesis of membrane or periplasmic proteins *in vivo* (see section 6.5.4). These RNA molecules would then aggregate during preparation to produce the appearance assumed by the nucleoid in photographs such as Fig. 4.1. Equally, it is not certain that the attachment of the chromosome to the membrane is specific; this could also occur during the isolation procedure (Meyer *et al.* 1976). The packaging of the chromosome *in vivo* involves the binding of a basic protein, HU. The protein is believed to be a prokaryotic counterpart of histones. HU has a subunit M_r 9000 and there are about 30 000 copies per *E. coli* chromosome (Haselkorn & Rouvière-Yaniz 1976).

4.1.2 Plasmids

In addition to the chromosome, many bacterial strains contain independent replicons called plasmids. These are named and classified in a variety of ways. Recently isolated plasmids receive a cryptogram consisting of a small p (for plasmid), two capital letters (for the person or laboratory of origin) and a number. Among the genes found on plasmids are those for self-transmissibility, resistance to drugs or toxic metal ions, catabolism of certain carbon sources, production of bacteriocins (a group of polypeptide antibiotics) and restriction-modification systems. Plasmids primarily recognized for their drug resistance genes are called resistance factors, R factors or R plasmids. Those with genes for catabolic enzymes are called catabolic plasmids. Colicinogenic factors are plasmids that encode particular *E. coli* bacteriocins called colicins.

If the plasmids are capable of transfer to a recipient cell the strains harbouring such plasmids (donors) are said to be 'fertile' or 'male'. The *E. coli* fertility factor, F, is the best-studied plasmid of this type but the majority of R factors and catabolic plasmids in Gram-negative bacteria have transfer genes. Transfer genes include the genes that specify sex pili. These are extracellular extensions of proteins involved in the conjugative event. The pili are also receptors for certain groups of phage, known as the pilus-specific or male-specific phages. Finally there are certain plasmids of no known function and lacking any identified genes. These are called cryptic plasmids.

Plasmids will not all coexist in the same cell. This is probably because they share sites on the plasma membrane for replication. Plasmids that cannot coexist in the same cell are said to belong to the same incompatability group (genetic designation, *inc*); for example, an R factor that is *incF* cannot be maintained in a cell that harbours the F factor. Incompatibility is partly associated with pilus type and pilus-specific phages can be used effectively to allocate plasmids to incompatability groups. Thus an F-specific phage, such as coliphage R17 will lyse *E. coli* strains that are F^+ or Hfr or, alternatively, harbour an *incF* R factor. However, certain different incompatability groups share the same pilus type.

So far we have seen that plasmids can be grouped either according to their functions or according to the nature of their Inc genes. There is a further way of classifying plasmids according to the regulation of their replication. For this the plasmids fall into one of two groups. Single copy number plasmids are maintained in cells so that the ratio of plasmids to chromosomes is approximately unity. Multi-copy number plasmids are present in larger numbers. Typically large plasmids, including the majority of R factors and catabolic plasmids belong to the single copy number group and small plasmids, including

Col E1, a well-studied colicinogenic factor, are multi-copy number plasmids. In part of the literature, single copy number and multi-copy number plasmids are referred to as 'stringent' and 'relaxed' plasmids respectively.

Certain plasmids can integrate into the chromosome, thus generating one replicon out of two. Such a plasmid is called an episome and the best accredited example is the F factor (Fig. 4.2). Episome integration and excision (the reverse process) requires the enzymes involved in genetic recombination and is indeed a single cross-over event. It follows, therefore, that there must be homology between the plasmid and the chromosome.

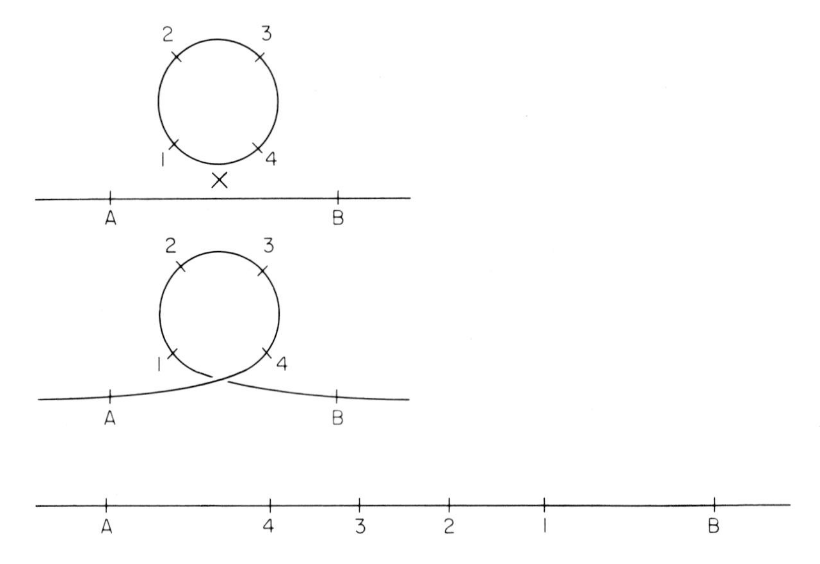

Fig. 4.2. Insertion of an episome, such as the F factor, into a bacterial chromosome. A and B are bacterial genes and 1, 2, 3 and 4 are plasmid genes.

Excision of the episome may involve cross-overs other than those that occurred during integration. In these cases bacterial chromosomal genes may end up on the plasmid. Such plasmids were first obtained with F and were termed F′ factors. However, other 'plasmid primes' can be constructed for genetic analysis. The construction of these plasmids *in vivo* has largely been superseded by recombinant DNA methods (see section 11.1).

For general reviews of plasmid biology see Lewin (1977), Bukhari *et al.* (1977), Williams (1981) and Willetts and Skurray (1980).

4.1.3 Transposable genetic elements

A transposable genetic element is a piece of DNA capable of moving from one replicon to another or from one part of a replicon to another part. The mechanisms of excision and insertion of transposable genetic elements (see section 4.3.7) do not involve the alignment of homologous DNA sequences and are, therefore, independent of the recombinational system of the bacterium. Although it is not true that transposable genetic elements can integrate absolutely anywhere with equal frequency and they certainly demonstrate some sort of site selectivity, nevertheless in a large replicon, such as the *E. coli* chromosome, the number of sites is so large that the event appears superficially to be random.

There are three classes of transposable genetic elements. The first class is represented by the insertion sequences (IS's). An IS is a piece of DNA of the order of 1000 base pairs in length; a total of seven such sequences are known. IS's were first discovered in the context of an unusual type of mutation in the galactose (Gal) operon in *E. coli*. The mutations were extremely polar and could not be reverted by any mutagen. These two findings alone would be characteristic of a deletion. However, the mutations did revert spontaneously with a very low frequency and it followed, therefore, that they must be insertions. The mutations arose because of the mobilization of an IS from one site in the chromosome to another site in the Gal operon. It is likely that all the roles of IS's in bacterial genetics have yet to be recognized. However, two functions can be ascribed to them. The homology between the F factor and the chromosome that allows integration of this episome (see section 4.1.2 and Fig. 4.2) is due to an identical IS in the plasmid and the chromosome. Secondly, one example (IS2) is known to function as a promoter in one orientation but not the other.

All IS's have a sequence feature called an inverted repeat. This means that the sequence at one end is a mirror image of the sequence at the other end. A consequence is that in a single strand of denatured IS DNA the two ends are complementary so that if such a molecule is annealed, loops are formed and these can be visualized in the electron microscope (see section 3.2.5).

The second class of transposable genetic elements are the transposons (abbreviated to Tn). Transposons are blocks of genes that can move around in a way analogous to the IS's. Drug resistance genes and catabolic genes are frequently found in transposons and the transposons undoubtedly contribute to the usefulness of plasmids as natural vehicles for trying out the effectiveness of various combinations of DNA sequence in the natural environment and in the evolution of R factors. Through the transposons drug resistance genes can be transferred from plasmid to chromosome or vice versa or between plasmids of different Inc groups. In this way bacterial populations can respond rapidly and effectively to new environmental stresses and requirements. All transposons have an inverted repeat structure characteristic of IS's. The apparent exception, Tn3, contains a direct repeat. However, the repeat sequence is itself an IS so that the very ends of the total transposon sequence are the inverted repeats of the IS. See Bukhari, *et al.* (1977) for detailed examples of the roles of IS's and transposons, and Calos and Miller (1980) for a more recent catalogue.

The third class of transposable genetic elements contains only one member, phage Mu (see section 4.1.4).

4.1.4 DNA phages

Cells infected with DNA phages contain a phage replicon. DNA phages are either class I or class II viruses. The single-stranded phages all contain a covalently continuous molecule and double-stranded replicative forms are produced in the infected cells. All the double-stranded *E. coli* DNA phages examined so far contain a linear molecule and this is true of the vast majority of such phages in other bacteria. The DNA molecules from many of these phages contain a characteristic type of sequence properly referred to as terminal redundancy, that is the two ends are identical. Terminal redundancy reflects the mechanism for the generation of new phage chromosomes from a replicative intermediate. The lambdoid phages (λ, ϕ80, 21 and 434) have a special type of terminal redundancy, 'sticky endedness'; the ends of the molecule are single-stranded extensions of complementary sequences that enable the molecule to anneal and form a continuous double-stranded structure, often misleadingly referred to as a circle. Another special type of terminal redundancy is reiteration. In such phages, which include the *E. coli* 'T-even' phages, T2, T4 and T6, the redundancy consists of several genes. Any one T4 phage, for example, contains double copies of certain genes. Another phage may contain a different set of duplicate genes. The different types of terminal redundancy can be diagnosed by electron microscopic examination of DNA which has been pretreated in a variety of ways (Fig. 4.3).

Several phages are temperate. The state of the prophage DNA is not known in the majority of cases. Of those phages in which the prophage state is known with certainty, *E. coli* phage P1 is unique in that its prophage is a plasmid. In the other cases, the prophage is integrated into the chromosome. The process is best understood for

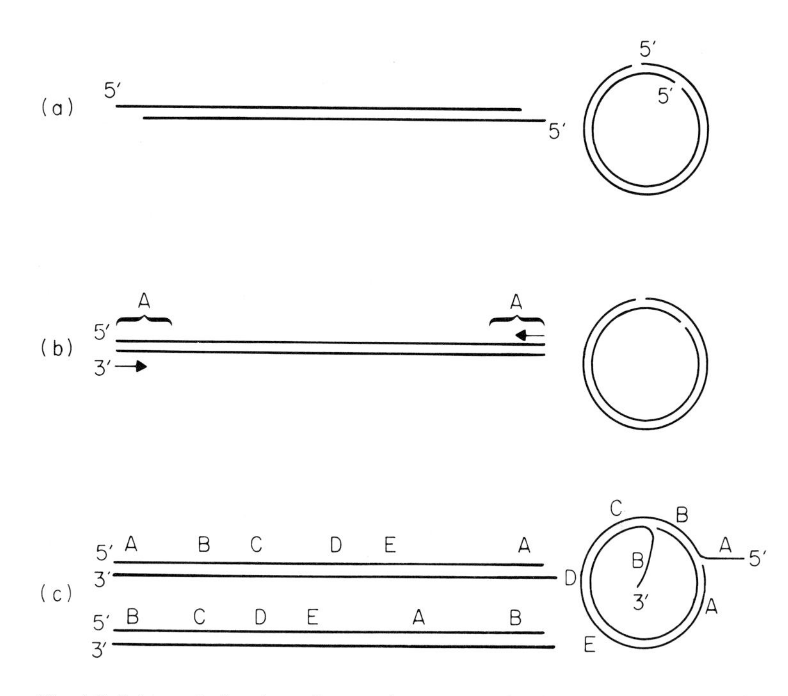

Fig. 4.3. Schematic drawings of types of sequence redundancy in phage DNA molecules. (a) A 'sticky-ended' genome, such as that in λ. The two single-strand extensions are complementary so that if the DNA is annealed, closed 'circles' are formed. (b) The two sequences, A, are identical. If the DNA is treated with a 3′-exonuclease, two stretches (represented by the short arrows) are eroded. If the resulting molecule is annealed, a circle is formed. (c) Reiterative sequence molecules form circles with single-strand 'tails' if the molecules are denatured and re-annealed. The reason can be seen from the very simple example shown. The molecule has been represented as consisting of six regions (each region, represented by a capital letter, corresponds to a set of genes) of which two are identical. A population of double-stranded DNA molecules will consist of a mixture of several forms, of which two are shown. Circles can be formed following denaturation and re-annealing. The one shown is formed by annealing the 'top strand' of the upper double-stranded molecule with the 'bottom strand' of the lower one.

the lambdoid phages. Each of these has a single attachment site in the *E. coli* chromosome (between *gal* and *bio* in the case of λ itself) and the integration process is summarized in Fig. 4.4. The process is analogous to the integration of the F factor (Fig. 4.2). However, the mechanism is different and does not require the recombinational system of the bacterium (see section 9.4.2). Certain temperate phages (for example P2) have a number of attachment sites with different chances of integration at each.

In certain cases, bacterial chromosomal DNA can become incorporated into the phage chromosome to produce specialized transducing phage. These are the phage counterparts of plasmid primes

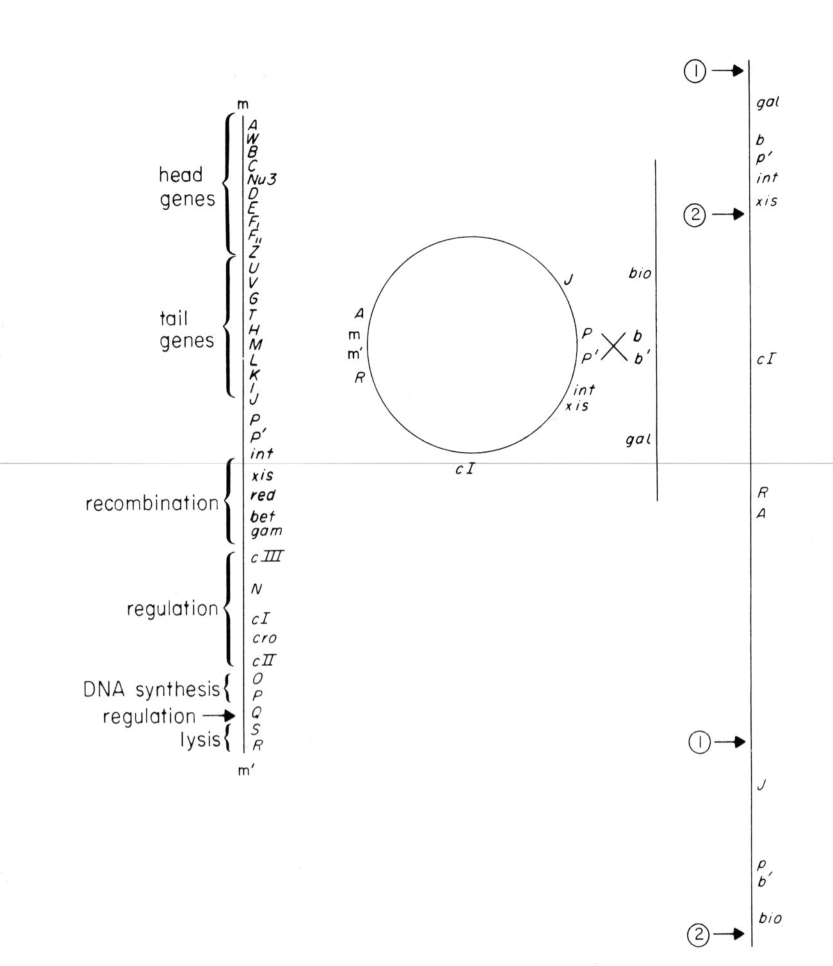

Fig. 4.4. The genome of phage λ. The double-stranded molecule is represented by a single line. The left-hand diagram shows the arrangement of genes in the molecule in the phage. This is referred to as the vegetative genetic map. Note that the genes are clustered according to their functions. The two ends, m and m′, are the cohesive ends (*cos*). Circularization generates a transient plasmid that integrates by a recombinational cross-over between the phage attachment site (*pp*′) and the bacterial attachment site (*bb*′). Induction results in a reversal of this process. Occasionally excision is faulty and this generates specialized transducing phages. If, for example, the excision event were to involve a cross-over at the points (1) (rather than the correct cross-over at the points *bp*′ and *pb*′) then the *E. coli* gal operon would be incorporated into the phage genome. The site *pb*′ and some tail genes would be left behind in the bacterial chromosome. Such a phage would be designated λ*dgal* (*d* for defective; because of the incomplete set of tail genes the phage will need a helper phage to complement this loss if it is to form plaques). Excision at points (2) generates another specialized transducing phage, λ*pbio* (*p* for plaques; this phage has all the genes needed for vegetative propagation).

(see section 4.1.2); the origin of specialized transducing λ phages is discussed in section 4.3.6.

Finally there is phage Mu. This is the only known phage that is also a transposable genetic element (see section 4.1.3). Its name, Mu, is short for mutator because the phage can integrate into an enormous number of sites and if it integrates into a structural gene, not only is that gene not expressed but the mutation is totally polar. Unlike the lambdoid phages and other temperate phages, integration is essential for Mu replication. Phage Mu is discussed further in section 4.3.7.

The foregoing paragraphs are no more than an elementary outline of DNA phages. Many phages, notably those of *E. coli, Salmonella typhimurium* and *Bacillus subtilis* have been the subject of detailed molecular biological study and have yielded important parts of our current understanding of nucleic acid biochemistry. A selection of examples of these studies occurs later in this chapter. For more comprehensive reviews see Kornberg (1979) and Lewin (1977).

The details of the molecular understanding of what must be only a tiny fraction of all the phages in the world have rather overshadowed the questions of the biological roles of phages and their evolution. Probably most groups of bacteria have many phages. The ways in which phage populations behave in the wild has been discussed by Campbell (1961). In the particular case of temperate phage, the prophage can be regarded as a piece of DNA adapted for survival as a passenger in its bacterial genomic host. The process of induction of a lysogen can be seen as a device whereby the phage genome can exploit its capacity to form a particle that can traverse the environment if its host bacterium is endangered. Induction is often triggered by treatments that put the host at risk in some way (such as inhibition of DNA synthesis) that would endanger the host. The molecular biology of a phage such as λ (see for example section 5.2.4) can be considered against this background.

4.2 DNA REPLICATION

4.2.1 Introduction

The replication of the *E. coli* chromosome consists of the initiation of the process at a unique site, the chromosome origin, the elongation phase during which two replication forks channel their way round the chromosome and termination to generate two new chromosomes (Fig. 4.5(a)). The two new chromosomes have to segregate into the daughter cells and the process must be regulated in such a way that DNA correlates with the other events that define the cell cycle. In the following sections we shall consider each of these stages and events but not in the most logical order; it is better to consider the elongation process before looking at the special problems associated with initiation.

It is necessary to be aware of some methodological difficulties. Although many of the enzymes involved in DNA replication, such as

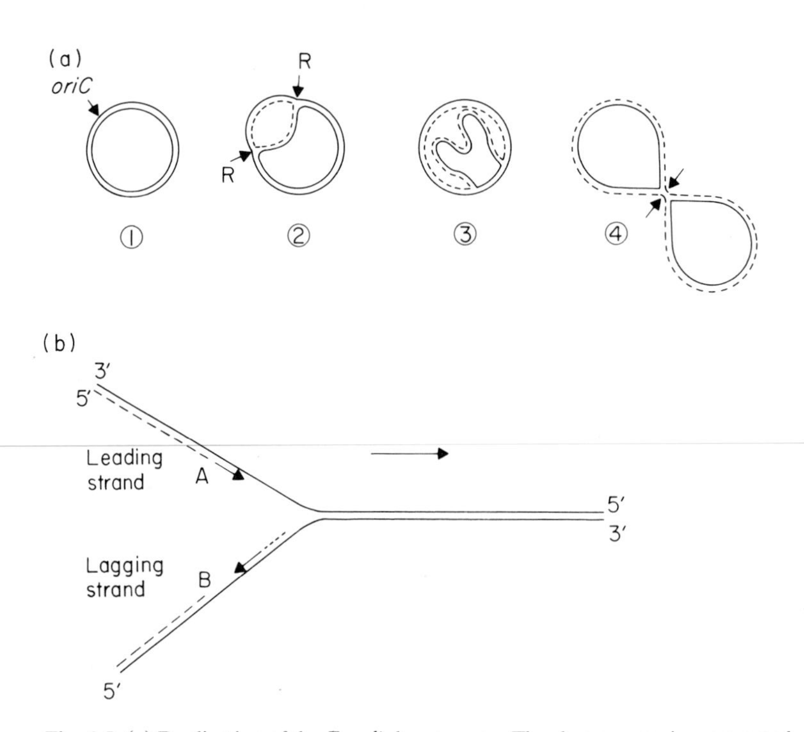

Fig. 4.5. (a) Replication of the *E. coli* chromosome. The chromosome is represented as the two strands of the double helix. Newly synthesized DNA is represented by a discontinuous line. The chromosome (1) has a unique origin of replication, *oriC*. In (2), replication has started and the two replication forks (R) are moving round the chromosome. In (3) replication is almost complete. In (4) the forks have met. The diagram has been twisted round to show that for the two daughter chromosomes to segregate, the newly synthesized strands must be cut (at the arrows). (b) A replication fork. The overall direction of movement of the fork is from left to right as shown by the longer arrow. The two parental strands form templates. The upper template (3′–5′) is of the appropriate polarity for continuous extension of the primer by DNA polymerase; there is a free 3′-OH at position A and this is a substrate for the enzyme. The small arrows show the direction of chain extension (5′–3′) The other template is of opposite polarity so that the corresponding position in the newly synthesized strand (B) is a 5′-terminus. In this case DNA polymerase requires a short primer (. . . .) and moves away from the overall direction of movement of the replication fork. The synthesis is necessarily discontinuous and the polymerase will shortly reach the 5′-terminus of the previously used primer (at B).

DNA polymerase, can be assayed satisfactorily, it is not possible to replicate the whole chromosome *in vitro*. There are two ways of establishing whether a suspected enzyme activity is essential for chromosome replication. The genetic method is to study *dna* mutations (conditional-lethal mutations which, when expressed, prevent DNA replication) and to identify the defective activity in the *dna* mutants. The alternative, biochemical approach is to study a smaller replicon, such as a phage chromosome, which can be replicated completely *in vitro* with defined mixtures of enzymes and other proteins. We shall consider the evidence from these two approaches in parallel in building up a picture of bacterial DNA replication. Finally (see section 4.2.5), the special features of phage and plasmid replication will be outlined.

4.2.2 Elongation

Chain extension

If we consider the relationships between the sequences in a replication fork, such as one of those illustrated in Fig. 4.5(a), the process of biosynthesis which accompanies the movement of the fork along the chromosome is seen to be an essentially asymmetric process. This follows from the facts that DNA polymerase can only extend a primer from its 3′-end and the two growing chains, each of which must act as primer during the reactions, are of opposed polarities (Fig. 4.5(b)). It follows, therefore, that there must be discontinuous chain synthesis in one strand. Thus, in addition to DNA polymerase, the process requires a polymerase that can put in the primers and an enzyme to join up the discontinuities. An additional activity is required to erode the primer sequences because the primer enzyme incorporates ribonucleotides, as we shall see shortly. Before putting the process together with biochemical reactions it is necessary first to introduce the enzymes. The number given in parenthesis after each enzyme is the polypeptide molecular weight $\times 10^{-3}$, the abbreviation in italics is the designation of the gene in question in *E. coli.*

E. coli has three DNA polymerases, often abbreviated to pols, pol I (108, *polA*), pol II (*polB*) and pol III (*polC*). The physiological role of pol II is obscure and it will not be considered further here. All DNA polymerases have an associated 3′–5′ exonuclease activity (see

Table 4.1. Summary of polypeptides in the *E. coli* DNA polymerase complex using the nomenclatures of Kornberg (1979) and Wickner (1978). See text for discussion. The genetic designations *polC* and *dnaE* are synonymous. The number is the number of subunits present in the holoenzyme according to Kornberg (1979); 'initiation' here refers to the introduction of the first deoxyribonucleotide residue following the RNA primer.

Synonyms			Subunit	M_r ($\times 10^{-3}$)	No.	Role	Gene
pol III*	pol III		α	140	1	polymerase	*dnaE/polC*
			ϵ	25	1	?	?
			θ	10	2	?	?
			τ	83	1	?	?
	EFII	DnaZ	γ	52	2	initiation	*dnaZ*
		EFIII	δ	32	1		?
co-polIII*	EFI		β	40	1	initiation	?

section 3.4); however, pol I also has an associated 5′–3′ exonuclease activity. It is this activity that constitutes the biochemical basis for the erosion process referred to in the previous paragraph. The pol III enzyme is the chain extension polymerase. It has several polypeptide co-factors. These used to be referred to as co-polymerases ('co-pols') but now either as elongation factors, EF's (Wickner 1978) or by Greek letters (Kornberg 1979; Table 4.1). There is controversy (discussed by Kornberg on p. 177 of his book) as to whether the co-factors are firmly bound to the enzyme and whether or not ϵ, θ and τ are genuine components of the DNA polymerase system.

The priming activity resides in a mixed polymerase (65, ***dnaG***) that introduces a very short RNA primer. The DnaG protein acts in concert with a mixture of proteins, DnaB, DnaC, n, n′, n″ and i, which together form a priming system referred to as a 'primosome' (Low *et al.* 1981). The primosome travels along the lagging strand template to generate the initiation sites for priming. The result is the synthesis of short pieces of DNA (the 'Okazaki fragments') which (following removal of the primers as will be explained) are sealed by DNA ligase (see section 3.4.1 and Fig. 3.27).

We can now write down the reactions involved in the extension of the DNA chain in the half of the replication fork in which continuous synthesis is impossible (Fig. 4.6). Although three polymerases, the DnaG enzyme, pol III and pol I, are required, only the first two are employed as chain-extending polymerases. The polymerase activity

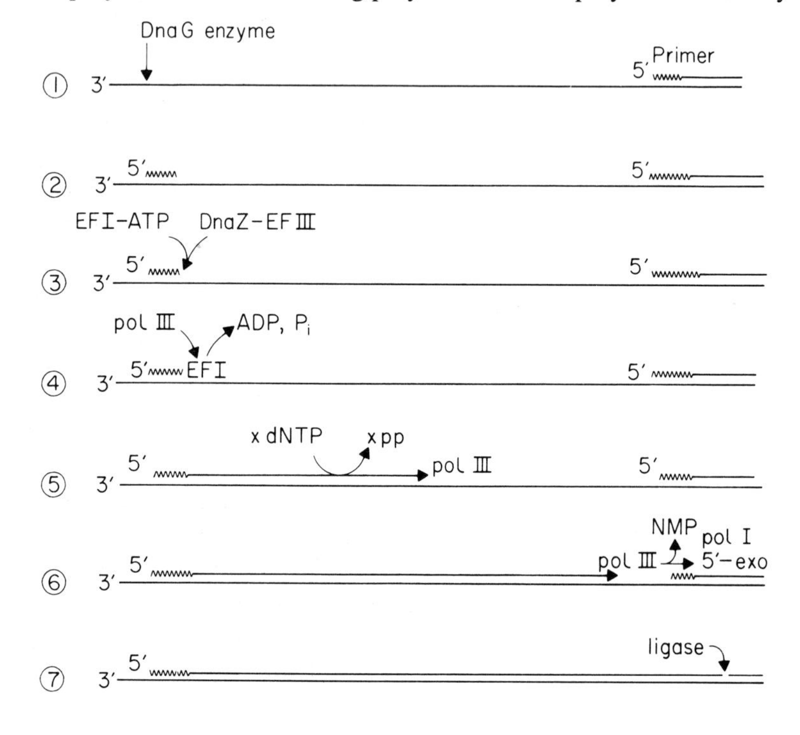

Fig. 4.6. Stages in DNA chain extension. (1 and 2) The primer enzyme, the product of the *dnaG* gene synthesizes a short RNA molecule (shown as a zigzag line). (3) In the presence of the *dnaZ* gene product and EFIII, EFII and ATP (or dATP) bind to the template. (In some literature, the complex DnaZ–EFIII is referred to as EFII; see Table 4.1.) Given EFI and ATP, pol III binds (4). Extension of the primer now occurs (5). In (6) the previous primer is being eroded by the 5′-exonuclease activity of pol I and the exposed template is used for further extension by pol III. The product is a newly synthesized strand with a covalent discontinuity that is sealed by DNA ligase (7). This scheme follows Wickner (1978). See Kornberg (1979) for an alternative interpretation of the results in which the co-factors, EFI, EFIII and DnaZ are not released from the holoenzyme.

of pol I has an important role in DNA repair (see section 4.3.4); however, *polA* mutants of *E. coli,* which have no detectable pol I polymerase activity, are able to grow satisfactorily provided that the defect in the pol I protein is of such a character that the 5′–3′ exonuclease activity is unimpaired.

The experimental evidence for this scheme is discussed by Wickner (1978). Two outstanding questions, which are not completely answered from the available data, are the length of the primer and the number of ribonucleotides in it, and the length of extended synthesis, without re-priming in the other strand. In principle, this strand does not require any priming once initiation has started.

Topological problems

There is obviously much more to DNA chain extension than the simple biochemistry represented by Figs 4.5 and 4.6. The templates have to be revealed by unwinding and denaturing of a double helical molecule; the chromosome is supercoiled so that the supercoiling must also be removed. The newly formed double-stranded molecules must re-assume the correct secondary and tertiary structures. Proteins that effect changes in DNA topology are called topoisomerases (see reviews by Champoux 1978, Kornberg 1979, and Cozzarelli 1980). A distinction must be made between proteins that alter the secondary structure of unconstrained molecules and proteins that can alter the topological winding number of covalently closed molecules. This last process necessarily involves the making and breaking of covalent bonds (see section 3.1.4).

In the following summary of the proteins of *E. coli* that affect DNA conformation, entries in brackets are molecular weights $\times 10^{-3}$, synonyms or older names and (italicized) genetic designations if these are known.

Helix destabilizing protein (HDP) (4 × 20, DNA binding protein) promotes denaturation of unconstrained, double helical DNA by binding to single-stranded DNA and hence lowering the T_m. However, in the presence of divalent cations HDP promotes the process of renaturation. The protein is probably essential for the separation of the template strands in the replication fork. Not only does the protein have appropriate properties but it is known to be an essential component of the system for replicating ϕX174 DNA *in vitro* (see section 4.2.5).

Unwinding enzyme (180) has the distinction of being the largest polypeptide chain to have been characterized from *E. coli*. It differs from HDP in that it unwinds the double helix actively (the enzyme is a DNA-dependent ATPase) starting at single-stranded ends. The role of the enzyme in replication of the bacterial chromosome has still to be established. However, the *E. coli* Rep protein (68, *rep*), which is required for ϕX174 DNA, seems to be a counterpart to the unwinding enzyme. Rep is not required for growth of *E. coli* but its requirement for phage growth suggests that active unwinding is a necessary component, together with HDP activity, for the unwinding of the double helix and maintenance of single-stranded templates.

DNA swivelase (ω, 56+31) alters the helical winding number to reduce the number of superhelical turns. This constitutes a change in the topological winding number (see section 3.1.4). The enzyme is, therefore, an endonuclease and also can rejoin single-strand discontinuities. The enzyme does not require ATP; it obtains its energy from the negative free energy charge that is associated with the loss of supercoiling. The role of swivelase in DNA replication is obscure.

DNA gyrase (2×110+another subunit, *nalA* and *cou*) is certainly required for DNA replication. One class of mutants resistant to the antibacterial drug, nalidixic acid, have mutations in the gene *(nalA)* for one of the subunit polypeptides; similarly coumaromycin resistance mutations lie in the gene *(cou)* for the other subunit. As these drugs inhibit DNA replication in sensitive strains, the requirement for the expression of the *nalA* and *cou* genes is established. Gyrase resembles swivelase in that it alters the topological winding number. Indeed in the absence of ATP the enzyme has the same activity as swivelase itself and the swivelase activity resides in the two identical *nalA* subunits. However, in the presence of ATP and the *cou* subunit, the enzyme becomes a DNA-dependent ATPase and increases the number of supercoils. Thus gyrase is the ATP-dependent generator of supercoils that is required to assemble the correct tertiary structure of newly formed chromosomes.

Other proteins that affect DNA conformation are RNA polymerase (see section 5.1.2) and the *recBC* nuclease (see section 4.3.5).

We can now produce a plausible scenario for the changes in conformation of DNA that accompany replication. In order to generate the replication fork, the supercoiled and double-stranded parental molecule loses its supercoiling (possibly following swivelase activity) and is denatured (unwinding protein and HDP) to reveal the two templates. Following replication, the newly synthesized double-stranded molecules are wound up into their superhelical conformation by DNA gyrase. The process is summarized with the current view of other details in Fig. 4.7.

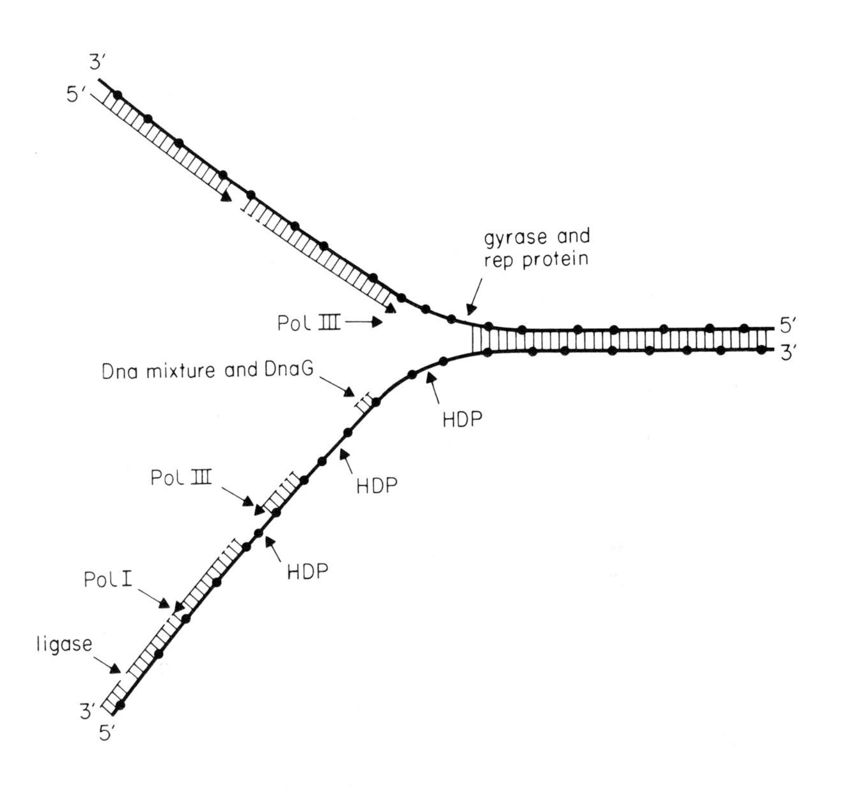

Fig. 4.7. Components of the chain elongation system according to Ogawa and Okazaki (1980). The template strands are shown as heavy lines. RNA primers are shown as discontinuous lines. Blobs in template strands (approximately every five residues) are potential initiation sites. The arrows point to the sites of action of enzymes and other proteins. The Dna mixture (required for priming) consists of the DnaB and DnaC proteins which alter the DNA in such a way as to form what the authors refer to as a 'mobile promoter' and proteins i, n, n′ and n″ which are involved in a 'pre-priming' process. Proteins DnaB and n′ both have DNA-dependent ATPase activity. The model differs from Fig. 4.5(b) in one detail: some discontinuity is believed to occur during biosynthesis of the 'leading strand'.

4.2.3 Initiation and termination

Initiation

Chromosome replication starts at the origin (*oriC*, Fig. 4.5). The sequences of this and the origins of several other replicons are known (Meyer *et al.* 1979, Rosen *et al.* 1979, Sugimoto *et al.* 1979). One of these sequences is shown in Fig. 4.8. Two features of it are common to all known origin sequences: the potential for assuming the type of elaborate alternative secondary structure shown in the sequence and the polypeptide coding region in the sequence. The former property is presumably associated with the recognition of the sequence by initiation proteins (see section 4.3.5). The significance of the coding sequence in the *E. coli* origin and in R factors (such as the one illustrated in Fig. 4.8) is uncertain but the corresponding polypeptide from certain phages has been characterized and is known to be required for the initiation process itself.

Initiation must consist of three events: recognition of *oriC*; some sort of denaturation event; and the synthesis of the first primers. With regard to the last process, the first RNA primers are not produced by the mixed polymerase *(dnaG)* but rather by RNA polymerase (*rpoB*,

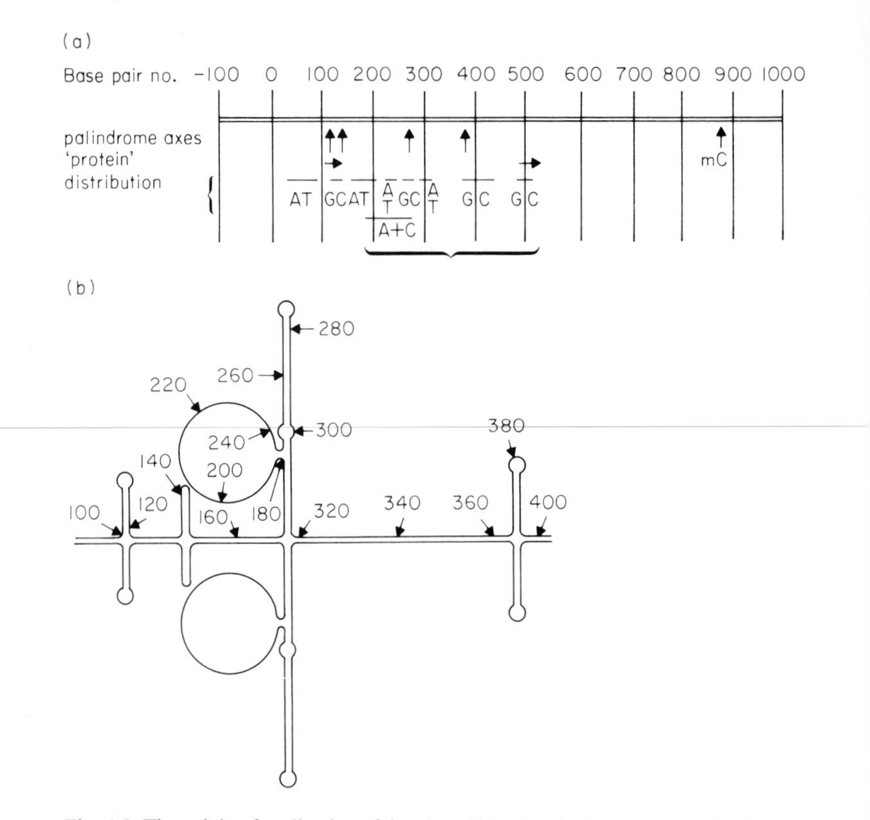

Fig. 4.8. The origin of replication of the plasmid R100. The figures summarize features revealed by the sequence of Rosen *et al.* (1979). The base pairs are numbered according to these authors' conventions; base pair 1 defines a junction between restriction fragments from two sets of derivative plasmids that contain the R100 origin. (a) The two horizontal lines represent the two strands of the sequence between base pairs −100 and +1000. The short vertical arrows show axes of palindromic symmetry; one of these palindromes (mC) contains 5-methylcytosine in the sequence 5′...C—mC—T—G—G...3′. Residues 124–127 in the 5′–3′ strand are ATG and residues 509–511 in the same strand are TGA. These are in the same reading frame so that if this were to be the non-transcribing strand for a part of an operon, these residues would appear in mRNA as an initiation and termination codon (AUG and UGA; see sections 5.2.3 and 5.2.4) and a protein may be coded for by this region (represented as the region between the two short horizontal arrows). The sequence contains some regions of extremely asymmetric residue distribution. Regions marked GC and AT contain very high proportions of G:C and A:T base pairs respectively and in the region marked 'A + C', 56 out of 70 A and C residues all appear in one strand. The curved bracket at the bottom contains, with a high degree of probability, the actual origin sequence. (b) A possible conformation of the sequence between base pairs 90 and 400 with all the palindromes drawn in their alternative, cruciform conformation (see also section 4.3.5). The numbers of base pairs from (a) are indicated as residue numbers in the 'upper' strand.

see section 4.3.1). Also protein synthesis *de novo* is required for initiation. These two facts are reflected in the findings that the initiation of chromosome replication is inhibited by rifampicin and other inhibitors of transcription and by chloramphenicol and other inhibitors of protein synthesis. From a study of the behaviour of Dna mutants, it is known that expression of *dnaA*, *dnaI*, *dnaP*, *dnaB*, *dnaC* and *dnaD* is required for initiation (Wickner 1978). The products of the first three genes are unknown and their modes of action are obscure. They are not required for elongation. The genes *dnaC* and *dnaD* are required for synthesis of a single protein. Moreover, *dnaB* and *dnaC/D* are required for elongation although it is not clear why. The product of *dnaB* is a nucleoside triphosphatase and the *dnaC/D* protein interacts with this enzyme in some way possibly as an allosteric effector. It is reasonable to suppose that these proteins are involved in the recognition of *oriC*, a change in its conformation and modification of the promoter-recognition properties of RNA polymerase so that it starts transcription at the origin.

Termination

Remarkably little is known about the termination of the replication process. The terminus lies in a relatively 'quiet' region of the *E. coli* genetic map lacking any high density of known genes. It is not clear whether this fact itself is significant or accidental. Termination is dependent on the gene *dnaT* (far removed from the terminus region) but neither the DnaT protein nor any other termination-specific proteins have been characterized so the mechanism that solves the intriguing topological problem of what happens when two replication forks meet one another is obscure. However, the intermediate needs to be processed in a way analogous to the recombinational intermediate discussed in Fig. 4.18 and section 4.3.5.

4.2.4 DNA replication and the bacterial cell cycle

Replication of the chromosome is tied into the cell cycle in a remarkably efficient way. *E. coli* can grow in different media with a wide variety of different doubling times and the cells can divide at different sizes (fast-growing cells are larger than slow-growing ones) but nevertheless the organism invariably makes the right amount of DNA and (with the exception of particular mutant strains) never makes a mistake, such as producing a daughter cell without any DNA. For a review of the subject and current ideas of the mechanism, see Donachie (1979). Here we only discuss the simplest aspects of the process.

The *E. coli* chromosome is replicated in 40 min; 20 min after completion of chromosome replication the cell septates and divides. Thus for a cell growing in a medium in which the doubling time is 1 h, a newly born cell initiates chromosome replication immediately; after 40 min the replication is complete and after a further 20 min the cell divides. If the doubling time is more or less than 1 h, in the case of slower-growing cells, the initiation of chromosome replication is delayed, in the case of fast-growing cells (doubling time of less than 1 h), initiation of chromosome replication becomes more frequent. The idea takes a little getting used to at first sight because in everyday life we do not normally think of this as a method of 'speeding things up'; nevertheless the process does work. To use an analogy first suggested by R.H. Pritchard, imagine the movement of loads of goods from a factory to a shop by lorries. If it is desired to increase the rate of arrival of goods at the shop, either the lorries can be made to go faster or, alternatively, lorries can be made to leave the factory at more frequent intervals. The replication fork is the analogue of a lorry load. The journey taken by the forks is not from factory to shop but rather from origin to terminus of the chromosome. We have said that the time taken for chromosome replication, i.e. for the 'journey from origin to terminus' is invariant so that in order to speed up the rate of arrival of forks at the terminus we must alter the rate of re-initiation. The consequence for the cell is that if the doubling time is less than 1 h, each daughter cell inherits a chromosome in the process of dividing; indeed very fast-growing cells inherit two such chromosomes. Of the many questions raised by the cell cycle in *E. coli,* the two that directly involve DNA replication are, How does the progress of DNA replication 'inform' the cell about the right time to separate? and, How does the cell 'inform' the enzymes and other components that it is time to initiate chromosome replication?

With regard to the first of these questions, it is known that the event of termination of DNA replication is a prerequisite for cell division. The most direct evidence for this comes from the observation that if a replicon whose replication cannot be terminated (e.g. a plasmid damaged with UV light) is introduced into *E. coli* the cells elongate but are unable to septate. However, the termination event is not sufficient to ensure septation. A period of termination protein synthesis is required. Whether or not specific termination proteins are made is not known.

The second question (how the cell knows it must initiate chromosome replication) is unresolved but there are several models for the process. Among the data that the models must accomodate is the fact that initiation is determined by the mass (and hence the volume) of the cell. When the cell mass divided by the number of chromosome origins reaches a characteristic value, the 'initiation mass', initiation occurs. Just two models are worthy of brief mention. A negative control model supposes that a repressor of initiation is produced during replication of the *oriC* region. The repressor is diluted during cell growth and when the cell has grown sufficiently for the concentration to be one half of that at initiation, re-initiation is allowed. Alternatively, a positive control could be affected if there were an 'initiator' of replication which accumulated during cell growth. Of the several ways in which this could work, possibly the most attractive is the idea that as the cell grows proteins are made that contribute to the building up of a replication complex, 'replisome'. Re-initiation would be possible once the replisome was complete.

4.2.5 Phages and plasmids

The replication of small replicons involves special mechanisms and variations.

Single-strand phages

A single-strand DNA phage is replicated in several stages, the first of which involves the production of a double-stranded replicative form. The main features of the replication of ϕX174 DNA are shown in outline in Fig. 4.9. The phage DNA (+ strand) contains internal secondary structure and is primed by the DnaG enzyme and several other *E. coli* proteins. The significance of these proteins in the process is obscured by the fact that the similar phage, G4, does not require them. The extension requires pol III and is similar to the process described in Fig. 4.7. Completion involves removal of the RNA primer and its replacement by DNA with pol I. The maturation of the replicative form (RF) involves ligase and DNA gyrase. RFI is transcribed, may be replicated as a plasmid and is the intermediate for synthesis of more + strand by rolling circle replication. For this a nick is introduced into RFI by a complex process involving the ϕX174 gene A protein which remains attached to the 5′-end and forms an anchor; the 3′-end is extended by DNA polymerase. The mode of replication derives its name from the fact that if Fig. 4.9 were animated, the circle would be seen to roll (in a clockwise direction). The newly synthesized molecule contains many copies of the + sequence and these are cut out and parcelled up as circular molecules in progeny phage. The process involved ϕX174 coat proteins and is tightly controlled; a cell during the late period of infection does not have an excess of unpackaged DNA molecules.

Rolling circle replication is an exception to the rule that DNA replication is semi-conservative. It occurs during conjugative replica-

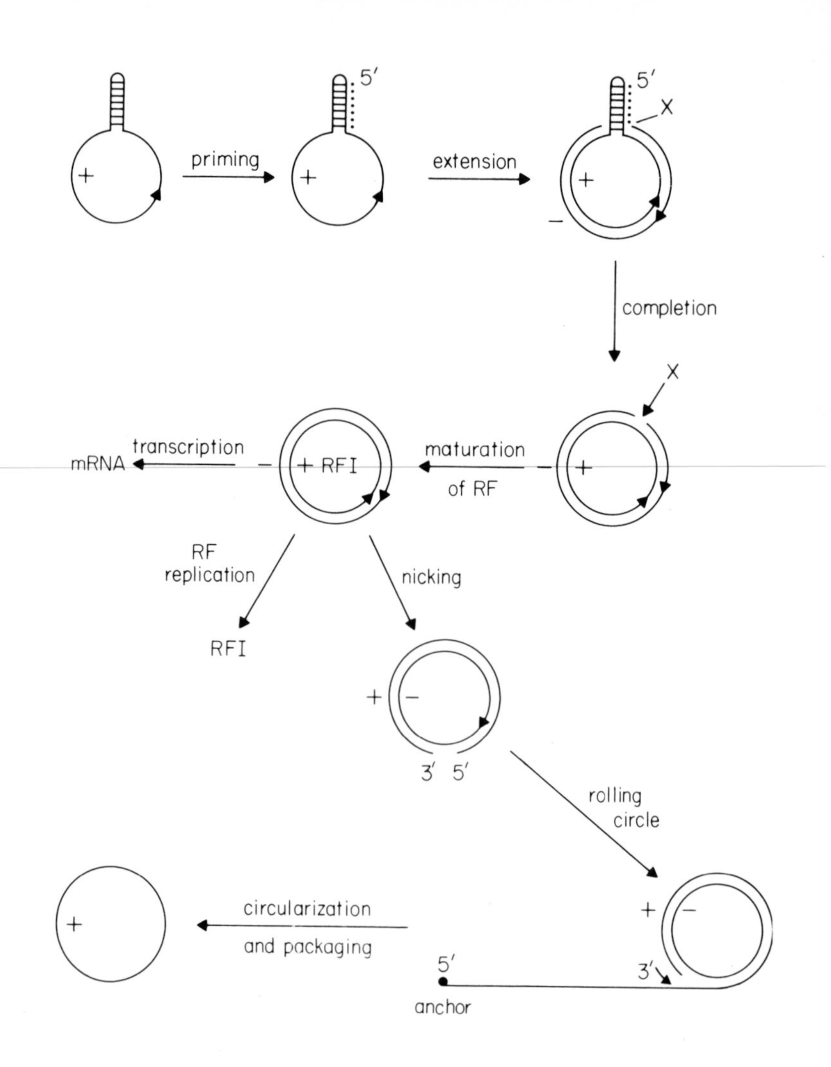

Fig. 4.9. Replication of the DNA of a single-stranded DNA phage. See text for discussion. A small arrow on a circle represents the 5′–3′ direction. X represents a particular point in the − sequence in two different stages of RF formation. The mature state of the replicative form (RFI) is supercoiled, although this is not shown here. The nicking of RFI is accompanied by a reversal of the positions of the + and − strands in the circular representation of the molecule; this is a purely diagrammatic device to make the illustrations of the later stages clearer.

tion of the F factor and also in the replication of double-stranded DNA phages. In these cases the '+ strand' of the lower part of Fig. 4.9 is a template for the biosynthesis of complementary strand and the ribbon that forms contains many copies of a double-stranded chromosome. Methods for recognizing the length of DNA to be cut out and packaged differ in different groups of phages and generate the different patterns of terminal redundancy (see section 4.1.4 and Fig. 4.3).

Other single-stranded DNA phages are replicated in a similar fashion to ϕX174 but with minor variations. In particular, the priming reaction requires RNA polymerase (rather than the DnaG enzyme) in the pilus-specific filamentous phages, such as M13, fd and their relatives.

Col E1

Studies on the replication of the small plasmid Col E1 have shed light on the initiation process (Wickner 1978). As with the *E. coli* chromosome, replication is initiated by RNA polymerase. Like many small DNA molecules, the two strands of Col E1 have significantly different base compositions and can thus be separated from one another in a CsCl gradient. The strands are named H (heavy) and L (light) from their relative densities. Initiation involves transcription of the origin region of the L strand before that of the H strand. Early DNA synthesis in Col E1 involves unidirectional extension (i.e. there is only one replication fork). The L strand template is in such an orientation that its primer has a free 3′-end, therefore the template/primer system that is first formed is like the upper part of Fig. 4.6 and does not require discontinuous synthesis.

4.3 DNA REPAIR AND RECOMBINATION

4.3.1 Introduction

DNA is involved in several different types of biochemical process. It acts as a template for the biosynthesis of DNA itself and for RNA. The covalent structure of DNA is altered in its own biosynthesis (a process in which DNA is a template, a substrate and a product) and in the process involved in DNA repair and genetic recombination. Although the roles of enzymes and other proteins and the nature of intermediates is less well understood for repair and recombination than for DNA replication and transcription, these events nevertheless form part of the subject of nucleic acid biochemistry. They are considered together in the same section because several enzymes and gene functions are common to the two processes.

Let us define some terms: DNA repair is the alteration of the structure of erroneous or damaged DNA; genetic recombination is the recognition of the consequences of the replacement of a piece of DNA by another piece carrying different alleles (such replacements may or may not be on a reciprocal basis between two chromosomes); additive recombination is a convenient phrase to describe the insertion of a piece of DNA into another piece or its reverse (the excision of a piece of DNA).

4.3.2 Types of repair process

DNA is one of the few macromolecules for which there exists a natural system for the correction of molecular errors. Two other examples of such macromolecules are aminoacyl-tRNA (section 6.4) and bacterial peptidoglycan (Donachie 1979). The following generalizations about DNA repair apply to *E. coli* and other enteric bacteria. Eukaryotic repair is discussed in section 7.4.1. For a concise review of DNA repair see Kimball (1980) and for a more comprehensive review see Hanawalt *et al.* (1979).

DNA is repaired in *E. coli* following exposure of the cells to a reagent that damages DNA. UV light can be regarded as a 'reagent' in this context; the photons are responsible for the dimerization of pyrimidines, especially thymines (see section 3.3.3).

Biochemically, DNA repair can be divided into two types of process. The first is the correction of damaged bases without re-synthesis or replacement of residues. This is referred to as 'pre-replication repair' (see section 4.3.3). The second type of process, often referred to simply as 'DNA repair', involves a more radical change in the DNA molecule. Several pathways for such template-dependent repair occur in *E. coli* and these pathways can be classified in a variety of ways (Clark & Volkert 1978). One type of template-dependent repair involves excision of a damaged region followed by re-synthesis (Fig. 4.10(a)). Alternatively, the repair may occur 'post replication', for example by genetic recombination (Fig. 4.10(b)). However, recombination is not the only possible mechanism for post-replication repair; the daughter molecule I in Fig. 4.10(b) could, in principle, be a substrate for a gap-filling process.

Repair pathways can be either inducible or constitutive. A simple experiment to detect inducible repair consists of inactivating a phage

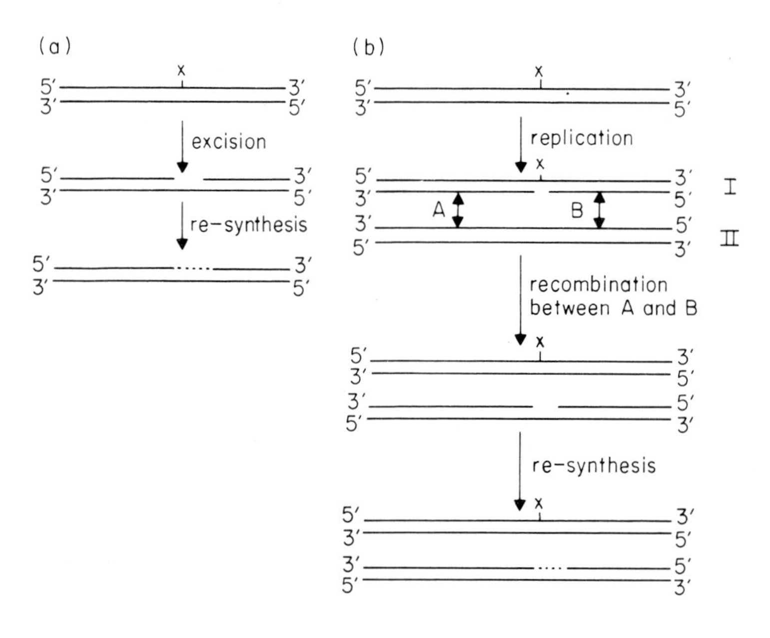

Fig. 4.10. (a) Outline of excision repair. x represents a damaged or modified residue (e.g. a thymine dimer or a methylated G-residue); the discontinuous line represents newly synthesized DNA. (b) Post-replication repair involving recombination. Semi-conservative DNA replication results in two daughter molecules (I and II). A recombinational event between points A and B between the 'lower' strand of I and the 'upper' strand of II separates the gap (caused by the replication polymerase being unable to use x as a template residue) and the x residue and the gap is subsequently filled using an undamaged template.

with UV light and plating the inactivated phage on control cells and also on cells that have themselves been exposed to UV light. If there is a UV-inducible repair system in the bacteria, the titre of phage will be higher on the UV-treated hosts than on the control. Such a finding is referred to as Weigle reactivation.

Certain repair pathways are error-prone; others are error-free. An error-prone pathway generates mutations. In *E. coli* the best-studied repair system is the so-called SOS pathway that involves error-prone, inducible repair. Thus a Weigle reactivation experiment involving a UV-treated *E. coli* phage results in a higher proportion of mutant phages among the plaques on the UV-treated cells than among the (much smaller) number of plaques on the control cells. However, in general, there is no reason *a priori* why error-prone pathways need to be inducible, or vice versa.

Errors during the repair process could, in principle, arise in a number of ways. One possibility is that the excised regions in an excision repair process might be longer than the spacing between the damaged bases (Fig. 4.11(a)). In this case a damaged base will appear in the template strand. Thus the chance of excision repair generating errors (given a certain average frequency of damaged bases) is a function of the length of the 'patch' of DNA excised and re-synthesized. Experimentally it is established that 'short-patch' (of the order of 10 nucleotides) and 'long-patch' (200–1000 nucleotides)

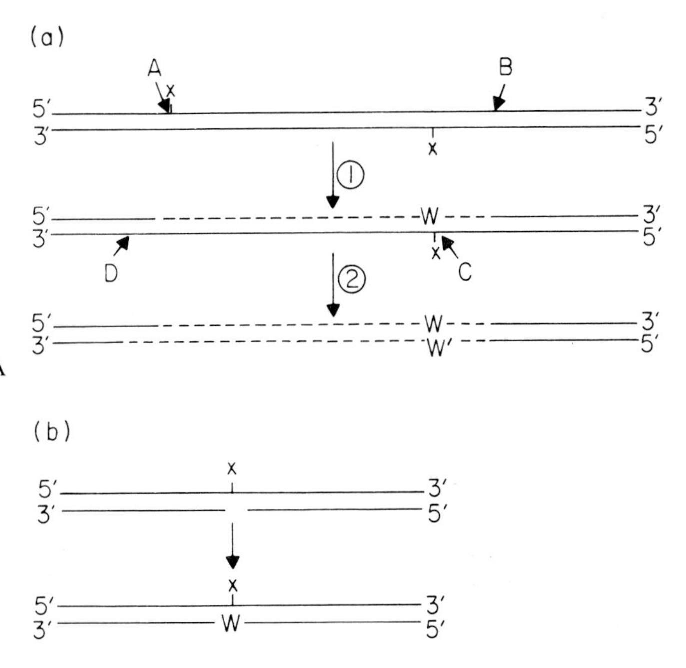

Fig. 4.11. Two theoretical ways in which DNA repair might generate errors. (a) Long-patch excision repair between A and B results in an x residue appearing in the template strand; this results in the incorporation of a 'wrong residue', W. When the other strand is repaired by long-patch excision between C and D, the residue complementary to W (W') is incorporated. (b) The daughter molecule I in Fig. 4.10(b) is repaired by a gap-filling process before recombination occurs. A wrong residue is incorporated.

repair systems operate in bacteria. An alternative source of errors occurs if a repair system attempts to fill in a gap in post-replication repair (Fig. 4.11(b)).

The Ames test

The conclusion from the last two paragraphs is that if one wishes to maximize the number of mutations produced by a mutagen, the error-prone system should be induced or made constitutive. For quantitative purposes, clearly the latter is an advantage. In *E. coli,* this may be done in a strain carrying a mutation called *tif.* We shall see later (see section 4.3.4) that *tif* is properly regarded as an allelic designation for *recA*. However, the effect of the *tif* mutation is to make the error-prone system constitutive at an elevated temperature. An alternative method is to introduce into the cells a plasmid that contains genes for a different, but constitutive, error-prone system.

Such a plasmid is employed in the most widely used set of strains for screening chemicals for their mutagenic properties, the Ames test (McCann & Ames 1976, Ames 1978). The interest in the sensitive and quantitative assay of mutagens derives from the correlation between mutagenicity and carcinogenicity. This correlation follows from the somatic mutation theory of chemical carcinogenesis which, in its simplest form, proposes that tumours derive from a mutant clone of somatic cells. In the Ames test, the strains all contain the plasmid with the error-prone system and all carry a mutation that affects the lipolysaccharide (LPS). The organism is not *E. coli* but the very closely related *Salmonella typhimurium.* The strains used for the test carry different defined mutations in the His operon. The mutations are known to be frame-shifts, transversions, transitions, etc. These strains are all unable to grow in the absence of histidine. The test consists of measuring the frequency of reversion to histidine prototrophy. Many carcinogens are either very poor mutagens in bacteria or do not produce mutations at all. It was suspected, correctly, by Ames that the reason for this is that these carcinogens need to be metabolized by mammalian enzymes to form the proximal mutagens and that endogenous bacterial metabolism is unable to effect the necessary conversions. Moreover, many of the proximal mutagens may be chemically very unstable and it is thus necessary to mix the compound with a mammalian enzyme system (the post-mitochondrial supernatant of liver) and the bacteria. The purpose of the LPS mutation is to allow the salmonellae to be in intimate contact with the membrane of the endoplasmic reticulum of the liver supernatant so that unstable metabolites of the carcinogen can react with bacterial DNA before they decay to inactive derivatives.

The importance of the Ames test lies not only in its substantiation of the somatic mutation theory of chemical carcinogenesis by improving enormously the correlation between carcinogenicity and mutagenicity, but also in affording a rapid, sensitive and inexpensive procedure for testing for the presence of putative carcinogens in pharmaceuticals, food additives, cosmetics and other products and chemicals in the environment.

4.3.3 Pre-replication repair

Photoreactivation

Photoreactivation is a mechanism for removing thymine dimers (see section 3.3.3 and Fig. 3.18); it requires radiation in the near UV (about 300 nm) and a photoreactivating enzyme. The mechanism involves tryptophan residues in the enzyme and is understood in considerable detail from studies with peptides that, *in vitro*, mimic the effect of the natural enzyme (Sutherland 1978).

Demethylation

This pathway, sometimes referred to as the adaptive response to alkylating agents, is the most recently discovered repair system in *E. coli*. The system is apparently specific for O^6-methyl G-residues, which are potentially highly mutagenic lesions (see section 3.3.2 and Fig. 3.15(d)). The system is independent of the genes *lexA* and *recA* (see section 4.3.4), is error-free and has the curious property of being saturatable so that only a certain number of methyl groups can be removed by this mechanism, following the challenge of the cell by a methylating agent. The reaction involves enzymic transfer of a methyl group from a G-residue in DNA to a cysteine residue in an acceptor protein. The phenomenon of saturatability presumably reflects the ability of the protein methyl-receptor molecules to function only once. Evidence for the production of residues of methyl cysteine, $MeSCH_2CH(NH_2)CO_2H$, and references to the discovery of the system and its biological implications are given by Olsson and Lindahl (1980).

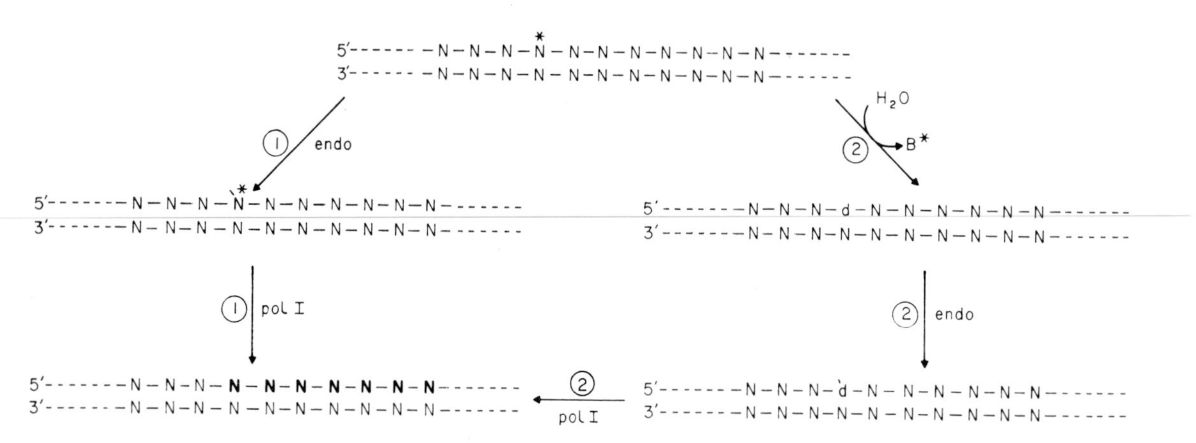

Fig. 4.12. Two possible biochemical pathways for excision repair. N represents a nucleoside and a hyphen represents a phosphate. N* is a damaged or incorrect nucleoside. In the case of UV-irradiated DNA, two adjacent residues (thymidines) would be damaged. B* is a damaged or incorrect base corresponding to N*; d represents deoxyribose. Endo's are endonucleases and pol I is DNA polymerase I. **N** represents a newly incorporated nucleoside residue. (1) refers to a pathway in which hydrolysis of a phosphodiester bond starts the process, and (2) refers to a pathway started by hydrolysis of an *N*-glycoside bond.

4.3.4 Excision and post-replication repair

Two theoretically possible pathways for excision repair are shown in Fig. 4.12.

Pathway (1) consists of hydrolysis of a phosphodiester bond to yield a 5′-phosphomonoester by a 'correction endonuclease', followed by erosion from the 5′-end and re-synthesis by the combined 5′-exonuclease and polymerase activities of DNA polymerase I (see section 4.2.1). When this synthesis is complete, there is a discontinuity (not shown in Fig. 4.12) that is sealed by DNA ligase. This pathway is well established as a mechanism for the repair of UV-induced pyrimidine dimers (see section 3.3.3). The pathway is defective in certain UV-sensitive (Uvr) strains of *E. coli*. The correction endonuclease is a multi-subunit complex determined by the genes *uvrA*, *uvrB* and *uvrC* (Seeberg 1978). The *uvrA* gene product is an ATP-dependent DNA binding protein that recognizes the distortion in the double helix; the expression of this gene is inducible by UV light (Kenyon & Walker 1981). The mechanism of such UV inductions is discussed later in this section.

Pathway (2) starts with removal of the inappropriate base by hydrolysis of the *N*-glycoside bond. This might occur spontaneously in the case of 7-alkyl purines (see section 3.3.2). Alternatively, there are enzymes (DNA glycosylases) that hydrolyse these bonds. The later part of the scheme involves nucleolysis by one of the two enzymes apurinic acid and apyrimidinic acid endonucleases followed by erosion and re-synthesis by DNA polymerase I (Friedberg *et al.* 1978). Although the physiological role of this pathway is not certain, it may be implicated in the removal of the lethal (but non-mutagenic) 3-methyl adenine lesions which can arise following exposure to

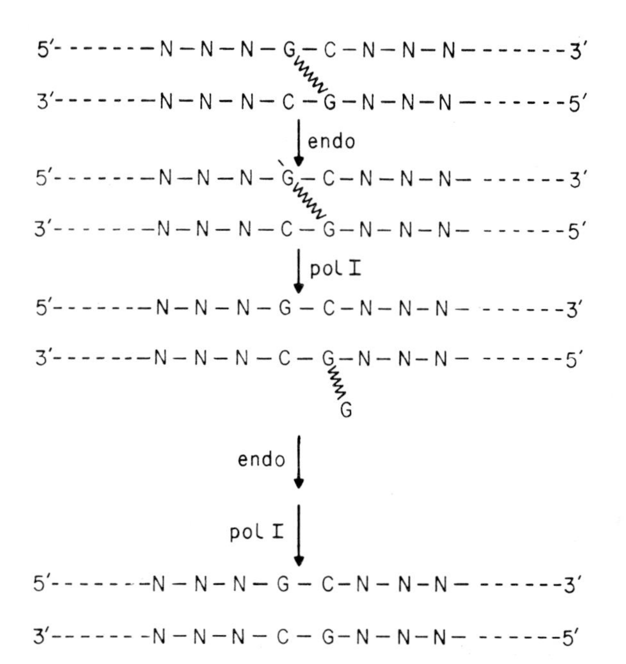

Fig. 4.13. Repair of cross-linked DNA. The process is illustrated according to pathway (1) of Fig. 4.12, although in principle, pathway (2) could operate. Cross-linked G-residues are represented as G ᴡᴡ G.

certain alkylating agents (see section 3.3.2). Additionally, there is speculation that the base B* (in Fig. 4.12) might be replaced by the correct base, by a template-dependent 'base insertase' (Linn 1978).

Excision repair can deal with cross-links, such as those introduced by the mustards (see section 3.3.2) by an extension of the earlier schemes (Fig. 4.13) and probably involves the *uvr* system.

The *uvr* repair system in *E. coli* is short patch and is error free. So *uvr* mutants are more UV-sensitive than wild type and are more UV-mutable (i.e. the proportion of mutants among the survivors is higher) since the cells rely more on error-prone SOS repair.

Involvement of recA

Genes that are defined by mutations that eliminate or reduce the frequency of recombination in an F^- recipient of a conjugal cross are designated *rec*. The gene *recA* is involved in several processes in *E. coli* as is evidenced by the fact that *recA* mutations are highly pleiotropic (that is they affect several phenotypes). For one thing *recA* mutants are more UV-sensitive than wild type but are not UV-mutable (cf. the *uvr* mutants). Thus *recA* is involved in the error-prone repair pathway.

The effects of UV light on *E. coli* include the induction of λ prophage, inhibition of septation, inhibition of exonuclease V (the *recB/C* gene product), induction of a long-patch repair system, mutagenesis, Weigle reactivation and the appearance in the cell of a new protein, referred to in the earlier literature as protein X. These effects all occur at an elevated temperature (without UV irradiation) in strains carrying the mutation *tif*; *tif* is in fact allelic for *recA*. Moreover, protein X is in reality the RecA gene product (Emmerson & West 1978). It seems feasible, therefore, to suppose that the gene *recA* is expressed constitutively at a low level, sufficient for its role in recombination but insufficient for 'protein X' to be detectable in mixtures of total *E. coli* proteins. Furthermore, it seems that UV induction results in over-expression of *recA* with a concomitant effect on phage induction and the other UV-inducible phenomena (this conclusion comes from the properties of *tif* mutants). There is, however, another element in the process: mutations in the gene *lexA* affect all the UV-inducible *recA* functions (including the appearance of protein X) but do not affect recombination. The implications of these conclusions are that *lexA* is the structural gene for a repressor of the *recA* operon but that the RecA protein itself and a UV-sensitive target are both components of the control system.

The UV-sensitive target is DNA. Part of the evidence for this comes from the fact that other DNA-damaging reagents, such as mitomycin c, also produce the *recA*-dependent inducible effects. Emmerson and West (1978) proposed that the inducers might be deoxyribonucleoside monophosphates. These might be expected to accumulate in cells with damaged DNA templates through the proofreading functions of DNA polymerase III. If the polymerase inserts a wrong residue, its 3′-exonuclease activity removes it and the polymerase tries again. If the template strand contains an aberrant residue, possibly there is no base that can fit. The effect will be a net conversion of dNTPs to dNMPs. This explanation may be correct in part; however, *in vitro,* the inducers appear to be single-stranded polynucleotides (Craig & Roberts 1980) which would also accumulate in cells with mismatched double helices. The control of *recA* expression and the mode of action of the RecA protein itself are believed to be due to the proteolytic activity of the protein in the presence of such DNA fragments. The suggestion that the RecA protein is a specific 'repressor protease' was made by Witkin (1976). The regulation of the *recA* system shown in Fig. 4.14 is based on Witkin's ideas but incorporating the role of damaged DNA fragments as inducers. The other difference between this figure and Witkin's views is that, rather than invoking the RecA protein as a protease, Witkin postulates that the LexA protein plays an additional

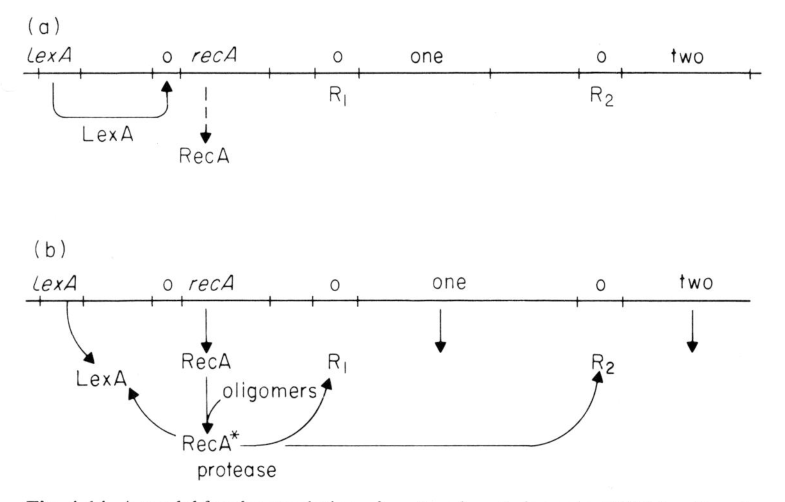

Fig. 4.14. A model for the regulation of *recA* and *recA*-dependent SOS functions in *E. coli.* In (a) the expression of *recA* is limited by the binding of a repressor (the LexA protein) to the *recA* operator (o). 'One' and 'two' are two operons regulated by *recA*. They might, for example, be operons required for the induction of a prophage and the septation inhibitor. These operons are not expressed because there are repressors (R_1 and R_2) bound to the appropriate operators. In (b), DNA has been damaged (for example by UV light). DNA oligomers (or alternatively mononucleotides—see text) convert the RecA protein into a protease ('RecA*'). This hydrolyses LexA so that expression of *recA* becomes unconstrained and also hydrolyses R_1 and R_2 so that the operons one and two are expressed. The original model comes from the review by Witkin (1976) and differs in ascribing to LexA a proteolytic protective role (see text).

role as a protector of the repressors (such as R_1 and R_2 in Fig. 4.14) from endogenous protease(s) and hence the effect of RecA is on LexA alone. The consequence is formally equivalent: in this scenario there is simply an extra component in the sequence of events that leads to proteolysis of the repressors for UV-inducible functions.

So far we have concentrated on known UV-inducible functions. However, with the advent of the construction of recombinant phage molecules (see Chapter 11) it is possible to identify the positions on the *E. coli* chromosome of *recA/lexA*-dependent UV-inducible operator/promoter combinations without prior knowledge of the gene functions. The method is based on specialized transducing derivatives of phage Mu. This phage is capable of inserting effectively at random in the *E. coli* chromosome (see section 4.3.7). The defective phage Mud (Ap, *lac*) contains the structural genes of the Lac operon but without the promoter. However, if the phage DNA inserts within a transcriptional unit, the expression of Lac (easily observed on indicator media for β-galactosidase) is a test for transcription or otherwise from the promoter of the transcriptional unit in question (Casadabadan & Cohen 1979; Fig. 4.15). Kenyon and

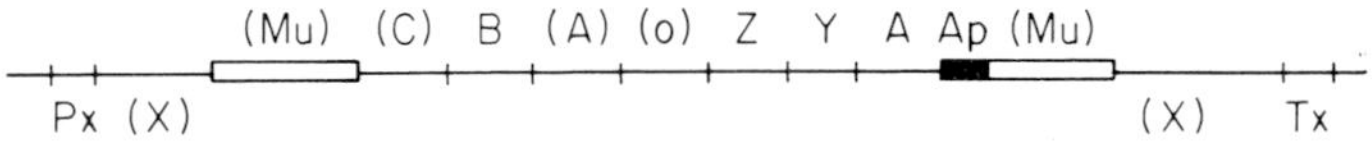

(Mu) (C) B (A) (o) Z Y A Ap (Mu)
Px (X) (X) Tx

Fig. 4.15. The Mud (Ap, *lac*) phage DNA inserted in a transcriptional unit (after Casadabadan & Cohen 1979). The genetic map is not to scale. Brackets surround genes or groups of genes that are either incomplete or split. The open bars represent Mu DNA; the thick black line is a part of the transposon Tn3 carrying the gene for ampicillin resistance (Ap); (C), B and (A) form a part of the Trp operon on to which a part of the operator (o) and the three structural genes, Z, Y and A of the Lac operon have been fused. X (split by the Mud insert) is an *E. coli* operon with its promoter (Px) and terminator (Tx). If the X operon is expressed, a giant mRNA transcript will be formed and the Lac operon will be expressed. This can be recognized because the LacZ protein, β-galactosidase, can readily be detected by indicator media that incorporate histological stains for the enzyme.

Walker (1980) used the Mud phage to identify damaged induced *(din)* loci in *E. coli.* Mud inserted in such a *din* locus gave rise to β-galactosidase expression if the cells were exposed to UV light or other DNA-damaging reagents and this induction was dependent on *recA* and *lexA*. One such Mud insertion led to greater UV sensitivity and it was this observation that led Kenyon and Walker (1981) to the discovery that *urvA* is a *din* locus.

These observations all add up to accounting for the way in which *recA*-mediated processes can lead to increased DNA repair: RecA protein itself is a recombinational protein required for post-replicative repair and *uvrA* is required for excision repair. An unresolved question is exactly how the SOS phenomenon of error-prone repair arises. The subject is reviewed by Witkin (1976) and the genetics of the system are reviewed by Gottesman (1981). One possibility is that the inactivation of the *recBC* nuclease (exonuclease V) and the over-expression of the gene *(recA)* for another recombinational protein lead to error-prone recombination. In other words it is the recombinational post-replication repair that is error-prone. On the other hand long-patch repair is induced by UV in a *recA*-dependent manner. This type of repair apparently relies on DNA polymerase III. However, the nature of the protein, presumptively induced like the proteins of 'operons one and two' of Fig. 4.14, and the mode of action of this protein in allowing long-patch processes are both unknown. There are certainly additional components, for example, genes *umuC* and *recF* are involved in the mutagenesis. The papers already cited by Kenyon and Walker (1980) and Craig and Roberts (1980) include references to the recent literature.

4.3.5 Recombination

General considerations

Genetic recombination involves replacement of one region of a double-stranded DNA molecule by another (Fig. 4.15). Reciprocal recombination of the kind illustrated here is not often observed in prokaryotic genetics; for example, in a cross between an Hfr and an F^- strain, one of the two molecules III and IV in Fig. 4.15 would not be a viable chromosome. In those cases where both products of the cross-overs can be recovered as viable progeny and subsequently analysed genetically, the pattern of recombination is sometimes found not to be reciprocal and this is important evidence in support of the kinds of model for recombination about to be discussed. This and other evidence for the formation of a heteroduplex intermediate during recombination (scheme 2 in Fig. 4.16) in fungal systems are discussed in section 7.4.2. However, in the context of prokaryotic recombination there is plenty of clear evidence for the existence of heteroduplex intermediates. One example comes from the analysis of rapid lysis (*r*) mutants of phage T4 whose plaques are large and continue growing and can readily be distinguished from wild-type (r^+) plaques. Plaques produced by the progeny from certain $r^+ \times r^-$

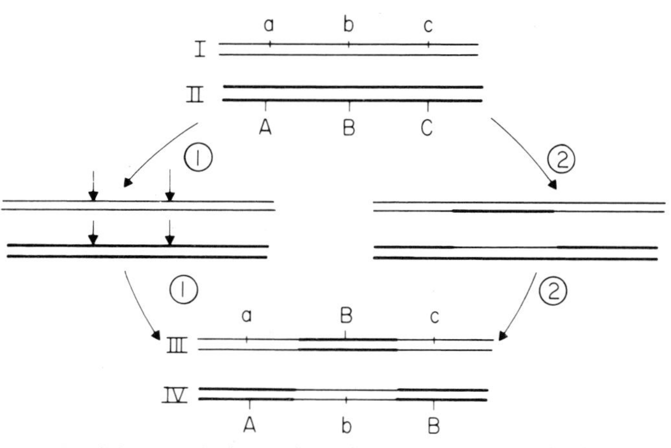

Fig. 4.16. Possible general schemes for reciprocal genetic recombination. I and II are two chromosomes shown diagrammatically as double-stranded DNA molecules; a/A, b/B and c/C are allelic variants of three genes. III and IV are the recombinant chromosomes. Scheme (1) is a hypothetical pathway in which double-stranded pieces (between the arrows) are exchanged between the chromosomes. Scheme (2) involves heteroduplexes, partly mismatched in the region including the gene b/B, as intermediates. See text for a discussion of genetic evidence that supports scheme (2).

crosses include some small (r^+) plaques with large flares of r^- phage. Phage isolated from the two parts of such plaques are found to be true (either pure r^- or pure r^+). The T4 particle from which the original plaque of this type was derived must have been heterozygous and contained both alternative r alleles. As the phage is haploid, it must have contained heteroduplex DNA.

For references to the literature on recombination and a discussion of the models, see Radding (1978).

How do the homologous sequences find one another?

Do two homologous double helices associate by some sort of H-bonding or other interaction different from the Watson–Crick base pairing schemes (such as a quartet of bases forming an association) or do the two molecules denature and re-anneal? And is the process dependent upon some sort of specific 'recognition sequence'? In the only carefully argued model for recognition, Sobell (1972) proposes that the process is essentially one of re-annealing of partially denatured structures and that the association does require specific sequences. The essentials of the Sobell model are illustrated in Fig. 4.17. There is no direct proof for this model but it is attractive for the following two reasons. It invokes regions of twofold symmetry, which have been established as being important in other regulatory regions of DNA (see section 4.1). The suggestion that

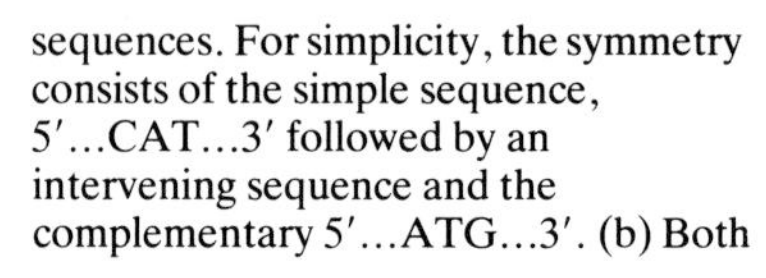

Fig. 4.17. Recognition of homologous DNA sequences containing partial palindromic symmetry. This is a much simplified version of the Sobell model (see text). (a) The two homologous sequences. For simplicity, the symmetry consists of the simple sequence, 5′...CAT...3′ followed by an intervening sequence and the complementary 5′...ATG...3′. (b) Both molecules adopt a cruciform conformation and are nicked by an endonuclease at the arrows. (c) Recognition by hybridization. The symbol ● represents a 5′-end.

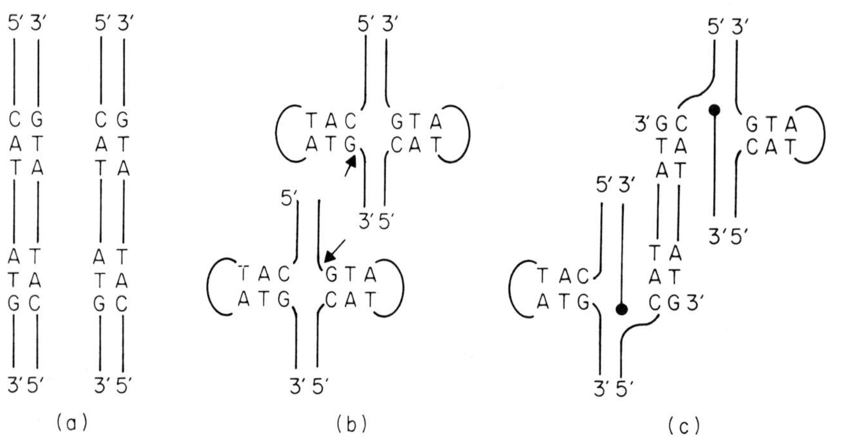

there are specific sequences involved in the initiation of recombination suggests that there ought to be 'recombinational hotspots' surrounding such sequences. There is evidence for such hotspots in several genetic systems; the most dramatic example is probably the mitochondrial chromosome of yeast. In particular parts of this chromosome recombination rates of 25% occur within 1000 base pairs (Borst & Grivell 1978).

How is the heteroduplex formed?

Of the many models proposed to account for heteroduplex formation, only two are considered here. The first is the Holliday model (Holliday 1964; Fig. 4.18). The Holliday model is particularly attractive because of its simplicity and the elegant way in which it gives rise to reciprocal and non-reciprocal recombination by alternative patterns of cleavage by nucleases. In fact the model is somewhat less simple than it appears in Fig. 4.18 because if three-dimensional models are used in place of the 'tramline' representations of the figure, the two alternative ways of cutting the intermediate (at the C's or D's in Fig. 4.18) are found to be non-equivalent. A structure similar to this must arise when two replication forks of the replicating bacterial chromosome come up against one another at the chromosome terminus (see section 4.2.3); in this case it is clearly important that the inherent asymmetry is recognized

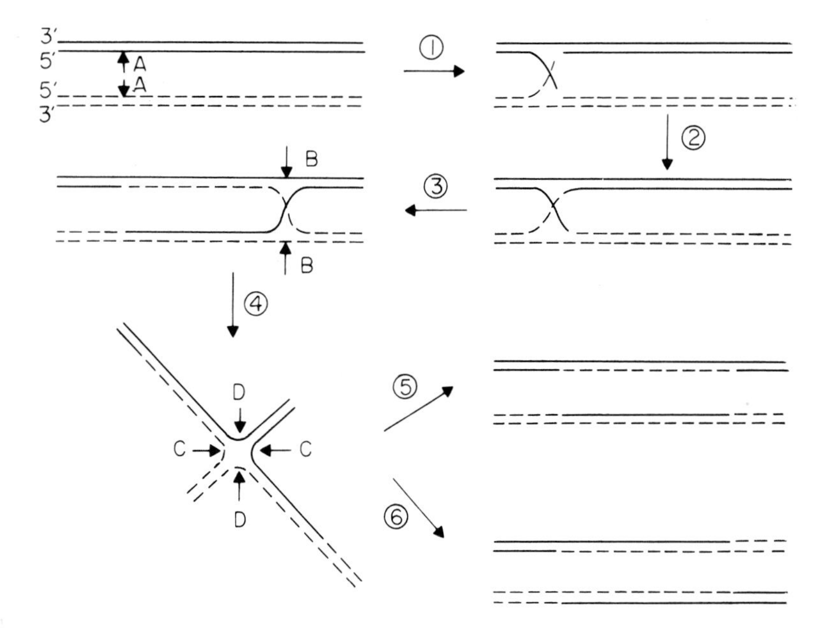

Fig. 4.18. The Holliday model for genetic recombination. Two homologous double-stranded sequences are nicked in one strand at A by an endonuclease (1). The ends associate with one another by a process for which Fig. 4.17 suggests one possible mechanism. The ends are ligated (2) and the branch so formed migrates (3). If the resulting structure is 'bent' at B and rotated about one of the strands in the cross-over, the molecule can be re-drawn as in (4). If an endonuclease cuts at points C and the ends are ligated (5), two heteroduplexes are formed. Following replication the reciprocal pattern of recombination would be produced. Cutting at D and ligating (6) yields a pattern of heteroduplex molecules that, following replication, would generate a non-reciprocal pattern of recombination.

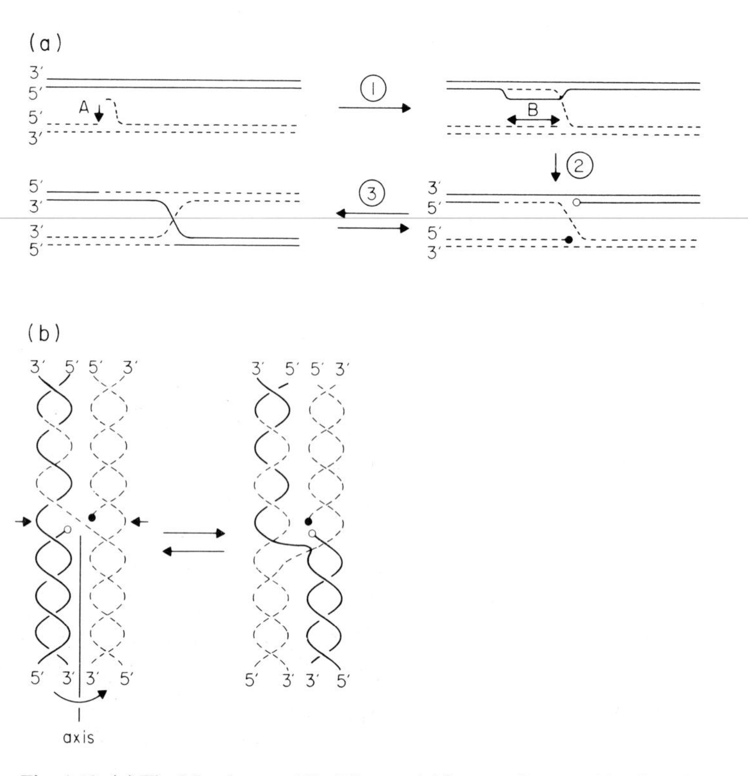

Fig. 4.19. (a) The Meselson and Radding model for genetic recombination. One strand has been nicked by an endonuclease at A. During stage (1), the 3′ terminus acts as a primer for resynthesis of segment B by a process analogous to repair (scheme 1 of Fig. 4.12). The displaced strand anneals to the complementary strand of the other double helix. Stage (2) consists of cleavage of the non-hybridized loop to leave a free 3′-end (●) in one double helix and a free 5′-end (O) in the other one. Stage (3) is 'isomerization' to generate a complete cross-over. Migration and cutting occurs as in the Holliday model (stages 4–6 in Fig. 4.18). (b) Isomerization explained. Stage 3 from (a) redrawn to show the helices. The lower part of the structure rotates around the axis as shown so that the two arrowed sequences generate the two strands in the cross-over. This leaves the ends (● and O) adjacent in the same helix and, therefore, capable of being joined by DNA ligase.

correctly because the correct pair of cuts will generate two daughter chromosomes whereas an incorrect pair of cuts would merely generate a closed molecule twice the size of a single chromosome.

This problem and also the second model are discussed by Radding (1978). The second model is that of Meselson and Radding (Fig. 4.19). The two main differences from the Holliday model are a requirement for DNA synthesis and a fundamental asymmetry of the initial stage with a requirement for an isomerization step to generate the single cross-over.

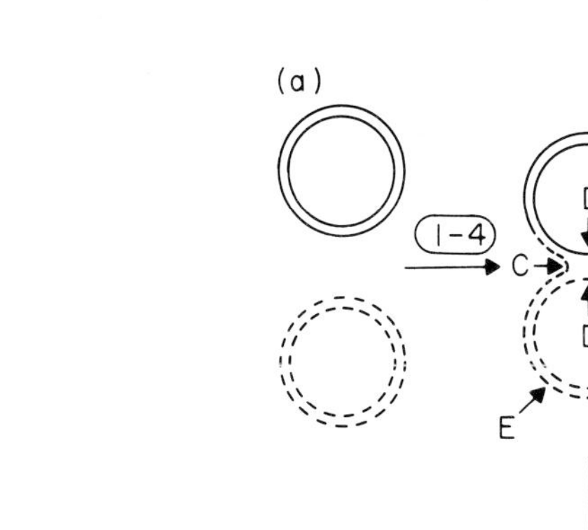

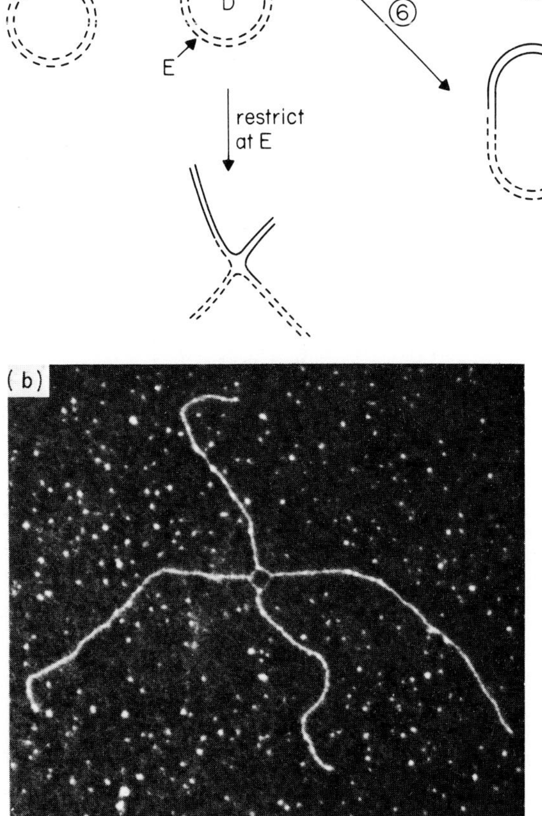

Fig. 4.20. Dimerization of the plasmid, Col E1 by recombination. (a) The Holliday model (Fig. 4.18) adapted to a closed molecule. The first four stages generate a figure of 8. Processing can either generate the reciprocal products (i.e. two Col E1 plasmids, step 5) or the non-reciprocal product (step 6), which is in this case a dimer. The letters C and D have the same significance as in Fig. 4.18. Col E1 has a single site of the restriction endonuclease, *Eco* RI. The figure of 8, therefore, contains two such sites (E). If a mixture, believed to contain such structures is restricted, some molecules shaped like a χ with a central single-stranded region, should be present. (b) A χ-shaped restricted intermediate of the sort shown as a diagram in (a) actually visualized by electron microscopy. (We are grateful to Professor D. Dressler for supplying this photograph.)

Direct experimental evidence for the formation of the penultimate structure in the recombinational models was obtained by Potter and Dressler (1977, 1978) who showed that the plasmid Col E1 forms dimers in *E. coli* cells or cell-free extracts by a recombinational mechanism (the dimerization did not occur in *recA*$^-$ cells or extracts). Electron micrographs of DNA extracted from these cells or produced in the cell-free system showed, among the dimers, perfect examples of Holliday intermediates (Fig. 4.20).

Recombinational proteins

The biochemical details of bacterial recombination are limited largely to the RecA and RecBC proteins. RecA protein (protein X) was first purified by McEntee (1977) who showed it to be a tetramer. Radding and his colleagues (Cunningham *et al.* 1979, Shibata *et al.* 1979a, b) have examined the biochemical properties of the protein. It is a DNA-dependent ATPase; in the presence of supercoiled double helical DNA, a homologous piece of single-stranded DNA and ATP, the double helix is locally denatured and the single-stranded DNA hybridizes to form a D-loop. This reaction could, therefore, account for the uptake step (1) in the model of Fig. 4.17. The stages in the RecA reaction can be resolved experimentally by using the ATP analogue, ATPγS, $^{-}O_3PSPO_2^{-}OPO_2^{-}OAdo$. This compound cannot be hydrolysed effectively by an ATPase. It was found that RecA protein effected the local denaturation of the double-stranded DNA in the presence of either ATP or ATPγS and any single-stranded DNA, whether homologous or not. Given homology between the single-stranded DNA and the double-stranded molecule, ATP hydrolysis was required for the establishment of the D-loop.

By using a different DNA substrate West *et al.* (1981) showed that strand separation as shown in the first panel of Fig. 4.18(a), is probably not required for RecA-mediated alignment of DNA molecules. In their experiment, the single-stranded molecule was annealed to a covalently closed single-stranded circle. As the RecA protein was shown not to function as a 'helicase' or HDP (see section 4.2.2) the authors concluded that homologous pairing can occur without prior strand separation of any sort. However, this experimental system would suggest that a cut such as 'A' in Fig. 4.19 would be required for the pairing to occur because there is no evidence that RecA can align two covalently continuous double helices.

The RecBC nuclease, otherwise exonuclease V, the product of two recombination-specific genes, *recB* and *recC*, has a bewildering range of specificities (Champoux 1978). The enzyme has a molecular weight of 270 000 and is inhibited by the RecA protein. It is both an ATP-dependent nuclease and a DNA-dependent ATPase and can degrade DNA chains from either 5′- or 3′-ends to produce pentanucleotides and other short oligomers. Shibata *et al.* (1979a, b) propose it might be involved in cleaving single-stranded remnants following the formation of the D-loop by the RecA protein.

4.3.6 Translocatable elements—F and λ

Introduction

This and the following two sections deal with processes variously referred to as additive, illegitimate or site-specific recombination. The mechanism of translocation and the consequences of translocation are of great current interest and the field is among the most exciting in current prokaryotic molecular biology; see Kleckner (1977), Shapiro (1977) and Calos and Miller (1980) for reviews.

Insertion of the F factor

The insertion of the episome, F, is a recombinational event consisting of alignment of homologous IS sequences in plasmid and chromosome followed by a single cross-over (see section 4.1.2 and Fig. 4.2). The process depends on the bacterial recombinational system and does not occur in Rec⁻ cells (Willetts & Skurray 1980).

Insertion and excision of λ

Although the insertion and excision of the λ chromosome is formally similar to episome insertion and excision (see section 4.1.4) it differs in two important respects: the process is independent of the

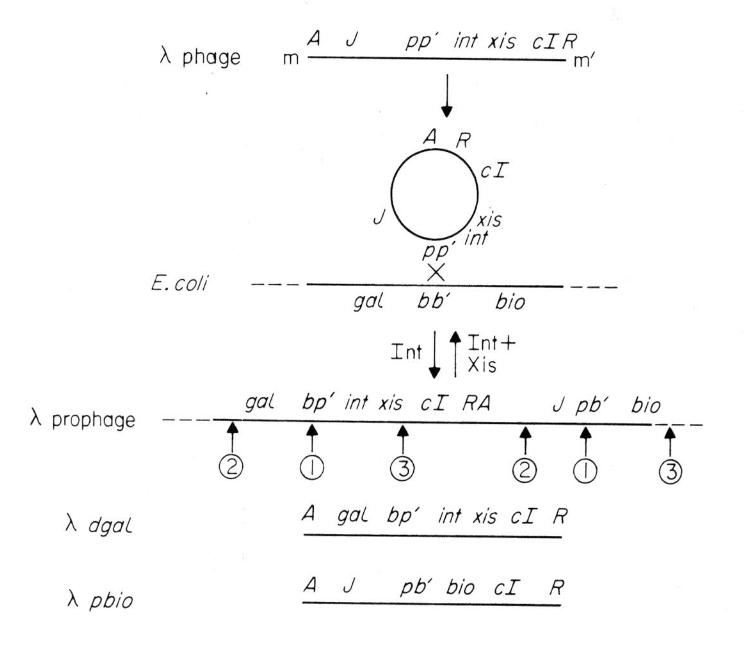

Fig. 4.21. Integration and excision of λ. The top line represents the vegetative map (the linear genetic map of the phage). Only a few genes are shown (see Fig. 4.4 for details); m and m′ are the cohesive ends (*cos* region) and *pp′* constitutes *attP*. Integration at *attB* (*bb′*) requires Int protein. A reversal of this process at (1) requires Int protein and Xis protein. Excision at (2) generates λ*dgal* and excision at (3) generates λ*pbio*. Both of these types of specialized transducing phages contain hybrid *attP*/*attB* sites.

Rec system of *E. coli* and it involves the alignment of two related but not homologous sequences. The λ genes involved in the process are *int* (integration) and *xis* (excision). Integration requires *int*; excision requires both *int* and *xis*.

The sequences involved in integration and excision are designated *attP* (attachment site in the phage) and *attB* (attachment site in the bacteria). Each site consists of two parts (p and p′ for *attP,* b and b′ for *attB*) and is split during the excision process. The excision process is occasionally erroneous and in these cases, specialized transducing phages are generated. Such a phage contains a recombinant attachment site (*attR*) of the type otherwise only found in a lysogen (Fig. 4.21). Note that the *int* gene is adjacent to *attP*. From the direction of transcription in this part of the λ chromosome, it is the distal or C-terminal part of the gene that abutts *attP*. Davis *et al.* (1977) and Landy and Ross (1977) determined the *attP* DNA sequence. The sequence extends to cover the distal region of the *int* gene (Fig. 4.22). The significance of the promoter sequences in *attP* remains unclear. Possibly transcription or the binding of RNA polymerases effects a change in the tertiary structure of the supercoiled DNA in this region. The high-AT regions would also favour strand separation. The palindromic regions would be required to generate Sobell intermediates (Fig. 4.17) and λ integration and excision may be properly

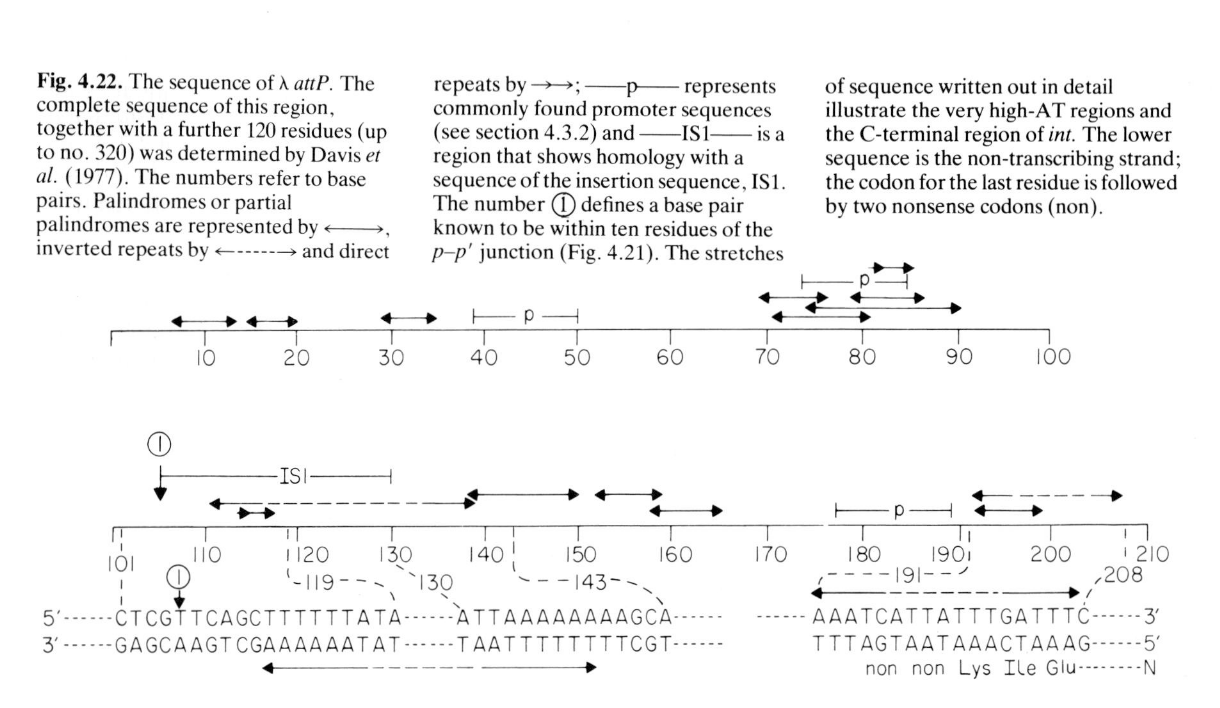

Fig. 4.22. The sequence of λ *attP*. The complete sequence of this region, together with a further 120 residues (up to no. 320) was determined by Davis *et al.* (1977). The numbers refer to base pairs. Palindromes or partial palindromes are represented by ←——→, inverted repeats by ←------→ and direct repeats by →→; ——p—— represents commonly found promoter sequences (see section 4.3.2) and ——IS1—— is a region that shows homology with a sequence of the insertion sequence, IS1. The number ① defines a base pair known to be within ten residues of the *p–p′* junction (Fig. 4.21). The stretches of sequence written out in detail illustrate the very high-AT regions and the C-terminal region of *int*. The lower sequence is the non-transcribing strand; the codon for the last residue is followed by two nonsense codons (non).

regarded as a variant on a recombinational cross-over in which partial homology between the sequences (*attP* and *attB*) does not allow integration and excision to proceed in a Rec-dependent system but rather is brought under control of two phage genes (*int* and *xis*).

4.3.7 Translocatable elements—insertion sequences, transposons and phage Mu

These are the transposable elements of section 4.1.3. The basic mechanism of transposition is probably common to all these elements; it is convenient to consider insertion sequences and transposons first and leave the special features of Mu to the end.

Insertion sequences and transposons

The elements all have an inverted repeat sequence at each end (see section 4.1.3; Fig. 4.23). Models for the mechanism of transposition have to account for several findings including the following. Following insertion, there is a short duplication of host DNA on either side of the transposable element. The length of the repeat is characteristic of the transposable element and is commonly of either five base pairs (e.g. the ampicillin-resistance transposon, Tn3 and Mu) or nine base pairs (e.g. IS1 and the kanamycin- and tetracycline-resistance transposons, Tn5 and Tn10). Transposable elements can cause deletions and inversions of host DNA. Transposition can also

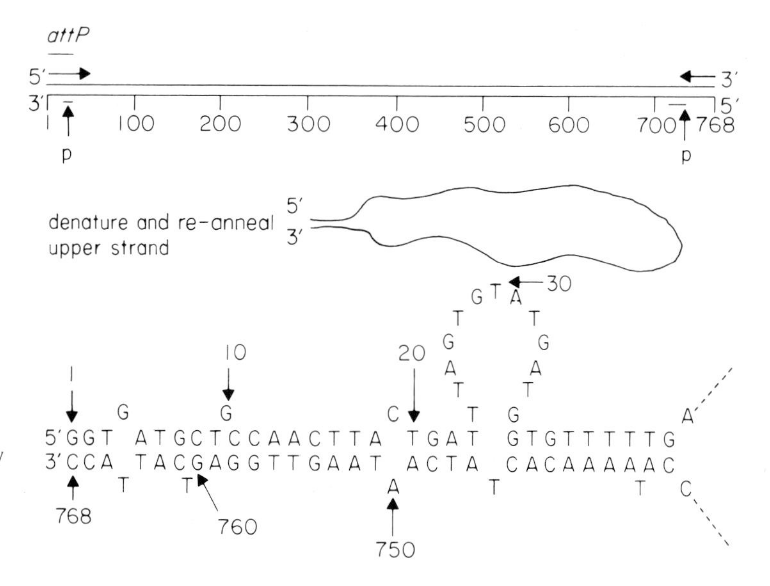

Fig. 4.23. Insertion sequence IS1. The upper part of the figure shows the positions of inverted repeats, promoter-like sequences (p) and a region of homology with part of the λ attachment site (*attP*, Fig. 4.22). Annealing of the upper strand reveals the complementarity inherent in the inverted repeats. The lower part of the figure shows the actual nature of the complementarity (Ohtsubo & Ohtsubo 1978).

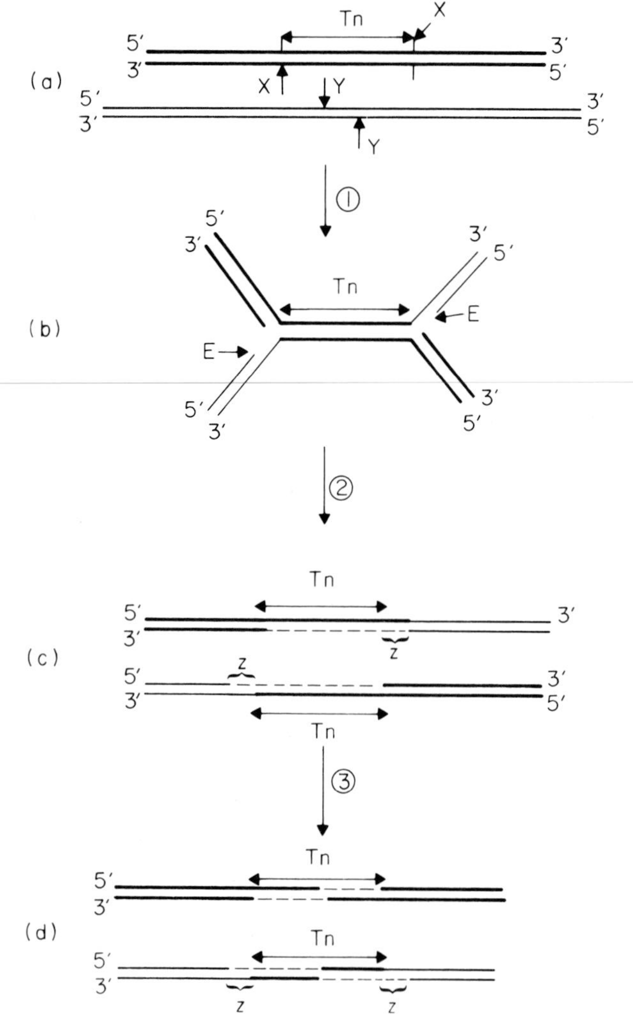

Fig. 4.24. Transposition according to Shapiro (1979). (a) A donor molecule (bold lines) containing a transposable element (Tn) aligned with a recipient molecule (thin lines); X and Y are sites where nicks are to be introduced. Process (1) is the nicking of the molecules in (a) followed by two ligation reactions. (b) The ligated intermediate; E's are unligated 3′-ends. Process (2) is the extension of the primers from E by DNA polymerase, accompanied by separation of the Tn double helix. (c) An intermediate formed by process (2). The dotted lines represent newly synthesized DNA. Process (3) is genetic recombination between the Tn sequence and (d) is the situation following this process. In (c) and (d), z is the region between the Y's in (a) and is the short repeat (usually 5 or 9 base pairs) that flank transposable elements (see text).

cause fusion of replicons to form 'co-integrates', for example, transposition of a drug resistance transposon from a plasmid to a chromosome can result in integration of the plasmid into the chromosome in an arrangement in which plasmid DNA is flanked by two direct repeat copies of the transposon. This is different from the recombinational insertion of the F factor. Finally, in certain cases, transposition can result in the generation of heterogeneous circles containing the transposon and pieces of host DNA (this is particularly important in the case of Mu, see below).

Several models have been proposed to account for transposition (reviewed by Bukhari 1981). The model of Shapiro (1979) is summarized in Fig. 4.24; the alternative models differ from this mainly by making the replication step precede the recombination step. The essence of the model is the creation of staggered breaks, ligation and DNA synthesis that lead to an intermediate arrangement (Fig. 4.24(c)) from which the transposed products are resolved by recombination.

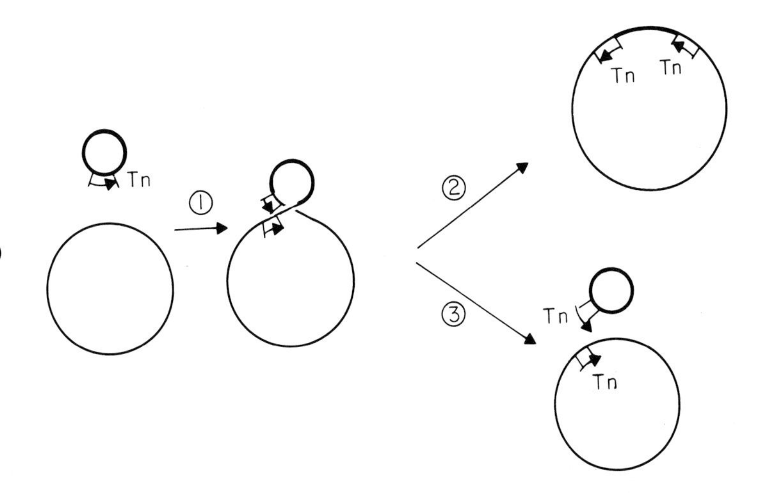

Fig. 4.25. Transposition of a transposable element (Tn) from a plasmid (smaller circle) and a chromosome. Process (1) corresponds to steps (1) and (2) in Fig. 4.24. If the co-integrate (equivalent to Fig. 4.24(c)) is unresolved by recombination (2), the plasmid DNA is inserted in the chromosome flanked by direct repeats (as shown by the directions of the arrows) of the transposable elements. Alternatively (3) recombination resolves the co-integrate to form two replicons again, corresponding to Fig. 4.24(d).

The model predicts that transposable elements never have an autonomous existence and that transposition invariably leads to amplification of the transposable element. It also accounts for the finding that transposable elements can effect fusion of replicons by forming a co-integrate (Fig. 4.25).

In the case of transposition within a replicon, the events depend critically upon the orientation at the ligation step (Fig. 4.26). In one

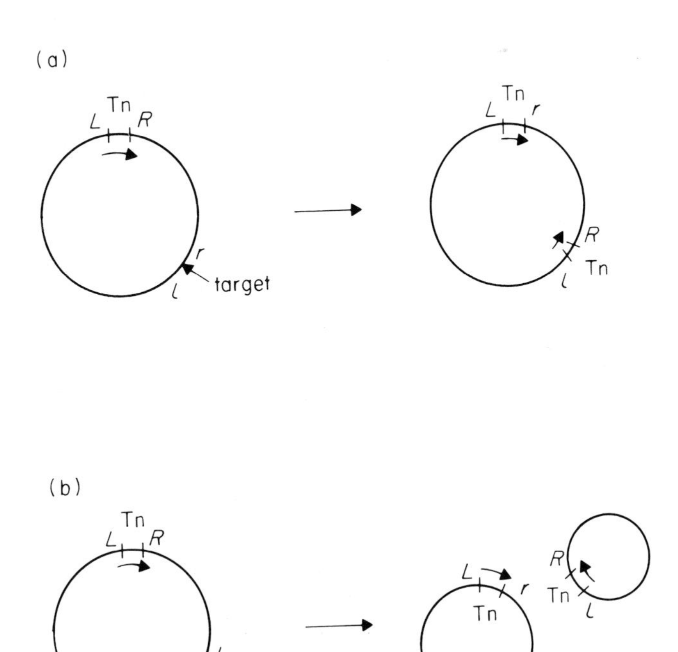

Fig. 4.26. Consequences of transposition within a replicon. *L* and *R* are regions flanking the transposable element corresponding to the upper molecule in Fig. 4.24(a); *l* and *r* are the corresponding sequences (on either side of the Y's) in the lower molecule in Fig. 4.24(a). (a) In one orientation of the target (the site to which transposition is to occur) the effect of the stages illustrated in detail in Fig. 4.24 is to generate an amplification of the transposon within the replicon. (b) In the opposite orientation, the resolution step fragments the replicon.

orientation the transposon is duplicated within the replicon; in the other orientation the replicon is split in two in a process that would normally be lethal.

The steps leading to the co-integrate (i.e. conversion of the structures in Fig. 4.24(a) to Fig. 4.24(c)) involve a transposase. In the case of certain transposons and the elements described in section 4.3.8 the structural genes for such proteins are known. The resolution of the co-integrates requires a second protein that also (at least in the case of certain transposons) functions as a repressor for the transposase gene. The evidence and references are summarized by Simon *et al.* (1980).

Phage Mu

The sequence at the ends of phage Mu DNA contains two distinct (left and right) attachment sites, *attL* and *attR*. There are no inverted repeats except for TG (at the very ends) and the rather dubious pair of 5′-ACGAAAAAAAC-3′ (*attL*) and 5′-ACGAAAAAC-3′ (*attR*). However, the sequences are such that the two ends of a single strand of Mu DNA could associate in a more complex structure (Fig. 4.27).

The Mu DNA molecule behaves as a transposable element. During establishment it integrates at one of an enormous number of potential sites in the *E. coli* chromosome. When integration occurs within an operon, highly polar mutations are generated and give the

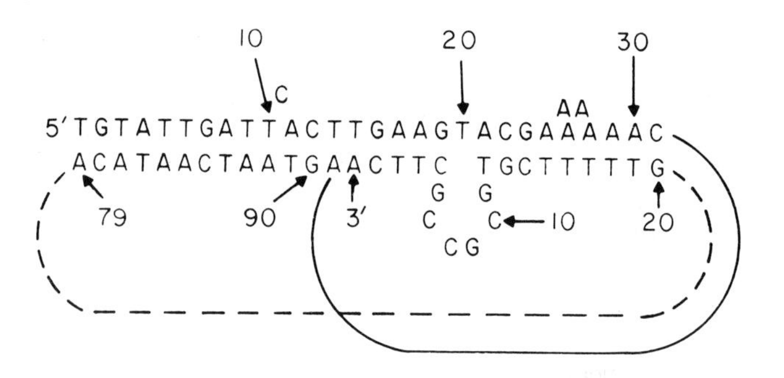

Fig. 4.27. The two termini of a single strand of Mu DNA showing a possible secondary structure. The numbers are the distances (in residues) from the junctions with *E. coli* DNA. The 5′-end is the first 31 residues of the *attL* sequence. The continuous line represents the whole of the rest of the Mu DNA sequence apart from *attR*. The 3′-end is the first 91 residues of *attR*; the discontinuous line represents the sequence of residues, 21–78 of *attR*, that are not involved in forming this structure (Kahman & Kamp 1979).

phage its name (see section 4.1.4). Moreover, the phage is replicated by repeated transpositions of the type shown in Fig. 4.26(b). The phage is quite unlike other known temperate phages (such as λ) in that integration is essential for replication as the phage genome is not a replicon. Clear plaque mutants of Mu necessarily carry mutations in the maintenance, not the establishment system. A consequence of this mode of replication is that the DNA in the Mu phage particle is flanked by *E. coli* DNA. These *E. coli* sequences (to the left of *attL* and the right of *attR*) are different in every phage particle and are the remnants of the last circles (such as those in Fig. 4.26(b)) to have been formed during replication. See Shapiro (1979) and Bukhari (1981) for references to the literature on Mu replication.

4.3.8 Phase shifts

Phase variation is a phenomenon found in prokaryotes in which clones of a single strain may express one of two alternative phenotypes. To take a familiar example, strains of *Salmonella* can express one of two alternative flagella antigens. The corresponding genes are *H1* and *H2*; when *H2* is 'on' the H2 form of flagellin (the protein subunit of bacterial flagella) is expressed and *H1* is repressed. Conversely when *H2* is 'off', H2 flagellin is not made and H1 is expressed. Cells can switch from one phase (H2) to the other (H1) and vice versa. The mechanism of phase variation in this case is well established (Simon *et al.* 1980; Fig. 4.28). Two 14 base pair inverted repeats, IRL and IRR, flank a sequence of 970 base pairs. This sequence includes a gene *hin* (H inversion), whose product, Hin protein, can invert the sequence between IRL and IRR. The promoter for the H2 operon also lies within the invertible sequence. In the orientation shown in Fig. 4.28 the operon is expressed to produce H2 flagellin and a repressor for the H1 operon. However,

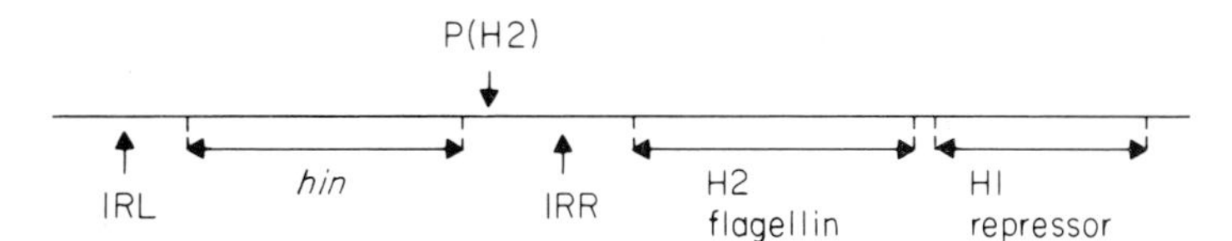

Fig. 4.28. The H2 region in *Salmonella* in the 'H2 on' phase. The elements and genes are not shown to scale. IRL and IRR are the left and right inverted repeats; P(H2) is the promoter for an operon with two structural genes, H2 flagellin and H1 repressor. See text for discussion and an explanation of *hin*.

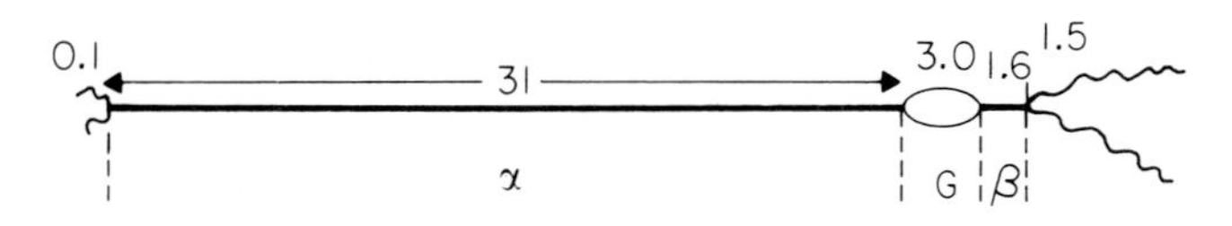

Fig. 4.29. Structure of the Mu genome as revealed by electron microscopy of reannealed DNA; represents denatured *E. coli* DNA. Numbers are lengths in kbp.

following inversion, reading of the H2 operon from this promoter, P(H2), is impossible and H1 is expressed.

A similar system operates in phage Mu. When Mu DNA, obtained from phage propagated from a lysogen, is denatured and re-annealed and examined by electron microscopy, the molecules are seen to have a structure shown schematically in Fig. 4.29 (Howe 1980). The G-region forms a loop owing to the fact that one half of the original molecules have the G-region in one orientation and half have it in the opposite orientation. The G-region is flanked by inverted repeats. The ragged ends are pieces of *E. coli* DNA that invariably flank Mu DNA (see section 4.3.7). The majority of Mu genes lie in the α region (Fig. 4.29). However, two tail-fibre genes lie in the G-region. The explanation for the behaviour of G lies in the fact that there are two sets of these genes. With G in one orientation one kind of tail-fibre is produced; these fibres are involved in the attachment of Mu to receptors in *E. coli.* In the other orientation, different fibres are produced; these are involved in the attachment to an alternative host, *Citrobacter freundii* (van de Putte *et al.* 1980). G-inversion is similar to H2 inversion in *Salmonella*; a gene *gin* (G inversion) is responsible. Moreover, *hin* can complement *gin* defects and, therefore, the Hin and Gin proteins must be similar. The Hin protein itself shows considerable sequence homology with the product of a gene *TnpR* in transposon Tn3 (Simon *et al.* 1980). The TnpR protein is responsible for the recombinational resolution of the co-integrate formed during transposition and also functions as a repressor for the transposase gene (see section 4.3.7 and Fig. 4.24). Thus it appears that the Hin, Gin and TnpR proteins are examples of a class of site-specific recombinational proteins, summarized in Fig. 4.30.

The transposition of ISs also requires (presumably) site-specific recombinational proteins. The question as to whether the ISs themselves have the coding capacity for a small protein with such a role is unresolved. However, the genes for such proteins have been discovered in the F factor (Hopkins *et al.* 1980) as part of the transfer

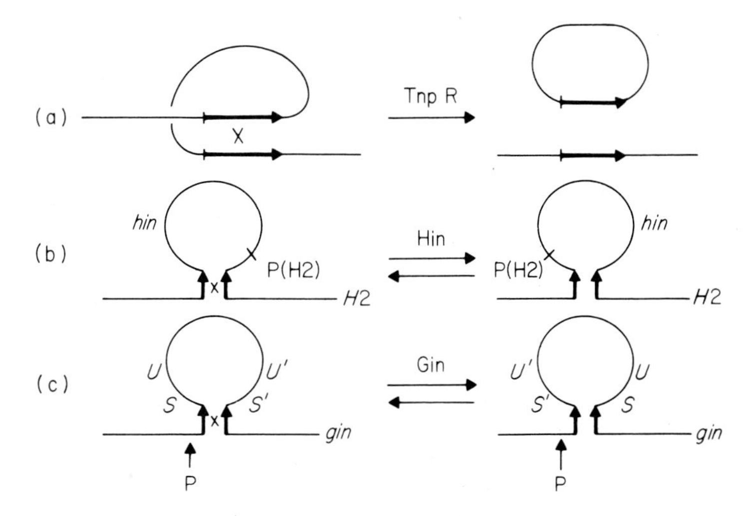

Fig. 4.30. Site-specific recombination between inverted repeat sequences, →. (a) Resolution of the co-integrate during Tn3 transposition (cf. Fig. 4.24 step (3) and Fig. 4.25). (b) Inversion of the H2 regulatory region in *Salmonella* (cf. Fig. 4.28). (c) Inversion of the G-loop. P is the promoter for the tail-fibre genes; *S* and *U* are the genes for attachment to *E. coli* and *S'* and *U'* are the corresponding genes for attachment to *Citrobacter freundii.* Thus the left-hand orientation defines Mu as an *E. coli* phage whereas the right-hand orientation defines it as a *Citrobacter* phage.

operon. The properties of the genes (*ferA* and *ferB*) suggest that they may be the counterparts of genes for transposase and the recombinational protein/transposase repressor.

Are there any other recombinational mechanisms in bacteria?

The preceding sections have shown that the covalent structure of DNA may be altered *in vivo* by at least three totally independent mechanisms: replication, recombination and transposition. Other types of event, repair and lysogenization by lambdoid phages can perhaps be regarded as variants on the first two of these processes. Is there another mechanism? The majority of bacteria contain restriction and modification systems. The role of these systems *in vivo* is unclear. Claims that restriction represents a sort of prokaryotic immunity should be regarded critically. For one thing, as an immune system restriction is not very efficient; for another it would be of rather dubious value to a bacterium living in a culture because there is a high probability that an infecting phage will have been propagated previously in an identical cell; the phage DNA will, therefore, be modified and the phage itself will be immune to the restriction system. On the other hand, as restriction enzymes make very specific cuts in DNA molecules, the ligation of restriction fragments *in vivo* would result in new arrangements of genes. Whatever the didactic force of this argument about the functions of restriction enzymes, the enzymes are employed *in vitro* for making recombinant DNA molecules (see section 11.2.1).

Chapter 5 Prokaryotic Biosynthesis of RNA

5.1 TRANSCRIPTIONAL COMPONENTS

5.1.1 Introduction

This chapter is concerned with the biosynthesis of RNA from a double-stranded DNA template by RNA polymerase. Synthesis of RNA from an RNA template during RNA phage replication is discussed in section 6.5.6.

The essential components of the transcriptional apparatus are DNA-dependent RNA polymerase, the promoter, the structural gene(s) and the terminator. In addition, most operons have regulatory devices that may involve modulation of the activities of these essential components or may involve additional protein molecules and regulatory sites in the DNA template.

5.1.2 RNA polymerase

The enzyme from *E. coli* has been well characterized and seems to be typical of bacterial RNA polymerases. Unlike DNA polymerase, the enzyme does not require a primer but incorporates the first ribonucleoside triphosphate to start the RNA chain. Thus all newly synthesized RNA molecules have a 5′-triphosphate terminus. The enzyme can be inhibited by several antibiotics of which rifampicin and streptolidigin are the most important in experimental studies on transcription. Rifampicin inhibits initiation of transcription; streptolidigin inhibits elongation. The enzyme consists of two parts, the σ factor (M_r 95×10^3), which is required only for initiation, and the core enzyme (M_r 360×10^3). The core enzyme consists of four subunits and has the formula $\alpha_2 \beta\beta'$ where the Greek letters represent polypeptide chains with the following molecular weights (all $\times 10^{-3}$): α, 40; β, 155; β', 165. The β subunit contains binding sites for both rifampicin and streptolidigin and is, therefore, involved in both initiation and elongation. Both β and β' subunits are involved in binding to DNA (Okada *et al.* 1978). The σ factor is an allosteric effector of the enzyme that makes it recognize the promoters. It has two effects on the affinity of the enzyme for sites in DNA. These are most readily seen from some actual data. The core enzyme alone binds to approximately 1300 sites in phage T7 DNA with binding constants (mol^{-1}) of around 2×10^{11}. The core enzyme plus σ factor binds to two classes of sites: there are approximately 1300 sites with binding constants in the range 10^8–10^9 and 8 sites with binding constants between 10^{12} and 10^{14} (Chamberlin 1974). Thus the σ

factor not only increases the affinity of the enzyme for promoters (the 8 sites in the T7 experiment) but also reduces the affinity for non-specific sites. Following binding to a recognition site the polymerase is registered for the initiation of transcription and unwinds the DNA double helix through 240° to separate the transcribing and non-transcribing strands (see section 5.1.3). It is the formation of this 'open promoter' complex that is inhibited by rifampicin.

Kinetic studies on RNA polymerase established two different nucleoside triphosphate binding sites. The initiation site will bind only ATP or GTP, does not require Mg^{2+} and has a relatively low affinity for triphosphates. In contrast, the elongation site will bind all four NTPs, requires Mg^{2+} and has a high affinity for the triphosphates (Minkley & Pribnow 1973).

In the remaining part of this section, the different stages in the transcription of an operon will be considered in turn, starting with the recognition of the promoter and the binding of RNA polymerase and proceeding past the operator to transcription of the structural genes and termination. Emphasis is given to the roles of these processes in transcriptional regulation. Finally, the post-transcriptional modification of RNA will be considered.

5.1.3 Promoters

This section concerns only the RNA polymerase binding functions of promoters. Certain promoters contain an additional site for the CAP protein (see section 5.2.3).

Conventionally, the first base pair to be transcribed is numbered +1 and consecutive base pairs are numbered +2, +3, etc. The pair before +1 is numbered −1 and the preceding pairs are numbered −2, −3, etc. From analysis of known promoter sequences, the regions of DNA protected by bound RNA polymerase from DNAase attack and by studies on the effects of promoter mutations, several generalizations can be made about *E. coli* promoters (Pribnow 1979, Siebenlist *et al.* 1980, Travers *et al.* 1981). The arrangement and functions of different promoter regions are summarized in Fig. 5.1. The first sequence to be recognized was a heptamer centred around position −9 (more rarely −10) and called the Pribnow box. This is now recognized as a site for strong binding of the core enzyme and appears in Fig. 5.1 as Rc. A second region around position −35 (often referred to as 'the −35 region') is the site for strong interactions with the σ factor. This region appears in Fig. 5.1 as Rσ. The base pair for the initiation of transcription (I in Fig. 5.1) is in a

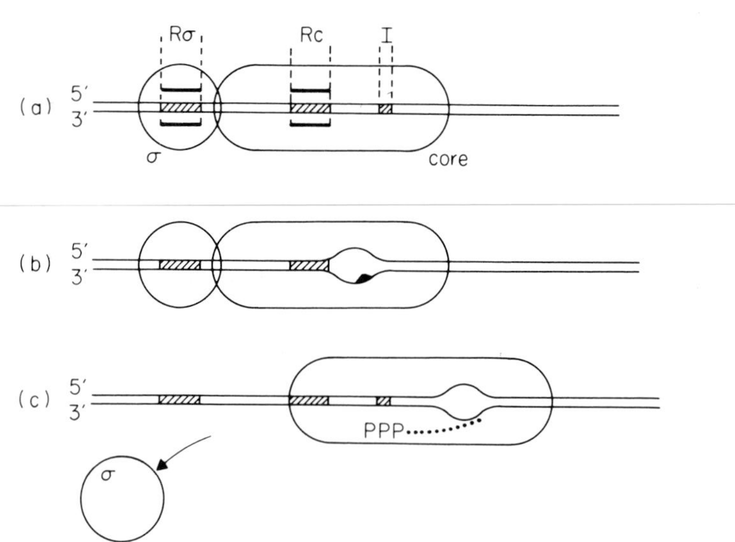

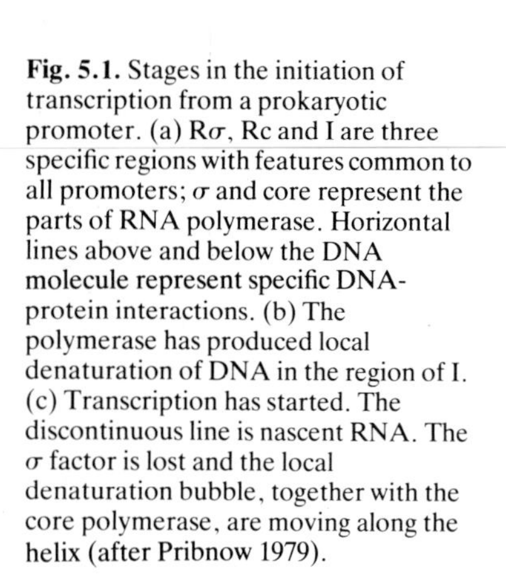
Fig. 5.1. Stages in the initiation of transcription from a prokaryotic promoter. (a) Rσ, Rc and I are three specific regions with features common to all promoters; σ and core represent the parts of RNA polymerase. Horizontal lines above and below the DNA molecule represent specific DNA-protein interactions. (b) The polymerase has produced local denaturation of DNA in the region of I. (c) Transcription has started. The discontinuous line is nascent RNA. The σ factor is lost and the local denaturation bubble, together with the core polymerase, are moving along the helix (after Pribnow 1979).

sequence that is denatured by the polymerase so that transcription can be initiated. Fig. 5.2 shows the actual sequence of a promoter. The current generalizations (following Pribnow 1979) about other promoters are as follows. A consensus sequence for the −35 (Rσ) region is TGTTGACAATTT (in this and the following the non-transcribing strand is shown in the conventional 5′–3′ direction). If different promoters are compared they tend to conform to the consensus more in the first half of this sequence than the second. The Trp promoter (Fig. 5.2) is an example of a promoter very close to the consensus in this region. There then follows a non-specific region which shows little homology between different promoters. This region varies from 12 to 14 base pairs (13 in the case of Trp). The consensus sequence for the Pribnow box (Rc) is TATPuATPu. The Trp promoter is less close to the consensus in this case. Another non-specific sequence of five or six base pairs (six in the case of Trp) separates this from the initiation pair (I) which invariably has a purine in the non-transcribing strand. Thus all nascent bacterial RNA molecules start with the sequence pppPu . . . .

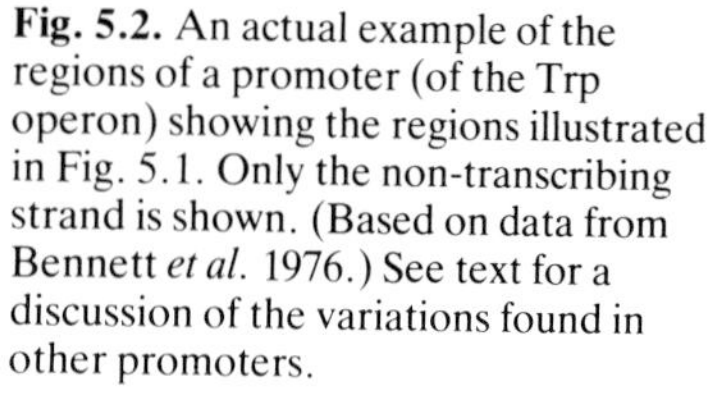
Fig. 5.2. An actual example of the regions of a promoter (of the Trp operon) showing the regions illustrated in Fig. 5.1. Only the non-transcribing strand is shown. (Based on data from Bennett *et al.* 1976.) See text for a discussion of the variations found in other promoters.

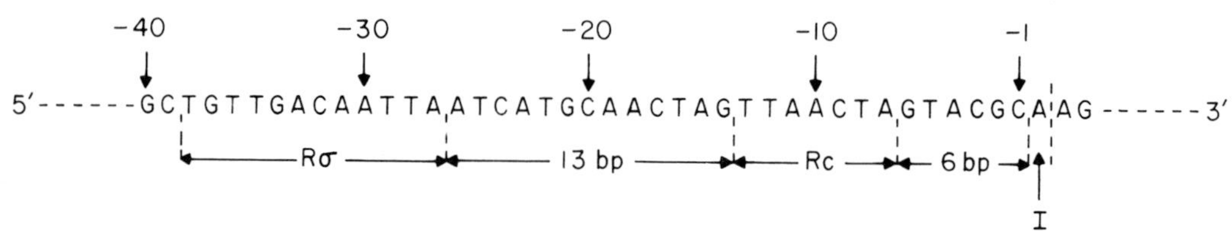

Operons for stable RNA that demonstrate the stringent effect (see section 5.2.2) have additionally another conserved region, 'the discriminator region' spanning −5 to +2. The sequence for this is of the form CggCNCC (g represents either C or G; Travers *et al.* 1981).

The affinity of RNA polymerase for different promoters varies (quite apart from the regulatory devices described in the following sections). Promoters for which the enzyme has a very high affinity are referred to as 'strong promoters'. The 'strength' of promoters is undoubtedly used as a device for achieving different levels of maximal expression of different operons. These differences can be visualized directly by electron microscopy of newly transcribed RNA (Stüber & Bujard 1981).

5.1.4 Terminators

A DNA sequence at which transcription terminates is called a terminator. Terminators are preceded by palindromes that, following transcription, result in a stem-and-loop structure in the RNA transcripts. Certain terminators are recognized by RNA polymerase in the presence of a protein termination factor rho, ρ, a tetramer of subunit M_r 50 000. Other terminators are ρ-independent; a third class are recognized weakly without ρ but the efficiency of

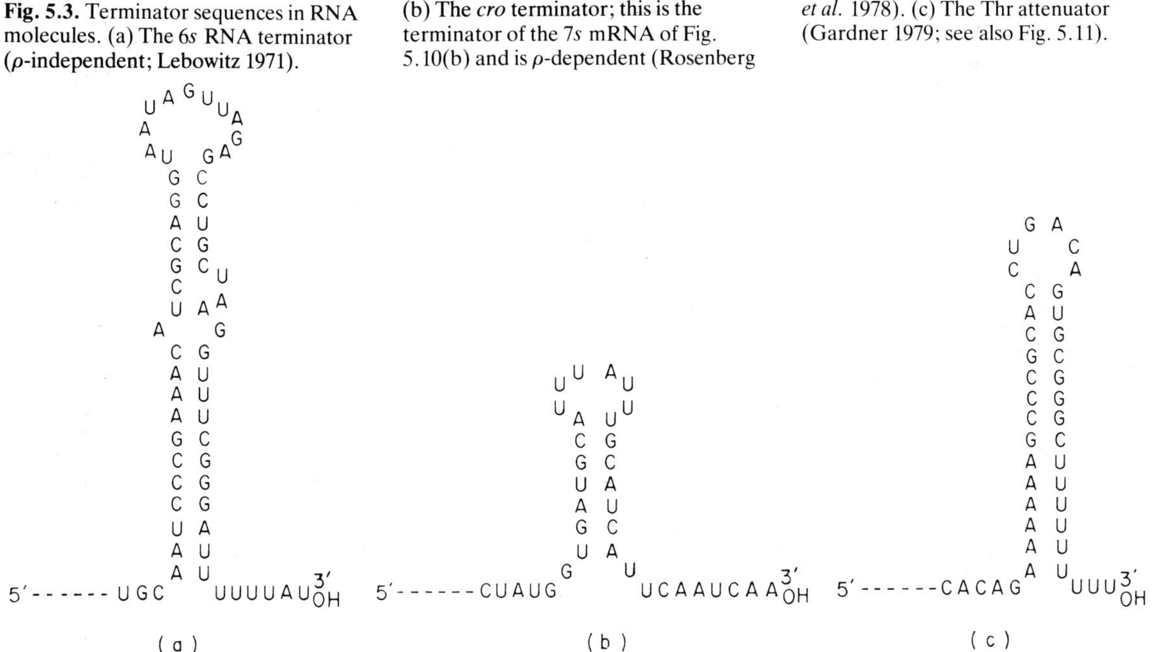

Fig. 5.3. Terminator sequences in RNA molecules. (a) The 6*s* RNA terminator (ρ-independent; Lebowitz 1971). (b) The *cro* terminator; this is the terminator of the 7*s* mRNA of Fig. 5.10(b) and is ρ-dependent (Rosenberg *et al.* 1978). (c) The Thr attenuator (Gardner 1979; see also Fig. 5.11).

termination is enhanced by ρ. In general, ρ-independent terminators have a sequence of U-residues at their 3′-termini; ρ-dependent ones do not necessarily have this feature and their stems can be shorter (Fig. 5.3; see Adhya & Gottesman (1978) for a review).

5.2 CONTROL OF TRANSCRIPTION

5.2.1 Introduction

The regulation of transcription constitutes the major method for the control of gene expression in bacteria. Several different mechanisms are employed and there are a few generalizations that can be made.

Coarse control of transcription is achieved by modulating the activity of RNA polymerase. Examples of this are discussed in section 5.2.2. 'Coarse control' refers to the switching between major families of operons. Examples are the transcription of genes for mRNA as opposed to rRNA or tRNA; the changes in transcription associated with cytodifferentiation (such as the formation of spores by bacilli); and the change in transcriptional pattern following infection by a phage (when transcription of phage genes takes place sometimes at the expense of transcription of host-cell genes).

Other examples of transcriptional control reflect responses to changes in substrate availability or requirements for biosynthesis of metabolites. There are several ways in which such 'fine control' mechanisms can be classified. One approach is to look at the nature of the nucleic acid sequence involved in the control process: this could be the promoter, an additional regulatory sequence such as an operator or attenuator or a terminator. We shall see in section 5.2.5 that these last two are related. An alternative approach is to ask about the nature of the control protein: this can either be an activator, which effects positive control, or a repressor, which effects negative control (Fig. 5.4). In either case there are theoretically two kinds of control circuit: 'classical' (in which case the regulatory protein is the product of a constitutive gene not under the control of the controlled operon) and 'autogenous' (in which case synthesis of the regulatory protein is so controlled). Applications of systems analysis lead to the conclusions that:

(1) activator systems are likely to be common in cases where there is likely to be a high demand for expression of an operon;

(2) autogenous control leads to unstable or unsatisfactory networks in activator-mediated control but confers advantages on

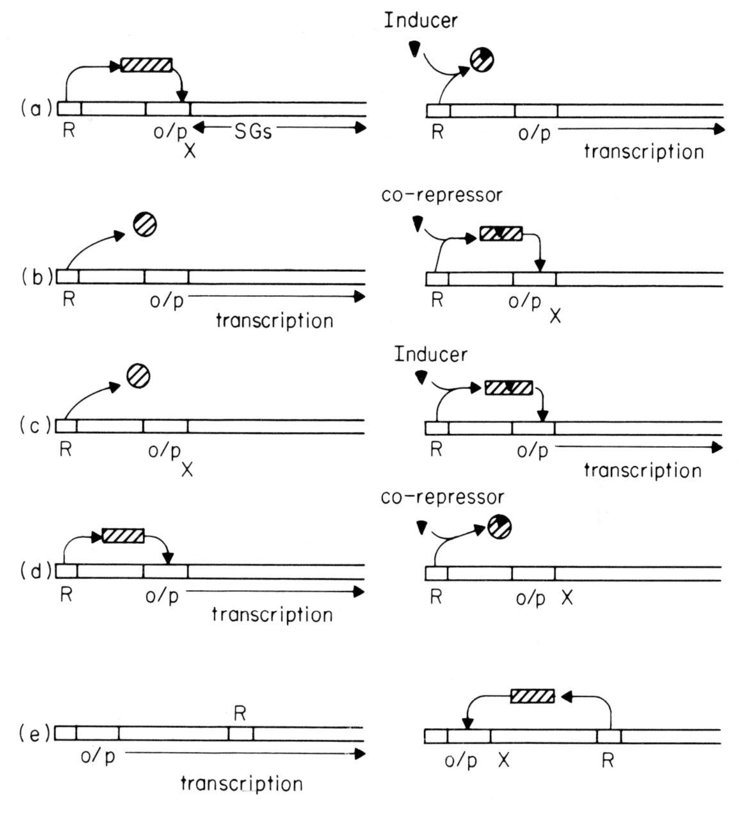

Fig. 5.4. Different types of regulatory circuits for the control of transcription. R is the structural gene for a regulatory protein (repressor in (a) and (b) and activator in (c) and (d) and the hatched shape is that protein. In (a)–(e) the regulatory protein is allosteric and is capable of binding to DNA in its rectangular conformation but not in its circular conformation. o/p (operator/promoter) is the regulatory DNA sequence to which the protein can bind. The allosteric modulator (either inducer or co-repressor) is shown as a black wedge. The structural genes (SGs) are to the right of o/p. An X indicates that transcription from o/p is prohibited. (a) An inducible repressor-controlled operon; (b) a repressible repressor-controlled operon; (c) an inducible activator-controlled operon; (d) a repressible activator-controlled operon. All of these are examples of classical regulation. (e) An example of an autogenously controlled operon with a repressor. A similar scheme could be drawn for autogenous control involving an activator but both theoretical and empirical considerations make this unlikely (see text; after Savageau 1979).

repressor-mediated control (Savageau 1976, 1979).

Experimental evidence with *E. coli* and *Salmonella typhimurium* tends to support conclusion (1): for example, in its natural habitat an enteric bacterium would have low demand for the expression of operons required for the utilization of galactose and lactose and a high demand for the expression of operons for the utilization of maltose and arabinose. The Lac and Gal operons are indeed regulated by repressors and the Mal and Ara operons by activators. The evidence for (2) is less clear. Certainly, no examples of autogenous activator-mediated regulation are known but examples of genuine autogenous regulation are never common. A rather special case operates in λ (see section 5.2.4 and Fig. 5.10; other examples are the histidine utilizing (Hut) operon of *S. typhimurium* (Fig. 5.5) and *recA* (see section 4.3.4 and Fig. 4.14).

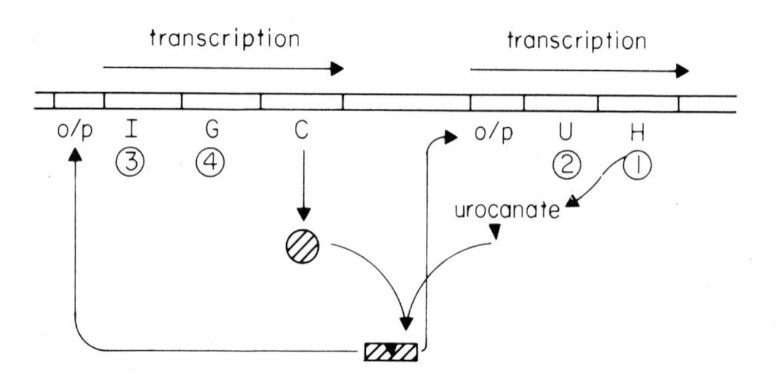

Fig. 5.5. Autogenous regulation in the Hut operon. The two transcriptional units contain genes (H, U, I and G) for enzymes for the four metabolic steps, (1)–(4), in the conversion of histidine to glutamate and formamamide. Urocanate is the product of step (1). The other symbols have the same meaning as in Fig. 5.4. The lines above the figure show the pattern of transcription. The lines below the figure show repression by urocanate (the co-repressor); gene C is the structural gene for the repressor (after Savageau 1976). Other features of the regulation of the Hut operon are discussed in section 5.2.3.

The most familiar example of an activator is the CAP protein (see section 5.2.3). The regulator of the Ara operon is more complex because the product of the regulatory gene (*araC*) plays a kind of dual role. In the presence of arabinose (the inducer) the AraC protein positively activates (as in Fig. 5.4(c)). However, in the absence of inducer AraC behaves as a repressor (as in Fig. 5.4(a)). Thus the inducer effects a change in a regulatory protein which affects the consequences of its binding to DNA rather than determining whether or not the protein will bind to DNA at all (see Savageau (1979) for details and references).

Several regulatory networks are complicated by the fact that a regulatory protein can affect the transcription of more than one transcriptional unit. CAP is a good example and the λ repressor is another. Both of these are considered in more detail later. We have already seen the role of repressors in the Hut operon. Another type of complex operon comprises a regulatory region flanked by

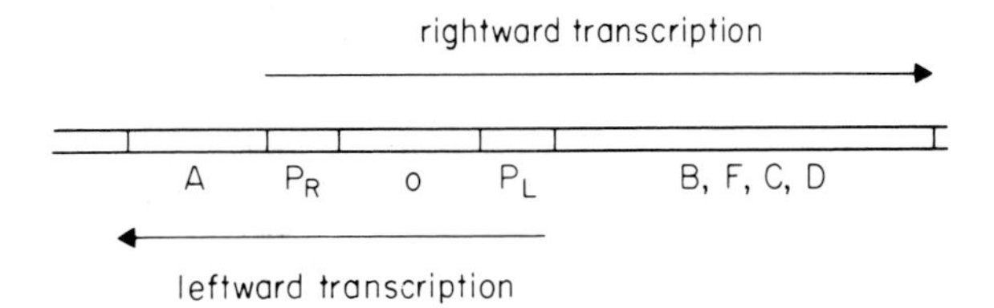

Fig. 5.6. Regulation of the cluster of biotin biosynthetic genes (after Campbell 1979). A–D are the structural genes; o is the operator and P_R and P_L are the rightward and leftward promoters; the operon is shown in its derepressed state.

structural genes. This type of system operates in the operons for the biosynthesis of arginine and biotin (Fig. 5.6).

The following sections contain examples of regulatory sequences, DNA-protein interactions and termination regulation. These examples are chosen, not as a systematic survey of transcriptional regulation (reviewed by Campbell (1979), Savageau (1979) and Travers *et al.* (1981)), but rather to illustrate the underlying principles behind the molecular interactions involved.

5.2.2 Modulation of RNA polymerase activity

Modification of σ

There are several examples of the modification of the σ factor to change the affinity of RNA polymerase for large groups of promoters. *Bacillus subtilis* is a spore-forming bacterium. During sporulation, several vegetative genes are shut off. This is achieved by a progressive inactivation of σ with the consequence that weak vegetative promoters are not used (see Chambliss (1979) for review).

Other modifications to RNA polymerase

Bacteriophage infection may cause modifications to the polymerase. Typically this is used as a device for delaying the expression of the 'late genes' of the phage (i.e. those genes whose products are required late in the infectious cycle but not in the early stages). *E. coli* phage T4 produces proteins that result in adenylation (substitution of AMP groups) in the subunits and alters the promoter specificity to allow expression of later genes. A more extreme case is that of T7; here the late genes are read by an unusual single-subunit RNA polymerase that is a phage gene product.

Guanine and adenine nucleotides

The biosynthesis of stable RNA (rRNA and tRNA) is under the so-called 'stringent control' (see section 6.3.3). Essentially the finding is that in wild-type cells, starvation of amino acids results in a diminution of stable RNA biosynthesis and this is accompanied by the appearance in the cells of two highly phosphorylated guanine nucleotides, ppGpp and pppGpp. The biosynthesis of these compounds is regulated by uncharged tRNA and free ribosomes and, as ppGpp appears to operate as an allosteric effector of RNA polymerase, the stringent system gives the organisms a way of 'noticing' that there is little protein biosynthesis going on and consequently no requirement for more ribosomes and tRNA. However, the stringent response (cessation of stable RNA synthesis) also occurs in the absence of amino acid starvation if the cells run out of ATP under certain circumstances (for example shifting a temperature-sensitive aldolase mutant to its non-permissive temperature). There is no accumulation of ppGpp under these circumstances. These observations, together with studies on the promoter specificities of separated forms of RNA polymerase, have led Travers *et al.* (1981) to propose that there are two interconvertible forms of RNA polymerase: A (that responds to adenine nucleotides) and G (that responds to guanine nucleotides). The A state is thought to be metastable and to convert slowly to the G state. However, high concentrations of polymerase result in a cycling of the enzyme between the two forms because this promotes dimerization and the G-dimer and A-dimer equilibrate through 'isomerization'. Dissociation of A-dimer generates the A form. The A form of the enzyme responds to ATP, or more exactly to the energy charge, i.e. $(ATP+\frac{1}{2}ADP)/(ATP+ADP+AMP)$. A high energy charge promotes transcription of stable RNA; a low one promotes transcription of certain mRNAs. The G form responds to ppGpp and GTP. In this case high GTP (or low ppGpp) promotes transcription of stable mRNA and high ppGpp prevents transcription of stable RNA and allows transcription of certain mRNAs. The nature of the allosteric changes made in RNA polymerase by these nucleotide couples is unknown but presumably they alter the response of the polymerase in its initiation complex to the discriminator region.

5.2.3 The catabolite effect

The catabolite effect is mediated by the protein CAP (a dimer of subunit M_r 22.3×10^3) and cyclic AMP. Within the terms of section

5.2.1 and Fig. 5.4, CAP is an activator and cyclic AMP is an inducer (as in Fig. 5.4(c)). The characteristic feature of the CAP system is that it is highly pleiotropic. Very many catabolic operons in *E. coli* are subject to the catabolite effect and thus require a high concentration of cyclic AMP (found during an absence of glucose) for expression, or at least for full expression. The regulation due to the catabolite effect is usually overlayed by operator–repressor control. The DNA sequence to which CAP binds is called the CAP site and it shows twofold rotational ('palindromic') symmetry. CAP sites lie within the promoter around position -60 (see section 5.1.3). Thus a CAP site would lie immediately to the left of the region covered by σ in Fig. 5.1(a).

The mode of action of CAP is believed to be to stabilize the RNA polymerase–DNA recognition complex (Pribnow 1979). Not all CAP sites are identical. Two actual examples are discussed in section 5.2.4 and Fig. 5.8. Just as there are weak and strong promoters with respect to affinities for RNA polymerase, so there are probably weak and strong CAP sites with different affinities for CAP.

The effect of the CAP system is to ensure that the operon is not expressed if there is a surplus of glucose, a readily metabolized carbon source, in the cell even if there is an inducer that would otherwise derepress the operator–repressor system. Obviously the system works well for such operons as Lac and Gal which are expressed to allow utilization of 'difficult' or 'exotic' carbon sources (lactose and galactose). However, there are certain circumstances in which this effect would work against an organism's best interest unless there was an alternative method of overriding it. The classical example is the Hut operon in *Salmonella typhimurium*. The Hut enzymes are required for the degradation of histidine and the purpose of the operon is to allow the organism to use histidine as a source of either carbon or nitrogen if other sources are not available. The Hut operon is regulated by the catabolite effect. Obviously from the point of view of carbon sources, it works well and efficiently. If there is glucose in the cell and consequently cyclic AMP is in sufficiently low concentration to allow CAP to bind to its site, there is no point in the cell making enzymes to use histidine as a carbon source. However, a cell with a surplus of glucose but lacking any available nitrogen source other than histidine faces a difficulty. This is overcome by the fact that there is an alternative to CAP that can bind to the CAP site and allow RNA polymerase to bind to its site in the Hut promoter. This alternative is the enzymically active form of glutamine synthase. The system is summarized in Fig. 5.7. Note that

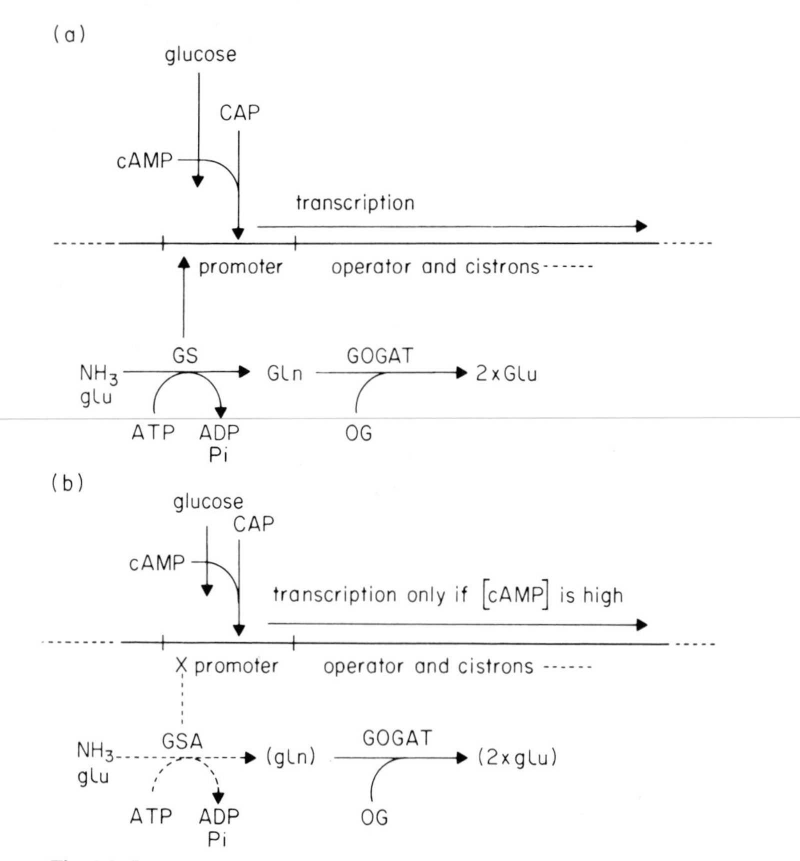

Fig. 5.7. Regulation of the Hut operon in *Salmonella typhimurium*. Abbreviations used: GS, glutamine synthetase; GOGAT, glutamate oxoglutarate amino transferase; OG, oxoglutarate. (a) In the presence of His and absence of NH_4^+, GS is an alternative effector to CAP so that transcription occurs whether or not cyclic AMP is present in high concentration (i.e. whether glucose is present or not). The small amount of NH_4^+ produced by histidine catabolism is efficiently converted to glutamine by the fully active form of the GS enzyme. The glutamine is converted to glutamate by GOGAT. Note that the overall reaction catalysed by GS and GOGAT is conversion of one NH_3 and one OG to yield one glutamate with concomitant hydrolysis of one ATP. (b) In the presence of His and excess NH_4^+: as in (a) the inducer (His) is present but the high concentration of NH_4^+ results in a high ratio of OG : Gln. This, in turn, results in adenylation of GS to its inactive form (GSA). The conversion of NH_3 to Gln is now minimal (represented by the broken arrow) and the GSA (actually a mixture of forms as there are 12 adenylation sites in GS) cannot act as a positive effector for RNA polymerase binding to the promoter. The expression of the operon is now under the control of CAP and the level of cyclic AMP. Thus the Hut genes will not be expressed if glucose is present. If the NH_4^+ salts become exhausted, the ratio of OG : Gln falls; this promotes deadenylation of GS and the situation in (a) is produced. For details of the Hut system see Tyler *et al.* (1974).

in the presence of NH_4^+, a readily metabolized form of nitrogen, glutamine synthetase is converted into its inactive form; this form does not bind to the CAP site and so the Hut operon is repressed.

5.2.4 Operators and repressors

The operator is the DNA sequence to which the repressor binds. The operator–repressor system constitutes fine control of transcription and is specific to the operon in question. From the sequences that have been determined, the operator is a region of palindromic symmetry and the repressor is an oligomeric protein (normally a tetramer). The way in which the two structures may recognize one another is discussed in section 5.2.2. Repressors are DNA binding proteins with a high affinity for the specific operator sequences. As an example, the repressor for the Lac operon forms its complex with the operator with an equilibrium constant of 10^{-13}, which is more than an order of magnitude less than for its non-specific binding to DNA (Riggs *et al.* 1970). Actual operator sequences are shown in Figs 5.8 and 5.9. The recognition of these sequences may involve the adoption of cruciform conformations (see section 5.2.6; Malcolm 1977).

Operator sequences may be separate sites between the promoter and the initiation codon for the first cistron; in these cases the operator is transcribed when the operon is induced. Alternatively, the operator may overlap wth promoter sequences so that repressor competes for binding with either CAP or RNA polymerase. Examples of these alternatives follow.

The Lac and Gal operons

The sequences of the regulatory regions of these operons are shown in Fig. 5.8. The Lac operon is unusual in that the structural gene for the repressor, *lacI,* is close to the Lac promoter. The promoters have many similar features including comparable, but not identical, CAP sites. However, the operons illustrate different modes of action of repressors. In the Lac operon, the operator follows the promoter so that repressor binding does not interfere with interactions between DNA and RNA polymerase. When the operon is expressed, the operator is transcribed as a part of the leader sequence of Lac mRNA. The Gal operon (Fig. 5.8(b)) illustrates an operator that overlaps with the CAP site so that repressor and CAP compete for binding. However, the Gal operon has palindromic sequences, reminiscent of operators such as that in the Lac operon, in the leader region. The Gal operon also has a weak alternative promoter for CAP-independent transcription.

The Lac system has been the subject of very intensive genetic and biochemical studies. The sequences of mutant operators and mutant repressors and measurements of binding parameters have led to a detailed thermodynamic model of the regulation of Lac expression (von Hippel 1979; Fig. 5.9).

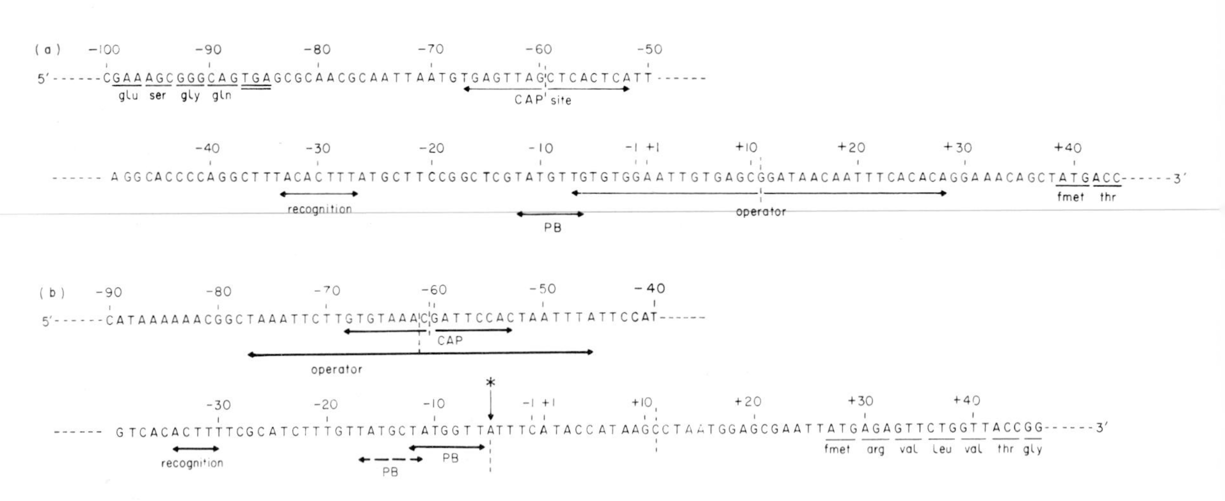

Fig. 5.8. Regulatory sequences of (a) the Lac operon and (b) the Gal operon of *E. coli* (after Dickson *et al.* 1975 and Musso *et al.* 1977, respectively). Only the non-transcribing strand is shown. Thus palindromic sequences can be recognized by complementarity between pairs of bases working away from the axis of symmetry. In the following, the sequences are followed from the 5′-end. (a) The sequence up to −85 is the *C*-terminus of *lacI*, the structural gene for the Lac repressor. The termination codon is marked by double underlining. A palindromic sequence with its symmetry axis between −60 and −59 forms the CAP site. Inspection of this, and other palindromes in the figure, shows that the symmetry is imperfect. The sequence from −33 to −27 is a part of the RNA polymerase recognition site; −11 to −5 is the Pribnow box (PB). The operator is the long palindrome whose symmetry axis goes through residue +11. Lac mRNA starts at +1. The first structural gene of the operon, *lacZ*, starts at +39. (b) The CAP site and the operator overlap. The CAP site and the residues flanking it are partly homologous with the corresponding site in the Lac operon (a). The part of the RNA polymerase recognition site between −36 and −30 is identical to the corresponding sequence in the Lac operon (a). Gal mRNA normally starts at +1 and its Pribnow box (PB) is at −12 to −6. However, there is an alternative (CAP-independent) transcriptional start at −5 (*) and this has its corresponding Pribnow box at −17 to −11. Although their significance is unclear, two other palindromes can be detected in the sequence (only the axes of symmetry are shown here): the palindrome whose axis is at −5 and stretches from −18 to +9; and the one whose axis is at +11 and stretches from +4 to +18. The first structural gene, *galE*, starts at +27.

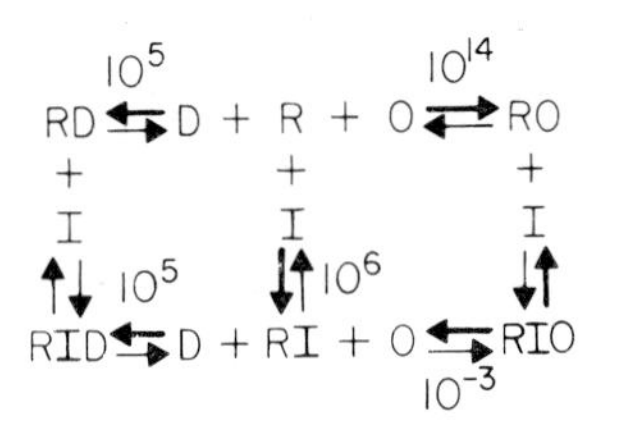

Fig. 5.9. Model for the interactions involved in regulation of the Lac operon by repressor–operator interactions (after von Hippel 1979). R, repressor; O, operator; I, inducer (the gratuitous inducer IPTG that, unlike lactose the natural inducer, is not a substrate for the enzymes of the operon); D, DNA sites other than the operator. Combinations of these symbols represent the various complexed species. Thus RO is repressor bound to operator and RD is repressor bound 'non-specifically' to DNA. Of the two arrows for each equilibrium, the bold one shows the direction in which the equilibrium lies. The numbers are equilibrium constants for the formation of complexes. Thus in the top right-hand corner, 10^{14} is the value of [RO]/[R].[O].

The λ immunity region

During maintenance of λ, gene *cI* is transcribed from a maintenance promoter, P_{RM}. Its gene product is the λ repressor which mediates immunity against incoming λ DNA and also prevents induction of the lysogen by binding to two operators to the left and right of *cI* (Fig. 5.10(a)). Expression of *cI* is autoregulated: when repressor reaches a certain level its binding to the right operator prevents transcription from P_{RM}. Following the inactivation of the loss of the repressor, the lysogen is induced and two short early transcripts are produced (Fig. 5.10(b)). The right transcript results in expression of gene *cro*; the Cro protein is also a repressor-like molecule that prevents *cI* transcription. The λ left and right operators overlap the promoters and hence repressor and RNA polymerase compete for binding to DNA. An unusual feature of the λ left and right operators is that they are both trios of palindromes (Fig. 5.10(c)). The λ repressor binds with highest affinity to O_L1 and O_R1 and this prevents transcription from the promoters P_L and P_R. The repressor has a lower affinity for O_L2 and O_R2 and even lower affinity for O_L3 and O_R3. As the concentration of repressor increases it is, therefore, titrated out by operator sites in the order 1, 2, 3. When O_R3 is filled, transcription from P_{RM} is cut off because the promoter lies within this operator (Fig. 5.10(c)) and autoregulation is thus effected. Following induction, Cro protein also binds to the operators but its order of affinities is different from those of repressor. Thus one of the first sites to be filled by Cro protein is O_R3; this masks P_{RM} and expression of *cI* is prevented. The induction process is thus irreversible: the removal or inactivation of the λ repressor needs only to be transient. We can thus regard the transcription of *cI* on the one hand and the left and right operons on the other as alternative 'phases' of gene expression in a λ lysogen and this is an example of a theoretical mechanism for phase shifts that, unlike the *hin* and *gin* systems (section 4.3.8), does not require DNA rearrangement (Parish 1979).

The maintenance and right transcripts are relatively unusual among mRNA molecules in having no leader sequence; the initiation codon for protein synthesis in both of these lies at the extreme 5′-terminus. More commonly a ribosome binding site lies within the leader sequence (see section 6.5.1). The use of a 5′-terminal AUG results in inefficient initiation and reflects the requirement for only small amounts of repressor and Cro protein. However, the *cI* gene is preceded by a ribosome binding site (Fig. 5.10(c)). During

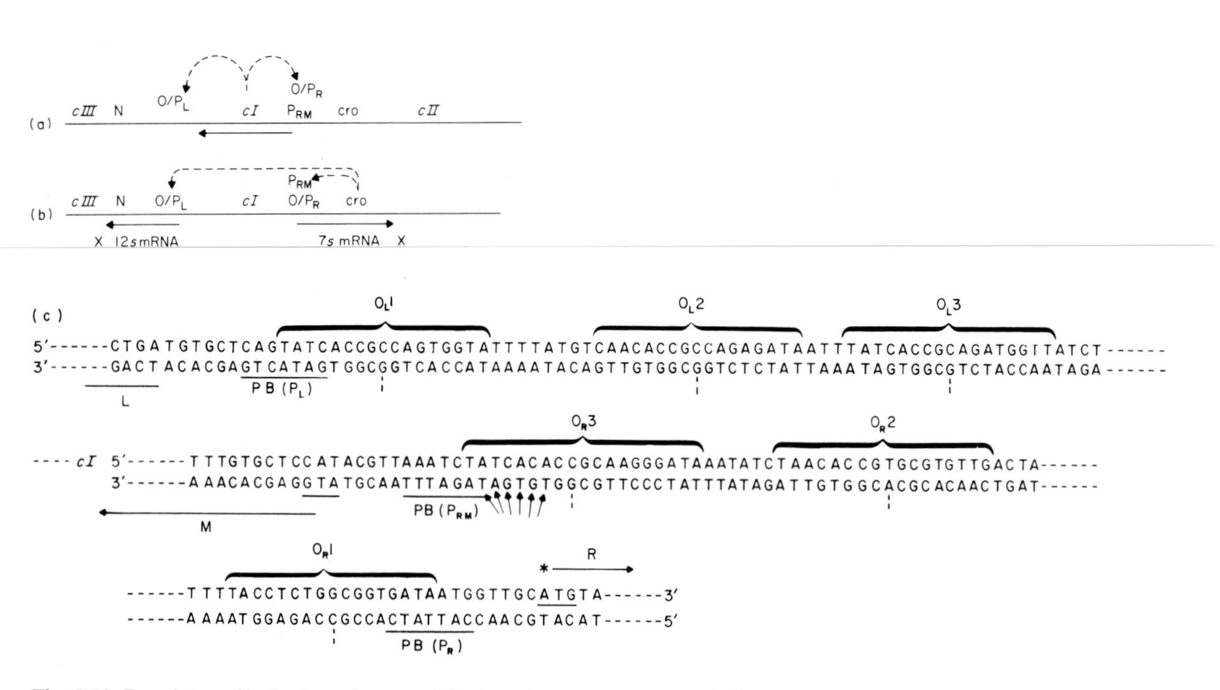

Fig. 5.10. Regulatory sites in phage λ. (a) Transcription from the maintenance promoter (P_{RM}) during the maintenance of lysogeny. A short transcript (continuous arrow) of *cI* produces λ repressor that binds to the left (O/P_L) and right (O/P_R) operator–promoter complexes and allows limited transcription from P_{RM} only.
(b) Immediately after induction (by inactivation of repressor), transcription from O/P_L and O/P_R proceeds. The Cro protein, like the repressor, binds to O/P_L and the O/P_R–P_{RM} regions but with different consequences: transcription from P_{RM} is now not allowed. The short transcripts (7*s* and 12*s*) will be extended beyond the points, X, once N protein has accumulated (see section 5.2.5).
(c) Sequence of DNA from part of the regulatory region of phage λ (Ptashne *et al.* 1976). The bulk of the *cI* gene is not included. O_L1, O_L2, O_L3, O_R1, O_R2 and O_R3 are the six operator palindromes; vertical lines show the axes of symmetry (see text for discussion). The arrow marked L shows the leftward transcript with an asterisk at the first transcribed residue (A in the non-transcribing strand). PB(P_L) is the Pribnow box for the leftward promoter. The arrow marked M shows the maintenance transcript with other conventions as above. The underlined 5′ATG3′ in the non-transcribing strand is the initiation codon for λ repressor. The six small arrows show a ribosome binding site. The arrow R and the preceding Pribnow box refer to the rightward transcript. The underlined ATG is the initiation codon for Cro protein.

establishment the *cI* gene is transcribed from a different promoter, lying to the right of *cro* (Fig. 5.10(b)) and the ribosome binding site lies within the very long leader sequence of the establishment transcript (see section 6.5.1).

5.2.5 Attenuation and regulation of termination

Termination of transcription can, in principle, constitute a regulatory device. If a transcript covers, for example, the first three cistrons of an operon followed by a ρ-dependent termination sequence, transcription of a fourth and subsequent cistrons could be effected by the removal or inactivation of the ρ factor. An elegant example of this regulatory device is achieved by phage λ during induction. The short left and right transcripts shown in Fig. 5.10(a) are limited by ρ-dependent termination sequences. The left transcript ends just beyond gene *N*. In time, N protein accumulates in the cell. This protein is an anti-ρ factor. It behaves as though antibodies were introduced into the cell so that these short transcripts become extended to allow expression of distal genes from the same promoters but after a short time delay. The structure of the ρ-dependent terminator of the right transcript is shown in Fig. 5.3(b).

Attenuation

Attenuation, a special case of the regulation of gene expression by termination, occurs in certain amino acid biosynthetic operons. It was first discovered in the Trp operon (reviewed by Crawford & Stauffer 1980) and has also been discovered in the Phe, His and Thr operons. An example of an attenuator sequence is shown in Fig. 5.11. The leader sequence has a coding capacity for a short polypeptide, 'the leader peptide'. The peptide includes a large proportion of residues of the amino acid whose biosynthesis is determined by the expression of the operon. Thus in the case of the Thr operon (Gardner 1979; Fig. 5.11) 8 of the 21 residues are threonine and a further 4 are the biosynthetically related amino acid,

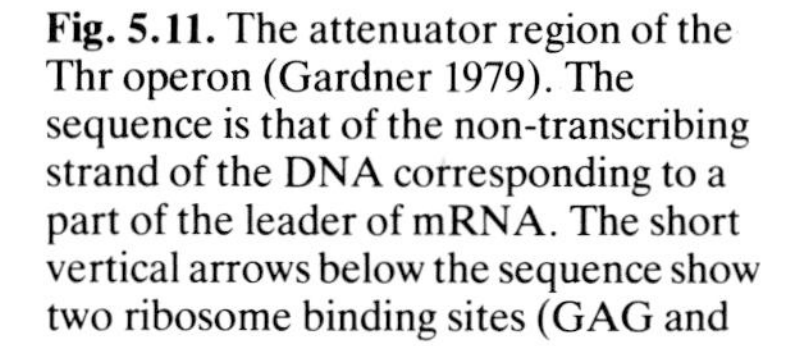

Fig. 5.11. The attenuator region of the Thr operon (Gardner 1979). The sequence is that of the non-transcribing strand of the DNA corresponding to a part of the leader of mRNA. The short vertical arrows below the sequence show two ribosome binding sites (GAG and GAGG). The sequence of the leader peptide is shown; the termination codon has double underlining. At the 3′-end of the sequence is the start of the first structural gene, *thrA1*. The attenuation terminator is at t; the sequence x → t corresponds to the mRNA sequence shown in Fig. 5.12(c); the two sequences, 2 and 2′ are complementary and form the stem shown in Fig. 5.12(c). The sequences 1 and 1′ are also complementary and can form an alternative secondary structure (see text for discussion).

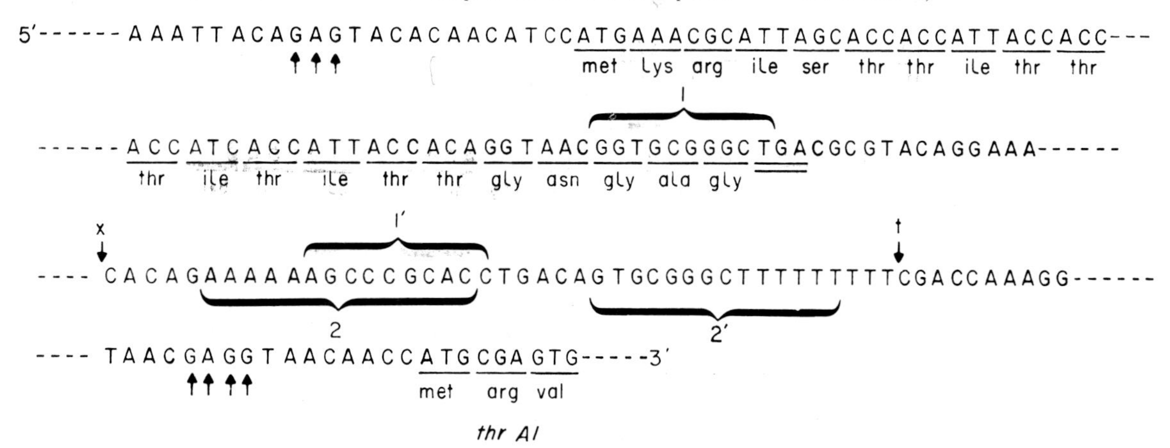

isoleucine. In the cases of the Thr, Phe and His operons, the leader peptides contain high proportions of threonine, phenylalanine and histidine respectively. The leader peptides themselves have no functions as proteins but their biosynthesis is regulatory. The sequence at the 3′-end of the leader is capable of forming one of two alternative secondary structures. In Fig. 5.11 either the complementary sequences, 1 and 1′, can form a stem or, alternatively, 2 and 2′ can do so. These two alternatives are mutually exclusive because the sequences 1′ and 2 have residues in common. The pairing of 2 and 2′ forms the terminator (Fig. 5.12(c)); the pairing of 1 and 1′ does not form a terminator. As RNA is transcribed from the 5′-end, one would expect the 1:1′ pairing to occur and preclude the formation of the terminator (2:2′) and the polymerase would read through the termination residue t (in Fig. 5.11). However, translation of the leader peptide results in the region 1 being covered by ribosomes; this prevents the formation of the 1:1′ structure and allows 2:2′ to form and termination to occur. Hence the recognition or otherwise of the terminator is dependent upon the efficiency of translation of the leader peptide, and because of the curious

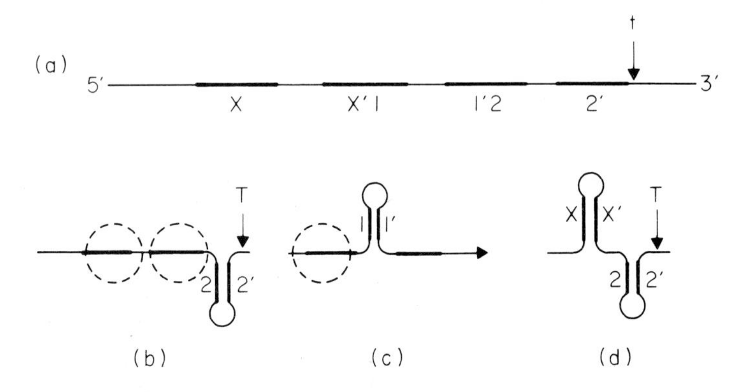

Fig. 5.12. Schematic drawing of the leader region of Trp mRNA (after Crawford & Stauffer (1980). (a) Distribution of regions with degrees of complementarity shown as a heavy line. The numbers 1, 1′, 2, 2′ have the same significance as the corresponding regions of the Thr leader (Fig. 5.11). X and X′ are also complementary. However, just as it is impossible to pair 2 with 2′ at the same time as 1 and 1′ (because 1′ and 2 overlap) so it is impossible to pair 1 and 1′ at the same time as X and X′. The potential transcriptional terminator is at t. (b) Under conditions of excess tryptophan, ribosomes (dashed circles) cover X and X′/1 to allow pairing of 2 and 2′. Termination occurs (T) and expression of the operon is attenuated. (c) In the absence of tryptophan the ribosome is held in the early part of the leader peptide template, 1 and 1′ pair up, preventing the formation of the terminator structure. (d) During starvation of all amino acids, ribosomes are unable even to initiate translation of the leader peptide. This allows pairing of X with X′ and also 2 with 2′ (termination).

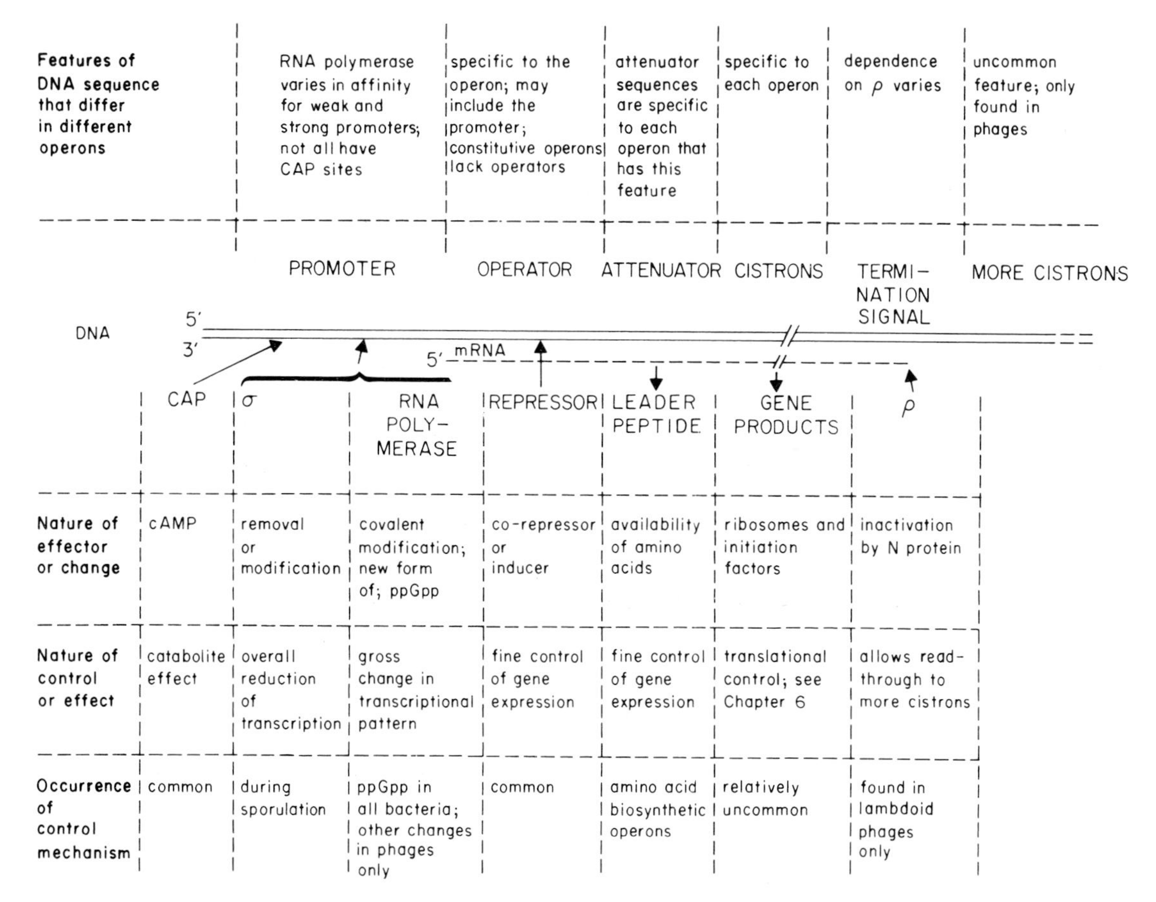

Fig. 5.13. Summary of nucleic acid sequences that can regulate prokaryotic gene expression.

composition of the leader peptide, this is dependent on the concentration of threonine. If threonine is lacking (or depleted), there will be a shortage of thr-tRNA in the cells and the ribosomes will get stuck in an early part of the leader peptide mRNA sequence. These conditions lead to expression of the operon. As threonine accumulates the efficiency of translation increases and transcription is terminated.

In certain cases, e.g. the His operon, attenuation is the only mechanism of regulation. In others there is a repressor–operator system. In the case of Trp, charged trp-tRNA^{Trp} is co-repressor, the repressor and RNA polymerase compete for overlapping operator/promoter sequences and the repressor works as in Fig. 5.4(b). It is also true of the Trp operon that the attenuator system is more sophisticated and ensures attenuation under conditions of starvation when there will be a lack of all charged tRNA molecules (Crawford & Stauffer 1980, Yanofsky 1981; Fig. 5.12).

5.2.6 Summary

The potential for the different components of the bacterial transcriptional apparatus to regulate transcription is summarized in Fig. 5.13.

The preceding sections have illustrated a variety of DNA sequences which are specific sites for the binding of proteins: CAP sites; RNA polymerase binding sites; operators; and terminators. Other examples found elsewhere in the book include restriction sites (see section 3.4.2), the origins of replicons (see section 4.2.3), the site for the cleavage of a concatenate phage chromosome (see section 4.2.5) and various types of recombinational events (see section 4.3). Some generalizations can be made about these sites. Non-palindromic sites could in principle, function in one of three ways:

(1) the primary structure of the DNA might be recognized as such by the protein;

(2) the primary structure of an RNA transcript might be recognized;

(3) the site might have an exceptional secondary structure or might readily be induced to adopt an exceptional secondary structure. The obvious example would be a region rich in A:T that would denature more readily than the rest of the DNA.

Likewise, palindromes could, in principle, be recognized in one of three ways. These are illustrated in Fig. 5.14. It is rather unlikely that all palindromes are recognized in the same way. The direct recognition of the symmetry by a multimeric protein (a) would be

Fig. 5.14. Three ways in which a palindromic region of DNA might, in principle, be recognized by a protein. (a) The symmetry of the DNA sequence complements the symmetry of a protein with an even number of subunits. The subunits each recognize the same sequence (5′AATTCCGG3′ in this simple case) and the protein binding relies also upon subunit–subunit interaction (e.g. between the two hatched areas on the picture). (b) The DNA adopts a cruciform conformation which can be recognized by a DNA-binding protein. (c) The palindrome is part of a transcribable region. In this example, the lower strand in (a) is imagined to be the transcribing strand and the transcript has a hairpin secondary structure that can be recognized by an RNA-binding protein.

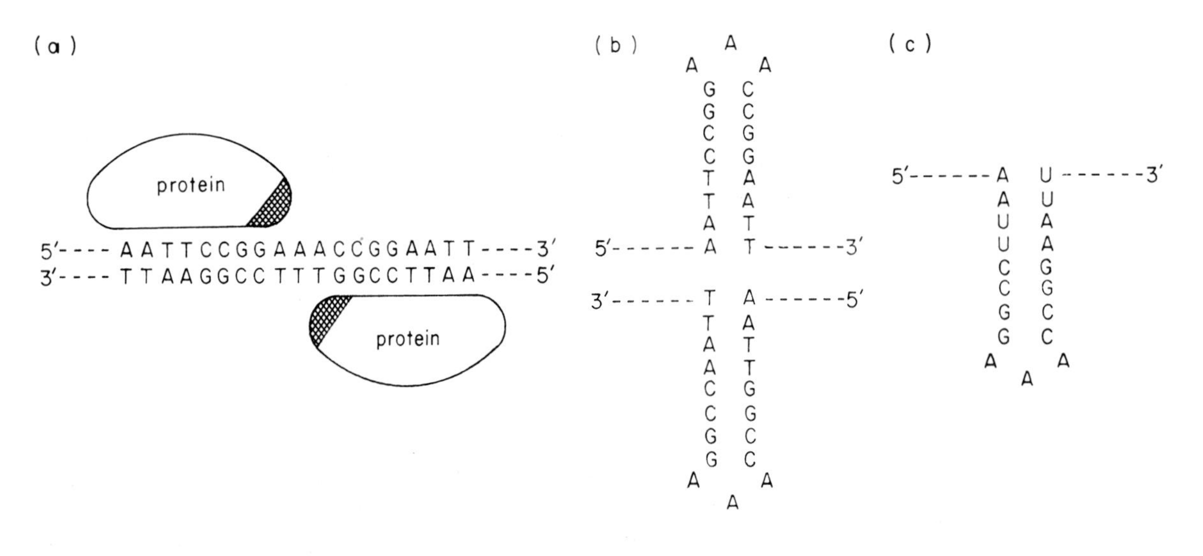

appropriate for a short palindrome, such as a restriction sequence. Very long palindromes are more likely to be recognized by their effects on the secondary structure of either the DNA or a transcript. Conversely the cruciform secondary structure (b) or the hairpin structure of RNA (c) are unlikely to represent sufficiently stable structures from a short palindrome. Direct experimental support for model (b) has come from studies on the single-strand nuclease-sensitive sites in supercoiled plasmids. Such sites lie in the centre of palindromes, corresponding to the loops in Fig. 5.14 (Lilley 1980, Panayotatos & Wells 1981). The types of sequence and their possible modes of recognition by proteins are summarized in Table 5.1.

Table 5.1. Types of DNA sequence recognized by proteins.

Names of process or site	Type of sequence	Possible mode of recognition process
Restriction site	Palindromic	Symmetry of multimer
Generation of λ sticky ends	Palindromic	Symmetry of multimer (?)
Origin of replicons	Complex palindromes	Unusual DNA secondary structure
CAP site	Palindromic	Symmetry of multimer or unusual DNA secondary structure
RNA polymerase binding site	Not palindromic	Ease of denaturation (?)
Operators	Palindromic	Unusual DNA secondary structure (?)
Transcription termination (ρ)	Palindromic	Secondary structure of RNA
Recombination	Not known	Unusual DNA secondary structure (?)
Insertion of λ	Complex; high A:T	Ease of denaturation (?)

The generation of λ sticky ends refers to the final stage in the processing of the DNA (see sections 4.1.4 and 4.2.5).

'Symmetry of multimer' refers to the model illustrated in Fig. 4.16(a). See text for discussion.

5.3 POST-TRANSCRIPTIONAL PROCESSING OF RNA

5.3.1 Messenger RNA

In bacteria, mRNA is typically the immediate product of transcription and the only processing that it receives is its degradation; this may occur while transcription is still proceeding.

There is one well established example of pretranslational processing of an mRNA molecule; this is the 'early' mRNA transcript of phage T7. Transcription starts from one of a cluster of three strong promoters and the transcript covers five cistrons. The RNA is cleaved at five sites by RNAase III to produce six molecules: a non-informational leader sequence and five mRNAs—one for each cistron. There is no known example of bacterial gene expression being regulated in this fashion. See section 6.5 for the usual modes of translational control in bacteria and the review by Perry (1976) for references to the literature on the maturation of prokaryotic RNA, including T7 early mRNA.

5.3.2 Modification of transfer and ribosomal RNA

Modification of tRNA and rRNA consists of the conversion of specific C, U, A and G residues in precursors to the modified nucleosides, such as those illustrated in Fig. 3.29 (see section 3.5.2). The modification reactions include acylations, methylations, thiolations, reductions and (in the case of ψ) rearrangement of the *n*-glycoside bond. The majority of the modification reactions involve mechanisms familiar from the biochemistry of the interconversion of intermediate metabolites. Thus, for example, acetylation of N^4 in a C-residue involves acetyl CoA as acetyl donor, I is derived from A by hydrolytic deamination and the majority of methylations involve *S*-adenosyl methionine as methyl donor. However, certain tRNA methylations in bacilli involve methylene FH_4 and thus resemble conversion of dUMP to dTMP (see section 2.2.5; Kersten *et al.* 1975). (See Altman (1981) for a review.)

A different type of mechanism is required for acquisition of the extraordinary nucleoside Q (see Fig. 3.29). Although this nucleoside (which occurs in certain tRNAs) behaves as a highly modified G-residue, it is not a purine (it contains a C in place of an N at position 7). In this case a Q precursor, which contains the 7-deazapurine ring system, displaces a Gua in the tRNA precursor in a reaction catalysed by a tRNA transglycosidase. The final modifications, including incorporation of the dihydroxycyclopentadine group occur in the tRNA (Okada *et al.* 1978).

5.3.3 Transcriptional units and tRNA and rRNA precursors

Transfer RNA

Transfer RNA genes are found in many parts of the *E. coli* chromosome, including at least one rather unexpected location (see the final part of this subsection). The general structure of a tRNA precursor is as follows: 5′ p-leader-(tRNA-spacer)$_n$-tail-OH 3′. In this formula, 'tRNA' represents the sequence that (following modification) ends up as a mature tRNA molecule. In different precursors, n may be one or another small number, that is, in many cases, several tRNA's are co-transcribed to form a sequence analogous with polycistronic mRNA. The 5′-end is represented as a phosphomonoester rather than a triphosphate because this is the structure that has been isolated and studied. These precursors cannot, therefore, be the immediate products of transcription. Presumably the extreme 5′-terminus of the transcript is removed at an early stage, perhaps very soon after the initiation of transcription.

As tRNA assumes a characteristic secondary structure (see section 3.1.5), the precursors must have an extensive secondary structure in the regions that include the tRNA sequence. Such a structure is shown in outline in Fig. 5.15. To pursue the analogy of the clover leaf, the 'plant' now has some roots in the form of spacer sequences. Note that the stem is joined to the roots that involve extra base pairs to those present in the stem of the mature molecule and that, therefore, one of the processing enzymes (RNAase P) is a double-strand nuclease.

Ribosomal RNA

A transcriptional unit of rRNA is as shown in Fig. 5.16. It differs from that of eukaryotes (see section 8.2) in the following respects.

(1) 5*s* rRNA is co-transcribed with the large rRNA's (16*s* and 23*s*).

(2) A tRNA sequence contributes to the spacer between the 16*s* and 23*s* genes.

(3) There is no counterpart to the 5.8*s* rRNA (see section 7.1) in bacteria.

(4) Cleavage of the precursor molecule begins before transcription is completed. This relatively minor point of distinction means that no counterpart of the 45*s* rRNA precursor of eukaryotic cells (see section 8.2.1) is normally found in bacteria. However, by suitable

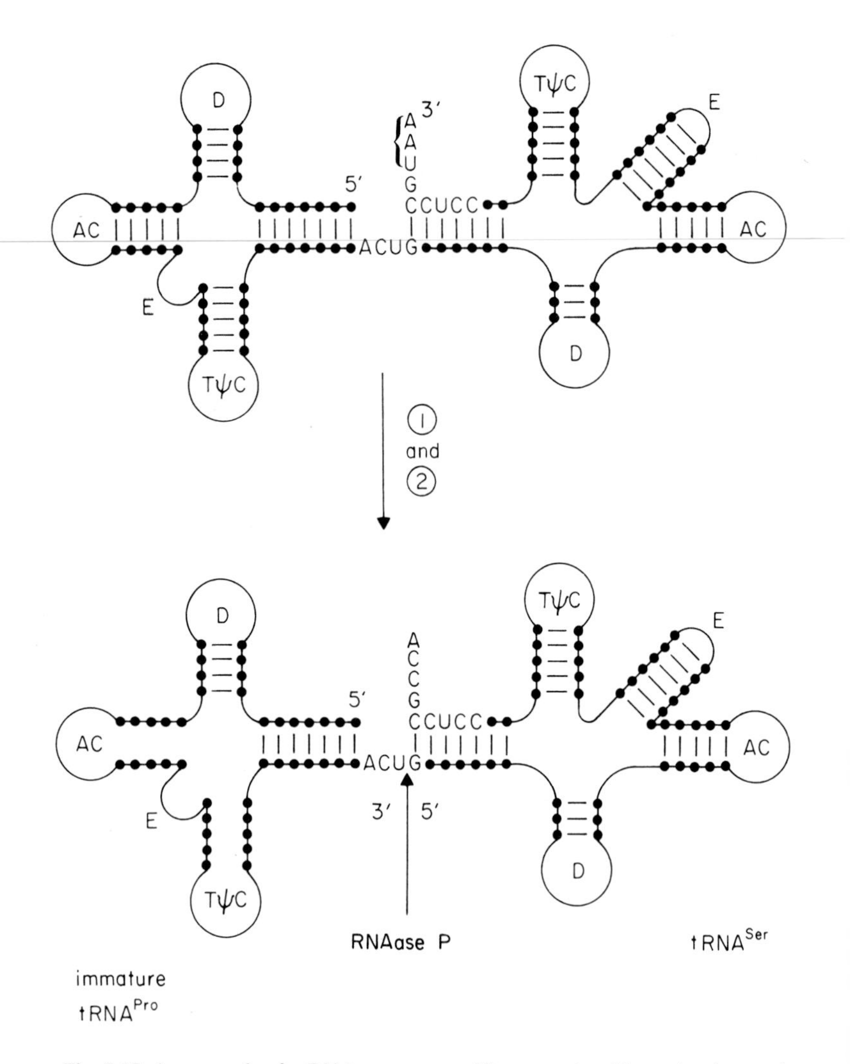

Fig. 5.15. An example of a tRNA precursor and its processing. The molecule contains sequences for two tRNAs and is a product of phage T4. The left-hand side of the precursor generates $tRNA^{Pro}$; the right-hand side generates $tRNA^{Ser}$. Only residues involved in forming stems are represented and are shown as either dots or letters. In order to help to orient the two sequences, the loops are lettered AC (anticodon), TψC, D and E (extra) (cf. Fig. 3.8, section 3.1.5). The precursor (top of figure) contains neither of the 3′-CCA termini. The stages of processing shown here are only concerned with breakage and formation of phosphodiester bonds; nucleoside modification is not included. The first two stages are (1) removal of the three 3′-terminal residues (represented by a bracket) and (2) incorporation of the CCA terminus for the $tRNA^{Ser}$ sequence by the template-independent polymerase, tRNA nucleotidyl transferase. The precursor is next split by RNAase P. This split generates the complete $tRNA^{Ser}$ sequence and an immature $tRNA^{Pro}$ which needs further processing by steps analogous to (1) and (2). (Based on data from Seidman & McClain (1975); the paper contains references to other examples of tRNA processing.)

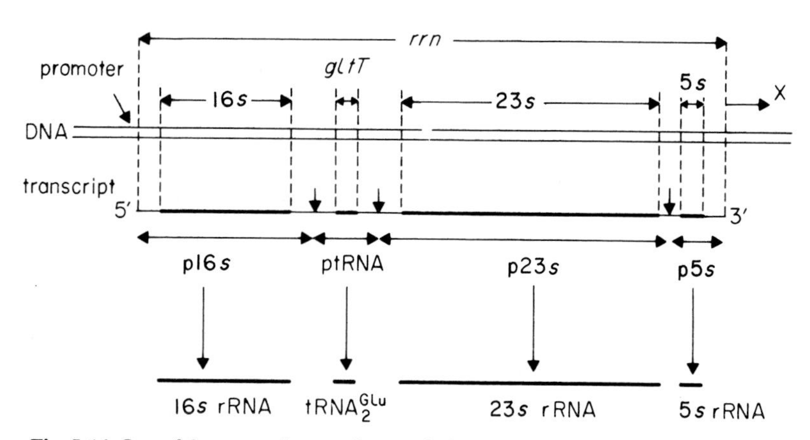

Fig. 5.16. One of the seven ribosomal transcriptional units (*rrn*) in *E. coli*. This unit is at approximately 71′ on the *E. coli* genetic map and the cluster of ribosomal and RNA polymerase genes discussed in section 5.3.2 is located in the direction marked X. The four structural genes are for rRNA molecules (16*s*, 23*s* and 5*s*) and tRNA$^{Glu}_2$ (*gltT*). The region is co-transcribed. The transcript does not occur in normal cells because it is cleaved at points indicated by short vertical arrows during transcription. The enzyme RNAase III is responsible for at least one of these cuts. The products are the four precursors (p16*s*, etc.). In the transcript, the thickened parts of the line represent sequences that end up in the mature molecules. The final stages consist of further cutting of the precursors and (with the exception of 5*s* RNA) modification of nucleoside residues. The other six *rrn* regions are similar. A tRNA$^{Glu}_2$ sequence has been located between 16*s* and 23*s* in three of them; the remaining three contain both tRNA$^{Ala}_{1B}$ and tRNA$^{Ile}_1$ genes. There is also some evidence to suggest that the distal ends of at least some *rrn* regions may contain tRNA genes (Morgan *et al.* 1977).

manipulation of strains of *E. coli* that carry a mutation which affects RNAase III, it is possible to isolate the long precursor corresponding to 45*s* RNA. The bacterial molecule has a sedimentation coefficient of 30*s*.

The regulation of the biosynthesis of compounds of the protein synthetic apparatus is discussed further in section 6.3.3.

NOTE ADDED IN PROOF

For further information on plasmid biology and on other material in Chapters 4 and 5 see R.E. Glass (1982) *Gene Function: E. coli and its heritable elements*. Croom Helm, London.

A review of transcriptional control can be found in S. Baumberg (1981) The evolution of metabolic regulation. *Symp. Soc. gen. Microbiol.* **32,** 229–72.

Chapter 6 Bacterial Protein Synthesis

6.1 INTRODUCTION

For those regions of the bacterial genome which are transcribed into molecules of mRNA, protein synthesis represents the final stage in the expression of genetic information. The transcript is translated, using the genetic code, to yield polypeptide products whose amino acid sequences are determined by the nucleotide sequence of their respective mRNAs. In bacterial cells the processes of transcription and translation are closely coupled (Fig. 6.1). Ribosomes form the sites of protein synthesis in all living cells and are themselves highly complex, being composed of at least 57 different macromolecules. Protein synthesis itself involves the coordinated action of over 130 different macromolecules some of which, e.g. aatRNAs, mRNA, ribosomes, are involved in each of the three major steps in translation whereas others, the protein factors, are specifically involved in only one.

In this chapter some features of the genetic code and bacterial ribosomes will be discussed first to assist the understanding of the detailed mechanisms and control of bacterial protein synthesis. Since the genetic code is universal and the basic mechanisms of protein synthesis are invariant in all living organisms much of the discussion in the following sections is also relevant to eukaryotic protein synthesis. The major differences between eukaryotic and prokaryotic translation are discussed in Chapter 9.

6.2 THE GENETIC CODE

6.2.1 General features of the code and coding rules

The 64 (4^3) permutations of the four bases of mRNA into triplets (codons) that constitute the genetic code are shown in Table 6.1. See vol. 31 (1966) of the *Cold Spring Harbor Symposia on Quantitative Biology* for a historical account of the elegant experiments performed in the early to mid 1960s which led to the elucidation of the meanings of the individual codons. Those coding assignments have subsequently been extensively confirmed by nucleic acid and protein sequencing studies. The 61 codons specifying amino acids are termed sense codons; the remaining three signal termination of translation and are collectively termed nonsense or chain termination codons although the trivial names, amber, ochre and umber, are frequently used. A total of 61 codons specify the 20 amino acids incorporated into proteins during translation (other

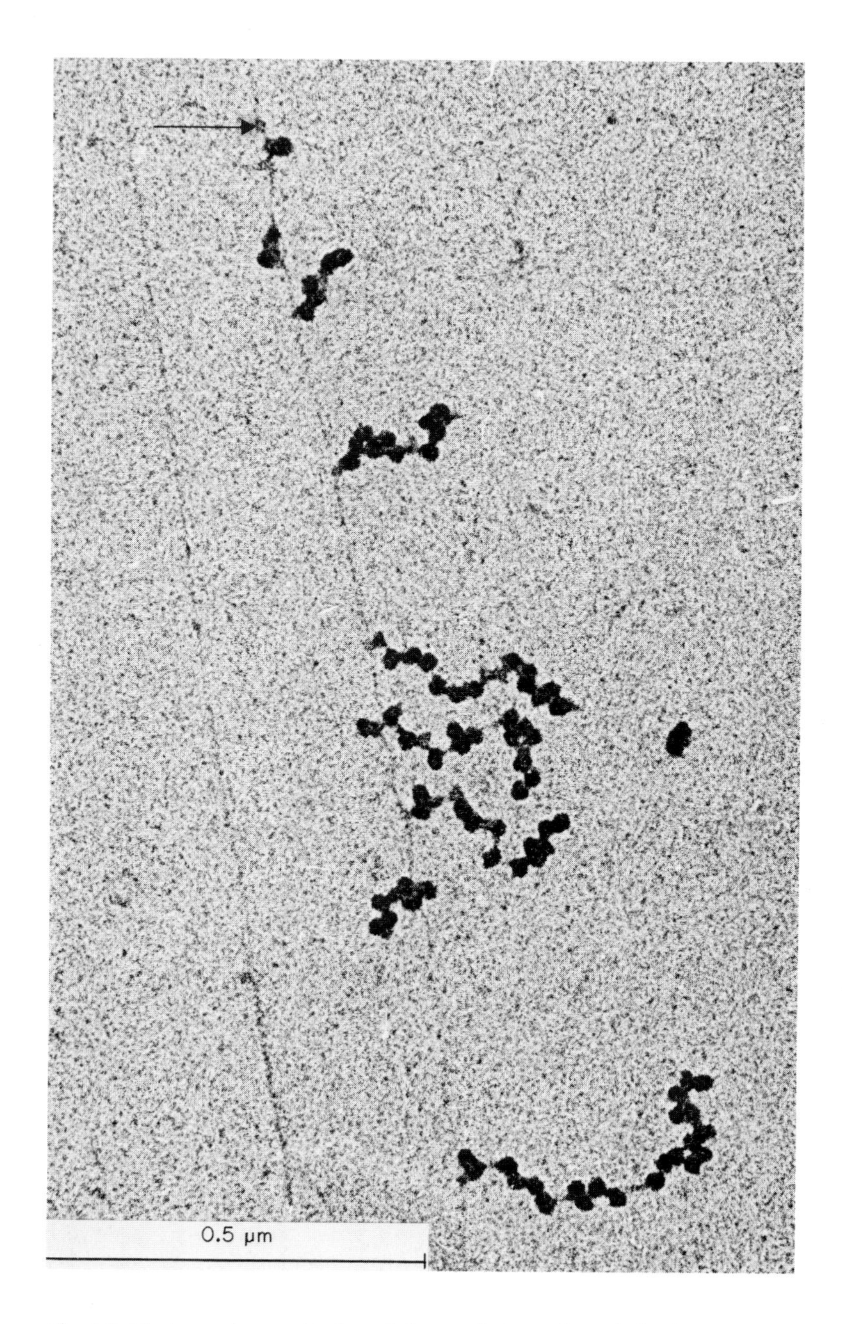

Fig. 6.1. Electron micrograph of coupled transcription and translation in *E. coli*. The arrow indicates a putative RNA polymerase molecule presumably on or very near a transcriptional initiation site. Ribosomes attach to the nascent mRNA and an incomplete gradient of increasing polysome length is seen. (Based on data from Miller *et al.* (1975). We are very grateful to Dr O.L. Miller for supplying a print; reproduced, with permission, © 1970, The American Association for the Advancement of Science.)

amino acids not present in the table may be found in proteins as a result of post-translational modifications; see Chapter 9), i.e. the code is highly degenerate. With the exception of Met and Trp, each amino acid is specified by more than one codon and the extent of coding degeneracy varies. Application of rapid DNA sequencing (see section 3.5.4), together with knowledge of many protein sequences, has led to the elucidation of different patterns of selective codon utilization in both prokaryotic and eukaryotic systems. Here we comment on codon utilization in the code of Table 6.1. The particular case of codon utilization in mitochondria, in which the code differs in significant details from this table, are discussed in section 10.4.1.

Table 6.1. The Genetic Code. Numbers in brackets are the total numbers of codons for each amino acid. Hatched boxes are 'codon families' (see text). *Indicates initiation codon and CT indicates chain termination codon.

First codon base ↑		Second codon base → U	C	A	G	Third codon base ↓
	U	UUU Phe (2)	UCU Ser (6)	UAU Tyr (2)	UGU Cys (2)	U
		UUC Phe	UCC Ser	UAC Tyr	UGC Cys	C
		UUA Leu (6)	UCA Ser	UAA (ochre) CT	UGA (umber) CT	A
		UUG Leu	UCG Ser	UAG (amber) CT	UGG Trp (1)	G
	C	CUU Leu	CCU Pro (4)	CAU His (2)	CGU Arg (6)	U
		CUC Leu	CCC Pro	CAC His	CGC Arg	C
		CUA Leu	CCA Pro	CAA Gln (2)	CGA Arg	A
		CUG Leu	CCG Pro	CAG Gln	CGG Arg	G
	A	AUU Ile (3)	ACU Thr (4)	AAU Asn (2)	AGU Ser	U
		AUC Ile	ACC Thr	AAC Asn	AGC Ser	C
		AUA Ile	ACA Thr	AAA Lys (2)	AGA Arg	A
		AUG* Met (1)	ACG Thr	AAG Lys	AGG Arg	G
	G	GUU Val (4)	GCU Ala (4)	GAU Asp (2)	GGU Gly (4)	U
		GUC Val	GCC Ala	GAC Asp	GGC Gly	C
		GUA Val	GCA Ala	GAA Glu (2)	GGA Gly	A
		GUG* Val	GCG Ala	GAG Glu	GGG Gly	G

Factors such as tRNA base composition, codon–anticodon interaction energy, modification of DNA bases and structural requirements of mRNA probably influence the use made of the inherent degeneracy of the code. With the exception of the amino acids, Arg, Ser and Leu, which are each specified by six codons, the degeneracy of the code is solely a feature of the third codon base. Provided that the first two codon bases are the same, the coding sense is unchanged with either of the two pyrimidine bases in the third position and any codon ending in C specifies the same amino acid if U is substituted. Likewise, unless the first two codon bases are AU or UG, the coding sense is preserved if the third position is filled with either of the purines A or G. The patterns of third base degeneracy are shown in

Table 6.2. Patterns of third base degeneracy.

No. of groups represented in code table	Permitted 3rd base(s) with no change in codon sense	Amino acid(s)/signal specified	Total No. of codons
8	U,C,A or G (codon family)	Leu, Val, Ser, Pro, Thr, Ala, Arg, Gly	32
1	U,C or A	Ile	3
7	U or C (pyrimidine restricted)	Phe, Tyr, His, Asn, Asp, Cys, Ser	14
6	A or G (purine restricted)	Leu, Gln, Lys, Glu, Arg, ochre, amber	12
2	G	Met, Trp	2
1	A	umber	1

Table 6.2; in order to explain these patterns it was proposed that the third (3′) base of a codon could form a non Watson–Crick base pair with the corresponding base at the first (5′) position in the anticodon (Crick 1966). As shown in Table 6.3 this 'wobble' pairing is permitted between U and G, G and U, I and U or A and involves minimum distortion of bond angles. However, pairings of U with U or C; C with U, C or A; G with A; or I with G are not permitted at the third (wobble) position of the codon and the classic G:C, A:G pairing rules persist at the first and second base positions. The wobble hypothesis has subsequently been confirmed and expanded through knowledge of the recognition properties of tRNAs of known anticodon sequences. Adenine has never been found at the first anticodon position; it is apparently always converted to hypoxanthine by the enzyme anticodon deaminase, to give an anticodon starting with I and, therefore, capable of recognizing synonymous codons with U, C or A in the third position. The absence of A as a first anticodon base explains the lack of discrimination between U and C in the third base position of pyrimidine restricted codons. In the case of Met and Trp, which are each specified by one codon only and depend on a third codon base of G, the first anticodon base of the respective tRNA

Table 6.3. Wobble hypothesis.

5′ Anticodon base	Base recognized on codon
C	G
A	U
U	A or G
G	C or U
I	C or U or A

molecules has been identified as C. Modification of the first anti-codon base can increase or decrease pairing specificity, e.g. 2-thiouridine (s^2U) pairs with A but not G whereas uridine-5-oxyacetic acid (oa^5U) pairs with A, G and U.

Although tRNA sequence determinations have shown that wobble pairing does occur *in vivo* there is evidence for selection against it in favour of classic base pairing, for example analysis of the RNA of the pilus-specific RNA phage MS2, and its three protein products revealed that for six out of the seven amino acids specified by pyrimidine-restricted codons (Table 6.2), there is preferential use of codons ending in C, i.e. there is selection for C:G codon–anticodon pairing and against U:G wobble pairs. It has been proposed that such selection may serve to minimize the occurrence of translational errors because of differences in binding energies between classic and 'wobble' pairs.

In the case of at least two of the eight groups (codon families) shown in Table 6.2, where codon sense is independent of the nature of the third codon base, there is evidence that only the first two codon bases are read *in vitro* with indiscriminate pairing at the third position. It has been proposed that this 'two out of three' method of codon reading may be used *in vivo* in the case of the eight codon families and that the genetic code is organized to permit such reading without loss of translational fidelity (Lagerkvist 1978).

In summary, translation of the genetic code depends on pairing between codons in mRNA and the complementary bases of anti-codons in the corresponding tRNA molecules. With rare exceptions (see section 6.2.3) pairing at the first two codon positions obeys classic pairing rules whereas pairing at the third position may follow the more flexible wobble rules or, in certain cases, be completely indiscriminate.

6.2.2 The origin and evolution of a universal genetic code

The genetic code appears to be so arranged as to minimize the effects of certain DNA, or translational, mutations. Thus all codons with a middle base of U specify hydrophobic amino acids, mutation to yield a middle base of C results in codons that still specify non-polar or neutral amino acids. The two pairs of codons for the acidic amino acids differ in only one base as do those for the aromatic amino acids, Phe and Tyr. It seem highly unlikely that these and other patterns are coincidental; they probably reflect either a mechanism of evolution or confer a selective advantage and thus determine the structure of the code. Although there are different patterns of codon usage and these may be still be evolving, and despite the special features of the code in eukaryotic organelles (see section 10.4.1), it is otherwise universal and, in the absence of primitive codes, the many models proposed to explain the codon assignments and their universality are necessarily speculative. These explanations range from stereochemical pairing between codons and amino acids, to assignment of codons by chance, or origination from an ancestral organism arrived from outer space. Investigations designed to evaluate the various theories include: analysis of relationships between protein composition and the number of codons assigned to each amino acid; analysis of correlations between amino acids and their anticodon nucleotides; nucleic acid:amino acid affinity studies; study of homologies within tRNA molecules and between different tRNA molecules, and molecular model building. It seems likely that the initial codons selected their respective amino acids through the stereochemical interactions of free amino acids with nucleotides of either the primitive message or a primitive tRNA molecule. There is much evidence that the current range of organism-specific tRNA molecules arose from a single ancestral gene through gene duplication followed by differentiation, and such multiplication of tRNAs probably played a key role in the evolution of the code. The proposal that the primitive code was a doublet code with the third codon base serving as a comma may find support from the proposal that 'two out of three' reading occurs at present (see section 6.2.1). The genetic code probably evolved together with the amino acids it specifies; an initially small number of amino acids—those formed most easily in the primordial soup—have been postulated as precursors for other amino acids with concomitant subdivision of their codons amongst the new products. The stages during the evolution of the code at which primitive ribosomes and activating enzymes emerged and translational fidelity was established are issues open to much speculation. There must have been some point at which the code was frozen so that slightly different codes did not emerge and new amino acids did not evolve and/or become incorporated into polypeptides (Jukes 1977).

6.2.3 The initiation codon(s)

For accurate translation of the genetic code there must be a fixed starting point in order to define the correct triplet reading frame and in the case of polycistronic mRNAs of prokaryotes each cistron has its own initiation signal. In all but a few recorded cases (see, for example, the later part of this section) the initiation codon has been found to be AUG which is the sole codon for methionine. However, the methionine residue incorporated in response to the initiator AUG is frequently removed from the nascent polypeptide by a specific aminopeptidase and Met is the N-terminal amino acid in only 45% of the total *E. coli* proteins. This cleavage is one example of post-translational modification in prokaryotes (see also section 6.5.5). AUG also codes for methionine at internal positions in the message and there are two types of tRNA for methionine, one of which specifically interacts with the initiator AUG ($tRNA^{fMet}$) and one which responds to all other AUG codons ($tRNA^{Met}$). Some structural and functional properties of these two isoaccepting tRNA species from *E. coli* are shown in Table 6.4. Functional roles have been proposed for some of the indicated structural differences between the two tRNA species; for example it has been proposed that the lack of base pairing with the first nucleotide of the initiator tRNA is important for initiation factor binding and that the substitution of A for G following the TψC sequence permits aminoacyl tRNA entry into the ribosomal P site. However, such structural features, although unusual, are not unique to $tRNA^{fMet}$ and it is likely that it is the overall conformation of fMet-$tRNA^{fMet}$ that enables it to bind to the ribosomal site otherwise available only to peptidyl-tRNA molecules and hence to fulfil its role in protein synthesis.

GUG has been identified as the initiation codon for the A protein of MS2 bacteriophage and the repressor protein (*I* gene product) of the Lac operon. GUG normally specifies the amino acid, valine (Table 6.1); however, in these two exceptional cases where it serves

Table 6.4. Some properties of the methionine tRNAs from *E. coli*.

Relative distribution	$tRNA^{fMet}$	$tRNA^{Met}$
Relative distribution	70%	30%
Anticodon	CAU	CAU
5′ terminal residue	Not base paired	Base paired
Sequence in ψ loop	TψCA	TψCG
Base on 3′ side of anticodon	A	Modified A
Recognized by:	Initiation factors	Elongation factors
Binds at:	Ribosomal P site	Ribosomal A site
Bound Met:	Formylated following aminoacylation; forms N terminal residue of nascent polypeptide	Not formylated; incorporated at internal positions of polypeptide

as an initation codon, it is recognized by $tRNA^{fMet}$ and, therefore, serves as a codon for methionine. Pairing of the anticodon CUA with the codon GUG involves an atypical wobble pairing at the *first* (5′) base of the codon. It has been suggested that the absence of a bulky modifying group on the A adjacent to the 3′ anticodon base of $tRNA^{fMet}$ (Table 6.4) is necessary to permit the required distortion of bond angles at this unusual codon position.

Efficient initiation of translation is not solely dependent on the initiator AUG (or GUG) codon; mRNAs frequently contain an untranslated region (leader sequence) preceding the initiation codon. The influence of a purine-rich sequence within the leader sequence on ribosome binding during initiation will be discussed in section 6.5.1. The two bases on either side of an initiator AUG can influence its efficiency as an initiation signal. Initiation efficiency is apparently increased if a pyrimidine rather than a purine base precedes AUG. Moreover, the initiation complex (see section 6.5.1) forms more readily if the base on the 3′ side of the initiation codon is a purine. It has been suggested that the base U_{33}, which precedes the anticodon in all tRNA molecules, can, during the process of initiation only, form a base pair with G or A. It appears likely that the most efficient initiation signal is a quadruplet, rather than a triplet, codon.

6.2.4 The termination codons and nonsense suppression

Any of the three codons UAA (ochre), UAG (amber) and UGA (umber) in phase in mRNA signal termination of protein synthesis. Under normal conditions these codons are not recognized by any aminoacyl tRNA species but, at the appropriate stage in translation, cause the binding of release factors to the ribosome with consequent release of a completed polypeptide chain from the tRNA–ribosome–mRNA complex (see section 6.5.3).

Nonsense mutations and nonsense suppressors

Point mutations which result in the conversion of a sense codon to any one of the three chain termination codons are termed nonsense mutations and result in premature termination of protein synthesis at the position of the mutation (see section 1.4.4); for example, in R17 (a phage related to MS2, section 6.2.1) substitution of U for C at the 5′ position of the seventh codon of coat protein mRNA converts a glutamine codon (CAG) to an amber codon (UAG). Translation of this message in wild-type, i.e. non-suppressing, strains of *E. coli* yields an acid-soluble hexapeptide composed of the first six residues of wild-type coat protein. However, translation of the mutant message in other, so-called suppressive or permissive, strains yields, in addition to small amounts of the hexapeptide fragment, large amounts of full-length coat protein. The suppressor activity of these strains resides in the tRNA fraction; thus a cell-free extract from a non-suppressing strain is capable of complete translation of coat protein message carrying the amber nonsense mutation in a protein-synthesizing system *in vitro* supplemented with tRNA from an amber suppressing strain. The particular tRNA species involved in the suppression is termed suppressor tRNA and results from mutation in a tRNA gene to give the anticodon, CUA, which, according to the pairing rules discussed in section 6.2.1, will only recognize the amber codon. Bacterial strains capable of suppressing ochre or UGA nonsense mutations also contain mutant suppressor tRNA species. The anticodon of ochre suppressor tRNAs from bacteria is UUA which, because of the wobble rules, can recognize both UAA and UAG codons. Consequently, bacterial strains containing ochre suppressor tRNAs can also suppress amber mutations. (In contrast, the anticodon of an ochre suppressor tRNA from yeast has been identified as IUA which can only suppress ochre mutations.)

In summary, the suppressor, *su,* allele directs the synthesis of a mutant suppressor tRNA which recognizes a nonsense codon. The nonsense suppressor strains of *E. coli* are numbered according to the map position of the mutant tRNA gene and by the amino acid inserted by the suppressor tRNA in response to the nonsense mutation. In all cases the codon for an amino acid inserted in response to the nonsense codon is related to the latter by a single base change, for example an *su*1 strain inserts serine at the position of an amber mutation because mutation in a $tRNA^{Ser}$ gene has caused a change in the anticodon from CGA to CUA whereas an *su*3 strain would insert tyrosine at the same position as a result of a GUA:CUA conversion. Only half of the 20 amino acids specified by the genetic code are one mutational step away from one of the chain termination codons. The codons for the remaining ten can change to a nonsense codon only by two or three mutational steps. Although suppression may prevent premature chain termination and yield a protein product of the correct length the amino acid inserted at the site of the nonsense mutation will frequently differ from that coded for in the wild-type message. Consequently, efficient suppression does not always result in an active protein product. Likewise, a nonsense mutation does not always result in inactive products; if the site of the mutation is close to the normal termination site the polypeptide synthesized will be only slightly shorter than the wild-type protein and may be partially, or fully, biologically active.

Suppressor mutations do not invariably involve a base change in the anticodon sequence; the suppressor tRNA from the *su*9 strain inserts a tryptophan in response to UGA mutations but also responds to the normal UGG Trp codon. Sequencing studies have revealed that there is no change in the anticodon (CCA) in the suppressor tRNA and the only difference from wild-type $tRNA^{Trp}$ is the substitution of an A for a G in the D stem (see Fig. 3.8). This substitution converts a G:U wobble pair to an A:U pair which, possibly by increasing the stability of the helix in that region of the molecule, may affect the total conformation so as to permit the pairing of C with A in the third codon position in violation of the wobble rules. Thus a base change in a tRNA molecule far from the anticodon region can affect codon:anticodon pairing. In most cases there is more than one tRNA species available for the recognition of synonymous codons for any particular amino acid. The ability of the *su*9 suppressor tRNA to recognize the sole Trp codon in addition to UGA is of the utmost importance to the host cell as there is only one gene for $tRNA^{Trp}$; for example, although one species of $tRNA^{Tyr}$ with the anticodon sequence GUA would serve to respond to both codons for tyrosine (UAU and UAC) there are known to be three species in *E. coli* in the relative proportions 60:25:15; only the minor species can be mutated to give an amber suppressor tRNA (anticodon CUA) or an ochre suppressor molecule (anticodon UUA).

The loss of a tRNA acceptor activity by mutation can be overcome by the construction of partial diploids which carry the information for synthesis of the wild-type tRNA species. The use of partial diploids led to the identification of the *su*7 locus, in which mutation in the sole $tRNA^{Trp}$ gene converts the anticodon from CCA to CUA to give an amber suppressor tRNA that can no longer respond to the UGG codon for Trp. Interestingly, the amino acid inserted by this suppressor molecule at the site of amber mutations is not Trp but Gln and, therefore, the change in anticodon results in mischarging of this particular tRNA. Although mutations such as *su*7 are recessive with respect to lethality in the presence of the appropriate wild-type tRNA, nonsense suppression is normally dominant. A rare exception is found with a type of UGA suppression apparently involving a defect in tRNA modification; in this case the tRNA species responsible for the suppressor activity seems to be an under-methylated tRNA species and suppression is recessive in nature.

It would obviously be disadvantageous, and possibly lethal, to the host cell if the presence of suppressor molecules resulted in the read-through of normal chain termination codons to produce elongated proteins. In fact the efficiency of suppression depends on the reading context of the nonsense codon and the surrounding nucleotide sequence plays an important role in chain termination as well as initiation. Consequently, it is possible for suppressor tRNAs (and/or release factors; see section 6.5.3) to distinguish between nonsense mutations and bona fide chain termination signals. Moreover, different types of suppressor molecules exhibit different efficiencies of suppression; in general, ochre suppressors are less efficient than amber suppressors and consequently UAA may be a more efficient chain termination signal than UAG. Chain termination at UGA codons is not always 100% efficient even in bacterial strains not known to harbour nonsense suppressors. The RNA bacteriophage Qβ takes advantage of the low level read-through of UGA codons exhibited by its *E. coli* host to enable the synthesis of small amounts of an extended coat protein required for the assembly of active phage particles (see section 6.5.6).

Some bacterial and phage cistrons terminate with double full stop signals (UAAUAG and UAGUAA) and these could have evolved as fail safe devices to prevent read-through resulting from inefficient termination codons or the presence of suppressor tRNAs. Chain termination codons are frequently found in the untranslated regions preceding the first initiation signal of a polycistronic message and in the intercistronic sequences between the coding sequences where they may serve to maintain the correct reading phase and/or prevent translation errors. The chain termination process is discussed in further detail in section 6.5.3; further details of the genetics of suppression and references are to be found in Caskey (1977) and Smith (1977).

6.2.5 Further mechanisms of suppression

Suppression of a nonsense mutation by alteration in the structure and specificity of a tRNA molecule is an example of *intergenic* or external suppression. The primary mutation still exists but its effect is suppressed by a second mutation in a different gene. Other examples of intergenic suppression by mutant tRNAs include suppressors for missense and frame-shift mutations; for example, a G to A base substitution that changes a glycine codon (GG_{G}^{A}) to an arginine codon (AG_{G}^{A}) can be suppressed by a mutant $tRNA^{Gly}$ that inserts glycine in response to the new arginine codon. The efficiency of suppression must be low to prevent misreading of bona fide arginine codons.

Suppressors specific for frame-shift mutations in the *Salmonella* His operon have been located genetically. Characterization of the tRNA specified by one of these suppressor loci revealed the presence of an additional C in the CCC anticodon loop of a $tRNA^{Gly}$ molecule. Such a molecule could suppress single base addition frame-shift mutations in a sequence of repeating Gs coding for glycine.

Intergenic suppression mediated by altered tRNA molecules has been termed informational or direct suppression since tRNA plays a vital role in the transfer of information from DNA or RNA into protein. An example of indirect intergenic suppression is found in *Salmonella* where mutation to arginine auxotrophy can suppress a requirement for proline (Fig. 6.2). In this case the second mutation circumvents the original mutant phenotype by providing an alternate metabolic pathway. Such a mechanism can, in principle, suppress a deletion and one example of this has been described (Tanemura & Bauerle 1979).

The suppression of polarity by mutation in the *suA* gene is another example of indirect intergenic suppression (see section 6.5.5).

Intragenic, or internal, suppression refers to suppression mediated by a second mutation within the original mutant gene. Because mutation is a chance event, spontaneous mutations leading to intragenic suppression occur less frequently than those resulting in intergenic suppression. Some examples of intragenic suppression are given on the next page.

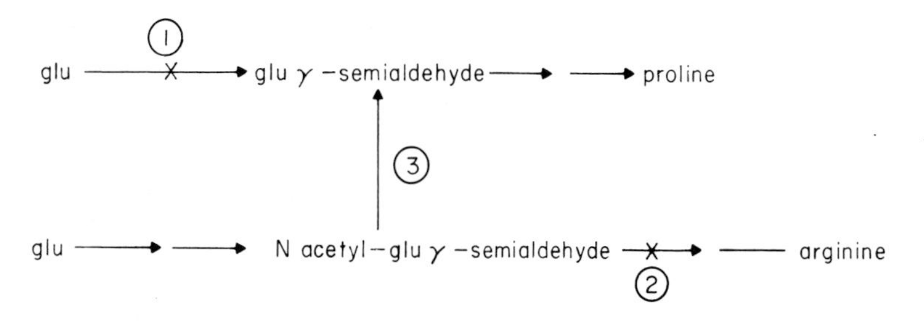

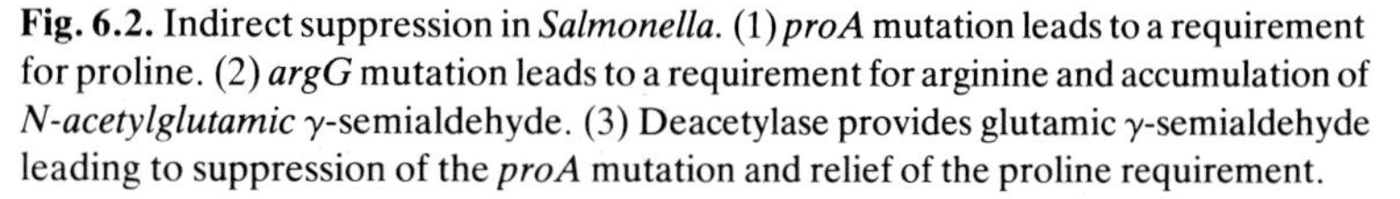

Fig. 6.2. Indirect suppression in *Salmonella*. (1) *proA* mutation leads to a requirement for proline. (2) *argG* mutation leads to a requirement for arginine and accumulation of *N-acetylglutamic* γ-semialdehyde. (3) Deacetylase provides glutamic γ-semialdehyde leading to suppression of the *proA* mutation and relief of the proline requirement.

(a) Suppression of a nonsense mutation by alteration in a nucleotide not involved in the original mutation:

AAA→UAA→UUA
Lys ochre Leu (suppression results if substitution of Leu for Lys does not affect biological activity).

(b) Suppression of a frame-shift mutation by a second compensating frame-shift within the same gene.

(c) Suppression by restoration of active configuration of: (i) a protein, e.g. the tryptophan synthetase of *E. coli* is inactive if a nonsense mutation results in the substitution of Glu for Gly at residue position 210 or Lys for Trp at position 174. However, if both nonsense mutations occur together biological activity is restored. (ii) tRNA, e.g. *su*3 tRNA loses its suppressor activity if the G in position 2 of the sequence is replaced by A or if C in position 80 is replaced by U. Since base 2 normally pairs with base 80 in the amino acid acceptor stem suppressor activity is restored if both base changes occur in the same molecule.

(d) Suppression of polarity by introduction of a new polypeptide initiation signal near and distal to the site of the polar mutation (see section 6.5.5).

6.2.6 Overlapping genes

Until recently, the genetic code was accepted to be a triplet, commaless, non-overlapping code which was translated using a single fixed reading frame. Accordingly, translation was assumed to begin at an initiation signal and proceed to the first in-phase chain termination signal producing a single polypeptide from each coding sequence contained in the message. It now appears that the bases of the message may occasionally be read as doublets or even quadruplets and also there are many examples of overlapping genes both in prokaryotes and eukaryotes.

Undoubtedly the best examples of overlapping genes in prokaryotic systems are found in the genome of the related bacteriophages ϕX174 and G4. In most bacteriophages, genome length is restricted to that which can be efficiently packaged into the phage head. In the case of the single-stranded DNA phage ϕX174, all the various functions of the genome must be contained within its 5375 nucleotides. The genome codes for at least nine proteins and in order to compress the required information into the small genome, two genes (B and E)

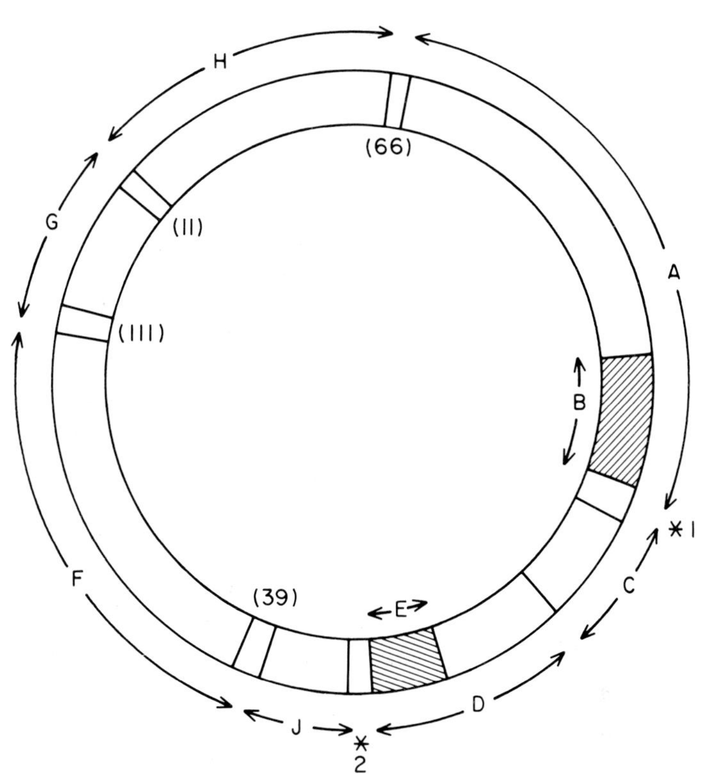

Fig. 6.3. The overlapping genes of ϕX174 (after Sanger *et al.* 1977a). The functions of the products of the genes shown are as follows: F, G, H, J, structural proteins; A, required for double-stranded DNA replication and single-stranded DNA synthesis; B, C, D, involved in production/packaging of single-stranded DNA; E, required for host lysis. The hatched areas indicate genes completely contained within other genes; the gene B coding sequence is one nucleotide to the left of the A protein reading phase and the gene E coding sequence is one nucleotide to the right of the D protein reading phase. Numbers inside the brackets indicate the length of intergenic regions. *Indicates an initiation/termination overlap between adjacent genes as follows:

		gene C initiation
*1 ATGA→	TACT→	AUGA
		gene A termination
genome sequence	cDNA sequence (replicative form)	mRNA sequence
		gene J initiation
*2 TAATG→	ATTAC→	UAAUG
		gene D termination

are completely contained within others (A and D respectively) and there are at least three further sites of overlap between the termination signal of one gene and the initiation signal of the adjacent gene (Fig. 6.3).

In addition to the overlapping genes found in ϕX174 the genome of G4 has been found to contain a region of overlap which shows even greater efficiency in the use of a single DNA sequence. In this phage, the gene for a tenth protein (gene K), of as yet unknown function, overlaps along its whole length with other genes (Shaw *et al.* 1978); the 5′ proximal region overlaps with the last 86 nucleotides of gene A and the 3′ region overlaps with the first 89 nucleotides of gene C, and at two sites the coding capacity of the DNA is used in all three reading frames (Fig. 6.4). Moreover, there is evidence that initiation or termination of translation at different sites in the same reading frame in a single gene serves to increase the variability of protein products encoded by these small bacteriophage genomes (Pollock *et al.* 1978).

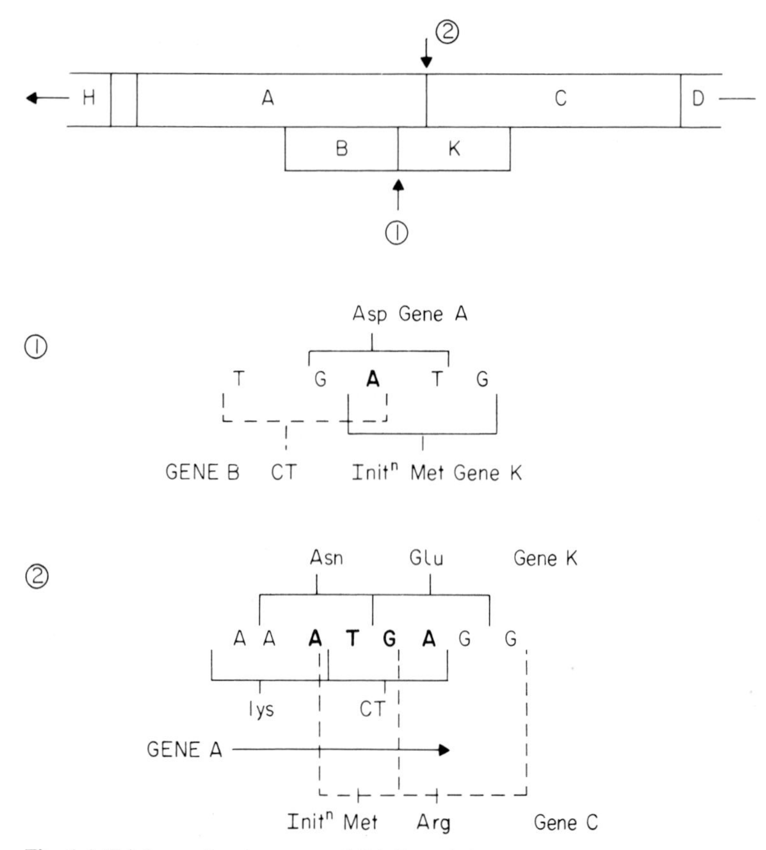

Fig. 6.4. Triple overlapping genes of G4. Bases in bold print are read in all three reading frames (after Shaw *et al.* 1978).

The read-through of a UGA codon at the end of the Qβ coat protein gene to give a protein with a different function was mentioned in section 6.2.4. Since the coat protein gene is completely contained within the so-called A_1 protein gene, this is another example of overlapping genes although there is no change of reading frame in this case.

Although selective pressure for overlapping genes is maximal in the case of bacteriophages and viruses, initiation/termination overlap has been reported in a messenger RNA transcribed from the *E. coli* Trp operon and the dual application of rapid DNA sequencing techniques and recombinant DNA technology may soon reveal more cases of overlap in the bacterial genome.

In summary, translation of the genetic code proceeds via certain codon:anticodon pairing rules, the code being translated usually, but not always, in steps of triplets of bases with a reading frame which is fixed for any one particular protein but which can be changed to produce different protein products.

6.3 BACTERIAL RIBOSOMES

6.3.1 Structure and function

Ribosomes are complex ribonucleoprotein particles that form the site of protein synthesis in all living cells. The importance of protein synthesis, and hence ribosome biogenesis, is indicated by the fact that there are $15–18 \times 10^3$ ribosomes per cell in a rapidly growing culture of *E. coli*. This means that up to 25% of the total cellular mass of an *E. coli* cell may be composed of ribosomes, and that ribosomal RNA (rRNA) and ribosomal proteins (r proteins) may respectively constitute up to 80% and 25% of the total cellular complement of RNA and protein. All ribosomes are composed of two subunits of unequal size which, in the case of bacterial ribosomes, are characterized by sedimentation coefficients of 50*s* and 30*s*. Protein and RNA constitute 95% of the dry weight of both subunits, the remainder being accounted for by cations, chiefly magnesium and polyamines, which are required for maintenance of structural integrity; in fact, a high proportion of the total cellular magnesium content may be contained within ribosomes. Unless otherwise stated, the remainder of this section refers to the ribosomes from *E. coli* since these have been far more widely studied than those from any other organism, and there is no reason to suppose that ribosomes from other bacterial species will show major differences in their basic structural and functional properties. Numerous reviews of the structure and function of bacterial ribosomes have been published, among the most recent are those by Kurland (1977), Brimacombe *et al.* (1978), Hardy (1979), and there is a collection of excellent essays in Chambliss *et al.* (1980).

Ribosomal subunits

Some details of the ribosomal subunits of *E. coli* are given in Table 6.5. A variety of techniques, including electron microscopy and small angle X-ray scattering, have been used to estimate the shape and dimensions of the two subunits. Although there is disagreement between different research groups as to the degree of structural asymmetry, the 30*s* subunit appears to have an elongated, slightly curved, prolate shape divided into two unequal regions by a transverse cleft. In contrast, the 50*s* subunit is more symmetrical with a curved base, large middle protuberance and two smaller lateral protuberances giving an 'armchair' appearance in some electron micrographs. The active unit of bacterial protein synthesis, the 70*s* ribosome, is formed by the association of the two subunits with mRNA and aa-tRNA molecules.

Table 6.5. Ribosomal subunits from *E. coli*.

	Large subunit	Small subunit
Sedimentation coefficient	50*s*	30*s*
Particle weight (M_r)	1.55–1.80 × 10^6	0.7–1.0 × 10^6
RNA component(s): sedimentation coefficient, approximate no. of nucleotides, (M_r)	23*s* 3200 1.1 × 10^6 5*s* 120 4.0 × 10^4	16*s* 1600 0.55 × 10^6
No. of individual protein species	34	21

Polyribosomes, formed by the association of several functional ribosomes with a single mRNA molecule during translation, represent an efficient and economical method of using mRNA molecules which, in the case of prokaryotes are unstable and constitute only 1–2% of the total cellular RNA. In the 70*s* ribosome the 30*s* subunit appears to rest on the 'two arms' of the 50*s* subunit (Fig. 6.5), the resultant tunnel region between the two subunits apparently affords protection from the action of RNAases in the case of (a region of) mRNA and from proteolysis in the case of the nascent polypeptide. The two subunits are superficially, functionally, as well as structurally, differentiated; the catalytic activities involved in protein synthesis appear to be confined to the 50*s* subunit whereas the 30*s* subunit appears to be mainly concerned with the specific binding of substrates during the course of protein synthesis. However, it is becoming increasingly apparent that some, if not all, ribosomal functions *in vivo* may be determined by 'functional domains' in which parts of several different ribosomal components contribute to a

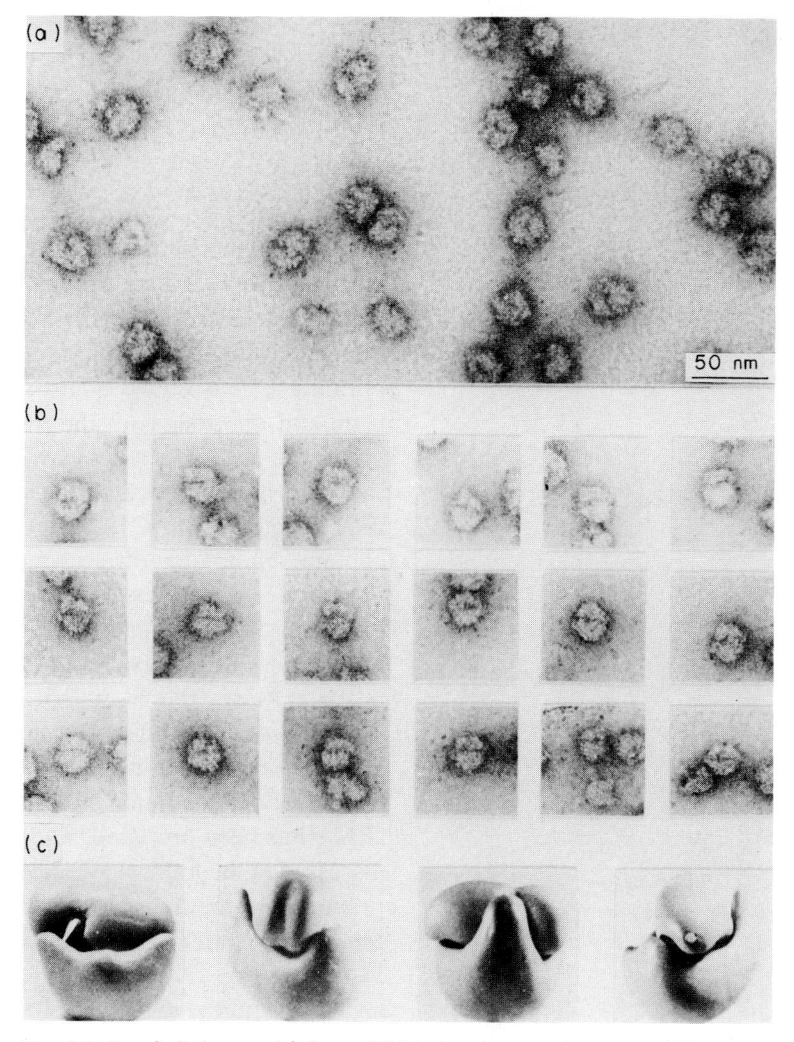

Fig. 6.5. *E. coli* ribosomes. (a) General field of an electron micrograph of 70*s* ribosomes. (b) Selected micrographs of particles. (c) A model of a ribosome successively rotated through 90° (G. Tischendorf, B. Tesche & G. Stöffler unpublished data). (We are very grateful to Dr Stöffler for supplying this photograph.)

particular functional site. Moreover, it is likely that many of the functional steps in protein synthesis require the cooperative activity of both 30*s* and 50*s* components.

The RNA and protein components and their interaction

Ribosomal RNA accounts for approximately 66% of the dry weight of both subunits, a fact which indicates its important structural role; in addition, evidence for a functional role for each of the three rRNA species has recently emerged (see section 6.5). *E. coli* 5*s* rRNA, which was the first RNA molecule to be sequenced by Sanger's group, contains no modified bases and can be divided into two halves with a high degree of sequence homology indicating an evolutionary past involving gene duplication and joining. The correspondence between 5*s* RNA and several tRNA sequences has led to a proposal for the evolution of tRNA and 5*s* RNA from a common ancestral gene. The 5′ and 3′ termini of 5*s* RNA can readily interact through base pairing and various structural models incorporating extensive secondary structure have been proposed. The complete sequences of 16*s* rRNA and 23*s* rRNA are published and these are further discussed on p. 247. There are no extensive sequence homologies between the two major rRNA species, i.e. 23*s* rRNA is not a 16*s* rRNA dimer. Both 16*s* and 23*s* rRNA molecules contain minor quantities of modified bases including methylated nucleosides and pseudouridine. These are distributed along the molecules in a non-random manner and may influence the properties of the molecule; for example, the presence of a methylated sequence $m^6_2Am^6_2A$ near the 3′-end of the 16*s* rRNA is responsible for sensitivity to the antibiotic kasugamycin. Physical studies of 16*s* and 23*s* rRNA molecules indicate that 60–70% of the nucleotides are base paired in double-stranded RNA and a secondary structure map of 16*s* rRNA has been proposed. The genetics and synthesis of rRNA are discussed in the following section.

Numerous methods have been developed for the isolation, purification and characterization of *E. coli* ribosomal proteins (Stöffler & Wittmann 1977). Comparison of electrophoretic, immunological, chemical and physical data obtained by several different laboratories has led to a common nomenclature S1–S21 for the 30*s* r proteins and L1–L34 for the proteins from the 50*s* subunit. Proteins S20 and L26 are indistinguishable immunologically and proteins L7 and L12 differ only in the presence of an N-terminal acetyl group in L7. With the exception of these two pairs, the remaining proteins are immunologically, physically and chemically distinct. The sequences of more than half of the 54 different *E. coli* r proteins have been determined and secondary structures have been proposed for many of these. The r proteins are rich in basic amino acids with a low content of aromatic and sulphur-containing residues; 60% of the proteins have isoelectric points of pH 10 or above; most are insoluble in dilute buffers at neutral pH and some have a strong tendency to aggregate. There is one copy of each ribosomal protein (including S20 = L26) per 70*s* ribosome and earlier reports of 'fractional' r proteins were due to artefactual loss during purification. The only exception is L7/L12 which is present in 2–3 copies per ribosome.

The roles of protein–protein and RNA–protein interactions in the maintenance of ribosome structure were first revealed by unfolding and disassembly experiments. Unfolding is accomplished by removal of Mg^{2+} ions from subunits and proceeds through discrete stages to produce less compact structures with the same RNA:protein ratio as the native subunits (Fig. 6.6). Disassembly results from treatment of the subunits with high concentrations of monovalent cations. In this case, r proteins are removed in discrete groups ('split proteins') to yield a series of protein-deficient 'core particles' and, eventually, protein-free rRNA (Fig. 6.7). Reassembly of split proteins and core particles yields active subunits without the formation of any new intermediates, i.e. the split proteins are reattached by a reversal of the process that detached them.

$$50s \underset{+Mg^{2+}}{\overset{-Mg^{2+}}{\rightleftharpoons}} 36s \underset{+Mg^{2+}}{\overset{-Mg^{2+}}{\rightleftharpoons}} 21s$$

$$30s \underset{+Mg^{2+}}{\overset{-Mg^{2+}}{\rightleftharpoons}} 26s \underset{+Mg^{2+}}{\overset{-Mg^{2+}}{\rightleftharpoons}} 16s$$

Fig. 6.6. Unfolding of *E. coli* ribosomal subunits.

increasing CsCl or LiCl concentration →

50*s* →	43*s* →	28*s* →	23*s* rRNA
↓ 20% total protein	↓ 30% protein	↓ remaining 50% protein	

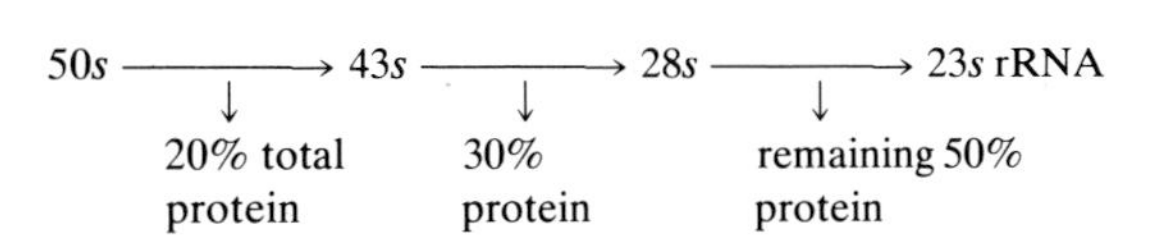

Fig. 6.7. Disassembly of 50*s* subunits.

Functionally active subunits can also be reconstituted *in vitro* from separately purified rRNA and r proteins (Fig. 6.8). Assembly occurs via the stepwise addition of specific groups of protein to form reconstitution intermediates (RIs) which are deficient in several protein components. The rate-limiting, temperature-dependent steps are thought to involve structural rearrangement of the respective, functionally inactive, RI particles. With the establishment of conditions for total, as well as partial, reconstitution of subunits

(a) *30s subunits* (Traub & Nomura 1969)

16*s* rRNA + 15 proteins → RI particles (21*s*)

$$\text{RI particles} \xrightarrow{40°C} \text{RI* particles}$$

RI*particles + 6 remaining proteins → 30*s* particles.

(b) *50s subunits* (Dohme & Nierhaus 1976)

$$23s\ \text{rRNA} + 5s\ \text{rRNA} + \text{proteins} \xrightarrow[4\ \text{mM}\ Mg^{2+}]{0°C} RI_{50(1)} \quad (33s)$$

$$RI_{50(1)} \xrightarrow[4\ \text{mM}\ Mg^{2+}]{44°C} RI^*_{50}$$

$$RI^*_{50} + 5s\ \text{rRNA} + \text{proteins} \xrightarrow[4\ \text{mM}\ Mg^{2+}]{44°C} RI_{50(2)} \quad (48s)$$

$$RI_{50(2)} \xrightarrow[20\ \text{mM}\ Mg^{2+}]{50°C} 50s\ \text{particles}$$

Fig. 6.8. Proposed schemes for assembly of ribosomal subunits *in vitro*.

further experiments for the study of ribosome structure and function became feasible. The number of proteins binding directly and specifically to 16*s*, 23*s* and 5*s* rRNA is estimated to be seven, ten and three respectively. Thus rRNA–r protein interactions are highly dependent on the method of preparation of both the RNA and the proteins and the above numbers may be underestimates. Protein S11 (and possibly also S12) binds specifically to 23*s* rRNA indicating that S11 may play a role in the stabilization of the 70*s* ribosome. It was initially hoped that reconstitution experiments omitting one protein at a time would enable identification of the structural and/or functional roles of individual proteins. Although these hopes were largely unrealized, the results of reconstitution experiments did lead to the construction of an assembly map showing the beneficial effect of the binding of certain proteins to RNA on the binding of other proteins, and also facilitated the identification of altered r proteins in mutants changed in their antibiotic sensitivity together with deduction of the functional roles of certain individual proteins. However, as the importance of cooperative protein–protein, RNA–protein and RNA–RNA interactions in ribosome structure and function became apparent, new techniques for the study of ribosome topology were developed. One group of techniques shows examples of affinity labelling (Pellegrini &

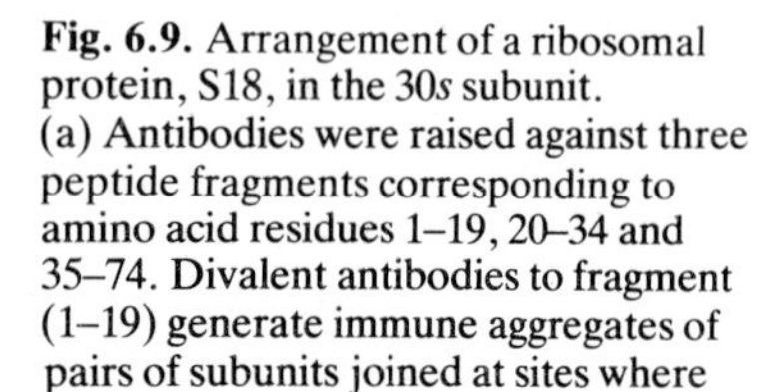

Fig. 6.9. Arrangement of a ribosomal protein, S18, in the 30*s* subunit. (a) Antibodies were raised against three peptide fragments corresponding to amino acid residues 1–19, 20–34 and 35–74. Divalent antibodies to fragment (1–19) generate immune aggregates of pairs of subunits joined at sites where this fragment is exposed. The top four pictures are electron micrographs of such aggregates. Antibodies to fragment (20–34) produced no such aggregates so this region of S18 must be buried in the subunit. The lower eight pictures are of aggregates produced by antibodies to fragment (35–74). (b) An interpretation of the data showing the accessible antigenic sites for S18 on the surface of a model of the subunit (based on unpublished data by G. Stöffler, K.H. Rak, G. Tischendorf & M. Yaguchi). (We are very grateful to Dr Stöffler for supplying this photograph.)

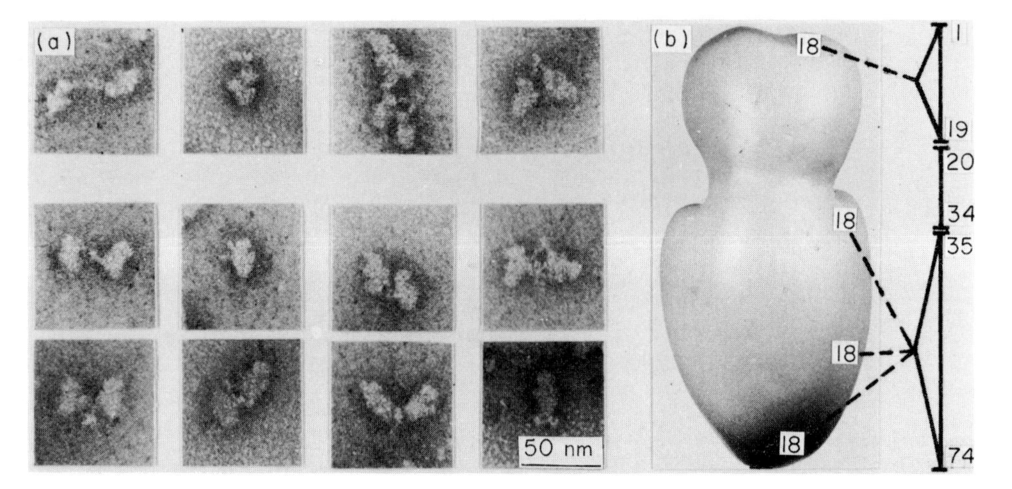

Cantor 1977). Examples of such labels are radioactive derivatives of tRNA and antibiotics containing highly reactive substituents capable of covalent attachment to ribosomes. Following incubation with an appropriate label, the ribosomes are dissociated to identify the component(s) to which the label was bound. Other techniques for the study of ribosome topology include the use of bifunctional cross-linking agents, neutron scattering and immune electron microscopy. The last technique has indicated that many r proteins are not globular but have possibly highly extended conformations which may, in some cases, stretch the whole length of a ribosomal subunit (Fig. 6.9); further studies using monoclonal antibodies are necessary to confirm these proposals. Although there is, as yet, no detailed model for the topology of the ribosomal subunits, all the evidence points to an extraordinary degree of structural and conformational cooperation between the ribosomal components—a feature which is also seen in the assembly and disassembly of ribosomes *in vitro* (this section) and *in vivo* (see section 6.3.3).

Finally, with regard to the functional role of ribosomes, mention must be made of the two different sites for the binding of tRNA to the larger subunit. These two sites are referred to as A (for aminoacyl or acceptor) and P (polypeptidyl); the P site is occasionally referred to as D (for donor). The A and P sites are essentially operational terms defined by the ability of bound aminoacyl or peptidyl tRNA to react with puromycin, an antibiotic which mimics the 3′-end of aa-tRNA (Fig. 6.10). If peptidyl (or initiator) tRNA is in the P site, peptidyl transferase catalyses the formation of a peptide bond between the activated carboxyl group of peptidyl-tRNA and the amino group of

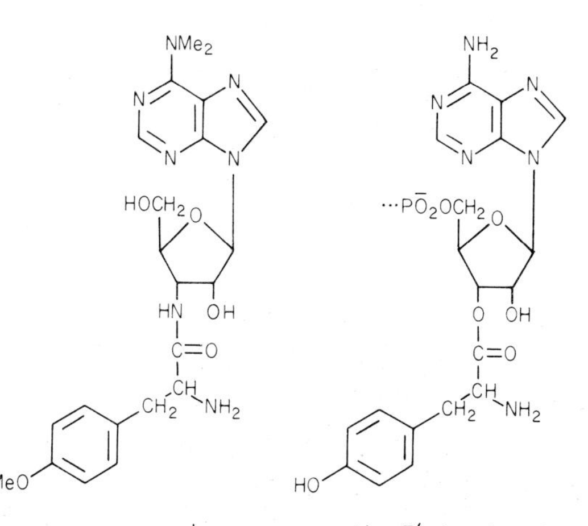

Fig. 6.10. Puromycin—an analogue of aminoacyl-tRNA.

puromycin. However, puromycin lacks both an activated carboxylic acid ester group and the tRNA structure, so no further peptide bond formation occurs and the resultant peptidyl-(or fMet-) puromycin dissociates from the ribosome leading to premature chain termination. Thus if charged tRNA reacts with puromycin it is bound to the P site of the ribosome; conversely, lack of reaction with puromycin indicates binding to the A site. The two sites could be structurally distinct with the A site composed of one set of ribosomal components and the P site of a constantly different set. Alternatively, the two sites could be equivalent so that following translocation (see section 6.5.2) a former A site becomes a new P site and vice versa. It is certain that the structural elements of both sites must be partially overlapping so that the bound substrates can react at the same peptidyl transferase centre to enable peptide bond formation between them. Moreover, each site must contain components of 30*s* and 50*s* subunits.

6.3.2 Genetics

Although rRNA can account for up to 80% of total cellular RNA, the results of DNA:RNA hybridization experiments indicate that the number of gene copies for each of the three rRNA species is seven. This means that rRNA genes comprise only 0.6–0.8% of the bacterial genome and the promotors for these genes (see Chapter 5) must, under optimal growth conditions, enable transcriptional initiation at a frequency an order of magnitude greater than that at a fully derepressed (mRNA-producing) structural gene. The genes for the three rRNA species are part of a transcriptional unit from which precursor rRNAs are produced by an RNAase before transcription is finished (see section 6.3.3). The precursors of 16*s* and 23*s* rRNA (p16*s* and p23*s* rRNA) are respectively 10% and 5% longer than the mature molecules and are submethylated with respect to the latter; p5*s* rRNA is usually identified as a molecule with 3 additional nucleotides at the 5′-end of the mature molecule. Protein-free rRNA is not normally detected *in vivo* and it is assumed that r proteins become attached to the precursor rRNA molecules before their transcription is complete (see section 6.3.3).

Most of the r protein genes of *E. coli* have been mapped and appear to be organized into at least 15 transcription units. Many of the genes are located in two clusters at 72′ and 79′ on the chromosome map. The 28 r protein genes mapping close to the 72′ region appear to be organized into five or more transcription units, the largest of which, the *S10* operon, contains the genes for four 30*s* r proteins intermixed with seven 50*s* r protein genes. The *spc* operon comprises seven 50*s* r protein genes and three 30*s* r protein genes; the *str* operon contains the genes for the elongation factors EF-Tu and EF-G and two 30*s* r proteins whilst a fourth transcription unit, the α operon, from the same gene cluster region, contains genes for one 50*s* and four 30*s* r proteins together with a gene for the α subunit of RNA polymerase (Fig. 6.11).

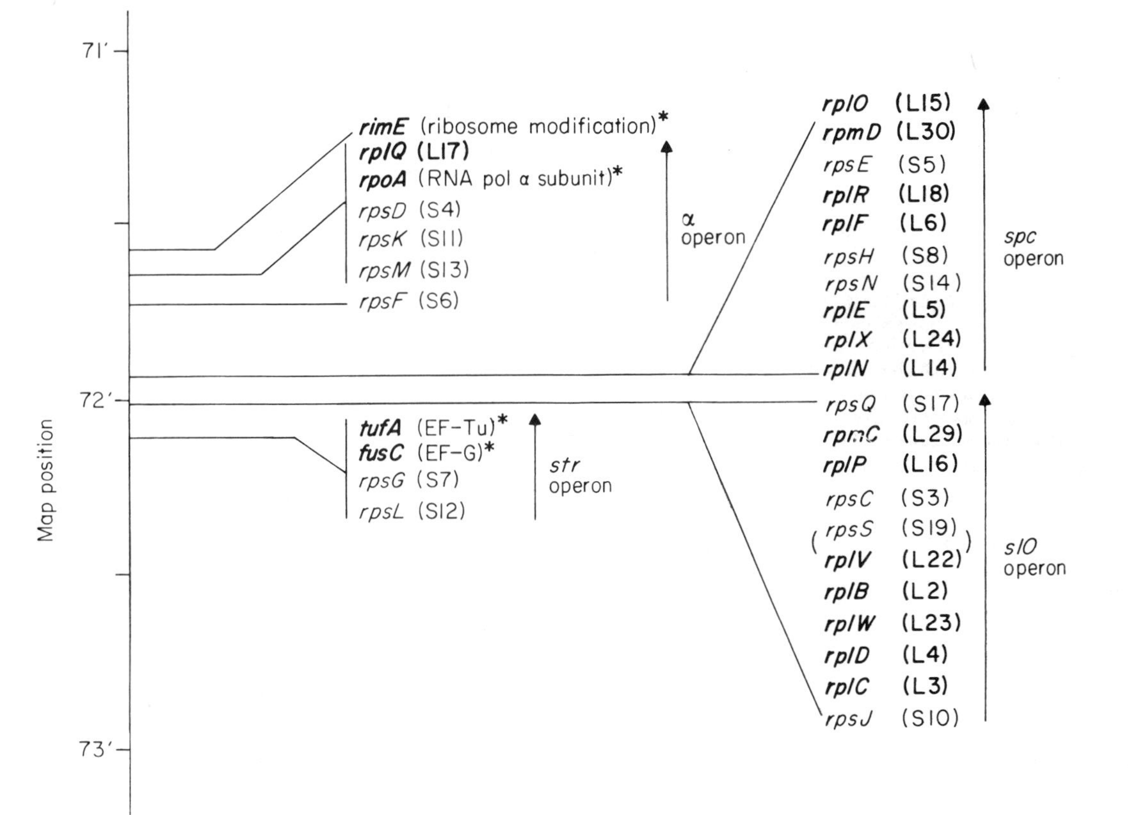

Fig. 6.11. Ribosomal gene clusters of *E. coli*. Vertical arrows indicate transcriptional units. Genes for 30*s* proteins are in normal type; those for 50*s* proteins and non-ribosomal proteins (*) are in heavy type. (Based on data from Champney & Kushner 1976 and Yates & Nomura 1980.) The order of L22 and S19 is not yet determined.

In the 79′ region, at least five 50*s* r protein genes, a second EF-Tu gene, genes for the β and β' subunits for RNA polymerase and an rRNA transcriptional unit are clustered. Moreover, genes for L7/12, L10 and the polymerase subunits are co-transcribed (the β operon). Thus genes for all the subunits of core RNA polymerase and genes for at least some of the factors required for protein synthesis are co-transcribed with ribosomal genes. Clearly, coordinate control of gene expression is facilitated by grouping genes of related functions into transcriptional units. However, not all the r protein genes are clustered; for example, genes for S2 and S18 map at 4′ and 94′

respectively. Indeed other control mechanisms are implied by studies on the regulation of RNA polymerase subunit gene expression and by studies of levels of mRNA for ribosomal proteins in cells containing relatively large numbers of r protein genes (see the following section).

6.3.3 Biogenesis and possible control mechanisms

Studies of the kinetics of ribosome assembly *in vivo* from labelled protein and RNA have revealed the following scheme for the processing of precursor particles:

$$P_1\ 30s\ (21s) \rightarrow P_2\ 30s\ (26s) \rightarrow 30s$$
$$P_1\ 50s\ (32s) \rightarrow P_2\ 50s\ (43s) \rightarrow 50s\ .$$

The designation P_1 30*s* refers to the first precursor of 30*s* subunits and the value in parenthesis (21*s*) is the sedimentation coefficient of that precursor. These precursors represent detectable kinetic hold-up points in subunit biogenesis. They contain precursor rRNA, indicating that rRNA maturation occurs at a late stage of ribosome biogenesis. As no protein-free RNA is normally detected *in vivo*, the first group of proteins becomes attached to a precursor rRNA during its transcription. The assembly process *in vivo* is related to assembly *in vitro* because the compositions of P_1 30*s* and P_2 30*s* correlate with those of RI particles.

Further studies on the assembly process have been facilitated by the isolation of mutants. With the discovery that the rate of assembly of subunits *in vitro* is temperature dependent, a survey was made of cold-sensitive mutants of *E. coli* with a view to obtaining organisms that could not assemble mature ribosomes at low temperatures. The genetics of such mutants are reviewed by Smith (1977). One group of such mutants are designated Sad (subunit assembly defective); these fall into three classes. At the non-permissive temperature (20 °C), mutants in classes (1) and (2) fail to make 50*s* subunits and accumulate P_1 50*s* (32*s*) and P_2 50*s* (43*s*) respectively. Class (3) mutants fail to make either 50*s* or 30*s* subunits and accumulate P_1 30*s* (21*s*) and P_1 50*s* (32*s*). From the finding that class (3) mutants each carry single mutations in genes for 30*s* r proteins, it follows that assembly of 50*s* subunits *in vivo* is dependent upon the simultaneous assembly of 30*s* subunits.

A further group of cold-sensitive mutants of *E. coli* which have defects in the maturation of 50*s* ribosomal subunits have been termed Rim (ribosome maturation). In contrast to the Sad mutants, Rim mutants map outside the known positions of r protein genes and no alterations in any of the r proteins have been demonstrated in these mutants. It has been proposed that Rim mutants are defective in ribosome maturation accessory factors rather than in ribosomal structural proteins. Such factors could be methylases or nucleases involved in rRNA maturation or enzymes responsible for the methylation or acetylation of r proteins.

Another approach to the study of ribosome biogenesis is the investigation of ribosome assembly under conditions of unbalanced growth. In the presence of metabolic inhibitors such as puromycin, streptomycin and chloramphenicol, protein synthesis is rapidly inhibited with little or no effect on RNA synthesis. Under these conditions ribonucleoprotein particles, which contain precursor rRNA molecules, and some, but not all, ribosomal proteins accumulate. Although the relationship of these particles to bona fide ribosome precursor particles remains to be established, their accumulation again points to an ordered stepwise addition of specific groups of proteins to precursor rRNAs *in vivo* (Sykes *et al.* 1977).

Studies of ribosome assembly under conditions of amino acid starvation led to the identification of the RNA control or *relA* locus in *E. coli*. This can exist in two allelic states—stringent *(relA$^+$)* or relaxed *(relA)*. In an amino acid auxotroph carrying the *relA$^+$* allele, removal of the required amino acid triggers a pleiotropic effect termed the stringent response. Synthesis of certain RNAs ceases (see section 5.2.2) and other effects include cessation of synthesis of specific proteins, decreased synthesis of phospholipids, altered pattern of mRNA synthesis, decreased rate of nucleoside transport and increased protein turnover. In contrast, in an amino acid auxotroph carrying the *relA* allele (relaxed strain) protein synthesis stops upon removal of the required amino acid but RNA synthesis continues at a normal, or near normal, rate for a considerable period of time and the other effects of the stringent response noted above are not seen. The continuation of RNA synthesis in the absence of protein synthesis results in the accumulation of protein-deficient 'relaxed particles', which are capable of being processed to form mature ribosomal subunits if the medium is supplemented with the required amino acid.

During amino acid starvation of stringent, but not relaxed, strains of *E. coli* two unusual nucleotides accumulate in large amounts. These were initially termed 'magic spot' nucleotides MS I and II and were subsequently identified as guanosine tetraphosphate (ppGpp) and guanosine pentaphosphate (pppGpp) respectively. The addition

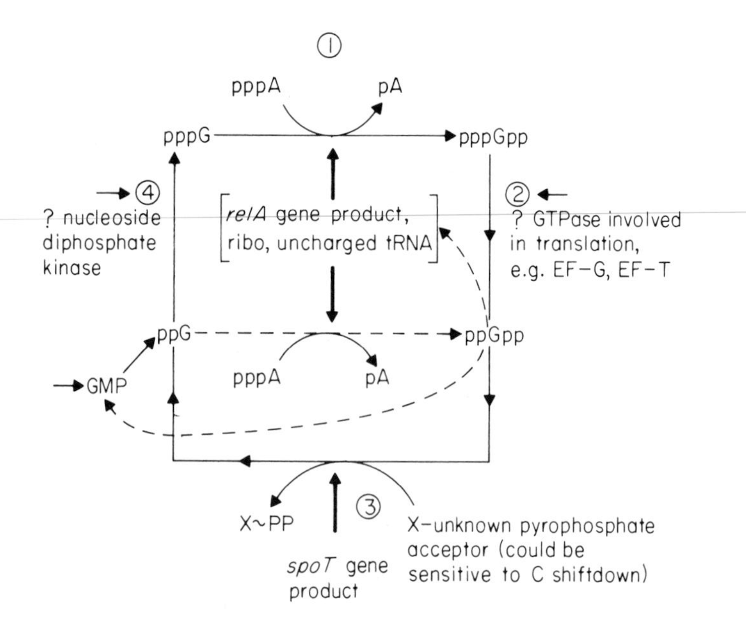

Fig. 6.12. The ppGpp cycle in *E. coli*. The broken line represents the possible feedback of ppGpp on its own synthesis. Step (1) The *relA* gene product has been identified as the stringent factor, a ribosome-bound pyrophosphotransferase which promotes the binding of uncharged tRNA to the A site of the ribosome and catalyses the formation of pppGpp from GTP and ATP in response to this binding. Stringent factors may also catalyse the direct production of ppGpp from GDP and ATP (both reactions can be demonstrated *in vitro*), however, this is likely to be a minor pathway for ppGpp synthesis *in vivo* as the intracellular concentration of GTP is usually 50-fold higher than that of GDP. Step (2) pppGpp is rapidly converted to ppGpp by an, as yet, unknown mechanism, presumably any GTPase could be responsible, including one of those involved in translation. Step (3) The breakdown of ppGpp is under the control of the *spoT* gene product and may not be a single-step reaction; no inorganic pyrophosphate accumulates at this step but the nature of the pyrophosphate acceptor X is unknown. The availability of such an acceptor would control the rate of ppGpp catabolism and if X were a glycolytic intermediate it would explain the accumulation of ppGpp in both stringent and relaxed strains during a shift from a rich to a poor C source. Step (4) Nucleoside diphosphate kinase is the only known bacterial enzyme which non-specifically converts nucleoside diphosphates to nucleoside triphosphates, although its involvement in the ppGpp cycle is not proven. (Based on data from Fiil *et al.* 1977 and Kari *et al.* 1977.)

of the deprived amino acid results in a rapid decay of these two nucleotides prior to the resumption of RNA and protein synthesis.

It is ppGpp that is the pleiotropic effector of the stringent response. The synthesis of ppGpp during amino acid starvation requires uncharged tRNA, ribosomes and a pool of mRNA, and forms part of a self-regulating cycle (Fig. 6.12). This cycle is over-simplified because it does not take into account three other Rel loci:

relC codes for r protein L11 implying that this protein may determine in part the affinity of the ribosomes for stringent factor; *relB* mutants have a delayed relaxed response but the RelB protein is not known; *relX* is a locus apparently involved in the basal synthesis of ppGpp during normal growth.

It is anticipated that further details of the ppGpp cycle will soon emerge together with an understanding of the mechanisms enabling ppGpp to control a wide variety of metabolic systems. The reduction of guanosine nucleotide pools as a result of ppGpp accumulation might be expected to exert a general effect on nucleic acid metabolism. Moreover, as discussed in section 5.2.2 ppGpp acts, at least *in vitro*, as an allosteric effector of RNA polymerase resulting in alteration of the initiation specificity of this enzyme. The effect of ppGpp on the transcription of structural genes varies according to the gene product and the net result of the various effects is to alter cellular metabolism so as to increase amino acid pools under conditions of amino acid starvation. There is little point in synthesizing ribosomes under conditions of limited protein synthesis, therefore RNA polymerase is 'directed away' from rRNA genes and transcription of at least some of the r protein gene units (see section 6.3.2) is inhibited. Also ppGpp effects a decrease in tRNA methyltransferase (the enzyme that introduces T-residues into tRNA molecules) and inhibits enzymes of lipid and nucleotide biosynthesis. In contrast, ppGpp promotes protein degradation and also amino acid biosynthesis. This last is achieved by increased expression of the His and Trp operons and of the phosphoenol pyruvate carboxykinase gene (thus increasing the amount of oxaloacetate available for transamination). (See the references cited in Fig. 6.12. and Nierlich (1978) for an introduction to the ppGpp literature.)

In summary, ppGpp is the central controlling element in the stringent response and may play a role in ensuring that the number of ribosomes per genome is proportional to growth rate. Also by the clustering of genes for rRNA and r protein into transcriptional units under the control of ppGpp, *E. coli* ensures that r proteins and rRNA are produced together in stoichiometric amounts. However, there is probably more to maintenance of the proper ratio of these macromolecules. There is no evidence that all ribosomal genes (especially the non-clustered genes for r proteins) are under the control of ppGpp. Moreover, although RNA polymerase subunits, β and β', and r proteins, L7/L12 and L10, are co-transcribed, the regulation of their synthesis is different, a suggestion that expression of genes in these clusters is modulated, in part, by translated factors.

Other devices used by the cell to ensure the proper balance between rRNA and r protein undoubtedly include the sequential addition of r proteins to nascent rRNA and the regulation of 50*s* assembly by 30*s* proteins revealed by the class (3) Sad mutants. These two factors could be related: the gene order in an *rrn* operon (Fig. 5.16) means that 30*s* r proteins bind before their 50*s* counterparts and the nascent 30*s* particle might influence the conformation of 50*s* precursors. Furthermore, the Rim mutants indicate a role for, as yet, unidentified factors for which putative regulatory roles can be designed.

More recently evidence has emerged for the regulation of some r protein operons by feedback inhibition at the level of translation (Yates & Nomura 1980; Yates *et al.* 1980; Zengel *et al.* 1980). Certain key regulatory proteins appear to inhibit the synthesis of specific r proteins whose genes are in the same operon as their own—so-called autogenous regulation. S4, for example, specifically inhibits the synthesis of r proteins S4, S11 and S13 (all components of the α operon); L1 inhibits the synthesis of L1 and L11 (the two protein products of the L11 operon); S8 inhibits the synthesis of L5, L24, S14, and S8 of the *spc* operon, and L4 inhibits the synthesis of S10, L3, L4, L23 and, possibly, L2 of the S10 operon (Fig. 6.11). Although the precise mechanism of these specific inhibitions is as yet unknown, the effect appears to take place at the level of translation rather than transcription. The four regulatory proteins identified so far are all proteins which bind directly to RNA during ribosome assembly. Thus a decrease in rRNA synthesis would result in a rapid accumulation of these proteins which, in view of the small pool size of free (unbound) r proteins in exponentially growing *E. coli* cells, would lead to a rapid decrease (by translational repression) in the synthesis of the aforementioned r proteins. Likewise, any increase in rRNA synthesis would lead to a rapid depletion of free r proteins with concomitant increase in r protein synthesis. Such regulatory r proteins may, therefore, not only regulate the expression of certain r protein operons but may also play a role in the regulation of the coordination of r protein and rRNA synthesis. It remains to be seen whether all r protein operons contain one (or more) regulatory proteins.

6.4 AMINO ACID ACTIVATION

Whereas the specificity of each of the three major steps in translation (see the following section) resides in codon–anticodon interactions, the enzymatic aminoacylation of tRNA molecules to form the activated precursors for protein synthesis represents an amino acid specific step. The enzymes responsible for charging tRNA with the correct amino acid (and simultaneously activating the amino acid by virtue of the resultant carboxylic ester bond) are termed aminoacyl tRNA synthetases. The aminoacylation reaction is a two step process. In the first step an aminoacyl adenylate enzyme complex and pyrophosphate are formed by reaction of the synthetase with the appropriate amino acid and ATP. The second step involves transfer of the aminoacyl group to the terminal A of an appropriate tRNA molecule (Fig. 6.13). Whether transfer occurs initially to the 2′ or 3′-OH of the ribose of the terminal A appears to be a function of the synthetase concerned; there is probably rapid and reversible transfer between these two positions. However, the aminoacyl group must apparently be esterified to the 3′-OH for aa-tRNA to fulfil its role in translation. The sum of the two reaction steps is:

$$\text{ATP} + \text{amino acid} + \text{tRNA} \rightleftharpoons \text{aa-tRNA} + \text{AMP} + \text{PP}_i.$$

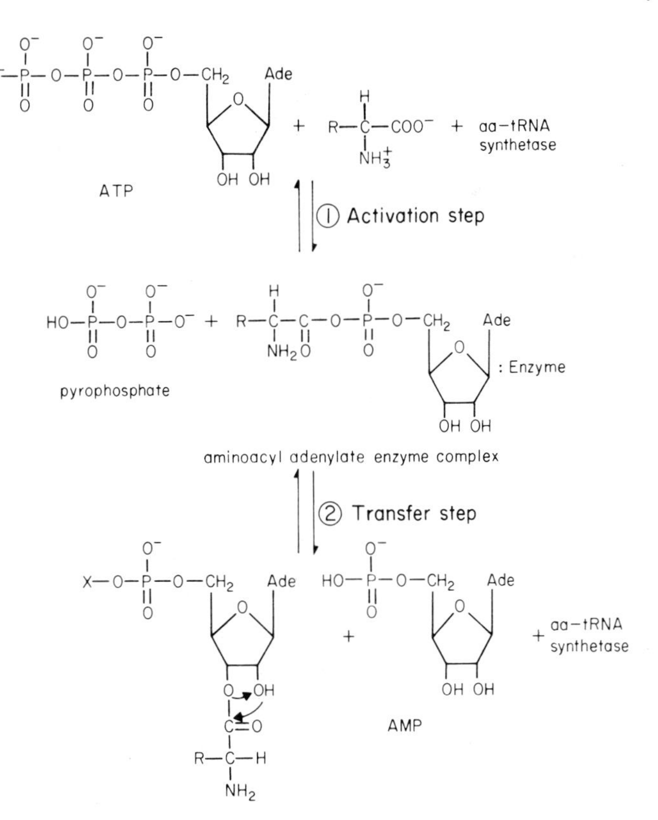

Fig. 6.13. Amino acid activation. R, amino acid side chain; X, remainder of tRNA molecule. For an explanation of the reversible transfer of the aminoacyl group in aa-tRNA see text.

Thus, activation of each amino acid residue consumes two high-energy bonds; hydrolysis of pyrophosphate drives the reaction to the right and the net reaction is analogous to the activation of fatty acids prior to β oxidation. In the latter case, the first step produces fatty acyl adenylate and the second step involves transfer of the fatty acyl group to the SH group of CoA with the formation of a thioester (fatty acyl CoA) instead of the ester (aminoacyl tRNA).

In prokaryotes there is only one synthetase for each of the 20 amino acids specified by the genetic code, regardless of the number of codons and/or cognate tRNA molecules for that amino acid. All aa-tRNA synthetases catalyse the same type of reaction with the same type of substrates but they differ greatly in size (M_r 50–200×10^3) and subunit structure (monomers, dimers, tetramers with identical or heterologous subunits). Some of the larger, single-subunit enzymes may be composed of two similar covalently linked peptides and may have arisen by gene duplication and fusion. The aa-tRNA synthetases are remarkedly specific, they must recognize not only the correct amino acid from the cellular amino acid pool(s) containing amino acids with very similar structures (e.g. Ile and Val; Phe and Tyr) but also the appropriate tRNA molecule(s) from the whole range of tRNA species, which are all of the same size and conformation. How activating enzymes recognize their cognate tRNA is not known and is the subject of intensive research; in some cases the recognition involves the anticodon sequence (e.g. *su*7 product, see section 6.2.4), whereas in other cases mutational change in the anticodon sequence does not appear to affect recognition (Schimmel 1979). It is likely that evolution of the various aa-tRNA synthetases has led to different classes of tRNA recognition features. Similarly, the formylase responsible for the reaction:

$$\text{Met-tRNA}^{\text{fMet}} + \text{formyl-FH}_4 \rightarrow \text{formyl-Met-tRNA}^{\text{fMet}} + \text{FH}_4,$$

shows a similar degree of specificity since it will not formylate Met-$\text{tRNA}^{\text{fMet}}$ or any other aa-tRNA species.

The process of amino acid activation is strikingly error-free. It has been estimated that the maximum frequencing of replacement of a correct amino acid (as encoded in the mRNA base sequence) by a sterically similar but incorrect amino acid is less than 1 in 10^4. Although an isoleucyl-tRNA synthetase will form valyl-adenylate (with a lower affinity of binding than the correct isoleucyl adenylate), fidelity of translation is maintained in the transfer reaction as the isoleucyl tRNA synthetase enzyme will not form valyl-tRNA^{Ile}. Moreover, some synthetases are able to deacylate tRNA charged with the wrong amino acid and, although a physiological role for deacylation reactions has been disputed, recent evidence indicates that certain synthetases possess a hydrolytic site which is distinct and separate from the activation site. The double sieve hypothesis (Fersht 1980) proposes that the activation site discriminates between amino acids on the basis of size (by steric hindrance) and that the hydrolytic site excludes the correct substrate but accepts amino acids sufficiently similar in size and chemical characteristics to the latter to be activated by the enzyme. Experimental evidence for the kinetic proof-reading of the tRNA and the amino acid by aa-tRNA synthetase has been presented (Yamane & Hopfield 1977). Although any error correction after step (1) of the aminoacylation reaction has occurred is energy expensive, fidelity of aa-tRNA synthesis is a prerequisite for translational fidelity and must be essential for the survival of the cell.

6.5 BACTERIAL PROTEIN SYNTHESIS

For a general review of the stages in protein synthesis see Grunberg-Manago *et al.* (1978).

6.5.1 Initiation

The initiation of protein synthesis has been reviewed by Hunt (1980) and Bosch and van der Hofstad (1979).

The end-product of the initiation step of protein synthesis is an initiation complex in which both ribosomal subunits are bound to an initiation signal of mRNA, charged initiator tRNA—with its anticodon paired with the initiation codon—is positioned in the ribosomal P site and the vacant A site is ready to receive the aa-tRNA specified by the next codon on the 3′ side of the initiation codon. Initiation is the most critical, and probably the rate-limiting, step in protein synthesis, it is then that the correct reading phase of the message must be defined, thereafter the elongation cycle (see section 6.5.2) operates, and it is not surprising that modulation of the efficiency of initiation is the basis of the majority of proposed translational control mechanisms.

Initiation factors

In addition to mRNA carrying an initiation signal, charged initiator tRNA and ribosomal subunits, initiation requires a supply of

GTP and the cooperative interaction of various initiation factors (IFs). Three types of initiation factor have been isolated and characterized from *E. coli.* They are all monomeric proteins which are distinct from any of the 30*s* r protein species but can be separated from native 30*s* subunits by centrifugation of the latter in high-salt buffer. The three factors are essential for translation of native mRNAs *in vitro* and it is now known that early studies employing synthetic polymers to direct polypeptide synthesis *in vitro* only succeeded in the absence of these factors because of the fortuitous use of high Mg^{2+} concentrations. If these experiments had been performed under physiological ionic conditions, the elucidation of the genetic code may well have been delayed by several years.

IF1 is a small basic protein (estimates range from M_r 8.5–9.5×10³) which appears to assist the remaining two types of initiation factor. Thus IF1 stabilizes the binding of IF2 to the 30*s* initiation complex and is required for IF2 release from the 70*s* initiation complex. IF1 also enhances the rate of dissociation of run-off 70*s* ribosomes into their component subunits, thereby cooperating with IF3 in supplying 30*s* 'initiation' subunits from the 70*s* ribosome pool. The fact that a factor analogous to IF1 has not been isolated from certain bacterial species emphasizes its role as a helper as opposed to a factor with an indispensable functional role.

IF2 exists in two forms, M_r 118×10³ and M_r 90×10³, although the smaller form may be derived from the larger form by proteolytic cleavage and individual roles for the two have not been described. IF2 recognizes and promotes the ribosomal binding of charged initiator tRNA and exhibits a ribosome-dependent GTPase activity.

IF3 (M_r 21–23×10³) binds to 30*s* subunits with a high affinity to form a complex (30*s*:IF3) which cannot reassociate with 50*s* subunits. Thus, in association with IF1, IF3 displaces the ribosome dissociation equilibrium:

$$\begin{array}{c} \text{IF1} \\ 70s \rightleftharpoons 50s + 30s \\ 30s + \text{IF3} \rightleftharpoons 30s\text{:IF3} \\ \hline 70s + \text{IF3(IF1)} \rightleftharpoons 50s + 30s\text{:IF3} \end{array}$$

IF3 is sometimes termed the dissociation factor.

IF3 also stabilizes the binding of mRNA to 30*s* subunits; the 30*s* subunits to which it is bound are earmarked for initiation. A functional role for IF3 in mRNA selection was proposed following the identification of two forms of IF3 from *E. coli*; the largest form, M_r 23×10³, was termed IF3α and was reported to be selective for

MS2 RNA as opposed to late T4 mRNA, whereas IF3β, M_r 21×10^3, exhibited the opposite specificity. However, IF3β may come from proteolytic cleavage of IF3α. Nevertheless, the appearance of a new initiation factor has been reported following T4 infection and during the switch from the early to late phase of T7 infection and a role for IF3 in translational control cannot be discounted (Revel 1977).

The steps leading to the formation of the initiation complex are shown in Fig. 6.14—note that the true shape of the 70*s* ribosome is not really that of a 'cottage loaf' (see Fig. 6.5) and that 30–40 nucleotides of mRNA, not 6 as indicated, are protected from RNAase action by ribosome binding.

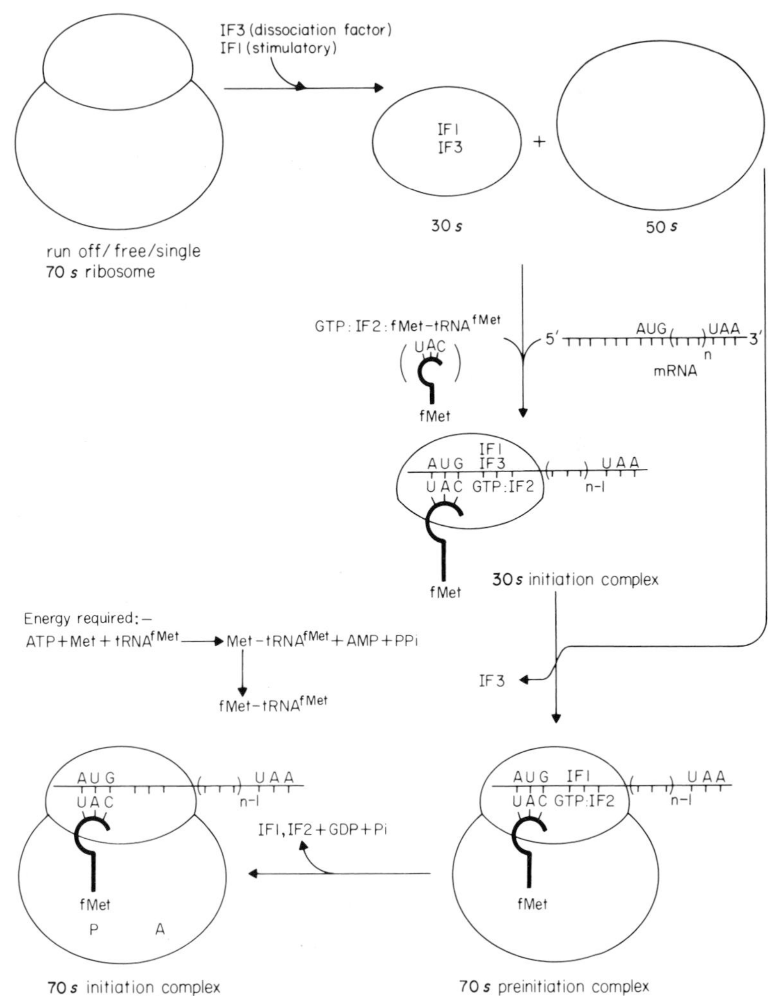

Fig. 6.14. Bacterial protein synthesis. Stage I, initiation.

The combined action of IF3 and IF1 results in a shift of equilibrium away from 70*s* ribosomes towards free subunits. The binding of mRNA and fMet-tRNA^{fMet}, under the influence of IF3 and IF2 respectively, to a 30*s*–IF3 complex, results in the formation of a 30*s* initiation complex. The order of binding of mRNA and initiator tRNA is controversial but may not be important *in vivo,* i.e. there may be two alternative routes for the formation of the 30*s* initiation complex. Although GTP is required for the binding of IF2 to ribosomes and IF2 is required for the binding of fMet-tRNA^{fMet}, the existence of the ternary complex, fMet-tRNA^{fMet}:IF2:GTP shown in Fig. 6.14 has not been conclusively established in the same way as the EF-Tu complex (see the following section).

Initiation sequences

Accurate translation requires the binding of ribosomes to mRNA at the correct initiation signal and selection of the correct binding site may be determined both by the sequence and conformation of the message and by the small ribosomal subunits. Much of our current knowledge of the mechanism and control of translational initiation in prokaryotes has come from studies on the translation of RNA phages. The demonstration that *E. coli* ribosomes translate all three cistrons of denatured phage f2 RNA whereas ribosomes from *Bacillus stearothermophilus* translate only one (the A protein cistron) regardless of the source of initiation factors, tRNA or supernatant enzymes, was the first clear evidence for an active ribosomal role in the selection of mRNA initiation signals. Subsequent studies showed that ribosome selectivity was a feature of the 30*s* subunits and that the particular 30*s* components involved were protein S12 and 16*s* rRNA (Gorini 1971). The specific interaction of mRNA with 30*s* subunits also requires r protein S1 and is stabilized by IF3 (Isono & Isono 1975). A 30–40 nucleotide-long stretch of mRNA is protected from the action of RNAase by ribosome binding. In the absence of elongation factors and charged tRNAs, other than fMet-tRNA^{fMet}, salt-washed ribosomes will, in the presence of IFs, bind specifically to mRNA initiation sequences. Subsequent RNAase treatment, followed by dissociation of the mRNA–ribosome complex, enables the isolation and sequence analysis of the ribosome binding sites.

In all the bacteriophage ribosome binding sites so far sequenced, the initiator AUG is approximately in the middle of the ribosome-protected sequence and there is a polypurine stretch of 3–9

R17 A protein	GAUUCCUAGGAGGUUUGACCU	AUG	GGA
QB A protein	UCACUGAGUAUAAGAGGACAU	AUG	CCU
R17 coat protein	CCCUCAACCGGAGUUUGAAGC	AUG	GCU
QB coat protein	AAACUUUGGGUCAAUUUGAUC	AUG	GCA
R17 replicase	AAACAUGAGGAUUACCC	AUG	UGG
QB replicase	AGUAACUAAGGAUGAAAUGC	AUG	UCU

E. coli 16*s* rRNA	3′ $_{HO}$ AUUCCUCCACUAG
Complementary sequence	GAAGGAGGUGAUC

Fig. 6.15. Sequences of some phage ribosome binding sites. The underlined bases show sequence complementarity with the 3′ terminal sequence of *E. coli* 16*s* rRNA. R17 coat protein initiation site contains two sequences of four bases complementary to overlapping regions of the rRNA. The boxes indicate translational initiation codons. Each binding site contains one or more nonsense codon either in or out of the translational reading frame. (Based on data from Steitz & Jakes 1975 and Steitz *et al.* 1977.)

nucleotides approximately 10 nucleotides from this codon on its 5′ side (Fig. 6.15). Shine and Dalgarno (1975) proposed that a polypyrimidine sequence AUCACCUCCUUA_{OH} at the 3′ terminus of *E. coli* 16*s* rRNA participates directly in translational initiation by forming a number of base pairs with the purine-rich sequence preceding the initiation codon. In direct support of this proposal, hybrid RNA complexes containing the 3′ terminal region of 16*s* rRNA and the A protein initiation region of R17 RNA or the initiation region of the λ P_R transcript have been isolated. Upon mRNA binding one set of RNA base pairs formed intramolecularly within 16*s* rRNA are thought to exchange for another set formed intermolecularly between 16*s* rRNA and mRNA (Fig. 6.16). The *E. coli* protein S1 interacts specifically with 16*s* rRNA at the 3′ pyrimidine-rich sequence ACCUCCU and can be cross-linked to this 3′ terminal region by periodate oxidation. This binding may open up the proposed double-stranded structure at the 3′ terminus (Fig. 6.16) to allow the base pairing with the complementary mRNA sequence. It is proposed that IF3 stabilizes this mRNA–16*s* RNA interaction. It is also possible that S1 recognizes and unfolds elements of mRNA structure (see below) and the RNA-binding properties of S1 are important in another context when this protein becomes a subunit of Q*β* replicase (see section 6.5.6). When the ribosome moves beyond the initiation site during elongation, rRNA base pairs would be expected to reform in the reverse reaction—this may happen following the dissociation of IF3 from the 30*s* initiation complex. There are a number of lines of evidence that proteins S1 and S12 are neighbours in the 30*s* subunit although the precise role of S12 in initiation is uncertain; protein–RNA interactions are undoubtedly required in order to stabilize the RNA–RNA interactions at initiation and other r proteins, in addition to S1 and S12, may contribute to this stabilization. The initial rRNA–mRNA interaction may occur some distance away from the initiation sequence (Backendorf *et al.* 1980). The final mRNA–rRNA interaction is thought to result in the correct positioning of the initiation codon on the ribosome and the extent of the region of sequence complementarity between the two RNA species could determine the efficiency of ribosome binding and hence the efficiency of initiation.

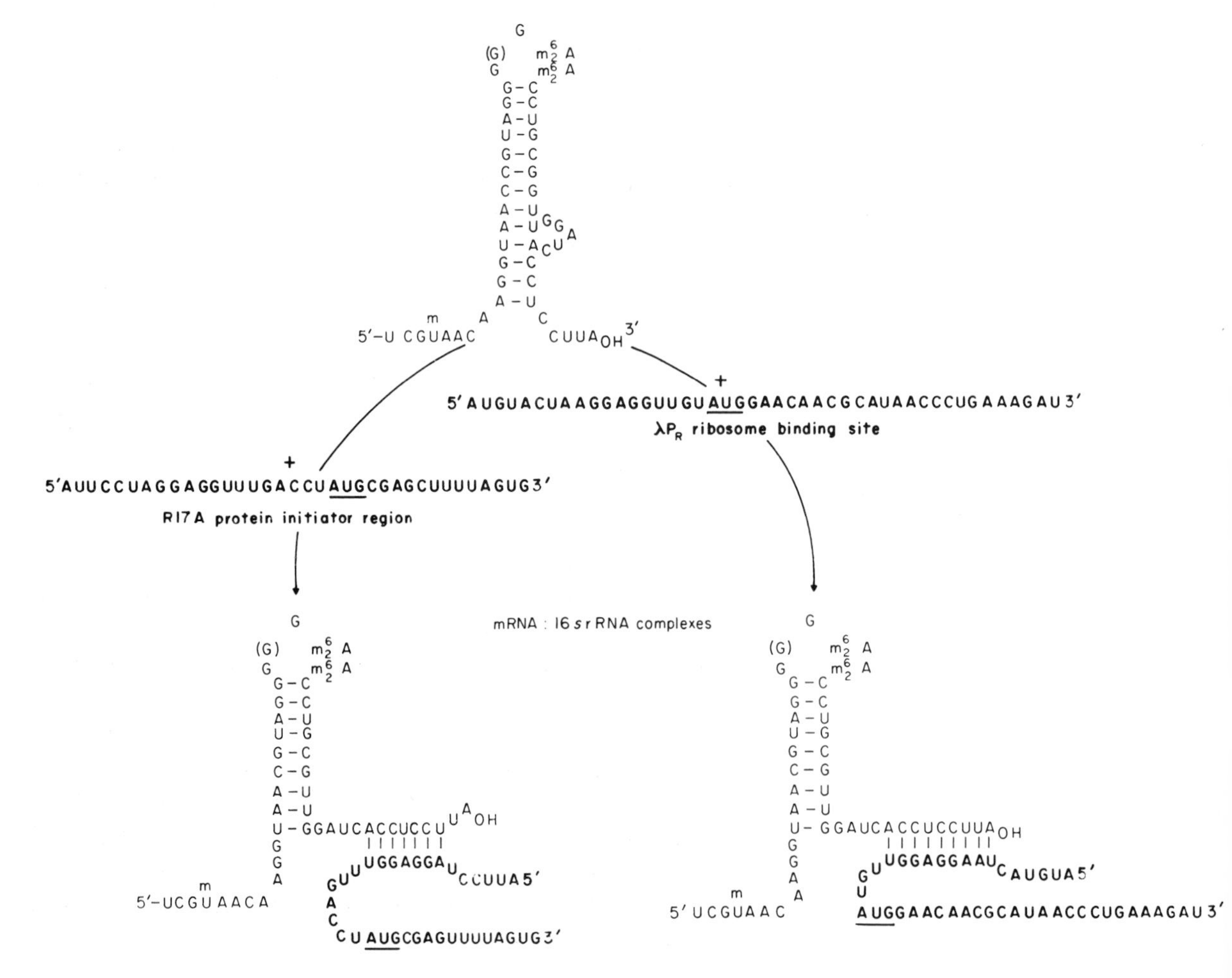

Fig. 6.16. A role for 16*s* rRNA in polypeptide chain initiation. mRNA-16*s* rRNA complexes were isolated by gel electrophoresis, following incubation with ^{32}P-labelled fragments of either R17 A protein initiator or the ribosome binding site of the λ P_R transcript; in each case the incubation was followed by treatment with colicin E3 which removes the 3′ terminal 16*s* rRNA fragment from 16*s* rRNA. It is proposed that the interaction involves partial melting of the 16*s* fragment and reannealing to the mRNA ribosome binding site. For experimental evidence, see Shine and Dalgarno (1975) and Steitz and Steege (1977). Initiation codons are underlined.

An example of translational control by ribosome affinity for an mRNA initiation site may be the synthesis of the *cI* repressor protein of bacteriophage λ (see section 5.2.4). For the establishment of lysogeny, large amounts of *cI* are required and the appropriate mRNA is transcribed from the establishment promoter P_{RE} and contains a leader sequence on the 5′ side of the initiator AUG. In contrast, maintenance of the lysogenic state following integration of λ DNA into the bacterial chromosome requires relatively low levels of *cI* and transcription of *cI* mRNA now occurs from the P_{RM} promoter. The absence of a leader sequence preceding the initiation AUG in the P_{RM} transcript is thought to result in less efficient ribosome binding with consequent limited utilization of the message and decreased synthesis of *cI* protein.

The patterns of translational specificity exhibited by ribosomes from several different bacterial strains with various phage RNAs can be correlated with the degree of complementarity of the respective 3′ terminal 16*s* rRNA sequences with the various ribosome binding site sequences.

In addition to the presence of many purine bases, certain other features of the mRNA sequence preceding the initiation codon are shown in Fig. 6.15. The presence of one or more AUG codons either in or out of phase emphasizes the importance of correct positioning of the bona fide initiation codon on the ribosome so that only the latter is available for pairing with the anticodon of the initiator tRNA. The role of the nonsense codon(s) found either in or out of phase in the nucleotide sequence preceding the initiation codon (Fig. 6.15) may be to prevent translation of intercistronic or leader sequences and/or to ensure recognition of the correct initiation codon.

Secondary structure

The secondary structure of the mRNA influences ribosome binding. Thus cell-free synthesis of phage f2 proteins by *E. coli* ribosomes and formaldehyde-denatured phage f2 RNA, involves independent initiation of all three cistrons and yields the three proteins in roughly equal amounts. However, *in vivo* and *in vitro,* using native phage RNA, the three phage proteins are formed in different amounts and at different times; initially ribosomes bind only at the coat protein initiation site because the initiator regions of the A protein and replicase cistrons are masked by the secondary structure of the RNA. As ribosomes translate the coat protein cistron, part of the RNA secondary structure is disrupted to expose the ribosome binding site of the replicase cistron, i.e. the synthesis of RNA replicase starts only when a part of the coat protein gene has already been translated. The initiation region of the A protein gene is only transiently accessible for ribosome binding during RNA replication; further details of the influence of RNA secondary structure on the control of translation of RNA phages are given in section 6.5.6. Comparison of the nucleotide sequence of several different phage RNAs has led to the conclusion that the presumed secondary-structured regions are more highly conserved than other unstructured regions, implying some conservation of overall conformation, possibly related to the packaging of such RNA molecules into phage heads. RNA secondary structure is probably also involved in the processes that give rise to the phenomenon of translational polarity of expression of bacterial operons (section 6.5.5).

*The 70*s *initiation complex*

The 30*s* initiation complex, probably still carrying all three IFs, binds a 50*s* subunit in a reaction which involves RNA–protein, protein–protein and RNA–RNA interactions. The latter may be due to pairing between complementary sequences in the 3′ terminal regions of 16*s* and 23*s* RNA and, moreover, IF3 could be concerned in the alignment of these sequences because it can be cross-linked to sites near the termini of both RNA species (reviewed by Grunberg-Manago *et al.* 1978). However, the presence of IF3 in the complex is only transitory and it is lost to leave the '70*s* pre-initiation complex' with fMet-tRNA^{fMet} in the P site and IF2 in the A site. Hydrolysis of the GTP, probably by the GTPase present in the 50*s* subunit, now changes the conformation of IF2 so that its affinity for the ribosome is reduced. The release of IF2 also requires IF1 and these two factors, GDP and P_i are probably all lost together. In those cases in which a fourth base pair is involved in the initiator codon–anticodon interaction (see section 6.2.3), the release of the factors must also be associated with the breaking of this base pair to allow the mRNA base to be recognized as the first letter in the next codon.

Addition of puromycin to the factor-free 70*s* initiation complex results in the formation of fMet-puromycin and uncharged tRNA^{fMet}, indicating that by this stage, the fMet-tRNA^{fMet} must be properly located in the P site and the A site is available for the start of the elongation cycle. The energy used in the formation of the 70*s* initiation complex amounts to two high-energy phosphates from ATP (to form fMet-tRNA^{fMet}) and one from GTP (to release IF2).

6.5.2 Elongation

See Clark (1980) for a review of the elongation process.

Elongation factors and fidelity of translation

Peptide chain elongation requires three elongation factors, EF-Tu (M_r 47×10^3), EF-Ts (M_r 34×10^3) and EF-G (M_r 80×10^3). Tu and Ts are named, rather arbitrarily, after the thermal lability of the factors from *Pseudomonas fluorescens.* It is not true of all bacteria that Tu ('u' for unstable) is more labile than the 'stable' Ts. The EFs together constitute a significant proportion of total soluble bacterial protein (Wurmbach & Nierhaus 1979).

The Tu and Ts genes are parts of ribosomal transcriptional units (see section 6.3.2). Both proteins are involved in RNA replication (see section 6.5.6). Another role for EF-Tu is suggested by the fact that it is present in fourfold excess over EF-Ts. Indeed EF-Tu is a major protein component in *E. coli* and may be important as either a membrane protein or (as a polymer) an internal structural protein (Beck 1979).

The elongation cycle is shown in Fig. 6.17. EF-Tu first forms a ternary complex with aa-tRNA and GTP. The factor has an affinity for any aa-tRNA other than fMet-tRNA^{fMet} or Met-tRNA^{fMet}. EF-Tu does not bind to uncharged tRNAs, which require stringent factor for binding to the ribosome.

The aa-tRNA.EF-Tu.GTP ternary complex binds to the A site of the ribosomes. Once the initiation system has defined the correct reading frame, and given the proof-reading by the aa-tRNA synthetases is accurate, it is up to the binding of the ternary complex to maintain translational fidelity. There is much evidence to implicate the ribosome itself in maintaining this fidelity. Loss of fidelity, which amounts to a misreading of the genetic code, can be induced by streptomycin. *In vitro,* poly(U) directs the synthesis of poly Phe. In the presence of streptomycin, Ile, Ser, Tyr or Leu (as well as Phe) are incorporated. This corresponds to a misreading of UUU as AUU, UCU, UAU or CUU and, therefore, (unlike the 'wobble pairings', see section 6.2.1) involves ambiguities at positions other than the 3′ residue of the codon. Moreover, even in the absence of streptomycin, the poly(U) system will erroneously incorporate some Ile residues if the incubation includes large amounts of Ile-tRNA^{Ile}, but no other aa-tRNAs. Streptomycin is, therefore, to be thought of as amplifying an inherent ambiguity rather than inducing new misreadings. *E. coli*

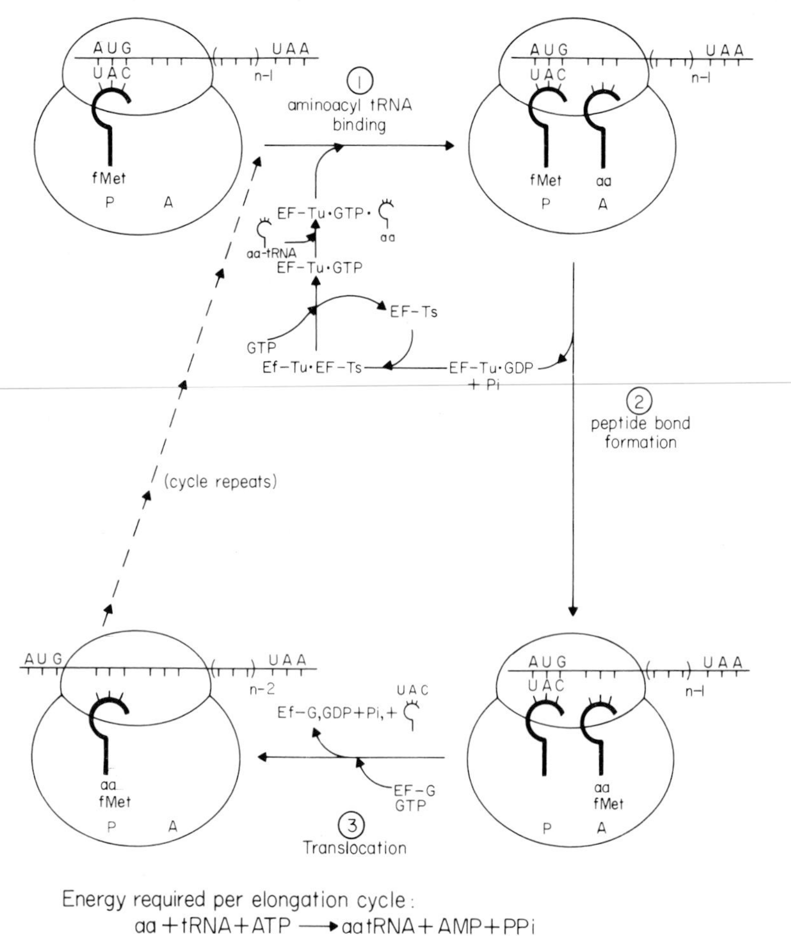

Fig. 6.17. Bacterial protein synthesis. Stage II, elongation.

strains resistant to high concentrations of streptomycin (Str^R) carry a mutation (originally designated *strA* but now *rspL*) in the gene for r protein S12. Another group of mutants (formerly *ramA* but now *rspD*) misread the code in the absence of streptomycin; *rspD* is the gene for r protein S4. Taken together, these data suggest proteins S12 and S4 are involved in the maintenance of translational fidelity. It is also likely that other r proteins (S5 and S17) are involved in the interactions that contribute to this maintenance.

Altered S12 in Str^R strains results in a more selective ribosomal 'screen' against misreading. In an Str^S strain, streptomycin lowers the screen. A final group of streptomycin mutants are those that are conditionally streptomycin dependent (CSD). The organisms carry mutations in an essential metabolic pathway that can only be relieved by adding streptomycin to lower the screen to allow sufficient ambiguity in translation for a wrong reading of a codon to correct an unrelated mutation. This phenomenon is termed 'phenotypic suppression'. Although it resembles informational suppression (see section 6.2.4) in operating at the level of translation, phenotypic suppression differs in that it does not require an altered tRNA. (See Gorini (1974) and Tai *et al.* (1978) for reference to the streptomycin effects.)

Lake (1977) proposes that the mechanism for maintaining translational fidelity involves the loose binding of aa-tRNA to a recognition (R) site on the 30*s* subunit. If the screen is satisfied with the fit, a conformational change in the anticodon loop is induced and this allows a switch from the R to the A site.

There are several lines of evidence to suggest that the 5*s* rRNA component of the 50*s* subunit also plays a role in the binding of aa-tRNA during elongation by hydrogen bonding between a specific, highly conserved, CGAA sequence of 5*s* RNA with the TψCG sequences common to all non-initiating tRNAs. This TψCG sequence is not exposed in tRNA in solution and it seems likely that codon–anticodon interactions result in a conformational change of the aa-tRNA positioned at the ribosomal A site with consequent exposure of the TψCG sequence (Erdmann 1976).

The binding of each molecule of aa-tRNA to the ribosome is accompanied by hydrolysis of the GTP component of the EF-Tu.GTP.aa-tRNA ternary complex. This GTP hydrolysis is not required for the correct positioning of the aa-tRNA in the A site but rather for the release of EF-Tu in order that the next step of the elongation cycle—peptide bond formation—can proceed. The resulting EF-Tu.GDP complex has a decreased affinity both for aa-tRNA and for the ribosome and is released. EF-Ts is involved in the phosphorylation of EF-Tu.GDP to regenerate EF-Tu.GTP which is now able to form a new ternary complex with aa-tRNA and thus re-enter the elongation cycle. Several workers have proposed that the energy of GTP hydrolysis in tRNA binding is used to 'proofread' the codon–anticodon interaction on ribosomes and evidence has been presented that r protein S12 influences tRNA selection both before and after GTP hydrolysis (Yates 1979).

Peptide bond formation

The second step of the elongation cycle involves the formation of a peptide bond between the activated carboxyl group of the peptidyl- (or formyl-methionyl-) tRNA bound at the ribosomal P site and the α-amino group of the aa-tRNA bound at the A site. This transpeptidation reaction is catalysed by peptidyl transferase which may be regarded as one of the 'functional domains' (see section 6.3.1;

Harris & Pestka 1977) of the 50*s* ribosome since several 50*s* proteins including L11, L18, L27, L16, and L2 have been implicated in this enzymatic activity. It has been proposed that the CCA termini of the two reacting, charged tRNA molecules may be positioned at the peptidyl transferase site by binding to complementary UGG sequences at the 3′- and 5′-ends of 23*s* rRNA. Energy in the form of two high-energy bonds is expended in the activation of each amino acid (see section 6.4) and the peptidyl transferase reaction itself does not require additional energy expenditure in the form of ATP or GTP and no specific soluble protein factors are required for this step.

Translocation

Following the peptidyl transferase reaction, uncharged tRNA occupies the P site of the ribosome and peptidyl-tRNA is present in the A site. During the third (translocation) step of the elongation cycle the uncharged tRNA is displaced from the P site, peptidyl tRNA moves from the A to the P site and the ribosome moves along the mRNA three nucleotides towards the 3′-end. These three events are all triggered by the binding of EF-G and GTP to the ribosome. Following translocation, the GTP is hydrolysed, EF-G leaves the ribosome in the form of an EF-G.GDP complex and the next elongation cycle can commence with the binding of the appropriate EF-Tu.GTP.aa-tRNA complex. The translocation-associated movement of the ribosome along the mRNA can be experimentally demonstrated by showing that the mRNA fragment protected from RNAase action by the associated ribosome extends three nucleotides further towards the 3′-end in the mRNA–ribosome complex following the addition of EF-G and GTP than in the same complex prior to their addition. Movement of the peptidyl-tRNA from the P to the A site of the ribosome is implied by the demonstration that this molecule is unable to react with puromycin in the pre-translocation complex and becomes puromycin-reactive following translocation. The mechanisms for movement of tRNA in the ribosome and the movement of the ribosome along the message are obscure and it is also not known whether the uncharged tRNA molecule vacates the A site before the peptidyl-tRNA movement or whether peptidyl-tRNA actively displaces the uncharged tRNA.

The 50*s* r proteins L7 and L12 have been implicated in the translocation event; these proteins are also required for the ribosome-dependent GTPase activities of IF2, EF-Tu and EF-G. As described in section 6.3.1, L7 and L12 differ only in the presence of an N-terminal acetyl group in L7 and are present together in multiple copies in the 70*s* ribosome; they are acidic proteins with a high content of Ala, Glu and Gln and some methylated amino acids and possess a high α-helix content (50–60%). The chemical and physical properties of L7 and L12 closely resemble those of contractile proteins such as myosin and flagellin. It was thought that the EF-G-dependent GTP hydrolysis provided energy necessary to drive the translocation mechanism. However, it now seems likely that GTP hydrolysis occurs after translocation in order to ensure the release of EF-G from the ribosome. Thus the GTP hydrolysis associated with the ribosomal binding of IF2, EF-Tu and EF-G occurs after each of these factors have fulfilled their functional role and in each case appears to cause a conformational change in the factor with consequent decrease of its affinity for the ribosome. Since there is much evidence that all three factors bind to the ribosome at a common or overlapping site (which presumably corresponds, or is very close, to the ribosomal A site) their rapid binding and release is essential. The sequential interaction of EF-Tu and EF-G with the ribosome at the beginning and end of each elongation cycle may be determined by the ribosome-bound tRNA molecules; ribosomes carrying peptidyl-tRNA at the P site will readily bind EF-Tu.GTP.aa-tRNA, whereas ribosomes in which peptidyl-tRNA occupies the A site will readily bind EF-G and GTP. Activation of the EF-G binding site may result from a conformational change in the ribosome and/or the peptidyl-tRNA following peptide bond formation. It is also possible that tRNA molecules play an active role in translocation, for example, a conformational change in the peptidyl-tRNA molecule with maintenance of its anticodon–codon interactions could effectively pull the mRNA molecule through the ribosome during the movement of the peptidyl-tRNA from the A to the P site. Moreover, codon–anticodon interaction at the P site is a prerequisite for aa-tRNA binding at the A site (Lührmann *et al.* 1979). Evidence to suggest that tRNA molecules could play a critical role in determining the distance along the message travelled by the ribosome during the translocation step is provided by the frame-shift suppressor tRNAs. One of these contains an extra residue in its anticodon loop (see section 6.2.5). Others might operate by causing the ribosome to advance by four, rather than three, residues following the reading of a '+1' frame-shift mutation.

Rate of synthesis and energetics

The rate-limiting step in prokaryotic protein synthesis is initiation. Following the formation of the 70*s* initiation complex elongation normally proceeds rapidly—in an *E. coli* cell in rich medium at 30 °C, approximately 10 amino acids per second can be incorporated into the growing peptide chain. Although, in principle, it would be possible to control the rate of chain elongation, and hence the rate of translation of different mRNAs, by the frequency of occurrence of codons whose translation is dependent on aa-tRNA molecules present in rate-limiting concentrations, there is little evidence that such control is of physiological importance. Two high-energy bonds of ATP are expended in the activation of each amino acid prior to its incorporation and, in addition, each elongation cycle involves the hydrolysis of two molecules of GTP—one following the EF-Tu-directed binding of each aa-tRNA and one following EF-G-dependent translocation.

6.5.3 Termination

Termination represents the final, and least well-defined, stage in translation (Grunberg-Manago *et al.* 1978; Caskey 1980) and normally occurs when the last amino acid specified by the message (i.e. the carboxyl-terminal amino acid of a polypeptide) has been incorporated and translocation results in the positioning of one of the three chain termination codons in the ribosomal A site. The events associated with the termination process are the cleavage of the completed polypeptide chain from its linkage to tRNA in the P site, consequent release of this polypeptide from the ribosome and dissociation of the residual ribosome complex (Fig. 6.18).

Three soluble protein release factors (RFs) are involved in termination. RF1 (M_r 44–49 $\times 10^3$; 500 molecules per cell) and RF2 (M_r 47–50 $\times 10^3$; 700 molecules per cell) both respond to the nonsense codon, UAA; RF1 also responds to UAG and RF2 also responds to UGA. The fact that, of the three codons, UAA is the only one recognized by both factors may correlate with the lower efficiency of informational suppression of UAA (see section 6.2.4). The third factor, RF3 (M_r 70 $\times 10^3$), is required for the binding of one of the other two factors when a nonsense codon comes into register over the P site although RF3 itself does not recognize these codons. The nature of the conformational change in the ribosome, effected by a nonsense codon, that allows binding of RFs is unknown. The 3′

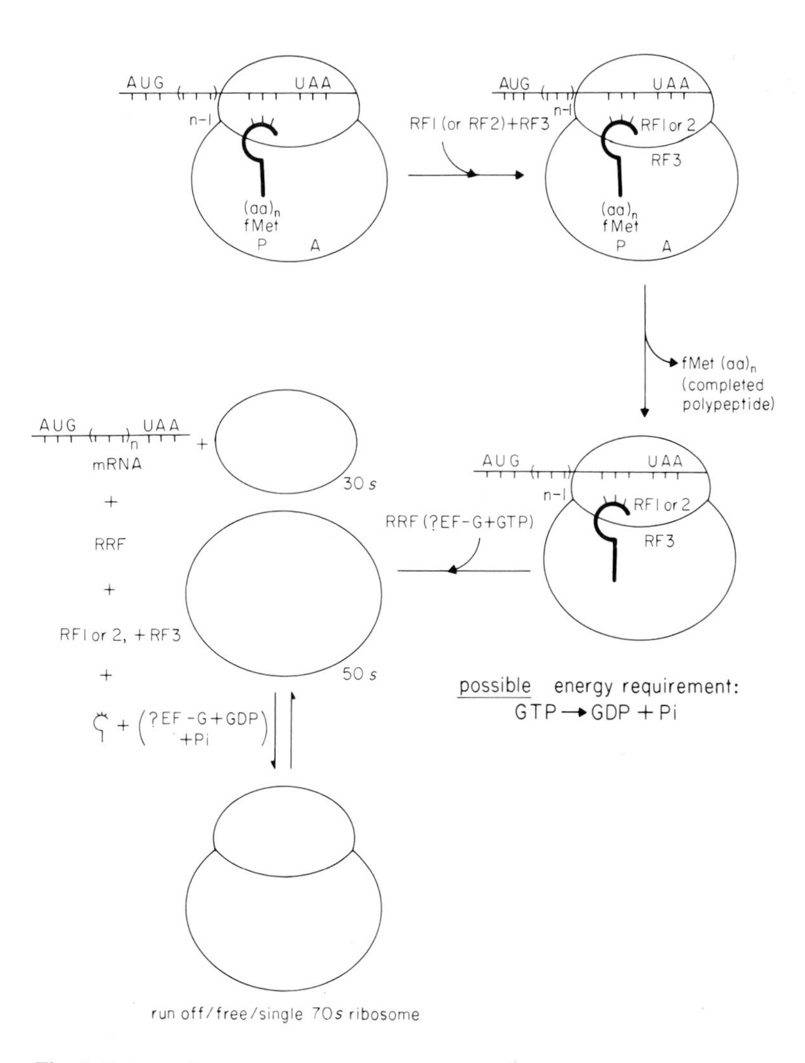

Fig. 6.18. Bacterial protein synthesis. Stage III, termination.

terminal sequence of 16*s* rRNA is implicated because if the 49 3′ terminal nucleotides are removed (by the bacteriocin, cloacin DF13), codon-directed binding of RF1 and RF2 is inhibited. As for the actual mechanism, it is reasonable to suppose that interaction between the 16*s* molecule and the mRNA sequence that includes the nonsense codon will involve H-bonding. Indeed, if we look at the 3′ terminus of 16*s* RNA, $3'_{HO}\underline{AUU}CCUCC\underline{ACU}A...5'$, the first of the underlined triplets could associate with UAA or (given a wobble) UAG and the other one could associate with UGA.

The release factors appear to function as allosteric effectors of peptidyl transferase, modulating the activity of this enzyme in such a way that the transpeptidation reaction, in the absence of an α-NH_2 group in the ribosomal A site, is catalysed between the peptidyl residue of the peptidyl-tRNA in the P site and water. Following the release of the completed polypeptide chain from the last codon-specified tRNA molecule (and hence from the ribosome), dissociation of the mRNA–ribosome–tRNA complex into the separate components occurs (but see also section 6.5.5). There is evidence that release of ribosomes from mRNA requires a protein factor of M_r 18×10^3 isolated from the salt wash of ribosomes and termed ribosome release factor (RRF). The details of action of RRF are not clear although there is evidence that EF-G and GTP may also be required for the release of the ribosome and uncharged tRNA from the mRNA. It is not yet clear whether the ribosome is released from the mRNA in the form of a 70*s* (run off/free/single) ribosome or whether disassembly of the termination complex directly yields the two component subunits. However, equilibrium between 70*s* ribosomes and 30*s*+50*s* subunits will be rapidly established prior to displacement of this equilibrium by initiation factors (see section 6.5.1).

In accordance with the roles of GTP in the initiation and elongation steps of prokaryotic translation, it might be anticipated that GTP cleavage may be required for the release of RFs following termination to permit subsequent binding of initiation factors. Alternatively, it is plausible that the dissociation of the ribosome into its component subunits (following termination) might obviate a requirement for an energy-dependent conformational change for dissociation of RFs. Present experimental evidence for and against a GTP requirement is equivocal but the case for such a requirement is substantiated in part by the fact that RF3 has a ribosome-dependent GTPase activity.

6.5.4 Biosynthesis of extracellular proteins

Although all bacterial proteins are synthesized in the cytoplasm by the mechanisms described above, the final destination of some of these proteins is exterior to the inner surface of the plasma membrane. Thus, approximately 25% of the total protein synthesized by *E. coli* is ultimately located either in the inner (plasma) membrane, outer membrane or in the periplasmic space located between these two membranes. Moreover, many bacteria are known to secrete extracellular enzymes into the external medium. For all these extracellular proteins, mechanisms must exist for their selective secretion across, or insertion into, the highly hydrophobic plasma membrane (Lin *et al.* 1978).

The signal hypothesis (see section 9.4.4) was developed by Blobel to account for the mechanism of the biosynthesis of eukaryotic secretory proteins. The original version of the hypothesis proposed that mRNAs for such proteins code for a signal peptide that constitutes a metabolically short-lived hydrophobic extension of 15–30 amino acids on the NH_2-terminus of the nascent polypeptide chain. This hydrophobic sequence binds to the endoplasmic reticulum membrane inducing the formation of a channel for the co-translational and unidirectional transfer of the nascent chain across the membrane and also causing binding to the membrane of the ribosome-mRNA complex. Having established the conditions for the vectorial transport for the nascent polypeptide, the signal peptide is cleaved from the growing chain by a microsomal membrane protein—the signal peptidase.

One example of a similar system in bacteria is the biosynthesis of penicillinases by *Bacillus licheniformis.* Although there is only one penicillinase gene there are two enzymes. One is extracellular and the other is membrane bound. The difference between the two is a signal peptide, present in the membrane-bound form of the enzyme and removed by a counterpart of signal peptidase, outside the cytoplasmic membrane, to produce the extracellular form. The signal in this case consists of 25 amino acids, the N-terminal residue being phosphatidyl serine.

Other examples of precursors with signal sequences are those for the major outer membrane lipoprotein and periplasmic alkaline phosphatase, both in *E. coli.* In the latter case, not only is the precursor well characterized, but so are the membrane-bound polysomes, engaged in its biosynthesis. See Davis and Tai (1980) for a review with references to the original literature (see also p. 247).

6.5.5 Polarity, polar mutants and suppression

Transcription and translation of polycistronic mRNA is a characteristically prokaryotic device (see Chapter 8 for the alternative systems used in eukaryotic cells). Polycistronic mRNA has the information for as many as 20 different proteins and consequently an equal number of translational initiation sites. Polarity of gene expression is a selective deficiency of gene products from cistrons more distal from the operator. A simple example of natural polarity is found in the three structural genes of the Lac operon (Fig. 6.19). Translation of Lac mRNA *in vivo* does not yield equimolar amounts of the three protein products: 3–5 times more β-galactosidase is synthesized from the *Z* gene transcript than galactoside acetyltransferase from the *A* gene transcript. Also natural polarity can be enhanced by polar mutations.

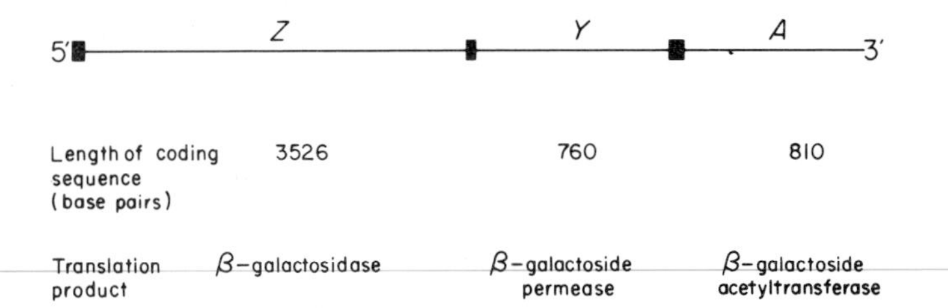

Fig. 6.19. Lac operon mRNA. Vertical bars indicate non-translated leader and intercistronic sequences; transcription starts within the proximal region of the operator gene.

In principle there could be three explanations for polarity:

(1) transcriptional termination—premature dissociation of RNA polymerase from the template could result in a lack of operator–distal mRNA sequences;

(2) translational initiation—different efficiencies of formation of translational initiation complexes (see section 6.5.1) could explain polarity;

(3) message stability—if, for some reason, the 3′ region of mRNA was unstable, there would be a deficiency of operator–distal products.

The natural polarity of the Lac operon is due to (2) although the effect is somewhat amplified by the relative metabolic instability of the acetyltransferase enzyme. However, in general, polarity represents a balance between the alternative mechanisms (Zipser 1970) and the interaction between them is revealed by an analysis of polar mutants. These mutants have also proved of great value in revealing detailed features of the reading of the genetic code.

Consider a nonsense mutation within the *lacZ* gene; such a mutation will, in the absence of an appropriate suppressor tRNA, cause premature chain termination during translation of the *Z* message and it is unlikely that the resultant β-galactosidase fragment will be enzymically active. However, the synthesis of Lac permease and acetyltransferase is frequently also decreased as a result of a mutation in the *Z* gene even though there are no mutations in *Y* and *A*. This enhancement of the natural polarity is typical of a polar mutation. The polar effect generally increases, the closer the nonsense mutation maps to the operator, i.e. the further away the mutation maps from the initiation signal for the next cistron. The explanation for this phenomenon lies in the coupling of transcription and translation in prokaryotes. A transcribing RNA polymerase is closely followed by a functioning ribosome (Fig. 6.1). The transcription–termination factor ρ (see section 5.1.4) recognizes and binds to unprotected (ribosome-free) mRNA behind the polymerase and, functioning as an RNA-dependent ATPase, induces premature termination of transcription at the next ρ-sensitive termination site. Consequently, although concomitant translation is obviously not a requirement for continued transcription of, for example, rRNA and tRNA genes, and transcription of natural templates can occur *in vitro* in the absence of simultaneous translation of the nascent mRNA, the necessity for close coupling of transcription and translation of structural genes *in vivo* can be explained by the action of ρ. Moreover, the interaction of ρ with unprotected nascent mRNA explains the basis for mutational polarity. Ribosome release at the site of a nonsense mutation in nascent β-galactosidase mRNA would result in ρ-dependent termination of transcription at a site *within* the *Z* gene so that the mRNA for the 3′ portion of *Z* and all of the *Y* and *A* sequences would not be synthesized. If, however, the nonsense mutation maps close to the initiation signal for the *Y* gene (or close to a potential reinitiation signal within *Z*—see (4) below), there is a chance that a ribosome may associate with the message before ρ can bind and exert its effect and the polar effect of the nonsense mutation will be decreased.

There are four possible mechanisms of suppression of polar mutations.

(1) Informational suppression. In the presence of an appropriate suppressor tRNA molecule (see section 6.2.4) an amino acid will be inserted at the site of the nonsense mutation, i.e. premature chain termination will not occur and, in the case of a polar mutation in the *Z* gene, complete transcription of the Lac operon will occur. Although this type of intergenic suppression will not always result in the synthesis of active β-galactosidase, the polar effect of the nonsense mutations will be suppressed and normal amounts of *Y* and *A* gene products will be synthesized.

(2) Intergenic suppression by mutation in the gene coding for ρ. In the absence of functionally active ρ, transcription can continue without concomitant translation. In the case of a *lacZ* polar mutant, suppression of this type would not result in the synthesis of active β-galactosidase but normal amounts of *Y* and *A* gene products would be synthesized. The first class of this type of polar suppressor mutants to be isolated were termed *suA* mutants and subsequent studies showed ρ to be the product of the *SuA* gene. Other apparently ρ-defective mutations, including those termed *psu, sun, rho* and *nitA,* have been isolated by widely different selection procedures and several exhibit suppression of polar mutations (Adhya & Gottesman 1978).

(3) Intergenic suppression by mutation in an RNA polymerase gene. The precise mechanism of action of ρ in premature transcription termination is obscure; for example, the role of ATP

hydrolysis in the release of nascent mRNA and/or the dissociation of RNA polymerase from the DNA template is unclear. Factor ρ recognizes (and probably binds to) ribosome-free nascent mRNA and also interacts with the β subunit of RNA polymerase during transcription termination at the end of an operon. If such an interaction also occurs during premature termination of transcription within an operon, it is possible that mutation within the structural gene for the β subunit of RNA polymerase (*rpo* B) will result in mutants defective in transcription termination within an operon in the presence of ρ.

(4) Intragenic suppression by translational reinitiation within the mutant gene. Fine structure studies in several *E. coli* operons have established that, following premature chain termination, translational reinitiation can occur not only at the initiation signal at the beginning of the next cistron but also at sites within the mutated gene proximal to the site of the nonsense mutation. If reinitiation of translation occurs before attachment of ρ to the nascent mRNA, the polar effect of the nonsense mutation will be suppressed. Two natural reinitiation sites within the *Z* gene were defined and mapped using a double mutant polarity test (Yanofsky & Ito 1966). The polar effects of two nonsense mutations existing together are multiplicative if the two mutations are separated by an initiation site, otherwise only the more operator-proximal mutation exerts an effect. The two alternative reinitiation sites in *Z* together with the initiation site at the start of *Y* result in three possible polarity gradients within *Y*.

Some of the most interesting data on reinitiation have come from experiments on *lac*I, the structural gene for Lac repressor. From studies of *E. coli lac*I amber mutants, alternative potential reinitiation sites have been identified within the mRNA sequence which encodes the first 62 amino acid residues of Lac repressor (Files *et al.* 1975). The nature of the sites that can be used for reinitiation is further revealed from the sequence of *lac*I mRNA (Steege 1977). The wild-type Lac repressor protein is initiated with GUG; translation reinitiation within *lac*I can, depending on the position of the nonsense mutation, occur at a GUG codon normally specifying the insertion of valine at amino acid position 23, at an AUG codon normally specifying methionine at position 42 or at a UUG (leucine) codon in position 62.

Analysis of such sequence data reveals the probable factors which determine whether a potential initiation codon (AUG, GUG or exceptionally UUG) can actually be recognized as such by the translational apparatus. In the case of wild-type *lac*I mRNA there is a cluster of three initiation codons near the 5′-end of the molecule, only one of which is recognized by fMet-tRNA^{fMet} and the initiation complex (see section 6.5.1). From the data of Steege (1977), the chosen codon is the one that fits into the most complete 'Shine and Dalgarno pairing scheme' (see section 6.5.1). In the cases of the *lac*I amber mutants, there are in principle a total of nine possible reinitiation codons. By examining the sequence surrounding the acceptable AUG or GUG alternatives, Steege concluded that the potential for acting as an initiation codon was only realized in those codons that appear in loops in one of the alternative secondary structures that can be adopted by the mRNA. Something of a mystery still surrounds the UUG codon. There is no evidence that UUG can act as an initiation codon at the start of a cistron and the UUG in *lac*I mRNA is the most distal of the possible reinitiation codons that have been identified in this sytem. Possibly this is representative of a different type of initiation event in which ribosomes may remain attached to mRNA, after having passed a termination codon, and 'drift' along mRNA until they find a possible initiation codon. Such a process would not involve a dissociation of ribosomes and would thus bypass the formation of the 30*s* initiation complex.

It has been proposed that the role of the formyl group of the initiator methionyl tRNA is to act as an anti-dissociating factor in polycistronic messenger translation and evidence has been presented that formylation is required *in vivo* for the coordinate expression of the *E. coli* Lac operon and that a strong polarity ('metabolic polarity') in the expression of this operon is induced in the presence of trimethoprim, an antagonist of folic acid coenzymes. It has also been suggested that initiation of translation of the first cistron of an operon requires dissociated ribosomal subunits (see section 6.5.1) but *no* formylation whereas 70*s* ribosomes initiate translation at intercistronic regions with a stringent requirement for formylated initiator tRNA (Danchin & Ullman 1980).

6.5.6 Translational control

The regulation of expression of structural genes of the bacterial genome and of DNA phages is largely exerted at the level of transcription although there is good evidence for translational control of the expression of genes for the translational apparatus itself (see section 6.3.3). Furthermore, as described in the preceding sections of this chapter, fine control of gene expression may be superimposed at the level of translation through differential mRNA stability and differential efficiencies of initiation. Attenuation (see section 5.2.5) also provides an example of the coupling of translation with transcriptional termination as a regulatory device.

Translational control in RNA phages

The single-stranded RNA bacteriophages contain a genome which not only serves as a template for its own replication but also directly constitutes the message for the synthesis of essential phage proteins. As this means that there can be no transcriptional control, these RNA phages exhibit a highly intricate, self-regulating system of translational control.

The single-stranded RNA phages of *E. coli* are classified into two serological groups: group I includes phages f2, R17 and MS2 whereas group II is represented by phage Qβ. The physiological properties of the phages belonging to the two groups are very similar with the exception of size—phage Qβ and its RNA are approximately 15% larger than the group I phages. As with all non-filamentous phages (and viruses) a limitation is imposed on the genome size of the RNA phages by virtue of the fact that the nucleic acid molecule must be packaged into the phage head. Consequently each RNA phage genome contains the information for only three proteins although in the case of Qβ limited amounts of a fourth protein species are synthesized as a result of read-through of a termination codon at the end of the coat protein gene (see the following discussion). The three phage genes are arranged in the order 5′-A protein–coat protein–replicase–3′; the Qβ read-through protein is termed the A_1 protein. In spite of the limited coding capacity resulting from the restricted genome size, the three phage genes are separated by non-translated intercistronic sequences which presumably function to ensure correct translational termination and initiation, and there are also highly conserved non-translated regions at the 5′- and 3′-ends of the genome which may be important for the initiation of replication of the genome. In the absence of any translational control mechanisms, it might be anticipated that approximately equal amounts of the three phage proteins would be synthesized following infection of a host cell. However, the amounts of the different proteins required for the assembly of new phage particles are in the order coat protein > replicase > A (and A_1) protein. Large amounts of coat protein—180 molecules per particle in the case of the group I phages—are required for the assembly of new phage heads. As the replicase functions catalytically and each phage particle contains only one single-stranded genome, smaller amounts of this protein are required. In the case of A protein, there is only one molecule per phage particle. The latter protein plays a role in phage adsorption and probably also in the penetration of the phage RNA into the host cell; in addition it may be required for the correct packaging of newly synthesized RNA into the phage head and is sometimes termed maturation protein. A role for the A_1 read-through protein which is found in variable amounts in Qβ particles, has yet to be defined. Translational control mechanisms operate to ensure the synthesis of the phage proteins in the required proportions.

Following infection of an *E. coli* cell by an RNA bacteriophage, host ribosomes bind almost exclusively to the initiation site of the coat protein gene in spite of the fact that by the criterion of mRNA:16*s* rRNA complementarity equally, or more, efficient ribosome binding would be expected to occur at the A and/or replicase initiation sites (see Fig. 6.15). It appears that the initiation sites for the A and replicase genes are initially inaccessible for ribosome binding because they are buried within the secondary structure of the RNA. Following disruption of RNA secondary structure, e.g. by heat or formaldehyde treatment, ribosomes can initiate translation at all three sites with an efficiency predicted from the known nucleotide sequences. In the cases of f2 and R17 RNA the sequence containing the replicase initiation site is complementary to a sequence of nucleotides early in the coat protein gene. During translation of the coat protein gene, any base pairs between these complementary regions will be disrupted and studies with phage carrying nonsense mutations in the first half of the coat protein gene have shown that the initiation site of the replicase gene becomes accessible to ribosome binding following incorporation of the first 50 amino acids of R17 coat protein and the first 70 amino acids of f2 coat protein.

This control by the secondary structure of the infecting phage RNA ensures that early in infection translation of the coat protein gene precedes that of the replicase gene. The product of translation of the replicase gene is the only phage-coded subunit of the functional RNA replicase, which is a tetrameric protein, the remaining three subunits being sequestered from the host cell following infection. These three host proteins are the translational elongation factors EF-Tu and EF-Ts and r protein S1, each of which plays an essential role in protein synthesis in the uninfected host cell. The phage-specified subunit is primarily responsible for the polymerization activity of the functional enzyme, whereas the subunits provided by the host are thought to be involved in template recognition. The synthesis of new copies of plus-strand phage RNA proceeds via the synthesis of complementary minus strands from the infecting plus-strand RNA template. The synthesis of minus strands occurs in the direction 5′–3′, i.e. the replicase initiates replication at the 3′-end of the template and moves towards the 5′-end. Ribosomes, however, translate the phage genes in the opposite, 5′–3′, direction and simultaneous translation and replication would set ribosome and replicase on a collision course. A role for the replicase in ensuring that such a collision does not occur has been implied from the demonstration that, *in vitro*, Qβ replicase binds strongly to two internal sites in the Qβ plus-strand RNA molecule (Blumenthal & Carmichael 1979; Fig. 6.20). One of these sites overlaps the ribosome

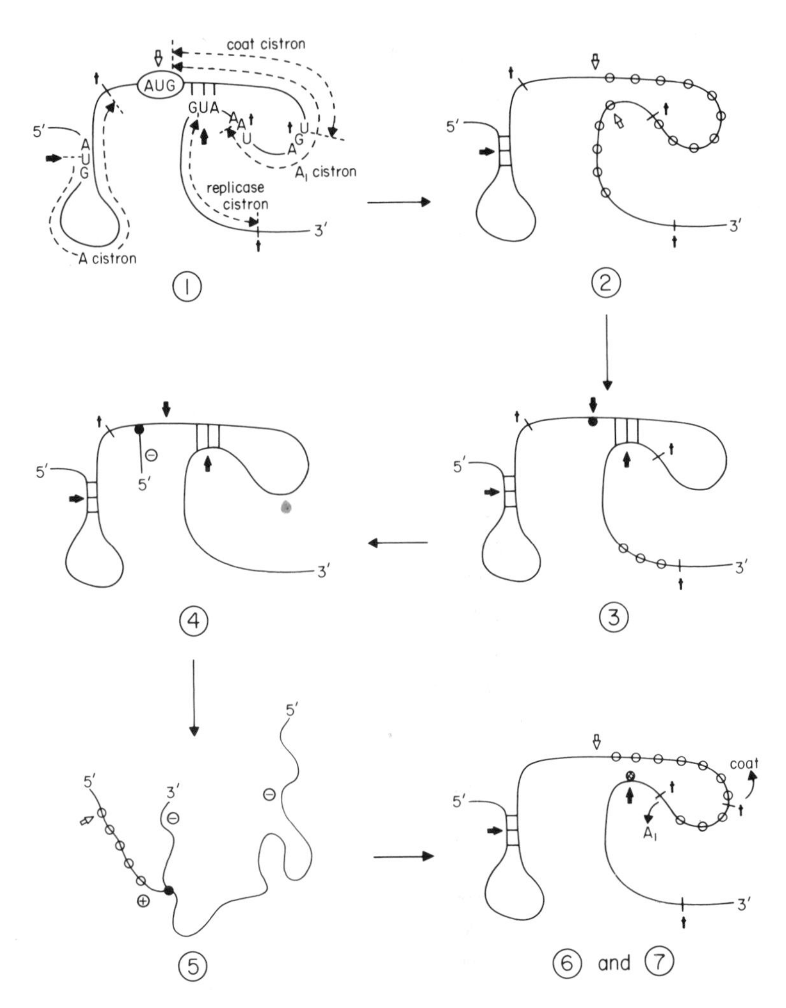

Fig. 6.20. Self-regulating translation and replication of Qβ RNA. ⇨ indicates initiation site accessible to ribosome binding. ➡ indicates initiation site inaccessible to ribosome binding. **t** indicates translational termination site. III indicates H-bonding for the maintenance of RNA secondary structure. • = replicase. ⊗ = coat protein. ○ = ribosome. Numbers in brackets refer to the following events after infection of the host cell. (1) Modulation of initiation by the secondary structure of phage RNA. (2) Translation of coat protein gene opens up RNA secondary structure thus enabling translation of the replicase gene. (3) Translational repression (i): replicase (one phage-coded subunit plus three host proteins) binds to plus strand ⊕ RNA—all translational initiation sites are inaccessible. (4) Replication (i): synthesis of RNA minus ⊖ strands. (5) Replication (ii): minus strands used as template for plus strands (no requirement for host factor protein unlike (4)); coupled translation and replication possible and A protein initiation site is transiently open before the nascent plus strands assume secondary structure. (6) Translational repression (ii): as its intracellular concentration increases, coat protein binds to plus strand RNA and prevents ribosome binding to replicase initiation site. (7) Translational read-through past coat protein termination codon results in synthesis of low level of A_1 protein.

binding site at the beginning of the coat protein gene and the second is located near the beginning of the replicase gene. Binding of the replicase to the site at the start of the coat protein gene precludes ribosome binding at the coat protein initiation site. Ribosomes already engaged in translation will dissociate at the termination codons of the coat or replicase cistrons and thereafter all translational initiation sites will be inaccessible for ribosome binding, either as a result of the secondary structure of the RNA or because of the presence of replicase at the coat (and possibly replicase) initiation site. The resultant ribosome-free RNA can then be used as a template for the synthesis of complementary minus strands.

The role of r protein S1 as an RNA-binding protein during the initiation of protein synthesis is described in section 6.5.1; it is absolutely required for replicase binding to both internal sites of the Qβ RNA molecule and, in addition, has been shown to bind to a pyrimidine-rich sequence close to the 3′-end of Qβ plus-strand RNA. An additional *E. coli* protein, termed host factor, is required for Qβ minus-strand RNA synthesis and has been shown to bind to the same site at the 3′-end of the plus-strand template as S1. Thus both S1 and host factor appear to be involved in the recognition of the replication initiation site at the 3′-end of plus strand Qβ RNA; host factor is not required for minus-strand-directed RNA synthesis and its role in the uninfected *E. coli* cell is not known. During the process of phage RNA replication, S1 may also function in unfolding the duplex between the nascent RNA and template RNA as the replicase moves along the template molecule. This would be analogous to the proposed role of S1 in unfolding the secondary structure of 16*s* rRNA, and possibly mRNA, during formation of the translational initiation complex (see section 6.5.1). The sole function of the newly synthesized minus-strand RNA molecule is to serve as a template for the synthesis of complementary plus strands. Both *in vitro* and *in vivo,* the product of Qβ replicase is predominantly plus-strand RNA, which is as expected since plus-strand RNA, in addition to serving as a template for minus-strand synthesis, is also translated and packaged into phage particles. There are a number of possible mechanisms to ensure the overproduction of plus-strand RNA. Host factor is only required for the synthesis of minus strands and this protein may be present in the host in limiting amounts. The Ef-Tu and EF-Ts subunits of Qβ replicase may enable more efficient recognition by the replicase of the 3′-end of minus-strand molecules than the 3′-end of plus-strand RNA. Some plus strands may be packaged into phage particles before they become available as templates for RNA synthesis.

Coupled translation and replication of nascent plus-strand RNA is possible since both ribosome and replicase will be moving in the same direction along the molecule. Indeed it appears that ribosomes can only bind to the A protein initiation site for the short period before the nascent strand is long enough to assume its secondary structure—during the replication of f2 plus-strand RNA the A protein initiation site becomes inaccessible for ribosome binding when the nascent strand length exceeds 500 nucleotides. Consequently, the synthesis of A protein only occurs for a short fraction of the phage RNAs intracellular lifetime.

At some stage following phage infection, sufficient RNA minus-strand molecules will have been synthesized to serve as templates for the synthesis of the required amount of plus-strand RNA which will, in turn, ensure synthesis of adequate amounts of A protein. The remaining requirement at this stage is for large amounts of coat protein; the intracellular concentration of this protein will have steadily increased throughout the subsequent infection period to a level sufficient to enable it to repress further replicase synthesis by binding to a site within the plus-strand RNA molecule which includes the replicase initiation site. The coat protein initiation site is now the only site accessible to ribosome binding and subsequent translation will yield the required large amounts of this protein. It is likely that significant amounts of the Qβ A_1 protein are also synthesized at this later stage of infection; approximately 3% of translating ribosomes read through the UGA codon at the end of the coat protein cistron and terminate translation at an ochre codon preceding the replicase initiation site to produce an A_1 protein considerably larger than the coat protein. As mentioned in section 6.2.6, the A_1 and coat proteins may be considered to be the products of two overlapping genes. The reason for the observed translational leakiness of a single UGA codon is not clear although it may reflect a low level of UGA suppressor tRNA in the wild-type *E. coli* cell. In contrast, the coat protein of MS2 RNA is terminated with the double termination sequence UAAUAG and no read-through protein corresponding to Qβ A_1 has been detected following infection by any of the group I RNA phages.

Although direct evidence for the translational repression of coat protein synthesis by replicase binding has only been obtained with Qβ replicase, and direct evidence for the translational repression of replicase synthesis by coat protein has only been obtained with the group I phages, it is highly likely that both mechanisms will be found to operate in both groups. Together with the control exerted by the secondary structure of the phage RNA these mechanisms operate to switch host ribosomes to the appropriate translational initiation sites at the correct time during the infectious cycle. Consequently the Qβ proteins are synthesized in the ratio 5 A protein molecules:100 coat protein molecules:3 A_1 protein molecules:25 replicase molecules. Some 40–60 min following infection of an *E. coli* cell with Qβ, the host cell lyses to release 10–40$\times 10^3$ Qβ particles, of which 10–15% are infectious. It is truly remarkable that the small, single-stranded RNA genome from a single phage particle can, with a little help from its host, direct the synthesis of 1–4$\times 10^3$ fully infectious phage particles in such a short period of time.

NOTE ADDED IN PROOF

Although we are still some way from having a complete topology of the ribosome (p. 211), comparison of the rRNA sequences from both small and large subunits of very diverse species has led C. Glotz, C. Zweib, R. Brimacombe and K. Edwards (*Nucl. Acids Res.* (1981) **9,** 3287–306; 3621–40) to describe the rRNA secondary structure domains and to show that these have been remarkably conserved.

Developments in the biosynthesis of extracellular proteins (p. 236) are reviewed by T.J. Shilvay, P.J. Bassford Jr. and J. Beckwith (1979) in *Bacterial Outer Membranes* (Ed. M. Inouye), pp. 203–54. John Wiley & Sons, New York.

Chapter 7

Replication of Eukaryotic Genomes

7.1 GENERAL INTRODUCTION AND DISTINCTIVE FEATURES OF GENE REPLICATION AND TRANSCRIPTION IN HIGHER ORGANISMS

The classical studies of Kornberg and Cairns (reviewed in Chapter 4) established the basic mechanisms of the replication of bacterial genes. The sheer complexity of the genome of eukaryotes, however, has raised critical problems in the design and interpretation of experiments on gene replication in higher organisms. After years of painstaking effort, investigators are only just beginning to unravel the processes of DNA replication and cell division in organisms other than bacteria.

For many years, it also looked as though studies on gene transcription in eukaryotes would be a slow, even painful, experience. Historically, the portents should have been favourable, because the first demonstration of a DNA-dependent RNA polymerase was made by Weiss in rat liver, and about this time Hurwitz, August and others were making considerable progress in the study of RNA replicases (RNA-dependent RNA polymerases) in viruses. Rather surprisingly, these innovative studies were generally overlooked and most researchers turned their attention to the relatively easier task of studying gene transcription in bacteria. However, technological advances in the last four years or so have brought about such a biochemical revolution that Chambon (1977) could entitle his erudite summary of a Cold Spring Harbor Symposium 'The Molecular Biology of the Eukaryotic Genome is Coming of Age.' His title was singularly apt, because extremely exciting and imaginative work has now established that gene transcription in higher organisms is fundamentally different from the corresponding process in prokaryotes. As Chambon puts it, 'the structure and function of higher genomes are not just those of a big *Escherichia coli*'.

Some of the distinctive features of gene replication and transcription in eukaryotes are presented in Table 7.1, and all aspects of this Table will be discussed in detail in the following sections of this chapter. The information in Table 7.1 is deliberately restricted to gene replication and transcription in the nucleus; DNA replication in mitochondria and chloroplasts has much in common with bacterial processes and Chapter 10 should be consulted for details.

Despite these marked differences, the fundamental similarities of macromolecular syntheses on a template of genomic DNA both in prokaryotes and eukaryotes should nonetheless not be overlooked. DNA polymerases and RNA polymerases of eukaryotic origin also

Table 7.1. Distinctive features of genetic replication and transcription in eukaryotes.

Property	Prokaryotes (e.g. *E. coli*)	Eukaryotes* (e.g. calf thymus, and liver, chick oviduct)
(A) DNA replication		
Number of replication forks	One only	Several thousand
Relationship to a cell cycle	Can be coupled	Strictly related; DNA synthesis exclusively in S-phase
Coupling to the synthesis of structural proteins	Not coupled†	Coupled to the synthesis of histones and tubulin
Coupling to the synthesis of components for DNA replication	Not coupled	Coupled to the synthesis of DNA polymerase α, thymidine kinase and DNA-unwinding proteins
Sensitivity of inhibitors		
(a) Colchicine	Insensitive	Sensitive
(b) Vinblastine	Insensitive	Sensitive
(c) Sulphanilamides	Sensitive	Insensitive
(B) DNA transcription		
Relationship to genetic translation	Coupled	Not coupled
Number of RNA polymerases	One	Three‡
Different subunit types in RNA polymerases	Present	Absent
Need for cAMP-binding protein in transcription	Needed	Not involved
Need for ppGpp in rRNA synthesis	Needed	Not involved
Processing of RNA transcripts	Minimal	Very extensive for all types of eukaryotic RNAs
Intervening sequences (introns) in genes	Absent	Present
Sensitivity to inhibitors		
(a) Rifamycin	Sensitive	Insensitive
(b) α-Amanitin	Insensitive	Sensitive (with the exception of RNA polymerase A)

*The properties listed may not be invariably true for all eukaryotes, e.g. certain fungi and yeasts.
†Coupling to structural components other than those of bacterial cell walls.
‡There are further distinctive types of RNA polymerase in chloroplasts and mitochondria.

initiate nucleic acid synthesis on 3′-OH ends of the DNA template in the 5′→3′ direction, using complementary nucleosides or 2′-deoxynucleoside 5′-triphosphates and with a mandatory requirement for Mg^{2+} or Mn^{2+} ions. DNA replication in eukaryotes is similarly a semi-conservative, symmetric and discontinuous process, where short, Okazaki fragments of replicated DNA are finally joined by DNA ligases and other factors. Likewise, RNA synthesis in higher organisms is a conservative, asymmetric continuous process, with stringent strand selection. Once a eukaryotic gene has been replicated or transcribed, however, the fundamental differences in macromolecular synthesis between prokaryotes and eukaryotes become very evident indeed. After replicating their long and complex genes, eukaryotes must segregate the newly synthesized chromosomes into the separate daughter cells and package them efficiently for enclosure within the spatially tight confines of the nucleus. Furthermore, after transcribing 'active' genes, eukaryotes subject all the primary RNA transcripts to elaborate processes of post-transcriptional modification. Additionally, mature species of RNA must be transported into the cytoplasm to enable them to fulfil their respective roles in protein synthesis (genetic translation). The emphasis on 'active' genes is important, because 90% or more of the genes in fully differentiated cells are never expressed, or put another way, 90% of eukaryotic genes are normally closed down and not transcribed by RNA polymerases. A central problem, then, is the selective activation of specific DNA sequences in the complex eukaryotic genome and further, how this activation is regulated in a precise and orderly manner. In this context, studies on eukaryotes have certain advantages: sensitivity to hormones, the influences of development, sexual maturation and differentiation and the identification of sex-linked attributes have all played an invaluable part in the elucidation of the complex structure and function of the eukaryotic genome. In addition, amplified (repeated) genes and chromosome abnormalities associated with clinical syndromes have also provided vital insights into the function and importance of individual genetic loci on specific eukaryotic chromosomes.

7.2 ORGANIZATION OF EUKARYOTIC GENOMES

7.2.1 The nucleus and its importance

The word, eukaryon, is taken from the Greek and simply means, with a nucleus. The English botanist Brown, in 1839, was the first to recognize the nucleus as a prominent structure of the cells of higher organisms. Subsequent investigators observed the division of living cells under the microscope and noted conspicuous structures or chromosomes, at certain stages of the division process. During division, morphological differences were noted in cells, leading to the formulation of the important concept of the cell cycle in eukaryotes. With the introduction of decolourized pararosaniline as a precise histochemical reagent by Feulgen and Rossenbeck in 1924, it was later shown that the nucleus and its associated chromosomes contained DNA.

The nucleus is the vital control centre of eukaryotic cells. Without it, the biochemical processes become totally disrupted and uncontrolled, resulting in cell involution and death. If the simple protozoan, *Amoeba proteus,* is briefly exposed to X-rays, the organism becomes quiescent and loses all the attributes of organized life, including its flowing or amoeboid form of locomotion. If a damaged cell is enucleated and receives a transplanted nucleus from a non-irradiated *Amoeba,* all cellular activities, including locomotion, are promptly restored. The nucleus is also the repository of most inherited characters. The unicellular alga *Acetabularia* exists in two subspecies, *A. mediterranea* and *A. crenulata,* both of which consist of a base, containing the nucleus, a long stalk and a morphologically distinctive cap. If the base region from one subspecies is grafted on to the stalk from either type, complete regeneration of the alga occurs, but the characteristic morphology of the cap is always dictated by the origin of the nuclear-containing base, never by the origin of the stalk. Ivarie *et al.* (1975) have demonstrated the importance of the nucleus in a more contemporary manner. Hepatoma tissue culture (HTC) cells respond to the synthetic steroid, dexamethasone, with the synthesis of a limited number of proteins, including tyrosine aminotransferase. HTC cells have a receptor mechanism for such glucocorticoids and the selective accumulation of dexamethasone in the nucleus stimulates the synthesis of the mRNA for tyrosine aminotransferase. This is a striking example of enzyme induction under conditions *in vitro.* However, the structural integrity of mammalian cells, including anchorage of the nucleus, is maintained by a fine network of actin-like microfilaments and this supporting substructure can be broken down by cytochalasin B, a drug obtained from the mould, *Helminthosporium dematoiderum.* Cytochalasin promotes enucleation of HTC cells, so that the induction of tyrosine aminotransferase by dexamethasone is lost.

It should be borne in mind that a functional interdependence exists between the nucleus and its surrounding cytoplasm. The nuclear DNA cannot be replicated in the absence of cytoplasm and nuclear activity is subject to stringent control by cytoplasmic factors. The continuous exchange of regulatory proteins between the cytoplasm and nucleus is firmly established and transplantation experiments between the amphibia, *Xenopus laevis* and *Discoglossus pictus,* illustrate the real significance of nuclear–cytoplasmic interactions. It is possible to produce nuclear-transplant hybrids, in which a nucleus of *X. laevis* is placed in the alien environment of *D. pictus* cytoplasm and vice versa. In the surrogate cytoplasm the normal pattern of nuclear genetic activity (notably transcription) is soon lost, leading to complete and early developmental arrest. Control nuclear-transplant hybrids, containing homologous nucleus and cytoplasm, develop in the normal manner.

Certain hereditary traits are not carried in the nucleus, or even in DNA. Both in mitochondria and chloroplasts, the DNA carries the genes for their unique species of rRNA and tRNA, plus a variety of proteins necessary for the function of these organelles. If yeasts are cultured for long periods of time in the presence of ethidium bromide, an interchelating agent for DNA, there is a complete breakdown of mitochondrial DNA and the production of small, mitochondria-depleted or 'petite' mutants. These survive only because yeasts can grow under totally anaerobic conditions. Other examples of cytoplasmic inheritance factors include the resistance of chloroplasts to streptomycin inhibition in the unicellular alga, *Chlamydomonas;* proteins for controlling pollination in corn and other cereals; and the lethal factors produced by certain rogue or 'killer' strains of the simple ciliate, *Paramoecium,* and yeasts. These lethal or κ-factors are proteinaceous toxins, surprisingly encoded in species of double-stranded RNA which are integral components of particles associated with cytoplasmic membranes. Normal strains of yeast and *Paramoecium* contain neither these particles nor these unusual 'RNA-type' genes (Vodkin & Fink 1973).

7.2.2 Eukaryotic chromosomes

Unlike bacteria, the more specialized and highly evolved eukaryotic organisms reproduce exclusively by sexual means. The remarkable work of T.H. Morgan on the fruit fly, *Drosophila melanogaster,* led to our understanding of the genetic basis of sexual reproduction. Except for the germ cells, each eukaryotic cell contains a constant and species-specific number of paired or *diploid* chromosomes, the total complement of individual chromosomes being termed the chromosome number. Man has a chromosome number of 46. Each homologous chromosome pair is made up of two daughter chromatids, firmly held together at a constricted region called the centromere. The chromosome pairs exhibit great differences in terms of length, shape and size, but on average they are between 1 and 15 μm long. The shape of each chromosome pair depends on the location of the centromere and four subclasses of chromosomes are recognized (Fig. 7.1); the range in structure is from telocentric (with the centromere at or very near the end of the chromosome pair) to metacentric (with a central centromere). The differences in chromosome structure are subtle, but nevertheless, so reproducible in a given species that they may be unequivocally identified by experienced cytologists, using modern histochemical techniques. Current practice favours either the fluorochrome, quinacrine mustard, or Geimsa strain, after limited digestion with trypsin to remove certain chromosome-associated proteins.

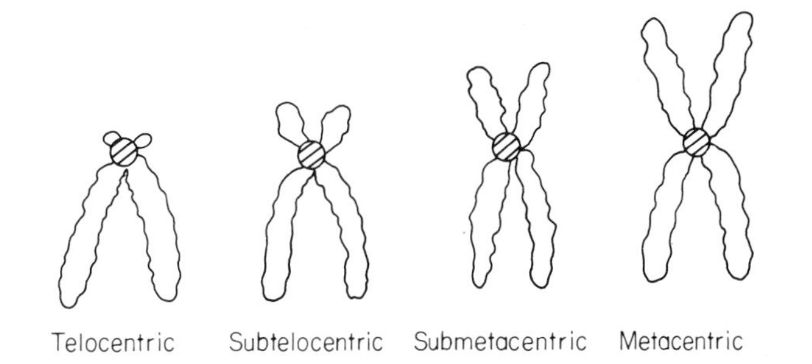

Fig. 7.1. The various shapes of eukaryotic chromosomes, dependent on the location of the centromere. The centromere in each case is represented by the hatched area.

One pair of chromosomes are unique in that they dictate the sex of the individual. In the human male, one unusual chromosome (Y sex chromosome) has a distinctive short, one-armed appearance and is paired with a submetacentric chromosome (X sex chromosome), so that the male-type pair of sex chromosomes is XY. In the human female, the sex chromosomes are alike, so that the female-type pair of sex chromosomes is XX. The remaining chromosomes, not associated with sex, are termed collectively the autosomes. The presence of distinctive sex chromosomes is the remarkable means by which evolution has initiated and maintained the differences between sexes. Rather than having a large battery of chromosomes or genes which are either male- or female-specific, the chromosomal difference in Man is but the single Y chromosome. However, the Y chromosome specifically directs the production of the male sex hormone, testosterone, and the somatic and psychic differences between the sexes are achieved by the hormonal activation of specific chromosomes.

The germ cells are unique in producing the male or female gametes necessary for sexual reproduction. The gametes vary enormously from one eukaryote to another, but essentially the female gonad produces a female gamete or egg and the male gonad a male gamete or spermatozoon. The gametes are formed by a special reductive means of cell division, *meiosis,* with the result that the gametes are unique in being *haploid* or contain single, unpaired chromosomes. During meiosis, only one chromosome from each diploid germ cell is passed randomly to the forming gamete, so that the complete gamete receives single chromosomes of either maternal or paternal origin. The female gamete derived from an XX germ cell can only carry a single X sex chromosome. By contrast, the male gamete derived from an XY germ cell can carry either a single X or Y chromosome; there are, therefore, two populations of spermatozoa, androsperm (X) or gynosperm (Y). In Man, each gamete will, in addition, contain 22 unpaired autosomes.

During reproduction, the haploid male and female gametes fuse, forming a single diploid cell, the zygote. During chromosome segregation and fusion, each chromosome pair of the zygote will contain a daughter chromatid (haploid chromosome) of either maternal or paternal origin. Fertilization of the female gamete (X) by androsperm (Y) or gynosperm (X) produce a male (XY) or female (XX) of the species, respectively. By successive rounds of cell division, *mitosis,* the zygote will proliferate sufficiently to form the completed offspring or sibling of the parents, each cell containing the identical complement of diploid chromosome pairs. Approximately 50 mitotic divisions will produce a human fetus. During these divisions, however, the cells acquire specialized functions by the process of differentiation. The actual basis for differentiation is still unclear, but chemical signals originating both from the environment and from adjacent cells are believed to interact with specific genes and thus dictate the pattern of development, and thus gene activation, within a given cell. The ultimate outcome is always remarkably accurate for a given species; for example, the simple nematode, *Oxyuris equi,* contains precisely 1 excretory cell, 18 midgut cells, 64 muscle cells and 251 nerve cells.

In view of the bewildering variety in the form and reproductive mechanisms in eukaryotic organisms, it is not surprising that the complement of autosomes and sex chromosomes is extremely variable. Representative data on this vast subject are summarized in Table 7.2. The remarkable differences in the sex chromosomes suggest that the fundamental basis for sexual reproduction is not always as simple as that described above for Man and the majority of other mammals.

Table 7.2. Chromosome variations in various species of animals.

Chromosome number		Sex chromosomes		
Species	Number	Species	Male	Female
Ascaris (roundworm)	2*	Man	XY	XX
Mosquito	6	Most Mammalia	XY	XX
Drosophila (fruit fly)	8	Insects (Orthoptera)	X	XX
Toad	22	Insects (Lepidoptera)	XX	XY
Pig	38	Birds	XX	XY
Mouse	40	Fish (Teleostea)	XX	XY
Rat	42			
Rabbit	44			
Man	46			
Monkey	54			
Horse	64			
Aulacantha (protozoan)	1600			

*The single chromosome pair does not refute the general concept of sex chromosomes as roundworms are hermaphrodites.

There are two biological advantages of sexual reproduction; extensive mixing of genes throughout a long line of ancestors and the possibility of rectifying defective genes. During gametic fusion, the zygote receives half its genes from either parent, but these in turn, will contain genes inherited from their own ancestors, and so on, right down the line of many generations. The result is a subtly mixed population of genes for a given species, a finding exploited in the initial breeding of new strains of animals and plants. Thorough mixing of DNA by repeated generations of breeding also permits the expression of evolutionary selection pressures. Further, if the gene on the haploid chromosome from one parent is defective, there is a reasonable chance that the corresponding gene donated by the other parent to the final chromatid pair will be sound. Despite the exploitation of sexual reproduction in creating animal strains with particular experimental advantages and in developing new hybrids of staple cereals and livestock, breeding between closely related humans is actively discouraged. Dangerous chromosomal defects could be perpetuated by such familial inbreeding.

Although eukaryotic cells initially have a uniform set of chromosomes, changes are known to occur. The presence of the Y chromosome is mandatory during early development of the male, but this sex chromosome is not always detectable in the cells of aged males. Although female cells have two X chromosomes, only one is active and the other becomes an inactive, densely staining Barr body

(Barr & Bertram 1949) at the periphery of the nucleus. The genes in this condensed X chromosome are not expressed, meaning that they are not transcribed into RNA. The random inactivation of one X chromosome is known as the Lyon effect (Lyon 1961) and explains why females heterozygous for glucose 6-phosphate dehydrogenase deficiency have two populations of erythrocytes; one has the normal amount of the enzyme, whereas the other is deficient. This enzyme deficiency is a sex-linked trait, the X chromosome specifying the enzyme in red cells but not other organs. The structure of the erythrocyte needs high levels of reduced glutathione and this, in turn, is maintained by the NADPH produced in the pentose phosphate cycle. As many as 10% of women are sensitive to haemolytic anaemia induced by antimalarial drugs because of this lack of reduced glutathione in their red cells. This gene mutation has not been bred out in the course of evolution because it provides protection against certain malarial parasites. The most important of

Table 7.3. The assignment of marker genes to specific human autosomes. The contents are largely taken from King (1975). Where known, assignment to the long or short chromosome arm is indicated by (l) or (s), respectively. Many enzyme names are in their trivial form, because these were used in naming specific loci.

Chromosome number	Marker (locus name)	Locus symbol
1 (s)	6-Phosphogluconate dehydrogenase	PGD
	Phosphoglucomutase	PGM-1
1 (l)	UDGP Pyrophosphorylase	UGPP
2 (l)	Isocitrate dehydrogenase	IDH-1
2 (s)	Malate dehydrogenase (cytoplasmic)	MDH-1
2	Interferon synthesis	If-1
4	Haemoglobin	—
5	Hexosaminidase B	Hex-B
	Interferon synthesis	If-2
6	Phosphoglucomutase	PGM-3
	Malic enzyme (cytoplasmic)	ME-1
7	Malate dehydrogenase (mitochondrial)	MDH-2
10	Glutamate-oxaloacetate transaminase	GOT-1
	Hexokinase (cytoplasmic)	HK-1
11 (s)	Lactate dehydrogenase (isoenzyme A)	LDH-A
12	Triosephosphate isomerase	TPI
	Citrate synthase	CS
14 (l)	Nucleoside phosphorylase	NP
15	Mannosephosphate isomerase	MPI
17 (l)	Galactokinase	GaK
	Thymidine kinase	TK
19	Glucosephosphate isomerase	GPI
20	Adenine deaminase	ADA
	Cholesterol synthesis	—

these, *Falciparum,* has a high requirement of reduced glutathione for optimal growth. Other X-linked disorders, particularly common in males, include neonatal jaundice, muscular dystrophy, haemophilia and colour blindness. The Y chromosome has only one known function, to promote sexual differentiation (see section 8.5.2); perhaps one other is to control testis size.

The ultimate objective in all genetic investigations is to assign specific gene function to individual chromosomes. Recent technological advances have enabled genetic markers to be assigned to specific chromosomes even in man (Table 7.3). In studies such as this, two approaches have proved invaluable, first, genetic hybridization of somatic human cells to lines of rodent cells and second, molecular hybridization of specific, ^{3}H-labelled RNAs to the DNA in mounted, cytological specimens. In fact, the mapping of human chromosomes is even more advanced than suggested in Table 7.3.

7.2.3 The complexity of eukaryotic DNA

The genome of eukaryotic cells is very much larger than that of a typical prokaryote, such as *E. coli* (Table 7.4). When fully extended, the DNA in a single chromosome from *Drosophila* and Man is 4.0 cm and 40.0 cm long, respectively. Conservative estimates suggest that the DNA in Man is long enough to represent 10^5 different genes. This extreme length makes eukaryotic DNA very prone to shearing forces during its isolation, yet by gentle means (Gross-Bellard *et al.* 1973), DNA of M_r 2.0×10^8 can be isolated. Such DNA preparations are free of even single-stranded breaks and are about 190 μm long (Fig. 7.2).

If calculations are made on the basis of the one gene–one enzyme concept of Beadle and Tatum, then eukaryotic DNA is very much longer than its calculated length for 10^5 genes of average size. An explanation of the complexity and length of the DNA of higher

Table 7.4. The relative size of haploid genomes in various organisms.

Organism	M_r	Number of nucleotide pairs in the genomic DNA
E. coli	2.5×10^{9}	3.8×10^{6}
Yeast, *Saccharomyces cerevisiae*	1.3×10^{10}	2.0×10^{7}
Fruit fly, *Drosophila melanogaster*	6.0×10^{10}	9.0×10^{17}
Man	1.5×10^{12}	2.3×10^{9}

Fig. 7.2. An electron micrograph of DNA from cultured CVI cells. The total length of the DNA is approximately 190 μm. The arrowhead points to one end of the molecule, the other being out of field. (Reproduced, with permission, from Gross–Bellard *et al.* 1973.)

organisms began with the illuminating work of Britten and Kohne (1968) on the rate of annealing (renaturation) of mammalian DNA. The rate of annealing of much of eukaryotic DNA was so rapid that certain base sequences were probably repeated many times over (see section 3.1.3). This was a vital clue. Subsequent work has established that the excessive length of eukaryotic DNA can be attributed to six factors which have little, if any, parallel in prokaryotes.

(1) Many genes are known to be highly reiterated or amplified.

(2) There are distinctive types of eukaryotic DNAs called satellite DNAs. These are readily demonstrated by equilibrium centrifugation in CsCl but they are not transcribed into RNA.

(3) There are spacer regions between genes which *may or may not* be transcribed. If transcribed, their contribution to the RNA transcript is removed by an excision process during the maturation of the RNA.

(4) The initial RNA transcripts of many eukaryotic genes contain intervening sequences of introns, which are subsequently excised during the process of post-transcriptional modification. A fundamental corollary is that eukaryotic genes are longer than their gene

products (RNA) and this clouds the unconditional application of the colinearity concept to higher organisms (Chambon 1977).

(5) Even in the active forms of mature eukaryotic mRNA, there are non-coding (non-translatable) regions at both the 5′- and 3′-ends; thus the mRNAs are actually longer than one would expect in terms of their function in protein synthesis.

(6) Finally, many RNA transcripts never leave the nucleus and play no part in cytoplasmic protein synthesis.

Therefore eukaryotic cells contain more and longer genes than anticipated. In addition, the long DNA of the genome contains regions which are replicated during cell division but never transcribed, e.g. satellite and certain spacer regions. Some examples of the variation in frequency of eukaryotic genes are given in Table 7.5.

Table 7.5. The relative frequency of eukaryotic genes.

	Class I	Class II	Class III
Description	Highly reiterated / Amplified	Reiterated	Single / Unique
Copies per diploid cell	10^2–10^3	10–10^2	1
% Total DNA	10	20	70
Examples	5, 18 and 28*s* rRNA	mRNAs for	mRNAs for
	mRNAs for	membrane proteins	α- and β-Globin
	histones	Proteins of seminal	Ovalbumin
	Proteins of prostrate secretion	vesicle secretion	Silk fibroin

Amplified genes and spacers

Oogenesis in Amphibia is accompanied by nucleolar proliferation and selective replication of the genes for 18 and 28*s* rRNA, to the extent that the mature oocyte remarkably contains 10^{12} ribosomes. About 75% of oocyte DNA is represented by amplified ribosomal genes. Without amplification, the assembly of this wealth of ribosomes would not be possible. After fertilization, there is an explosion of growth, and the ribosomes are essential to meet this enormous demand for protein synthesis. Skilful electron microscopy (Miller & Beatty 1969), combined with CsCl gradient ultracentrifugation, established the arrangement of these amplified genes (Fig. 7.3(a,b)). The genes are tandemly repeated, hundreds of times, each repeating unit of 18 and 28*s* rRNA genes being separated by a non-transcribed spacer region, about 5000 base pairs long. The tandem genes are about 3 μm long, which is sufficient DNA to

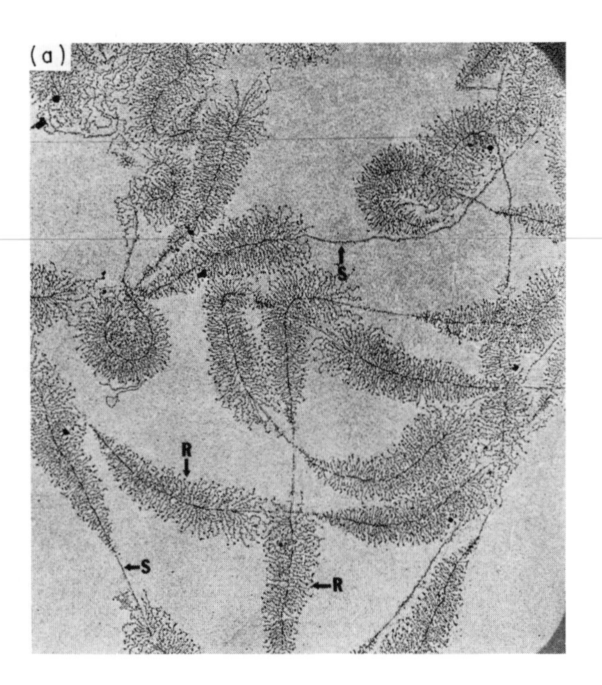

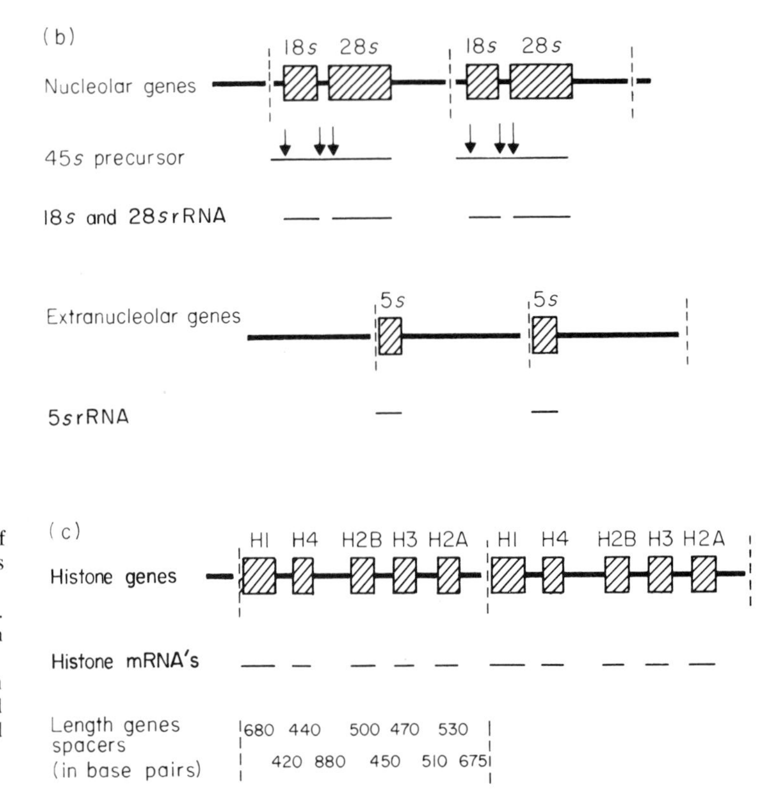

Fig. 7.3. The arrangement of some amplified genes in eukaryotes. (a) The tandem array of rRNA genes in the nucleolus of *Xenopus laevis* oocytes. The delicate filaments, displayed like arrowheads, represent the synthesis of 40*s* rRNA precursor, R; the tip of each arrowhead is the initiation point of transcription. The filamentous regions, S, are non-transcribed spacers. (Reproduced, with permission, from Miller & Beatty 1969.) (b) The arrangement of nucleolar 18*s* and 28*s* rRNA genes and the extranucleolar 5*s* rRNA gene in *Xenopus;* the presence of transcribed and non-transcribed spacers in the nucleolar genes should be noted. Cleavage points are indicated by arrows. (c) The arrangement of histone genes in the sea-urchin, *Psammachinus miliaris;* the length of genes and spacers are given in numbers of base pairs. In both (b) and (c) the genes are represented by hatched boxes, spacers by solid bars and RNA transcripts as solid lines.

encode for these two species of rRNA. The initial 45*s* transcript is subsequently cleaved to produce 18 and 28*s* rRNA. The genes for 5*s* rRNA are not in the nucleolus, but distributed on many of the extranucleolar chromosomes; again, these genes are tandemly repeated yet the spacer in this case is much longer than the actual gene (Fig. 7.3(b)).

Histone genes are also amplified in sea-urchin oocytes (for a review see Kedes 1979) to meet the problem of DNA packaging during the rapid cycles of DNA replication after fertilization. Again, the genes are tandemly repeated, one gene for each of the five histones per unit and separated by spacers of varying length (Fig. 7.3(c)). The spacers are AT-rich and the histone genes GC-rich and thus readily separated when sheared sea-urchin DNA is subjected to ultracentrifugation in CsCl gradients. A similar picture applies to the histone genes in *Drosophila,* where they are located on chromosome 2.

Satellite DNA

Britten and Kohne (1968) established that certain sections of eukaryotic DNA contained repeated or repetitive sequences of DNA. A distinction should be made here between the use of 'reiterated' and 'repetitive' in terms of DNA sequences. Reiterated is generally used for DNA of unique sequence which is repeated or amplified many times over in the haploid genome. Repetitive is generally reserved for a base sequence which is repeated many times *within a given DNA molecule.* This is simply illustrated in the following diagram.

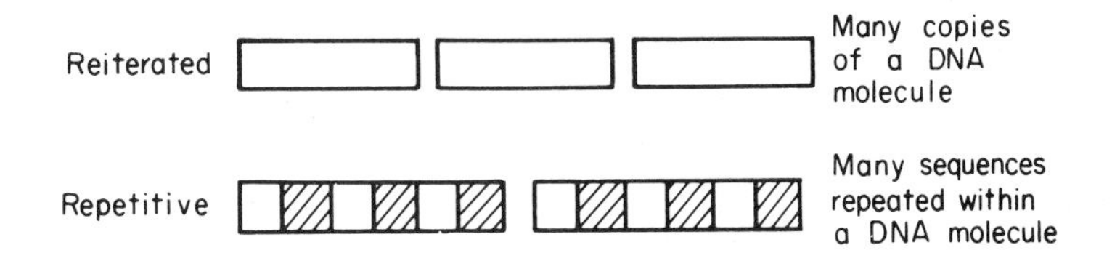

Many investigators examined sheared eukaryotic DNA by centrifugation to equilibrium in CsCl gradients and identified a series of satellite DNAs, distinctive in buoyant density to the main bulk of the DNA; for example, Gall *et al.* (1973) found three satellite DNAs in *Drosophila virilis* (Fig. 7.4(a)). Satellite DNA is found in varying amounts in different species, but can account for up to 30% of eukaryotic DNA. Without exception, satellite DNA is composed of

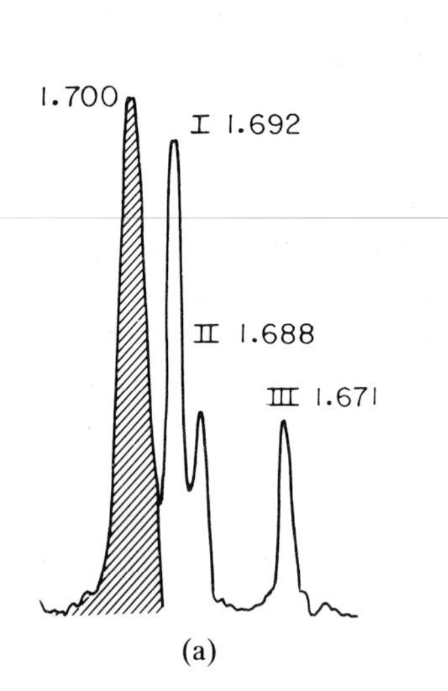

(a)

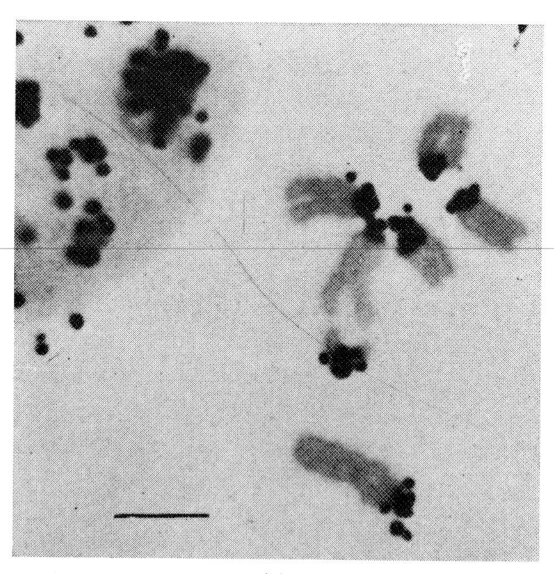

(b)

Fig. 7.4. Satellite DNA of eukaryotes. (a) A densitometer tracing showing the three satellite DNAs in *Drosophila virilis,* demonstrated by Gall *et al.* (1973) in a CsCl gradient. Non-satellite DNA is indicated by hatching. (b) Radiograph showing the deposition of silver grains over the centromeres after hybridization *in situ* with a ^{3}H-labelled RNA probe made from satellite DNA and *E. coli* RNA polymerase. Silver grains are also evident in the interphase nucleus (top left field). (Reproduced, with permission, from Pardue & Gall 1970.) The length of the bar is 5 μm.

repeated sequences and is thus highly repetitive DNA. Examples of repeated sequences include the following:

Crab	A-T-A-T-A-T	repeat every two bases
Rat	G-C-A-C-A-C-G-C-A-C-A-C	repeat every six bases
Drosophila	A-C-A-A-A-T-T-A-C-A-A-A-T-T	repeat every seven bases

The three satellites isolated by Gall *et al.* (1973) in *Drosophila* are composed of repeating units of three hepanucleotides:

I	II	III
5′ ACAAACT 3′	5′ ATAAACT 3′	5′ACAAATT 3′
3′ TGTTTGA 5′	3′ TATTTGA 5′	3′TGTTTAA 5′

Using in-situ nucleic acid hybridization on mounted specimens of separated chromosomes, it has been established that satellite DNA is concentrated largely in the centromeres (Pardue & Gall 1970). The chromosomal DNA in the specimens was heat denatured and then exposed to either satellite DNA labelled with [^{3}H]thymidine *in vivo* or to [^{3}H]RNA made with satellite DNA as template *in vitro.* After washing the specimens, radioactive hybrids were located by radioautography; the resultant silver grains were deposited over the centromeres (Fig. 7.4(b)). Centromeric DNA is not transcribed, but the repetitive sequence in satellite DNA may play a role in the separation of replicated, daughter chromosomes during mitosis, and also maintain the structure of chromosome pairs throughout interphase. The real necessity for satellite DNA is still not clear, for it is absent from certain eukaryotes. Additional roles for satellite DNA have been proposed (for a review see Bostok 1980). Its wide variation both in amount and nucleotide composition precludes any regular role in coding terms for satellite DNA. Furthermore, satellite DNA is not evenly distributed between chromosome pairs and can vary in amount rapidly. Based on these observations, it has been suggested that these repeated DNA sequences may provide an additional means of altering the overall genotype of an organism during evolution. The presence of satellite DNA may, therefore, be advantageous to eukaryotes, particularly those with long generation times, enabling an organism to adapt to changing environments.

Pseudogenes

New and unexpected insights into the complexity of eukaryotic genomes have come from molecular analysis of the genes coding for the globin polypeptide chains of haemoglobins. From studies on a wide range of mammalian species it is now evident that there are more globin-related genes in a given genome than are necessary to encode the known globin polypeptides (for a review see Proudfoot 1980). Sequence analysis of these extra globin-like genes shows that they have extensive changes in nucleotide sequence relative to the normal, expressed globin genes. These nucleotide changes are such that either gene expression is prevented or the globin polypeptide would be biologically inactive (i.e. aberrant gene expression). These additional, gene-like elements have been termed pseudogenes, meaning tracts of DNA exhibiting significant homology to a functional gene but containing mutations which prevent their expression. Pseudogenes are best considered as relics of evolution. Extrapolating from the present findings on the globin genes, Proudfoot (1980) has suggested that one in four DNA sequences may be pseudogenes, or as he neatly puts it, 'one quarter of our genes might be dead'. Pseudogenes could, therefore, contribute significantly to the size and length of eukaryotic genomes, without contributing to their store of expressable genetic information.

Selfish or functionless DNA

Independently, in back-to-back papers, Doolittle and Sapienza (1980) and Orgel and Crick (1980) drew attention to the fact that the operation of natural selection within genomes leads to the appearance and persistence of DNA sequences which play no part in phenotypic expression and have no known function. Nonetheless, these tracts of 'selfish' DNA would be replicated along with functional (expressed) genes during cell division and mitosis. These stimulating papers aroused widespread interest and debate (for example, see four consecutive papers in *Nature (London)* (1980) **285,** 617 *et seq.).* Together with the identification of pseudogenes, the concept of selfish DNA raises important aspects of the organization and complexity of eukaryotic genomes and their evolution; these issues will be reviewed later. It is now clear that eukaryotic DNA has a unique length, containing long tracts with no known function; this situation has no strict parallel in prokaryotes.

Transposable elements or transposons in eukaryotes

As shown by the classical studies of McClintock on maize, it seems that eukaryotes contain transposable genetic elements, capable of movement from one site on a chromosome to another site on the same or a different chromosome. These transposable elements of DNA, or transposons, can promote chromosomal rearrangements and thus modify gene expression. Transposons introduce an element of genetic flexibility into eukaryotic genomes which runs counter to the traditional views of genetic stability and consistency in the association of specific genes with particular chromosomes. There is a lively interest in transposons at the present time because of their relevance to both heredity and development (for a review see McKay 1980). Transposons have been studied in detail in prokaryotes, especially enterobacteria (see section 4.3.7). While studied in great detail in molecular terms, the biological importance of certain transposons remains somewhat obscure; for example, the transposons *copia,* in *Drosophila,* and *Tyl* in yeasts, share considerable homology and both are widely scattered throughout the chromosomes of both organisms. One outcome of the transposition of *copia* and *Tyl* is the formation of unstable mutants which have changes in the expression of structural genes for histidine-metabolizing enzymes. The biological relevance of other eukaryotic transposons is clearer. The mating type of many yeasts is controlled by a different site-specific recombinational system (see section 7.4.3). The protozoon, *Trypanosoma brucei,* causes sleeping sickness in humans by being able to escape from the immune response of the host. The infective organism achieves this by having remarkable flexibility in the protein components (or antigens) of its surface proteins. A battery of transposable genes code for individual surface proteins and these are moved to special sites in the genome for their expression. Transposons raise interesting questions about eukaryotic DNA. First, the traditional view of conservative gene arrangements and unchanging sequences of nucleotides in DNA may not be consistent with the findings of contemporary research. Second, differentiation or other changes in gene expression may be partly achieved by transposition of genetic elements within the genome. As in prokaryotes, therefore, there is a dynamic or flexible aspect to the base sequence of eukaryotic DNA.

Palindromic sequences in eukaryotic DNA

Palindromes are regions of double-stranded DNA which contain an axis of twofold rotational symmetry; such structures are unique in that they are 'read' identically from either end during DNA replication or transcription.

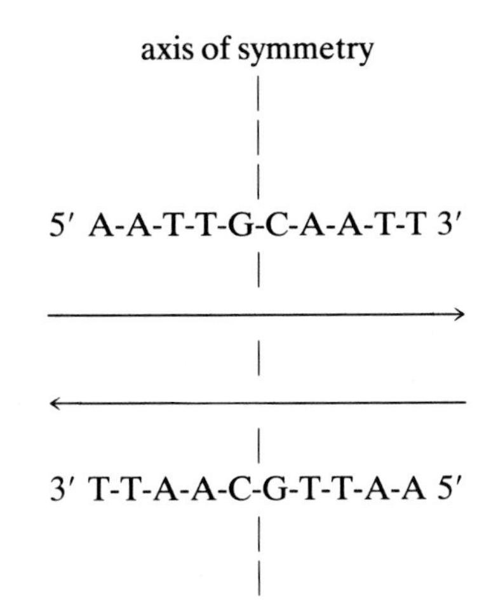

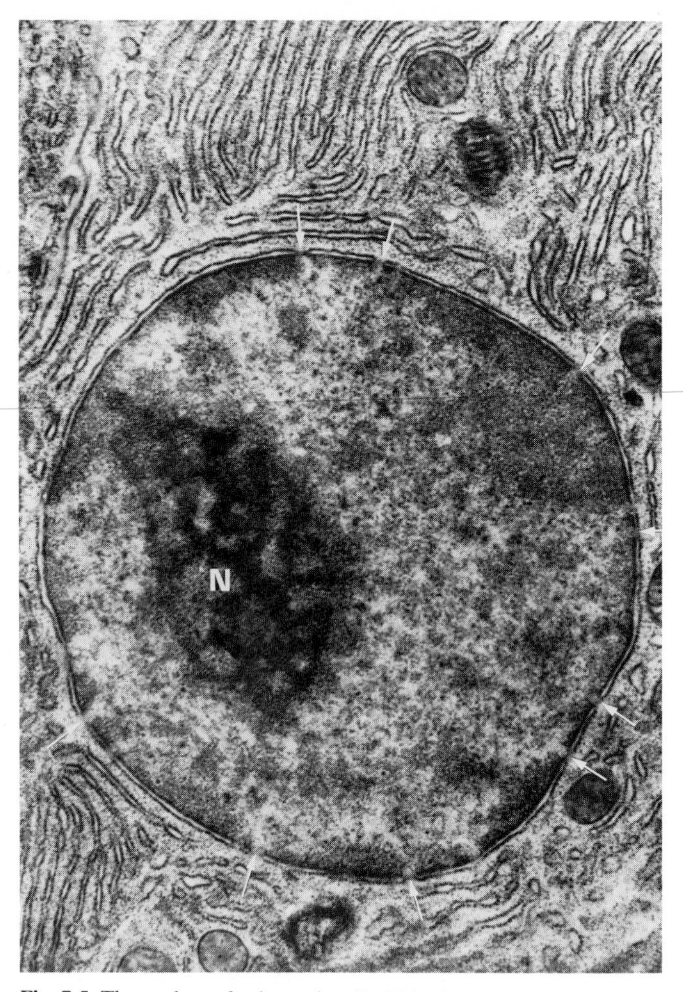

Fig. 7.5. The nucleus of eukaryotic cells. This electron micrograph shows the nucleus in a section of rat pancreas. The two membranes of the nuclear envelope contain pores, indicated by the arrows. The nucleolus, N, is particularly prominent. (Reproduced, with permission, from Fawcett 1966.)

Combined sedimentation and electron microscopic studies were used by Wilson and Thomas (1974) to identify palindromes as fold-back or hairpin structures in HeLa DNA. Palindromes are found every 10–80 μm in eukaryotic DNA, the regularity being dependent on the species. The rRNA genes in particular have been described as giant palindromes and contain large numbers of hairpin structures.

7.2.4 The structure of the nucleus

Very soon after the discovery of DNA by Miescher, microscopists examined histological sections with basic dyes, and in 1882 Flemming described the presence of densely staining chromatin in the nucleus. He further distinguished condensed, deeply staining *heterochromatin* from extended, weakly staining *euchromatin*. Chromatin is a complex of DNA and nuclear proteins. Of the latter, the basic histones were first isolated by Kossel in 1884 and non-histones were first studied by Stedman and Stedman in 1949. The average nucleus has a diameter of 5μm and a volume of 65 μm^3. Even with the electron microscope, little detail is seen of the nuclear interior, except for differences in particulate structures, equivalent to heterochromatin and euchromatin. However, the nucleolus, one or more per cell, is a conspicuous structure (Fig. 7.5). The nuclear periphery is delineated by the nuclear envelope, a flattened sack, composed of two membranes and between which is the perinuclear space. The nuclear envelope is an extension of the membranes of the endoplasmic reticulum and its outer surface is associated with ribosomes. Nuclear pores occur at intervals along the envelope as roughly circular or polygonal structures; they are vital in facilitating nuclear–cytoplasmic exchanges, particularly in the translocation of RNA. The centrioles of the cell are paired structures which are often found outside the nucleus. During cell division, they migrate to the edge of the nucleus and play a significant part in forming the mitotic spindle (see also Fig. 7.10.).

The organization of the nuclear interior

For many years, the structure of chromatin was an elusive enigma but a dramatic change has occurred over the last five years. Plausible explanations have now been found for the structural importance of histones and for the efficient packaging of eukaryotic DNA. The true importance of non-histone proteins and nuclear RNA in regulating chromatin function is also becoming far clearer.

Nucleosomes or ν bodies

The concerted attack by biochemists and electron microscopists has led to revolutionary new insights into chromatin structure (for a review see Kornberg 1977). Three approaches proved particularly important: limited digestion of chromatin with DNAases; studies on histone–histone interactions and reconstitution experiments between histones and DNA. Many contributions were outstanding, but particular mention should be made of the work of Olins and Olins (1974), Kornberg (1974) and Oudet *et al.* (1975). Their work laid the foundations for the nucleosome concept. Two molecules each of histone 2A, 2B, 3 and 4 interact and form a spherical particle or nucleosome, around the outside of which is

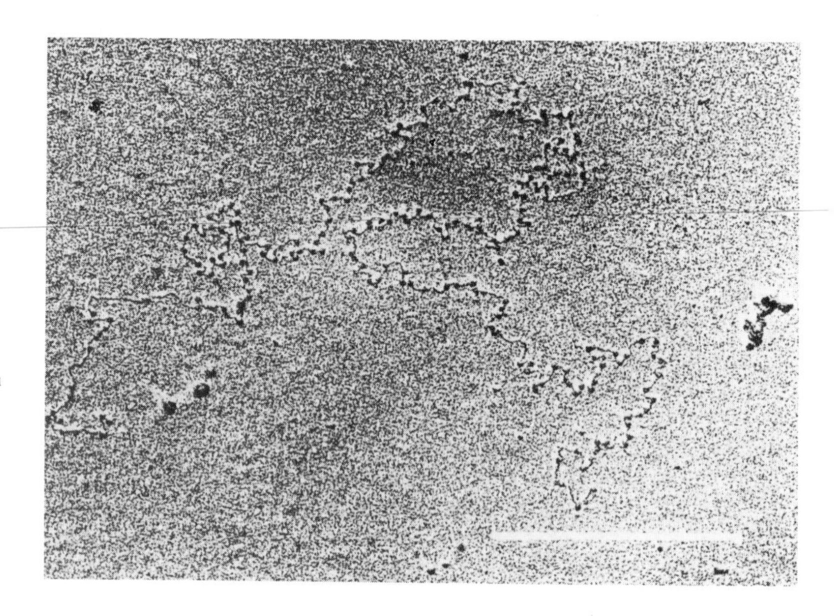

Fig. 7.6. The presence of nucleosomes in eukaryotic chromatin. The electron micrograph shows the appearance of chick liver chromatin, after removal of histone 1, resembling beads on a string; each bead is a nucleosome, composed of histones (except histone 1). The bar represents 0.5 μm. (Reproduced, with permission, from Oudet *et al.* 1975.)

wound the DNA. The nucleosomes are distributed along the DNA like 'beads on a string' (Fig. 7.6) and separated by bridges or linker regions of largely protein-free DNA.

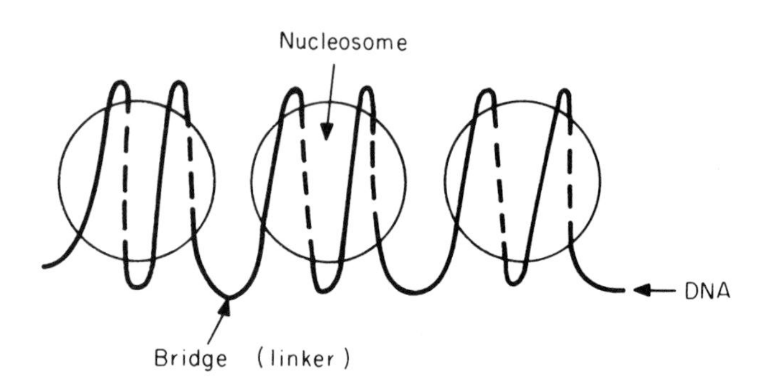

Exposure of chromatin to many DNAases, but particularly micrococcal nuclease (*Staphylococcus aureus*), leads to selective digestion of the linker DNA, releasing the individual nucleosomes or core particles. These have a diameter of approximately 10 nm and contain 200 base pairs of nuclease-resistant DNA. At high concentration of nuclease, the DNA of the core particles reaches a lower limit of 140 base pairs. The linker DNA is of variable length, depending on the location of the nucleosomes, but generally in the range 10–100 base pairs long.

Removal of histone 1 facilitates the conversion of chromatin into the looser, beaded structure seen in electron micrographs and this histone is believed to fulfil an important condensing role in maintaining chromatin structure. Experiments by Finch and Klug (1976) indicate a further level of organization in chromatin, in which the beaded structure is further condensed by packaging into a solenoid, with a pitch of 11 nM. Again, histone H1 is responsible for the stabilization of the solenoid. The mean number of nucleosomes per solenoid turn is six or seven.

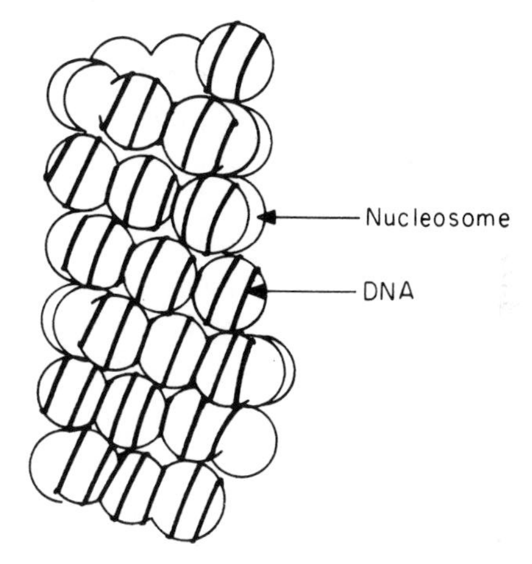

The structure of chromatin, particularly in the solenoid form, depends heavily on the involvement of histone 1. This lysine-rich histone is composed of three distinct functional regions or domains, namely N-terminal, central and C-terminal domains. Of these, only the globular or central domain is in a folded configuration and is essential for packaging and condensing the DNA in chromatin (Allan et al. 1980). Even without the flanking regions, the central domain is able to impose supercoiling in DNA to the extent of two full turns and to locate the histone 1 molecule as a whole within the nucleosome. In keeping with its important packaging role, the central domain shows the highest conservation of amino acid sequence during evolution. Histone 1 can also be readily phosphorylated and this promotes condensation of chromosomes as a necessary prelude to cell division (Matsumoto *et al.* 1980).

The precise location of DNA in chromatin is not known. Certainly it is wound round the outside of the nucleosomes and the conformation of the DNA, especially in terms of strand separation, is not influenced by interaction with the core of histone tetramers. Coiling of the DNA within each nucleosome does introduce one negative (or left-handed) superhelical turn. The packing or compaction ratio for

DNA in the solenoid is about 40:1 in the direction of the solenoid axis. The development of the nucleosome concept has done much to solve the problem of packaging the extremely long DNA in eukaryote chromosomes.

The nucleosome is an invariant feature of eukaryotic cells, from primitive to advanced organisms. This applies even to yeasts, which lack histone H1; their nucleosomes may thus be 'looser'. Beaded nucleosome-like structures can also be generated by the interaction of histones 2A, 2B, 3 and 4 with the closed, superhelical circles of DNA from viruses, bacteria and mitochondria; again, histone 1 is not required. In summary, it seems that the nucleosome is the invariable outcome of DNA–histone interactions under appropriate conditions.

As reviewed by Laskey and Earnshaw (1980), proteins other than histones are involved in nucleosome assembly within the intact cell. Topoisomerase I (a nicking-closing enzyme) is an abundant protein in chromatin of all eukaryotes which enhances DNA–histone interactions. The nomenclature of topoisomerases is discussed in section 7.3.3; an alternative name for this enzyme is DNA swivelase (see section 4.2.2). Another acidic protein, nucleoplasmin (M_r 29 000), plays a part in nucleosome assembly. It does not bind appreciably to either DNA or nucleohistone (DNA–histone) complexes, but it will interact with histones directly and is capable of mediating nucleosome assembly even in the presence of a vast excess of histones. The latter conditions strongly inhibit topoisomerase I.

Early studies indicated that nucleosome formation was random with respect to DNA sequence, and that nucleosomes did not restrict transcription of the DNA by hampering the access of RNA polymerases. Recent work suggests that this may be an over-simplification (e.g. Levy-Wilson *et al.* 1979). After limited digestion of nuclei from trout testis with micrococcal nuclease, these investigators isolated two subclasses of nucleosome monomers or core particles, termed MN1 and MN2. These differ in their content of high-mobility group (HMG) proteins, especially one called type 6. HMG-6 was present in MN1 but absent from MN2. The DNA from MN1 and MN2 nucleosomes had considerable overlap in terms of sequence, except that each nucleosome subset contained unique sequences of DNA not present in the other. On the basis of these studies, nucleosomes may not necessarily have an identical structure and, moreover, the distribution of nucleosomes may not be random with respect to the base sequence of DNA. Later work on the alignment of nucleosomes on the 5*s* rRNA genes in *Xenopus* confirms a *non-random* distribution of nucleosomes with respect to the base sequence of DNA (Gottesfeld & Bloomer 1980). This careful study was aided by the availability of restriction fragments of cloned 5*s* rRNA genes.

The histones have been intensively studied and many of their amino acid sequences are now known (Isenberg 1979). Essentially five types are known, H1, H2A, H2B, H3 and H4, with the nucleated erythrocytes of birds, amphibia and reptiles containing a unique histone, H5. The histones are ubiquitous in eukaryotic organisms, but certain protozoa and fungi may not contain H1. Although some variations in amino acid sequence have been detected, the histones have been highly conserved during evolution.

The other group of nuclear proteins are the non-histone proteins. For many years they resisted detailed analysis. Despite early difficulties presented by their relative insolubility and their tendency to aggregate, methods have now been developed for the reproducible separation and analysis of non-histone proteins. The number of different types of non-histone cannot be assessed accurately, because they show remarkable species and even organ diversity. In addition, many are present in trace amounts and can only be identified by tagging with radioactive tracers, e.g. phosphorylation by [γ–^{32}P]ATP. As a guide, HeLa cell nuclei contain about 450 different non-histone proteins, present in 10^2–10^6 copies per haploid genome (Peterson & McConkey 1976). Between 10 and 20 non-histones, however, may well form the bulk of these nuclear components. Establishing the location of non-histone proteins in chromatin is a matter of pressing urgency, but it is proving to be a daunting task. Striking success in locating the non-histones in Drosophila has recently been achieved (Elgin *et al.* 1977) using antibodies raised against homogeneous, non-histone proteins. Another approach is to 'lock' the chromatin in living cells by cross-linking reactions (Cech *et al.* 1977). Living cells can be exposed to derivatives of psoralen, which are activated by light (wavelength, 360 nm) and cross-link the DNA via the pyrimidine bases.

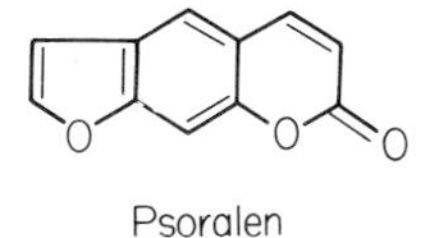

Psoralen

Methods such as these may soon reveal the precise location of non-histone proteins in active chromatin. A vast body of evidence suggests that non-histones play a vital role in controlling the expression (transcription) of eukaryotic genes during differentiation, development and hormonal stimulation (for a review see Stein *et al.* 1974). Current evidence suggests that histones play little part in regulating the transcription of template DNA in chromatin. Some examples of the chemical composition of eukaryotic chromatin and the properties of nuclear proteins are presented in Table 7.6.

Table 7.6. The properties of chromatin and nuclear proteins.
(a) The chemical composition of chromatin.*

Cell type	DNA	RNA	Histone	Non-histone
Rat liver	1.0	0.04	1.15	0.95
Pig cerebellum	1.0	0.13	1.61	0.52
Chick erythrocyte	1.0	0.02	1.08	0.54
Calf thymus	1.0	0.05	0.89	0.21
HeLa cell	1.0	0.05	1.08	0.78

(b) The properties of histones and non-histones.

Property	Histones	Non-histones
Number	Five†	300–600‡
Distribution	Universal; highly conserved	Species- and tissue-specific
M_r	11 000–21 000	7000–80 000
pI	7.5–10.5	3.9–9.2
Amino acid sequence	Known	Not known
Function in:		
structure of chromatin	Formation of nucleosomes, DNA supercoiling	Formation of nuclear matrix
transcription	Probably little	Both activators and inhibitors

*Relative to DNA as 1.0.
†Certain erythrocytes and yeasts may be different (see text).
‡Probably a conservative underestimate.

Many species of small (sn) RNA are present in the nucleus. While their functions are not fully understood, they are distributed throughout the nucleus, in the nucleolus, nucleoplasm and the nuclear membrane. Many are associated with chromatin and are believed to help in the maintenance of nuclear structure. Certain sn RNAs (193, 171 and 127 nucleotides long) are covalently linked to the RNA of chromatin in HeLa cells (Pederson & Bhorjee 1979). DNA-linked RNA may be involved in the tertiary structure of chromatin, particularly during the S-phase of the cell cycle. These structural and regulatory roles of sn RNA may be the attachment of chromatin to the nuclear skeleton or nuclear protein matrix. There is growing evidence that sn RNAs may play a part in directing splicing enzymes to their correct cleavage sites during post-transcriptional processing (Roberts 1980b, Lerner *et al.* 1980).

Nuclear protein matrix

The size and morphology of the nucleus are not static. Classical experiments by Graham and Harris showed that the nucleus enlarges significantly as a necessary prerequisite for the initiation of RNA, and particularly, DNA synthesis. The migration of cytoplasmic proteins into the nucleus was later found to be the trigger for nuclear swelling but structural components of the nucleus responsible for these dynamic changes in shape have only recently been identified. If nuclei are exposed sequentially to hypotonic swelling, extraction with 2M NaCl and Triton X-100 and digestion with nucleases, the final structure is a delicate meshwork of fibres, called the nuclear protein matrix (Berezney & Coffey 1974). It is this subnuclear structure that gives the nucleus its dynamic properties. The nuclear protein matrix (Fig. 7.7) contains no histones, but has small amounts of metabolically active DNA and RNA, associated with essentially three non-histone proteins (M_r 62 000, 67 000 and 69 000; Berezney & Coffey 1977). The network of protein fibres is particularly prominent around the nucleolus and the periphery of the nuclear matrix contains the structures of the nuclear pores. The nuclear matrix clearly plays a vital part in the initiation of the synthesis of DNA and RNA (for a review see Berezney & Coffey 1977). When cells are exposed to pulses of radioactive precursors of nucleic acids, the specific activity of the DNA and RNA associated

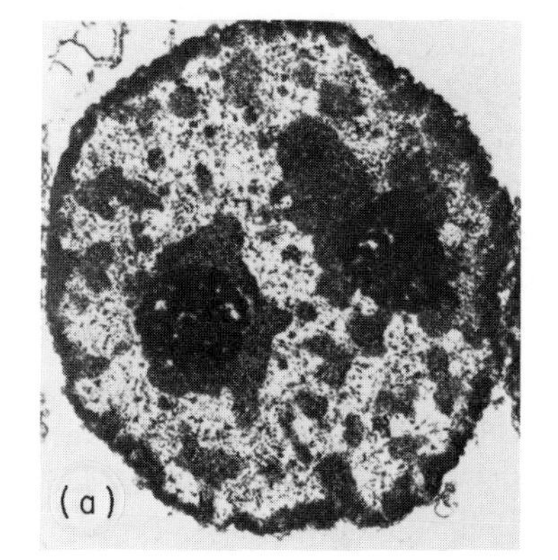

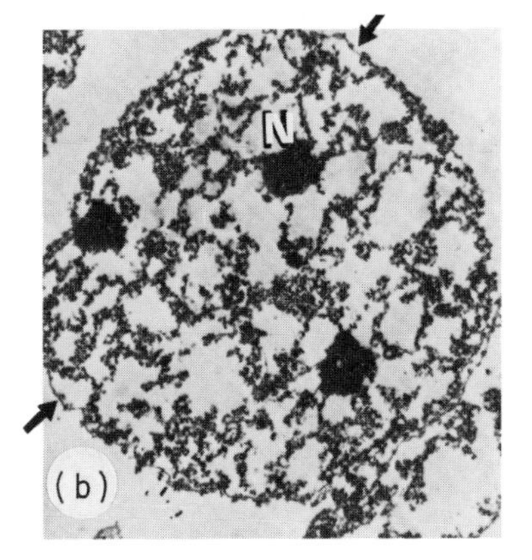

Fig. 7.7. The nuclear protein matrix in eukaryotic cells. (a) A nucleus from rat liver, showing two nucleoli and dense chromatin close to the nuclear envelope. (b) The residual nuclear protein matrix, composed of >98% protein and virtually no DNA or RNA. The matrix is particularly prominent around residual nucleolar structures, N, and also contains remnants of the nuclear envelope, marked by arrows. (Reproduced, with permission, from Berezney and Coffey 1974.)

with the matrix is vastly higher than that of the bulk of nucleic acids in the remainder of the nucleus.

Chromatin reconstitution

Despite its complexity, it is possible to reconstruct chromatin from DNA, histones and non-histones into a form resembling that in the intact cell. Prompted by the early studies of Gilmour and Paul (1971), erythropoietic cells have been particularly useful in this context because the fidelity of reconstitution can be checked by the accurate transcription of the single α- and β-globin genes by exogenous RNA polymerase. It was in reconstitution experiments of this type that the crucial role of non-histone proteins in the regulation of gene transcription was established. Non-histones from erythropoietic cells promoted the expression of the globin genes, whereas non-histones from other sources could not. Gadski and Chae (1976, 1978) have demonstrated that non-histones must be reassociated with the DNA before the histones for the globin genes to be faithfully transcribed. These important experiments underline both the structural and functional importance of non-histone proteins. In the subsequent histone-DNA interactions, histones 3 and 4 probably attach first, followed by histones 2A and 2B.

Covalent modification of nuclear proteins

It has been known for some time that both histones and non-histones can be modified by a variety of enzyme systems present in the nucleus; particular attention has been given to methylation, acetylation and phosphorylation reactions (for reviews see Sheimin *et al.* 1978, Isenberg 1979). The modified amino acid residues include the N-terminal residue and the frequent ϵ-amino groups of lysine. Many modification reactions are rapid and occur in both the nucleolus and the extranucleolar nucleoplasm (for a review see Jungman & Kranias 1977). The subtle weakening of DNA interactions by this means has been widely accepted as an early step in the activation of genes.

Another interesting mechanism involved in the covalent modification of nuclear proteins involves the transfer of moieties of ADP-ribose from NAD^+, so that the protein finally contains a long tract of poly (ADP-ribose). This process is carried out by ubiquitous eukaryotic enzymes, the poly (ADP-ribose) synthetases, according to the following reaction:

$$xNAD^+ + Protein \rightarrow (ADP\text{-}ribose)_x - Protein + xNMN + xH^+.$$

These enzymes have been widely studied (for a review see Hayaishio Ueda 1977) and completely purified from many sources. Current evidence suggests that poly ADP-ribosylation of nuclear proteins is significant in maintaining chromatin structure and in promoting the early phase of DNA replication (this point is discussed in section 7.3.4). In its structural role, poly (ADP-ribose) can cross-link two histone H1 molecules together into a dimer–polymer complex, histone 1-poly (ADP-ribose)-histone 1. The location of poly (ADP-ribose) synthetase in the linker region between nucleosomes suggests that this form of covalent modification of histone 1 could be important in the supercoiling of DNA into the final solenoid structure of chromatin.

7.3 DNA REPLICATION

7.3.1 Introduction

DNA replication in eukaryotes is a complex process. The presence of many chromosomes and the sheer length of their DNA introduce difficulties not encountered in bacteria. Nonetheless, the replication process is relatively efficient; for example, the helical DNA in the largest chromosome of *Drosophila* is 2.1 cm long and contains no less than 6.2×10^7 H-bonded base pairs. Replication in *Drosophila* proceeds at about 2600 base pairs min^{-1} and if there were only one origin of replication, the synthesis of complete daughter strands of DNA would take some 16 days to complete. That this is achieved in less than 3 min indicates that DNA replication proceeds in a bidirectional fashion and requires the cooperative activity of more than 6000 replication forks or origins. This important concept is illustrated in Fig. 7.8. As reviewed by Sheimin *et al.* (1978), considerable efforts have been made to characterize the replication units in eukaryotes and the units vary considerably in size from one system to another. The entire process by which a diploid eukaryotic cell divides is called the cell cycle and this begins and ends with the division event itself, or mitosis. However, whereas DNA synthesis takes place almost continuously in rapidly dividing bacteria, eukaryotic cells duplicate their DNA only during a restricted phase of the cell cycle, the S(DNA synthetic)phase.

Many of the components necessary for DNA replication in higher organisms have now been characterized but the overall process is still

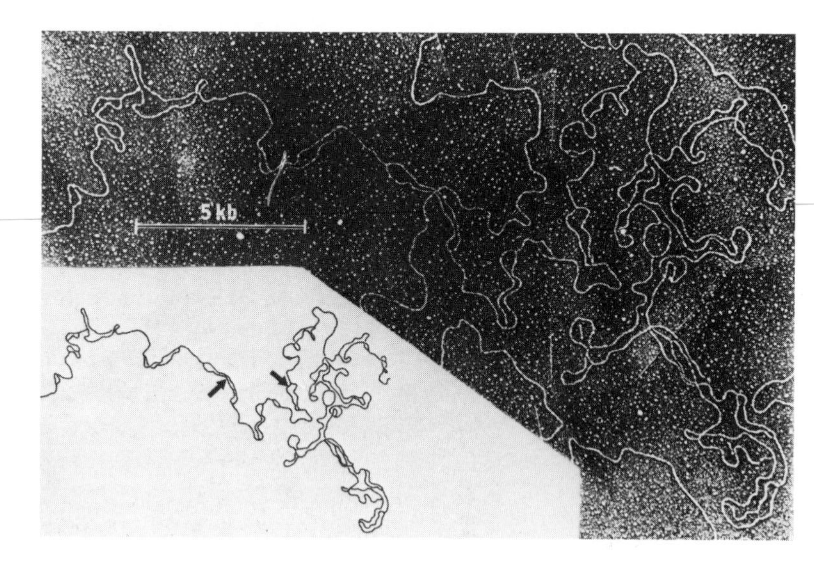

Fig. 7.8. The multiple replication forks in eukaryotic DNA. This electron micrograph shows the replication of DNA in the eggs of *Drosophila melanogaster*. The long tract of DNA contains many centres of replication, or areas of DNA destabilization. Selected replication centres or 'eyes' are marked by arrows on the line drawing (inset). The bar indicates the length of 5 kb of DNA. (Reproduced, with permission, from Kreigstein and Hogness 1974.)

not fully understood. Several critically important facts have emerged from these recent endeavours. First, the components for DNA replication in eukaryotes demonstrate considerable inter-species and evolutionary variation. Second, the most important enzyme for replication, DNA polymerase α, has little, if any, activity towards a template of helical DNA and the necessary unwinding or denaturation of DNA, prior to replication, is the subject of widespread interest. Third, few DNA polymerases of eukaryotic origin have associated $3' \rightarrow 5'$ and $5' \rightarrow 3'$ exonuclease activities like their bacterial counterparts. Fourth, explanations must be found for the extensive coupling of the S-phase of the cell cycle to the synthesis of many proteins engaged in DNA replication. In addition, more information is needed on the assembly of the mitotic apparatus which effects the separation of replicated chromosomes. Finally, an insight is only just emerging into the necessary involvement of RNA polymerase in DNA replication in higher organisms. Clearly, much still remains to be learned, but in his summary of a Cold Spring Harbor Symposium, Stahl (1978) is optimistic that an explanation of the mechanism of eukaryotic gene replication may soon be within reach.

7.3.2 The cell cycle and mitosis

The concept of the cell cycle was launched by the pioneering work of many cell biologists, notably Howard and Pelc; the cycle

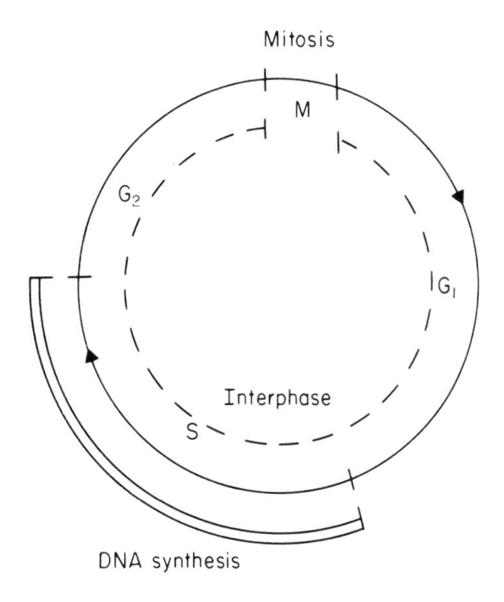

Fig. 7.9. The cell cycle in a mammalian cell.

is presented in Fig. 7.9. Most of the recent work on the cell cycle has been conducted on mammalian cells in culture under conditions where their rate of division can be experimentally controlled (for a review see Pardee *et al.* 1978). In most cell lines, the duration of the three phases S (6–8 h), G_2 (2–6 h) and M (1 h) is fairly constant, the variation in cell cycle time being mainly due to differences in the length of the G_1 phase. There is also considerable variation in the duration of the G_1 phase between the individual cells of a single population. Certain evidence suggests that this variability can be explained by phenotypic variations in cells imposed at birth. Eukaryotic cells can exist in a quiescent or non-proliferating state for long periods; this may be days under conditions of cell culture or even months, if ever, in the intact animal. Clearly, in most differentiated cells, division is highly suppressed. Several explanations have been forwarded to explain this quiescence in the G_1 phase. The first possibility, discussed at length by Prescott (1976), is that cells can enter a novel state, G_0 which is actually outside the conventional cell cycle and from which they can only re-enter the G_1 phase with considerable difficulty. The second alternative, proposed by Dell'Orco *et al.* (1975) is that cells are never really quiescent and are passing, albeit very slowly, through a greatly extended G_1 phase. The final alternative stems from the imaginative paper by Smith and Martin (1973) in which they question whether cells are actually cyclic at all. Smith and Martin suggest that once mitosis is complete, cells enter an indeterminate A state in which they are not specifically preparing for replication. The probability that a cell in a given population will leave the A state per unit time and make the crucial transition into the determinate B state is in fact *random*. The B state is essentially equivalent to the S, G_2, M and G_1 phases of the traditional cell cycle and here cells are committed to divide. The probability of transition from state A to B is determined by both the cell type and by environmental conditions. As a corollary, changes in the *rate of proliferation* (cell division) are thus accounted for by changes in *transition probability*. An impressive body of evidence supports the transition probability model, including kinetic analyses of cell division on changing the environmental conditions of culture with respect to cell density, hormones and serum concentration. A modified probability model, based on the necessity for two random transitions, has recently been proposed (Brooks *et al.* 1980).

Since eukaryotic cells divide within the confines of a cell cycle whereas prokaryotic cells do not, there has been considerable debate on whether a common mechanism determines the rate at which all cells undergo division (for a review see Nurse 1980). Deterministic models have been proposed to explain the proliferation of prokaryotic cells; the fundamental premise is that the rate of division is determined by the rate of growth of the cell. Further, it is envisaged that some key event, or trigger, for division can only occur when a cell has attained a critical size. Deterministic models accurately describe the division of certain eukaryotic cells, including bud emergence in budding yeasts and division in fission yeasts. The alternative, transition probability models were proposed to explain division in higher organisms. In attempting to reconcile these contrasting models, it could be argued that the controls for division are fundamentally different in prokaryotes and eukaryotes, but considerable evidence suggests that this is not true. It is possible that deterministic and probabilistic models describe different aspects of a common mechanism for controlling the rate of division of all types of cells. Nurse (1980) concludes that only further experimentation can resolve this issue, and in particular, kinetic investigations on the duration of phases of the cell cycle in conjunction with measurements of cell size. He also argues that the behaviour of centrioles, or their equivalents, during these further experiments, would be valuable.

Brooks *et al.* (1980) emphasize a striking similarity between their modified probabilistic model (or cycle) for cell division and the centriole cycle in eukaryotes. The centrioles undergo ultrastructural changes during cell division, and in particular, they duplicate close to the entry of mammalian cells into the S phase. Accordingly, centrioles are convenient markers for monitoring cell division (see also Fig. 7.10).

There is considerable evidence now to support the idea of Pardee (1974) that for cells to begin to grow they must pass a point of previous restriction (or a halt) and then be fully committed to the process of DNA replication. The division of yeasts has been investigated in detail and in these cells, the point of restriction is colloquially termed 'start'. Two genes, *whi 1* and *whi 2* have been unequivocally identified in *Saccharomyces cerevisiae,* whose function is to enable the yeast cell to traverse through 'start' into cell division. If this point of restriction is not overcome, cells remain in the A state (or alternatively, in more conventional terms, the G_1 phase) and fail to proliferate. Cells can be stimulated to divide by artificial means, including the manipulation of nutritional and growth factors in the media for cells in culture, surgically induced regeneration (such as partial hepatectomy) or by the administration of steroid hormones (notably sex hormones) to immature or castrated animals. Studies on such systems indicate that extensive biochemical changes must occur as a necessary prerequisite to the initiation of DNA synthesis. Unquestionably, the driving forces for DNA replication are protein and RNA synthesis. No single biochemical response can be said to be the critical trigger for DNA replication, rather the initiation of cell division requires the concerted synthesis of many vital components and regulatory factors. As reviewed by Sheimin *et al.* (1978), a broad spectrum of inhibitors of eukaryotic DNA replication are available and inhibition of protein synthesis by any mechanism results almost immediately in the shut-down of cell division and entry into the S phase is prevented. There are lines of cultured cells with temperature-sensitive defects in amino acid activation or the synthesis of single ribosomal proteins in which DNA replication can only occur at permissive temperatures.

It has been known for some time that histones are synthesized only during the S phase of the cell cycle. The regulation of histone synthesis is further discussed in sections 8.4.6 and 9.3.5, but these observations have wider implications. Studies by Chiu and Baril (1975) and Rennie *et al.* (1975) have established that many necessary components for DNA replication are active only in the S phase. In particular, DNA polymerase α, DNA ligase, DNA-unwinding proteins and thymidine kinase are conspicuously active during the phase of DNA replication and decline to barely detectable levels thereafter. These fluctuations in activity are a reflection of accelerated synthesis of these replication components during the

S phase, after which their synthesis abruptly ceases. It would appear, therefore, that the prerequisite for protein synthesis in DNA replication is partially explained by the synthesis of the replication machinery itself.

During the last phase of the cell cycle, or mitosis, cell division occurs and the chromosomes replicated during the S phase are equally distributed between the two daughter cells. Just prior to mitosis, the centrioles move away from the periphery of the nucleus and take up positions at opposite sides of the cell as discrete structures, the mitotic poles (Fig. 7.10). Each centriole has a smaller counterpart, disposed at right angles, and microtubules are laid down around each centriole pair in a radiating fashion, forming two star-like structures, the asters. Plants do not contain centrioles and their mitotic apparatus has a more diffuse structure altogether. The successive phases of mitosis are prophase, metaphase, anaphase and

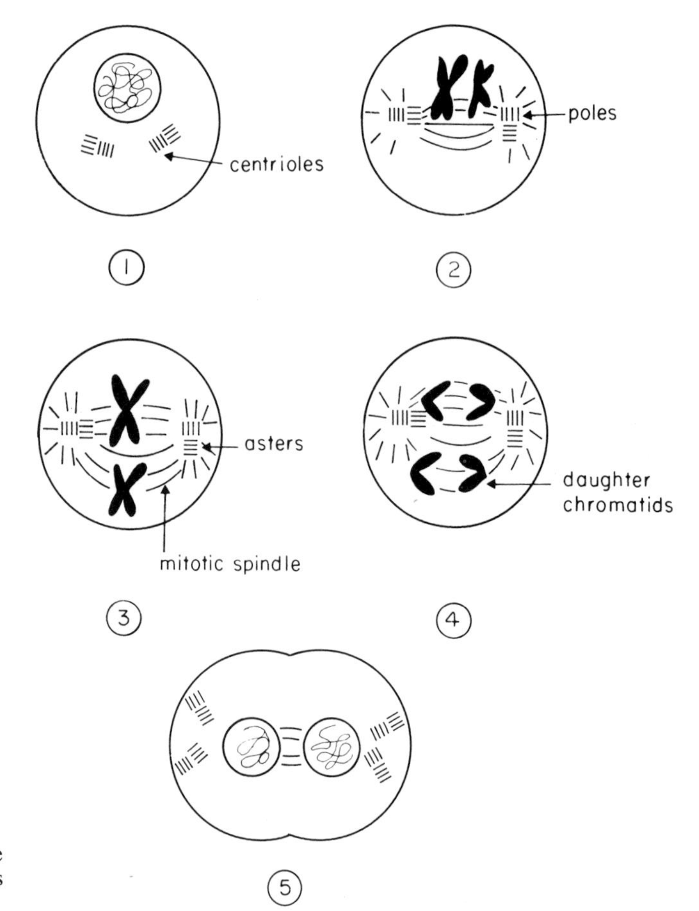

Fig. 7.10. The stages of mitosis: (1) interphase, (2) prophase, (3) metaphase, (4) anaphase and (5) telophase. For simplicity, only one pair of homologous chromosomes, plus their replication copies, is shown.

telophase. During prophase, more microtubules are deposited between the poles, forming the mitotic spindle, an important structure with an inherent ability to bind the centromeres on the replicated chromosomes. The chromosomes condense during prophase and become far more prominent in appearance. The chromosomes are present as daughter chromatids, each of which was formed from one of the two identical, helical DNA molecules formed during the S phase. Dissolution of the nuclear membranes and loss of the nucleolus end prophase. At metaphase, the chromosomes are lined up on the mitotic spindle in the centre of the cell, forming a distinctive structure, the metaphase plate. The centromere now divides, enabling the daughter chromatids to separate. During anaphase, the separated chromatids or daughter chromosomes move towards opposite poles, seemingly by contraction of the microtubules in the mitotic spindle. The actual mechanism of chromosome migration remains uncertain. In the concluding telophase, new nuclear membranes are formed around each set of daughter chromosomes and the cell membrane pinches in, or invaginates, producing the two daughter cells. In plants, telophase is characterized by the construction of a new cell wall right across the centre of the parent cell under division. Many stages of mitosis are sensitive to inhibition (see section 7.3.3 for details), but as an

Table 7.7. Cell cycle mutants in eukaryotic organisms.

Cell type	Mutant nomenclature	Site of block*	Function affected
Yeast (*Saccharomyces cerevisiae*)	cdc 6 cdc 7	Nuclear division	Entry into S phase
	whi 1 whi 2	G_1	Entry into S phase
Chinese hamster	ts K/34C	G_1	Synthesis membrane glycoproteins
	ts 41	G_1	Assembly 60*s* ribosome subunit
	K-12	G_1	Entry into S phase
	cs3-D4	G_0†	Progressive $G_0 \rightarrow G_1$
	ts-H1	G_1	Leucyl-tRNA synthetase
Syrian hamster	ts 546	M	Transition, metaphase → anaphase
	ts 665	M	Progression of prophase
	BF 113	G_1	Entry into S phase
	ts AF8	G_1	Entry into S phase
Mouse	ts 2	M	Unspecified as yet
	ts B54	G_1	Entry into S phase

* The conventional cell cycle is used here, as this was the nomenclature employed by most authors.
† This conclusion may finally need reassessment because of the uncertainty surrounding the G_0 phase, but this was the explanation given in the literature.

example, colchicine arrests cells in metaphase, a fact exploited during microscopic analysis of chromosomes.

In recent years, several mutations in eukaryotic organisms have been reported with specific blocks either in mitosis or other stages of the cell cycle (Table 7.7). Mutants such as these will play an invaluable part in future studies.

7.3.3 Enzymes and other components for DNA replication

Proteins for destabilizing double-stranded DNA

It was forcibly emphasized by Weissbach (1977) that eukaryotic DNA polymerases cannot use native, helical DNA as a template for DNA synthesis directly. A prerequisite for DNA synthesis is thus a destabilization or change in the conformation in the DNA template. The current literature is full of reports on proteins which can bind to DNA, thereby destabilizing it, but confusion in their nomenclature has tended to obscure their real significance in DNA synthesis. Fortunately, the excellent review by Champoux (1978) has helped to clarify the situation. By definition, helix-destabilizing proteins bind tightly and preferentially to single-stranded DNA, often lowering transition temperature (T_m) for the helix→coil transition, lowering the optimum temperature for strand renaturation and inducing a conformation suitable for replication. There are also a group of proteins that alter the conformation of DNA. First, topoisomerases, which have the combined action of an endonuclease and a polynucleotide ligase, and relax closed, circular superhelical DNA; alternative names include DNA-untwisting or nick-and-joining enzymes, DNA swivelases or relaxing proteins. Second, DNA-unwinding enzymes catalyse strand separation in helical DNA and derive the necessary energy from the concomitant hydrolysis of ATP or other high-energy triphosphate. Third, DNA gyrase enzymes couple the hydrolysis of ATP to the introduction of negative superhelical turns in closed circular DNA. As yet, DNA gyrase activity has not been found in eukaryotic cells.

Helix destabilizing proteins have been identified in a diversity of eukaryotic cells, including plants, fungi and many mammalian sources. These stimulate the homologous DNA polymerase α and bind also to double-stranded DNA to some extent. This suggests that these proteins may bind to and denature DNA, rather than simply extending and stabilizing single-stranded regions already present. The role of such proteins in DNA replication is clearly important. It should also be remembered that RNA polymerase would fulfil such a role.

Topoisomerases have been found in many eukaryotic cells and specifically in the nucleus. From all sources, the enzyme is of M_r 60 000–70 000 and is distinguishable from its bacterial counterpart in being able to remove both negative and positive superhelical turns in DNA. The direct involvement of these enzymes in the replication of eukaryotic DNA, other than that in viruses, seems extremely limited. Certainly, topoisomerase activity remains essentially constant throughout the cell cycle; its activity is not modulated by cell division. One role of topoisomerase is important in all cells, namely, the attachment of histones to newly synthesized DNA (Laskey *et al.* 1977).

Many eukaryotic proteins have been isolated by their selective retention on affinity columns containing immobilized, double-stranded DNA. These have been termed DNA-unwinding proteins on the basis of two criteria:

(a) to effect the retention of [^{3}H]-labelled helical and superhelical DNA on nitrocellulose membranes;

(b) to stimulate DNA synthesis by DNA polymerase α on its 'non-preferred' template of native, helical DNA.

The unwinding of double-stranded DNA needs concomitant hydrolysis of ATP as the driving force, and the *rep* protein from *E. coli*, for example, has the dual properties of a DNA-unwinding protein and a DNA-dependent ATPase. The eukaryotic DNA-unwinding proteins do not have an associated ATPase activity, but separate DNA-dependent ATPases have been widely identified in higher organisms.

In summary, the formation of the many replication forks in eukaryotic DNA during the late G_1- and early S-phases of the cell cycle requires the concerted action of at least four distinct types of proteins; DNA-destabilizing proteins, RNA polymerases, DNA-unwinding proteins and DNA-dependent ATPase. Topoisomerases are needed only in the later attachment of histones to the replicated DNA (for a review see Falaschi *et al.* 1980).

DNA polymerases

For many years, the status of eukaryotic DNA polymerases was rather chaotic. Several factors contributed to this uncertainty; the enzymes were poorly characterized and anomalously found in cytoplasmic rather than nuclear fractions of tissue homogenates, and no consistent nomenclature was observed. The acceptance of a standard nomenclature (Weissbach 1977) and the purification of enzymes to a state approaching homogeneity has greatly improved matters and a summary of the physicochemical properties of the principal DNA polymerases in higher eukaryotes is presented in Table 7.8.

Table 7.8. The properties of mammalian DNA polymerases.

	Enzyme type			
Property	α	β	γ	mt (mitochondrial)
M_r	120 000–300 000	30 000–50 000	150 000–300 000	150 000
Effect of N-ethyl maleimide	Inhibited	No effect	Inhibited	Inhibited
Preferred template	Destabilized DNA with RNA primer	Any DNA	Any DNA; RNA preferred	Any, even super-helical DNA
Associated activities				
exonucleases	No*	N.D.†	N.D.	N.D.
DNA-dependent ATPase	Yes	N.D.	N.D.	N.D.
Optimal KCl or NaCl concentration	<25 mM	100–200 mM	100–250 mM	100–250 mM
Nuclear location	Yes	Yes	Uncertain	No

*There is only one report of such an association.
†N.D. = no reliable data.

Great difficulties have been encountered in the purification and characterization of the DNA polymerase α, because of its tendency to aggregate and to adsorb other proteins non-specifically. Despite earlier reports, it is unquestionably located within the nucleus, although it is readily released into the cytoplasm during homogenization in media of low ionic strength. If nuclei are released by treatment with cytochalasin B, or prepared in organic solvents or glycerol, over 90% of the DNA polymerase α is retained in the nucleus. The molecular architecture of DNA polymerase α remains uncertain. Most reports suggest that the enzyme moiety is a single polypeptide chain, M_r 155 000–170 000, but alternatively, it may be a dimer composed of two dissimilar subunits, M_r 76 000 and 66 000. The enzyme has other proteins associated with it in a tight, functional complex. The possible association of exonuclease activity, like bacterial polymerases, has not been substantiated. However, the enzyme is probably complexed in the nucleus with a DNA-dependent ATPase, M_r 60 000. The template specificity of DNA polymerase α

has been examined in detail by Spadari and Weissbach (1975). The enzyme cannot utilize double-stranded DNA, and works best on nuclease-treated DNA containing hairpin structures and many 3′-OH ends. Whereas it can utilize the copolymer poly[d(A-T).d(A-T)], it cannot use synthetic ribohomopolymers or RNA as templates. Importantly, DNA polymerase α catalyses the addition of polydeoxyribonucleotides to an RNA primer synthesized on the DNA template with *E. coli* RNA polymerase.

The low molecular weight DNA polymerase β is not ubiquitous and is absent from protozoa, plants and fungi. DNA polymerase β has been purified to homogeneity from many sources and is immunologically distinct from other eukaryotic DNA polymerases. This enzyme can use a wide variety of templates, including the ribohomopolymer, poly(A), but generally works better on polydeoxyribonucleotide polymers such as poly(dA) and poly(dC) than the corresponding polyribonucleotides, poly(A) and poly(C).

DNA polymerase γ is present in only small quantities in most eukaryotic cells, but is not yet available in a highly purified form. It may not even be a distinct species of eukaryotic DNA polymerase, since the important paper by Bolden *et al.* (1977) suggests that mitochondrial (mt) and type γ DNA polymerases are probably identical.

Despite the uniformity of enzymes for DNA replication in highly developed organisms (Weissbach 1977), lower eukaryotes contain a diversity of DNA polymerases, the evolutionary significance of which remains to be established; for example, the fungus, *Ustilago maydis,* contains two DNA polymerases, both of M_r 1 000 000. One polymerase is associated with a 3′→5′ exonuclease and a range of temperature-sensitive mutants with specific defects in DNA replication are available. The cilated protozoan, *Tetrahymena pyriformis,* contains a single DNA polymerase, distinctive from any other eukaryotic enzyme. The yeast, *Saccharomyces cerevisiae,* contains two DNA polymerases, types I and II; type I is similar but not identical to DNA polymerase α, and type II is like the bacterial DNA polymerases II and III, and carries a 3′→5′ exonuclease activity. From these limited examples, it would seem that DNA replication in eukaryotes is lacking in overall uniformity and that extreme evolutionary variation exists. In this context, DNA replication is unique, because the processes of gene transcription and translation are very similar in all eukaryotic organisms.

To summarize, as far as higher eukaryotes are concerned, DNA polymerase α must be considered the critical enzyme for DNA replication. This conclusion essentially rests on two lines of evidence. First, a DNA template with RNA primer regions is believed to be the natural template during DNA replication and only DNA polymerase α can utilize such a template. Second, only the activity of DNA polymerase α changes in a wide range of cells when they are stimulated to divide. DNA polymerase β remains constant in most cells, irrespective of the state of cell division, but nonetheless, plays a part in the concluding phases of DNA replication and DNA repair. For a review of eukaryotic DNA polymerases, see Scovassi *et al.* (1980).

RNA polymerases and ribonuclease H

RNA polymerases play both indirect and direct roles in DNA replication. Their indirect role is to provide the necessary RNA to meet the demand on the machinery for protein synthesis, particularly rRNA (Burke & Fangman 1975). Their direct role is to insert primer segments of RNA into the template DNA as an essential prerequisite for DNA synthesis by DNA polymerase α (for a review see Sheimin *et al.* 1978). In a few systems, it has been possible to detect nascent, covalently linked RNA–DNA intermediates during the early phase of DNA replication in the yeast, *S. cerevisiae* and the broad bean, *Vicia faba.* Other evidence for a primer function for RNA in DNA replication has been provided by nearest neighbour frequency analyses of nascent RNA–DNA intermediates in *S. cerevisiae* and the malarial parasite, *Physarum polycephalum.* Collectively, these results indicate that oligoribonucleotide segments, 8–15 bases long, are joined via their 3′-end to the 5′-end of the growing DNA strand. The RNA primer does not have a unique sequence and all the complementary ribo- and 2′-deoxyribo-nucleotides are present in these RNA–DNA intermediates. The critical question is, Which RNA polymerase inserts these primer sequences? In a thorough study, based on HeLa cell nuclei, Brun and Weissbach (1978) suggest it is specifically RNA polymerase A. In their experiments, DNA replication was insensitive to α-amanitin but abolished by an antiserum raised against purified RNA polymerase A. Studies on other eukaryotic systems, however, indicate a less stringent requirement for RNA polymerase A and it may be that insertion of primers into template DNA is a general property of all forms of eukaryotic RNA polymerase.

Once the nascent chains of DNA have been formed on the short RNA primers, the latter must be excised. Selective digestion of RNA in RNA–DNA hybrids requires a special type of nuclease, ribonuclease H. The most intensively studied RNAase H activity in eukaryotes is in yeast, where it exists as two types, both of which are basic proteins. Type H_1 (M_r 48000) hydrolyses RNA–DNA hybrids to produce ribonucleoside 5′ monophosphates and DNA, whereas type H_2 (M_r 21000) produces a mixture of oligoribonucleotides, with free 3′-ends, and DNA. This duality of ribonuclease H activity in eukaryotes now seems a general finding. Huet *et al.* (1977) suggest that ribonucleases H_1 and H_2 are associated specifically with RNA polymerase A in yeast, again indicating a very important role for this particular RNA polymerase in DNA replication (see also Brun & Weissbach 1978).

DNA ligase

Eukaryotic DNA ligase is a notoriously labile enzyme which has resisted all efforts at complete purification. Nonetheless, it has been demonstrated in all higher organisms. It is usually a dimer, M_r 180000–220000 and sedimentation coefficient 12.4*s*, with a tendency to gradually dissociate into constituent monomers of M_r 90000–100000. In all systems investigated it is a nuclear enzyme, but like DNA polymerase α, readily released into the cytoplasm during conventional methods of homogenization. Importantly, its activity rises dramatically during the S phase of the cell cycle. The enzyme joins up vicinal 3′-OH and 5′-phosphate groups in double-stranded DNA in the presence of Mg^{2+} ions, but unlike its bacterial counterpart, the energy for phosphodiester bond formation is met by the covalent activation of the enzyme by ATP rather than NAD^+. In keeping with this mechanism DNA ligase–AMP and DNA–AMP intermediates have been identified during the course of the reaction with eukaryotic DNA ligase. Eukaryotes may contain two forms of DNA ligase, type I and II; both have identical reaction mechanisms, but type II is more thermolabile than type I and does not cross-react with antibodies raised specifically against type I DNA ligase.

Other factors

There are complex fluctuations in deoxyribonuclease activities in mammalian cells during the S-phase, presumably to increase the number of 'start points' or free 3′-OH groups in the DNA template, but the enzymes specifically associated with DNA replication remain to be characterized.

Calf thymus contains a terminal addition DNA polymerase which adds single deoxyribonucleoside 5′-triphosphates to the 3′-ends of intact DNA as homopolymers, 20–50 bases long. While useful in genetic engineering (see section 3.4.1), such enzymes play no proven part in DNA replication.

A wide search has been carried out for proteins which accelerate DNA replication and cell division of cells maintained in culture *in vitro*. Two such proteins, epidermal growth factor (EGF) and nerve growth factor (NGF) have been identified and widely investigated. The structure of EGF has been completely determined (Fig. 7.11). EGF has been isolated from a wide range of sources, including mouse submaxillary glands, human urine and snake venoms (for a review see Carpenter & Cohen 1979). EGF stimulates DNA synthesis in a wide variety of cultured eukaryotic cells and vastly accelerates their rate of proliferation. Randomly distributed binding sites for EGF have been identified on the plasma membranes of most eukaryotic cells and from which the mitogen is internalized by endocytosis of vesicles containing both EGF and membrane fragments (Hopkins 1980). However, its mode of action in stimulating cell division remains enigmatic. NGF has also been identified in many eukaryotes (for a review see Bradshaw 1978). As its name suggests, NGF promotes the division and maturation of neural cells, but many other cells of non-neural origin are also sensitive to this mitogen. NGF has an M_r of 130000 and is composed of three different subunits, designated α, β and γ, with the overall stoichiometry of $\alpha_2\beta\gamma_2$. As yet, no individual protein can be said to trigger DNA replication alone, but EGF and NGF indicate the types of molecules whose presence helps to promote the entry of cells into the S phase.

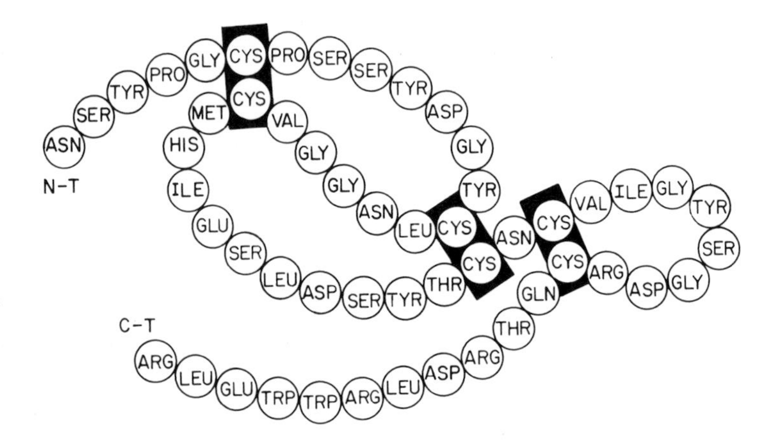

Fig. 7.11. The amino acid sequence of epidermal growth factor. EGF is a single polypeptide chain, containing three disulphide bridges, shown here in black. N-T is the N-terminal group, C-T is the C-terminal. (Reproduced, with permission, from Savage *et al.* 1973.)

Tubulin and assembly of the mitotic spindle

Separation of the daughter chromosomes on the mitotic apparatus constitutes the last stage of DNA replication in eukaryotes. Although recognized by cell biologists for over a century, the biochemistry of spindle formation is still far from understood (for a review see Timasheff 1979). Electron microscopy revealed that the early mitotic spindle is composed of long, hollow cylinders or microtubules, of inside and outside diameters of 15 and 25 nm, respectively. Microtubules are formed of discrete units or protofilaments, running parallel to the axis of the cylinder and with 13 protofilaments arranged around the circumference (Fig. 7.12(a)). Viewed from the side (Fig. 7.12(b)) the protofilaments are ordered axially with respect to each other in such a way that each is displaced from horizontal juxtaposition by an angle of 10°. The microtubule is thus a helix with 13 protofilaments per turn and a pitch of 8 nm. The protofilaments, in biochemical terms, are composed of α- or β-tubulin subunits (both of M_r 55 000) and of similar but not identical amino acid sequence. The stable form of extracted tubulin is a dimer, M_r 110 000 and sedimentation coefficient 5.8*s*, containing one unit each of α- and β-tubulin. The helical arrangement of protofilaments is actually an alternate, repeating sequence of α- and β-tubulin subunits (Fig. 7.12(b)).

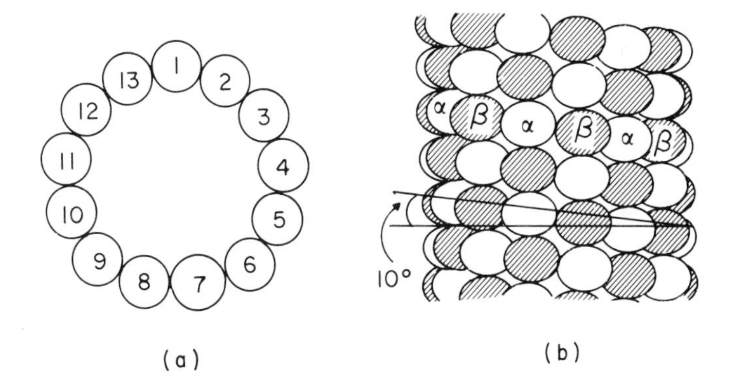

Fig. 7.12. The structure of microtubules of polymerized tubulin. (a) Top view. (b) Side view, in which the alternating α- and β-monomers of tubulin are indicated by hatching.

If the Ca^{2+} ions in tubulin solutions are sequestered, the subunits spontaneously assemble axially and laterally, forming microtubules. The assembly mechanism is not clear, but each subunit of α- and β-tubulin bears one molecule of tightly bound GTP and microtubule formation proceeds with the generation of inorganic phosphate, possibly according to the following reaction sequence:

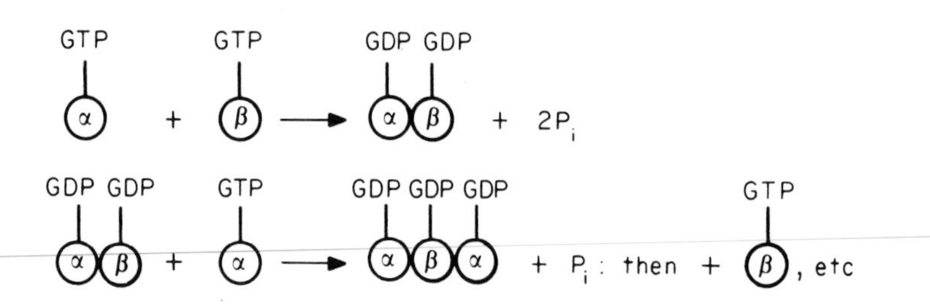

It seems likely (Timasheff 1979) that other factors are needed for microtubule formation. The absence of Ca^{2+} ions and the presence of Mg^{2+} ions seem critical and proteins of both high and low molecular weight have been implicated; these are termed HMW proteins (M_r 350000) and tau factors (M_r 60000), respectively. Both types of proteins are detectable by immunofluorescence procedures in the final microtubule structure. Certain HMW proteins are substrates for cAMP-dependent protein kinase and phosphorylation plays a part in microtubule assembly. Overall, the assembly process is best described as nucleated, cooperative polymerization, in which growth of the helix is driven by changes in entropy.

Both α- and β-tubulin appear to have been highly conserved during evolution. From studies on synchronized Chinese Hamster ovary (CHO) cells, it is known that tubulin synthesis accelerates during the late S- and early G_2-phases of the cell cycle. This enhanced availability of microtubule precursors can be accounted for in two ways; first, synthesis of tubulin monomers *de novo,* and second, activation of pre-existing monomers by phosphorylation of the associated HMW proteins. Significantly, mammalian cells attain their maximal potential for protein phosphorylation towards the end of the S phase. Furthermore, addition of bibutyryl cAMP to cultures of CHO cells increases the number and length of their microtubules, probably by stimulating a cAMP-dependent protein kinase.

7.3.4 DNA replication in eukaryotes and other related processes

Units of replication

In bacteria, it is possible on genetic and molecular grounds to define the single chromosome as a replicon during its replication. This is not possible in eukaryotic systems and the use of the term replication unit is preferable.

Many techniques have been used to size eukaryotic replication units, the range being 4 μm long (1.3×10^4 base pairs) to 280 μm long (for a review see Sheimin *et al.* 1978). Many replication units are known to be arranged in tandem, as is the case for satellite DNA and for the genes coding for rRNA and histones. The rate of replication on all units is roughly similar, at about 0.1–2 μm, or 1–15$\times10^3$ nucleotide residues per min.

Chromosome replication

Taken overall, DNA replication in eukaryotes probably proceeds in a series of stages, none of which are fully defined. The first stage requires the initiation of DNA replication during the early S phase and produces single-stranded 4–7*s* DNA (220–280 nucleotide residues), akin to the Okazaki fragments identified in bacteria. The second stage involves elongation of the replicated DNA, first as 6–26*s* DNA and later, into longer 30–100*s* sequences. The third stage involved the maturation of the DNA by joining together all the replicated pieces. The next stage involves the association of the replicated DNA with nuclear proteins to form the chromosome replicas prior to mitosis. Finally, the replicated chromosomes must be separated, and packaged as nucleosomes in the nuclei of the two daughter cells. Evidence for this staged process comes from work with inhibitors and from cells derived from patients with certain clinical syndromes. Fibroblasts derived from individuals with Bloom's syndrome initiate DNA synthesis poorly and only short pieces of replicated DNA are found. Lymphocytes from patients with megaloblastic anaemia have serious interruptions in the elongation phase of DNA replication (for a review see Sheimin *et al.* 1978).

The initiation of DNA replication is not understood in detail, but initiation points or origins of replication are possibly located in the hairpin structures created by the frequent palindromic sequences in eukaryotic DNA. The formation of the multiple replication forks needs the concerted action of DNA-destabilizing proteins, DNA-unwinding proteins, DNA-dependent ATPases and RNA polymerase (probably type A). The formation of short RNA primer sequences ends the initiation stage.

During the elongation stage, DNA polymerase α forms short tracts of DNA on the RNA primers. The template specificity of DNA polymerase α is in keeping with its vital role in DNA replication; it is the only mammalian enzyme capable of utilizing hairpin structures and RNA-primed DNA as templates (Spadari & Weissbach 1975).

DNA polymerase α is only partially processive, (McKune & Holmes 1979) meaning that it detaches from the template after polymerizing a few bases and does not copy the entire replication unit in one fell swoop. Other factors may be necessary to assist in the continuous reassociation of DNA polymerase α with the template. Such factors have been described (Cobianchi *et al.* 1978). DNA polymerase α will proceed to synthesize Okazaki fragments on both strands of the replication forks and when this is complete, RNAase H, and other enzymes, remove the RNA primers. DNA polymerase β then fills the gaps in the tracts of replicated DNA and completes the elongation phase.

Finally, DNA ligases join up the replicated DNA sequences and the two parent DNA strands in each replication fork have now been fully replicated. The entire process to this stage is represented in Fig. 7.13. For a review of the DNA replication in eukaryotes, see de Pamphilis and Wassarman (1980).

The nascent DNA is next packaged into chromatin, in the form of nucleosomes, by its association with histone octamers (or core

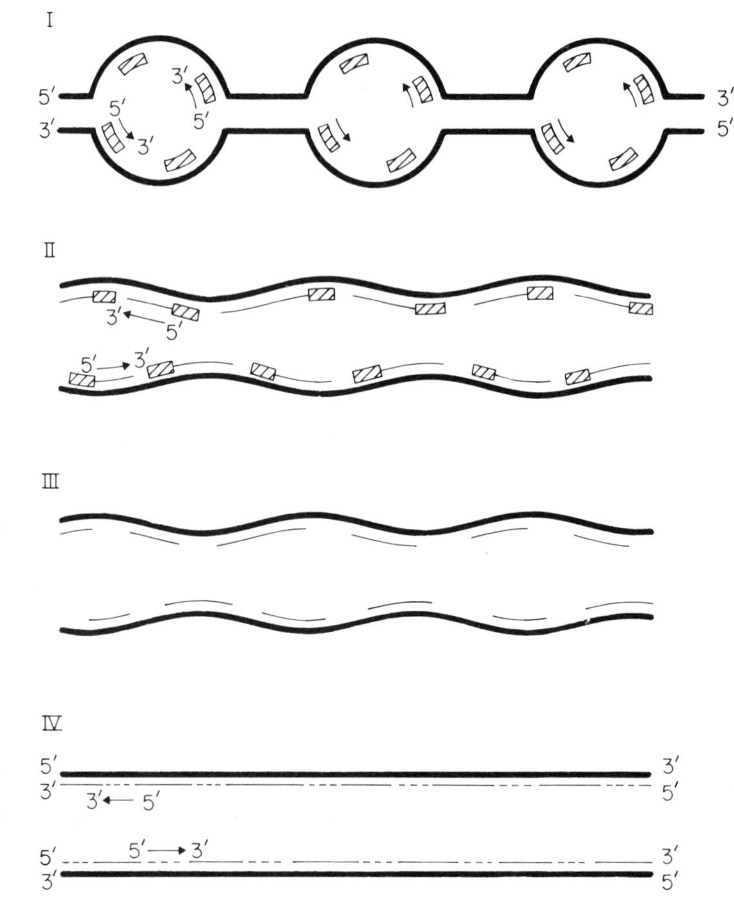

Fig. 7.13. A diagram of eukaryotic DNA replication. Throughout the figure, the following symbols are used; ▬▬, parental DNA; ▨, RNA primer; ——, daughter DNA inserted by DNA polymerase α and - - -, daughter DNA inserted by DNA polymerase β. I. Unwinding of the extremely long DNA template into many replication forks and the formation of RNA primers on each fork. II. Insertion of Okazaki fragments by DNA polymerase α. III. Excision of the RNA primers. IV. Filling of gaps and joining up of replicated daughter DNA strands.

particles). Most evidence suggests that newly synthesized histones, themselves made in the S phase, are not preferentially bound to the nascent DNA. It is now generally believed that replicated DNA acquires preformed nucleosome core particles, both old and newly synthesized, and that these are distributed in a cooperative and random manner on both daughter strands. An alternative model, based on non-random association of core particles, has less experimental support (for a review see Sheimin *et al.* 1978). Supercoiling of DNA must occur in this packaging process and this is accomplished by topoisomerases (Laskey *et al.* 1977). The distribution of histones round the replication fork is probably as indicated in Fig. 7.14 (see also de Pamphilis & Wassarman 1980).

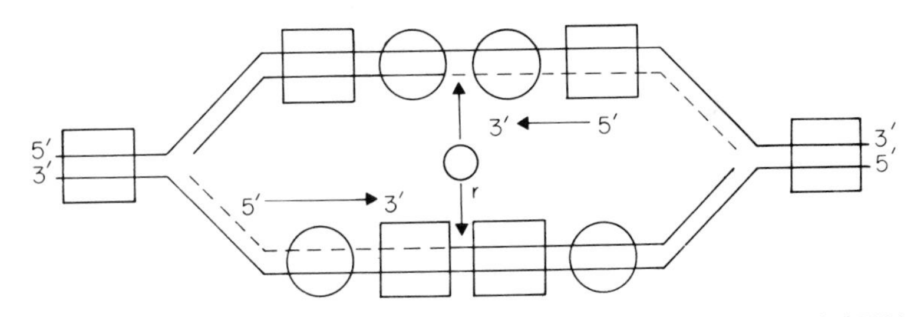

Fig. 7.14. The probable appearance of a eukaryotic replication fork at the end of DNA replication. For convenience, the origin of replication, O, is placed centrally; this need not be so *in vivo*. Since DNA synthesis occurs in the $5' \rightarrow 3'$ direction, indicated by the arrows, the process will be continuous (———), and discontinuous (- - - - -), beginning at the central origin of replication. Newly synthesized histone octamers are given as circles, and old octamers as squares.

The biochemical events necessary for completion of the replicated chromosomes and their separation during mitosis are simply not understood. Poly(ADP) ribosylation, particularly of histone 1, may be vitally important in completing the condensed structure of the chromosome replicas. Poly(ADP)-ribose synthetase activity is maximal in cultures of HeLa cells during the G_2 phase of the cell cycle. The critical involvement of microtubular polymers of tubulin in the separation of chromosomes has already been discussed.

The above is an account of the general mechanism for replicating eukaryotic chromosomes in the S phase. It should not be forgotten, however, that unique sequences of DNA can be replicated at specific stages of growth and development. Examples include the amplification of rRNA genes in *Xenopus laevis,* selective DNA replication during larval development and metamorphosis in the mosquito, *Aedes aegypti,* and the selective replication of giant (polytene) chromosomes in many plants. Where DNA replication can proceed in discrete organelles, such as nuclei, mitochondria, chloroplasts and even symbiotic organisms, the S phases in all are remarkably *synchronized* (Klein & Bonhoeffer 1972). The regulatory mechanisms underlying all these interesting phenomena are not understood.

Origins of replication in higher organisms

Origins of replication have been unequivocally identified in the closed circular DNA of bacterial chromosomes, bacterial plasmids, mitochondria and certain viruses. By contrast, the size and complexity of eukaryotic chromosomes have made it very difficult to isolate the origins of replication in the long, linear chromosomes of higher organisms. The development of techniques for the genetic transformation of yeasts by autonomously replicating, cytoplasmic plasmids provided vital opportunity to search for eukaryotic origins of replication by the techniques previously used for bacteria. Beach *et al.* (1980) have constructed a plasmid which contains no effective replication origin but whose efficiency of replication in the yeast, *S. cerevisiae,* is greatly enhanced by the insertion of specific fragments of yeast chromosomal DNA. Stinchcomb *et al.* (1980) have extended this work and prepared a series of plasmids which will only replicate autonomously in *S. cerevisiae* in the presence of additional sequences of chromosomal DNA from a wide range of eukaryotes, including the slime mould, *Dictyostelium discoideum,* fruit fly, *Drosophila melanogaster,* and maize, *Zea mays.* By this novel means, it may soon prove possible to identify and purify the origins of replication of many eukaryotic chromosomes. This should provide the means to analyse DNA replication in higher organisms in greater detail than has been possible hitherto.

Model systems

Studies on DNA replication in the future will centre on simpler systems, where the genes are simpler or amenable to cloning and genetic manipulation; DNA from the chromosomes, mitochondria, chloroplasts and kinetoplasts of lower eukaryotes fall into this category.

Since the nucleus is the prime site for DNA replication, studies of DNA synthesis in nuclear-derived systems *in vitro* are important. Brun and Weissbach (1978) have pioneered such work in HeLa nuclei. These authors separated two subnuclear fractions on the basis of their solubility in 0.32M NaCl. The soluble fraction contained DNA polymerases, including α, and regulatory factors; the insoluble fraction contained DNA and RNA polymerases. Neither fraction alone carried out DNA replication extensively, but there was impressive DNA synthesis when mixed together. There are two important features of this work. First, there was authentic initiation of DNA synthesis, requiring RNA primers made by RNA polymerase A; RNA–DNA hybrids were also identified as obligatory intermediates of the replication process. Second, the requirement for the salt-soluble factors could not be simulated by the addition of exogenous, highly purified DNA polymerases. Brun and Weissbach (1978) have, therefore, developed a sophisticated yet simple means of studying the regulatory elements of DNA replication. Their method could be invaluable in future experiments on the mechanisms for regulating DNA replication during the hormonal stimulation, viral infection or the course of normal developmental change.

Benbow *et al.* (1978) have pioneered the development of multi-enzyme systems capable of faithfully replicating DNA in its entirety. Their work is based on two principles. First, the templates are simple and well defined, being the single-stranded and double-stranded circular DNAs from the phages, fD and ϕX174, and the plasmid pX1r11. Second, soluble extracts of unfertilized eggs of *Xenopus laevis* induce DNA synthesis in the nuclei of the resting (non-dividing) liver of the adult. Fractionation of egg proteins has helped in their search for the proteins which are crucial for DNA replication. With these admittedly artificial templates, the necessary components include DNA polymerase α (and γ in certain experiments), RNA polymerase B, DNA-binding proteins, RNAase H and DNA-swivelase. Even the authors admit certain anomalies in their work (Benbow *et al.* 1978), but nonetheless, their commendable approach will be invaluable in future studies on the replication of more complex DNA templates, such as an interphase mammalian chromosome.

Crews *et al.* (1979) have applied the most advanced methods of molecular biology to sequence the origin of DNA replication in mitochondrial DNA; this is a remarkable step forward. It had been known for some time that this closed circular DNA was replicated according to the rolling circle model of Cairns, and that a loop structure, (D-loop), near one of the 12*s* mitochondrial rRNA genes, was the origin of replication. In addition, a short-lived, 7*s* species of DNA was an early intermediate in mitochondrial DNA replication. Mitochondrial DNA was selectively cleaved with restriction endo-

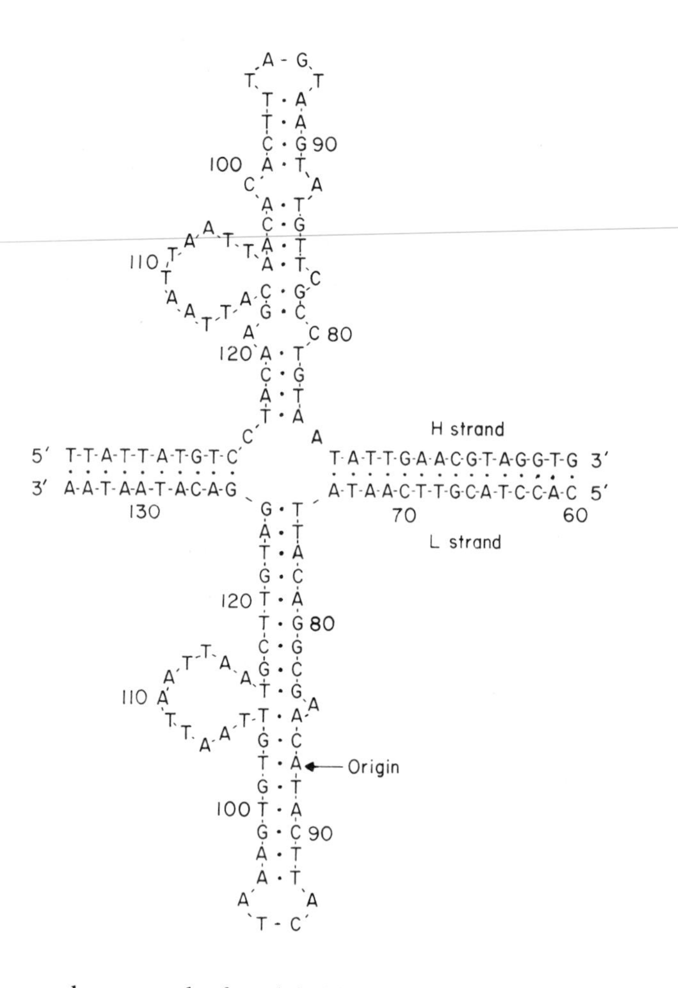

Fig. 7. 15. The origin of replication of mitochondrial DNA. The two strands of mitochondrial DNA, H and L (heavy and light in CsCl gradients) are indicated. The numbers indicate the order of base in the restriction endonuclease fragment $\Delta 86^{Hae}$, not shown in its entirety here. The arrow marks the origin itself. (Reproduced, with permission, from Crews *et al.* 1979.)

nuclease, and after labelling the 5′-ends of the fragments with nucleotide kinase and [γ-^{32}P]ATP, each labelled fragment was checked for its ability to hybridize to 7*s* DNA. Base complementarity was finally narrowed down to a restriction fragment, termed $\Delta 86^{Hae}$ (approximately 200 base pairs). Crews *et al.* (1979) then determined the base sequence of fragment $\Delta 86^{Hae}$ and the initiating (5′) end of 7*s* DNA; strict complementarity was established and the initiator region had been unequivocally identified (Fig. 7.15). A possible secondary structure was proposed, and a loop structure is clearly found near the origin of replication; other important aspects in the structure relative to DNA initiation remain to be determined. Two points are of particular interest, however. First, palindromes are found in this restriction fragment, but they are not in the immediate vicinity of the initiator region. Second, although 7*s* DNA is the putative precursor of mitochondrial DNA, no covalently bound RNA primer was detected. These findings run counter to accepted aspects of the replication of nuclear DNA in eukaryotes; it could be that the mechanism for the replication of closed, circular DNA, as in mitochondria, is fundamentally different.

Nonetheless, progress such as this indicates the true merit of studying simpler model systems. Based on such work, the replication of larger and linear chromosomes can now be tackled with more confidence (see, for example, Beach *et al.* 1980, Stinchcomb *et al.* 1980).

The importance of the nuclear matrix in DNA replication

Berezney and Coffey (1977) have performed some elegant work which highlights the crucial importance of the nuclear matrix in DNA replication. When either regenerating liver or cultured 3T3 fibroblasts were pulse-labelled with [^{3}H]thymidine for 30 s, and the nuclear matrix isolated, the small fraction of DNA retained in the matrix contained the vast majority of the newly synthesized, [^{3}H]-labelled DNA (Pardoll *et al.* 1980). Furthermore, the ^{3}H label could be displaced from the matrix-associated DNA by a 'chase' of an excess on non-radioactive thymidine. These and other kinetic data strongly suggest that the initiation of DNA replication occurs on the nuclear matrix. This important conclusion is substantiated by the work of Buckler-White *et al.* (1980) on the infection of cultured 3T6 fibroblasts by polyoma virus. As reviewed by Crawford (1980), the small, circular DNA of the genomes of polyoma and SV40 viruses codes for only a few proteins. One protein, the T-antigen, appears during infections by both viruses and is necessary for the transformation of the host cells and also for stimulation of viral (and perhaps host) DNA replication. During polyoma infection, both newly synthesized T-antigen and viral DNA were particularly associated with the nuclear matrix of the host cells. A key role for the nuclear matrix in DNA replication is thus confirmed.

An alternative, fixed point model for DNA replication, based on immobilized replication complexes

A new model has recently been proposed by Coffey and his co-workers for nuclear DNA replication in eukaryotes, which differs significantly from the more traditional model (see Fig. 7.13 and de Pamphilis & Wassarman 1980). Recent evidence suggests that if interphase nuclei are depleted of histones and other nuclear proteins, the DNA is found to be organized in extremely long, supercoiled structures, without any breaks (Vogelstein *et al.* 1980). These supercoils can be visualized as a halo surrounding the nuclear matrix. Furthermore, the DNA loops can be reversibly wound using various concentrations of ethidium bromide yet irreversibly unwound when nicks are introduced by low concentrations of DNAases. The extreme length of these DNA supercoils raises serious problems in their replication, for the long tracts of daughter (replicated) DNA and parent (template) DNA could easily become entangled and disorganized. This problem is averted by the tight association of small regions of the DNA supercoils with the nuclear matrix. Pardoll *et al.* (1980) proposes that DNA replication complexes or replisomes, containing the necessary enzymes and factors for the DNA synthesis, are distributed across the nuclear matrix and the DNA is reeled through these immobilized complexes as it is replicated. This highly imaginative concept is visualized in Fig. 7.16 and provides a most plausible explanation of how precise order may be maintained during the replication of long eukaryotic chromosomes. The only limitation of the fixed-point model is that the enzymes required for DNA replication must be restricted exclusively to the replisomes. This stricture is well illustrated by DNAases; on the one hand they are needed for replication, probably to provide start points or free 3′ OH groups, yet on the other hand, they irreversibly unwind the supercoils

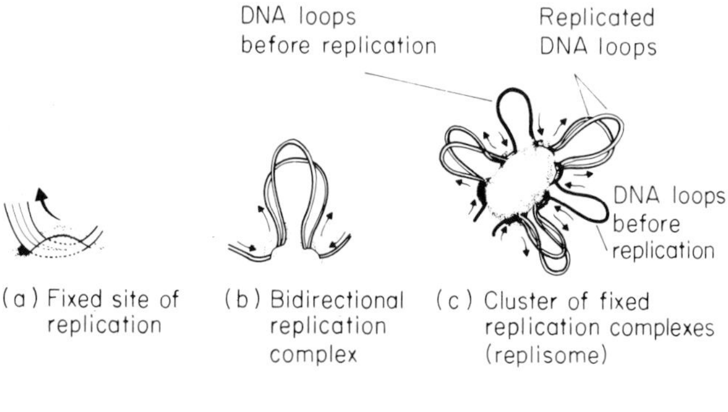

Fig. 7.16. A fixed-site model for the replication of DNA in eukaryotic nuclei. Nuclear DNA is in long supercoils and the model predicts that the replication machinery is present as replisomes, scattered throughout the nuclear matrix. (a) A single replisome, represented as an arch, the DNA being reeled through as it is replicated. (b) A pair of adjacent replisomes, immobilized on the nuclear matrix, permitting bidirectional DNA replication. The DNA template is reeled in from both sides and nascent, replicated DNA loops out between. (c) A cluster of immobilized replisomes. DNA is bound by the replisomes in alternating replicated and non-replicated (template) loops. In all figures, arrows represent the direction of movement of DNA. (Reproduced, with permission, from Pardoll *et al.* 1980.)

in both template and replicated DNA (Vogelstein *et al.* 1980). If randomly distributed, therefore, DNAases could completely disorganize DNA replication.

Replication of the genomes of DNA viruses

The small, closed circular genomes of polyoma and SV40 viruses, some 5200 base pairs, are ideal for experimentation in all aspects of molecular biology, including DNA replication. In particular, the genomes have been fully mapped (see Fig. 7.17 and Crawford 1980), the origins and termini for DNA replication are known and the genomes have been fully sequenced (e.g. Soeda *et al.* 1980). The genomes are so small that they can encode only a few proteins. The virus particles themselves (or virions) contain the genomic DNA, viral coat proteins, a few small polypeptides and four histones (histone 1 is absent). During the early and late phases of viral infection, virus-induced proteins appear in the host cells; these are called viral antigens, namely T and U (early) and V (late). T-antigen (M_r 90000) is found on the nuclear matrix of lytically infected cells and is necessary for the initiation of viral DNA replication. V-antigen represents viral coat protein, but the function

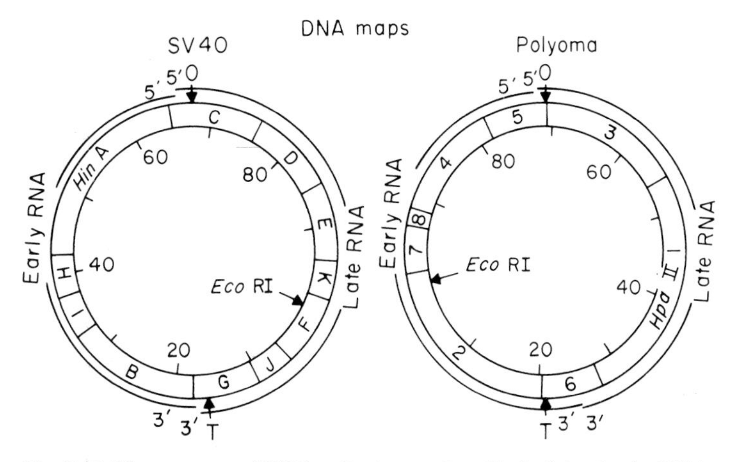

Fig. 7.17. The genomes of SV40 and polyoma virus. Each of the circular DNA molecules is presented with the origin of DNA replication (0) at the top and the terminus of replication (T) at the bottom. Regions transcribed for RNA, both early and late in the course of infection, are indicated on the left and right sides, respectively. The restriction enzyme fragments used to produce the physical maps are indicated by the letters A–I (SV40; *Hind* II+III) and the numbers 1–8 (polyoma; *Hpa* II). The single cleavage site in each genome for the restriction enzyme, *Eco* RI, is shown because this is conventionally used as the reference point, 0/100, on the map of each genome. (Reproduced, with permission, from Crawford 1980.)

of U-antigen is unknown. These viral genomes are replicated using exclusively the biosynthetic machinery of the host cells (as reviewed by Fareed & Davoli 1977). Replication begins at an origin (often referred to as map position, *ori*) in a semiconservative, bidirectional manner, towards the terminus of replication. Electron microscopic studies of replication intermediates indicate that replication proceeds according to the rolling circle model (see section 4.2.5). The unreplicated parts of the viral genomes contain superhelical turns because the parental strands remain covalently closed during replication. Unwinding of the template strands occurs close to, or precisely at, the two replication forks; DNA-swivelases are engaged in this destabilization of the template. Importantly, the virus-induced T-antigen binds to the genomes, close to the origin of replication, and signals the onset of replication. Initiation occurs on RNA primers and the two replication forks proceed in opposite directions, until they meet at a position 180° from the origin. Once replication is completed by host DNA ligases, packaging of the viral genomes is expedited by their association with host-derived histones 2A, 2B, 3 and 4.

The adenovirus genome is a linear duplex DNA molecule, containing about 35 000 base pairs. Replication begins at the 5′ terminus of each strand and a protein (M_r 55 000) appears to act as a primer for replication *in vitro,* in soluble extracts derived from viral-infected HeLa cells (Challberg *et al.* 1980). The linkage is a covalent, phosphodiester bond, between the serine residue and the 5′ terminus of the DNA template. Presumably, the initiation of DNA replication at the 5′ termini of the genome requires the presence of RNA primers.

DNA inhibitors

DNA replication in eukaryotes can be inhibited by a wide variety of chemicals. Because of the dependence of replication on RNA and protein synthesis, many inhibitors are not very specific, e.g. cycloheximide. Other inhibitors, however, attack more specific aspects of DNA replication, such as the supply of vital precursors or assembly of the mitotic spindle (Table 7.9). Many of these drugs are widely used in the chemotherapy of cancer, either alone or in combination (adjuvant therapy). Colchicine, from the crocus, *Colchicum autumnale,* binds slowly, yet non-covalently, to $\alpha\beta$ tubulin dimers, 1 mole:1 mole dimer. Colchicine need only bind to the end of a growing microtubule to prevent further attachment of $\alpha\beta$ dimers. Vinblastine and vinchristine are alkaloids from the periwinkle, *Vinca rosea,* and they also bind to tubulin dimers in an

Table 7.9. Some selective inhibitors of eukaryotic DNA replication.

Drug	Site of action	Effect on DNA replication
Methotrexate	Inhibits dihydrofolate reductase	Replication stops in elongation step (mid-S-phase)
5′-Fluoro-deoxyuridine*	Inhibits thymidylate synthetase and DNA polymerase	Blocks initiation step (very early S-phase)
Cytosine arabinoside†	Inhibits DNA polymerase	Blocks initiation step
Hydroxyurea	Inhibits ribonucleoside diphosphate reductase	Curtails the supply of precursors; S-phase suspended
5-Methylnicotinamide	Inhibits poly (ADP-ribose) synthetase	Cells held in G_2 phase
Chlorambucil, Melphalan	Alkylating agents, cross-link the template DNA	Replication stops in elongation step (early and mid-S-phase)
Vinblastine, Colchicine, Podophyllotoxin	Inhibit the assembly of microtubules	Cells are held in metaphase

*The active inhibitor is a metabolite, 5-FdUMP.
†The active metabolite is araCTP.

equimolar ratio. Both act remarkably quickly, and precipitate microtubules in anomalous coils, defective helices and macrotubules which cannot assemble into a functional microtubule. Podophyllotoxin, from the berberis, *Podophyllum peltatum,* inhibits tubulin assembly by as yet unknown means.

DNA repair

DNA polymerases also have an important function in DNA repair (see section 7.4.1). Several inheritable and serious human syndromes, including progeria, ataxia telangiectasia, xeroderma pigmentosum and Fanconi's anaemia, are all typified by defects in DNA repair mechanisms. DNA polymerase β and many other factors, yet to be fully characterized, are involved in repair processes (for a review see Hanawalt *et al.* 1979). Interestingly, DNA repair is not coupled to histone synthesis (Sheimin & Guttman 1977).

7.4 REPAIR AND RECOMBINATION

7.4.1 Repair

The repair of eukaryotic DNA is of particular importance in cancer research. Chemical carcinogenesis involves molecular lesions in DNA (see section 3.3.2); moreover, many antineoplastic drugs are general antimitotic cytotoxic compounds whose use is based on the premise that many tumour cells divide rapidly and might be selectively sensitive to inhibition of DNA replication. Clearly the repair of DNA can, in principle, eliminate the lesions during exposure to carcinogens and could render less effective the antineoplastic drugs. Also if eukaryotic repair can be error prone (see section 4.3.2) somatic mutagenesis and its consequent carcinogenic risk could be enhanced and antineoplastic drugs could themselves prove to be carcinogenic. For a convenient summary of the extensive literature on eukaryotic repair processes, see Generoso *et al.* (1980).

Several experimental systems have been exploited for the study of repair processes. In the case of *Neurospora crassa* and yeast, several mutagen-sensitive mutants have been analysed both genetically and biochemically (see papers by Schroeder & Olson, Schroeder *et al.* and Lemontt, in, Generoso *et al.* 1980). As with *E. coli*, these fungi have several repair pathways, certain of which are error prone and others error-free excision processes. The *E. coli* systems seem to have their fungal counterparts. In particular the *uvs-3* mutation in *Neurospora* is in many ways analogous to *recA* (see section 4.3.4) in that the organisms are mutagen-sensitive, relatively non-mutable and also show aberrant mitotic recombination. Moreover, the *Neurospora* repair systems rely, in part, on a family of DNAases that are derived from a common precursor by proteolysis and *uvs-3* and certain other repair-defective alleles appear to be lesions in this proteolytic system. Proteolysis is similarly implicated in the inducible *recA* functions of *E. coli* (Fig. 4.14). Among higher eukaryotes, mutagen-sensitive mutants have been characterized in *Drosophila* (see the paper by Green, in, Generoso *et al.* 1980).

In the case of mammalian systems, there is also evidence for systems comparable with those in *E. coli* although evidence for error-prone repair is still equivocal because of a lack of sensitive mutagenesis systems. There are mutagen-sensitive mutant cell lines available. In humans a well-studied mutation of this type occurs in patients suffering from xeroderma pigmentosum, a familial ultrasensitivity of skin to the UV radiation in sunlight. Cells cultured from such patients have a defective long-patch repair system (responsible for repair of lesions caused by UV and certain aromatic hydrocarbon carcinogens) but short-patch repair (responsible for repair to lesions caused by alkylating agents and ionizing radiation) is unimpaired. Nevertheless, different mammalian cells differ in their sensitivity to alkylating agents; for example, O^6-alkyl G-residues are removed efficiently in liver and inefficiently in brain. In the particular case of O^6-methyl G-residues, the removal appears not to involve base removal and may be a demethylation (see section 4.3.3). Different xeroderma cell lines are either effective or non-effective in this process and are designated mex$^+$ and mex$^-$ respectively. Although the genetic basis of mex is still unclear it is possible that tissue differences may reflect differences in mex (Sklar & Strauss 1981).

Post-replication repair has been implicated in the repair of methylation damage in Chinese hamster kidney cells (Friedman & Huberman 1980). Post-replication repair occurs in *E. coli* (see section 4.3.4). Post-replication repair can either be recombinational or non-recombinational. In this latter case the process is error-prone (Fig. 4.10(b)). This might represent an important repair hazard in animal cells.

7.4.2 Recombination

The mechanisms of recombination are discussed in section 4.3.5. There is no reason to believe that the process is fundamentally different in eukaryotes. Recombination is usually associated with meiosis although mitotic recombination can be analysed in certain fungi. Historically fungal genetics is particularly important for recombinational theory because it provided the first evidence for heteroduplex intermediates (see Radding 1978 for references to the original literature).

The first clear evidence for heteroduplex intermediates came from the studies of Kitane and Olive on the ascomycete fungus, *Sordaria*. The products of meiosis in *Sordaria, Neurospora* and other ascomycetes are eight haploid spores. This is because there are two (rather than the more usual one) mitotic divisions following meiosis. Moreover, the spores are produced in a row in a structure called the ascus, rather like a tiny pod of peas. Kitane and Olive studied recombination between strains that differed in a gene that determines spore colour (either brown or white). Asci from recombinants can be recognized and analysed microscopically. A small proportion were found to have brown and white spores in the ratio of 5:3 (shown schematically in Fig. 7.18). This means that brown and white alleles must be present in a haploid cell from which the mature spores segregate. The only way a haploid cell can behave in this way (as a heterozygote) requires a mismatched heteroduplex DNA structure. Post-meiotic segregation can equally be observed in *Neurospora*; as there is no convenient spore colour system to

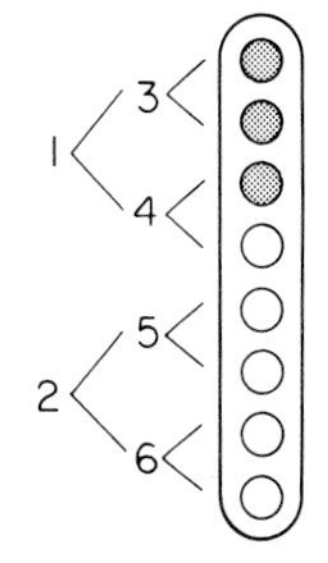

Fig. 7.18. Post-meiotic segregation. The figure shows a *Sordaria* ascus containing three brown spores (stippled circles) and five white ones. (1) and (2) represent the immediate products of meiosis and the first post-meiotic mitotic division produces (3)–(6); the second post-meiotic mitosis produces the eight spores. Numbers (1) and (4) must be heterozygotes, even though all the cells in the diagram are haploid.

exploit, the experiment involves more labour. The individual spores are dissected out of the ascus. They are then grown and biochemical markers (such as nutritional requirements) are analysed. Again segregation ratios of 5:3 are, rarely, observed. A prokaryotic counterpart to the study of post-meiotic segregation is the analysis of *r* (rapid lysis) mutants of phage T4 (see section 4.3.5).

The other discovery in fungal genetics, crucial to the development of an understanding or recombination was the finding that recombination is frequently non-reciprocal, particularly for closely linked markers. The non-reciprocal pattern results in an allele (e.g. C in Fig. 7.19(a)) being 'lost'. The phenomenon is referred to as post-meiotic gene conversion and is genetically a useful method of linkage. Of the various explanations for the phenomenon, all involve heteroduplex intermediates. The Holliday Model (see section 4.3.5 and Fig. 4.18) suggests a simple explanation for non-reciprocal patterns of recombination. However, gene conversion is believed usually to be due to the repair of mismatched heteroduplex intermediates (Fig. 7.19(b)).

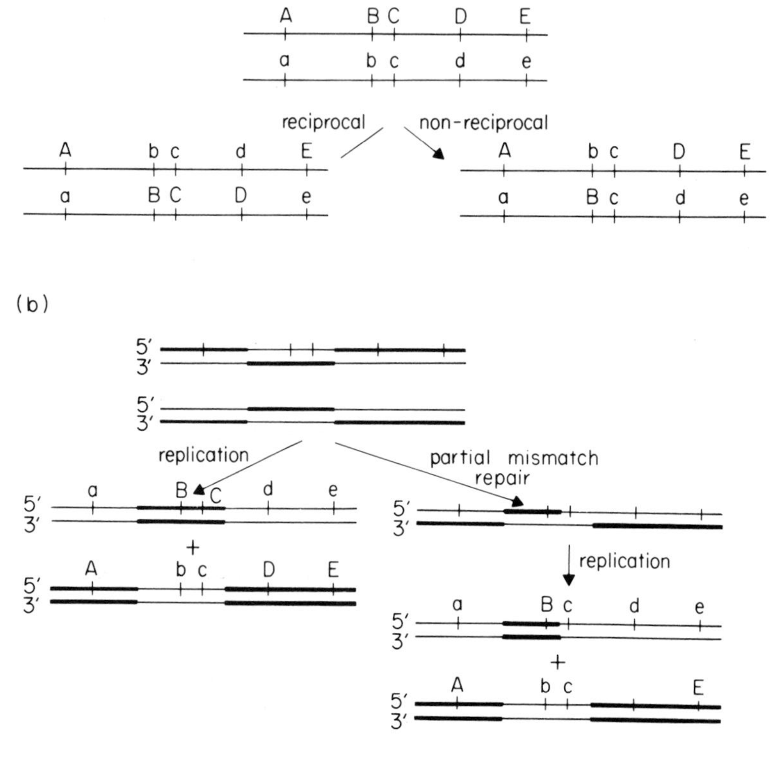

Fig. 7.19. (a) Reciprocal and non-reciprocal recombination. A and a, B and b, etc. are different alleles for five genes, A–E and the single horizontal lines represent double-stranded DNA molecules. In the reciprocal pattern of recombination, two cross-overs have occurred (between A and B and between D and E). The two recombinant chromosomes, AbcdE and aBCDe, show reciprocal exchange of the alleles. In the non-reciprocal example, one cross-over is between A and B and there is another to the left of D. However, the C allele is 'lost' and the exchange is non-reciprocal.
(b) Heteroduplex intermediates formed during recombination. If the mismatched region (e.g. that including the C/c alleles) is repaired before segregation of the heteroduplex, one allele is lost. The 'replication' and 'repair' processes only refer to the lower heteroduplex.

7.4.3 Insertions and transpositions

Just as eukaryotic genetics first suggested mechanisms of recombination so the first evidence for transposable genetic elements came from the formal genetic analysis of maize (McClintock 1958). Similar evidence for transposable genes has since been obtained for *Drosophila* (Green 1977). In this section we just refer to two systems in yeast which have been analysed in detail. The special case of DNA rearrangements in the immune system will be discussed in section 8.7.1.

The his4-912 element

The *his4* gene in yeast encodes a complex protein with three enzymic domains, A, B and C. The gene is transcribed in the order A–B–C. The mutation *his4-912* arose spontaneously; the mutation lies towards the start of *his4* and obliterates all A, B and C functions. The mutation is non-reversible by mutagens, is non-suppressible but can revert spontaneously. These data are suggestive of an insertion sequence (see section 4.1.3). The revertants are mostly not wild type. Many have a cold-sensitive requirement for histidine, thus implying a difference from the wild-type *his4* allele. Other revertants contain a transposition of *his4* from its normal site on chromosome III to chromosome VIII and others have extensive chromosomal translocations and an extensive inversion in chromosome III (Chaleff & Fink 1980).

The *his4-912* insertion consists of 6200 base pairs and is homologous to a class of independently discovered repetitive yeast DNA sequences called Ty1. In the majority of *his4-912* His^+ revertants, excision of Ty1 is incomplete and leaves behind a 300 base pair sequence, δ. The excision is supposed to involve site-specific recombination between direct repeats at the ends of Ty1 (Fig. 7.20(a)). Other reversions and rearrangements could be due either to recombination between Ty1 (or δ) sequences in different parts of the yeast genome. Alternatively, they might be generated by transposition as described in the Shapiro model (Fig. 4.24). The inversion of a part of chromosome III could, in principle, be the counterpart in a linear eukaryotic chromosome of the co-integrate intermediate of transposition (Fig. 7.20(b)). If this latter explanation is correct, there are detailed differences between the mechanisms of bacterial and fungal transposition as revealed by the analysis of deletions generated in the two cases (Roeder & Fink 1980).

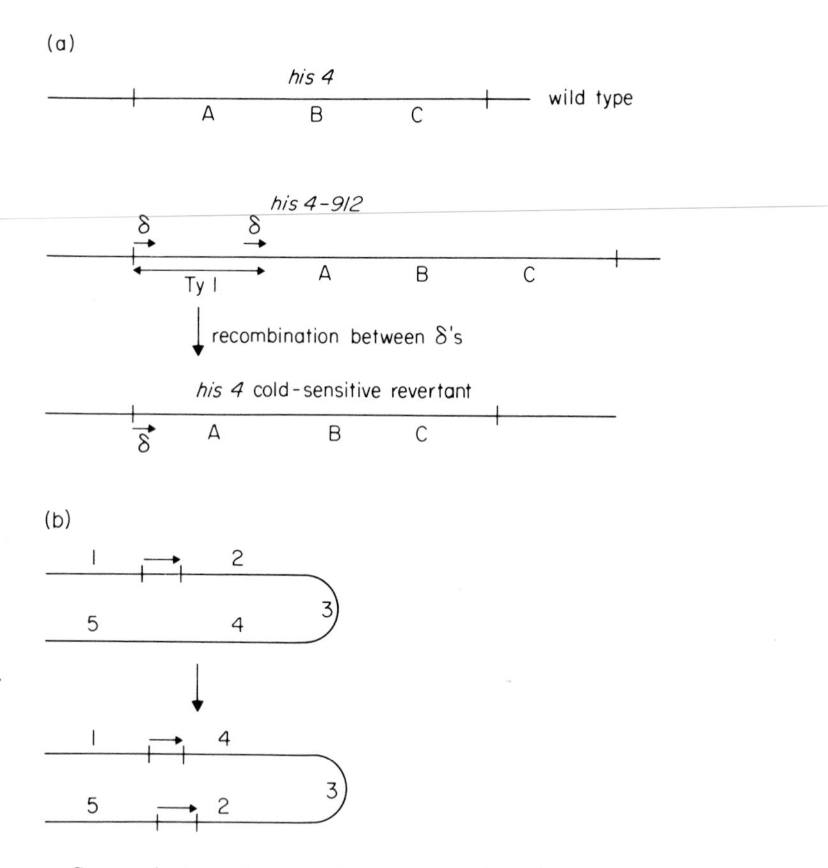

Fig. 7.20. (a) The *his4* gene in wild-type yeast (upper line), the insertion of the Ty1 element in the *his4-912* mutation and excision of the Ty1 element by recombination between the direct repeats (δ). (b) An example of how the transposition of a transposable element (such as Ty1 or δ) could affect the inversion of a part of a linear chromosome. The example is comparable with the co-integrate in Fig. 4.24. Numbers 1–5 are genes or markers. Note that the same effect could, in principle, be generated by recombination if there is a second Ty1 (or δ) element present between genes 4 and 5 in the upper figure.

Somewhat analogous insertion-mediated mutations have been characterized in a yeast alcohol dehydrogenase gene (Williamson *et al.* 1981).

Mating type conversion in yeast

The two mating types in yeast, a and α, are determined by alternative alleles at the mating type locus, *MAT* in chromosome III. Strains are capable of switching mating type from $MAT\alpha$ to $MATa$ and vice versa. The process is determined by two other loci, $HML\alpha$ (to the left of the centromere in chromosome III) and $HMRa$ (to the right of the centromere in chromosome III). The sequences of $HML\alpha$ and $MAT\alpha$ are identical, likewise $HMRa$ and $MATa$ are identical. The α and a alleles have identical terminal sequences but differ in a unique central region of 700 base pairs (in a) or 900 base pairs (in α). These findings have led to the 'cassette model' for yeast mating type conversions (Hicks *et al.* 1977; Fig. 7.21(a)). The essentials of the model are that *HML* and *HMR* are quiescent

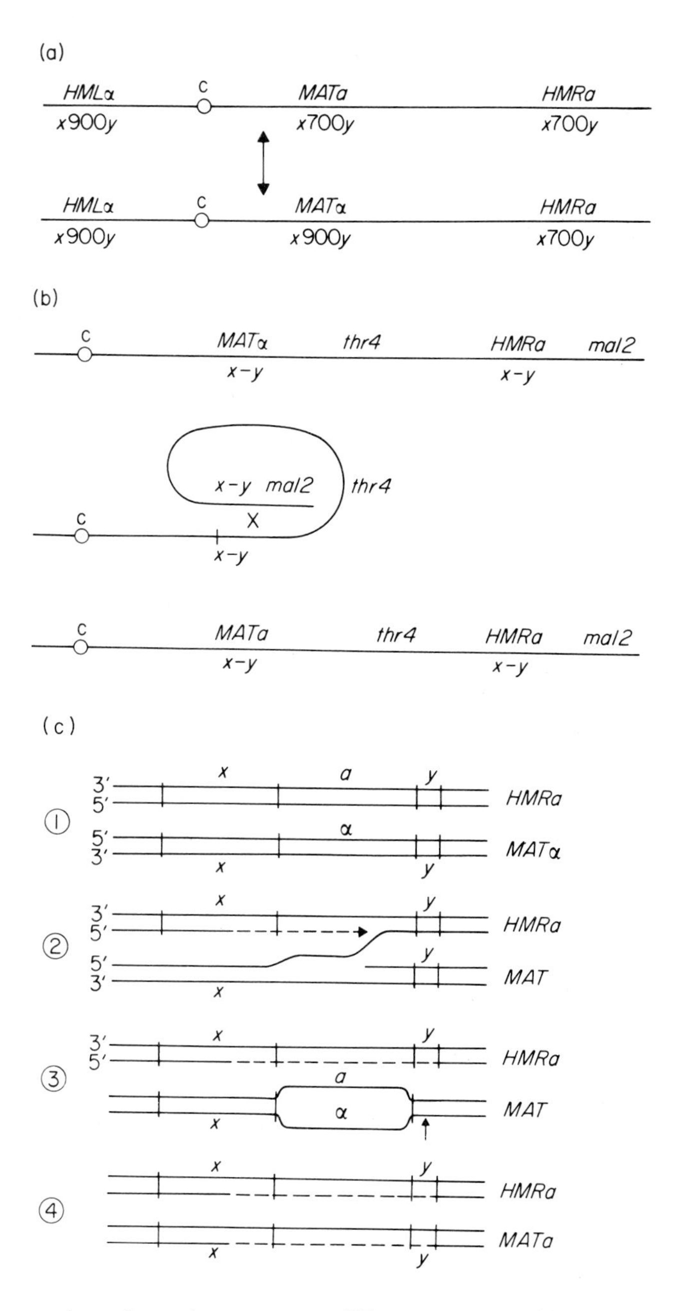

Fig. 7.21. Mating-type conversion in yeast. (a) Elements of the cassette model. The line represents chromosome III; c is the centromere. The *HML*, *HMR* and *MAT* alleles are explained in the text. The symbols x700y and x900y represent the 700 and 900 base pair sequences flanked by the two sequences (represented here by *x* and *y*) that are common to all these genes.
(b) Recombination between *MATα* and *HMRa*. Only the right-hand part of chromosome III is shown; *thr4* and *mal2* are other markers. The final stage is a non-reciprocal recombination (c.f. Fig. 7.19b). If the recombination were to be reciprocal, two chromosomes would be formed: a shortened chromosome III ending with *MAT mal2* and a non-visible plasmid containing *HMR* and *thr4*.
(c) Model for the conversion of *MAT* to *MATa* (Klar *et al.* 1980). (1) The *HMRa* and *MATα* genes are aligned as in (b); *x* and *y* are the regions of homology. Not shown is the region to the left of *x*, homologous to the corresponding region in *HMLα* but missing from *MAT*. A nick is made at the small arrow in *HMRa* and DNA polymerase starts from the 3′-ends. The displaced strand is taken up and ligated to the corresponding place in the *MAT x* sequence. (2) This process is now under way. Newly synthesized DNA is shown as a broken line. The strand taken up by the *MAT* gene cannot pair in the *a*/*α* region because these are not homologous. However, the process started within a region of homology (*x*) and resulted in the erosion of the former 'upper strand' of the *MAT* gene. (3) The process of replication has finished at a point in *y*. The *MAT* gene is now a heteroduplex, its ***a*** strand being derived from the displaced strand from *HMRa*. Mismatch repair now commences from the short arrow. (4) Completion of the conversion of the *MAT* allele.

'library copies' or the mating type alleles and that the mating type is determined by the nature of the allele located at *MAT*. Thus in the upper drawing in Fig. 7.21(a) the cell is of mating type *a*. In the conversion to α (lower figure), *HML*α has effected replacement of *MATa* by *MAT*α but has retained its own configuration so that it can be kept as a library copy for future conversions.

The cassette model implies an event different from bacterial transpositions (see section 4.3.7) in that following the transposition event, the gene at the point of insertion (the former *MAT* allele) is lost. This has led to a proposal, supported by genetic and biochemical evidence, that the transposition is due to non-reciprocal recombination (Haber *et al.* 1980; Fig. 7.21(b)). A mechanism for this involving a special type of gene conversion (see section 7.4.2) is suggested by Klar *et al.* (1980) based on evidence for association between *HML*'s and *MAT*. The model (Fig. 7.21(c)) is based on displacement of a DNA strand in *HMR* by synthesis. The displaced strand is picked up by a strand in *MAT*. Thus in the early stages of the process a new *HML* strand is synthesized but a *MAT* strand is destroyed. The resulting mismatched *MAT* gene is repaired.

Chapter 8 Eukaryotic Biosynthesis of RNA

8.1 TRANSCRIPTION

8.1.1 Introduction

More than any other aspect of biochemistry or molecular biology, genetic transcription highlights the fundamental differences of macromolecular synthesis between prokaryotes and eukaryotes. Early studies raised several problems of eukaryotic transcription to the surface and solutions to these have now been found. In particular, means have been found to separate the multiple forms of RNA polymerase reproducibly and in high yield; plausible explanations have been found for the extreme length of RNA transcripts in higher organisms; it has proved possible to clone single eukaryotic genes in bacteria and thus aid the interpretation of experiments conducted on transcription *in vitro;* finally, many model systems for studying transcription have been developed in recent years. These systems have led to a detailed understanding of the extensive process of post-transcriptional modification.

Before describing eukaryotic transcription in detail, some preliminary remarks seem warranted. Unlike their bacterial counterparts, eukaryotic genes are essentially monocistronic and there is no substantial evidence that operator genes, repressor proteins or a cyclic AMP-binding protein (CAP) are present in higher organisms. Binding sites for RNA polymerases (promoter regions) are present in eukaryotic DNA, and indeed, their structure is now beginning to be determined. Eukaryotic RNA polymerases have no subunit equivalent to the bacterial σ factor and neither have transcription terminators, akin to ρ factor, been found in higher organisms. Taken collectively, the facts indicate that the classical Jacob-Monod model of gene operons is not applicable to eukaryotes.

The unique features of eukaryotic genes, however, are their intrinsic complexity and their arrangement. First, genes are often separated by spacer regions of DNA which may or may not be transcribed. Second, the genes necessary for the control of a given biochemical process may be located on different diploid chromosomes and hence gene activity must be highly coordinated. Third, and critically important, eukaryotic genes are complex and contain *intervening sequences* of DNA which are transcribed, but whose transcription products are *excised* during the maturation or modification of the initial RNA transcript. On the suggestion of Gilbert (1978), the coding regions of eukaryotic genes are called *exons* while the regions which interrupt the coding exons are called *introns* (short

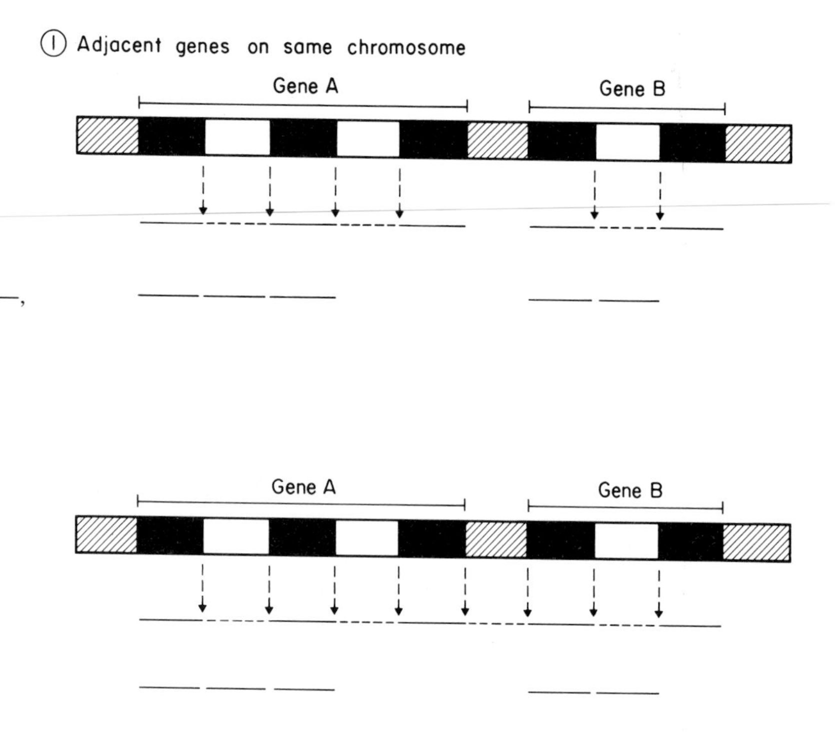

Fig. 8.1. The implications of gene structure and function for transcription. In all structures: ■, exons (coding region); □, introns (non-coding region); ▨, spacer (sometimes transcribed); ——, exon transcript; ---, intron or spacer transcript; and ↓, cleavage site at exon–intron or exon–spacer boundaries. In case (1) the genes are adjacent on the same chromosome; the complexity of transcription depends on whether the spacers between genes are transcribed or not. In case (2) (see facing page) two examples are given of genes on separate chromosomes which must work in harmony (be transcribed in a coordinated manner) to produce complex, oligomeric proteins. In all cases, the number of exons are arbitrary, but the final structures under case (2) are accurate.

for intervening sequences). The presence of introns and spacers supports the well-known fact that eukaryotic RNA transcripts are very long (for a review see Perry 1976). Excision of the introns and joining of the separated exons must be performed with remarkable accuracy. The differences between spacers and introns should be emphasized; spacers are between genes, and may or may not be transcribed. For the implication of these various properties of eukaryotic genes in terms of transcription, and its regulation, see Fig. 8.1.

The importance of the discovery of introns in eukaryotic genes cannot be over-emphasized. Indeed, it has probably been the most revolutionary finding in cell and molecular biology for over a decade. It provides the culminating evidence that the transfer of genetic information in prokaryotes and eukaryotes is fundamentally different. The discovery of introns has crucial importance in terms of evolution, and plausible explanations for the need for introns are emerging; these important points are discussed in more detail later.

8.1.2 The structure and function of eukaryotic RNA polymerases

Two incisive breakthroughs by Roeder and Stirpe helped to resolve the complex pattern of RNA polymerases in eukaryotes.

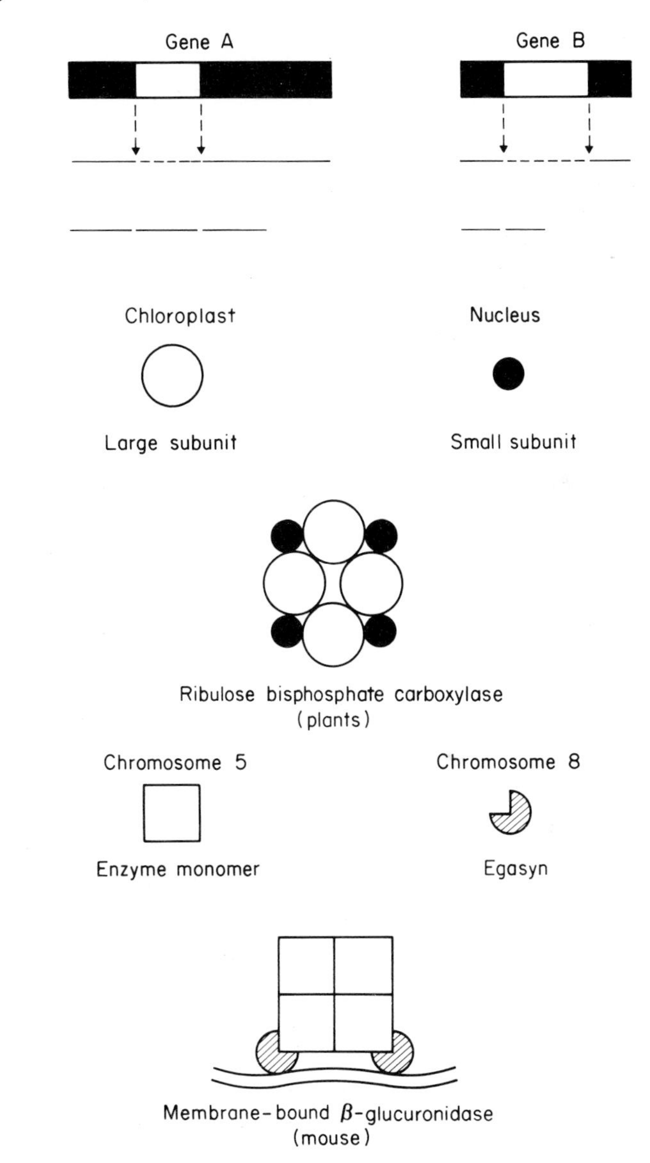

First, DEAE-cellulose was found to resolve several peaks of enzyme activity and these could be distinguished by their relative sensitivity to α-amanitin, the toxin from the toadstool, *Amanita phalloides.* All forms of eukaryotic RNA polymerase have now been purified to homogeneity, particularly useful sources being calf thymus, yeasts, rat liver and wheat germ.

Two forms of nomenclature have been proposed for the three principal types of nuclear RNA polymerases; the classification A, B and C (Chambon 1975) will be followed here, rather than the alternative classification of I, II and III. The enzymes show chromatographic heterogeneity in certain cells and this may only be partially explained on the basis of isoenzymes. The properties of eukaryotic RNA polymerases are summarized in Table 8.1; additional information may be found in the excellent review by Chambon (1975). It should be added that mitochondria contain a single RNA polymerase, which transcribes all the species of mitochondrial RNA; in contrast to the nuclear enzymes, mitochondrial RNA polymerase is of low molecular weight. All the enzymes have an absolute requirement for Mg^{2+} or Mn^{2+} ions, and, with a few exceptions, have a unique molecular architecture and especially a quaternary structure of distinctive subunits. These structures are totally different from bacterial RNA polymerases.

Table 8.1. The properties of eukaryotic RNA polymerases.

	Enzyme type			
Property	A	B	C	mt*
M_r	550 000	600 000	600 000	64 000–68 000†
Location	Nucleolus	Nucleoplasm	Nucleoplasm	Mitochondria
Role in RNA synthesis	rRNA (except 5*s* rRNA)	hnRNA (mRNA precursor)	tRNA 5*s* rRNA viral RNA	Mitochondrial RNAs‡
Sensitivity to α-amanitin	Refractory	Extremely sensitive	Sensitive, high concentrations	Refractory
Optimum ionic strength	Low	High	High	Low

*Not included in the classification of Chambon (1975).
†Pure mitochondrial enzyme from *Neurospora crassa, Xenopus laevis* and rat liver.
‡4*s* tRNA, 7*s* and 12*s* rRNAs.

Unlike the single bacterial enzyme, all eukaryotic RNA polymerases are resistant to rifampicin. However, all the eukaryotic enzymes are inhibited by the semisynthetic antibiotic, rifamycin AF/013; initiation, and to a lesser extent, elongation, are sensitive to this inhibitor.

In the main, RNA polymerases appear to have been highly conserved during evolution. Antibodies raised against a purified type A enzyme cross-react extensively with the amanitin-insensitive RNA polymerases from virtually all eukaryotes, but not with enzymes of type B and C; similarly, antibodies against type B enzyme are specific to enzymes of that type and not others.

There is no convincing evidence that changes in the amounts or activities of RNA polymerases play a significant part in the regulation of eukaryotic transcription (Chambon 1975). During the development of *Xenopus laevis* and the sea-urchin, *Strongylocentrotus perpuratus,* there are marked fluctuations in the pattern of RNA synthesis; these changes are not matched by marked fluctuations in RNA polymerase activities. RNA polymerase A activity is present even in cells which are not actively synthesizing ribosomes, surprisingly including the anucleolate mutant of *X. laevis,* where there is absolutely no ribosome synthesis. Furthermore, there is no correlation between developmental changes in mRNA synthesis and the complement of RNA polymerase B. Based on the accurate assay of this enzyme by binding of [^{14}C-methyl]-γ-amanitin (Meilhac *et al.* 1970), most eukaryotic cells have an excess of enzyme to meet the demands of mRNA synthesis.

The mechanism of eukaryotic transcription

This is best described in three phases; initiation, elongation and termination (Chambon 1975). Whether specific elongation or termination factors are involved remains in some doubt; certainly, studies on transcription in model systems suggest that the RNA polymerases can recognize an 'end' or terminator sequence in eukaryotic genes (see section 8.3.3 for details). By contrast, the initiation of eukaryotic transcription is not readily achieved. In early work, DNA templates containing either many breaks (DNAase treatment) or areas of denaturation (usually achieved by heating) were found to be readily transcribed by eukaryotic RNA polymerases. However, when eukaryotic DNA was prepared carefully to avoid any breaks, such a helical native template was poorly transcribed by all types of RNA polymerase.

An imaginative approach by Mandel and Chambon (1974) was to use Simian virus (SV) 40 DNA as the template in transcriptional studies. This viral DNA is a closed, superhelical circle of double-stranded DNA; the superhelix imposes a strain on the overall structure such that a small region of separated (unpaired) bases is present. Furthermore, the superhelical form can be converted to relaxed circular and linear forms by treatment with appropriate enzymes and in which base pairing is very effectively preserved.

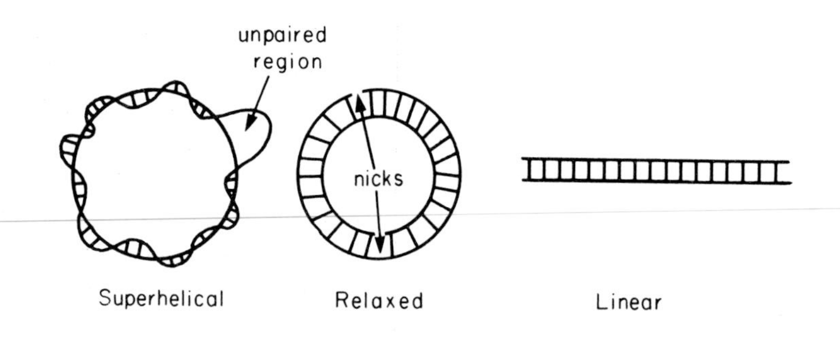

Forms of SV40 DNA

Significantly, only the superhelical form of DNA, containing the unpaired region, was transcribed by RNA polymerases A and B. These observations raise the important point that initiation of eukaryotic transcription needs destabilization of the template DNA, as a local region of denaturation (unpaired bases). Unless the RNA polymerases are present in vast excess, as in experiments *in vitro,* then initiation needs the help of other factors. DNA-destabilizing proteins and DNA-unwinding proteins may well be involved, as in DNA replication (see section 7.3.3), but, surprisingly, this possibility has not been investigated in depth. In the intact cell, the template DNA is associated with non-histone proteins and their involvement in the initiation of transcription is vital (O'Malley *et al.* 1977). There are also stimulatory proteins which enhance the binding of eukaryotic RNA polymerases and promote the initiation of transcription (for a review see Chambon 1975).

Studies on small, relatively simple eukaryotic genes indicate that rigorous strand selection occurs during their transcription. During the early and late phases of infection of cells in culture by SV40, there is a pronounced *switch* in the transcription of the viral DNA strands. The small but amplified genes for 5*s* rRNA in *X. laevis* and *X. borealis* (described in the original literature as *X. mulleri*) can be separated in CsCl gradients into H (heavy) and L (light) strands; only the L strand is transcribed *in vivo,* indicating the *strict asymmetry* of eukaryotic transcription. Some insights into the mechanisms and importance of strand selection are emerging. Strand selection during the early and late phases of SV40 viral infection cannot be simply explained by the RNA polymerases moving between a large number of promoter sites distributed differently between the two DNA template strands. Regulatory proteins are clearly involved in directing and controlling faithful transcription from a small number of promoter sites (Nevins & Winkler 1980). Faithful transcription also demands a double-stranded DNA template and thus rigorous strand selection. From elegant studies on the transcription of cloned tRNA genes, faithful transcription can only occur from double-stranded DNA and not its separated, single-stranded counterparts. The non-transcribed strand clearly plays an important part in maintaining the accuracy and efficiency of transcription.

The problems of initiation, strand selection and genome complexity illustrate the difficulties in unravelling the basis of eukaryotic transcription and its regulation. The need to develop model systems and to devise sophisticated methods for the analysis of RNA transcripts was widely recognized; indeed, concerted effort on these topics has resulted in some of the most exciting findings in contemporary biochemistry.

8.1.3 The development of new technologies for investigations on transcription

Selective gene activation

The selective activation of eukaryotic genes has been of vital importance to recent progress; only by this means can the function of a particular gene be seen against a complex background of transcriptional activity. There is a vast literature on the modulation or transformation of eukaryotic cells by viruses. Exposure to viruses provokes profound changes in the transcription of the host genome, but the mechanisms are truly complex and rather beyond the scope of the present text (for reviews see Soeda *et al.* 1980, Ziff 1980). Many simpler alternatives are now available, in which hormones and simple chemicals can activate specific gene loci.

In a classical study, Karlson and Clever (1960) exposed larvae of the midge, *Chironomus tentans,* to the insect moulting hormone, ecdysone, and observed microscopically the uncoiling or 'puffing' of specific genes in the giant, polytene chromosomes in the salivary glands. These prominent puffs or Balbiani rings were found to be centres of intense RNA synthesis. Sekeris and Lang (1964) demonstrated that ecdysone promoted the induction (or synthesis *de novo*) of the enzyme, L-dopamine decarboxylase, needed for hardening on the skin of pupae; importantly, enzyme induction was dependent on the appearance of the mRNA for DOPA-decarboxylase. Clearly, hormones regulate the activity of specific genes and the provision of specific species of RNA. Work on insect hormones foundered for some time, because the interaction between ecdysone and the insect nucleus could not be explained; this impasse has recently been resolved and ecdysone binds to and activates selective insect genes (Gronenmyer & Pongs 1980). It is now clear that the imaginative prediction of Jensen & Jacobsen (1962) is correct; many hormones, including those in plants and insects, work via receptor mechanisms. In essence, the following chain of events is envisaged:

Hormone (H) + cytoplasmic receptor (R) $\rightarrow$ high affinity, H–R complex

H–R complex $\xrightarrow{\text{activation}}$ H–R$^{\circledast}$ complex, changed physicochemical configuration

H–R$^{\circledast}$ complex $\xrightarrow{\text{translocation}}$ nucleus, occupation of specific acceptor sites

Nucleus + H–R$^{\circledast}$ complex $\xrightarrow[\text{modification}]{\text{transcriptional}}$ increased synthesis specific mRNAs, increase rRNAs and even tRNAs

H–R$^{\circledast}$ complex $\xrightarrow{\text{inactivation}}$ end of hormonal response or breakdown.

By contrast, polypeptide hormones work via membrane receptors and cyclic nucleotides (second messengers; see section 8.4.1).

Support for the receptor concept is overwhelming (for a review see King & Mainwaring 1974) and a representative sample of hormones acting by this mechanism is presented in Fig. 8.2. The specificity of hormonal response is controlled by the distribution of the receptor proteins; hormone-responsive (or target) cells contain receptors whereas non-target cells do not. Target cells contain the receptors appropriate to their needs in terms of transcriptional control and this pattern is established by means unknown during the course of differentiation. Most hormones act directly, but some must be metabolized to their biologically active form; examples include 5α-dihydrotestosterone (from testosterone) and 1α,25-dihydroxy-cholecalciferol (from cholecalciferol, formerly called vitamin D_3). Many of the hormones in Fig. 8.2, markedly accelerate the synthesis of specific species of mRNA (for reviews see Yamamoto & Alberts 1976, O'Malley *et al.* 1977). These transcriptional changes can be enhanced further by the use of synthetic analogues of higher biological activity than the naturally occurring hormones. Powerful analogues in current use include diethylstilboestrol (for oestradiol), ponasterone (for ecdysone) and dexamethasone or triamcinolone acetonide (for cortisol). Examples of genes which are subject to stringent control by hormones are given in Table 8.2. The only drawback to hormonal manipulation is that it can sometimes activate hormone-sensitive, nuclear ribonucleases.

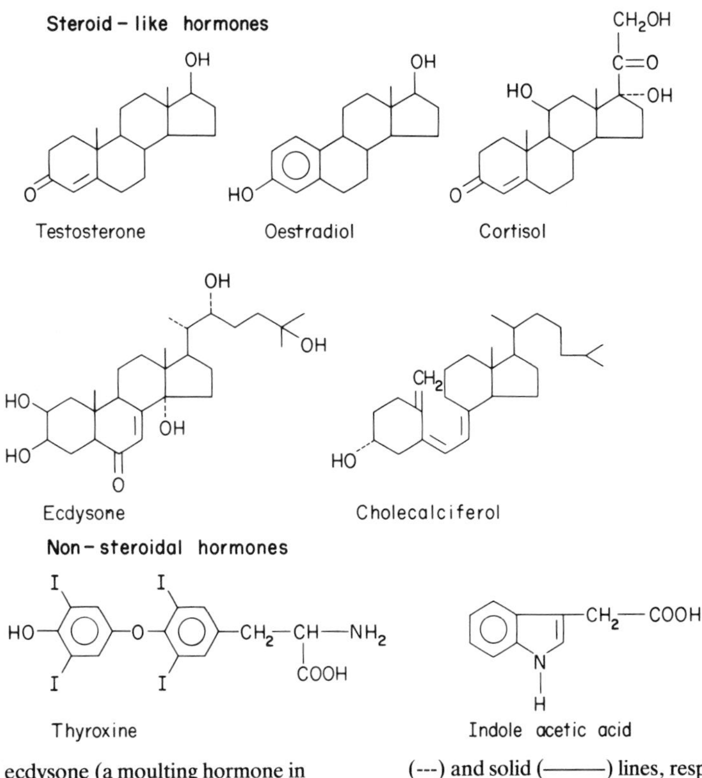

Fig. 8.2. Hormones which selectively activate eukaryotic genes. The steroid-related hormones include testosterone (androgen or male sex hormone), oestradiol (oestrogen or female sex hormone), cortisol (glucocorticoid), ecdysone (a moulting hormone in insects) and cholecalciferol (formerly vitamin D_3) needed for the absorption of Ca^{2+} ions. In these hormones, bonds in the α (lower face) or β (upper face) configurations are indicated as broken (---) and solid (———) lines, respectively; by convention, angular methyl groups are not indicated in detail. The two non-steroidal hormones are thyroxine (a regulator of basal metabolic rate) and indole acetic acid (a plant auxin).

Table 8.2. Genes whose activities are regulated by hormones. In most of the systems below, genes for rRNA are activated as well.

Cell type	Hormone (or analogue)	Responsive gene
Chick oviduct	Diethylstilboestrol	Ovalbumin Ovomucoid Transferrin (conalbumin)
Rabbit uterus	Progesterone	Uteroglobin
Rat uterus	Oestradiol	Enolase
Rat liver	Cortisol	Tryptophan oxygenase Tyrosine aminotransferase
Hepatoma (HTC) cells	Dexamethasone	Tyrosine aminotransferase
Rat prostate, seminal vesicle	Testosterone*	Secretory proteins Aldolase
Mouse kidney (both sexes)	Testosterone	β-Glucuronidase
Avian, Amphibian } liver (both sexes)	Oestradiol	Vitellogenin

*The active metabolite is 5α-dihydrotestosterone.

Studies on the synthesis of haemoglobin have featured prominently in recent work for several reasons. Human haemoglobin is perhaps the most intensely studied of all proteins and many clinical syndromes are available for genetic investigation of its biosynthesis, structure and regulation. Erythropoiesis involves the progression of undifferentiated, nucleated stem cells along an irreversible course of cell division and maturation, resulting in the anucleated, terminally differentiated erythrocyte.

Haemocytoblast→proerythroblast→early, intermediate→reticulocyte→erythrocyte.
(stem cell) and late normoblast

Erythropoiesis involves a massive synthesis of haemoglobin, such that the erythrocyte is effectively gorged with haemoglobin, and contains few other proteins, except the enzymes and factors needed for glycolysis, the pentose phosphate cycle and osmotic balance. Erythropoietic systems have vital advantages in terms of studies on transcription. First, the entire process is regulated by erythropoietin, a glycoprotein hormone secreted by the kidney. Fetal spleen and liver cells can be maintained in culture and respond to erythropoietin *in vitro* with the selective activation of the globin genes, present in only one copy each per genome. Second, phenylhydrazine terminates erythropoiesis at the reticulocyte stage, producing cells conspicuously active in haemoglobin synthesis and rich in the mRNAs for globin chains. Third, certain oncogenetic viruses can transform mouse haemopoietic cells in such a way that erythropoiesis is suppressed; however, such cells can be encouraged into the normal route of differentiation and haemoglobin synthesis by exposure to low concentrations of dimethyl sulphoxide (Friend *et al.* 1971). These lines of Friend cells provide elegant systems for studying gene activation by the simplest experimental manipulation.

Certain poisons and carcinogens can be used to enhance genetic activity; for example, low doses of thiocetamide cause considerable enlargement of the nucleolus in many cells, and this has been exploited in studies on the activation of rRNA genes and the increase in their transcription by RNA polymerase A.

Unlike normal cells, tumour cells thrive under conditions of culture *in vitro*. Furthermore, the malignant process is accompanied by profound changes in gene transcription and stable, tumour sublines can produce vast quantities of a single protein; for example, myeloma cells (tumours of lymphocyte-producing cells of bone marrow) synthesize only light or heavy chains of immunoglobulins and pituitary cells secrete polypeptide hormones. Such cells provide opportunities for the isolation of specific eukaryotic mRNAs, for the purification of RNA polymerases and for general investigations on transcription.

The nature of active genes

Despite a great deal of research, the precise changes in the structure of a gene to enable it to be an 'active' initiation site for transcription still remain unclear. Many disparate theories have been proposed, but in view of the complexity of eukaryotic genomes, it is unlikely that gene activation can be attributed to one mechanism alone. Almost certainly, gene activation is the result of several molecular events, working in a coordinated manner. On present evidence, the best criterion for the activation of a gene is that it becomes acutely sensitive to digestion by pancreatic deoxyribonuclease, type I (DNAase I). This important observation, first made by Garel and Axel (1976) on the ovalbumin gene in hormonally stimulated chick oviduct, has since been found to be a general phenomenon of gene activation, including the genes for α- and β-globins, fibroin and many proteins in *Drosophila.* It is important to stress, however, that DNAase I sensitivity reflects a gene conformation which is clearly necessary to permit transcription, but not sufficient in itself to explain completely the actual mechanism of gene activation in molecular terms; for example, the mature, anucleate erythrocyte no longer synthesizes haemoglobin, yet its globin genes are still very sensitive to digestion by DNAase I.

In a more extended study, Garel and Axel (1977) reported that the entire ovalbumin gene was readily detected in nucleosome core particles and clearly the presence of histones about a specific gene does not prevent its transcription. The general outcome of this work is that the genes are organized by non-histone proteins and other nuclear components in such a way that they are in an extended, transcriptionally active and DNAase I-sensitive state. Histone–DNA interactions can be weakened by covalent modifications, including methylation, acetylation and phosphorylation. In addition, the nuclear complement of non-histone proteins is changed qualitatively and quantitatively as a prerequisite for both gene transcription and gene replication. Striking examples of the selective synthesis of non-histone proteins have been demonstrated during the responses of the prostate to androgens and of the uterus to oestrogens. Wider evidence supporting the role of non-histone proteins in the initiation of gene transcription can be found elsewhere (Stein *et al.* 1974; O'Malley *et al.* 1977). Once a gene has been activated, its efficiency in initiation rises markedly. Studies on the ovalbumin gene in chick oviduct indicate that its rate of initiation increases several orders of magnitude after hormonal stimulation (Schwartz *et al.* 1975), to the extent that the content of ovalbumin mRNA rises from baseline of 10 copies per cell to 100 000 copies per stimulated cell.

The original concept that active genes have a different configuration has been extended in an imaginative way by Wu *et al.* (1979a,b). Their new concept is based on the idea of domains, as originally formulated to describe the structure–function relationships in immunoglobulins, but also applicable to more complex structures, such as eukaryotic chromosomes. Accordingly, Wu *et al.* (1979a) found evidence for preferential cleavage sites for DNAase I in *Drosophila* chromosomes which they suggest delineate higher orders of structure or domains. Expressed another way, chromatin DNA is seen as a series of discrete functional (transcriptional) units which are demarcated by regions of defined base sequence and which may be activated selectively. The latter possibility was confirmed in an elegant manner by Wu *et al.* (1979b). If *Drosophila* cells in culture are exposed to an elevated temperature of 35 °C rather than their usual 25 °C environment, the cells go into 'heat shock'. As a consequence, a set of nine genes become puffed or activated and virtually all of the biosynthetic capability of the cells is directed to the synthesis of a few proteins, characteristic of cell survival at 35 °C. Importantly, when these nine genes are activated, their cleavage by DNAase I changes, yet the normal pattern of cleavage is restored when the cells are returned to an environment of 25 °C and the heat shock genes become inactivated. Wu *et al.* (1979b) interpret their observations as a novel manipulation of the domains of *Drosophila* chromatin via changes in environmental temperature. This interesting work underlines the close correlation between gene structure and function in a very novel way; gene activation clearly involves subtle changes in structure extending beyond their sensitivity to DNAase I.

Using the expression of the α- and β-globin genes in chicken erythrocyte nuclei as their model system, Gazit *et al.* (1980) have greatly advanced our understanding of active genes. These investigators identified the active globin genes by their sensitivity to DNAase I and radioactively labelled them by the nick translation procedure; after introducing nicks with DNAase I, the breaks were repaired by *E. coli* DNA polymerase (type I) in the presence of dATP, dCTP, dGTP and [^{3}H]TTP. DNAase-sensitivity of the globin was lost if erythrocyte nuclei were extracted with 0.35 M-NaCl, but nuclease sensitivity was restored in reconstituted nuclear systems supplemented with the NaCl extract. Gene activation, as monitored by DNAase sensitivity of the ^{3}H-labelled globin genes, was controlled by two HMG (high-mobility group) proteins, types 14 and 17. The expression of selected genes, therefore, needs their association with specific non-histone proteins and the synthesis and distribution of HMG proteins is clearly a powerful means of regulating gene transcription in eukaryotes. This extends the significance of this interesting group of nuclear proteins, since HMG protein (type 6) has an important role in the assembly of nucleosomes (see section 7.2.4).

The concept of chromatin domains helps to explain one paradox of hormonal stimulation. The number of hormone–receptor protein complexes which may be translocated into the chromatin of hormone-sensitive cells is extremely high, up to 50 000 molecules per nucleus (King & Mainwaring 1974); this vastly exceeds the number of regulatory proteins, such as Lac repressor, in *E. coli*. If, however, the receptor complexes have to inactivate many chromosome domains, in addition to activating a select few, then large numbers of molecules could be envisaged (Yamamoto & Alberts 1976). Certainly, there is evidence that purified receptor complexes can rapidly and specifically activate specific genes in chromatin under conditions *in vitro,* as reported for the ovalbumin gene (Schwartz *et al.* 1976). While the specificity of this response is extremely gratifying, the speed, essentially instantaneous, is rather problematical. Other investigators have reported a significant lag in the induction of ovalbumin mRNA in the intact oviduct cell. The lag can be explained by the time needed for the translocation of receptor complex to key sites in chromatin adjacent to the ovalbumin gene itself or to the necessary synthesis of other nuclear components (including HMG proteins)

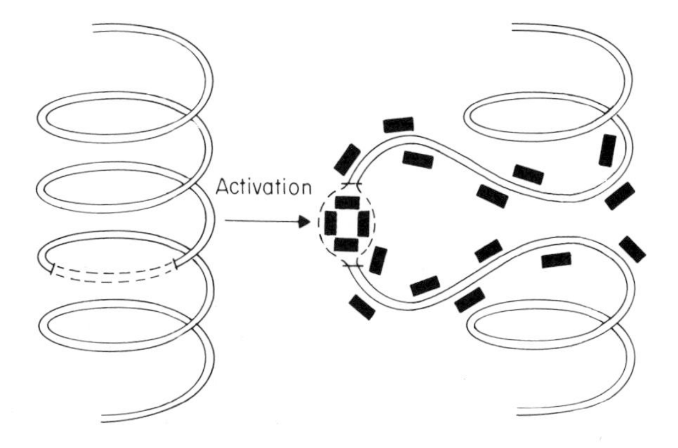

Fig. 8.3. The activation of genes by hormones. Many molecules of the hormone–receptor complex (■) may be needed to activate even a single hormone-sensitive gene (= = =) because the domains of chomatin may need to be changed markedly in order to initiate a change in transcription.

necessary for maximal synthesis of ovalbumin mRNA. Another possibility is illustrated in Fig. 8.3; there, it is envisaged that the hormonal activation of even a single gene needs a major change in the overall structure of a eukaryotic chromosome, involving modifications in the configuration of many chromosome domains by the hormone–receptor complex.

8.1.4 The analysis of transcripts from eukaryotic genes

Probes for nucleic acid hybridization

Only nucleic acid hybridization provides the necessary specificity for accurately monitoring the expression of specific eukaryotic genes. Other methods have been used in the past, but their value, apart from simplicity, is debatable. Prompted by the work of Gilmour and Paul (1971) and many others, the consensus view is that 90% or more of the genes in their natural environment, as nuclear chromatin, are closed down or repressed in terms of transcription.

Hybridization can be used in two ways in monitoring gene expression. First, hybridization is the ultimate test for correct strand selection during transcription. The strands of small genes, obtained from amplified sources in nature or by cloning, can be separated in CsCl gradients. The separate DNA strands can then be used in hybridization experiments against RNA species labelled in intact cells, either *in vitro* or *in vivo*. By such hybridization, the active gene sequence for the transcription of 5*s* rRNA in *Xenopus* was found to be in the L (light) strand. Second, and far more commonly, hybridization is used to analyse RNA transcripts made on templates of purified DNA, chromatin or even intact nuclei. In these cases, the non-radioactive transcript is analysed by hybridization to a highly radioactive (^{32}P- or ^{3}H-labelled) complementary DNA (cDNA) to the gene product (RNA transcript) under study. In a restricted number of cases (e.g. amplified rRNA genes in amphibia), it is possible to label the appropriate DNA strand directly *in vivo*. In the vast majority of cases, the RNA species in question is purified to homogeneity and the radioactive cDNA made with reverse transcriptase and precursors such as [α-^{32}P]dCTP, [^{3}H]dCTP or even [^{125}I]dCTP (Bhalla *et al.* 1976).

The identification of poly(A) tracts at the 3′-end of many eukaryotic mRNAs was a momentous discovery (Lim & Canellakis 1970, Darnell *et al.* 1971) and opened up new opportunities for the purification of poly(A)$^+$ mRNAs by chromatography on oligo(dT)-cellulose or poly(U)-cellulose. Reverse transcriptase lacks template specificity and in the presence of a short oligo(dT) primer to form an initiation site or hairpin with the poly(A)$^+$ tail, the enzyme makes a cDNA to most eukaryotic and viral mRNAs. Messenger RNAs with a 3′ poly(A) tail are termed poly(A)$^+$; mRNAs without such a tail are poly(A)$^-$. Under appropriate conditions, *E. coli* DNA polymerase I can also utilize mRNA and rRNA as templates and synthesize the corresponding cDNA. Certain eukaryotic mRNAs do not possess a poly(A) tract at the 3′ terminus, the best-studied poly(A)$^-$ mRNAs being those for histones. Nonetheless, if a poly(A) tail is added to histone mRNA using the ATP: polynucleotidylexotransferase from maize, cDNAs to histones can be made by reverse transcriptase. Because of its high degree of secondary structure and base pairing, tRNA is not a suitable template for the synthesis of a cDNA. This has not proved to be a problem, for, using techniques described later, investigators have found alternative means for studying the faithful transcription of tRNA genes.

The template for transcription in vitro

In attempts at the faithful transcription of specific eukaryotic genes *in vitro*, many alternative forms of template DNA can be used. In certain cases, the gene under investigation can be isolated by exploitation of certain unusual or distinctive properties. The silk fibroin gene has only one copy per haploid genome in the silk moth, *Bombyx mori*, yet it may be readily isolated in CsCl gradients on the basis of its anomalously high G+C content. Many genes can be amplified by natural or artificial means, thus providing ideal templates for transcriptional studies; a fine example of artificial gene amplification is provided by the work of Schimke *et al.* (1977) on the genes for dihydrofolate reductase, where resistance of cultured mouse cells to methotrexate was judiciously exploited. Many eukaryotic genes have now been successfully cloned in bacterial vectors, particularly impressive examples being the human genes for growth hormone (Goeddell *et al.* 1979a), interferon (Derynck *et al.* 1980) and insulin (Bell *et al.* 1980). In all these cases, the necessary presence of pre- or signal-gene sequences posed particular problems in the manipulation and correct expression of the genes. Of potentially even greater importance to future applications of genetic engineering in the clinical context has been the successful cloning of the mouse β-globin gene in a simple eukaryotic genome (Hamer &

Leder 1979). By such means as these, many purified, mammalian genes are now available in considerable quantity.

Less sophisticated templates for gene transcription obviously include isolated nuclei and chromatin. From the historical standpoint, studies on chromatin have been particularly important, but many problems need to be overcome to simulate faithfully the transcription of a specific gene in chromatin. Artefacts are only too easy to create unwittingly and the elegant papers of Gadski and Chae (1976, 1978) have laid down the fundamental sequence of experimental manipulations needed to reconstruct accurately the transcriptional activity of chromatin *in vitro*. Non-histone proteins must be able to reassociate with purified DNA before histones, if a specific gene (e.g. a globin gene) is to be transcribed with fidelity. To circumvent the problems associated with transcription in chromatin, investigators are now turning increasingly to using nuclei as a source of template. Not only are nuclei easy to prepare and purify, transcription is generally retained in a highly precise and correctly ordered manner. The complete fidelity of selective gene transcription in nuclei is well illustrated by the work of Orkin (1978) on the synthesis of α- and β-globin mRNAs in nuclei from erythroleukaemic cells and by the work of Hamada *et al.* (1979) on the synthesis of 5*s* rRNA in nuclei from rat liver and HeLa cells. In nuclei, both endogenous template DNA and RNA polymerase are used in transcription; in the two cases cited, transcription was asymmetric and from the correct gene strand, and the transcripts were of precisely the length found *in vivo*.

When hybridization to a cDNA is used to monitor transcription, a fundamental problem is the presence of the endogenous RNA in chromatin and nuclei; clearly, this must be accounted for in quantitative assessments of gene expression after, for example, hormonal stimulation. The transcription of cloned genes does not suffer from this practical limitation. A novel means of avoiding the problem of endogenous RNAs was proposed by Dale *et al.* (1974). The trick was to include 5′-mercuricytidine triphosphate in the assay medium for transcription, *in vitro*. There are two advantages to this approach. First, the intact cell does not make Hg-CTP and hence endogenous and exogenous transcripts (the latter being made *in vitro*, under controlled conditions) can be readily distinguished. Second, the exogenous transcript can be effectively separated by its selective adsorption on Sepharose beads containing immobilized sulphydryl groups, as follows:

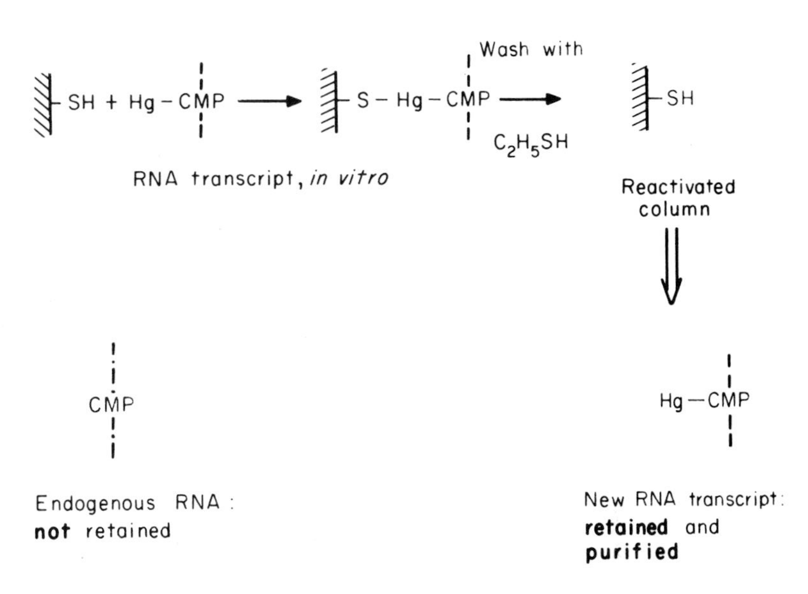

This approach has greatly improved the accuracy with which RNA transcripts made *in vitro* can be analysed by specific cDNA probes. The use of mercurated CTP (sometimes as [^{203}Hg]CTP) has featured in the better studies on eukaryotic gene expression.

The choice of RNA polymerase

This has turned out to be a controversial topic. For many years there has been an understandable preference for using the readily prepared and highly active RNA polymerase from bacteria for the transcription of eukaryotic genes *in vitro,* rather than utilizing the less stable and singularly less active eukaryotic RNA polymerases, either endogenous (present in nuclei) or exogenous (added RNA polymerase in purified form). In his innovative work on the transcription of liver chromatin from the mouse and *X. laevis,* Reeder (1973) suggested that the *E. coli* RNA polymerase did not necessarily work very accurately. In particular, inaccurate strand selection and the failure to recognize normally suppressed genes raised problems in the interpretation of results from such heterologous or 'mixed' systems (i.e. eukaryotic genes plus prokaryotic RNA polymerase). These problems came to a head with the studies of Zasloff and Felsenfeld (1977) on the transcription of globin genes *in vitro.* Under the usual conditions for transcription, *E. coli* RNA polymerase has the serious limitation of using endogenous RNA as a template and catalyses RNA-dependent as well as DNA-dependent RNA synthesis. Thus, with chromatin from erythropoietic cells, the bacterial enzyme made antisense RNAs (or RNAs complementary in base sequence) to the endogenous globin mRNAs. Using the mercurated CTP method of Dale *et al.* (1974) serious experimental artefacts were created because the antisense RNA formed a stable duplex with the endogenous mRNA. This association enabled the antisense RNA to be spuriously retained on SH-Sepharose. Specific [^{3}H]cDNA probes for antisense globin RNAs confirmed the difficulties surrounding the use of *E. coli* RNA polymerase (Orkin 1977). These problems are obviated if mercurated transcripts are subjected to thermal denaturation prior to adsorption on to SH-Sepharose (Zasloff & Felsenfeld 1977); this manoeuvre destroys the casual association between Hg-free endogenous RNA and Hg-containing RNA transcripts.

In summary, accurate transcription can be monitored efficiently using mercurated nucleotides, but unless special precautions are taken, eukaryotic genes are best transcribed *in vitro* by homologous (eukaryotic) RNA polymerases.

Choice of analytical approach

With the adoption of mercurated nucleotides (either Hg-CTP or Hg-UTP), non-radioactive transcripts made *in vitro* are generally analysed by a highly radioactive cDNA probe; to make the latter, the reverse transcriptase reaction contains deoxyribonucleoside 5′-triphosphates generally labelled with [^{3}H] or specifically labelled in the α-phosphate group with [^{32}P]. Since only the initial purine ribonucleoside 5′-triphosphate retains its full complement of three phosphate groups at the 5′ terminus of the RNA transcript, the initiation of transcription *de novo* can be readily monitored by the incorporation of [γ-^{32}P]ATP or GTP into RNA. Spurious results can be obtained, however, with impure RNA polymerases.

8.1.5 Model systems for studying the transcription of specific eukaryotic genes

The inherent difficulties associated with eukaryotic transcription have placed heavy reliance on model systems, each of which has certain favourable attributes for studying selective gene expression.

Transcription of hormonally activated chromatin in vitro *and* in vivo

Starting from the pioneering work of O'Malley and McGuire (1968) the chick oviduct is perhaps the best-known system for activating genes by steroid hormones and their analogues. Several proteins are only made in the oviduct when hormones are present, including ovomucoid, transferrin (formerly conalbumin), lysozyme, and particularly, ovalbumin. In a remarkable series of papers, culminating in O'Malley *et al.* (1977), the induction of ovalbumin synthesis by stilboestrol has been the archetype of how hormones act at the transcriptional level. The principal findings of O'Malley and his collaborators can be summarized as follows.

(a) The selective activation of the ovalbumin gene depends on hormonally mediated changes in the oviduct non-histone proteins.

(b) During this activation process, there is an increase in the number of binding sites for RNA polymerase. The number of initiation sites for transcription also increases dramatically and there is a precise correlation between the extent of the initiation of transcription of the ovalbumin gene and the degree of nuclear occupancy by steroid hormone–receptor complex (Tsai *et al.* 1975).

(c) In reconstitution experiments, the transcription of the ovalbumin gene was found to be asymmetric and only from the correct or coding DNA strand (Towle *et al.* 1977).

The only criticism that can be levelled at this work is that RNA transcription was largely measured by *E. coli* RNA polymerase in the presence of rifampicin. The antibiotic was included to eliminate the elongation of pre-existing RNA chains and thus only the initiation of transcription, *de novo,* was measured. As described elsewhere in this section, other investigators report aberrant transcription of eukaryotic genes by the bacterial RNA polymerase.

Other investigators, notably Palmiter and Schütz, have followed the induction of oviduct proteins under conditions *in vivo*. There is a large measure of agreement between these powerful research groups in their work on the activation of the ovalbumin and other oviduct genes. The only disparity is the lag in ovalbumin induction reported from studies *in vivo* but not the reconstitution work *in vitro*. Furthermore, all investigators primarily use stilboestrol for hormonal activation, yet the activation of the ovalbumin gene *in vitro* was effected with a progesterone–receptor protein complex, not an oestrogen–receptor protein complex.

The transcription of amplified or cloned genes

The genes for 5*s* rRNA are particularly suitable for transcription *in vitro*. In *Xenopus* oocytes, these genes are amplified and this increases the likelihood of their transcription, even in chromatin. Indeed, Parker and Roeder (1977) found that these genes were particularly well transcribed from *Xenopus* chromatin by RNA polymerase C, which unlike other eukaryotic polymerases, readily initiates RNA synthesis under conditions *in vitro*. In later experiments, it was demonstrated that only eukaryotic RNA polymerase C could effect this transcription. The 5*s* rRNA genes have been successfully cloned and they are faithfully transcribed in soluble, DNA-free extracts of *X. laevis* oocytes which are rich in RNA polymerase C (Ng *et al.* 1979). By all criteria, including strand selection and transcript size, the 5*s* rRNA genes were transcribed precisely in these cell-free systems. These are vitally important experiments, because they illustrate that eukaryotic RNA polymerases do have the intrinsic ability to recognize accurately initiation (promoter) and termination signals in eukaryotic DNA, aided perhaps by a few soluble, regulatory proteins.

Transcription in nuclei

The faithful transcription of specific genes by endogenous RNA polymerases has been unequivocally established by the work of Orkin (1978) and Hamada *et al.* (1979). A novel manipulation developed by Tekamp *et al.* (1979) maintains the fidelity of transcription in nuclei yet makes the process more efficient by accurately involving highly purified, exogenous but homologous (eukaryotic) RNA polymerases. When briefly exposed to a pH of 4.5, nuclei lose their endogenous RNA polymerase activity, but the template properties of nuclear DNA are fully preserved when the pH is restored to the physiological range and the nuclei are supplemented with purified, eukaryotic but not prokaryotic RNA polymerases. By this simple manipulation of pH, the highly specific biosynthesis of yeast rRNA by exogenous RNA polymerase A and of 4*s* tRNA and 5*s* rRNA by polymerase C is enhanced. This technique could have great potential in future experiments.

Giant chromosomes in the salivary glands of insects

The giant (or polytene) chromosomes of *Drosophila melanogaster* and *Chironomus tentans* have received a great deal of attention for over thirty years because of their extreme size and the ease with which detailed microscopic examination of them may be made. The transcriptional activity of a few genes in *C. tentans* can be dramatically and selectively controlled by simple experimental means; for example, galactose stimulates the activity of just two transcriptional units, the Balbiani rings 1 and 2 (BR1 and BR2), on chromosome IV. The transcripts from BR1 and BR2 are long, 75*s* mRNAs, which are finally located in the cytoplasm and direct the translation of the salivary polypeptides. This interesting system was elegantly exploited by Lamb and Daneholt (1979) in their visualization of the active BR2 genes during their selective

Fig. 8.4. The appearance of active transcription units. (a) A single chromosome, number IV, of the salivary gland of *Chironomus tentans*, showing a vast number of transcription units, as filamentous structures, and marked by arrows. The bar indicates 1 μm. (b) An active transcription unit in greater detail. The bar indicates 0.5 μm. (Reproduced, with permission, from Lamb & Daneholt 1979.)

transcription. Active transcription units for BR2 75*s* mRNA were about 7.7 μm long, and approximately 6 molecules of RNA polymerase B joined the active gene per second and the rate of RNA elongation was estimated at approximately 31 nucleotides per second at 18°C. The nascent 75*s* mRNA was seen as a series of lateral, ribonucleoprotein fibres, the maximum being about 120 per transcription unit. Lamb and Daneholt (1979) suggest that proteins are associated with growing RNA transcripts, forming organized, granular structures (50 nm diameter) along the complete length of the transcription unit. The most striking aspect of this beautiful study is the precise description of gene activation. Inactive genes exhibited a uniform, regularly beaded appearance, whereas active genes were irregularly or sparsely beaded. Gene activation was preceded by a marked extension of the gene fibre and a smooth, non-beaded region, 0.18 μm long, and possibly representing an RNA polymerase promoter region, preceded the active region of gene transcription, with its lateral, ribonucleoprotein fibres containing the 75*s* mRNA. Examples of the exquisite electron micrographs of Lamb and Daneholt (1979) are presented in Fig. 8.4.

Transcriptional systems derived from Xenopus *oocytes*

The transcription of tRNA genes was difficult to follow by conventional means, because reverse transcriptase and other polymerases cannot utilize tRNA as a template and cDNAs to tRNA are not readily available. Using genetic engineering, however, it has proved possible to clone certain tRNA genes in bacterial plasmids. Mertz and Gurdon (1977) found that DNA injected into *Xenopus* oocytes could be transcribed; furthermore, the cloned genes for $tRNA^{fMet}$ (Kressman *et al.* 1978) and ovalbumin (Trendelenburg *et al.* 1980) are precisely transcribed in intact oocytes. Schmidt *et al.* (1978) demonstrated that even the soluble extracts of *Xenopus* oocytes are capable of faithfully transcribing cloned genes *in vitro*; recent examples include the genes for 5*s* rRNA (Ng *et al.* 1979) and various tRNAs (Schmidt *et al.* 1978). Manley *et al.* (1980) have reported that soluble extracts of HeLa cells are also efficient in the transcription of purified exogenous genes. These soluble systems provide exciting possibilities for transcriptional studies in the future.

8.2 THE INITIAL RNA TRANSCRIPTS OF EUKARYOTIC GENES

In his excellent review, Perry (1976) points out how early studies on the synthesis of tRNA and rRNA in eukaryotes led to the idea that the initial transcripts of eukaryotic genes were much larger than the final, mature species of RNA involved in genetic translation. The nature of the initial RNA transcripts of eukaryotic genes has now become much clearer with the advent of new methods for the purification and analysis of all species of RNA.

8.2.1 Ribosomal RNA

Classical experiments by Perry and Scherrer established that the initial transcript of nucleolar rRNA genes by RNA polymerase A in mammals was a single RNA transcript of sedimentation coefficient, 45*s*, and M_r 4.5×10^6. In lower eukaryotes, the rRNA precursor is somewhat smaller, 36–38*s*, and even smaller in poikilotherms which have simpler transcriptional units still (Perry 1976).

Nucleolar rRNA genes have a complex structure and contain both transcribed and non-transcribed spacer regions. Their complexity is increased in some species by the presence of introns; indeed introns were first identified in cloned 26*s* rRNA genes of *Drosophila* (Glover & Hogness 1977, White & Hogness 1977). The primary 45*s* transcript of mammalian cells is the sole precursor for 5.8, 18 and 28*s* rRNA; the 5.8*s* species was formerly (and incorrectly) called 7*s* rRNA. Species such as *Drosophila* are even more complex in that they have a 2*s* rRNA as well. In the simpler cases, where introns are not present, the arrangement of the genes for the long rRNA precursor is arranged as follows:

5′- :	non-	— transcribed	— 18*s* rRNA	— transcribed	— 5.8*s* rRNA	— transcribed	— 28*s* rRNA	: 3′-
end	transcribed spacer	spacer	gene	spacer	gene	spacer	gene	end

This structure forms a transcriptional unit which is repeated possibly thousands of times in some eukaryotes.

The 5*s* rRNA is not made in the nucleolus, but is transcribed from genes in extranucleolar chromatin by RNA polymerase C. Again, these genes are separated by long, non-transcribed spacer regions and these transcriptional units are often repeated thousands of times per nucleus. Until recently, it was generally considered that the initial transcript of 5*s* rRNA genes was identical in size to the mature species of 5*s* rRNA. In some species, however, transcribed spacers are also present in this transcriptional unit. The careful study by Hamada *et al.* (1979) indicates that primary transcripts are longer by just 8 nucleotides at their 3′-ends than mature 5*s* rRNA. The structures of all rRNA genes are discussed in more detail in section 8.3.2.

In all eukaryotes, synthesis of the long rRNA precursor occurs exclusively in the nucleolus. The number of nucleoli per nucleus depends on the dispersion and fusion of the ribosomal rRNA genes with discrete components of eukaryotic chromosomes called the nucleolar organizers. Many eukaryotes have two nucleoli per nucleus during early embryonic development and are generally designated the genetic constitution, +/+ ν. The diploid, homozygous recessive condition (o/o ν) is lethal and results in the complete deletion of rRNA genes from the nucleolar organizers.

The nucleolar rRNA genes in *Xenopus* have been studied by Reeves (1977) in an effort to delineate the structural changes in chromatin which precede gene activation. This study suggests that the association between DNA and histones is a dynamic, flexible process, which is ultimately related to gene activation. As the rRNA genes become activated, they tend to lose their complement of tightly associated nucleosomes, but even when maximally activated, genes are never completely free of histones. By contrast, transcriptionally inactive spacer DNA has a constant structure, always rich in nucleosomes. As the rRNA genes shed their nucleosomes, they become more sensitive to digestion by micrococcal nuclease (*S. aureus*), a fact discussed earlier (see section 8.1.3) with respect to the increased sensitivity of hormonally activated genes to digestion by pancreatic DNAase I. From this study, it is clear that gene activation is always accompanied by structural changes in chromatin, irrespective of the trigger for activation. Foe (1977) extended this concept by elegant electron microscopic analysis of the nucleolar genes for rRNA in the milkweed bug, *Oncopeltus fasciatus,* where the rate of rRNA synthesis is extensively modulated during embryonic development. Foe (1977) concluded that inactive rRNA genes are present in dormant or ν chromatin, which has a beaded appearance, rich in nucleosomes, and which cannot be transcribed by RNA polymerase A. Activation of the rRNA genes necessitates a marked change in the structure of nucleolar chromatin. Active rRNA genes are found in transcribable or ρ chromatin, which has a non-beaded appearance, depleted of nucleosomes, along the entire length of the active transcriptional unit. The interconversion of ν to ρ chromatin is reversible and is regulated in *Oncopeltus* by factors associated with development. The findings of Foe (1977) may be schematically represented as follows:

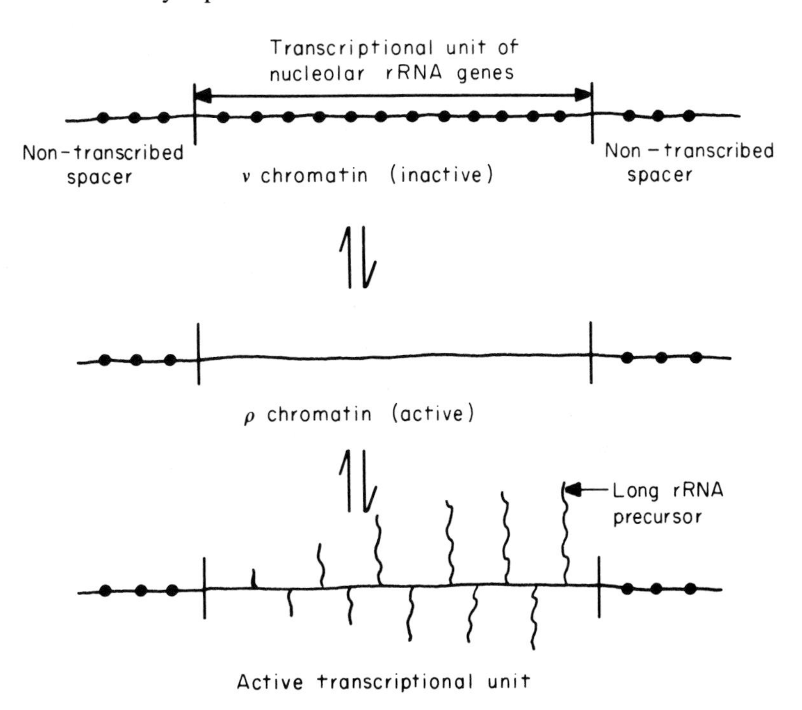

The important contributions of Reeves and Foe place heavy, even exclusive, emphasis on the loss of histones during gene activation. This conclusion is somewhat contrary to findings in other systems, where a fundamental role for non-histone proteins is envisaged in the modulation of gene expression. However, Reeder *et al.* (1977) have begun to investigate the qualitative and quantitative changes in all nucleolar proteins during the activation of rRNA genes and this may clarify the present impasse.

8.2.2 Transfer RNA

There is abundant evidence (for a review see Perry 1976) that the initial transcripts of tRNA genes are longer than the mature species of tRNA found in the cytoplasm of eukaryotes. As judged by sedimentation or electrophoretic criteria, the tRNA precursors are about 15–35 nucleotides longer than their ultimate form. The tRNA genes are transcribed by RNA polymerase C and the initial transcripts usually have a tract of uridine residues at their 3′ terminus. Many tRNA genes, including those in *Xenopus*, are separated by non-transcribed spacers, which are up to 10 times the length of the tRNA gene. Furthermore, it is now clear that tRNA genes often contain introns.

Eukaryotic tRNA genes are present in multiple copies. In mouse myeloma and *Xenopus* cells, each unique tRNA gene may be reiterated 200 times, but in most other eukaryotes, the average frequency of tRNA gene reiteration is only about 10. From the work of Hosbach *et al.* (1980) on the tRNA genes in *Drosophila,* clustering is found for the genes coding for a specific species of tRNA, e.g. clusters of five $tRNA^{Gln}$ genes, arranged in tandem, with the same polarity. In addition, some clusters contain tandem repeats of different tRNA genes; one repeat unit in *Xenopus,* some 3.18 kb long, contains two $tRNA^{Met}$ genes and one gene each for $tRNA^{Tyr}$ and $tRNA^{Phe}$.

As reviewed by Perry (1976), radioactive tRNA precursors often appear more abundantly in the cytoplasm of eukaryotic cells during pulse-labelling experiments. This could indicate their rapid exit from the nucleus, their site of transcription, or more probably their leakage from the nucleus during cell-fractionation procedures. Despite some rather tenuous evidence to the contrary, the initial tRNA precursor is made and processed in the nucleus.

8.2.3 Messenger RNA

Establishing the structure of the initial transcripts of mRNA genes made by RNA polymerase B has proved to be one of the most daunting problems of contemporary biochemistry. A vast body of evidence, ably reviewed by Darnell (1975), suggested that eukaryotic nuclei contained large molecules of heterogeneous (hn) RNA, and considerable circumstantial evidence supported the view that hn RNA was a precursor of mRNA. From studies on a variety of eukaryotic cells, hn RNA had certain unique properties:

(a) relative insensitivity to inhibition by actinomycin D;

(b) rapid labelling with RNA precursors;

(c) extreme heterogeneity and extreme length, often 5000–50 000 nucleotides long (sedimentation coefficients 20–100*s*).

It has taken a lot of work to dispel the doubt, originally expressed by Davidson and Britten (1973), that hn RNA really is the genuine precursor of eukaryotic mRNA.

Beginning about 1973, investigators recognized the potential of nucleic acid hybridization in the analysis of hn RNA; many, for example, raised [^{3}H]cDNA probes to the mature, 10*s* α- and β-globin mRNAs found in the polyribosomes of rabbit and avian erythroblasts. From these studies, putative globin mRNA sequences were identified in hn RNA of size 20*s* and greater. Using similar [^{3}H]cDNA probes raised against pure ovalbumin mRNA, other researchers failed to identify ovalbumin mRNA precursors in hn RNA greater than 18*s*, and suggested that mRNA precursors are not invariably very long and may even be the same size as the mature, cytoplasmic mRNA. MacNaughton *et al.* (1974) reconciled these differences by establishing that hn RNAs had a marked tendency to aggregate, and that if sedimentation analyses were carried out in denaturing conditions in the presence of formamide, the hn RNA sedimented as a broad peak, 10–40*s*, but rarely larger. Even in denaturing conditions, however, the hn RNAs of cells undergoing regeneration or malignant transformation were genuinely very long. Nonetheless, MacNaughton *et al.* (1974) provided convincing evidence that the nuclear precursors of globin mRNAs were 15*s* hn RNA molecules.

When poly(A) was identified specifically at the 3′-end of many eukaryotic mRNAs, there were justifiable hopes that this innovation could help in resolving the relationship between hn RNA and mRNA. In actual fact, although many hn RNAs can be readily polyadenylated, their precursor status remained somewhat enigmatic, even for globin mRNAs.

The resolution of the authentic role of hn RNA in eukaryotic mRNA transcription came from perceptive studies with cloned globin genes and from studies on the transcription of adenovirus type 2 (Ad2) DNA during the early and late phases of viral infection.

Maniatis *et al.* (1976) succeeded in cloning the gene for rabbit β-globin and Curtis *et al.* (1977) exploited the DNA of this plasmid to develop a very precise method for studying the synthesis of 15*s* globin mRNA precursor. From the kinetics of its synthesis in erythroleukaemic cells, Curtis *et al.* (1977) claimed a direct precursor relationship between 15*s* RNA and β-globin mRNA. This suggestion was supported by structural comparisons. First, RNAase T_1 digestion and fingerprinting revealed extensive base homology between 15*s* RNA and mature, β-globin mRNA; not surprisingly, in view of its length, the 15*s* RNA contained additional sequences which had no counterpart in 10*s* globin mRNA. Second, the 15*s* and 10*s* RNAs had identical 5′ and 3′ termini, $m^7Gppp^6A^mpC$ and poly(A), respectively.

Konkel *et al.* (1978) established that the gene for mouse β-globin had the following structure, containing two introns:

5′-end: globin exon	—	small intron (116 bp)	—	globin exon	—	large intron (646 bp)	—	globin: exon	3′-end

Tilghman *et al.* (1978) studied the hybridization of both 15*s* and 10*s* RNA from erythroleukaemic cells with plasmid DNA containing the β-globin gene, using the R-loop method devised by White and Hogness (1977). When mature, 10*s* β-globin mRNA was hybridized to the cloned β-globin gene, characteristic R-loop structures equivalent to the globin gene introns were found. By contrast, when the 15*s* RNA was hybridized to the cloned β-globin gene, the resulting structure was complete and showed *no interruptions,* confirming that both structural (exon) and intervening (intron) gene sequences are represented in the 15*s* RNA (Fig. 8.5). Clearly, the precursor role of 15*s* RNA in the synthesis of β-globin mRNA had been demonstrated unequivocally. Based on these important experiments, introns in eukaryotes genes will be transcribed in hn RNA and must then be excised during the subsequent formation of the mature, translatable species of mRNA.

Studies on viral genomes have substantiated the precursor relationship between hn RNA and eukaryotic mRNA. In a very favourable experimental system, there is a marked change in the expression of Ad2 genes during the early and late stages of infection

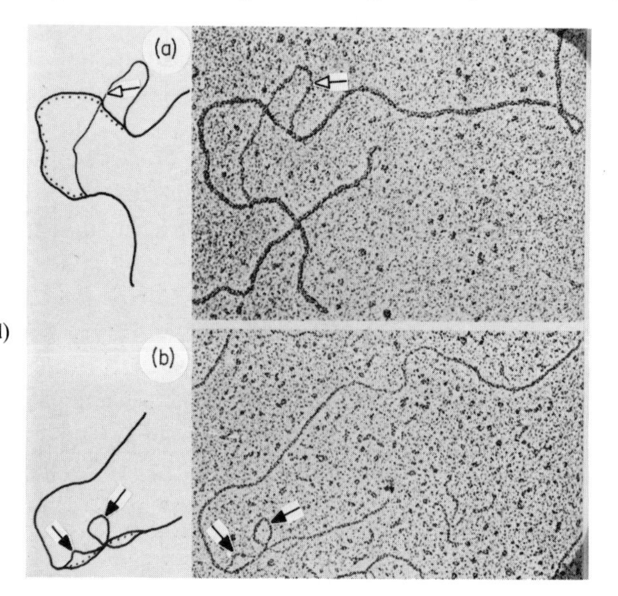

Fig. 8.5. Hybrid structures formed between a cloned, mouse β-globin gene and 10*s* and 15*s* β-globin mRNAs. (a) The hybrid between the globin gene and 15*s*, hn β-globin RNA was a continuous structure, along one strand of DNA; the other DNA (non-coding strand) is displaced and shown by the open arrows, →. (b) The hybrid between the globin gene and mature, 10*s* β-globin mRNA shows two displaced structures or R-loops, shown by the solid arrows, →. Line drawings to the left of both electron micrographs show DNA as solid lines, ———, and RNA as dotted lines, ······. (Reproduced, with permission, from Tilghman *et al.* 1978).

of cultured HeLa cells. These 'early' and 'late' viral mRNAs have been carefully characterized and assigned to specific regions on the map of the circular Ad2 genome. Importantly, as shown in Fig. 8.6, early and late mRNAs are transcribed from different strands of Ad2 DNA, from either the r (rightward reading) or l (leftward reading) strands. Additional work has defined the sizes of the hn RNA precursors for early and late viral mRNAs, and also the initiation sites for the transcription of these RNA precursors relative to the map coordinates of Ad2 DNA. Using restriction enzymes, Darnell *et al.* (1977) and Evans *et al.* (1977), selectively cut Ad2 DNA into a

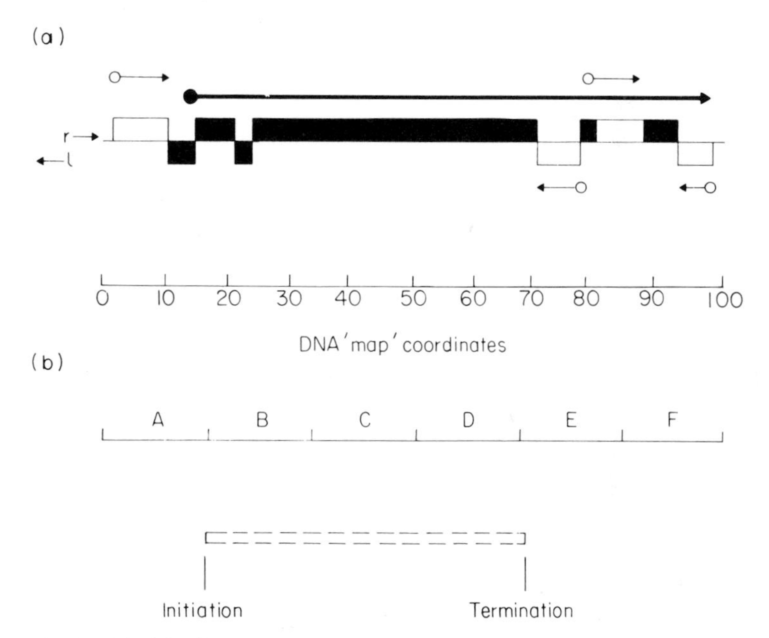

Fig. 8.6. The identification of the transcriptional units in Ad2 DNA for early and late viral mRNAs. This composite figure is based largely on the work of Petterson *et al.* (1976) and Evans *et al.* (1977). (a) A comparison of the known positions of the early and late mRNA genes in Ad2 DNA and their transcriptional units. The upper section shows the r (rightward reading) and 1 (leftward reading) strands, and the location of early, ■, and late, □, mRNAs. The initiation sites and directions of transcription units are shown by arrows for early, O→, and late, ●→, mRNA precursors. The lower section shows the map units, 0–100, for Ad2 DNA, used in monitoring the hybridization of nascent hn RNA to fragments of Ad2 DNA produced by digestion with restriction enzymes. (b) The rationale behind the hybridization of nascent hn RNA to DNA restriction fragments. A transcription unit is depicted as six sections of DNA, A to F, which may be separated by cleavage with restriction enzymes. If pulse-labelled [^{3}H]hn RNA, [- - -], hybridizes to separated segments B, C and D, but not to A, E or F, then this transcriptional unit is initiated near the cleavage site between A and B, and terminated near the cleavage site between D and E. In such an approach, precautions must be taken to prevent hn RNA degradation during its isolation.

series of defined fragments and determined their positions within the whole viral genome. Conventionally, the Ad2 genome is mapped in the rightward reading direction, 0 to 100, such that there are a total of 100 map coordinates, each being 1% of the total Ad2 DNA or 3500 base pairs (3.5 kb) long. Using pulse-labelling techniques (i.e. brief exposure of virus-infected HeLa cells to [^{3}H]uridine) a series of initial or nascent hn RNA precursors were isolated during early and late phases of viral infection and analysed by hybridization to the separate restriction fragments of Ad2 DNA. If a tract of Ad2 DNA is envisaged as a series of interconnected restriction fragments, A—B—C—D—E—F, as in Fig. 8.6(b), positive hybridization of a [^{3}H]hn RNA to fragments B, C and D, but not A, E and F, indicates clearly that the transcription unit for that hn RNA is contained in the DNA sequence, B—C—D. Furthermore, initiation of the transcription of that hn RNA must begin close to the restriction cleavage site between fragments A and B. Using this approach, six initiation sites in Ad2 DNA were identified. One initiation site at map position 16 was for an extremely long transcriptional unit, extending almost to the end of the Ad2 genome and serving as the sole precursor for at least four late mRNAs. Four additional initiation sites were revealed at other map positions, each of which represented the transcriptional unit for at least one early mRNA. The correlation between these various transcriptional units for hn RNA precursors and the topographical distribution of early and late mRNAs, as described by Petterson *et al.* (1976), is remarkable. Unquestionably, hn RNA represents the authentic, primary transcript for subsequent conversion into eukaryotic mRNAs. On the evidence presented by Evans *et al.* (1977), each species of hn RNA may be the precursor of one, or many, discrete species of mRNA; by inference, post-transcriptional processing can be extremely extensive or negligible. In the absence of infection by viruses, each hn RNA is the precursor for a single mRNA.

Radiological techniques have been widely used to define the size of transcriptional units. Such methods depend on the introduction of randomly spaced lesions in DNA by ultraviolet irradiation, which serve to terminate transcription in discrete transcriptional units. Using this approach, it was confirmed that the transcriptional unit for late Ad2 mRNAs was indeed very long, some 28 kb, and adequate to serve as the transcriptional template for at least four species of mRNA. Further, the ultraviolet target size of mRNA in uninfected HeLa and mouse cells was entirely consistent with hn RNA being the primary transcript in the formation of mRNA.

8.2.4 Initial gene transcripts in more detail

From everything that has been said so far, the initial transcripts of all eukaryotic genes are longer than the final, active species of RNA. By universal agreement (for a review see Perry 1976), eukaryotic transcripts often contain additional *leader* sequences at the 5′ P terminus and *trailer* sequences at the 3′ OH terminus; the leader and trailer sequences *may* or *may not* be subsequently removed by post-transcriptional processing, depending on the type of RNA. The gene sequence itself is long in eukaryotic cells, because it almost inevitably contains intron sequences and occasionally contains transcribed spacer sequences. By analogy with prokaryotes, there are *promoter* sequences where the initial binding of the appropriate RNA polymerase occurs; these are often 'upstream' of the *initiation* sequence at the 5′ P end of the gene where transcription actually begins. Similarly, *terminator* sequences, where transcription ends, may often be 'downstream' of the 3′ OH end of the coding gene sequence. The DNA strand actually transcribed by

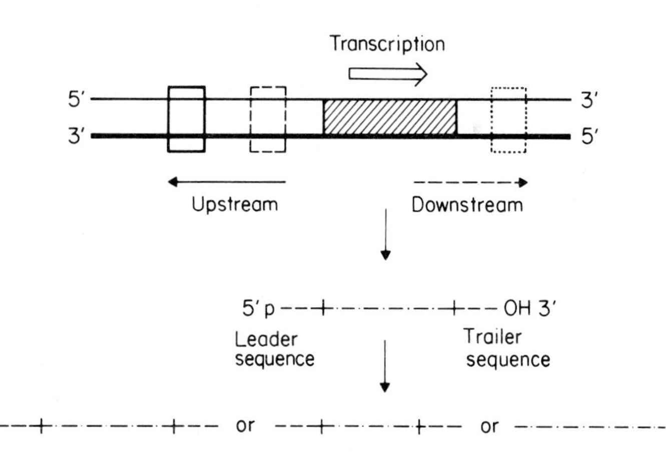

Fig. 8.7. The transcription of a eukaryotic gene. The transcribing strand is given as a heavier line, ▬, than the non-transcribing strand. Also included is the promoter region, ▭, where RNA polymerase is first bound, the initiator region, ⌜- - -⌝, where transcription begins and the termination region ⋮⋯⋮, where the transcription ends. The coding sequence of hypothetical mRNA is shown by hatching, ▨. The gene sequence between the initiation and termination points may contain introns and transcribed spacers. Features of the gene towards its 5′-end are said to be upstream, those towards its 3′-end are downstream. The initial transcripts often contain leader and trailer sequences at their 5′-and 3′-ends, respectively; depending on the complexity of the gene (presence of introns) post-transcriptional processing produces the mature (often shorter) RNA. Processing shows great variety and only three possibilities are shown here.

complementary base pairing in the $5' \rightarrow 3'$ direction is called the *sense* strand, whereas the non-transcribed DNA is called the *antisense* strand. As seen earlier with respect to Ad2 DNA, the sense strand in terms of transcription depends on the species of RNA being transcribed and can change from one DNA strand to another. The molecular basis for this crucial strand selection by eukaryotic RNA polymerases remains to be established. Cortese *et al.* (1980) have provided evidence that the transcription of cloned tRNA genes in *Xenopus* oocytes absolutely requires the double-stranded gene, and not its single-stranded components alone. Strand selection clearly involves more than simply the recognition of the sense strand by the RNA polymerase; the nonsense strand plays a vital part in strand selection by means as yet unknown. These current ideas on transcription are represented in Fig. 8.7 for a single, hypothetical gene sequence, less complex than that ascribed to Ad2 DNA in Fig. 8.6. The terms 'upstream' and 'downstream' are really scientific jargon, but, nonetheless, they have become familiar terms in the molecular biology of eukaryotic genes, particularly with reference to leader and trailer sequences.

Promoter sequences for RNA transcription

As more eukaryotic genes were cloned and their base sequences determined, some important and general features of gene structure began to emerge. It was first pointed out by Ziff and Evans (1978) that many eukaryotic genes contain an AT-rich region of seven bases upstream from the start of the 5′ leader sequences. Even more upstream, yet adjacent is a GC-rich region (Fig. 8.8). Together, these AT- and GC-rich regions are believed to constitute critical promoter sequences for the binding of RNA polymerase B (Ziff & Evans 1978, Gannon *et al.* 1979). Although not known in such detail yet, most histone mRNAs have these characteristic AT- and GC-rich regions at their 5′ termini (Kedes 1979). The sequence, 5′ T A T A A T A 3′, appears upstream of the initiation point of transcription at the 5′-end of nearly all mRNAs, although its precise location varies a little; in Fig. 8.8, for example, this AT-rich structure is more upstream in the histone 2A mRNA of sea-urchin, *Psammechinus miliaris,* than in other mRNAs of higher organisms. Present sequence data suggest that the AT-rich structure, referred to as either the TATA or Hogness box, is an essential recognition signal in all promoters read by RNA polymerase B (Wickens & Laskey 1981). Structures akin to the Hogness box have been identified in

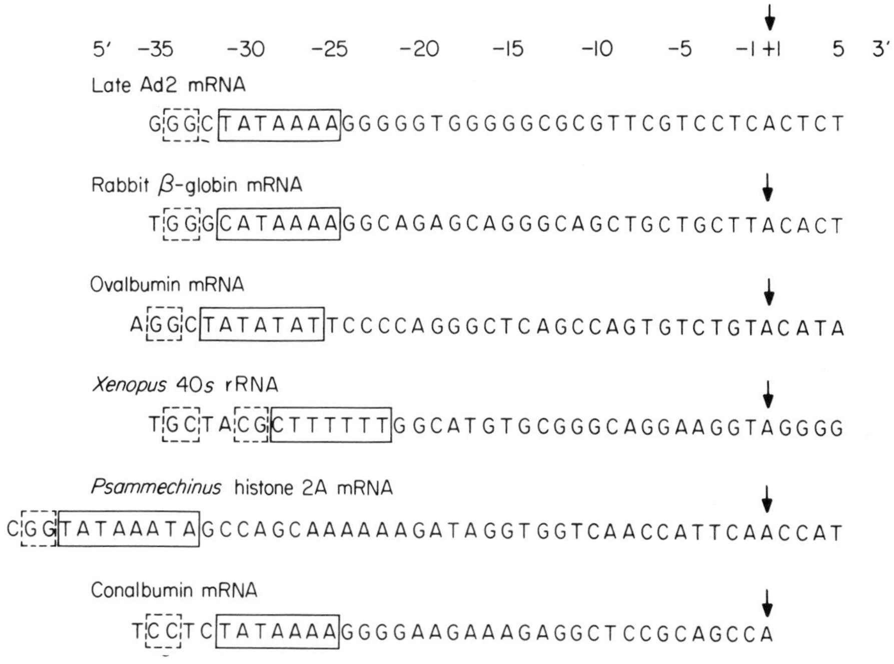

5′ −35 −30 −25 −20 −15 −10 −5 −1 +1 5 3′

Late Ad2 mRNA

G G G C T A T A A A A G G G G G T G G G G G C G C G T T C G T C C T C A C T C T

Rabbit β-globin mRNA

T G G G C A T A A A A G G C A G A G C A G G G C A G C T G C T G C T T A C A C T

Ovalbumin mRNA

A G G C T A T A T A T T C C C C A G G G C T C A G C C A G T G T C T G T A C A T A

Xenopus 40s rRNA

T G C T A C G C T T T T T T G G C A T G T G C G G G C A G G A A G G T A G G G G

Psammechinus histone 2A mRNA

C G G T A T A A A T A G C C A G C A A A A A A G A T A G G T G G T C A A C C A T T C A A C C A T

Conalbumin mRNA

T C C T C T A T A A A A G G G G A A G A A A G A G G C T C C G C A G C C A

Fig. 8.8. The base sequences at the 5′ termini of many eukaryotic genes. All sequences are written 5′→3′, left to right. The initiation point of transcription is at base +1. Bases upstream from this point (arrows) are given negative values; possible AT-rich promoter regions are enclosed in solid boxes. Other GC-rich regions, which may also be important in the binding of RNA polymerases, are enclosed in dashed boxes.

various cloned genes from the yeast, *S. cerevisiae,* including the mRNAs for histone 2B, cytochrome c and glyceraldehyde-3-phosphate dehydrogenase (Wallis *et al.* 1980). In all cases, the Hogness box was very upstream, some 140 nucleotides up from the 5′-end of these mRNAs. The location of these promoters is, therefore, variable in different genes, although relatively constant, or conserved, in terms of sequence. Viral mRNAs, such as adenovirus and Rous sarcoma, contain similar promoter sequences.

Promoter regions have been identified in the genes for tRNAs, but these differ from the promoters described above. Telford *et al.* (1979) cloned the *Xenopus* gene for $tRNA^{Met}$, and successively trimmed the sequence flanking the gene until it could no longer be transcribed on injection into frog oocytes. They concluded that the promoter region was retained within a region of 22 bases at the 5′ side of the tRNA gene, as follows:

−20 −10 −1+1

5′ A A T C A C C C A A A G G A C G G C A A T C A 3′

Since the genetic code is degenerate, many different gene loci exist for the various tRNAs for a given amino acid. De Robertis and Olson (1979) reported the base sequence of distinctive, cloned genes coding for four species of $tRNA^{Tyr}$ in the yeast, *Saccharomyces cerevisiae.* With but a single exception, the base sequences of the various $tRNA^{Tyr}$ species processed in *Xenopus* oocytes were *identical* and 92 bases long. However, each particular $tRNA^{Tyr}$ gene dictated the early synthesis of a unique pre-$tRNA^{Tyr}$, each of which had a different 5′ leader extension, both in terms of length and base sequence. From these interesting findings, it would appear that the promoter sequences for eukaryotic tRNA genes may turn out to be surprisingly subtle and complex.

5*s* rRNA genes provide a further insight into the different eukaryotic promoter sequences. Careful analyses by Hamada *et al.* (1979) show that in *Xenopus* nuclei, RNA polymerase C starts the transcription of 5*s* rRNA genes precisely at the same site as the 5′-end of the mature form of 5*s* rRNA. Clearly, no 5′ leader sequence need be subsequently removed by post-transcriptional processing and the initiation site for transcription is synonymous with the promoter sequence. The 5*s* rRNA genes contain extremely long, non-transcribed and AT-rich spacer regions, but only sequences close to the 5′-end of the gene feature as promoter sites; deletion mutants of *Xenopus,* lacking up to 80% of AT-rich spacer sequences, can still synthesize 5*s* rRNA in the normal manner (Federoff 1979a).

Recognition of promoter sites by RNA polymerases

A variety of cloned genes can be faithfully and efficiently transcribed in cell-free systems containing RNA polymerases and other regulatory factors. Examples include the 5*s* rRNA genes by RNA polymerase C (Ng *et al.* 1979), conalbumin and ovalbumin genes by RNA polymerase B (Wasylyk *et al.* 1980) and chloroplast genes by the distinctive, chloroplast-specific RNA polymerase (Jolly & Borgorad 1980). In every case, the most appropriate RNA polymerase (e.g. type B for mRNA-encoding genes) transcribed the genes far more accurately than inappropriate polymerases, indicating that eukaryotic polymerases do have an inherent ability to recognize correct promoter sites with great accuracy. However, stimulatory factors for eukaryotic RNA polymerases are also needed and they are now beginning to be purified, e.g. a regulatory factor for RNA polymerase C (Honda & Roeder 1980).

Despite the clear involvement of the TATAATA (Hogness) box in the binding of RNA polymerase B, other regions surrounding genes for mRNAs also play their part as transcription initiators. In many genes, the sequence

C
5′ G G or C A A T C T 3′
T

is more upstream by 40 nucleotides than the Hogness box and yet somehow assists in the accurate initiation of transcription (Benoist *et al.* 1980). The elegant studies of Grosschedl and Birnsteil (1980) on cloned genes for sea-urchin histones confirm this view and further suggest that accurate transcription of genes by RNA polymerase B requires at least three tracts of DNA, termed modulator, selector and initiator elements. The modulator elements control the rate of synthesis of the mRNA; these modulators are AT-rich, with inverted repeat sequences, even some 340 nucleotides upstream of the 5′-end of messenger mRNA coding sequence. The selector elements contain the Hogness box and base sequences which together direct the RNA polymerase B to select a unique initiation point for the 5′-end of the mRNA transcript. The initiator element contains the initiation point itself, and contains the crucial sequence 5′ C A T T C 3′. Clearly the promoter sites for the accurate initiation of eukaryotic transcription are complex and require the concerted action of several sequences of DNA, located even far upstream of the 5′-end of the coding sequence for the mRNA.

DNA sequences controlling the initiation of transcription of *Xenopus* 5*s* rRNA genes by RNA polymerase C have been identified; surprisingly, these regulatory sequences are situated in the centre of th 5*s* rRNA gene itself, not in the 5′ leader sequence, as described above (Sakonju *et al.* 1980). This interesting finding emphasizes the subtlety of the initiation of transcription in higher organisms and contrasts with prokaryotic promoters (see section 5.1.3).

Promoter sequences for mRNA capping

As described in detail in section 8.3.4, eukaryotic mRNAs often have a characteristic cap of 7-methylguanosine at their 5′ terminus. Since the first nucleotide incorporated into RNA retains its 5′ triphosphate moiety, capping enzymes provide an opportunity to identify unequivocally the initial 5′-end of all nascent RNA transcripts. Capping enzymes have been identified in vaccinia (cow pox) virus which catalyse the following reaction:

$$\text{pppG} + \text{S-adenosyl methionine} + \text{pppXpYpZp} \rightarrow {}^{7m}\text{GpppX}^{m}\text{pYpZp}.$$
$$(\text{or ppXpYpZp})$$

Using this reaction, it was confirmed that a 40*s* RNA is the sole precursor for rRNA in *Xenopus*. The vaccinia enzymes were also used to identify the initiation site for transcription of late Ad2 virus mRNA precursor and established that the initiating residues of the long, primary hn RNA transcript are the direct precursors of the capped, 5′ terminus of late Ad2 mRNA. Furthermore, Ziff and Evans (1978) made the perceptive observation that the 5′ leader sequences in β-globin mRNA and late Ad2 mRNA were virtually identical, suggesting that they probably represent distinct promoter regions for eukaryotic capping enzymes. These homologous sequences of DNA are as follows:

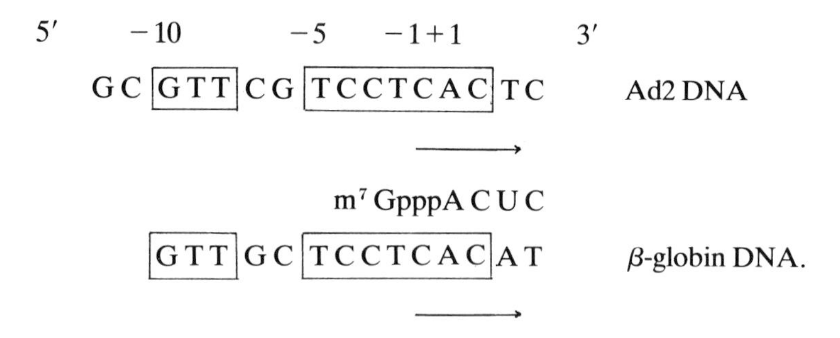

This abbreviated way of writing template and transcript sequences should be clarified, as it appears widely in the current literature. For brevity, only one DNA strand is written, from 5′→3′, the strand presented is in fact the *left-reading* strand. This shorthand has the advantage that the sequence of the RNA transcript is readily found, being the same as the DNA sequence, save that U replaces T.

Terminator sequences in eukaryotic transcription

As in bacteria, the terminator sequences are AT- and GC-rich regions near the 3′-end of genes. Such terminator sequences have been well documented by Korn and Brown (1978) for the 5*s* rRNA genes in several species (Fig. 8.9), where they form part of the non-transcribed spacer regions immediately downstream from the 3′-end of the coding sequence. The 3′-end of these 5*s* rRNA and viral RNA genes also contain GC-rich regions, as distinct centres of dyad symmetry; they are also seemingly part of the terminator

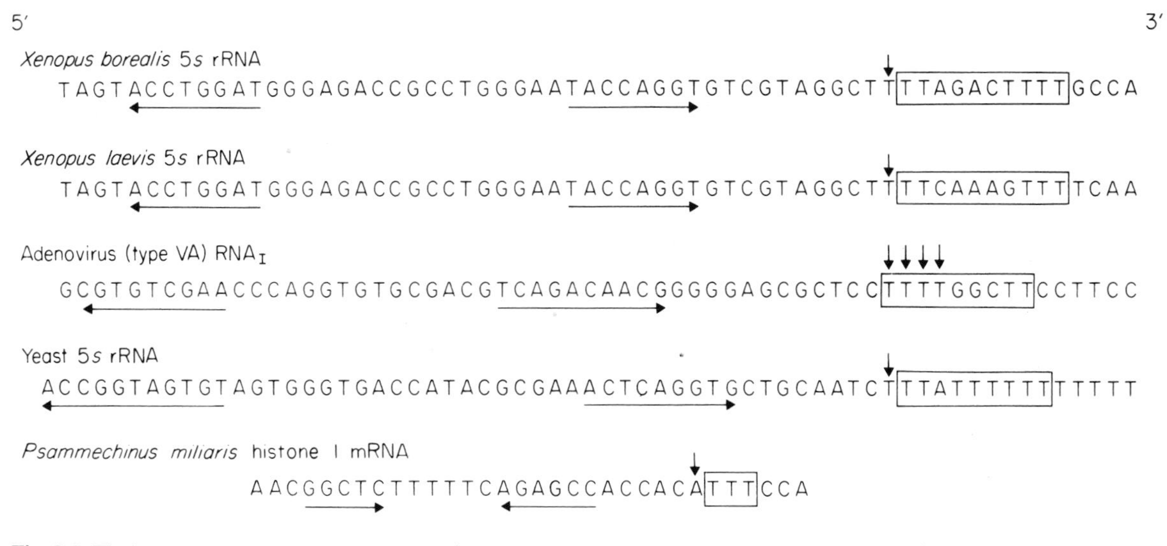

Fig. 8.9. The base sequences at the 3′ termini of several eukaryotic genes. All sequences are written 5′→3′, left to right. The 3′ terminus is indicated by a solid arrow, →. The probable AT-rich terminator sequences are enclosed in solid boxes. There are four alternative 3′ termini in the viral gene sequence. The two arms of dyad symmetric regions are indicated by arrows below each sequence; these, too, may feature in termination reactions. Note that the dyad symmetry runs in the opposite direction in histone mRNA.

signals. Similar AT-rich terminator sequences have also been identified at the 3′ termini of the genes for 40*s* rRNA precursor in *Xenopus,* histone mRNA in *Psammechinus* and mouse β-globin mRNA. In all of these cases, the terminator sequences were identified in cloned genes; without such technological innovations, it is difficult to envisage how the terminator sequences could have been clearly identified.

8.3 POST-TRANSCRIPTIONAL PROCESSING OF EUKARYOTIC TRANSCRIPTS

8.3.1 Introduction

All forms of eukaryotic transcripts are processed to form shorter, native forms of RNA with precise biological functions. As reviewed by Perry (1976), several types of reactions are involved:

(a) *nucleolytic reactions,* including cleavage and trimming of large precursors;

(b) *excision and splicing (joining) reactions,* where base sequences are removed from RNA precursors and the residual fragments are selectively recombined;

(c) *nucleoside modifications,* involving both ribose and base moieties;

(d) *terminal addition of nucleotides,* including the 5′ capping of mRNA, the 3′ polyadenylation of mRNA and the 3′ addition of the trinucleotide, CCA, to tRNA.

Before discussing these processes in more detail, however, the presence of introns in eukaryotic genes and consequently, the need for removing non-coding regions of eukaryotic transcripts must first be considered. Based on the excellent review of Abelson (1979), introns are a common feature of eukaryotic genes, but they are not ubiquitous (Table 8.3).

Table 8.3. The presence or absence of introns in eukaryotic genes.

Species	Gene	Number of introns per gene
(a) *The presence of introns*		
Yeast	$tRNA^{Tyr}$	1
	$tRNA^{Trp}$	
	$tRNA^{Leu}$†	
	$tRNA^{Ser}$†	
Drosophila	28*s* rRNA	1
Yeast (mitochondria)	21*s* rRNA	
Chlamydomonas (chloroplasts)	23*s* rRNA	
Mouse	α-globin mRNA	2
	β-globin mRNA	
	α-fetoprotein	
Rabbit	β-globin mRNA	2
Human	α-globin mRNA	2
	β-globin mRNA	
	insulin mRNA	
Mouse	immunoglobulin (L chain)	3*
Chicken	lysozyme	
Yeast (mitochondria)	cytochrome b	4*
Chicken	ovalbumin	7
	ovomucoid	
Chicken	conalbumin (transferrin)	16
Adenovirus (type 2)	late mRNA	uncertain
(b) *The absence of introns*		
Sea-urchin	histone mRNA (all types)	—
Yeast	$tRNA^{Asp}$	—
	$tRNA^{Arg}$	
Drosophila	5*s* rRNA	—

†Introns may not be present in all the degenerate genes for these species of tRNA.
*These are minimum values; small introns cannot be identified readily by R-loop methods.

The discovery of introns has been one of the most exciting events in contemporary molecular biology. From the historical standpoint, particular credit must be given to the innovative work of Glover and Hogness (1977) and White and Hogness (1977) in which they identified introns by the R-loop method in the cloned rRNA genes of *Drosophila*. The remarkable work of Cochet *et al.* (1979) on the chicken conalbumin gene reveals the true complexity of the genomes of higher organisms, since it contains no less than sixteen introns. The complexity of the chicken ovalbumin gene, with seven introns, is illustrated in Fig. 8.10.

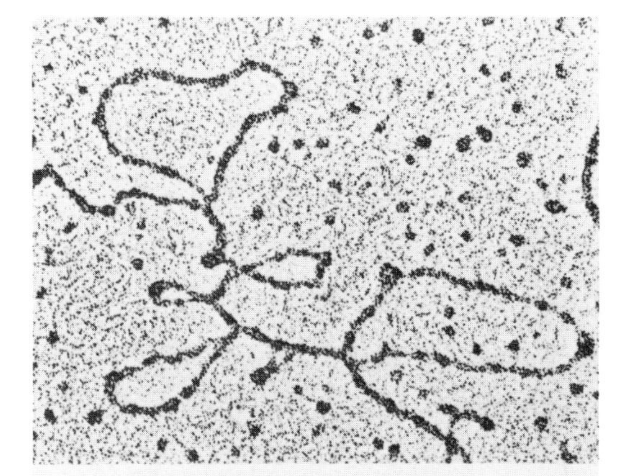

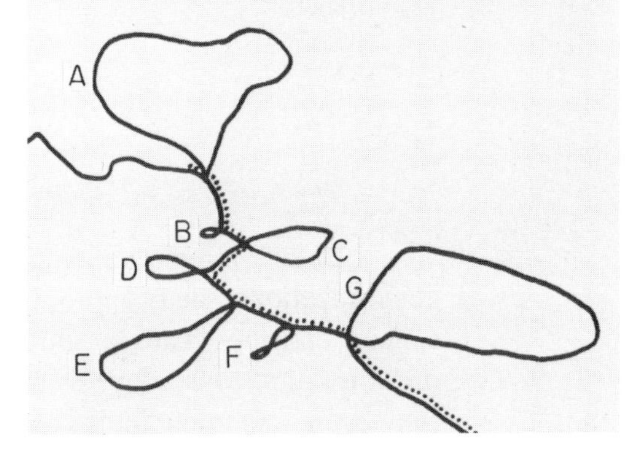

Fig. 8.10. The structure of the ovalbumin gene from chick oviduct. The electron micrograph shows the complex hybrid structure between the cloned ovalbumin gene and ovalbumin mRNA, with many R-loop structures. The line drawing illustrates the location of the seven introns, A–G; DNA is shown as a solid line, ———, and mRNA as a dotted line, ·······. (Reproduced, with permission, from Duglaiczyk *et al.* 1979.)

The structure of introns

In terms of gene organization, introns are generally found bounded within exons and never at the extreme 5′ or 3′ termini of genes (for a review see Abelson 1979). Some introns can be found, however, in non-coding regions, e.g. in the 5′ non-coding, leader sequence of ovalbumin mRNA. On current evidence, there is no common length or base sequence to introns. The only common feature in the structure of certain introns is found at the intron–exon boundaries, indicating that these are important recognition points for splicing enzymes (Table 8.4). From this information, it would appear that two ends of the 5′ ↓ GU———AG ↓ 3′ sequence are the landmarks recognized in the splicing of most hn RNAs; the situation appears more complex with respect to the post-transcriptional modification of tRNA precursors.

Table 8.4. Common features of exon–intron boundaries or splice points in purified eukaryotic genes. The important features in common are underlined. Note that these are the base sequences in the *genes;* the splice points in the processing of RNA transcripts will be the same, except that *U will replace T.* Consequently, the splicing of mRNA precursors (hn RNAs) may centre on the common sequences, 5′ GU---AG 3′, at the ends of introns. This generalization appears not to hold for the processing of tRNA and rRNA precursors.

5′ Exon		Intron		Exon 3′
Ovalbumin mRNA	AAG	GTG	CAG	GTT
	ATC	GTA	AAG	GAA
	CAG	GTA	CAG	CTT
SV40 (Early and Late) mRNAs	AAG	GTA	CAG	GTC
	GAG	GTA	TAG	ATT
Mouse α- and β-globin mRNA	CAG	GTT	TAG	GCT
	AGG	GTG	CAG	TCT
Mouse immunoglobulin mRNA	CAG	GTC	CAG	GGG
Yeast tRNA^{Tyr}	TAA	TTT	GAA	ATC
tRNA^{Phe}	AAG	AAA	GTT	ATC
tRNA^{Trp}	CAA	TTA	GCA	ATC

The function and significance of introns

Before attempting to discuss the function of introns, some generalizations seem appropriate on splicing in general and on splicing enzymes. Thus far split genes, or expressed another way, the presence of introns, seems a unique feature of eukaryotes. Indeed, if they were common in bacteria and blue-green algae, the most intensively studied prokaryotes, introns would surely have been discovered earlier. Interestingly, most introns have been reported to date in proteins which represent the terminal stages of differentiation, but this does not preclude the presence of introns in eukaryotic genes, say for enzymes, in general. On present evidence the order of exons in eukaryotic genes appears to be the same as the order in which they are present in the mature, spliced RNA. Splicing enzymes must be ubiquitous among eukaryotes and current evidence suggests this is the case. Accurate RNA processing has been reported in the transcription of many cloned genes in cell-free extracts and whole cells, ranging from the ciliated protozoan, *Tetrahymena pyriformis,* the frog, *Xenopus laevis,* to cultured mammalian cells (e.g. mouse L cells). Most splicing enzymes are present in the nucleus and released on cell damage into a soluble form. The number and mechanism of catalysis of splicing enzymes must await their complete purification. It would appear that specific splicing enzymes are needed for the processing of different types of RNA, since no structural homology is found between the introns in various RNA transcripts (e.g. compare mRNA and tRNA; Table 8.4). Even in mRNA precursors, the common ends of intron transcripts, 5′ GU ——— AG 3′, are probably inadequate alone to direct the splicing enzymes to the correct site. Perhaps the secondary structure of internal regions of introns is also important in terms of the recognition of cleavage sites.

Splicing enzymes must be remarkably accurate. The slightest inaccuracy in the excision or splicing reactions would wreak havoc in the precise transfer of genetic information in higher organisms. Since introns are present in mitochondrial and chloroplast genes, another intriguing question is whether these organelles contain their own complement of enzymes for processing their RNA transcripts. Even in the most up-to-date reviews (e.g. Sharp 1981), the conclusions are that splicing mechanisms are complex and not yet fully understood.

The functions of introns are even less clear. Certainly the work of Jeffreys and Flavell (1977) suggests that the presence of introns is not related to whether a gene is being expressed or not; rabbit DNA contains the long intron in the β-globin sequence, irrespective of the cells from which it is isolated, either erythropoietic or non-erythropoietic cells.

Three functions for introns have been proposed. First, it has been suggested that exon–intron boundaries may ultimately reflect subtle, functional domains within the protein encoded in the processed gene sequence. This proposal finds support from studies on the genes for mouse immunoglobulin H chain (Sakano *et al.* 1979b) and chicken ovomucoid (Stein *et al.* 1980).

Two other alternative functions for introns stem from genetic and biochemical studies on the sequence of intron-interrupted genes, termed *box* (or *cob*), in the yeast mitochondrial genome (see sections 10.2.1 and 10.3) which code for the apoprotein of yeast cytochrome b. Research from several research groups (for a review see Lewin 1980) indicates that, although the introns do not appear in the mature, coding sequence for cytochrome b, they do, nonetheless, have coding functions of their own. There is a general belief that the introns encode regulatory proteins or even RNA-splicing enzymes, necessary for their own removal from the initial transcripts of the cytochrome b gene complex. In this concept, introns and exons would constitute a genetic system of stringent self-regulation. An alternative proposal, first made by Church *et al.* (1979), is that the successive processing of different introns could provide an important mechanism for the regulation of gene aggression in eukaryotes (for a review see Borst & Grivell 1981). The critical feature of this work, notably by Lazowska *et al.* (1980), is that as introns are progressively removed and new exon–exon boundaries or fusions are created, this permits the synthesis of proteins engaged in the splicing out of

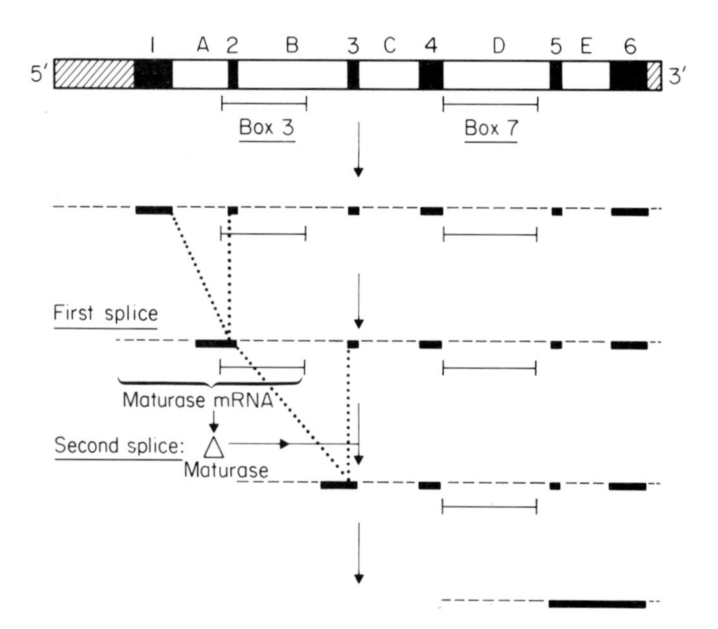

Fig. 8.11. The splicing of the mitochondrial gene for the apoprotein of cytochrome b. the cytochrome b gene in yeast mitochondria is made up of six exons 1–6, ▬, five introns A–E, ▭, and non-coding regions at both 5′ and 3′ termini, ▨. Two areas of the gene, called *box* 3 and *box* 7, can be identified by genetic rather than biochemical criteria, and mutants of both *box* regions are known. The primary transcript, 8 kilobases (kb) long, is first spliced to remove intron A by an enzyme coded by a *nuclear* gene. The newly spliced exons 1 and 2, together with the non-coding 5′ leader sequence constitute an active gene for the enzyme, *box* 3 maturase, necessary for performing the second splicing reaction. The maturase is clearly coded by a *mitochondrial* gene and excises intron B. Subsequent splicing steps in the 5′→3′ direction, by enzymes yet unknown, produces the mature, fully spliced mRNA for cytochrome b (2 kb) (after Borst & Grivell 1981).

subsequent introns. One such protein, M_r 42 000 and the product of very early exon–exon fusion, is called *box 3* RNA maturase. This maturase facilitates the next RNA splicing reaction. This concept of internally regulated, ordered splicing of long, primary transcripts is independently suggested by the processing of ovomucoid gene transcripts (Stein *et al.* 1980). It remains to be seen how far this novel system for the processing of transcripts of mitochondrial genes can be applied to the splicing and maturation of nuclear transcripts. A simplified picture of the early stages of processing of yeast cytochrome b transcripts is presented in Fig. 8.11.

The identification of introns in eukaryotic genes has tremendous importance in terms of evolution. Darnell (1978), for example, stated that non-contiguous sequences in eukaryotic DNA may reflect an ancient rather than a new form in the distribution of genetic information; accordingly, eukaryotes may have evolved independently of prokaryotes. This idea was contrary to the previous belief that eukaryotes evolved directly from prokaryotes. This and other aspects of evolution are considered further in the excellent review by Doolittle (1980). Living organisms are made up of three, distinct lineages which diverged very early in evolution, some 3.5 billion years ago, when genetic organization and genetic expression were undergoing rapid Darwinian or selective evolution, in the direction of accuracy and efficiency. The profound differences between these lineages are best seen as independently achieved solutions to the problem of gene organization.

The three lineages evolved independently to give the organisms known today: lineage 1, archaebacteria (prokaryotes); lineage 2, eubacteria (prokaryotes) and both chloroplasts and mitochondria (eukaryotes); lineage 3, nuclear organisms (eukaryotes). Eukaryotes resulted from repeated fusions of lineages 2 and 3. Direct evolution of eukaryotes from prokaryotes is, therefore, discounted by these contemporary ideas. Other evolutionary problems raised by comparisons of genome complexity include the similarities of mitochondrial, chloroplast and bacterial genomes, yet the presence of nuclear, mitochondrial and chloroplast genomes (in plants) in a single eukaryotic cell. Doolittle (1980) considers that the earliest progenitor of eukaryotes was a protoeukaryote, formed by the genomes of once free-living, aerobic bacteria being tapped within the cytoplasm of eubacteria. Such a fusion, or endosymbiosis, could ultimately lead to animal cells containing both nuclear and mitochondrial genomes. By similar fusions between free-living, photosynthetic blue-green algae and eubacteria, photosynthetic plants could be produced, containing independent nuclear and chloroplast genomes.

Comparisons of gene complexity have led, therefore, to revolutionary insights into the evolutionary origins of animals and plants. Evolutionary questions raised by the strict limitation of introns to eukaryotes and indeed, their significance to the genetic efficiency and accuracy of higher organisms, are also of great current interest. RNA splicing may well have had an important role in evolution, long before the division and separate evolution of prokaryotes and eukaryotes. There would be advantages to primitive organisms of making and testing new combinations of RNA sequences for translation into novel proteins, without having to modify the structure of their genomes. RNA splicing would, therefore, be an ancient process in the context of evolutionary time. Doolittle (1978) argues that pure prokaryotes, such as *E. coli,* devised mechanisms for streamlining their genomes by removing the ancient relics of introns. Indeed, *E. coli* can remove functional DNA sequences, for example identical tRNA genes, in the absence of any evolutionary pressure for their retention. By contrast, eukaryotes failed to evolve the streamlining mechanisms and continued to utilize introns in the regulation of gene expression. Certainly, introns in eukaryotic genes have persisted for a long time; the introns in globin genes are believed to be 100 million years old. While not completely understood, the discovery of introns and RNA splicing mechanisms has added new dimensions to molecular biology, particularly in terms of evolution, the organization of genomes and the control of gene expression.

8.3.2 Ribosomal RNA

Nucleolar rRNA species

The initial transcript of nucleolar rRNA genes in higher organisms is a long, 45*s* RNA of M_r 4.6×10^6. Since the review by Perry (1976) was published, the real complexity of rRNA genes has been fully revealed in *Xenopus laevis* and *Drosophila melanogaster* (Fig. 8.12). The arrangement of these genes in *Xenopus* is similar to the classical model described by Perry, but the nucleolar rRNA genes in *Drosophila* are more complex. First, there is a tiny 2*s* rRNA, which, like the 5.8s rRNA, is found associated with 26*s* rRNA; the location of the 2*s* rRNA gene is still unclear, but it is certainly adjacent to the 5.8*s* rRNA gene. Second, the 26*s* rRNA gene contains at least one intron. The initial rRNA transcript in *Xenopus* is

(a) Nucleolar rRNA genes

Xenopus laevis

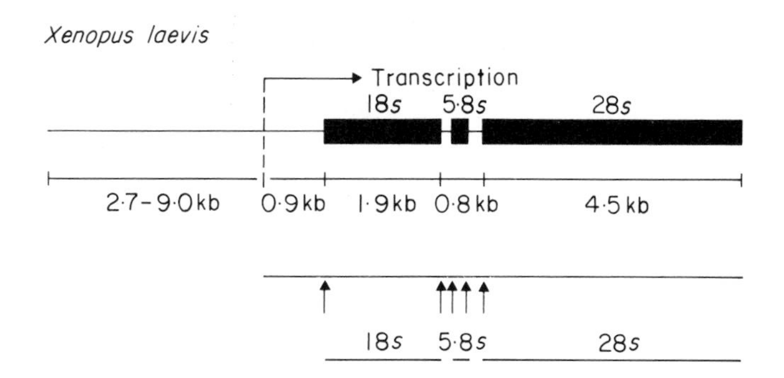

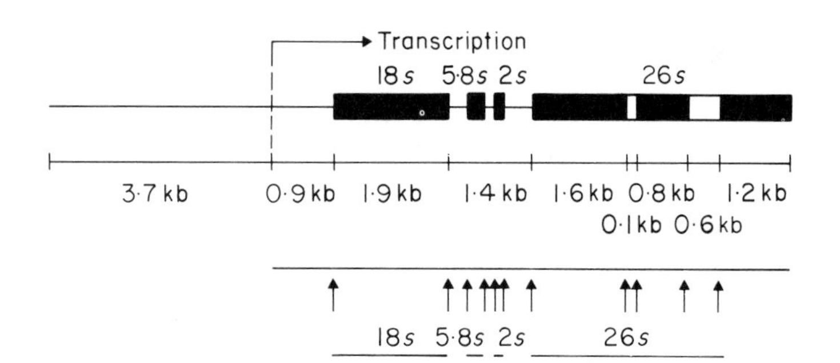

(b) Extranucleolar 5s rRNA genes

Xenopus borealis

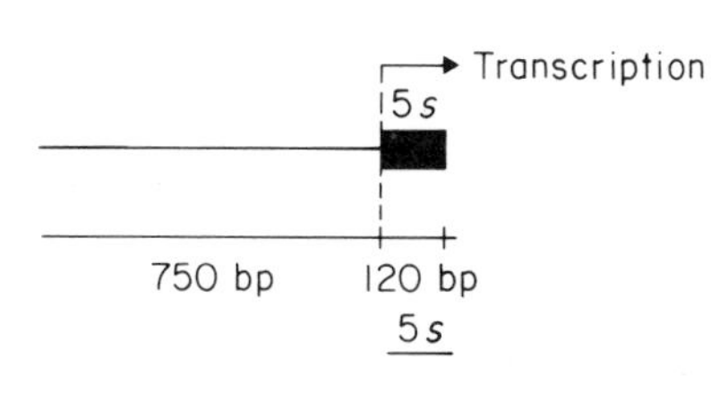

Drosophila melanogaster

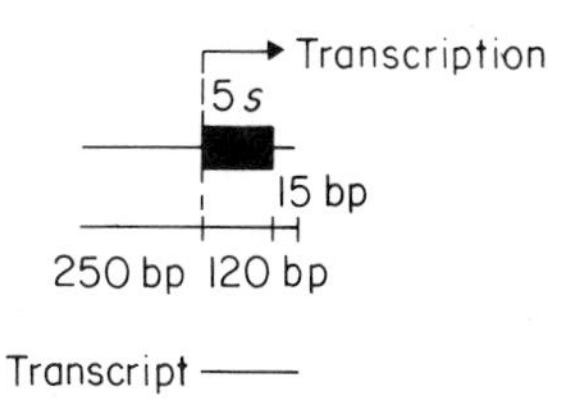

Transcript ———

Fig. 8.12. The repeating unit structure of the genes for eukaryotic rRNAs. Both transcribed and non-transcribed spacers are shown as solid lines, ———, structural genes (exons) as solid boxes, ■ and introns as open boxes, □. DNA lengths are shown as base pairs, bp, or kilobase pairs, kb. The genes are presented 5′→3′ (left to right) and cleavage points for processing enzymes are shown by arrows. (a) Nucleolar rRNA genes. In *Xenopus laevis,* transcription produces a 40*s* precursor, with a short 5′ leader but no 3′ trailer sequence; this is processed to produce 5.8, 18 and 28*s* rRNA. In *Drosophila melanogaster* (with two introns in the 26*s* rRNA gene), transcription produces a 43*s* precursor, with a short 5′ leader but no 3′ trailer sequence; this is processed to produce 2, 5.8, 18 and 26*s* rRNA. (b) Extranucleolar 5*s* rRNA genes. In *Xenopus borealis,* transcription produces 5*s* rRNA directly, without the need for any processing. In *Drosophila melanogaster,* the initial transcript has a short 3′ trailer and minimal processing produces 5*s* rRNA.

40*s*, and 43*s* in *Drosophila.* The non-transcribed spacers in both species are extremely variable in length and contain frequently repeated base sequences. The repeated sequences, however, are species of rRNA is as follows.

The complete sequence of reactions for producing nucleolar species of rRNA is as follows.

(1) RNA polymerase A binds to the extreme 3′-end of the non-transcribed spacer region, and after initiating transcription, proceeds along the complete length of the gene complex, including transcribed spacer regions and introns, until it approaches the terminator sequence at the extreme 3′-end of the 26 or 28*s* rRNA gene, or at the extreme 5′-end of the non-transcribed spacer in the next repeat of the gene sequence. Initiation of transcription proceeds upstream of the 5′-end of the 18*s* rRNA gene, which is transcribed first. The rRNA precursor thus has a 5′ leader sequence but no 3′ trailer sequence. In *Xenopus,* the initial transcript has the base sequence 5′ AGGGGAAGAC 3′.

(2) The transcript is then methylated extensively, particularly in areas destined to become the final sequences of nucleolar rRNA. The methyl donor is 5-adenosyl methionine and predominantly D-ribose moieties, rather than nitrogenous bases, are methylated. Interestingly, the converse is true in the post-transcriptional processing of bacterial rRNA; the significance of this evolutionary difference in the pattern of methylation remains uncertain. There is considerable evidence that methylation of the 2′ OH groups in ribose occurs earlier in processing than the limited methylation of bases.

(3) The long, methylated precursor is then subjected to extensive cleavage, such that the mature species of nucleolar rRNA are released. This process also involves splicing of the 26–28*s* rRNA if this gene contains introns and excision of the initial 5′ leader sequence. It is generally believed that the processing reactions proceed from one end of the methylated precursor to the other, rather than by multiple attack at all potential cleavage sites. The abundant methyl groups may well protect the mature species of rRNA from indiscriminate and incorrect cleavage. Since the long, initial transcript contains double-stranded regions (or hairpins), RNAases recognizing both single- and double-stranded templates are probably engaged in processing.

(4) Ribosomal proteins, with both structural and regulatory functions, are not made in the nucleolus, but on cytoplasmic polyribosomes. These, plus extranucleolar 5*s* rRNA, are transported into the nucleolus, to enable nascent ribosomes to be formed. There is some

evidence that these ribosomal proteins associate with the long rRNA precursor, and may even aid the specificity of the processing reactions or stabilize the mature species of rRNA. Processing cannot proceed in the absence of a continuous supply of ribosomal proteins, but there is a progressive decrease in the protein:RNA ratio of nascent ribosomes during maturation, perhaps as a result of compaction of these structures just prior to their release from the nucleus.

The synthesis of polyamines is tightly coupled to nucleolar rRNA synthesis. In the anucleolate mutant of *X. laevis*, the synthesis of both nucleolar rRNA and polyamines is absent (Russell 1971). Kuehn *et al.* (1979) have suggested that the phosphorylation of an acidic nucleolar protein (M_r 78 000) by a polyamine-activated protein kinase plays a vital role in the initiation of nucleolar rRNA synthesis. Once phosphorylated, this protein binds avidly to rRNA genes and greatly enhances their transcription by RNA polymerase A.

Many stages of the maturation process of eukaryotic rRNA have been simulated successfully *in vitro*. However, the accurate transcription of the long rRNA precursor by RNA polymerase A has only been found in intact cells, never in cell-free systems.

Extranucleolar 5s rRNA

For several reasons, studies on this particular species of rRNA are very advanced indeed. First, the genes are amplified in many oocytes and most have now been cloned (Fig. 7.3(b)). Second, RNA polymerase C performs the transcription of these genes very accurately in a variety of systems *in vitro* (see section 8.1.5). The genes are tandemly repeated in most species, but show considerable inter-species variation. The repeating unit of 5*s* rRNA genes in *X. borealis* consists only of a long, non-transcribed but GC-rich spacer and the gene itself; in *D. melanogaster,* the non-transcribed spacer is short and AT-rich, but the gene is bounded at its 3′-end by a very short, transcribed spacer.

From these two examples, it can be seen that the transcription of 5*s* rRNA in the adult forms of most species is very simple (it is far more complex in *Xenopus* oocytes, described later in section 8.4.6). In *X. borealis,* the promoter and terminator sites are located precisely in the spacer regions at the 5′- and 3′-ends of the 5*s* gene, so that the transcript is precisely of the right length and no post-transcriptional processing is needed. In many other 5*s* rRNA genes, there is a short spacer (15 bases or less) at the 3′-end which is transcribed. In these cases, just one cleavage produces the mature, 5*s* rRNA. The trailer sequence removed from the 3′-end of the initial transcript is the octanucleotide, 5′ GAUGCUUU 3′.

The function of spacers

Federoff (1979b) has reviewed the structure and putative functions of spacers. Unlike their associated genes, non-transcribed spacers have not been strongly conserved during evolution, even between closely related species. It is extremely interesting, however, that while non-transcribed spacers display extreme variation in terms of length and base sequence, the transcribed spacers are more strongly conserved (Tartof 1979). Stringent sequence conservation has only been maintained at non-transcribed spacer–gene boundaries, where crucial initiation and terminator sequences are known to reside. The 5*s* rRNA genes in *X. borealis* forcibly illustrate this point, where transcription from gene to precise, active gene product is accomplished in one step. The other, internal regions of non-transcribed spacers may play a part in the correct timing of gene expression during development or feature in the replication of the gene during growth and cell division. Transcribed spacers could play a part in providing the identification signals for the enzymes of RNA processing. While these are all plausible arguments, they do not alter the fact that spacers occur far less frequently than introns. For these and other reasons, the necessity and even the presence of spacers in most other eukaryotic genes still remains to be demonstrated. Indeed, it could transpire that spacers are virtually a *unique* feature of rRNA genes, and perhaps a few others only, including certain tRNA genes.

8.3.3 Transfer RNA

For some time, the post-transcriptional processing of the initial transcripts of tRNA genes was confused and contentious (Perry 1976). There was even debatable evidence that processing of tRNA precursors occurred in the cytoplasm, rather than the nucleus, and the precise order of the processing reactions was extremely uncertain. Genes are transcribed with complete accuracy in cell-free extracts of *Xenopus* oocytes and the processing of transcripts from yeast $tRNA^{Tyr}$ genes is now known in remarkable detail (de Robertis & Olson 1979, Melton *et al.* 1980). The intact oocyte, used by Melton and co-workers, has one significant advantage over the cell-free system; the nucleus is sufficiently large that it may be readily dissected out. Accordingly, the involvement of the nucleus or cytoplasm in the processing of the tRNA transcripts can be delineated precisely.

These types of $tRNA^{Tyr}$ genes contain a single intron (Table 8.3) and the production of mature species of eukaryotic $tRNA^{Tyr}$ is as follows.

(1) The tRNA genes are transcribed by RNA polymerase C with great efficiency, producing an initial transcript longer than the tRNA. In particular there is a 5′ leader sequence, 19 nucleotides long, a transcribed sequence of 14 nucleotides, equivalent to the gene intron, and a short 3′ trailer sequence, UU or UUU. The initial transcript is 108 nucleotides long.

(2) The 5′ leader sequence is then shortened (or trimmed) and certain residues at the 3′-end of the transcript are modified. The transcript is now 104 nucleotides long.

(3) The 5′ leader sequence is trimmed further, so that the intermediate tRNA precursor is now 97 nucleotides long. The limited base modifications in the previous step are seemingly necessary for specifying the correct cleavage site on the leader sequence.

(4) The 5′ leader and 3′ trailer sequences are now cleaved completely and the trinucleotide sequence, 5′ CCA 3′ is added to the newly exposed 3′ terminus. Further bases are modified, both at the 5′- and 3′-ends of the tRNA intermediate, which is now 92 nucleotides long.

(5) The 14 nucleotide sequence equivalent to the intron is excised and the two fragments of tRNA are spliced together. The 78 nucleotide product is the correct length of mature $tRNA^{Tyr}$, but it is only now that the tRNA leaves the nucleus and enters the cytoplasm.

(6) A single base at the 5′-end of the 78 nucleotide intermediate is now modified by cytoplasmic processing enzymes, forming mature, active tRNA.

Several points should be emphasized about these elegant studies. First, the sequence of processing is probably similar in principle for all tRNA precursors, but not all tRNA genes contain an intron (Table 8.3). Second, the remarkable specificity of RNA-processing enzymes is beautifully demonstrated in this work. Third, the cytoplasmic processing of tRNA precursors is minimal; post-transcriptional modifications are essentially conducted by nuclear enzymes. Fourth, excision of the transcribed sequence equivalent to the intron must be carried out with absolute accuracy, because it is immediately adjacent to the anticodon triplet (Fig. 8.13). Finally, much of the

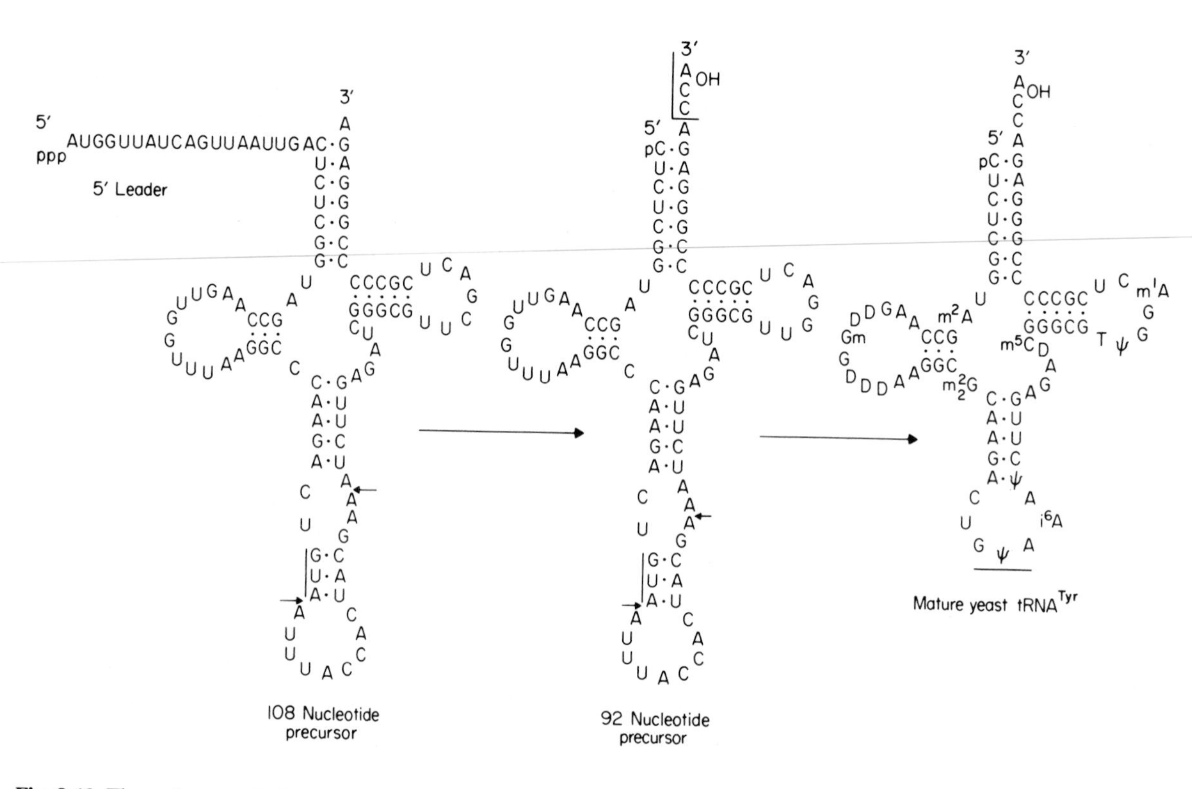

Fig. 8.13. The main stages in the processing of the primary transcript of tRNA^{Tyr} genes in *Xenopus* oocytes. The primary transcript, 108 nucleotides long, has a 5′ leader sequence, 19 nucleotides long, and a transcribed sequence equivalent to a single gene intron, demarcated within the arrows. The leader sequence is removed, yielding a 92 nucleotide precursor; this contains some modified bases (not shown here) and the sequence, CCA, added during processing at the 3′ terminus. The transcript equivalent to the intron is excised; splicing and further base modifications yield mature tRNA^{Tyr} (with the final modification shown). The anticodon triplet is *underlined* in all structures. (Reproduced, with permission, from de Robertis & Olson 1979.) Transcription of certain tRNA^{Tyr} genes can also produce a 2- or 3-base 3′ trailer sequence (see text); this was not observed here.

clover-leaf (H-bonded) structure of tRNA is evident, even in the initial transcript of the tRNA gene.

In the intact cell, many tRNA genes are preceded upstream by long, non-transcribed spacers with the next spacer immediately downstream of the 3′ terminus of the gene, so that the repeating unit of tRNA genes is probably:

5′...non-transcribed spacer...tRNA gene...non-transcribed spacer...3′

Clearly, promoter and terminator sequences must be present in the spacer regions adjoining the 5′- and 3′-ends of the tRNA gene. The promoter region is itself upstream of the 5′-end of the coding sequence, to account for the 19 nucleotide leader sequence which is subsequently cleaved off. The terminator signal is close to the 3′-end of the coding sequence because the 3′ trailer sequence is very short. Sprague *et al.* (1980) developed a cell-free system from *Bombyx mori* for the transcription of cloned $tRNA^{Ala}$ genes and truncated gene derivatives lacking 11 nucleotides from the flanking region upstream of the 5′-end of the gene. The truncated derivatives were poorly transcribed, indicating that the 5′ flanking region of tRNA genes constitutes part of the promoter sites for RNA polymerase C.

8.3.4 Messenger RNA

Eukaryotic mRNA has a unique complexity, not just accounted for by the frequent presence of gene introns. First, the 5′-end is frequently capped by 7-methylguanosine and the 3′-end is often a long tail of poly(A) (Perry 1976). Second, extensive 5′ (leader) and 3′ (trailer) sequences in the primary hn RNA precursor are retained as distinctive, non-coding sequences in the mature mRNA. Post-transcriptional processing of hn RNA is very extensive, yet distinctive in that it does not entail extensive modification of bases or ribose in internal positions of the mRNA (contrast this with nucleolar rRNA and tRNA).

The structure of eukaryotic mRNAs can be determined in two ways and elegant work on mouse β-globin mRNA illustrates this point well. The mature mRNA can itself be sequenced (Baralle & Brownlee 1978) or the cloned gene can be sequenced (Konkel *et al.* 1978). The agreement between the two methods is remarkable and the gene–mRNA–protein relationships for β-globin are presented in Fig. 8.14; this serves as a beautiful example of all aspects of gene structure, transcription and post-transcriptional processing in eukaryotic cells.

Abelson (1979) has carefully reviewed the evidence for the various biosynthetic steps in the synthesis of mouse β-globin mRNA.

(1) RNA polymerase B binds to a promoter site some 30 bases upstream of where transcription is initiated, and the initiator region is itself upstream of the initiating AUG codon in the coding sequence. This system enables a 5′ leader or non-coding sequence to be present in the mature mRNA. Transcription proceeds downstream, including the two introns; the first of these is 116 nucleotides long and the second, 646 nucleotides long. Transcription continues past the translational terminator codon, UAA, and the terminator sequence is considerably downstream of this point, so that the mature mRNA can also have a 3′ trailer or non-coding sequence. The transcript

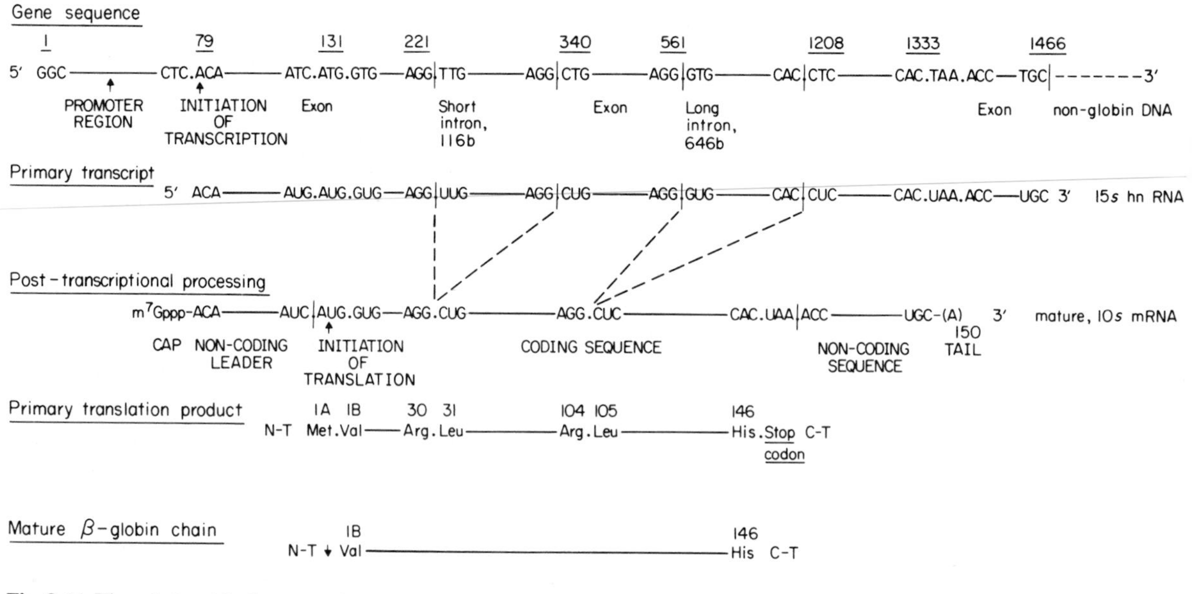

Fig. 8.14. The relationships between the mouse β-globin gene, β-globin mRNA and β-globin. The cloned gene is represented $5' \rightarrow 3'$ (left to right) with important bases numbered and *underlined*; the gene contains two introns, one short and one long. The initial transcript is a 15*s* hn RNA precursor, which in turn, is processed into mature, 10*s* β-globin mRNA. This is translated, $5' \rightarrow 3'$, into a primary translation product, N-terminal (N–T) Met. Important amino acids (relative to the location of the excised introns) are numbered with the N-terminal as one. The met residue, 1A, is removed by post-translational processing, producing β-globin (146 amino acids; N–T Val, 1B to C–T His).

contains 1500 bases and represents the 15*s*, hn RNA precursor to β-globin mRNA.

Talkington *et al.* (1980) monitored the transcription *in vitro* of cloned α-globin genes and truncated derivatives containing deletions of nucleotide sequences upstream of the 5′-end of the gene. They conclude that sequences up to 148 nucleotides upstream of the initiation point for transcription contribute to the promoter sites for RNA polymerase B. Furthermore, transcription of mutant globin genes with deletions in this 5′ flanking region could not be initiated.

(2) Immediately on completion of the hn RNA transcript, proteins are added to this long RNA intermediate. Importantly, these proteins associate with the hn RNA in a non-random, sequence-specific manner (Beyer *et al.* 1980). These proteins may help to direct RNA-splicing enzymes to the correct cleavage sites.

(3) The 5′- and 3′-ends of the primary transcript are then modified. A cap of 7-methylguanosine is added precisely at the initial base of the 5′ terminus, perhaps even while transcription of the rest of the gene and its trailer region is in progress. On completion of the

ribonucleoprotein hn RNA intermediate, a long tail of poly(A), at least 150 residues long, is added at the 3′ terminus. These distinctive features at the 5′ and 3′ termini have important roles in the ultimate translation of the mature mRNA (see section 9.2.2), and they certainly help at this stage to stabilize the hn RNA intermediate.

(4) The two sequences equivalent to the introns (both within the coding region) are excised and the RNA fragments spliced together to form the shorter, mature mRNA.

(5) Two further proteins, M_r 52000 and 78000, associate preferentially with the 3′ poly(A) tail (Blobel 1973) and the final ribonucleoprotein mRNA complex, sometimes referred to as an informosome, is translocated from the nucleus into the cytoplasm.

Available evidence suggests that, with minor modification, this scheme of mRNA synthesis and processing is universal to eukaryotic mRNAs (Darnell 1975). First, extremely long hn RNA precursors have been found in eukaryotic nuclei, including the primary transcripts for the mRNAs for ovomucoid, ovalbumin and the late adenovirus 2 genes. Second, modification of 5′ and 3′ termini of hn RNA precursors occurs on most primary transcripts. Third, even poly(A)$^-$ mRNAs often have 5′ cap structures. This finding suggests that capping has a vital role in the function of most eukaryotic mRNAs and further, that capping probably precedes, and is independent of, modifications at the 3′ terminus. Fourth, the introns in β-globin genes occur exclusively in the coding (or structural gene) sequence; this is not the case with other eukaryotic genes, where introns can occur in the gene sequences equivalent to the 5′ leader (non-coding regions) of mRNA (Breathnath *et al.* 1978). Nonetheless, the uniformity of splicing sequences (Table 8.4) suggests that gene arrangements such as these present no inherent difficulties in processing. Finally, although certain details remain to be clarified about the transport of nascent mRNA molecules into the cytoplasm, certain proteins are ubiquitously associated with eukaryotic mRNAs, both in the nucleus and in cytoplasmic polyribosomes.

Many aspects of eukaryotic mRNA processing have been accurately reproduced *in vitro*. The faithful initiation of hn RNA synthesis under artifical conditions has been difficult to achieve. Painstaking efforts have been rewarded in the case of the hn RNA precursor for the late adenovirus 2 mRNAs; accurate transcription has now been reported in the nuclei of viral-infected HeLa cells (Manley *et al.* 1979) and even in soluble extracts derived from viral-infected KB cells (Weil *et al.* 1979). After years of uncertainty and frustration, mRNA synthesis in eukaryotic cells is now beginning to be understood.

The 5′ capping of eukaryotic mRNA

Most eukaryotic and viral mRNAs have the characteristic cap of 7-methylguanosine at their 5′ termini. As reviewed by Shatkin (1976), the cap itself can have various forms. In addition to the 7-methyl guanosine moiety itself, methylation can extend to the 2′ OH groups on both the first and second ribose units at the 5′ terminus of the primary transcript and even the 6 position of the initiating base if this is adenine. It is, therefore, possible to have a structural hierarchy of caps, as follows.

(a) Cap zero: m^7GpppX, contains only the 7-methylguanosine residue.

(b) Cap one: m^7GpppX^m, has an additional methyl group at the 2′

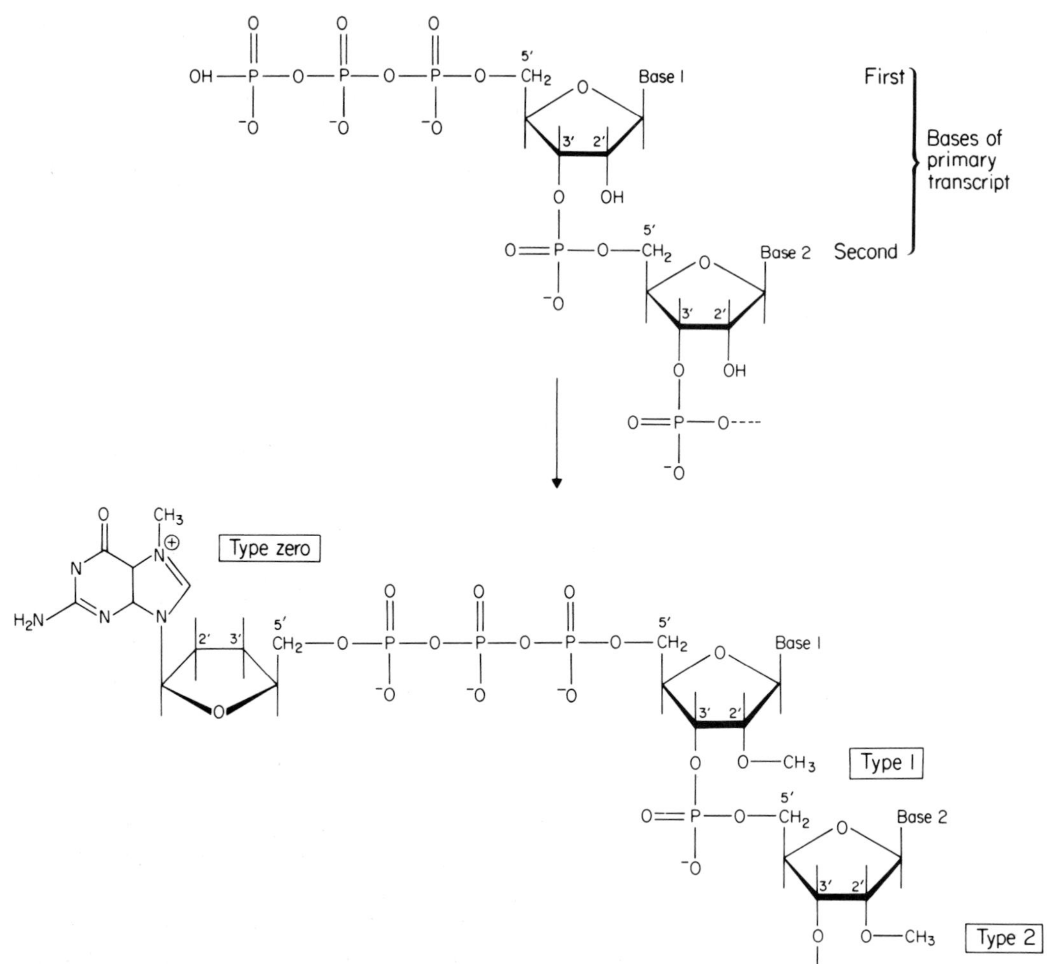

Fig. 8.15. Variations in the cap at the 5′ terminus of eukaryotic mRNAs. A guanosine residue is added to the initiating nucleotide of the primary transcript in all cases. The addition of methyl groups, to various positions (shown in boxes), then produces caps of types zero, 1 and 2.

OH of the original nucleotide at the 5′-end of the primary transcript (i.e. the *original* site of the initiation of transcription).

Another version of cap one is $m^7Gpppm^6A^m$, where yet another methyl group is inserted, provided that the initiating base of transcription is adenine.

(c) Cap two: $m^7GpppX^mY^m$, has yet a further 2-0-methyl group, inserted in the ribose moiety of the second nucleotide of the primary transcript.

Examples of these structures are found among the diversity of eukaryotic mRNAs (Fig. 8.15). In all cases, the guanosine moiety is added in a unique and *inverted* 5′→5′ linkage. The cap has two functions; first, it is essential for the correct initiation of translation of the mature mRNA and second, it protects the initial 5′ terminus from degradation by phosphatases.

Capping enzymes have been isolated from a variety of sources, especially virus particles. These enzymes operate in a stepwise manner to produce the cap:

Phosphodiester bond formation

$$\text{pppX} + \text{pppY} \xrightarrow{\text{RNA polymerase}} \text{pppX-Y} + \text{pp}_i$$

Removal of one 5′ phosphate

$$\text{ppX-Y} \xrightarrow{\text{nucleotide phosphohydrolase}} \text{ppX-Y} + \text{p}_i$$

Capping

$$\text{ppX-Y} + \underline{\text{pp}\overset{*}{\text{p}}}\text{G} \xrightarrow{\text{guanyl transferase}} \text{G}\overset{*}{\text{p}}\text{ppX-Y} + \underline{\text{pp}_i}$$

N^7 methylation

$$\text{G}\overset{*}{\text{p}}\text{ppX-Y} + \text{SAM} \xrightarrow{\text{N}^7\text{G-methylase}} \text{m}^7\text{G}\overset{*}{\text{p}}\text{ppX-Y} + \text{SAH}$$

(cap zero)

2′-0-methylation

$$\text{m}^7\text{G}\overset{*}{\text{p}}\text{ppX-Y} + \text{SAM} \xrightarrow{2'\text{-0-methylase}} \text{m}^7\text{G}\overset{*}{\text{p}}\text{ppX}^m\text{-Y} + \text{SAH}, \textit{etc.}$$

(cap one)

The methyl donor is exclusively S-adenosylmethionine.

The 3′ poly(A) tail of eukaryotic mRNAs

Polyadenylation of eukaryotic mRNAs can occur at various stages of processing, certainly during and after transcription of the gene, and even after the mRNA has entered cytoplasmic polyribosomes (Perry 1976). The enzyme responsible, RNA-dependent poly(A) polymerase, is ubiquitous. Cordecepin (3′-deoxyadenosine) concurrently blocks polyadenylation and the appearance of newly synthesized mRNA in cytoplasmic polyribosomes, indicating the probable importance of the 3′ tail sequence in mRNA transport. Other evidence strongly supports this view (Blobel 1973). The poly(A) tail also plays a major part in ensuring the stability of eukaryotic mRNAs during repetitive translation (see section 9.3.6).

8.4 ADDITIONAL ASPECTS OF EUKARYOTIC TRANSCRIPTION

8.4.1 The crucial importance of eukaryotic regulation at the transcriptional level

All currently available evidence supports the view that the critical regulation of macromolecular synthesis in higher organisms resides at the level of transcription rather than translation. Once the nucleus had become an invariant structure of higher organisms and the repository of the majority of the genes, then strong evolutionary forces would tend to emphasize the crucial role of the nucleus in cellular regulation. These evolutionary pressures are really not known, but energy expenditure and conservation could provide the key. The synthesis both of nucleic acids and proteins makes heavy demands on the energy reserves of the cell. For each base or amino acid incorporated into macromolecular linkage, at least two high-energy bonds (from ATP or GTP) must be used, as a minimum. Consequently, it is more efficient for a cell to avoid wasting nucleoside triphosphates in synthesizing an RNA molecule, only to degrade it or to fail to utilize it in translation. In addition, selective control at the translational level would require a horrific complexity of cytoplasmic, regulatory factors. There is absolutely no evidence for the latter. Of necessity, therefore, there is stringent regulation at the level of transcription, right at the beginning of the flow of genetic information in the cell. An analogy could be drawn to metabolic control, where overall regulation is generally exerted via a key enzyme, engaged in the first reaction of the metabolic pathway. Evidence supporting the crucial importance of transcriptional regulation is overwhelming. Results from just three model systems will be cited.

The type of haemoglobin in a given species is always subject to marked developmental change, from the embryo, to the fetus to the adult; in addition, the site of synthesis changes. For clinical reasons, the genes involved in the synthesis of human haemoglobins have

Table 8.5. The genetic control of human haemoglobins. All genes appear in a single copy, except for the α-globin gene, which is duplicated. There are two γ-globin genes, ${}^{G}\gamma$ and ${}^{A}\gamma$, depending on whether Gly or Ala occupies position 136 of the γ-globin chain, respectively.

Chromosome linkage	Both unknown		chromosome 11		chromosome 16	
Gene loci	ξ	ϵ	${}^{G}\gamma$—${}^{A}\gamma$—σ—β		—α——α—	
Globin chains	ξ	ϵ	${}^{G}\gamma$ ${}^{A}\gamma$ σ β		α	
Haemoglobin tetramer	$\xi_2\epsilon_2$	$\zeta_2\gamma_2^{G\ or\ A}$	$\alpha_2\epsilon^2$	$\alpha_2\gamma_2^{G\ or\ A}$	$\alpha_2\sigma_2$	$\alpha_2\beta_2$
	Hb Gower 1	Hb Portland	Hb Gower 2	HB F	HBA_2	HbA
Stage of appearance and site of synthesis	Embryo, Primitive yolk sac			Fetus, Liver, spleen, Marrow	Adult, Marrow	

been thoroughly documented (Weatherall & Clegg 1979; Table 8.5). If the switch from one type of haemoglobin to another does not appear at the appropriate developmental stage, the clinical consequences are severe. In particular, HbF is not compatible with adult life because its oxygen affinity is far too high; in contrast, it is vital in fetal development, as it ensures satisfactory oxygenation of the fetus. Similar developmental switches occur in animals, and although the trigger(s) for switching remains unclear, the process has been studied in great detail at the molecular level (for a review see Stalder *et al.* 1980). In sheep, there is a close correlation between the quantities of individual globin mRNAs in the cytoplasm of erythropoietic cells and the developmental pattern of globin synthesis, suggesting convincingly that the differential expression of globin genes is regulated in the nucleus (Young *et al.* 1978). These switches occur synchronously in all sites of erythropoiesis. In late fetal development, all sites produce the mRNAs for HbA rather than HbF, but as the marrow takes over the exclusive synthesis of HbA, the appropriate mRNAs are no longer detectable in liver or spleen. Translational changes cannot account for these phenomena.

Certain fish, such as the winter flounder, *Pseudopleuronectes americanus,* have to survive in environments of extreme temperature variation, including winter seas as cold as −2.7 °C. To prevent the freezing of their blood, such fish synthesize antifreeze proteins, one of which is a small protein, M_r 3300, and containing 66% Ala. Work by Lin (1979) has shown that the liver of the flounder produces antifreeze mRNA in response to environmental conditions. In winter, antifreeze mRNA production is at a maximum and the blood concentration of this vital protein approaches 30 mg ml^{-1}; in temperate seasons, antifreeze mRNA production is shut down. This is an interesting example of essentially a transcriptional rather than a translational response to provocative environmental cirumstances. The trigger for activating the antifreeze protein gene is not understood, but almost certainly it will prove to be hormonal.

During lactation, the mammary gland produces two unique constituents of milk, lactose and casein, in vast amounts; these two constituents of milk are not produced in other organs or indeed, in the mammary glands of non-pregnant females. The enzyme, galactosyl transferase, is present in many tissues, where it synthesized the carbohydrate moieties of certain glycoproteins, for example:

UDP-galactose+N-acetylglucosamine→UDP+N-acetyllactosamine.

During pregnancy, galactosyl transferase is synthesized and stored in the mammary gland, but it has little ability to utilize glucose, a component of the dissacharide, lactose. Immediately after parturition, there are profound changes in the hormonal milieu, including a pronounced secretion of prolactin and significant surges in the secretion of glucocorticoids and insulin. Acting in concert, these hormones promote specialized and differentiated functions in the mammary gland, leading to the dramatic synthesis of both casein and lactose. Both of these responses require the activation of selected genes. First, there is a marked acceleration in the synthesis of casein mRNA (Matusik & Rosen 1978). Second, there is expression of the gene for a modifier unit of galactosyl transferase, α-lactalbumin. This protein is only found in the lactating mammary gland and again, pregnancy evokes a dramatic activation of the synthesis of α-lactalbumin mRNA (Nakhasi & Aasba 1979). The modifier unit interacts with galactosyl transferase, forming a new enzyme complex, lactose synthetase, and the synthesis of lactose is initiated:

$$\text{UDP-galactose} + \text{glucose} \xrightarrow{\text{lactose synthetase}} \text{UDP} + \text{lactose}.$$
$$(\text{galactosyl transferase} + \alpha\text{-lactalbumin})$$

During pregnancy, an abundance of membrane-bound polyribosomes appears in the mammary gland, but the marked changes in lactose and casein synthesis, post partum, can only be satisfactorily explained by transcriptional rather than translational mechanisms, set in train by fundamental changes in gene expression. These changes in gene expression are reversible; once lactation is complete, the genes for α-lactalbumin and casein return to a quiescent, inactive state.

8.4.2 Additional control of eukaryotic gene expression at the level of processing (RNA splicing)

Since post-transcriptional modification of primary transcripts is so extensive in eukaryotes, there is considerable evidence now to support the original view of Church *et al.* (1979) that RNA processing and, particularly, RNA splicing, provides a means of regulating gene expression in higher organisms. Normal adult haemoglobin is a tetramer, $\alpha_2\beta_2$, containing 2α and 2β globin chains. There are a series of inherited diseases, α- and β-thalassaemias, which are characterized by lethal defects in the synthesis of α- and β-globin chains, respectively. In homozygous β^0 thalassaemia, no β-globin is produced, although such patients have measurable amounts of β-globin mRNA; in homozygous β^+ thalassaemia, there is quantitative deficiency in β-globin mRNA but such patients can synthesize small amounts of normal β-globin. Kantor *et al.* (1980) have presented convincing evidence that β^+ thalassaemia is an inherited disease, typified by a defective mechanism for the processing of β-globin mRNA. Hopper *et al.* (1980) have now isolated mutants in yeast which are incapable of excising the transcript equivalent to the intron in certain tRNA genes. The defect is specifically in enzymes for RNA splicing of tRNA intermediates; the processing of rRNA and mRNA in these mutants is normal. The non-spliced tRNA intermediates are biologically inactive (non-functional). Finally, as discussed in more detail in section 8.7, RNA splicing is vital for producing active mRNAs for the synthesis of the heavy and light chains of immunoglobulins, both of which contain constant and variable polypeptide sequences. The extreme specificity of a given antibody depends completely on the correct combination of constant and variable regions, and this in turn, depends on the correct combination of constant and variable gene sequences and ultimately, the correct splicing of the gene transcripts. Mutants have been identified in which anomalous splicing of the mRNAs for both light and heavy immunoglobin chains occur (Dunnick *et al.* 1980, Seidman & Leder 1980). These examples are sufficient to illustrate that RNA processing is another potential step for the regulation of gene expression in eukaryotes, although the detail of such control mechanisms remains to be elucidated.

8.4.3 The selective activation of genes and the selective degradation of gene transcripts

Throughout this chapter, heavy emphasis has been placed on the fundamental importance of the activation of selected genes and thus the transcription of genes, *de novo*. The alternative is the relatively indiscriminate transcription of eukaryotic genes, followed by selective degradation of all but a few transcripts. Considerations based on the waste of cellular energy militate against this second possibility and furthermore, experimental evidence is also strongly against the concept of selective degradation of RNA transcripts. Bellard *et al.* (1977) monitored the expression of the genes for ovalbumin and globin in chick oviduct, before and after oestrogenic stimulation. Using specific cDNAs to ovalbumin and globin mRNAs, they concluded that only the ovalbumin gene was activated and became sensitive to DNAase I digestion. The globin genes were dormant in the oviduct, a non-erythropoietic organ, at all times. Identical conclusions have been drawn by Schimke *et al.* (1977) and McKnight and Palmiter (1979) on the activation of the genes for dihydrofolate reductase and transferrin, respectively. The selective degradation of histone mRNAs during all but the S-phase of the cell cycle is perhaps the only example of the intentional or controlled degradation of gene transcripts (Melli *et al.* 1977).

As mentioned earlier, the nucleus contains large numbers of sn RNAs, which are clearly involved in regulating the structure and function of the nucleus. They are strictly confined to the nucleus and are rapidly degraded if they leak into the cytoplasm; the method of their degradation, however, remains unclear.

However, there is more to gene expression than simply activating the gene and producing a mature, active transcript. Examples are known where an mRNA appears to be preferentially stabilized as well, by as yet unknown post-transcriptional mechanisms. A striking example is the selective increase in the half-life of casein mRNA evoked by prolactin in the mammary gland (Guyette *et al.* 1979).

8.4.4 The relative abundance of different mRNAs in eukaryotic cells

As the result of the complexity of eukaryotic genomes, particularly with respect to gene frequency, RNA polymerase B cannot readily produce specific mRNAs in the same quantity. Bishop *et al.* (1974) provided evidence that there are three abundance classes of mRNA in mammalian cells. This work was extended by Hastie and

Bishop (1976) who carefully examined the poly(A)$^+$ mRNAs expressed from the genomes of three organs in the mouse, namely liver, kidney and brain. Mouse DNA is long enough to encode potentially for about 1×10^6 genes, although only 5×10^4–1×10^5 genes (or a maximum of 10%) are expressed during the normal lifespan of the mouse, and indeed all other mammals. Current dogma has attributed development and differentiation to the selective expression of different genes, but the basis for this assumption was far from satisfactory. Relying on mRNA–cDNA hybridization techniques, Hastie and Bishop (1976) established that between 11500 and 12500 different mRNAs are expressed in mouse kidney, liver and spleen.

These mRNAs are organized in each organ into three abundance classes, rather than as a continuum of concentration; a possible corollary is that the relative frequency of transcription of different genes is not necessary identical either. The three abundance classes for mRNAs were described as abundant, middle frequency and low frequency. The results indicated that the *abundant* class of mRNAs in each organ was *unique* or organ-specific, but the vast proportion of *total mRNAs* expressed was *common* to all three organs. Certain abundant mRNAs were present in all organs, but in much lower amounts, indicating that the genetic basis of differentiation is expressed by differences in (a) the *relative abundance* and (b) the *qualitative pattern* of gene transcription. About 550 mRNA sequences of middle frequency and between 9500–105000 of low frequency were *identical* in all three organs, suggesting that these were expressed from essential 'housekeeping' genes, necessary for the basic function, but not the individual character, of all three organs.

The cDNAs equivalent to each of the abundance classes of mRNA in each organ were selectively purified and hybridized individually against total mouse DNA, to investigate the relationship between mRNA abundance and the frequency (or repetition) of their template genes in the mouse genome. The important conclusions were as follows. First, the large number of housekeeping genes and the vast majority of the genes encoding the abundant, tissue-specific mRNAs were present in *single* copies only. Second, only a *minor* proportion ($<15\%$) of the abundant mRNAs were transcribed from repetitive genes, with up to 100 copies per haploid mouse genome. Making the reasonable assumption that RNA polymerase B will initiate and complete transcription from all active genes with a similar efficiency, the mRNA abundance is not simply attributable to the number of copies (or the repetition) of individual genes. Clearly, the organ-specific nature of gene expression, which is the basis of cellular differentiation, is ultimately controlled in a very subtle manner at the molecular level. These important experiments extend the original observation of Bishop *et al.* (1974) that three abundance classes of mRNA are present in mammalian cells.

Most investigators would agree with the findings of Bishop *et al.* (1974) and Hastie and Bishop (1976), but this is technically a demanding area of molecular biology. In particular, the frequency of the distribution (or abundance) of mRNAs are derived from kinetic data on two assumptions: first, that the cDNA faithfully reflects the distribution of mRNA sequences, and second, that the mRNAs of each kinetic class have the same average size as that of the overall population of mRNA species. Meynhas and Perry (1979) have rightly pointed out that these two assumptions may not be entirely valid during conventional mRNA–cDNA hybridization experiments.

However, despite technical limitations, the consensus view is that the mRNAs of eukaryotic cells are transcribed in an organ-specific manner. This cannot be simply explained by the number of copies of the genes being expressed, for most are unique (single copy) genes.

8.4.5 Initiation but premature termination of eukaryotic transcription

Initiation of transcription is difficult to achieve in most eukaryotes and thus it is likely to be a critical site of gene regulation. Indeed, the importance of the increased number of initiation sites after hormonal stimulation has been strongly stressed (Tsai *et al.* 1975b; Towle *et al.* 1977). There has been a widespread belief that once transcription of eukaryotic genes was initiated, completion of the transcript was inevitable. However, premature termination of transcription is an important mechanism of gene regulation in temperate bacteriophages and in the attenuation of amino acid operons of bacteria (see section 5.2.5). Premature termination during the transcription of adenovirus type 2 DNA has recently been reported by Evans *et al.* (1979). The mechanism of this termination remains to be established, but this work indicates that initiation need not lead to the completion of transcription in eukaryotic cells.

8.4.6 Untranslated and maternal mRNAs

Not all of the mRNA in the cytoplasm of eukaryotic cells is translated (Revell & Groner 1978). Globin mRNAs, for example, can exist in free, 20*s* ribonucleoprotein particles in immature duck erythroblasts, as well as in bound, 15*s* ribonucleoprotein particles which are actively engaged in the translation of globin chains. The protein constituents of these particles are not identical, and some proteins clearly suppress translation. Interestingly, the protein (M_r 75 000) normally associated with the 3′ poly(A) tail of eukaryotic mRNAs (Blobel 1973) is not present in the inactive, 20*s* particles. It appears, therefore, that some of the mRNA transcribed by RNA polymerase B is not for immediate use in protein synthesis, but rather to provide a cytoplasmic reserve of potentially translatable mRNA.

The eggs and zygotes of Amphibia, and a few other Phyla, are enormous structures, containing only a single nucleus. After fertilization, the enormous demand for protein synthesis cannot be met by transcription of new RNA. During development to the early gastrula stage, protein synthesis is largely dependent on stores of maternal mRNAs and rRNAs laid down during the protracted process of oogenesis (for a review see Davidson 1976). Two maternally derived RNAs have been studied in great detail.

Many investigators have shown that the synthesis of histone 1 in *X. laevis* is made at a slow rate during oogenesis and throughout development to the late cleavage stage, when it is markedly accelerated. Woodland *et al.* (1979) have subsequently demonstrated the expression of histone 1 mRNA in a very elegant manner. The early phase of histone 1 synthesis occurs entirely by mobilization of stores of maternal histone 1 mRNA. This source of mRNA is limited yet it lasts to the early gastrula stage. At this point, transcription of the histone 1 gene in the developing embryo is initiated by RNA polymerase B and continues throughout embryogenesis. The subtle manoeuvre used by Woodland *et al.* (1979) was to produce viable hybrids between *X. laevis* and *X. borealis,* two subspecies which have distinctive forms of histone 1. It proved possible to create androgenetic haploid hybrids, in which the maternal but not the paternal nucleus is destroyed, thereby enabling the depletion of maternal stores of histone 1 mRNA to be followed precisely.

The genes for 5*s* rRNA in *Xenopus* are complex; there are two distinct gene families, comprising separate tandem arrays, and encoding slightly different 5*s* rRNA sequences. The arrangement of 5*s* rRNA genes described previously (see Fig. 8.12(b)) is the *somatic* type, which is present and expressed throughout development, but not during oogenesis. The second arrangement of these 5*s* RNA genes is the *oocyte* type, which is present in the female ovary during *oogenesis only* and is the source of the stored (maternal) 5*s* rRNA.

When these oocyte genes were cloned, some surprising findings emerged. In *X. laevis,* the repeat unit contains an AT-rich spacer of variable length, the 5*s* RNA gene, a spacer and at the 3′ terminus, a gene-related sequence called the pseudogene, as follows:

AT– rich spacer	Gene	Spacer	Pseudogene
360 – 570 bp	120 bp	73 bp	106 bp

In *X. borealis,* the structure of the repeat unit is even more complex, namely:

AT– rich spacer	Gene 1	Spacer	Gene 2	Spacer	Gene 3
199 bp	120 bp	79 bp	120 bp	78 bp	120 bp

In both *X. laevis* and *X. borealis* oocytes, these 5*s* rRNA gene units are repeated hundreds of times, yet the pseudogene and gene 3 are not transcribed. Clearly, these maternal (oocyte) genes have a complexity all of their own and are quite different to the somatic type of these genes present in the later developing embryo. The basis for the switch in 5*s* RNA genes during oogenesis is completely unknown.

8.5 CHANGES IN GENE EXPRESSION DURING DEVELOPMENT AND DIFFERENTIATION

8.5.1 Introduction

Beginning immediately after fertilization, a series of discrete, integrated changes in gene expression occur during the successive rounds of cell division which ultimately produce the immature sibling of higher organisms at birth. During this remarkably complex process of development and differentiation, cells acquire their relatively small number of highly specialized functions, along with the maintenance or 'housekeeping' functions required by all cells, for example for providing energy and maintaining crucial ion and nutrient gradients (for a review see Rutter *et al.* 1973). At the time of birth, over 90% of the genes of higher organisms have been briefly activated at critical developmental stages, then remain quiescent, but not permanently shut down. Certain changes in differentiation and development, such as the establishment of sex, must be irreversible and hence require permanent changes in gene expression. These irreversible changes represent, in a sense, the 'coarse' or fundamental tuning of development. Throughout life, however, the organism must remain metabolically flexible in order to maintain homeostasis, to adapt to changes in the environment, and, in the female of higher organisms, to adapt to the changes imposed by pregnancy and lactation. By definition, therefore, some of the changes in gene expression during differentiation must remain essentially reversible, thus representing the 'fine tuning' of development.

In Mammalia, four stages of development are particularly important:

(a) early embryo;
(b) fetus;
(c) neonate;
(d) immature adult.

Taking the latter first, the attainment of sexual maturity depends on the onset of oogenesis (♀) or spermatogenesis (♂), and the growth of the sexual reproductive systems. These changes are promoted by the secretion of oestrogens (♀) and androgens (♂) (the detail of these processes is beyond the objectives of this book). The neonatal stage is one of extreme risk, because the newborn is precipitately deprived of the vital support systems of the mother, especially in terms of the supply of oxygen, the provision of nutrients and the removal of waste products. Major changes of gene expression occur at the neonatal stage to meet the harsh demands of life outside.

Development was originally studied by morphologists, who considered that development was staged, with abrupt changes occurring at precise times and in a coordinated manner. Biochemical investigations, beginning with the pioneering work of Flexner on brain development, are in accord with the concept of development by stages. In terms of gene expression, groups of genes are believed to be activated together at different stages of development. In developmental terms, for example, there are believed to be *clusters* of structural genes for various enzymes which are activated in unison to meet a particular demand imposed by the environment or to establish a fundamental function of the organism as a whole. There is also clinical evidence for developmental staging. The teratogenic effects of drugs (e.g. thalidomide) and viruses (e.g. *Rubella* or german measles virus), for example, are evident only at certain stages of early fetal development; such agents are essentially harmless to the late fetus, the neonate or the adult. The biology of development, particularly in molecular terms, is extremely complex; a limited number of examples will be discussed here, to illustrate certain important aspects only.

8.5.2 Developmental changes in organs, structural proteins and enzymes

Early embryo

The establishment of sex is vital at the early stage of embryonic development. In mammals, the female sex is inevitable; to establish the male, a series of important genetic diversions is needed, involving the integrated action of the Y and X sex chromosomes, plus an autosomal gene locus, not yet defined precisely, but usually referred to as gene Z. In the embryo of both sexes, there are common, primordial structures which can develop as a functional unit of gonad plus urogenital tract, typical of either sex (Fig. 8.16). There is a common pool of uncommitted cells, which can develop

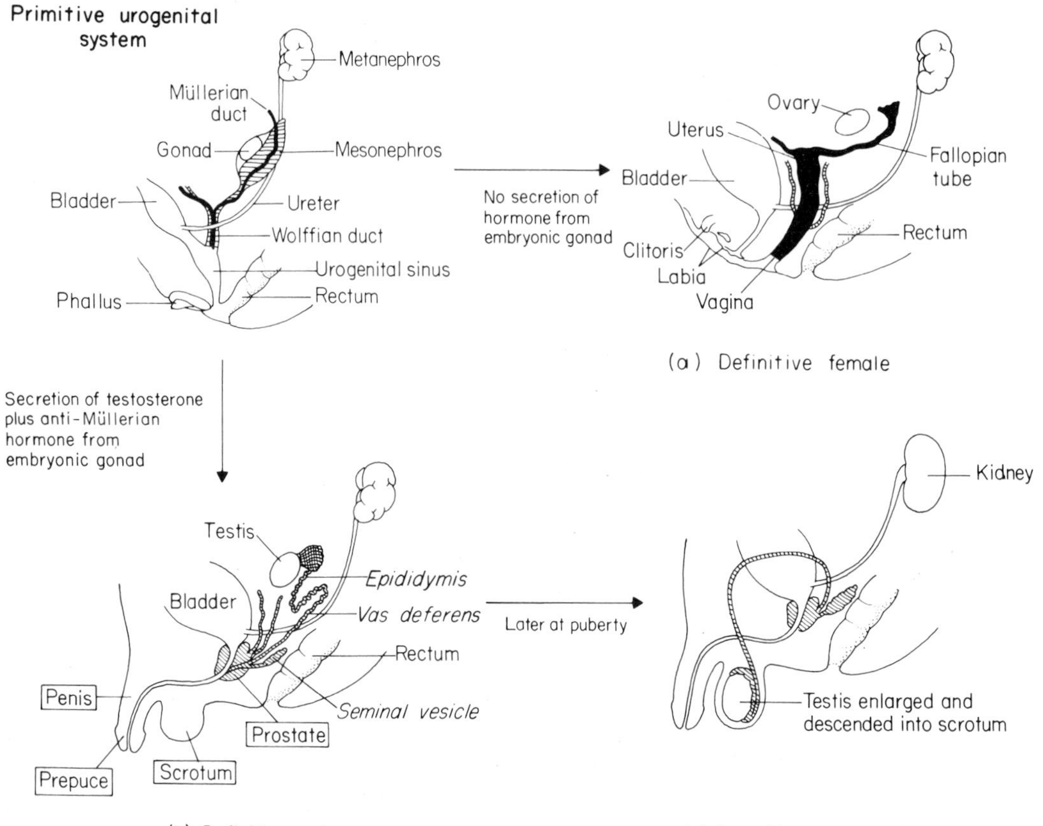

Fig. 8.16. The differentiation of the urogenital tract in mammalian males and females. In differentiation of the male structures in (b), structures requiring testosterone for differentiation are in *italics*, those requiring 5α-dihydrotestosterone are *boxed*. (Reproduced, with permission, from Mainwaring 1979.)

into either the testis (♂) or ovary (♀). In addition, some of the male and female accessory sex organs arise from common, primitive structures (or, from the German, anlagen), derived from primitive Müllerian and Wolffian ducts. Of these, the Wolffian ducts contain some form of 'suicide' message, which unless checked, leads to dissolution of Wolffian structures during development. Finally, the male and female external genitalia originate from common anlagen situated outside the primitive ducts.

The presence or absence of the male Y chromosome is critical in the development of sex. In its absence, the uncommitted (or primitive) gonad cells develop into the embryonic ovary; this is quiescent, and does not secrete any hormones at this time. Consequently, the female urogenital tract develops inevitably from the Müllerian structures; the Wolffian structures disappear also, because their suicide message remains active. In the presence of the Y chromosome, a glycoprotein called H-Y antigen appears and this directs the uncommitted gonad cells to differentiate irreversibly into the embryonic testis. Almost immediately, the embryonic testis secretes two hormones in small, but biologically active amounts; the steroid sex hormone, testosterone, and a small polypeptide hormone, the anti-Müllerian hormone. Provided that the Y chromosome is present, two other gene loci are activated. From the X chromosome (Tfm locus), specific androgen-binding proteins or receptors are formed, capable of binding either testosterone itself or a metabolite, 5α-dihydrotestosterone. The expression of an unknown autosomal gene (Z locus), produces an enzyme, 5α-reductase, necessary for the conversion of testosterone into 5α-dihydrotestosterone. The secretion of the anti-Müllerian hormone promotes the regression of the Müllerian duct structures and testosterone, either directly or indirectly (as 5α-dihydrotestosterone), and in both cases, aided essentially by receptor proteins, leads to the differentiation of the urogenital tract into the male form. Testosterone itself inhibits the regression of the Wolffian duct structures, and these develop into the embryonic epididymis, vas deferens and seminal vesicle; these structures do not contain 5α-reductase at this stage of development. The remaining anlagen contain an active 5α-reductase and the binding of 5α-dihydrotestosterone stimulates their development into the prostate and male external genitalia. Gene deletions profoundly disturb sexual differentiation. Mutations at a specific (Tfm) locus on the X chromosome present the appearance of androgen receptors; such individuals have a male (XY) genotype and testes secreting testosterone, but their internal and external organs are female rather than male. This is the clinical syndrome of testicular feminization, hence the name, Tfm mutants. Mutations at the Z locus (linkage unknown), impair the normal differentiation of the prostate and male external genitalia; their development specifically requires 5α-dihydrotestosterone.

Late fetus and early neonate

The most traumatic transition which mammals have to survive is that from the warm, moist and protected life *in utero* to the dry, intemperate, hostile life of our normal environment. In particular, the structure of the lungs must be changed for this transition to be successful. As reviewed by Rodney (1979), the lung must be differentiated further and produce a protein-phospholipid complex known as pulmonary surfactant. In its absence, newborn infants suffer the respiratory distress syndrome; fatalities from this are colloquially known as 'cot deaths'. Pulmonary surfactant contains only 30% protein (or even less) and is largely composed of dipalmitoyl-phosphatidylcholine. The biosynthesis of pulmonary surfactant is under multiple hormonal control, but certainly glucocorticoids and thyroxine are involved. From late fetal life onwards, both the adrenal cortex and the thyroid are active, thus providing the hormonal stimuli for many developmental changes, including this crucial change in the lung. Many enzymes must be induced in the fetal and neonatal lung to meet this acute demand for phosphatidylcholine, e.g. choline phosphotransferase, choline phosphate cytidyltransferase (Rodney 1979). These changes must be irreversible and are set in motion, at least in part, by the appearance of glucocorticoid receptors in the lung.

Enzymes

The developmental period, late fetus to early neonate, sees remarkable changes in the enzyme complement of many organs. A representative and admittedly incomplete summary of these changes in rat liver, kidney and brain is presented in Table 8.6. It is important to emphasize that these developmental changes fall into three groups or clusters of enzymes, namely, late fetal, neonatal and late suckling clusters, although in some cases, these divisions are somewhat arbitrary. In all of these organs, the appearance of these enzymes is vital for the acquisition of organ function at the right time. Taking the developing liver as an example, the rat is deprived of maternal

Table 8.6. The appearances of various enzymes in the rat at particular stages of development. The majority of these changes are best described in terms of three groups or clusters of enzymes.

	Clusters		
Organ	Late fetal	Neonatal	Late suckling
Liver	UDPG glycogen glucosyl transferase	Phosphoenolpyruvate carboxykinase (GTP)	Tryptophan oxygenase
	Glycogen phosphorylase	Tyrosine aminotransferase	Alanine aminotransferase
	Phosphoglucomutase	Cytochrome oxidase	Ornithine aminotransferase
	Glucose 6-phosphatase	UDP glucuronyl-transferase	Glutamine synthetase
	NADPH-cytochrome c dehydrogenase		Pyruvate kinase
	Ornithine carbamyltransferase		Xanthine oxidase
Kidney	Phosphofructokinase	Aldose-1-epimerase	Ornithine aminotransferase
		Phosphoenolpyruvate carboxykinase (GTP)	Glutaminase
			Arginase
			Alkaline phosphatase
			Carbonic anhydrase
Brain	Phosphofructokinase	Glucose 6-phosphatase	Cytochrome oxidase
	Monoamine oxidase		Fructosebiphosphate aldolase
			Glutamate decarboxylase

systems for bile excretion immediately after birth; hence the enzyme, UDP glucuronyltransferase, necessary for the excretion of bilirubin, emerges with the neonatal cluster of enzymes. The newborn must have its own liver glycogen reserves for survival, and so many enzymes emerging in the late fetal cluster of enzymes include those associated with glycogen breakdown and biosynthesis (e.g. glycogen phosphorylase).

The triggers for the appearance of these enzymes *de novo* are complex and perhaps no single stimulus is involved. As indicated in Table 8.7, thyroxine, glucagon and glucocorticoids (corticosterone) are all involved in the induction of the enzymes in rat liver. On comparing Tables 8.5 and 8.6, the real complexity of enzyme induction during development becomes apparent. Certain enzymes

Table 8.7. The hormones responsible for the induction of enzymes during the development of the rat liver.

	Enzyme cluster		
Hormone responsible	Late fetal	Neonatal	Late suckling
Glucagon	—	Phosphoenolpyruvate carboxykinase (GTP)	—
		Glucose 6-phosphatase	
		Serine dehydratase	
Corticosterone	Enzymes of glycogen synthesis	Tyrosine aminotransferase	Arginase
			Ornithine aminotransferase
Thyroxine	Arginase	—	Pyruvate kinase
	Glucose 6-phosphatase		Malate dehydrogenase ($NADP^+$)
	NADPH-cytochrome c reductase		

are activated by different stimuli at different stages of development; for example, arginase is made in the late fetal stage in response to thyroxine stimulation, but to glucocorticoid stimulation in the late suckling stage. In most instances, the induction of an enzyme is attributable almost entirely to the activation of its structural gene at the correct developmental stage by the appropriate signal or hormonal stimulus. A typical example of this induction process is seen in the synthesis, *de novo,* of the mRNA for phospho-enolpyruvate carboxykinase (GTP) in liver in response to stimulation by glucagon (Garcia Ruiz *et al.* 1978).

8.5.3 Developmental changes in organelles

It would be incorrect to see developmental changes in either enzymes or structural proteins as isolated, disconnected events. In reality, differentiation does have an underlying harmony, which includes organelles as well as individual enzymes (or proteins in general); for example, Pollack and Sutton (1980) have adroitly discussed the complex changes occurring in mitochondria during mammalian development. In some ways, these developmental changes in mitochondria are not too surprising, but they are now documented in detail. The synthesis of some of the mitochondrial constituents, both as RNA and protein, is determined by nuclear genes; on the other hand, mitochondria are central to energy metabolism and as such, must play a vital part in providing the energy for differentiation. To cite but one example from the paper of Pollack and Sutton (1980), there is a pronounced increase in the efficiency of ATP production in mitochondria throughout the entire body of mammals to meet the demand of neonatal protein synthesis. This general change in mitochondrial efficiency is promoted by the unusual combination of adrenalin and glucagon. Studies on development are beginning to underline the wider and perhaps surprising functions of glucagon in higher organisms.

8.5.4 Executive or master genes

Since differentiation is so complex, it is difficult to envisage the basis of its fundamental control. Recent work by Morata and Lawrence (1977) and Garcia-Bellido *et al.* (1979) on metamorphosis in insects has produced some very provocative ideas indeed. In essence, insects such as *Drosophila,* including the structure and spatial arrangement of all their composite organs, are laid down during metamorphosis from groups of rudimentary cells called imaginal discs. The concept emerging from this recent work is that higher organisms may well be derived from discrete compartments, each of which contains clones of cells committed to a common and integrated course of development, both temporally and spatially. Compartments, as units, may be connected to the enzyme clusters described earlier with respect to development. The crucial, and intrinsically exciting, aspect of this work on insect development, is that each fundamental compartment (or, possibly, each cluster of interrelated genes) is under the control of a few executive or master genes. This concept is very attractive, because it envisages a hierarchical basis for the integrated control of gene expression during development; differentiation is a precisely ordered, non-random process, and an underlying theme of control must surely be present. In addition, mutations or regulatory failure in these executive genes provide a very plausible explanation of many diseases, including cancer, which are basically a breakdown (or reversal) of differentiation. Opinion on the scientific acceptability of executive genes is currently divided (e.g. Slack 1978) but they remain an exciting aspect of contemporary developmental biology.

8.6 FUNCTIONAL COOPERATION BETWEEN INDIVIDUAL EUKARYOTIC GENES

8.6.1 Examples of gene cooperation

The distribution of genetic loci on disparate eukaryotic chromosomes leads to the situation where distant genes have to cooperate together in order to complete the synthesis of proteins and other macromolecules. Examples of gene cooperation have received passing attention already; for example, there is functional interdependence of the X and Y sex chromosomes in the development of the male urogenital tract and the α- and β-globin genes are on different chromosomes. In the latter case, haemin plays an important role in coordinating the synthesis of globin chains. In its absence, α-globin mRNA cannot bind extensively to the 40*s* ribosomal subunit, whereas β-globin mRNA can (Lodish 1976). However, the real complexity of gene cooperation is best illustrated by further examples.

Elegant studies by Paigen and co-workers (reviewed in Paigen *et al.* 1975) have shown that no less than six genetic loci are necessary for regulating the biosynthesis, molecular size, intracellular location and secretion of the enzyme, β-glucuronidase, in mouse kidney. Further, this work suggests that eukaryotic genes may best be classified into four groups:

(a) structural;
(b) regulatory;
(c) temporal;
(d) processing genes.

In mice of similar age, the kidneys of the male are larger and structurally distinct from those in the female; these differences are attributable to circulating androgens, particularly testosterone itself, rather than its metabolites. The basic subunit of β-glucuronidase is a polypeptide, M_r 70 000, but the six functional forms of the enzyme are tetramers, M_r range 260 000–470 000, which may be found in the lysosomes, or cell soluble fraction or bound to membranes of the endoplasmic reticulum. These six enzyme types may be summarized as follows: type L (lysosomes, M_r 260 000), type M (four species: microsomes, M_r 310 000–470 000) and type X (soluble fraction and microsomes, inducible by androgens, M_r 260 000, but distinct from L). The four species of type M enzyme contain varying amounts of another protein, egasyn (M_r 50 000), which enables the enzyme tetramer to bind to membranes of the endoplasmic reticulum. The six genes involved, together with their function and classification, are now known.

(a) *Gus*: structural gene controlling the synthesis of the mRNA for the enzyme monomer.

(b) *Tfm*: structural gene controlling the synthesis of the mRNA for the androgen receptor protein. The X form of the enzyme cannot be induced by androgens in receptor-deficient, Tfm mutants.

(c) *Gur* regulatory gene controlling the androgenic induction of the enzyme; the androgen receptor protein–testosterone complex is believed to bind to (and thus activate) this specific gene.

(d) *Gut*: temporal gene controlling the appearance of the enzyme during development.

(e) *Eg*: processing gene regulating the synthesis of egasyn, the 'cement' for the enzyme.

(f) *Bg*: processing gene controlling the entry of the enzyme into the lysosome and its secretion from the kidney. These gene interrelationships are summarized in Fig. 8.17. It can be seen that apart from the cluster of three genes on chromosome 5, the remainder are individual loci on different chromosomes.

The synthesis of the soluble enzyme, ribulose bisphosphate carboxylase (M_r>500 000), in the stroma of plant chloroplasts has been

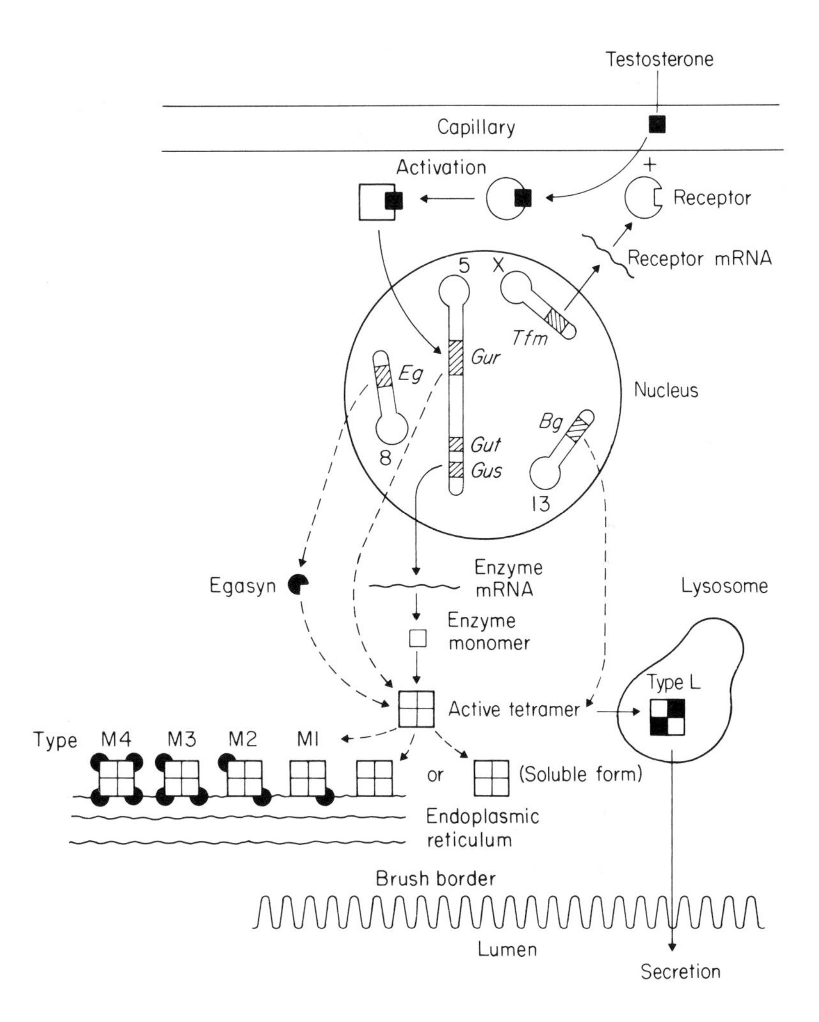

Fig. 8.17. The genetic control of β-glucuronidase in mouse kidney. Important genetic loci are indicated on the chromosomes by hatching. (Based, with permission, on Mainwaring & Irving 1979.)

widely studied (for a review see Ellis 1979). This enzyme is present in plants in massive amounts and it catalyses two vital reactions:

(1) the fixation of carbon dioxide during photosynthesis;

(2) the first reaction in the process of photorespiration:

$$(1)\ CO_2 + \text{D-ribulose-1,5-bisphosphate} + H_2O \xrightarrow{Mg^{2+}\text{ ions}} \text{2,3-phospho-D-glycerate}$$

$$(2)\ O_2 + \text{D-ribulose-1,5-bisphosphate} \xrightarrow{Mg^{2+}\text{ ions}} \text{3-phospho-D-glycerate} + \text{2-phosphoglycolate.}$$

Ribulose bisphosphate carboxylase has a complex quaternary structure, comprised of eight large (catalytic) subunits, M_r 52000–60000, and eight small subunits, M_r 12000–18000, of currently uncertain function. The novel feature of the synthesis of this enzyme is that it depends on nucleus–chloroplast interactions. The large subunit is encoded in the chloroplast genome and is synthesized within the chloroplast. The small subunit is encoded in the nuclear genome and first appears as a larger precursor, M_r 20000, which crosses the chloroplast envelope. Once within the chloroplast, the removal of the extra N-terminal (or signal) sequence (see section 9.4.4) releases the small unit of the enzyme in its final form and assembly of the enzyme follows. These closely integrated events are shown in Fig. 8.18. All of the structural genes involved have now been successfully cloned in bacteria.

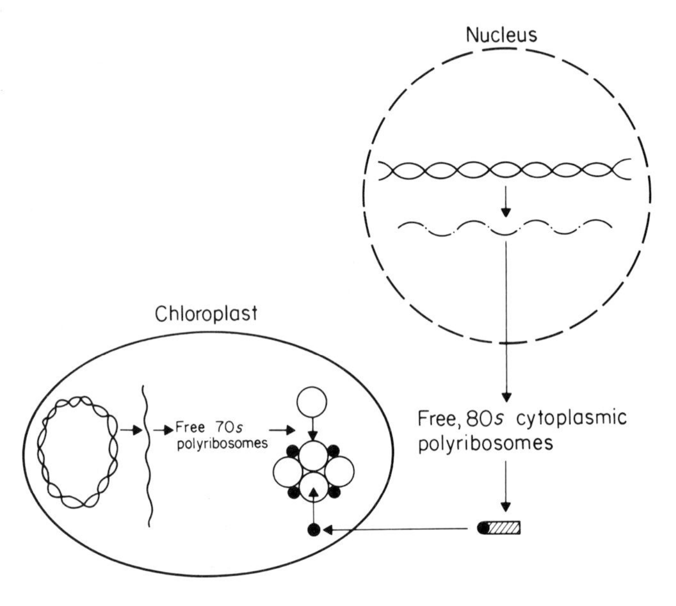

Fig. 8.18. Cooperation between nuclear and chloroplast genes in the synthesis of ribulose-1,5-bisphosphate carboxylase. This oligomeric protein is composed of large, O, and small subunits, ●. The nuclear gene is responsible for the synthesis of a precursor of the small subunit, ●▨. After translocation into the chloroplast, this precursor is converted in the final form of small subunit. The large subunit is produced exclusively within the chloroplast, from a gene in the circular chloroplast genome.

8.6.2 Experimental manipulation of gene expression

Cell hybridization has been a remarkably powerful technique for determining the chromosome linkage of mammalian genes (for a review see Kutherlapti *et al.* 1974). It is now being widely used to study the regulation of gene expression as well. A particularly striking example is the work of Deisseroth and Hendrick (1979) on the activation of human globin genes from non-erythroid cells. The basis for their work is that mouse and human globin chains can be readily separated by ion-exchange chromatography and the human α-globin gene is located on chromosome 16. Hybrid cells were prepared by fusing diploid human mononuclear lymphocytes with tetraploid mouse erythroleukaemia cells. The resultant hybrids retained nearly all of the tetraploid complement of mouse chromosomes but had lost 80% or more of the human chromosomes. Only hybrids containing human chromosome 16 were able to make human α-globin and yet this gene was not expressed in the original human cells, the non-erythroid lymphocytes. The important conclusion from this study is that the mechanisms responsible for restricting the expression of human globin genes in non-erythroid cells are *reversible*. The novel environment provided by the mouse chromosomes enables this genetic restriction to be overruled. These experiments provide considerable hope that the dysfunction of human genes in certain diseases may be precisely defined and perhaps even rectified.

The most dramatic reversal of gene activity, however, was not achieved by cell fusion techniques. In a truly remarkable paper, Mintz and Illmensee (1975) describe the conversion of highly malignant, dedifferentiated mouse teratoma cells back to perfectly normal, differentiated cells. This normalization of gene expression was achieved by exposing the tumour cells to unusual environments, first in the early blastocyst of a pregnant mouse and second, in the uterus of a pseudopregnant mouse.

8.6.3 Chemical signals between distant genes

We do not know how the functions of distant genes are integrated. One very fashionable model, proposed by Britten and Davidson (1969) and subsequently extended, has appeared in many textbooks (see also Fig. 8.19). In essence, this model proposes the existence of sensor sites, integrator genes and receptor sites in the eukaryotic genome, apart from structural genes. It is envisaged that the sensor site and integrator gene are situated together on one

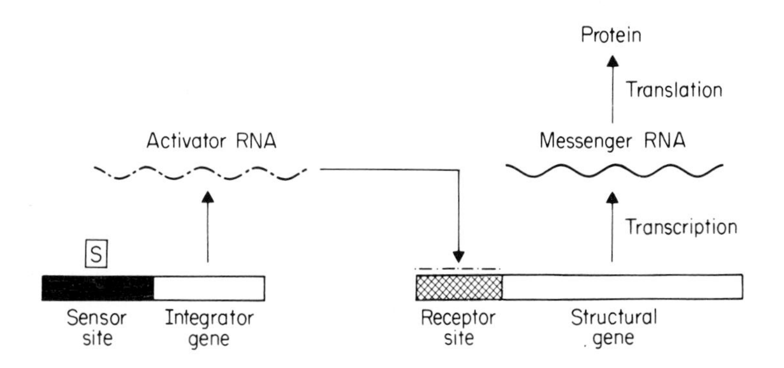

Fig. 8.19. The model of Britten and Davidson (1969) for the regulation of gene expression in eukaryotes. The structural genes are inactive in the absence of a hypothetical activator RNA. The latter is formed from a separate integrator gene in response to an appropriate signal, S, as a steroid hormone–receptor protein complex.

chromosome, while the receptor site and structural gene may be on the same or a different chromosome. Functional coordination between the two distant gene structures is maintained by the transcription of a hypothetical activator RNA from the integrator gene in response to a signal recognized by its adjacent sensor site. The activator RNA then translocates to the distant receptor site and promotes the transcription of messenger RNA from the structural gene. The fact that much of the RNA made in the nucleus never reaches the cytoplasm is in keeping with the model (for a review see Davidson & Britten 1973) but direct experimental support is seriously wanting. It may perhaps be argued that the sn RNAs described elsewhere (see section 7.2.4) could serve as activator RNAs, but this remains unproven. Certainly, there are reports that nuclear RNA fractions can regulate gene expression and even promote the cytolysis of tumour cells, but the molecular basis of the Britten–Davidson model needs wider verification.

8.7 GENE EXPRESSION AND IMMUNOGLOBULIN SYNTHESIS

8.7.1 Introduction

Mammals are capable of synthesizing a large number of different antibodies, perhaps as many as 1×10^6, each one being specific for a particular type of antigen. Antibodies are synthesized by lymphocytes present in bone marrow. Normally, lymphocytes do not divide actively, but in the presence of a foreign antigen (generally a protein), they differentiate and proliferate into plasma cells, which synthesize the antibody to the antigen in very large amounts. The principal antibodies are immunoglobulin (IgG) molecules, M_r 150 000. Each IgG molecule consists of four polypeptide chains, namely two identical light (L) chains and two identical heavy (H) chains. The IgG molecule has the overall shape of a Y (Fig. 8.20), with the two flexible arms meeting the tail at the hinge region. The N-terminal regions of both H and L chains serve together in recognizing the antigen and the remaining C-terminal regions carry out effector functions, such as the transport of antibodies across membranes. The relationship between structure and function in various parts of the IgG molecule is the key to their specialized role in defence against foreign proteins. The L and H chains are unique among proteins in possessing variable (V) and constant (C) regions of sequence at their N- and C-terminal ends, respectively. Each

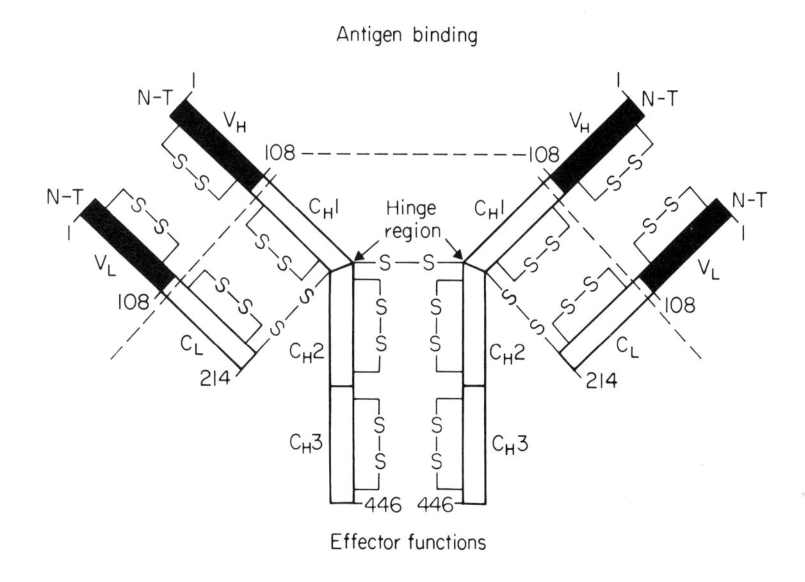

Fig. 8.20. The basic structure of an immunoglobulin (IgG). Areas of variable (V) and constant(C) sequences are indicated as solid boxes ■, and open boxes, □, to show regions of homology. Each half of the molecule contains six domains, as indicated. The N-terminal residues (N-T) are numbered as one. V_H and C_H1, etc., demarcate the domains of the IgG molecule; for details, see the text.

L chain (M_r 25000) contains approximately 214 residues, the variable region being residues 1–108; each H chain (M_r 50000) is composed of approximately 446 residues, the variable region again being residues 1–108. There are various regions of functional and structural homology or domains; each L chain has two domains and each H chain four. By combining the abbreviations for their location (H or L) and properties of their sequence (C or V), each IgG molecule has six defined domains (Fig. 8.20). Homology has been proved between domains V_L and V_H, and between C_L and C_H1, C_H2 and C_H3.

The synthesis of such a wide spectrum of specific antibodies raises some of the most profound problems in biology. Differences between antibodies occur in the variable region, particularly at three hypervariable regions or 'hot spots' around residues 30, 55 and 90 of the N-terminal sequences of H and L chains (for a review see Williamson 1976). There are minor or allotypic variations in the C regions of H and L chains; for example, residue 191 in the L chain can be Leu or Val. From studies on the normal Mendelian segregation of these allotypic markers, it was concluded that the gene coding for the C region was unique (one copy per haploid genome). By contrast, the enormous diversity of V regions suggested that multiple genes must exist for the antibody-specific regions at the N-terminal ends of the H and L chains. Further studies revealed the absorbing complexity of immunoglobulin biosynthesis (for a review see Williamson 1976). First, the mRNAs for IgG molecules have a contiguous coding sequence for both the V and C regions. Second, in the primitive germ cells of the embryo, the genes for the V and C regions are some distance apart in the genomic DNA, yet are brought close together during the process of differentiation and the formation of committed, functional lymphocytes in the bone marrow. To explain these observations, integration of genetic information must occur during differentiation by one or both of the following procedures:

(a) joining (or splicing) of the genes coding for the V and C regions;

(b) joining of separate transcripts of V and C genes into a single, contiguous mRNA.

There are many types of multiple myeloma in mice, where the malignant disease is manifest in the production of massive amounts of a single species of IgG. It has proved possible, therefore, to isolate and purify large amounts of the specific mRNA for a given mouse IgG molecule. Together with advances in mRNA–cDNA hybridization and gene cloning, the molecular biology of immunoglobin synthesis in mouse myeloma cells is now known in remarkable detail. In the mouse, there are three classes of C regions, namely λ_1, λ_2 and κ in the L chain and distinctive clones of myeloma cells are available, producing each class of L chain in abundance. The production of myeloma κ chains was studied in critically important experiments by Hozumi and Tonegawa (1976) and Rabbits and Forster (1978). Restriction fragments of total genomic DNA from mouse embryos and myeloma cells were separated and hybridized individually to κ chain mRNA. These studies proved unequivocally that the separate V_κ and C_κ genes in the embryo had been rearranged or spliced together during the course of differentiation:

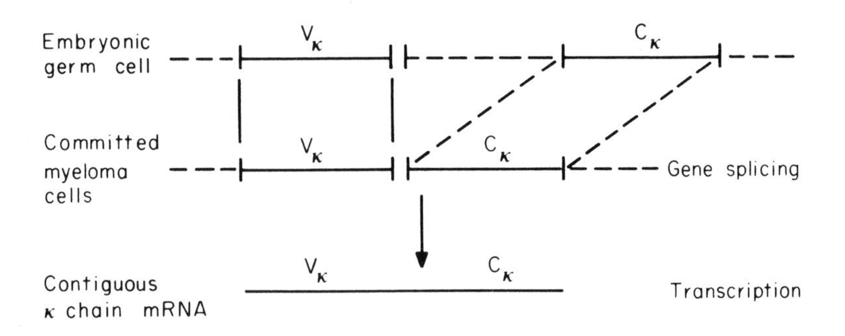

When introns were discovered, yet another level of complexity was found in IgG biosynthesis. Immunoglobulin production by lymphocytes is now known to be quite unique in that it involves gene (DNA) splicing during development and transcript (RNA) splicing during expression of the combined V plus C genes (for a review see Abelson 1979).

8.7.2 Synthesis of L chains

Studies on the base sequence of the mRNA for mouse λ_1 light chain have been in progress for some time, beginning with the work of Milstein *et al.* (1974); this study actually heralded the sequencing of eukaryotic mRNAs in earnest. Adult mouse λ_1 mRNA is now known to be very similar in structure to globin mRNAs, namely:

5′ 7-methylguanosine .	non-coding .	coding sequence .	non-coding .	poly (A) 3′
cap	leader sequence	*uninterrupted* for V and C regions	trailer sequence	

When the λ_1 genes from mouse embryos and the adult myeloma H2020 were successfully cloned and sequenced (for example Tonegawa *et al.* 1978), the real complexity of the recombination and expression of IgG genes became apparent. Several features of the embryonic and adult λ_1 genes warrant particular mention.

(a) The coding region of the $V_{\lambda 1}$ gene is not continuous. The first part of this gene (L region) encodes only the non-coding leader sequence of λ_1 mRNA and is interrupted by a short intron (93bp).

(b) The next section of the $V_{\lambda 1}$ gene encodes the V region as far as residue 97 (His), only. This region is followed by the large tract of DNA, length and sequence unknown, which is excised during the somatic rearrangement of V and C genes in the transition of the embryonic stem cell to the committed, adult lymphocyte.

(c) The next section of the $V_{\lambda 1}$ gene is usually referred to specifically as the J region (39bp) and it plays a pivotal role in the arrangement and expression of the entire L chain gene. The J sequence encodes the remainder of the V region, from residue 98 (Trp) to 110 (Gly) and is followed by a second, very long intron (1250bp).

(d) The $C_{\lambda 1}$ sequence follows thereafter, encoding the remainder of the C region from residue 111 (Gln) to the C-terminal residue (Cys), *plus* a 3′ non-coding trailer sequence.

The final λ_1 gene still has two introns, one in the sequence encoding the N-terminal signal peptide and one separating the classical V and C regions of the λ_1 L chain. Expression of this L chain gene is best visualized by examination of Fig. 8.21. The cleavage sites at exon–intron boundaries are between –G$\downarrow$G– residues, in keeping with the processing of other mRNAs (Table 8.4).

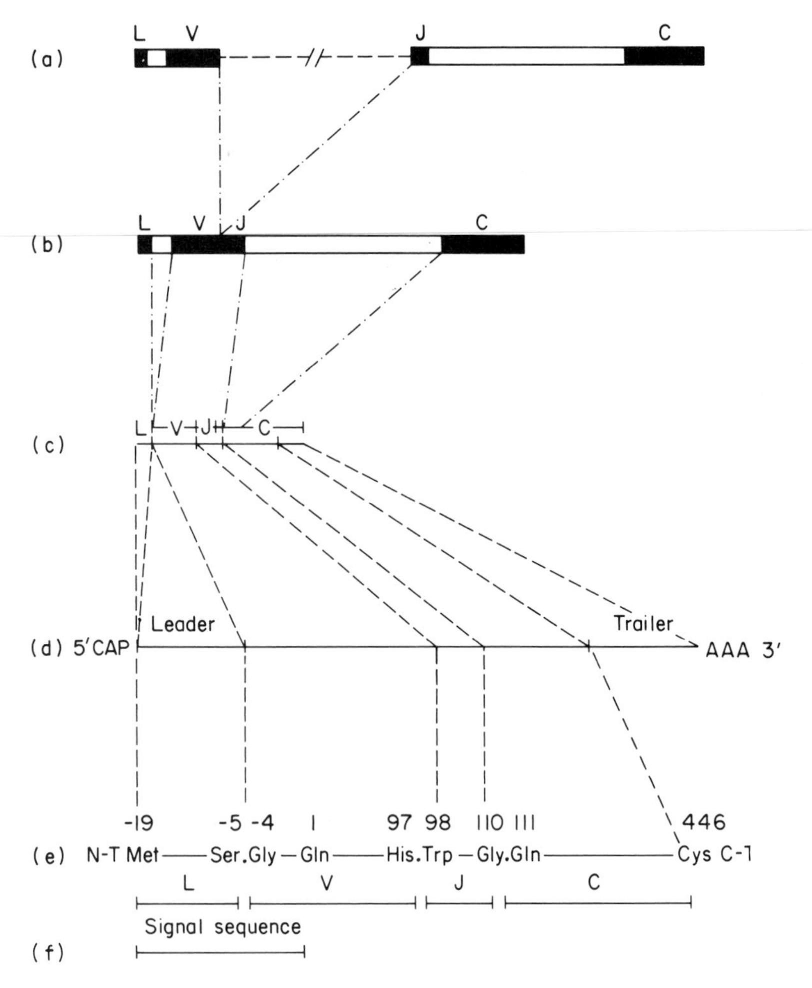

Fig. 8.21. Recombination and expression of the mouse genes for immunoglobulin light chain (type λ_1). Introns are shown as open boxes, □, and exons as solid boxes, ■. The N-terminal residue (N–T) of the final L chain is numbered as one; negative numbers include the original N-terminal signal peptide which is subsequently cleaved, forming the final λ_1 chain. (a) Gene arrangement in embryo germ cells, with the exons for the leader (L) and variable (V) regions separated by a long tract of DNA from the exons for the join (J) and constant (C) regions. (b) Gene rearrangement occurring in committed, adult stem cells, with juxtaposition of the V and J exons. (c) Primary gene transcript. (d) Mature, processed mRNA. (e) Primary translation product. (f) Processing, producing the final λ_1 L chain, by cleavage of the signal sequence.

The J region of light chain IgG genes is a crucial tract of DNA, with the quite unique feature of having both DNA- and RNA-recombination sites at its left- and right-hand ends (Abelson 1979):

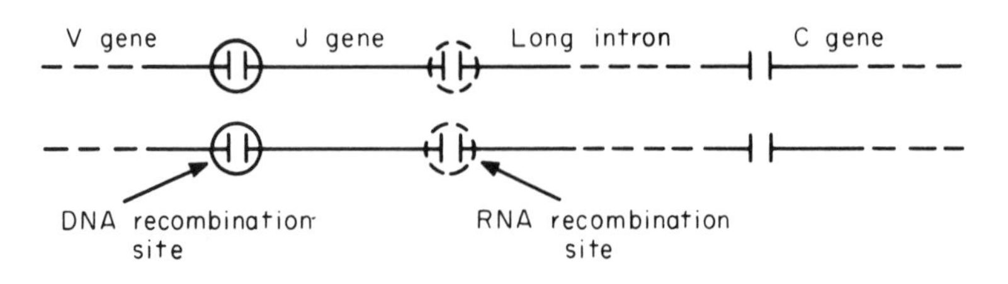

Clearly, its importance belies its relatively short length (39bp, only) and the fact that it encodes only 13 amino acids in the variable

N-terminal sequence. Nonetheless, any recombination in this region must be extremely precise for one of the antibody-specific or hypervariable 'hot spots' is in this sequence, at residue 95 or 96.

Sakano *et al.* (1979a) have determined the nucleotide sequences of five genes coding for specific myeloma κ light chains, together with their flanking sequences. As in the λ gene system, the κ chain secreted by the myeloma cells is encoded in three separate DNA segments in embryonic cells, namely V_κ, J_κ and C_κ and somatic recombination occurs between embryonic V_κ and J_κ genes. However, unlike the arrangement of the λ genes, the mouse genome contains *multiple* copies of V_κ and J_κ genes, the latter being clustered in the close vicinity of the *single* copy of the C_κ gene. Each of five J segments encodes amino acids, 96–108, of the κ chains. Sakano *et al.* (1979a) determined the sequences of the J_κ cluster and also the 3′ terminus of embryonic V_κ DNA. From these studies, an insight into the structure of the DNA recombination sites has been gained. The sequences of the five J segments were highly conserved in the coding region and in the region of the RNA recombination site; by contrast, the sequences in the flanking regions were rather diverse. Nonetheless, there was a short, conserved segment in the 5′ non-coding sequence of the V_κ gene, a palindromic septamer, immediately preceding the five J genes. This palindrome often had the sequence CACTGTG, interrupted by an AT base pair at the centre of symmetry. The presence of these palindromic sequences could permit the formation of an inverted, repeated stem sequence, between the 3′ end of an embryonic V_κ and the 5′-end of a J_κ sequence. A typical example of such a structure in the somatic recombination area is as follows:

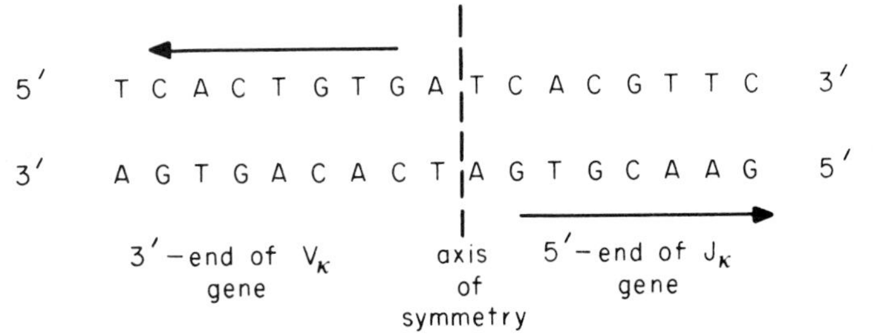

```
               <-------------       |
5′   T C A C T G T G A | T C A C G T T C   3′

3′   A G T G A C A C T | A G T G C A A G   5′
                        |  ------------->
   3′-end of Vκ       axis      5′-end of Jκ
      gene             of          gene
                    symmetry
```

Many theories of V–J splicing have been proposed (Williamson 1976), including the three theories of copy-insertion, excision-insertion and DNA deletion. Examination of the base sequences of these cloned genes by Sakano *et al.* (1979a) leads them to suggest that the DNA deletion model is probably correct. Accordingly, somatic recombination may be accompanied by excision of an entire DNA segment between a V gene and a J gene. Presumably, the inverted, repeated sequences described earlier serve as important recognition signals in the process of DNA excision and splicing.

8.7.3 Synthesis of H chains

The H chains of IgG molecules present some horrifying difficulties in terms of the arrangement and expression of their genes. The C regions of the H chain in the mouse can be classified as five distinctive types, namely μ, δ, γ, ϵ and α. Again, the application of genetic engineering to the H genes of experimental mouse myelomas has reaped a rich harvest.

As shown in Fig. 8.20, the C region of the H chains contains three functional domains (C_H1, C_H2 and C_H3) plus a hinge region. Sakano *et al.* (1979b) succeeded in cloning the mouse gene for the immunoglobulin γ_1 heavy chain. Extremely careful electron microscopic analyses, combined with sequence determinations, established that three introns were present in the gene, two of approximately 320bp and a short one, approximately 120bp. The exciting finding from the work of Sakano *et al.* (1979b) is that each of the three domains and the hinge region are encoded in separate DNA segments. Furthermore, the correct alignment of the three domains is seemingly dic-

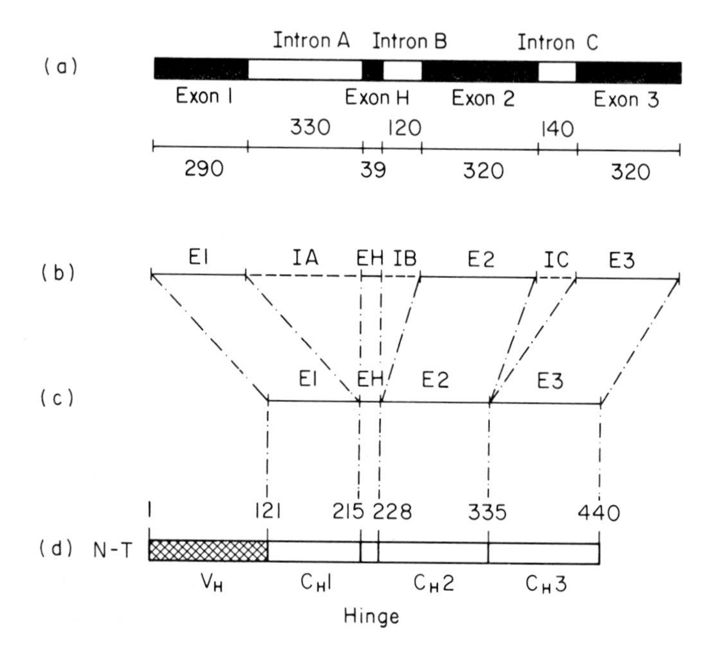

Fig. 8.22. The implication of the presence of introns in the mouse genes for immunoglobulin heavy chain (type γ_1). Introns are shown as open boxes, □, and exons as solid boxes, ■. The length of each DNA sequence is given in base pairs. The four domains of the final H chain are shown, namely the variable domain (in hatching), the three constant domains, plus the short hinge region. the numbering of amino acid residues is from the N-terminal end (N–T) as one. (a) Gene arrangement. (b) Initial RNA transcript. (c) Processed transcript. (d) Translation product of the three constant gene exons, indicating that each possibly encodes for an individual domain in the immunoglobulin H chain. For clarity, the presence of the variable gene sequence is not included (see also Fig. 8.21).

tated by the exon–intron boundaries. The splicing mechanisms involved in the post-transcriptional modifications of primary transcripts may thus create a mature mRNA coding for a polypeptide chain in which defined areas of function (or domains) are predetermined. This elegant work puts the function of introns into an entirely new context (Fig. 8.22).

So far, immunoglobulins have been referred to almost synonymously as IgG. There are, in fact, several classes of immunoglobulin, each of which has a distinctive structure and function (Table 8.8). IgM is the first class of antibody to appear in the serum after the injection of an antigen, whereas IgG is the principal antibody and appears later in response to an antigen. IgA is the major class of antibodies in external secretions, such as saliva, intestinal mucus, bronchial mucus and tears; this class of antibody is the first line of defence in viral and bacterial infections, along with interferon (see section 9.3). The real need for IgD and IgE is not yet known; however, the harmful effects of IgE in mediating allergic reactions are clearly established.

Table 8.8. The properties of the various classes of imunoglobulin.

Class	Serum concentration ($mg\ ml^{-1}$)	M_r	Light chains	Heavy chains	Quaternary structure
IgG	12	150 000	κ or λ	γ	$\kappa_2\gamma_2$ or $\lambda_2\gamma_2$
IgA	3	180 000 500 000	κ or λ	α	$(\kappa_2\alpha_2)_n$ or $(\lambda_2\alpha_2)_n$
IgM	1	950 000	κ or λ	μ	$(\kappa_2\mu_2)_5$ or $(\lambda_2\mu_2)_5$
IgD	0.1	175 000	κ or λ	δ	$\kappa_2\delta_2$ or $\lambda_2\delta_2$
IgE	0.001	200 000	κ or λ	ϵ	$\kappa_2\epsilon_2$ or $\lambda_2\epsilon_2$

In the response to an antigen, the expression of the H chain genes is complex, since it involves a series of switches, first from IgM (μH chain) to IgG (γH chain) and later, in some cases, to IgA (αH chain). Several hypotheses have been forwarded to explain these switches.

(a) Sequential insertion model: this suggests that the V_H gene is initially integrated with the C_μ gene, then subsequently integrated with the C_γ gene and finally with a C_α gene.

(b) Multiple insertion model: in this model, genetic integration occurs all at once, between several copies of V_H genes with individual C_H genes.

(c) Post-transcriptional model: this affirms that the V_H gene plus the combined array of C_H genes is transcribed as a single unit into a long precursor RNA which yields various V_H–C_H combinations in discrete mRNAs as the result of complex excision and splicing reactions.

(d) Deletion model: this envisages that the initial integration occurs between a V_H gene and the C_μ gene, and subsequent switching is achieved by deletion of the genetic material between the integrated V_H gene and the next C_γ gene—a final deletion would permit the functional association of the V_H gene and the C_α gene.

Evidence from two research groups (Honjo & Kataoka 1978, Rabbits *et al.* 1980) using quite different experimental approaches, suggests strongly that *deletion* mechanisms provide the most plausible explanation for the switches in H-chain synthesis. The order of C_H genes established by both groups, as $V_H:C_\mu:C_\gamma:C_\alpha$, is certainly consistent with the deletion model. Clearly, as more is learned about DNA–DNA and RNA–RNA recombination in eukaryotic cells, one can only marvel at the variety, accuracy and flexibility of these remarkable reactions. Certainly, excision and splicing enzymes have reached a zenith of sophistication in the molecular processing needed to synthesize immunoglobulin L- and H-chains. This is a most absorbing yet complex field and the current status of immunoglobulin synthesis may be found in the reviews by Davis *et al.* (1980) and Molgaard (1980).

8.8 METHYLATION OF EUKARYOTIC DNA

Throughout this chapter, considerable attention has been devoted to the mechanisms and importance of modifications to all types of eukaryotic RNA. Eukaryotic DNA is not subjected to particularly extensive modification, but 5-methylcytosine (m^5C) is found in the DNA of all eukaryotic organisms, right up the phylogenetic tree and the DNA of unicellular eukaryotes, such as *Tetrahymena* and *Paramoecium,* contains traces of 6-methyladenine as well. There has been considerable speculation on the significance of DNA methylation (for a review see Burdon & Adams 1980). The appropriate DNA methylases are located in the nucleus and utilize S-adenosyl methionine as the methyl donor. Most of the m^5C is found in the dinucleotide, m^5CG, and less commonly, m^5CC and m^5CT. There has been speculation that methylation of DNA may change its properties as a template for RNA polymerase, and most active or expressed genes are non-methylated. It remains to be seen whether the methylation of purified genes changes their rate of transcription *in vitro*, for example in *Xenopus* oocytes. Methylation of DNA could also change the activity and specificity of enzymes engaged in DNA repair; again, there is no conclusive evidence one way or the other. The true significance of the methylation of eukaryotic DNA remains enigmatic.

Chapter 9 Translation in Eukaryotes

9.1 INTRODUCTION AND DISTINCTIVE FEATURES OF EUKARYOTIC TRANSLATION

During the late 1950s, classical studies on eukaryotic cells by Littlefield, Zamecnik, Hoagland and others helped to lay the foundations of the process of genetic translation. These incisive studies led to an understanding of the form and function of rRNA, tRNA and the aminoacyl tRNA synthetases. For reasons of technical simplicity, the availability of mutants and high biosynthetic activity, later investigations on translation were conducted almost exclusively on bacteria. As described in Chapter 6, this work on prokaryotic organisms resulted in two of the most exciting achievements in biochemistry, namely the identification of mRNA and the cracking of the genetic code. Work on the mechanisms of eukaryotic translation was not entirely neglected and amino acid sequence analyses of abnormal haemoglobins provided the culminating evidence that the genetic code was truly universal (Beale & Lehmann 1965). Only mitochondrial translation shows any variation in the genetic code (see section 10.4.1).

While it remains true that the machinery for protein synthesis is essentially similar in all forms of life, evidence is rapidly accumulating now which indicates that translation in higher organisms has a complexity and subtlety all of its own. Indeed, one of the basic tenets of contemporary molecular biology, the colinearity concept of a structural gene and its polypeptide translational product may not be strictly applicable to eukaryotes (Yanofsky *et al.* 1964). Some of the distinctive features of genetic translation in higher organisms are presented in Table 9.1; it should be emphasized that this information applies only to cytoplasmic protein synthesis in eukaryotes, because genetic translation in mitochondria and chloroplasts resembles that in bacteria in certain respects (see sections 10.4 and 10.5). This functional similarity is reflected in structural homology; for example, the base sequence of the genes coding for 16*s* rRNA of *Zea mays* chloroplasts is homologous to the base sequence of *E. coli* 16*s* rRNA (Schwartz & Kössel 1980). Such evidence confirms the prokaryotic nature of the chloroplast ribosomes in plants.

Having emphasized the differences in translational processes between prokaryotes and eukaryotes, the fundamental similarities should not be overlooked. First of all, eukaryotic protein synthesis is best described in terms of four discrete stages, namely activation of amino acids, initiation, elongation and termination. The direction of translation of mRNA is $5' \rightarrow 3'$, beginning with the N-terminal

Table 9.1. Distinctive features of cytoplasmic translation in eukaryotes.

Property	Prokaryotes (e.g. *E. coli*)	Eukaryotes* (e.g. rabbit reticulocyte, wheat germ, rat liver)
Relationship between transcription and translation	Coupled	Not coupled
Nature of mRNA	Polycistronic	Monocistronic
Lifetime of mRNA	Short (h)	Extremely long (d)
Structure of mRNA	Uncapped	Poly(A) sequence at 3′, 7-methyl-guanosine at 5′†
Size of ribosomes:		
Ribosome monomer	70*s*	80*s*
Ribosome subunits	30*s*; 50*s*	40*s*; 60*s*
Nature of polyribosomes	Essentially free	Largely membrane-bound§
Sensitivity to inhibitors		
Streptomycin	Sensitive	Insensitive
Bacterial toxins	Insensitive	Sensitive
Ricin	Insensitive	Sensitive

*The properties listed may not invariably be true for all eukaryotic cells; the table is an attempt at generalization.
†Certain cells have an abundance of poly (U)-containing RNA and poly (C) tracts are present in certain viral RNAs. Histone mRNAs are poly (A)-deficient (poly A^-).
§Not true for certain tumour cells.

residue of the polypeptide product. Likewise, eukaryotic mRNAs are similar in having a non-coding or leader sequence at the 5′-end, with AUG as the initiating codon and UAA, UAG or UGA as alternative termination codons. Similarly, puromycin distinguishes two functional sites in eukaryotic ribosomes, the P or peptidyl site and the A or aminoacyl site, both of which are present in the large ribosome unit. Finally, the activation of amino acids is essentially similar in both eukaryotes and prokaryotes.

A contentious aspect of recent work on higher organisms has been the extent and importance of protein synthesis within the nucleus itself. The consensus view is that although mature and active polyribosomes may be present in the nucleus (Chatterjee *et al.* 1977), the cytoplasmic compartment bears the full brunt of protein synthesis. Nevertheless, the modest extent of nuclear protein synthesis may belie its true biological significance. There is evidence that important regulatory proteins, such as tubulin and certain non-histone, chromosomal proteins may be synthesized almost exclusively in the nucleus (Gozes *et al.* 1977). Accordingly, while small in quantitative terms, nuclear protein synthesis should not be dismissed as unimportant. More information is needed on this important point.

Certain bacteria have an unusual mechanism for the synthesis of short polypeptide chains which does not involve ribosomes, tRNA or mRNA. A good example is the synthesis of the cyclic antibiotic, gramicidin S, in certain strains of *Bacillus brevis.* This synthesis is of considerable interest because it involves unusual *D*-amino acids and ornithine, and is accomplished by activation of amino acids by just two enzymes, E_I and E_{II}, at the expense of ATP, yet with the formation of distinctive enzyme-bound thioesters,

$$\text{Enzyme S} \sim \overset{\displaystyle O}{\overset{\|}{C}} - \overset{\displaystyle H}{\overset{|}{\underset{\displaystyle R}{\underset{|}{C}}}} - NH_2$$

As illustrated in Fig. 9.1, the nearest mammalian peptides in terms of size and structure are thyrotropin-releasing hormone, oxytocin and vasopressin. However, there is absolutely no evidence that these eukaryotic hormones are made by other than conventional mechanisms, based on ribosomes, tRNA and mRNA. It seems that the thioester system of peptide synthesis is an uneconomical and primitive mechanism which was abandoned during the evolution of higher organisms.

Although the amino acid sequences of eukaryotic proteins show infinite variation and diversity, sequences can be repeated frequently

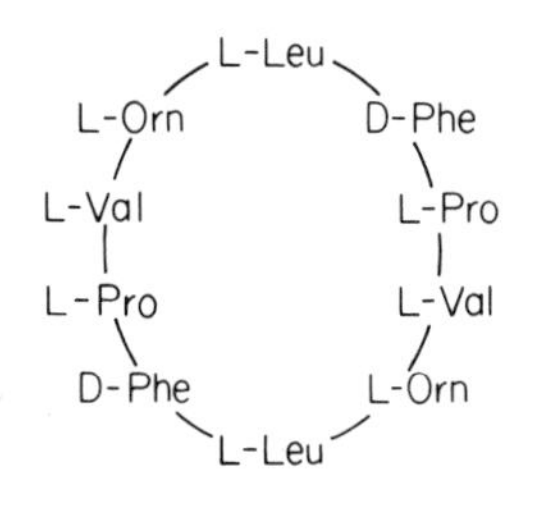

Gramicidin S

PyrGlu-His-ProNH$_2$

Thyrotropin-releasing hormone

Cys-Tyr-Ile-Gln-Asn-Cys-Pro-Leu-GlyNH$_2$
└── S ── S ──┘

Oxytocin

Cys-Tyr-Phe-Gln-Asn-Cys-Pro-Arg-GlyNH$_2$
└── S ── S ──┘

Vasopressin

Fig. 9.1. The structures of small or cyclic polypeptides. In all but gramicidin S, the amino acids are L-stereoisomers and the amino terminal residue, by convention, is to the *left*. PyrGlu is pyroglutamic acid and ProNH$_2$ is prolineamide. The N-terminal residues are underlined.

in certain proteins. Presumably, this uniformity of sequence contributes to the characteristic properties of the proteins. Examples include the sequences Gly-X-Pro(OH) and Gly-Pro(OH)-X in collagen [Pro(OH) is either proline or hydroxyproline and X is an undefined amino acid] and Ala-Ala-ThrG(Ac) in the 'anti-freeze' plasma protein of certain Antarctic fish. ThrG(Ac) is either aminodeoxygalactosyl threonine or its *N*-acetyl derivative.

The sublime accuracy of genetic translation in eukaryotes is well illustrated by recent studies on the regulatory hormones secreted by the hypothalamus and anterior pituitary. Errors in the synthesis of polypeptides in this vital area of the brain would have disastrous consequences for the well-being of all higher animals. The amino acid sequences of two of these regulatory hormones are presented in Fig. 9.2. Lutotropin-releasing hormone (formerly luteinizing hormone, LH) is devoid of biological activity until both terminal residues are modified, forming N-terminal pyroglutamic acid and C-terminal glycinamide. This finding illustrates the extreme accuracy with which peptide hormones must be synthesized. Even more striking is the case of corticotropin (formerly adrenocorticotrophic hormone, ACTH), where part of the identical amino acid sequence is the natural hormone antagonist or anti-corticotropin. A serious breakdown in the regulation of corticotropin synthesis would, in fact, be fatal.

Pyr.Gln.His.Trp.Ser.Tyr.Gly.Leu.Arg.Pro.Gly NH_2

Lutotropin-releasing hormone

Ser.Tyr.Ser.Met.Glu.His.[Phe.Arg.Trp.Gly.Lys.Pro.Val.Gly.Lys
1 5 10 15

Lys.Arg.Arg.Pro.Val.Lys.Val.Tyr.Pro.Asn.Gly.Ala.Glu.Asp.Glu
20 25 30

Ser.Ala.Glu.Ala.Phe.Pro.Leu.Glu.]Phe
35 39

Corticotropin

Fig. 9.2. The structures of two regulatory hormones isolated from human hypothalamus and anterior pituitary. Lutotropin-releasing hormone is inactive until both terminal residues are modified. In corticotropin, residues 7–38 (in brackets) constitute the natural antagonist (anti-corticotropin) to the hormone.

9.2 TRANSLATIONAL EVENTS IN EUKARYOTES

9.2.1 Ribosomes

In functional terms, no fundamental differences have been noted between the ribosomes of prokaryotes and eukaryotes. Nevertheless, mammalian ribosomes have a more complicated architecture than their bacterial counterparts. It is now generally accepted that translation in higher organisms is more sophisticated and demands the acquisition of a more complex type of ribosome during the course of evolution. The principal features which distinguish eukaryotic ribosomes are as follows. First, the ribosome monomer is larger. Second, both ribosome subunits contain more ribosome-associated proteins. Third, the large ribosome subunit contains two smaller species of rRNA rather than only one. Fourth, eukaryotic ribosomes are essentially refractory to streptomycin and other antibacterial agents, yet acutely sensitive to abrin and ricin.

Ribosome subunits

The functional unit for protein synthesis in eukaryotes is the 80*s* ribosome monomer, resulting from the union of the two ribosome subunits. With the further addition of mRNA and aminoacyl tRNAs, the ultimate structure is an active polyribosome cluster, in which all the ribosomes are synthesizing the polypeptide sequence dictated by the mRNA. The number of ribosome monomers attached to the mRNA can vary enormously, depending on the length of the amino acid sequence of the polypeptide under synthesis, and consequently the length of the genetic information present in the coding sequence of the mRNA. The α- and β-globin chains in rabbit consist of 121–146 amino acid residues and are synthesized on polyribosomes containing 5 or 6 ribosome monomers; by contrast, the heavy chain of muscle myosin is composed of 1800 amino acids, and synthesized in remarkable translational clusters containing 60–70 ribosome monomers. The structural integrity of each ribosome monomer is maintained by cations, especially polyamines and Mg^{2+} ions. The demand for the latter can be extremely high during intense periods of protein synthesis, for example during growth, regeneration and hormonal stimulation.

The two ribosome subunits have different functions, and as we shall see later, this is reflected in the different proteins and RNAs which contribute to their assembly and structure. The large (60*s*) subunit is responsible for performing many of the enzymic reactions associated with protein synthesis and additionally, binds the polyribosome clusters to the membranes of the endoplasmic reticulum. Membrane-bound polyribosomes are a very distinctive feature of eukaryotic cells, notably in those engaged in the secretion or 'export' of proteins, for example pancreas, endocrine glands and liver. By contrast, many ascites tumour cells are synthesizing proteins almost exclusively to meet internal demands imposed by a high and sustained rate of cell division; these cells are devoid of membrane-bound polyribosomes. The absence of bound polyribosomes (or 'rough' endoplasmic reticulum) is similarly a feature of rabbit reticulocytes; again, the translational product, almost entirely haemoglobin, is not for export. The small (40*s*) subunit is responsible for the specific binding of substrates, including mRNA, required for genetic translation. This simplistic division of labour between the large and small ribosome subunits may require modification as more becomes known of the fine detail of protein synthesis. There is increasing evidence that ribosomal functions may not be conducted by distinct or exclusive components, but by functional 'domains' to which many components in both subunits contribute, both functionally and structurally (Bielka 1980).

The RNAs and proteins associated with eukaryotic ribosomes are presented in Table 9.2; the structural elements of bacterial ribosomes are also included for comparison. Significant differences are readily apparent. High resolution electron microscopy has revealed a morphological similarity between prokaryotic and eukaryotic ribosomes; thus Fig. 6.5 equally illustrates the spatial relationships between the larger 40*s* and 60*s* ribosome subunits of higher organisms.

Many aspects of ribosome assembly are now understood. Ribosomal proteins are synthesized in the cytoplasm on free poly-

Table 9.2. The structural components of 80*s* ribosomes from rat liver. Certain properties of 70*s* (*E. coli*) ribosomes are included, in parentheses, for comparison.

	Large subunit	Small subunit
Physical properties of subunits		
Sedimentation coefficient, $S^0_{20,w}$	56.3 (50*s*)	36.9 (30*s*)
Diffusion coefficient, $D^0_{20,w}$ ($\times 10^7$ cm^2 s^{-1})	2.0	1.2
M_r	2.9×10^6	1.4×10^6
Buoyant density (g cm^{-3})	1.600	1.515
RNA (%)	59.4	44.9
Species of rRNA	28*s*, 5.8*s*, 5*s* (23*s*. 5*s*)	18*s* (16*s*)
Species of ribosomal proteins		
Number	49 (34)*	33 (21)*
M_r range	11 000–38 000	11 000–42 000
Average M_r	21 200	21 400
Basic amino acids (mol %)	16–30	15–28
Basic : acidic amino acids	0.73–1.83	0.68–1.43

*These figures are taken from the review by Wool (1979); other papers give other figures. All are significantly higher than in *E. coli*, but an unequivocal determination is awaited.

ribosomes ('smooth' endoplasmic reticulum), transported into the nucleolus and assembled there principally on the 45*s* rRNA precursor. Some ribosomal proteins may be added later in the cytoplasm to partially assembled ribosomes, rather than in the nucleus. It should also be remembered (see section 8.3.2) that eukaryotic 5*s* rRNA is made from its own gene, present outside the nucleolus in the nucleoplasm. Furthermore, in contrast to prokaryotic ribosomes, 80*s*-type ribosomes do not possess the ability to synthesize 'magic spot', ppGpp, and thus have quite unique mechanisms for regulating the supply of ribosomal components. Certainly in secretory cells, the synthesis of membrane components, such as phospholipids, and of polyamines, is intimately coupled to rRNA synthesis. This functional interrelationship is beautifully illustrated in the anucleolate mutant of *Xenopus laevis*. Despite the fact that rRNAs are made essentially in the nucleolus while polyamines are synthesized by purely soluble (cytoplasmic) enzymes, both functions are lost in the anucleolate mutant (Russell 1971).

Trying to elucidate the structure–function relationships within the eukaryotic ribosome has proved to be a daunting task. Certainly progress has not been as impressive as in the case of prokaryotic ribosomes; in particular, sequence analyses on both eukaryotic rRNAs and ribosomal proteins are not as advanced. Nonetheless, concerted investigations by chemical, physical and immunological means are now beginning to throw some light on the spatial arrangement and function of the components of 80*s*-type ribosomes. Treatment of ribosomal particles with chelating agents or media of gradually increasing ionic strength (especially LiCl) results in the orderly release of ribosomal proteins. It has proved possible by this means to separate most of the components of the eukaryote ribosome, which can then be analysed and characterized in detail.

Ribosomal RNAs

Three species of rRNA, types 5.8*s*, 18*s* and 28*s*, are all produced by selective cleavage of large transcript of the nucleolar ribosomal genes. Analysis of $5' \rightarrow 3'$ polarity of 45*s* rRNA precursor suggests that the 18*s* rRNA is derived from the $5'$-end, whereas 28*s* rRNA is derived from the $3'$-end. The two larger rRNAs are both (G+C) enriched, some 57% (18*s*) and 64% (28*s*). Both the 18*s* and 28*s* rRNAs have considerable secondary structure. Certainly in solution and also probably within the ribosome subunits, 60–80% of these rRNAs is composed of hairpin loops or regions of base-pairing, separated by tracts of unpaired RNA sequences. Analyses by low-angle, X-ray scattering suggests that both the 18*s* and 28*s* rRNAs are present as highly organized, tightly packaged entities at the centres of their respective ribosomal subunits. In neither, however, has much of the base sequence been determined. However, the 18*s* rRNA from all sources studied thus far has the invariant 3′-terminal sequence 5′ pGpApUpCpApUpUpA_{OH} 3′ and this may point to its possible biological roles within the ribosome monomer. Much of the base sequence of β-globin mRNA is now known and the highly conserved sequence at the 3′-end of 18*s* rRNA shows base complementary with only three regions of the mRNA, namely positions 12–14 or 34–36 in the non-coding 5′ region or positions 53–58 in the region of the initiator codon, AUG. On such indirect evidence, it has been argued that 18*s* rRNA–mRNA interactions may play some part in the initiation process in eukaryotes. More recently, Nakashima *et al.* (1980) covalently linked mRNA to 18*s* rRNA in eukaryotic 40*s* and 80*s* initiation complexes (see section 9.2.2) by photochemical reaction in the presence of an RNA cross-linking agent, psoralen. Covalent interaction between the 5′ mRNA cap and the 3′-ends of 18*s* rRNA was claimed. There is a diametrically opposed alternative in that the conserved, 3′ sequence G AUC AUU A contains two triplets, AUC and AUU, complementary to the terminator codons, UAG and UAA. Accordingly, it has been suggested that the 3′ sequence of 18*s* rRNA may be involved in the termination step of eukaryotic translation. Frankly, far more evidence is needed to strengthen belief in either of these alternative roles of 18*s* rRNA. The functional role of 28*s* rRNA is similarly uncertain.

A complete sequence analysis of yeast 5.8*s* rRNA is available, including the precise positions of methylated bases and pseudo-uridine. Despite the fact that models of its detailed secondary structure have been proposed (Erdmann 1976) its precise role in translation surprisingly remains enigmatic. In certain respects, it is analogous to the 5*s* rRNA in prokaryotes and thus may play some part in the interaction of the ribosome monomer with amino acyl tRNA, via the A-site of the large ribosome subunit, during the elongation phase of protein synthesis.

The 5*s* rRNA is distinctive in that it has a specific gene of its own and is exclusively restricted to eukaryotes. Its base sequence is known and it appears to have been highly conserved during evolution (Erdmann 1976). Models incorporating its secondary structure suggest a prolate shape, extensive base pairing and asymmetric folding. All eukaryotic 5*s* rRNAs contain an invariant sequence,

PypGpApU, which is complementary to a sequence pGpCpUpA found only in the eukaryotic initiator tRNA, $tRNA_f^{Met}$. It is now agreed (Bielka 1978) that 5*s* rRNA plays a part in the initiation step of translation in eukaryotes by interacting with met-$tRNA_f^{Met}$ in the vicinity of the P-site on the large ribosome subunit. This view is strengthened by the finding that 5*s* rRNA is intimately associated with a specific ribosomal protein, forming an rRNA–protein complex which has ATPase activity; the release of energy from ATP is a vital requirement for the initiation step of genetic translation.

Ribosomal proteins

Considerable attention has recently been devoted to the ribosomal proteins of eukaryotic ribosomes (Wool 1979). Although the total number of these proteins remains debatable, many have been characterized in terms of amino acid sequence. In addition, monospecific antibodies are now available for the pure proteins from a diversity of eukaryotic ribosomes. There is a prevalent confusion in this area of contemporary research in that a unified nomenclature has not been adopted for eukaryotic ribosomal proteins. This stricture is attributable to the use of different electrophoretic methods for their separation, each of which produces slightly different results in the hands of different investigators. The nomenclature of Welfle *et al.* (1972) will be used here; an alternative nomenclature may be found in the review by Wool (1979).

The ribosomal proteins associated with both subunits are freely accessible to a variety of reagents, including affinity labelling agents, antibodies and SH-specific blockers (N-ethylmaleimide and iodoacetamide). Collectively, these results suggest that the ribosomal proteins are situated largely on the external surfaces of each ribosomal subunit, surrounding the central core composed of 18*s* or 28*s* rRNA.

Several lines of evidence suggest that, with few exceptions, ribosomal proteins have not been conserved during evolution; for example, studies on the amino acid sequences of prokaryotic and eukaryotic ribosomal proteins reveal few homologies. Further, antibodies raised against pure proteins from *E. coli* do not cross-react with pure proteins from the ribosomes of rat liver and vice versa. However, immunological studies of this type indicated clearly that a few ribosomal proteins had been remarkably conserved throughout evolution. This has been confirmed by detailed electrophoresis of ribosomal proteins from a broad spectrum of biological sources.

Similar patterns were found in the ribosomal proteins of mammals, birds and reptiles, which are closely related in evolutionary terms. By contrast, little similarity was found in the ribosomal proteins of mammals, plants, molluscs and crustaceans which are distant in the context of evolutionary classification. A few ribosomal proteins, however, were consistently found in ribosomal particles, irrespective of their cells of origin, indicating a common ancestry of a few ribosomal proteins.

Although detailed information is not available at present, an attempt to classify the location and function of some of the ribosomal proteins in eukaryotes is presented in Table 9.3. In the nomenclature of Welfle *et al.* (1972), ribosomal proteins associated with the large (60*s*) and small (40*s*) subunits have the prefixes L and S, respectively. As indicated earlier in this chapter, the evidence presented in Table 9.3 indicates that ribosomal proteins on both subunits are involved in the performance of a particular function in the eukaryotic ribosome; this interdependence between subunits suggests the importance of composite, functional domains in the ribosome, rather than attributing functions exclusively to a particular ribosomal protein.

Considerable ingenuity and skill have been needed by investigators in providing the information in Table 9.3; its limitations are due to the inherent complexity of the eukaryotic ribosomes rather than to a lack of effort or imagination. Specific antibodies for purified ribosomal proteins have been particularly useful; for example,

Table 9.3. The provisional location and function of some of the proteins associated with rat liver ribosomes.

Property	Large subunit	Small subunit
Organization of the A-site	L2, L24, L27, L29	S13
Organization of the P-site	L7	S2, S9, S15, S18, S20, S21
ATPase activity	L3	
Peptidyltransferase activity	L2, L22, L25, L27, L29, L30, L36	
Binding of initiation factor* (eIF2)		S2, S9, S15
Binding of elongation factors*		
EF1	L19, L26, L38	S8, S12, S13
EF2	L1, L31	S13, S17
Binding of synthetic mRNA†		S12, S18, S19, S24
Site of phosphorylation		S9

*By accepted convention, eukaryotic initiation factors have the prefix, e, whereas under present practice, the elongation factors do not.
†Principally poly(U).

antibodies raised against proteins S2, S9 and S15 blocked the binding of the ternary initiation complex (eIF2.Met-$tRNA_i^{Met}$.GTP) to the small ribosomal subunit, indicating the crucial involvement of these proteins in initiation and their location near the P site of the functional ribosome complex. Furthermore, antibodies are now being used to investigate the precise location of all ribosomal proteins; this work is in a preliminary stage, but already the close proximity of groups of proteins in the active ribosome has been established; e.g. S8, S12 and S13, with L19 and L26. This observation fits well with their functional interrelationship, suggested in Table 9.3, in the binding of the elongation factor, EF1.

Biochemical 'dissection' of the ribosome has also provided useful information. Based on the sequential extraction of ribosomal proteins with increasing concentrations of LiCl, it has been possible to correlate the composition and function of ribosomes depleted with respect to particular proteins. By such means, it was established that proteins L22, L25 and L30 are needed for elongation and, more specifically, the peptidyltransferase reaction. Puromycin is a structural analogue of tRNA and is an essentially irreversible inhibitor of translation by promoting the formation of the anomalous translational intermediate, *N*-peptidylpuromycin (see section 6.3.1, and Fig. 1.9). It has proved possible to synthesize an analogue of N-peptidylpuromycin, N-bromoacetylpuromycin, which interacts covalently with a restricted number of ribosomal proteins; in modern parlance, N-bromoacetylpuromycin is a powerful 'affinity label'. It interacts only with ribosomal proteins L22, L27, L29 and L36 in eukaryotic ribosomes. Additional studies with N-bromoacetylpuromycin indicate that this group of proteins are located near the A site on the large subunit, and in particular, near the centre of peptidyltransferase activity.

Although current information is scarce, the enzymic functions of certain ribosomal proteins warrant some comment. When the large subunit is extracted with the chelating agent, ethylenediaminetetraacetic acid (EDTA), a ribonucleoprotein complex of sedimentation coefficient 7*s* is released. This complex is composed only of 5*s* rRNA and a single ribosomal protein, L3. Together, but not separately, these ribosomal components have ATPase activity, and hence they may play an important part in providing the energy required for the initiation step of protein synthesis. The eukaryotic ribosome possesses considerable protein phosphokinase activity, and when labelled *in vivo* with [^{32}P]phosphate, only one ribosomal protein, S9, is heavily labelled with radioactivity. S9, M_r 31 000, contains 15 serine residues and is one of the few ribosomal proteins conserved throughout evolution. Its high serine content explains its penchant for phosphorylation, as all the associated radioactivity is recovered as serine [^{32}P]phosphate. The extent of S9 phosphorylation in rat liver is influenced by a wide variety of agents. Phosphorylation of S9 is enhanced by glucagon, thyroxine, adrenalin, dibutyryl cyclic AMP and certain viruses yet reduced by insulin and the methionine analogue, ethionine. Despite extensive studies, the precise necessity for the phosphorylation of ribosomal protein S9 remains an enigma.

9.2.2 Initiation

Activation of amino acids

Before the initiation step of protein synthesis can occur, it is necessary for all the necessary amino acid precursors to be both activated and attached to the 3′ terminal sequence–CCA_{OH}3′ of their cognate tRNA, according to the general equation:

$$\text{amino acid} + \text{ATP} + \text{tRNA} \rightarrow \text{aminoacyl-tRNA} + \text{AMP} + \text{PP}_i.$$

These reactions are catalysed by aminoacyl-tRNA synthetases, a vitally important group of enzymes which have remarkable precision in recognizing their substrate, the tRNAs for a given amino acid. It is essential for the accuracy of genetic translation that the correct amino acid is linked to the correct tRNA. If a surrogate amino acid is stably linked to a tRNA, then the final amino acid sequence of the protein under synthesis will be incorrect.

Despite their functional uniformity in all forms of life, the aminoacyl-tRNA synthetases in higher animals are more complex than their counterparts in prokaryotic organisms. There is but a single aminoacyl-tRNA synthetase for each of the 'magic twenty' amino acids in prokaryotes, but there are distinctive amino acid activating enzymes in the cytoplasm, mitochondria and chloroplasts of higher organisms, and in certain cases, even multiple forms of a given aminoacyl-tRNA synthetase are present. As in bacteria, the aminoacyl-tRNA synthetases show a remarkable variation in terms of their molecular weight and subunit structure. Eukaryotes contain four distinctive types: α (monomer), α_2 (dimer) and α_4 or $\alpha_2\beta_2$ (tetramers). The physical properties of a selected number of aminoacyl-tRNA synthetases are presented in Table 9.4; the enzymes were chosen to illustrate their variety with respect to molecular weight, quaternary (subunit) structure and multiplicity. A

Table 9.4. The physicochemical properties of certain eukaryotic aminoacyl-tRNA synthetases. The enzymes from mitochondria and chloroplasts are not included; all those cited are cytoplasmic enzymes.

Synthetase	Source	M_r	Subunits	Type	Multiplicity (number of enzyme moieties)
Alanyl	Yeast	128 000	None	α	1
Aspartyl	Yeast	106 000	None	α	Unknown
Isoleucyl	Yeast	124 000	None	α	1
Lysyl	Yeast	138 000	2	α_2	1
	Reticulocytes	122 000	2	α_2	1
Methionyl	Wheat germ	70 000	None	α	2
		105 000	None	α	
Phenylalanyl	Liver	287 000	4	$\alpha_2\beta_2$	Unknown
	Yeast	220 000	4	$\alpha_2\beta_2$	3
		262 000	4	$\alpha_2\beta_2$	
		286 000	4	$\alpha_2\beta_2$	
Tyrosyl	Yeast	80 000	2	α_2	2
		116 000	4	α_4	
	Liver	124 000	2	$\alpha\beta$*	
	Soybean	122 000	2	α_2	

*The quaternary structure, $\alpha\beta$, is extremely unusual and does not fit into the usual four groups of enzyme subunits.

very distinctive feature of eukaryotic aminoacyl-tRNA synthetases is that they are frequently located within complexes of very high molecular weight; these complexes contain other important activities, including elongation factors and enzymes for the modification of tRNA bases. Many distant parts of a tRNA are implicated in the interaction with its aminoacyl-tRNA synthetase. By careful manipulation of tRNA structure (by chemical or enzymic means), the crucial regions of base sequences needed for recognition by its

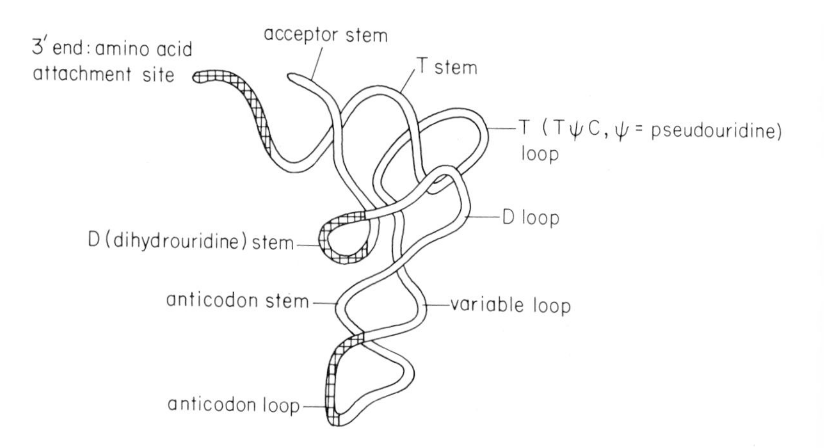

Fig. 9.3. The likely regions of eukaryotic tRNA involved in recognition by its cognate aminoacyl-tRNA synthetase. The important regions are shown by hatching. (Modified from Schimmel & Söll 1979; see also Fig. 3.8.)

aminoacyl-tRNA synthetase have been determined. As shown in Fig. 9.3, these include the amino acid attachment site (3′-end of tRNA), the D-stem region and part of the anticodon loop (see also Fig. 3.8).

The marked contribution to the overall fidelity of genetic translation by the aminoacyl-tRNA synthetases makes them the subject of extreme research interest. For further details on these enzymes, the excellent review by Schimmel and Söll (1979) is recommended.

Plague toxin

Plague, the historical scourge of man is caused by the bacterium, *Yersinia pestis*. It is now clear that this pathogenic bacterium synthesizes two forms of plague toxin, types A, M_r 240 000, and B, M_r 120 000. Both types contain the same polypeptide subunit, M_r 24 000 and type A is thus a decamer and type B a pentamer. These toxins are very selective inhibitors of protein synthesis in eukaryotes, because they are powerful mitochondrial poisons. They bind to a structural protein in the inner mitochondrial membrane, believed to be within Complex I of the electron transport chain (Fig. 9.4), and the regeneration of ATP via the oxidation of NADH is completely switched off. The plague toxins thus inactivate a similar stage of the electron transport chain as amytal, a powerful barbiturate, and rotenone, a poison from the root of *Derris elliptica* and other legumes. Plague toxins directly inhibit protein synthesis in mitochondria and indirectly prevent cytoplasmic protein synthesis as well; plague-infected cells cannot meet the extremely high demand for ATP needed for amino acid activation in the cytoplasm.

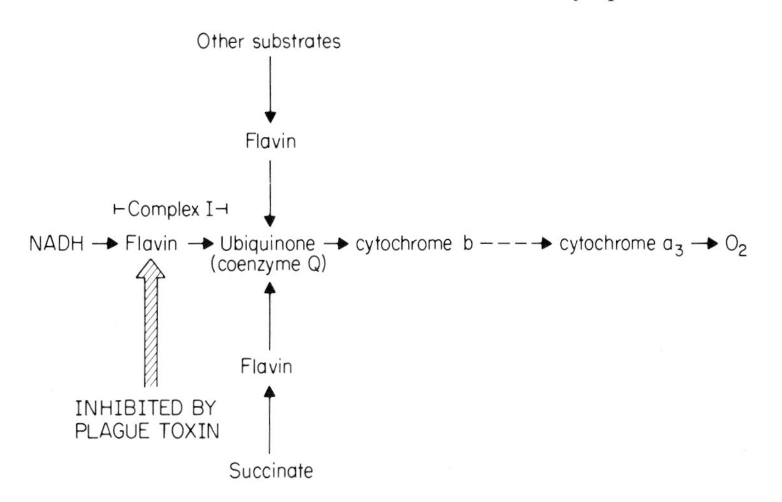

Fig. 9.4. The site of inhibition of the mitochondrial electron transport chain by the toxin from *Yersinia pestis*. The hatched arrow indicates the location of several intervening steps in the chain.

The initiation step of eukaryotic translation

The objectives of this crucial step of protein synthesis are twofold. First, to form a complex between ribosomal subunits and mRNA in which one ribosome monomer must be bound in close proximity to the initiator codon, AUG, in the mRNA. Second, that the charged initiator tRNA, Met-$tRNA_f^{Met}$, is positioned in the P site of the large ribosomal subunit, with its anticodon loop base-paired with the initiator codon. This leaves the ribosomal A site empty, and ready to receive the next species of aminoacyl-tRNA, as specified by the next codon in the mRNA in the 3′ direction.

The process of initiation in eukaryotes is more complicated than in prokaryotic organisms (for a review see Hunt 1980). First of all, far more initiation factors are involved; possibly as many as nine, as against only three in bacteria. Second, the mRNAs of eukaryotes have unique structural features, including the cap of 7-methylguanosine at the 5′-end, and frequently, but not always, poly (A) at the 3′-end; the importance of these distinctive ends is only partially understood at present. It should also be remembered that the initiating amino acid in higher organisms is methionine itself, rather than N-formylmethionine; as in bacteria, the methionine is attached to the initiator tRNA, $tRNA_i^{Met}$.

The protein factors necessary for the initiation step of translation in eukaryotes are an integral part of the ribosomal subunits, and may be readily extracted in 0.5M KCl. Many eukaryotic initiation factors have now been purified to homogeneity from such sources as rabbit

Table 9.5. The properties and functions of the initiation factors from rabbit reticulocytes.

Factor	M_r	Subunits	Function
eIF-1	15 000	None	Unclear; perhaps stabilizes the 40*s* complex
eIF-2	86 000	48 000, 38 000	Binding of both GTP and Met-$tRNA_f^{Met}$
eIF-2A	65 000	None	Binding of Met-$tRNA_f^{Met}$ to 40*s* subunit
eIF-3	750 000*	Many†	Binding of mRNA
eIF-4A	50 000	None	Binding of mRNA
eIF-4B	80 000	None	Recognition of 5′ cap on mRNA
eIF-4C	19 000	None	Stabilizes 40*s* complex
eIF-4D	17 000	None	Assembly of subunits and, perhaps, elongation
eIF-5	150 000	None	Formation of 80*s* complex and breakdown of GTP; it is a GTPase

* An approximation only.
† May be as high as nine.

reticulocytes, rat liver and wheat germ. For many years, there was considerable confusion in the literature caused by rival nomenclatures for eukaryotic initiation factors. A standard nomenclature has now been universally accepted, based on the original recommendations of Anderson *et al.* (1977); in this scheme, eukaryotic factors are prefixed by e, to distinguish them from bacterial initiation factors, from which they differ both in terms of structure and function.

The rabbit reticulocyte system has provided the most information on the initiation stage of genetic translation in eukaryotes. The properties and functions of reticulocyte initiation factors are presented in Table 9.5.

It is far easier to envisage the process of eukaryotic initiation by looking at a 'flow sheet' of the steps involved. Such a diagram, based on the work of Benne and Hershey (1978), is presented in Fig. 9.5. Each stage of the process will now be briefly described.

Stage I. Two distinct complexes are formed at this first stage; one is a binary complex between the small 40*s* ribosomal subunit and eIF-3 and the other is a ternary complex containing the initiator tRNA, GTP and eIF-2.

Stage II. Here, the binary and ternary complexes are brought together in the presence of eIF-4C, which somehow stabilizes this temporary union; eIF-2 and eIF-3 remain integral parts of this intermediary complex.

Stage III. This is an important stage where mRNA is added to the complex, forming the 40*s* initiation complex. Energy for this step is

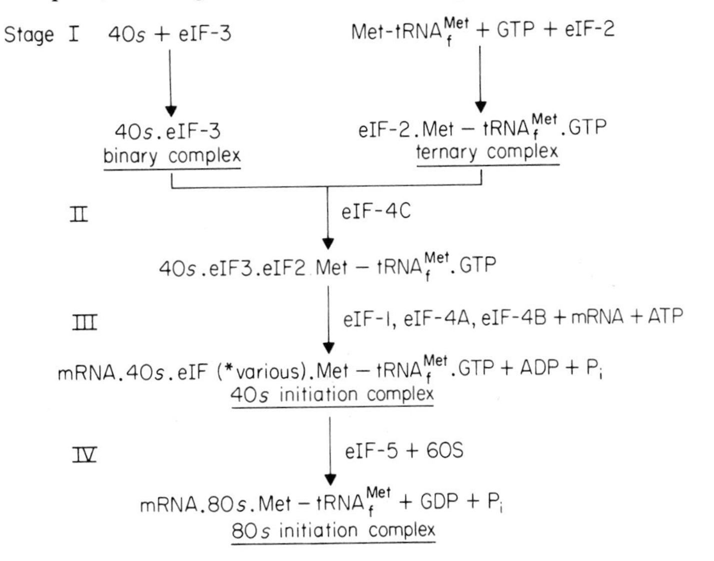

Fig. 9.5. A diagram of the formation of the active initiation complex in rabbit reticulocytes. (*There is some uncertainty concerning the type of initiation factors in the complex.)

provided by the breakdown of ATP. This stage requires eIF-1, eIF-4A and eIF-4B, but there is some uncertainty which initiation factors are a consistent feature of the 40*s* initiation complex.

Kozak (1980a) modified the secondary structure of reovirus mRNA either by reaction with bisulphite or by incorporating IMP in place of AMP; both procedures led to irreversible unfolding of the mRNA. Her studies indicate that an ordered secondary structure is not required for mRNA–40*s* ribosome subunit interactions, whereas the fidelity and efficiency of translation as a whole do depend on the precise folding, or secondary structure, of mRNA.

Stage IV. Finally, the large ribosome subunit is now added to the 40*s* initiation complex in the presence of eIF-5, forming the 80*s* initiation complex. The energy for this step is generated from bound GTP in the 40*s* complex, by the GTPase associated with eIF-5.

Kozak (1980b) has proposed a 'scanning model' for the initiation of translation in eukaryotes. In essence, this model suggests that a 40*s* ribosomal subunit binds at the 5′ terminus of an mRNA and migrates along the mRNA, searching for and stopping at the first, initiating AUG codon. Only at this time, does a 60*s* ribosomal subunit join the initiation complex and can protein synthesis begin. As described in more detail later, and particularly with respect to the function of the 5′ leader sequence of eukaryotic mRNA, there is now overwhelming experimental support for the scanning model.

With the formation of the 80*s* complex, the initiation step of eukaryotic translation is complete. A ribosome monomer is located in the region of the initiating codon, AUG, with the initiator tRNA, Met-tRNA$_i^{Met}$, precisely in the A-site of the large subunit, and temporarily anchored by anticodon–codon interactions (base-pairing).

The sequence of initiation reactions just described applies strictly to the processes necessary for the inclusion of natural mRNAs into a stable, active initiation complex, sedimentation coefficient 80*s*. At high concentrations of Mg^{2+}, eukaryotic ribosomes can bid Met-tRNA$_f^{Met}$ in the absence of initiation factors, provided that the synthetic mRNA, poly(AUG), is present. This reaction has little biological significance.

Inhibitors of eukaryotic initiation

Aurintricarboxylic acid is a powerful inhibitor of the formation of the ternary complex between eIF-2, GTP and Met-tRNA$_f^{Met}$; this is not specific to eukaryotes, as bacterial initiation factors are similarly inhibited. The antibiotic, pactamycin, is a selective inhibitor of eukaryotic initiation; it prevents the association of the eukaryotic (60*s*) ribosomal subunit with the 40*s* initiation complex.

The unique function and structure of tRNA$_f^{Met}$

The charged initiator, Met-tRNA$_f^{Met}$, is unique in being the only species of tRNA to place an activated amino acid directly into the ribosomal P site; all other charged tRNAs bring their associated amino acid into the A site. While the precise mechanism still remains to be fully established, elegant work by Wrede *et al.* (1979) has shown that initiator tRNA$_f^{Met}$ has a unique structure adjacent to its anticodon loop. tRNA$_f^{Met}$ is distinctive in having a short conformation of G:C base pairs in the anticodon stem (Fig. 9.6); this is not found in tRNA^{Met} or any other tRNA. It seems unlikely that this distinctive structure is a matter of serendipity and this short G-C sequence, therefore, appears vital for the unique biological role of tRNA$_f^{Met}$ in initiation.

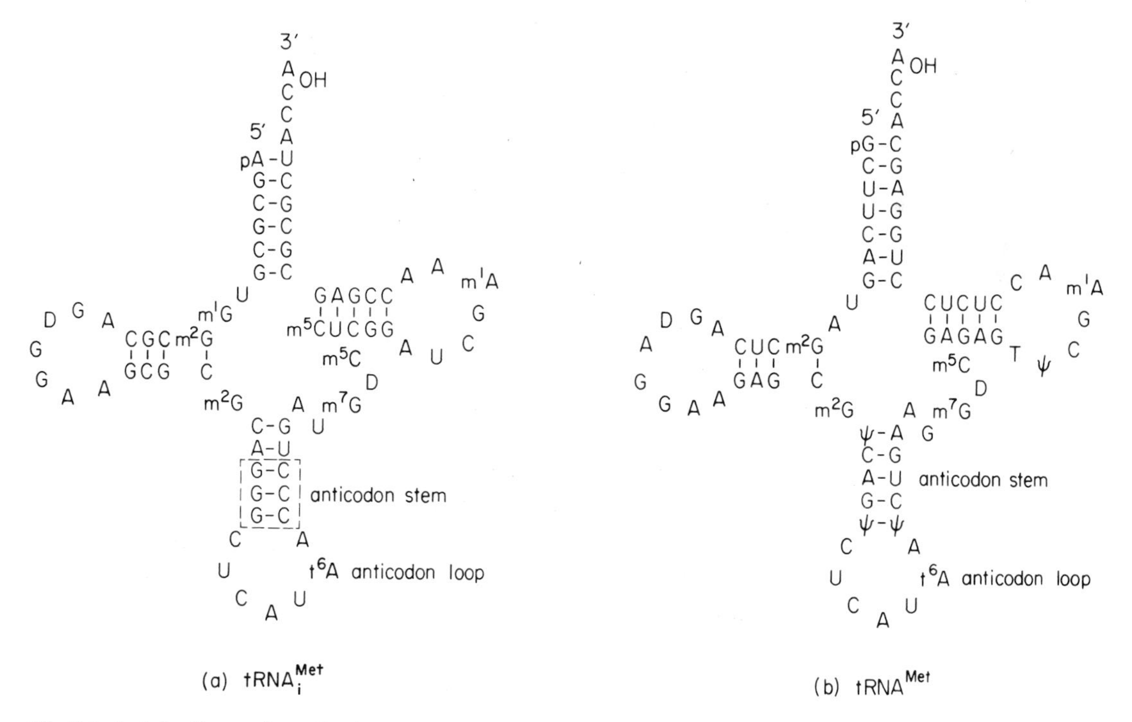

Fig. 9.6. A distinctive conformation in the anticodon stem of initiator, tRNA$_f^{Met}$. The examples used here are from yeast. Base pairing is indicated, with the unique structure enclosed within the box.

The 5′ terminal or leader sequences in eukaryotic mRNAs

Nearly all eukaryotic and viral mRNAs examined to date contain a sequence of bases at the 5′ terminus and preceding the

m⁷GpppAmCACUUGCUUUUGACACAACUGUGUUAUCUUGCAAUCCCCCAAAACAGACAGA.AUG

rabbit β-globin

m⁷GpppGmCUAAAGUCACGCCUGUCGUCGUCACU.AUG

reovirus mRNA (strain S54)

m⁷GpppGUAUUAAUA.AUG

brome mosaic virus (strain 4)

Fig. 9.7. The 5′ terminal or leader sequences of certain eukaryotic mRNAs. The initiating codon, AUG, has been lined up to emphasize the diversity of the leader sequences. In all, written 5′→3′.

initiating codon, AUG. As indicated in Fig. 9.7, there is no uniformity in these 5′ terminal or leader sequences with respect to either length or base sequence. Most eukaryotic mRNAs also have a distinctive cap of 7-methylguanosine. Since these leader sequences have no coding function and, therefore, are not translated into protein, their function remained obscure for some time. There is considerable evidence from studies by Kozak and Shatkin (1978b) and Filipowicz and Haenni (1979) that the ribosome 40*s* subunit binds in the first instance to the leader sequence, not directly to the region immediately surrounding the initiator codon. Once attached, the 40*s* ribosome subunit then moves in the 3′ direction to the initiating codon, AUG; at this time, the 60*s* subunit joins up with mRNA-bound 40*s* unit and translation is initiated. The molecular basis for recognition between the 40*s* subunit and the leader sequence is not yet known precisely, but there is now considerable support for the migration of the 40*s* ribosomal subunit 5′→3′ along eukaryotic mRNAs, as embodied in the 'scanning model' for protein initiation (Kozak 1980b).

The role of the eukaryotic mRNA cap at 5′ terminus

There is considerable evidence to support the contention that the 7-methylguanosine cap at the 5′-end of many eukaryotic mRNAs also plays a vital part in the attachment of the 40*s* ribosome subunit. In many systems for protein synthesis *in vitro*, derived from wheat germ, reticulocytes, L-cells and *Artemia salina,* removal of the 5′ cap of exogenous mRNA by pyrophosphatases prevents translation; in particular, the uncapped mRNA fails to bind to the 40*s* subunit. In addition, analogues of the m⁷GpppPu cap, such as m⁷pG, m⁷GTP and m⁷GpppXm, virtually stop the translation of many intact eukaryotic mRNAs in cell-free systems. Shafritz *et al.* (1976) have shown that m^7pG specifically inhibits the interaction of capped mRNAs to eIF-4B, indicating the importance of this initiation factor in the binding of the 40*s* subunit. What is particularly interesting is that short fragments derived from capped mRNAs on incubation with T1 RNAase, such as $m^7GpppGmpUp$ or $m^7GpppGp$, do not bind to wheat germ ribosomes. Clearly, the cap region alone is unable to promote the formation of a stable 40*s* initiation complex; other regions of the 5′ leader sequence are also implicated, as mentioned above. A cap-binding protein (CBP), M_r 24000, has now been identified in many eukaryotic cells which specifically aids the formation of stable initiation complexes for protein synthesis (Sonenberg *et al.* 1979); in particular, CBP enhances the interactions between the two initiation factors, eIF-3 and eIF-4B, and mRNA.

Certain observations tend to demote the importance of the 5′ terminal cap in eukaryotic translation. First, Bergmann and Lodish (1979) reported that uncapped reovirus and vesicular stomatitis virus mRNAs could be translated in cell-free preparations from wheat germ and reticulocytes, admittedly with a lower efficiency than their intact (capped) counterparts, provided that the ionic conditions of the incubations were grossly modified. The trick in these experiments was to conduct translation at very low concentrations of K^+ ions. Second, certain viral mRNAs do not have a 5′ cap at all, yet this does not preclude their faithful and efficient translation. The lack of a cap was first noted in poliovirus mRNA, a finding later confirmed in the mRNAs of encephalomyocarditis virus and satellite tobacco necrosis virus. In summary, the 5′ cap of 7-methylguanosine is required for the accurate translation of most, but certainly not all, eukaryotic mRNAs.

9.2.3 Elongation and termination

The elongation cycle in eukaryotic translation

The elongation phase of protein synthesis in higher organisms is similar in principle to that in prokaryotes, but subtle differences exist in the protein factors involved (for a review see Clark 1980). Eukaryotic translation requires two elongation factors, EF-1 and EF-2, which correspond in functional terms to the bacterial factors, EF-T and EF-G. It should be noted that eukaryotic elongation factors do not conventionally have the prefix, e, but present nomenclature nevertheless distinguishes clearly between eukaryotic and prokaryotic elongation factors. Both EF-1 and EF-2 have been purified to homogeneity from a variety of sources. In contrast to initiation factors, EF-1 and EF-2 are present in the soluble hyaloplasm or cytosol of eukaryotic cells, rather than being integral components of the 80*s* ribosome monomer. Mitochondria and chloroplasts contain distinctive elongation factors, rather than EF-1 and EF-2.

The elongation factor, EF-1

As indicated in Fig. 9.8, eukaryotic translation proceeds in stages. In the first stage, EF-1 is required to bring the appropriate aminoacyl-tRNA into the A site on the large (60*s*) ribosome subunit, as dictated by the codon of the mRNA at the A site. Conflicting values of the molecular weight of purified EF-1 have been published, but it seems that a fundamental unit, M_r approximately 50 000, is present in all eukaryotes, but with a tendency in some cells to aggregate into dimers and tetramers. The binding of EF-1 to the 80*s* ribosome requires both energy in the form of GTP and certain proteins associated with the large subunit (see Table 9.3). There is also unequivocal evidence that the two discrete complexes, EF-1.GTP and EF-1.GTP.aminoacyl-tRNA, are formed during eukaryotic elongation. During the entry of the incoming aminoacyl-tRNA into the A site, a binary complex of low energy is formed, EF-1.GDP, which must be regenerated at the expense of GTP hydrolysis by insoluble, ribosome-associated GTPases. EF-1 is continuously recycled throughout elongation (Fig. 9.8(a)), a process with a very high utilization of GTP.

Peptidyl transferase reaction

This is a complex reaction, the detailed steps of which are unclear (Fig. 9.8(b)). Certainly six or seven proteins associated with the large subunit (Table 9.3) and perhaps eIF-4D (Table 9.5) together constitute the machinery for the peptidyl transfer reaction. In this stage of elongation, a new peptide bond is formed by linkage between the amino group of the aminoacyl-tRNA in the A site with the carboxyl group of the aminoacyl-tRNA in the P site. This leaves an uncharged tRNA in the P site. The peptidyl transferase reaction is unique among the many steps of translation in not requiring the breakdown of either ATP or GTP. The new peptide bond is made at the expense of the bond energy of the ester linkage in the aminoacyl-tRNA originally located in the P site at the beginning of a particular round of elongation.

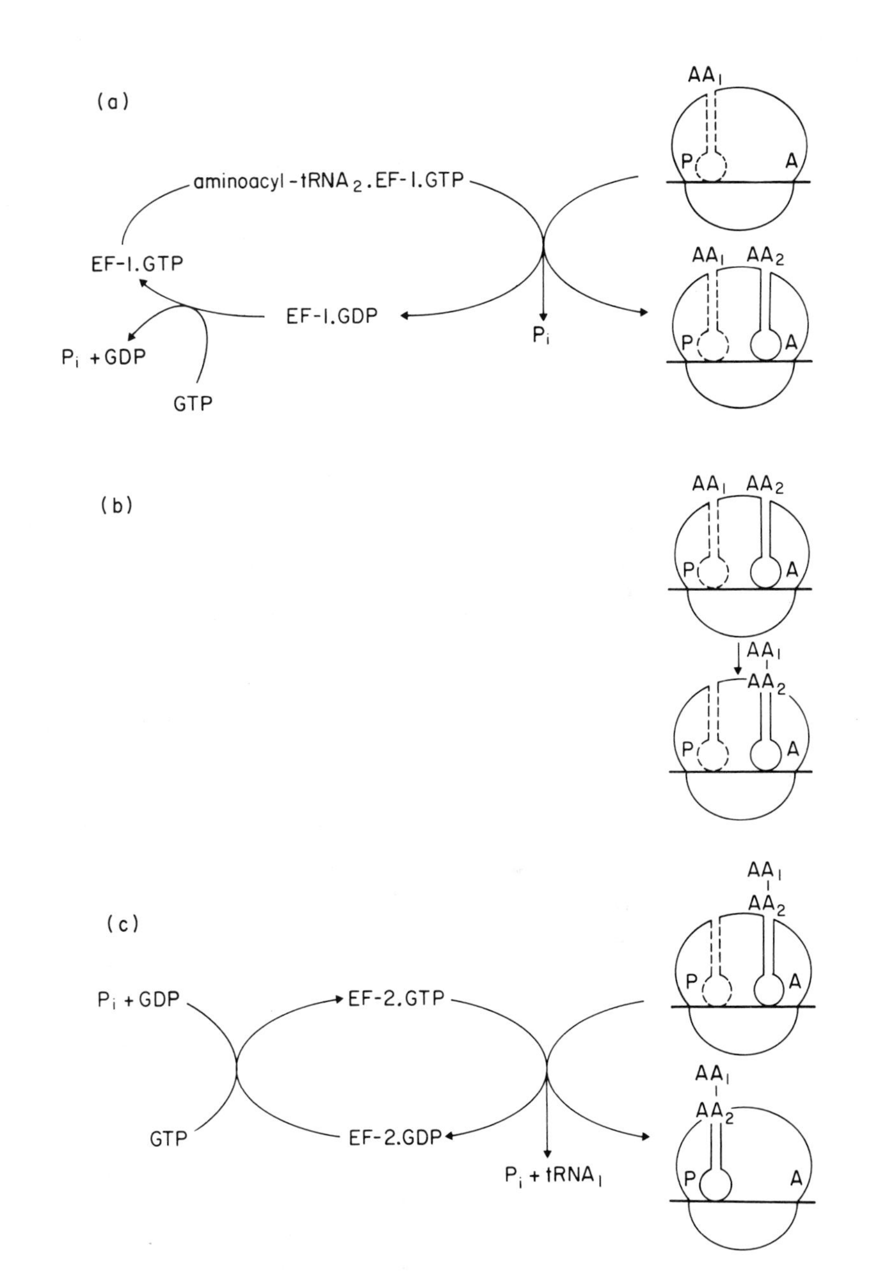

Fig. 9.8. Stages of the elongation phase of protein synthesis in eukaryotes. Elongation proceeds in three stages. Stage (a) involves the binding of incoming aminoacyl-tRNA and needs GTP and a specific soluble factor, EF-1. A low energy binary complex (EF-1.GDP) is converted into a high energy binary complex (EF-1.GTP) which forms a ternary complex with the aminoacyl-tRNA. Stage (b) involves the peptidyl transferase reaction; this complex process needs specific, ribosome-bound components, but full details are not yet available. Stage (c) involves the translocation reaction and needs GTP and a specific soluble factor, EF-2. A low energy binary complex (EF-2.GDP) and a high energy binary complex (EF-2.GTP) are involved.

Elongation factor, EF-2

The last stage of elongation involves a translocase reaction, in which a soluble factor, EF-2, promotes the expulsion of the uncharged tRNA from the ribosomal P site. The growing peptide chain is transferred into the vacant P site, leaving the A site empty for the next incoming aminoacyl-tRNA (Fig. 9.8(c)). During this translocation, the ribosome moves along the mRNA in the $5' \rightarrow 3'$ direction, resulting in the location of a new mRNA codon in the A site. EF-2 isolated from a variety of sources has a molecular weight in the range 90 000–110 000 and a remarkably high number of essential −SH groups, no less than 19 CySH mol^{-1} polypeptide chain. The translocase reaction has a high demand for GTP, necessary for the formation of a high-energy, binary complex, EF-2.GTP. The binding of this complex to the 80*s* ribosome also requires specific proteins associated with the large subunit (Table 9.3). During the overall performance of translocation, the binary complex is degraded and EF-2 is continuously recycled at the expense of more GTP.

These three stages occur for every round of elongation in which a further amino acid is added to the C-terminal end of the growing polypeptide chain. Elongation cycles continue until a termination codon in the mRNA (UAA, UAG or UGA) enters the ribosomal A site; at this point, elongation ceases.

Inhibition of eukaryotic translation

Several inhibitors of the translocase reaction in eukaryotes are known. Fusidic acid inhibits the translocase reaction in both prokaryotes and eukaryotes; in the latter, it forms a stable complex with EF-2.GDP and thus prevents the necessary recycling of EF-2.

However, there are two extremely interesting types of inhibitors in nature which specifically inhibit the translocase reaction in eukaryotes. The first is the toxin produced in *Corynebacterium diphtheriae* by a phage, β, associated with this pathogenic bacterium. The structure of diphtheria toxin is given in Fig. 9.9(a). It is a single polypeptide chain (M_r 62 000) containing two disulphide bonds and three structural domains, each of which has different functions in the pathogenic process. One disulphide bond subtends an exposed loop, rich in arginine residues, which demarcates the N-terminal portion of the molecule. The second disulphide bond demarcates a hydrophilic, C-terminal region, from an internal, hydrophobic region. The hydrophilic, C-terminal region is responsible for the attachment of

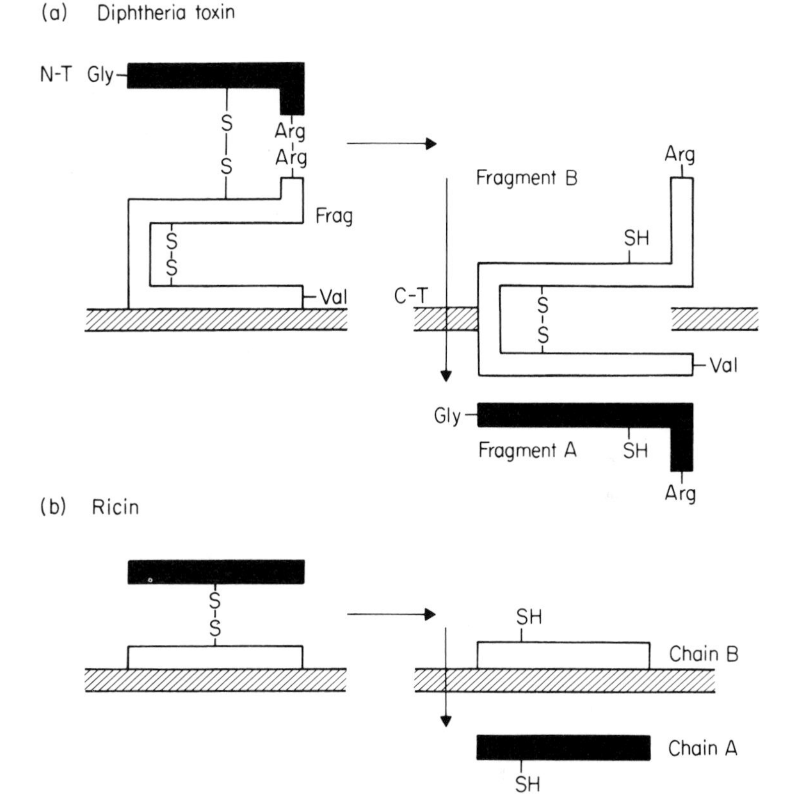

Fig. 9.9. The structures and entry mechanisms of powerful inhibitors of eukaryotic elongation. N-T is amino-terminal, C-T is carboxyl-terminal residue. The inhibitory portions of the molecules are shown as solid areas.

the toxin to as yet unknown constituents of the plasma membrane. The hydrophobic, internal region then helps the toxin to penetrate deeper into the membrane of the cell under attack. Proteolytic cleavage of the basic sequence, together with reduction of the adjacent disulphide bond, by enzymes in the plasma membrane, releases the N-terminal region (fragment A; M_r 21 000) for entry by endocytosis. The residual portion of the toxin, containing a disulphide bond more resistant to reduction, remains within the plasma membrane (fragment B; M_r 40 000), and somehow helps the uptake of fragment A. Fragment A is a potent enzyme which catalyses the transfer of the ADP-ribose moiety from NAD^+ to EF-2 and completely inactivates the elongation factor:

$$\underset{\text{(active)}}{NAD^+ + \text{EF-2}} \rightarrow \underset{\text{(inactive)}}{\text{ADP-ribose.EF-2}} + \text{nicotinamide} + H^+$$

The ADP-ribose is transferred to a single basic amino acid in EF-2 whose constitution remains to be established. Diphtheria toxin does not inactivate the elongation factors in either mitochondria or bacteria, clearly indicating that they have different structures to that of EF-2.

The second type of inhibitors are the powerful toxins found in the seeds of certain plants. The most widely studied is ricin, present in the seeds of the castor bean, *Ricinus communis*. In certain animals, the lethal dose of ricin is as low as 0.1 $\mu g\ kg^{-1}$ body weight. As shown in Fig. 9.9(b), ricin is made up of two subunits, chain A M_r 32 000 and chain B M_r 34 000, joined by a single disulphide bridge. Chains A and B have distinct functions. Chain B is necessary for the attachment of ricin to plasma membranes, after which reduction of the disulphide bridge releases chain A and enables it to enter the cell. Chain A alone cannot inhibit intact cells and essentially needs the presence of chain B in all but cell-free systems. Chain A interacts with a ribosomal protein on the eukaryotic-specific (60*s*) subunit, and thereby inactivates the GTPase activity of the ribosome. As a consequence, the attachment of EF-2 to the ribosome is totally prevented and the process of elongation is brought to a halt.

Termination factor, RF

In bacteria, two or even three soluble factors seem necessary for the termination of translation and release of the polypeptide chain. In eukaryotes, only one releasing factor, RF, seems necessary (for a review see Caskey 1980). RF has been purified to homogeneity from several sources, and has a high molecular weight, M_r 250 000 or even greater, which may reflect the tendency of a protomer, M_r 65 000, to aggregate. The release of peptidyl-tRNA from the 80*s* ribosome is promoted by RF provided that both GTP and tetranucleotides such as UAAA or UAGA are present; the latter sequences contain the termination codons, present in the ribosome A-site during the release phase of protein synthesis. Certainly, the binding of RF to the 80*s* ribosomes needs GTP and the subsequent activation of ribosome-associated GTPase activity is vital for release of the completed polypeptide chain. There are also indications that peptidyl transferase activity plays a part in the release process in eukaryotes.

9.3 MECHANISMS OF TRANSLATIONAL CONTROL

9.3.1 Introduction

In all organisms, irrespective of their complexity or evolutionary status, the major control resides at the level of transcription and RNA processing rather than translation. The presence of the nucleus in eukaryotes prevents the direct coupling of transcriptional and translational events, and consequently, higher organisms tend to have a lower rate of protein synthesis than bacteria, for example. Nevertheless, the very complexity of the structure, organization and function of eukaryotes provides opportunities for translational control which have no strict counterpart in prokaryotes. The cell cycle, sensitivity to hormones, sexual reproduction, differentiation and development all permit a certain measure of translational control. In addition, the unique structure and properties of eukaryotic mRNAs also provide the foundations for subtle means of translational regulation. These various aspects of control will now be briefly reviewed.

Perhaps the key to translational controls is the long half-life of eukaryotic mRNA. There is only one copy of the fibroin gene in the genome of *Bombyx mori* (silk moth), so that the vast secretion of fibroin from the silk gland can only be satisfactorily explained if each fibroin mRNA is translated thousands of times. Furthermore, when exogenous mRNA is injected into the oocyte of *Xenopus laevis,* it is translated up to 100–200 times per hour for several days. It is clear, therefore, that eukaryotic mRNAs are very stable and may be translated repeatedly.

It is generally assumed that elongation and termination of all eukaryotic polypeptide chains occur at the same rate. Certainly this is true for the translation of endogenous α- and β-globin mRNAs in rabbit reticulocytes and for exogenous mRNAs added to reticulocyte lysates depleted in globin mRNAs (for a review see Lodish 1976). Therefore, if translational control mechanisms exist in eukaryotes, they are most likely to be expressed at the initiation stage of protein synthesis.

Eukaryotic mRNAs, almost without exception, are monocistronic (Jacobson & Baltimore 1968), and contain a single functional site for the initiation of protein synthesis, the initiating codon, AUG, being a variable distance from the 5′-end (Fig. 9.7). In addition, eukaryotic mRNAs code for a single polypeptide chain, even if this is subsequently cleaved by post-translational processing to yield a number of different proteins, as is the case for vitellogenin mRNA and the RNAs from polio and encephalomyocarditis viruses. Studies on certain plant and animal viruses, including SV40, tobacco mosaic, Semliki forest and Sindbis viruses, indicate the presence of more than one initiating codon in these viral RNAs. Taking Sindbis virus as an example, the complete RNA of the virion (virus particle), sedimentation coefficient 42*s*, is recoverable from the polyribosomes of host cells during the early stages of viral infection. When translated in cell-free systems, the 42*s* RNA directs the synthesis of non-structural viral proteins, particularly the enzymes needed for replication of the RNA virus genes. Later in infection, the predominant form of Sindbis virus RNA is only 26*s*, this being derived from the 3′-end of the virion 42*s* RNA. When translated in cell-free systems, the shorter 26*s* RNA directs the synthesis of three structural proteins of the virus almost exclusively. This change is consistent with the need of the Sindbis virus to assemble new infective virions during the later stages of the infection process in the host cells; the viral genes are replicated in the early stages of infection and must be encapsulated within structural viral proteins to provide new virions. Thus there is a high demand for structural proteins late in the course of Sindbis virus infection.

Both the 42*s* and 26*s* forms of Sindbis virus RNA use a single site for initiating the synthesis of their respective proteins, but importantly, these initiation sites are different. Clearly, Sindbis and a few other viruses have the ability to change the site of initiation on their RNAs during the course of infection. The explanation of this change is a good example of fine control by both transcriptional and translational means working in harmony. Early in infection, when the entire virion RNA is present, the active initiation codon is near the 5′-end and the second, inactive initiation codon, nearer the 3′-end is used little, if at all. Later in infection, there is a marked change in the replication or transcription of the virion RNA, so that the replicating enzymes now make copies of the 3′-end of the virion RNA almost exclusively. By this means, the viral genes coding for structural proteins are amplified just when needed for virion assembly. How this subtle switch in transcription is made remains to be determined, but in certain RNA phages, there are masked or inactive initiation sites buried deep within secondary structure of the viral RNA and hence inaccessible to replicating enzymes (see section 6.5.6). If such a situation is germane to the transcription of Sindbis virus RNA, then late in the infection process, certain viral-induced proteins may modify the secondary structure of the virion RNA in such a way that

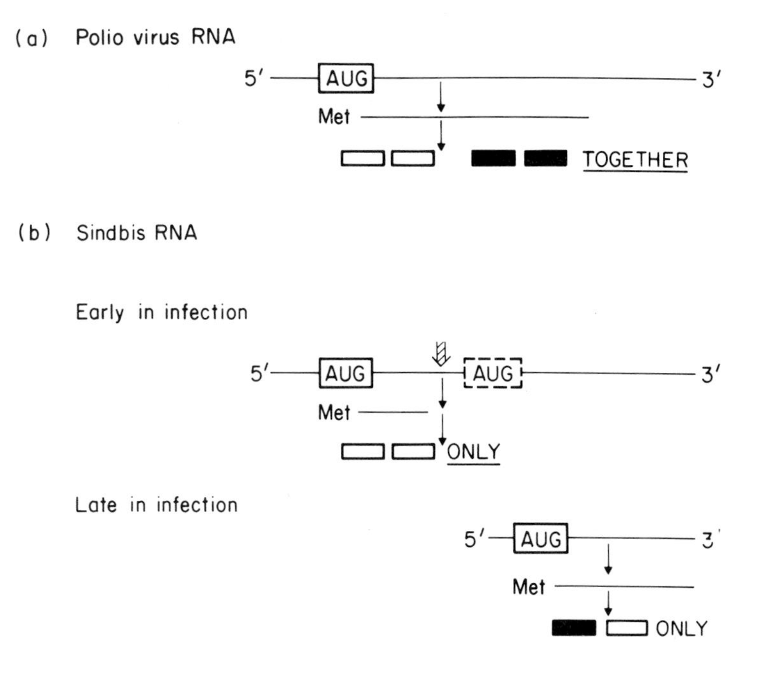

Fig. 9.10. Differences in the translation of viral RNAs in the course of viral infections. Active and inactive initiation codons are enclosed in solid and dashed lines, and viral enzymes and structural viral proteins are represented as open or enclosed boxes, respectively. Polio virus RNA has only one initiation site and throughout infection, produces viral enzymes and structural proteins at the same time. Sindbis virus RNA potentially has two initiation codons. Early in infection, only the initiation codon near the 5′ terminus is active and produces viral enzymes only. Later in infection, the viral RNA is shortened to contain only the 3′ terminus and this produces structural viral proteins only. There is, therefore, a marked switch in the pattern of translation of Sindbis RNA during the course of infection.

26*s* rather than 42*s* is preferentially transcribed (for a review see Steitz & Jakes 1975). The contrasting mechanisms for the translation of viral RNAs are best seen by comparing protein synthesis in polio and Sindbis virus infection (Fig. 9.10).

Since the most likely site for translational control mechanisms is the initiation stage of protein synthesis and there is generally only one active initiation codon in eukaryotic mRNAs, two forms of translational regulation can be envisaged (Lodish 1976). First, eukaryotic cells may be able to regulate the rate of polypeptide chain initiation, either by changing the rate at which the small 40*s* subunit binds mRNA or by changing the availability of the mRNA in the cytoplasm. Second, eukaryotic cells may possess initiation factors specific for a given species of mRNA. These and additional possibilities of translational control in eukaryotes will now be discussed.

9.3.2 The specificity of eukaryotic factors in initiating translation of different mRNAs

About five years ago, several publications indicated that initiation factors from a given eukaryotic cell always translated homologous (like) mRNAs better than heterologous (foreign) mRNAs in cell-free systems. Further, the translational product of one mRNA was claimed to inhibit the translation of other mRNAs *in vitro*. These observations raised considerable excitement and controversy. There was certainly excitement because these results suggested a novel means of translational control in eukaryotes. On the other hand, these results were singularly controversial because they ran counter to the observations of the vast majority of other investigators. In general terms, the ribosomes and initiation factors from one source translated the mRNA from another source to some extent, and often, very well (for a review see Revel & Groner 1978).

The controversy was finally settled by two simple but elegant studies. First, Gurdon *et al.* (1974) injected 'foreign' rabbit globin mRNA into fertilized eggs of *Xenopus laevis,* and let the normal processes of development and metamorphosis occur. The tadpoles derived from these eggs synthesized rabbit globin at precisely the same rate as the original fertilized egg, indicating that the translational machinery of *Xenopus* had by no means recognized the globin mRNA as foreign and degraded it; rather the contrary, the heterologous mRNA was efficiently translated over a long period of time. Second, Woodland *et al.* (1974) injected globin mRNA into fertilized eggs and showed that the developing muscle and nerve cells in *Xenopus laevis* tadpoles were synthesizing rabbit globin as efficiently as erythropoietic cells. This ingenious experiment shows that eukaryotic cells will translate heterologous mRNAs extensively, even if this is not in keeping with their normal biological activity. These telling experiments dismissed the possibility of mRNA-specific initiation factors once and for all. The translational machinery in eukaryotic cells has the same specificity towards all mRNAs, irrespective of their origins. An important corollary to these studies on *Xenopus* is that the synthesis of tissue-specific proteins during differentiation cannot be plausibly explained by changes in initiation factors or other translational components; the key to differentiation must, therefore, lie at the level of transcription (and RNA processing) in providing mRNAs in a tissue-specific manner.

9.3.3 Sequestering of eukaryotic mRNAs

The transcription of a given mRNA and its transfer into the cytoplasm of eukaryotic cells does not necessarily mean that it will be translated into protein. Several examples are now known where eukaryotic mRNA can be sequestered into ribonucleoprotein complexes in the cytoplasm and thus be rendered unavailable for translation.

The unfertilized eggs of sea-urchins and amphibia have a low rate of protein synthesis, yet they contain many mRNAs, including those for histones and microtubule proteins, including tubulin. These mRNAs are of maternal origin, being produced during the latter stages of oogenesis, yet they are not translated in the unfertilized egg where they are present as inactive ribonucleoprotein complexes. This store of inactive mRNAs is necessary to meet the extreme demand for protein synthesis during the early cleavage stages of cell division in the embryo. Immediately after fertilization, the mRNAs are released from the store of ribonucleoprotein complexes and protein synthesis increases five- or sixfold. Unfertilized eggs contain an inhibitor, believed to be a protein, which prevents the binding of aminoacyl-tRNAs to the cytoplasmic ribosomes in the egg (Gambino *et al.* 1973). By means as yet unknown, fertilization relieves the inhibition of translation and the amount of inhibitor declines dramatically.

The brine shrimp, *Artemia salina,* lives in salt flats and a dormant, encysted form can persist in this difficult environment indefinitely, provided they remain essentially dehydrated. The cysts contain inactive mRNAs which are not translated until hydration and the chance to proliferate return. The dehydrated cysts contain a powerful inhibitor of aminoacyl-tRNA binding to ribosomes, but in this case, the inhibitor is a short RNA, 20 nucleotides long, and rich in U and C (Lee-Huang *et al.* 1977). In addition, rehydration of *Artemia* cysts is accompanied by a 20-fold increase in both the activity and synthesis of eIF-2 (Lee-Huang *et al.* 1977). In immature erythrocytes, globin mRNA is present in two forms of ribonucleoprotein particle, a 15*s* form associated with polyribosomes and active in translation, and a 20*s* form, free in the cytoplasm. Active globin mRNA can be released from both 15*s* and 20*s* forms by deproteinization, suggesting that the 20*s* form is a reserve of globin mRNA. This reserve is prevented from being translated by an inhibitor associated with only the free, 20*s* ribonucleoprotein particles. These examples help to underline the fact that eukaryotic cells often contain sequestered and translationally inactive mRNA.

9.3.4 Control of the binding of eukaryotic mRNAs to eIF-2 and the 40*s* subunit

As shown earlier in Fig. 9.5, a crucial step in eukaryotic initiation is the formation of the ternary complex, eIF-2.Met-$tRNA_i^{Met}$.GTP, essential for the attachment of the initiator tRNA to the 40*s* ribosomal subunit. Evidence is now being accumulated to show that modification of eIF-2 may provide the basis for an interesting means of translational control. When deprived of iron, necessary for formation of the haemoglobin prosthetic group, haem, the synthesis of globin chains in intact reticulocytes rapidly comes to a halt. Similarly, when globin mRNAs are translated in reticulocyte lysates, the system must be supplemented with the protoporphyrin-Fe^{3+} complex, haemin, if translation is to continue for any appreciable time. In the absence of haemin, it was thought that a translational inhibitor accumulated which blocked the attachment of Met-$tRNA_i^{Met}$ to the 40*s* subunit; this idea gained recognition when it was found that a deficiency in haemin could be largely offset by addition of purified eIF-2 to reticulocyte lysates. The haemin effect in reticulocyte lysates aroused considerable interest, as it seemed to provide evidence for a subtle means of translational regulation. The molecular mechanisms for this interesting means of translational regulation are now largely known (for a review see Ochoa & de Haro 1979). The translational inhibitor, M_r 95 000, is a cAMP-independent protein kinase which is capable of inactivating eIF-2 by phosphorylation of its subunit, M_r 38 000 (see Table 9.5) for the quaternary structure of eIF-2). The phosphorylated form of eIF-2 (eIF-2[P]) is unable to participate in the usual formation of the ternary complex, eIF-2.Met-$tRNA_i^{Met}$. GTP, and consequently, attachment of globin mRNAs to the reticulocyte ribosomes is impaired. Haemin offsets the influence of the eIF-2 kinase (or translational inhibitor) by preventing the activation of a cAMP-dependent protein kinase in reticulocytes, necessary, in turn, for phosphorylating, and thereby activating, the eIF-2 kinase itself. Protein phosphatases in reticulocyte lysates must also play some part in the overall scheme of regulation which is represented in Fig. 9.11. The phosphatases would prevent both eIF-2 and eIF-2 kinase being permanently phosphorylated. In fact, when isolated from reticulocytes, eIF-2 is always in the non-phosphorylated (active) form.

The protective influence of haemin may have wider implications than in reticulocytes alone. Haemin enhances and prolongs protein synthesis in many cells which normally do not synthesize haemoglobin. Cell-free preparations from hepatoma, HeLa and ascites tumour cells all respond positively to haemin, and translational inhibitors have now been isolated and purified from Friend leukaemia cells and rat liver. In both cases, the inhibitors were classified as eIF-2 kinases. Clearly, the phosphorylation of eIF-2 is a novel means of regulating translation in eukaryotic cells, but how the phosphoryla-

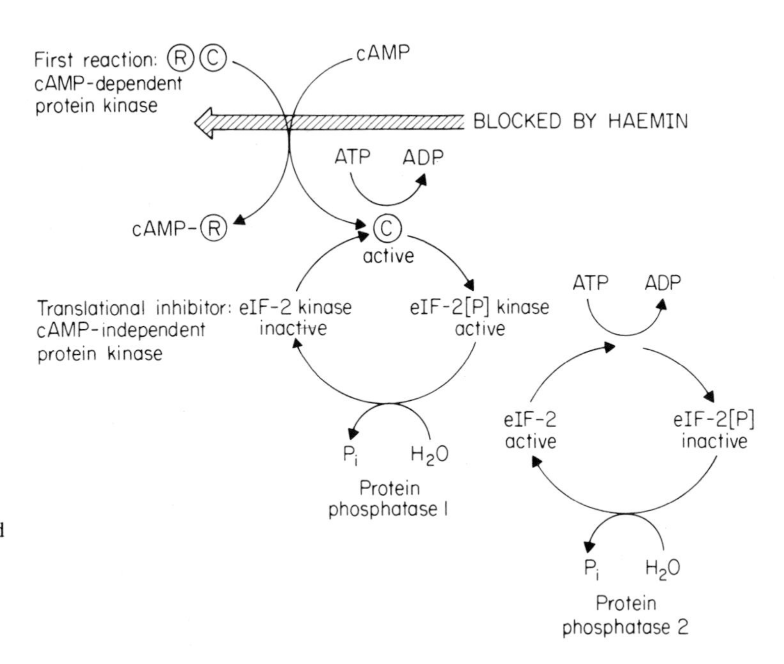

Fig. 9.11. The inactivation of eIF-2 in reticulocytes by phosphorylation and its prevention by haemin. The cascade of reactions is begun by a cAMP-dependent protein kinase, known to exist in regulatory (R) and catalytic (C) subunits. The protein phosphatases 1 and 2 are believed to have extreme substrate specificity; non-specific phosphatases could not perform the regulatory role envisaged here.

tion mechanism is kept under control in other than erythropoietic cells remains to be established. Most cells must adopt alternatives to haemin as their means of regulating the state of phosphorylation of eIF-2. Since all mRNAs are translated in an identical manner, it should be emphasized that blockade of the function of eIF-2 is a non-specific means of inhibiting eukaryotic translation.

9.3.5 Changes in eukaryotic translation promoted by differentiation, the cell cycle and hormones

The cell cycle

The cell cycle imposes considerable influence over eukaryotic translation. During the mitotic phase itself, the rate of initiation falls dramatically. The average size of the polypeptides made and the time needed to elongate those chains are the same during mitosis as in other phases of the cell cycle. During mitosis, however, the average size of polyribosomes, meaning the number of attached ribosome monomers per mRNA strand, falls to one-third of that of cells in the G1 phase. This means that the average mRNA initiates protein synthesis at a much lower rate in the M phase than in the G1 phase. Interestingly, most cytoplasmic mRNAs remain intact throughout mitosis and are translated at the normal rate during the subsequent

G1 phase. The shut-down of translation during mitosis may be a means whereby the eukaryotic cell can temporarily curtail biosynthetic activity in order to meet the high demand for energy during cell division.

Many proteins related to DNA replication and nucleosome formation are synthesized only during, or just prior to, the S phase of the cell cycle. Histone synthesis has been intensively studied from this standpoint. If DNA synthesis is arrested by thymine deprivation or by the selective inhibitor, cytosine arabinoside, histone synthesis is stopped completely. For many years, it was believed that histone mRNAs were transcribed only during the S phase (e.g. Stein *et al.* 1977) but recent works suggest this is an oversimplification of histone regulation. Even at the height of histone synthesis during the S-phase, a proportion of histone mRNA is found in free ribonucleoprotein particles rather than bound within active polyribosomes. On the other hand, when DNA synthesis is inhibited, histone mRNA is not accumulated in the free ribonucleoprotein complexes; rather the contrary, it is rapidly degraded. In an important study, Melli *et al.* (1977) reported that histone mRNAs are transcribed in the nucleus throughout the cell cycle, but are found in an active form in the cytoplasm only during the S phase. It is now thought that a specific and controlled degradation of histone mRNA takes place in the cytoplasm when DNA synthesis is not in progress. Current evidence suggests that histone synthesis is regulated by several integrated mechanisms, including transcription of histone gene, histone mRNA degradation and even histone turnover (for a review see Anderson & Lengyel 1980).

Many other proteins associated with DNA replication, including thymidine kinase, dihydrofolate reductase and DNA polymerase α, are similarly synthesized just prior to and during the S phase. The regulation of these enzymes appears to differ from that described for the histones. In the case of dihydrofolate reductase, regulation appears to be exclusively at the level of gene transcription with little, if any, translational control (Hendrickson *et al.* 1980).

Differentiation and development

In the slime mould, *Dictyostelium discoideum,* differentiation leads to a decrease in the rate of initiation of protein synthesis and the selective reduction in the translation of some species of mRNA (Alton & Lodish 1977). Both transcriptional and translational regulation are probably involved in these changes in protein synthesis.

During the development of many eukaryotes, protein synthesis is found to alter, both qualitatively and quantitatively. An explanation of these changes has yet to be found and lively controversy exists at present. Eukaryotes are distinctive in having several species of aminoacyl-tRNA synthetases (see section 9.2.2) and several groups of investigators have reported a qualitative change in these important enzymes during development, e.g. during the switch from embryonic to adult erythropoiesis and during development of silk moth, *Bombyx mori,* sea-urchin and mealworm, *Tenebrio molitor.* Improvements in technology have aroused serious doubts about these earlier findings, particularly those from *Tenebrio.* Developmental shifts in the complement of eukaryotic aminoacyl-tRNA synthetases should now be viewed with caution.

Hormones and translation

Many steroid hormones and thyroxine increase translational activity in a wide variety of cells. As reported in section 8.1.3, the stimulation of the production of rRNA and specific species of mRNA is now established beyond any doubt. However, as ably reviewed by Palmiter *et al.* (1976), there is nearly always a delay or latent period between the onset of mRNA synthesis and maximal synthesis of the encoded protein after hormonal stimulation. This is particularly evident in the androgenic induction of aldolase mRNA and the oestrogenic induction of vitellogenin mRNA. It is clear that some measure of translational control is exerted by steroid hormones, particularly in the provision of active initiation factors, ribosomal proteins or membrane components. All of these could be activated or synthesized at a slower rate than the mRNA itself. Many of the translational changes evoked by hormones are intimately related to a growth response, but this is not invariably the case; for example, the induction of vitellogenin in amphibian liver and of tyrosine aminotransferase in cultured hepatoma cells. Irrespective of a growth response or not, all of these effects are set in motion by the nuclear accumulation of the steroid hormone, as a hormone-receptor complex, and the subsequent transcription of hormone-sensitive mRNAs (see section 8.1.3).

There are several examples in the literature now which clearly indicate that steroids can influence protein synthesis by translational mechanisms alone. Metabolites of testosterone, the 5α-androstanediols, cause an instant but transient induction of β-glucuronidase in mouse kidney, yet this seems to be a translational response, centred on the microsomes. This enzyme response occurs even when all nuclear and cytoplasmic receptor sites are selectively blocked and 5α-androstanediols are bound to microsomes, never the nucleus. In a particularly interesting example of translational control, Baulieu *et al.* (1978) have reported that progesterone promotes the production of fertile eggs in dormant oocytes of *Xenopus laevis*. The hormone-induced changes are profound, including a stimulation of protein synthesis, breakdown of the germinal vesicle and ejection of the polar body; all of these changes are insensitive to actinomycin D and may be evoked even in anucleate oocytes. It is suggested that progesterone acts at the cell surface only, changing the intracellular distribution of Ca^{2+} ions such that a cascade of reactions occurs in the cytoplasm, leading ultimately to the formation of the mature egg. Organ cultures of chick blastoderm respond to 5β-reduced steroids with an enhanced rate of synthesis of fetal globins. These inducing steroids are not bound within the nucleus, and the amount of fetal globin mRNAs per blastoderm cell remains constant, even after stimulation. The inducing steroids selectively stimulate initiation factors for protein synthesis; elongation and termination factors are not affected (Mainwaring & Irving 1979).

9.3.6 The importance of the poly(A) sequence at the 3′-end of eukaryotic mRNAs

With the exception of histone mRNAs and perhaps a few others (Milcarek *et al.* 1974), eukaryotic mRNAs have a distinctive poly(A) tail at their 3′-end. This unique sequence has been an invaluable asset in the purification of the mRNA from higher organisms. In most instances, two proteins, M_r 52 000 and 78 000 are associated with the poly(A) tract (Blobel 1973). Removal of the 3′ tail by polynucleotide phosphorylase or 3′ exonuclease does not prevent the translation of eukaryotic mRNA, but there is growing evidence that deadenylation profoundly influences the stability of the mRNA. During long periods of incubation, either in cell-free systems or in intact *Xenopus* oocytes, histone mRNAs and deadenylated globin mRNAs are soon degraded; conversely, polyadenylation of histone mRNAs increases their half-life within oocytes very significantly (Chantrenne 1977). The sensitivity of histone mRNA to degradation was mentioned earlier with respect to the cell cycle. It would seem, therefore, that the poly(A) 3′ tail is not required for translation but makes a significant contribution to the stability and long half-life of eukaryotic mRNAs. Perhaps the only contradiction to these conclusions is the report that deadenylated ovalbumin mRNA is poorly translated.

9.3.7 Model systems for translation *in vitro*

Many post-mitochondrial fractions of tissue homogenates and cell lysates have been found useful for the translation of exogenous mRNA *in vitro*. Suitable sources include ascites tumour cells, rabbit reticulocytes, wheat germ, to name but a few. The most refined system, however, is undoubtedly to inject the mRNA into the oocytes of *Xenopus laevis* (Gurdon *et al.* 1971). The efficiency and accuracy of this system is truly remarkable and complete post-translational processing of many mRNAs has been reported; for example, immunoglobulin precursors are cleaved at the correct places, the proline residues of procollagen are hydroxylated and the N-terminal methionine of α-crystallin is acetylated.

9.3.8 Unusual controls of eukaryotic translation

Interferons

When exposed to viruses or certain chemical inducers, animal cells produce interferons or glycoproteins which strongly inhibit eukaryotic translation. Many interferons have now been fully characterized, and while they strongly inhibit translation of viral mRNAs, their influence on the translation of host cell mRNAs is contentious (e.g. contrast the reviews of Revel & Groner 1978 and Ochoa & de Haro 1979). The best inducers of interferon are double-stranded RNAs, either reovirus RNA or the synthetic polymer, poly(I): poly(C). Although double-stranded RNA is formed during the replication of certain viruses, it may not be the only stimulus for interferon induction; the induction process is complex and not fully understood (Friedman 1977). When many eukaryotic cells are exposed to interferons, a translational inhibitor is synthesized and may be recovered a few hours later in cell lysates, cell sap or ribosomal extracts (proteins solubilized or released in 0.5M KCl). The appearance of the inhibitor, which is not an interferon itself, requires RNA and protein synthesis in the host cell. The antiviral effects of interferons are expressed at both initiation and elongation phases of viral mRNA translation and depend on a considerable expenditure of host cell ATP. A summary of these unusual and complex reactions is presented in Fig. 9.12. Once produced in viral-

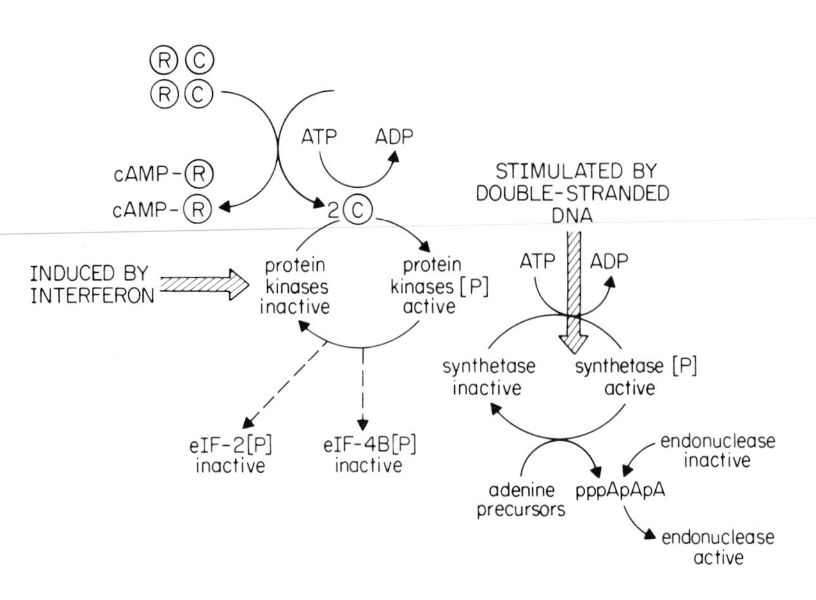

Fig. 9.12. A possible explanation of the inhibition of translation in eukaryotic cells by interferon. The cascade of reactions is begun by a cAMP-dependent protein kinase. Interferon induces the synthesis of one or more cAMP-independent protein kinases, which phosphorylate (and inactivate) eIF-2 and eIF-4B. One protein kinase is selectively stimulated by double-stranded RNA and activates the synthetase for a short nucleotide sequence, pppApApA. This, by means as yet unknown, activates an RNA endonuclease.

infected or chemically induced cells, interferons promote the synthesis of at least one cAMP-independent protein kinase (and probably more); the protein kinase(s) is in turn activated by the cAMP-dependent protein kinase described earlier with respect to the effects of haemin on eukaryotic translation. In essence, interferon triggers a complex cascade of phosphorylation reactions, leading to the activation of a sequence of enzymes and the production of translational inhibitors. In more detail, the responses to inferferons are as follows. First, the subunit of eIF-2 (M_r 38000) is phosphorylated and inactivated, so that the binding of initiator $tRNA_i^{Met}$ is prevented. Second, a ribosomal protein (M_r 67000) is phosphorylated and inactivated; this could be either eIF-2A or eIF-4B, but in any event, the binding of viral mRNA to host ribosomes would be severely impaired. Third, double-stranded RNAs specifically activate one protein kinase, leading to the phosphorylation and activation of a synthetase, and thus the appearance of a small nucleotide inhibitor with most unusual 2′–5′ phosphodiester bonds, pppA(2′)p(5′)A(2′)p(5′)A. There is now evidence that this small inhibitor somehow activates an endonuclease, leading to the degradation of viral mRNAs (for a review see Williams & Kerr 1980). By these mechanisms, viral mRNA translation is severely curtailed by the presence of interferon.

Inhibition of host protein synthesis by viruses

In many viral infections, particularly in their late or lytic phases, there is often a marked inhibition of the translational machinery of the host cell. Several possible explanations of this inhibition have been forwarded. First, it has been suggested that viral mRNAs bind more strongly to host ribosomes and are thus preferentially translated; certainly, relief of inhibition can be achieved *in vitro* by the addition of purified eIF-4B. Second, there are marked changes in host cell membranes during viral infections which lead to an influx of Na^+ ions. Protein synthesis from viral and host mRNAs has distinctive requirements for cations, and the influx of Na^+ ions could result in optimal translation of the viral mRNAs. Third, there is striking evidence that one viral-induced protein is a selective inhibitor of an initiation factor associated with eukaryotic ribosomes. Overall, the inhibition of host translation by viruses is likely to be a complex process in which many different mechanisms are involved.

Competition between viruses within a host cell

Poliovirus mRNA, which is uncapped, is translated in extracts from both uninfected and infected HeLa cells, whereas vesicular stomatitis virus mRNA, which has a 5′ 7-methylguanosine cap, is translated only in extracts of uninfected cells. Translation of the capped viral mRNA can occur in infected cell extracts, provided that they are supplemented with eIF-4B. This initiation factor is necessary for the recognition of the 5′ cap and it appears that poliovirus directs the synthesis of a specific inhibitor of eIF-4B. By this subtle means of inhibition, poliovirus can achieve translational dominance over viruses with capped mRNAs if they are competing within the same host cell.

Eukaryotic suppressor tRNAs

The termination of protein synthesis in eukaryotes is signalled by one of the three stop codons in the mRNA (UAG, UAA and UGA), which are recognized by the single releasing factor, RF.

However, termination can be circumvented, particularly during viral infection of both prokaryotic and eukaryotic host cells, by suppressor tRNAs; a well-known example is provided by translational suppression occurring during the infection of *E. coli* by the RNA bacteriophage, Qβ (see section 6.5.6). In such cases, the suppressor tRNA recognizes one of the stop codons, but actually overrides the termination signal and inserts an amino acid into the nascent polypeptide chain instead. The ribosome then continues moving along the mRNA until it encounters the next stop codon. By this mechanism, the suppressor tRNAs can promote the biosynthesis of 'read-through' proteins, namely those related to other proteins in terms of general structural homology, but with additional C-terminal sequences and thus distinctive functions. Geller and Rich (1980) have made the important identification of a suppressor tRNA in rabbit reticulocytes; accordingly, suppressor tRNAs may be a ubiquitous feature of eukaryotic cells, even in the absence of viral infection. This reticulocyte suppressor is a $tRNA^{Trp}$ moiety, capable of suppressing the UGA stop codon in β-globin mRNA and thus enabling small quantities of read-through proteins of β-globin to be produced.

This finding is of extreme importance to eukaryotic translation in general, but particularly in the context of the biosynthesis of biologically active polypeptides. There is growing recognition that polypeptide hormones, for example, may be grouped into families of related structure (for a review see Blundell & Humbel 1980). The insulin-like family encompasses relaxin (synthesized in females prior to parturition) and insulin-like growth factors or somatomedins; the glucagon-like family includes secretin and many other hormones needed for the control of gastric function. It is extremely interesting that many somatomedins are identical in sequence to insulin, but contain an additional C-terminal sequence. It seems possible, therefore, that some somatomedins may be insulin read-through proteins.

9.4 POST-TRANSLATIONAL MODIFICATIONS

In all eukaryotic cells, and particularly among the most advanced organisms, extensive post-translational modification of proteins occurs for several reasons. Although eukaryotic mRNAs are monocistronic and encode for only a single polypeptide chain, the initial translational product is often the precursor of many proteins which are produced by specific cleavage reactions. In addition, many translational products are for secretion and this makes important demands on the nature of nascent polypeptide chains to enable them to penetrate into intracisternal channels during the initial phase of the secretory process. Furthermore, many components of eukaryotic cells are glycoproteins or lipoproteins and these modifications must occur after the synthesis of the polypeptide chain. Many eukaryotic proteins also have a very complex final structure, as in elastin and collagen; these structures essentially require modified amino acids which do not possess codons of their own and so must be made by post-translational processing. The complexity of collagen, for example, is such that it could never be achieved in one biosynthetic step; the final structure is necessarily built up stage by stage. Lastly, eukaryotic cells rely heavily on hormones for metabolic regulation and in certain cases these biologically active proteins are temporarily stored in the cells of their origin. To protect these cells making polypeptide hormones, other proteins are often synthesized concomitantly which act as 'dampers' of hormonal activity during storage. These facets of eukaryotic cells and their translational machinery have no strict parallels in prokaryotic organisms. The alternative system is when polypeptide hormones are stored as inactive precursors or zymogens which are activated, on demand, by proteolytic cleavage.

The first graphic example came from the work of Steiner *et al.* (1967) who reported the synthesis of proinsulin, a larger precursor of insulin, in cultures of a human adenoma of the islet (β) cells of the pancreas. In this early example, a sequence of thirty amino acids in a connecting peptide, not present in active insulin, served to inactivate the hormone during storage (Fig. 9.13). As we shall see, there are now even earlier precursors of insulin known than proinsulin. Nevertheless, Steiner's work helped to establish the importance of post-translational modification in eukaryotic cells.

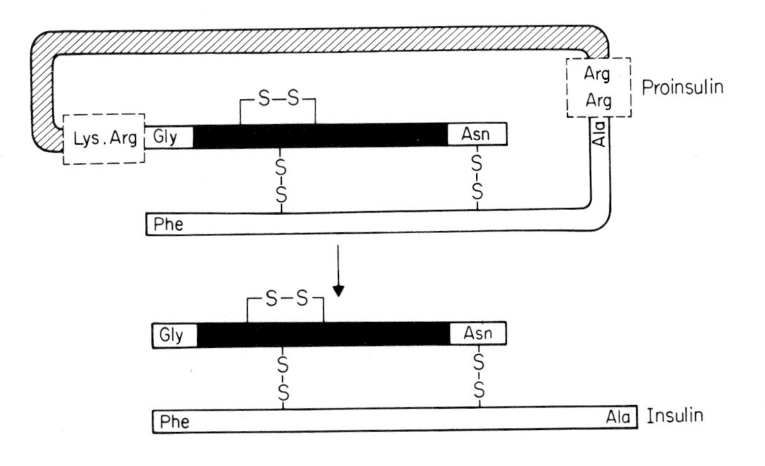

Fig. 9.13. The activation of proinsulin to insulin. This change is mediated by proteases acting on proinsulin in the storage granules of pancreatic β cells. The A and B chains of insulin are shown in solid and open areas, respectively; the connecting peptide is indicated by the hatched area, with the basic amino acids excised by proteolytic activation within the dashed boxes.

As reviewed by Schally *et al.* (1978) and Brestlow (1979), studies on hypothalamic and pituitary hormones help to illustrate three important aspects of protein processing in higher animals.

(1) They help to underline the extreme specificity needed in post-translational cleavage reactions. Oxytocin and vasopressin are two of the smallest known peptides with biological activity, yet both are derived by cleavage of very much larger precursors made by ribosomes and specific mRNAs. The proteases involved must have remarkably stringent requirements in terms of specificity to recognize the correct cleavage sites.

(2) They help to illustrate the need for rigorous control of protease activities; for example, the short sequence, Gly.Leu.Pro can be released from the N-terminal end of oxytocin by proteases and this fragment is a powerful inhibitor of the secretion of another pituitary hormone, α-melanotropin. Clearly, protease activities must be well controlled otherwise indiscriminate cleavage of polypeptide hormones could wreak havoc with delicately balanced endocrinological systems.

(3) They help to underline the necessary and concomitant synthesis of certain eukaryotic proteins. Neurophysins are acidic proteins, M_r about 10000, abundant in glycine, proline and disulphide bridges, which are synthesized at the same time as oxytocin and vasopressin. The neurophysins help to stabilize and contain the hormones inside the neurosecretory granules of the posterior pituitary until they are needed. Without the neurophysins, the small hormones would rapidly be degraded.

9.4.1 Simple mechanisms

With a few exceptions, the initiating N-terminal methionine residue is cleaved off after synthesis. It is sometimes necessary for the attainment of a completed and active protein to modify the N- or C-terminal residues. With hypothalamic pituitary hormones, both terminals must be blocked as a prerequisite for biological activity (Schally *et al.* 1978). In the case of many viral proteins, ovalbumin and histone 4, the N-terminal residue is acetylated.

Disulphide bridges must also be formed from appropriate cysteine residues in the nascent polypeptide sequence. As first demonstrated by Anfinsen (1964), there is considerable order in the formation of disulphide bridges. When denatured and reduced (—SH) forms of ribonuclease were allowed to slowly reoxidize using atmospheric oxygen and reform the —S—S— linkages, they always reformed accurately to the stable and active form of the enzyme. These elegant experiments indicated that the primary structure directs the correct tertiary structure of most proteins and also that the native form of proteins is thermodynamically the most stable structure. The correct linking of cysteine residues into disulphide bridges can only occur in the early biosynthetic precursors of many proteins, existing as single polypeptide chains; for example, proinsulin and chymotrypsinogen refold correctly, whereas their activated forms, insulin and chymotrypsin, with two and three polypeptide chains respectively, cannot.

Many eukaryotic proteins contain modified amino acids and since there are no codons for these in the genetic code, they must be made by post-translational processing. Collagen contains considerable amounts of hydroxyproline and hydroxylysine, formed by distinct mixed-function oxidases. Proline hydroxylation has been widely investigated and the enzyme involved, proline 4-monooxygenase, needs non-haem iron (Fe^{3+}) and ascorbic acid (vitamin C) as cofactors. Free proline is not a substrate as the enzyme essentially requires glycine on the C-terminal side of the proline residue. The reaction is as follows:

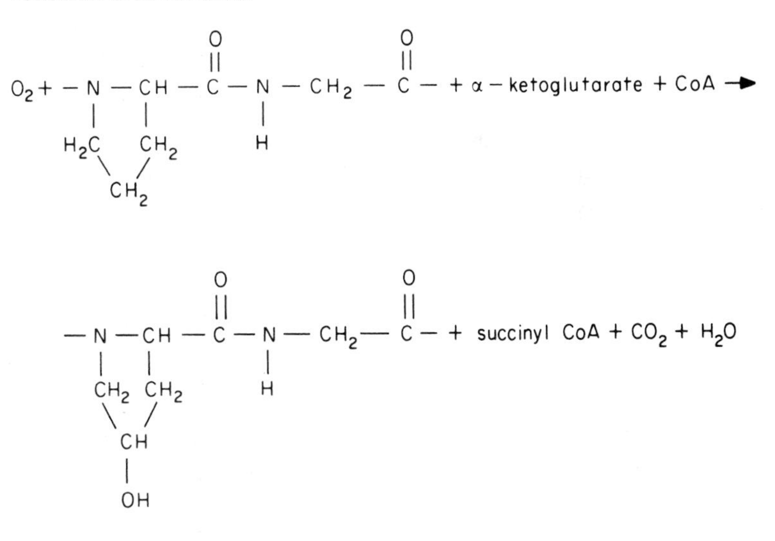

During adolescent growth, the demand for vitamin C for this reaction is very high, and a diet lacking citrous fruit results in scurvy.

9.4.2 Sophisticated mechanisms and polyproteins

It is now becoming clear that many eukaryotic mRNAs are translated into a single polypeptide chain which is subsequently cleaved to yield a number of discrete proteins. This is a generally accepted phenomenon of eukaryotic translation now, but only four very striking examples will be discussed in detail here.

Vitellogenin and yolk proteins

In oviparous vertebrates, including the South African clawed frog *Xenopus laevis*, and the domestic fowl, *Gallus gallus*, the precursor for certain protein constituents of egg yolk is vitellogenin, M_r 480 000 (for a review see Tata 1976). The synthesis of vitellogenin in females is part of their normal reproductive cycle, but it can also occur in males, often with fatal consequences. Vitellogenin is a complex protein, rich in lipids, sugars (sialic acid, mannose and galactose) and phosphorylated serine residues. Vitellogenin is composed of two polypeptide chains, M_r 240 000, each of which contains the amino acid sequences for three proteins, lipovitellin, M_r 135 000, and two forms of phosvitin, M_r 28 000 and 34 000. All evidence suggests that vitellogenin is a polyprotein, translated from a monocistronic mRNA.

Vitellogenin synthesis occurs in the liver of both sexes, even in cultures *in vitro* under the stimulus of trace amounts of oestrogens, especially the synthetic oestrogen, diethylstilboestrol. Oestrogen receptors are involved in this response (Tata 1976) and a particularly striking correlation between the ability to synthesize vitellogenin and the developmental appearance of oestrogen receptors has been reported (Lazier 1978). Translationally active vitellogenin mRNA has been isolated, directing the synthesis of a polypeptide of the correct molecular weight, M_r 240 000. Furthermore, there is a precise relationship between the extent of vitellogenin synthesis and the enhanced transcription of the vitellogenin gene. The mRNA for vitellogenin is one of the largest known, M_r 2.3×10^6, and having a high sedimentation coefficient, 30*s*. It has been calculated that this is 10% (or 600 nucleotides) longer than needed for the synthesis of lipovitellin and the phosvitins. Clearly, segments of vitellogenin are excised during post-translational processing. The number of copies of vitellogenin mRNA per liver cell rises from a baseline level of 5 to over 5000 after oestrogenic stimulation.

The translational response to oestrogens is dramatic. Protein synthesis in the liver is directed almost exclusively to vitellogenin production and normally expressed liver genes, including the albumin gene, are virtually closed down. Indeed, the serum of oestrogen-stimulated birds becomes engorged with vitellogenin, such that the male often

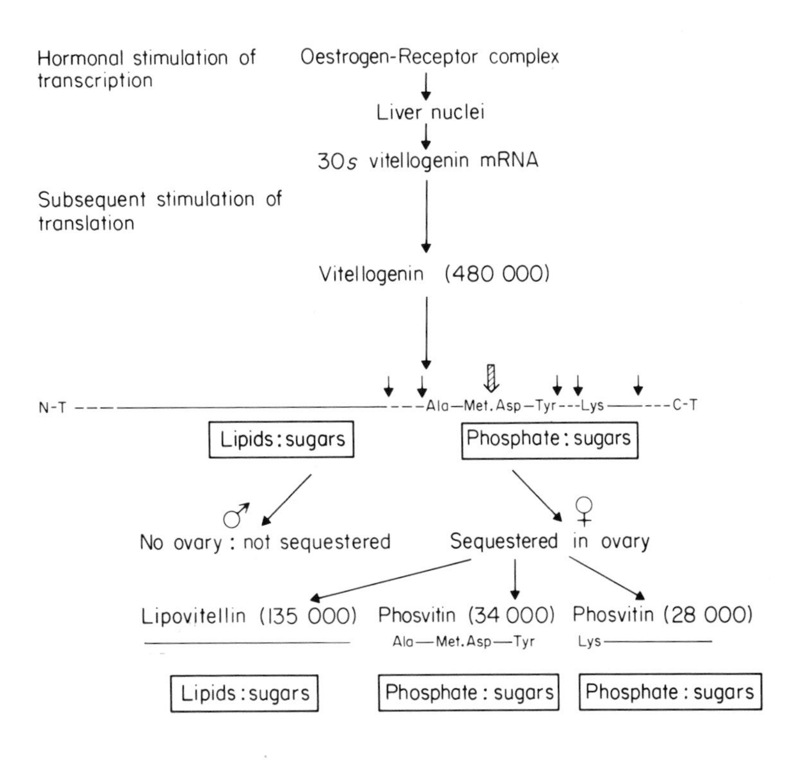

Fig. 9.14. The synthesis of lipovitellin and phosvitin in the liver of oestrogen-stimulated birds. Synthetic oestrogens, such as stilboestrol, saturate an oestrogen-specific receptor system in liver nuclei and enhance the expression of 30*s* vitellogenin mRNA. Vitellogenin (M_r 480 000) is composed of two chains (M_r 240 000 each). Lipovitellin and two forms of phosvitin are formed in the female ovary from sequestered vitellogenin; the male has no ovary and cannot sequester or process vitellogenin. The high blood levels of vitellogenin in male birds can cause death from anoxia. N-T is amino terminal end, C-T is carboxy terminal end. Proteolytic cleavage sites are indicated by arrows; an experimental cleavage site for CNBr is indicated by the hatched arrow. Sequences excised during post-translational processing are indicated by dashes. Components in the boxes are also added after completion of the polypeptide chain of vitellogenin.

dies of anoxia because, unlike the female, it cannot sequester vitellogenin. In the female, vitellogenin is sequestered into the ovary.

Information is scanty on the amino acid sequences of lipovitellin and the phosvitins, but CNBr cleavage of a single methionine residue in vitellogenin produces a fragment, M_r 90 000, containing both forms of phosvitin. These have now been assigned to the C-terminal region of vitellogenin. The stages in the synthesis of lipovitellin and the phosvitins are shown in Fig. 9.14. The phosvitins contain up to 56% serine and are the principal site of phosphorylation; by contrast, the lipovitellin moiety becomes exclusively associated with lipids. Carbohydrate moieties are distributed almost evenly among the post-translational cleavage products of vitellogenin.

Wahli *et al.* (1979) have now succeeded in cloning the vitellogenin gene and sequence analyses should provide supporting information on the location of lipovitellin and the phosvitins in vitellogenin, as well as indications of the nature of the proteolytic cleavage sites.

Corticotropin, melanotropins and natural analgesics

About a decade ago, the sequences of many pituitary hormones were known yet their mode of synthesis was not understood. In addition, a protein called β-lipotropin was isolated from the brain and its amino acid sequence was determined. At first, it was difficult to ascribe a precise function to β-lipotropin, but subsequent investigations, in rather remarkable directions, have led to one of the most exciting stories of contemporary biochemistry (for reviews see Chrétien *et al.* 1979, Nakanishi *et al.* 1979). There had been a wide search for pain-killing compounds, or analgesics, other than the opiate, morphine. This search was aided by pharmacological studies which suggested that potential opiate-like compounds should be blocked by a morphine antagonist, naloxone. Investigations by many research groups suggested that natural analgesics were present in the brain. Some were short pentapeptides and one was identical in sequence to part of β-lipotropin; it was named met-enkephalin. Later, investigators isolated further polypeptides with analgesic properties (see Chrétien *et al.* 1979); these were also structurally related to β-lipotropin, and like met-enkephalin, their opiate activity was selectively blocked by naloxone. These natural analgesics are now collectively called endorphins, and α, β and γ-endorphins have now been fully characterized. Next, cells of a mouse pituitary tumour were found to proliferate in culture and produce large amounts of corticotropin. The mRNA in these tumour cells was isolated from membrane-bound polyribosomes and translated *in vitro*. The principal translation product was a protein, pro-opiocortin, M_r 29500–31000, which quite remarkably, was precipitated equally well by specific antisera raised against either β-endorphins, β-lipoprotein or corticotropin. This incisive breakthrough suggested that all of these proteins, with a diverse range of biological activities, were all derived by cleavage of a common polypeptide precursor. This is now established unequivocally by work from a number of laboratories. Finally, the culminating evidence came in a quite remarkable paper by Nakanishi *et al.* (1979). Until this paper appeared, it was clear that specific cleavage sites must be present in the protein precursor but their nature was unknown. Nakanishi and his co-workers cloned the gene coding for the precursor and determination of its base sequence enabled the cleavage sites to be identified. Generally, these were basic sequences, Lys.Lys, Lys.Arg, or Arg.Lys, precisely as Steiner *et al.* (1967) had found in proinsulin over ten years before. This work, remarkable for the accuracy and speed with which it was performed, is summarized in Fig. 9.15. Extension of current work on the cloned vitellogenin gene is likely to make a similar impact.

The analgesic pentapeptides, enkephalins, have now been identified in many tissues other than brain, especially in chromaffin

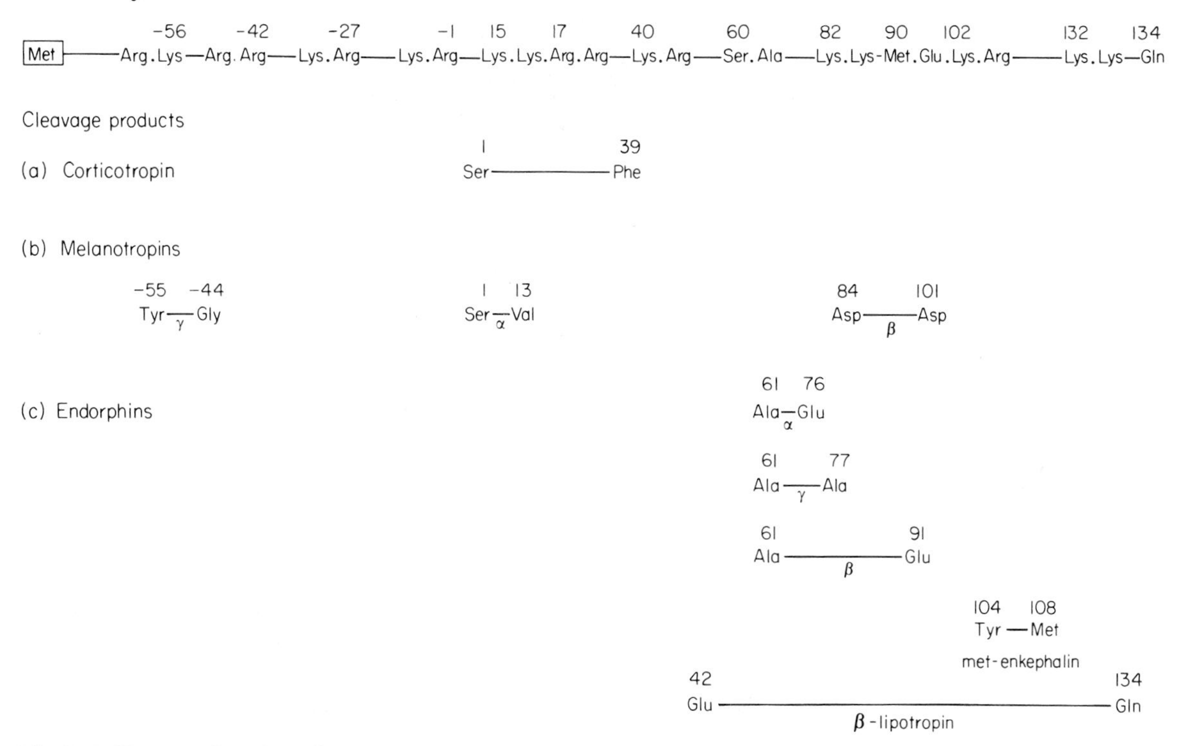

Fig. 9.15. Cleavage of a polypeptide precursor in brain to corticotropin, melanotropins and endorphins. This summary is based on amino acid sequences of the active polypeptides, but substantiated by the base sequence of the gene for the precursor (M_r 31 000, 265 residues). Cleavage points are mostly basic sequences and many proteins are derived from the C-terminal end. Amino acids are numbered with the N-terminal of corticotropin as 1. (Modified from Nakanishi *et al.* 1979.) The initiator is enclosed in a box.

granules of the adrenal cortex. While these pentapeptide sequences do occur in long polypeptide precursors, such as pro-opiocortin and β-lipotropin, these are not considered now to be the actual precursors of enkephalins; certainly in the adrenal medulla, enkephalins have distinctive precursor polypeptides (Kimura *et al.* 1980).

Collagen and elastin

Collagen is the most abundant protein in higher animals, constituting a quarter of most body protein in the form of teeth, cartilage, tendon and bone. It is a fine example of post-translational modification, since the overall process requires several novel steps. During collagen synthesis, soluble precursors are gradually converted into insoluble fibres, with high tensile strength. Collagen is secreted from an abundant cell type, the fibroblast. As reviewed by Fessler and Fessler (1978), there are four types of collagen, each with slightly different properties. All are composed of three polypeptide chains joined together; in certain collagens, these chains are identical, but in the commonest or type I collagen, there are two identical $\alpha 1$ chains plus a third $\alpha 2$ chain. The following account is exclusively on the biosynthesis of this type I collagen, found in bone, tendon and skin. The process is best described in a series of steps.

(a) Synthesis of procollagen. The initial translation products of the fibroblast are long precursors of the $\alpha 1$ and $\alpha 2$ chains; these are called pro-$\alpha 1$, M_r 120 000, and pro-$\alpha 2$, M_r 95 000. The mRNAs coding for these chains are necessarily long, M_r 1.7×10^6, and of high sedimentation coefficient, 28*s*. These chains are then hydroxylated, by mixed function oxidases, whereby many proline and lysine residues are converted in 4-hydroxyproline and 5-hydroxylysine, respectively. These reactions are conducted by different hydroxylases. Once hydroxylated, sugars may then be attached to the hydroxylysine residues, usually as a disaccharide unit, forming

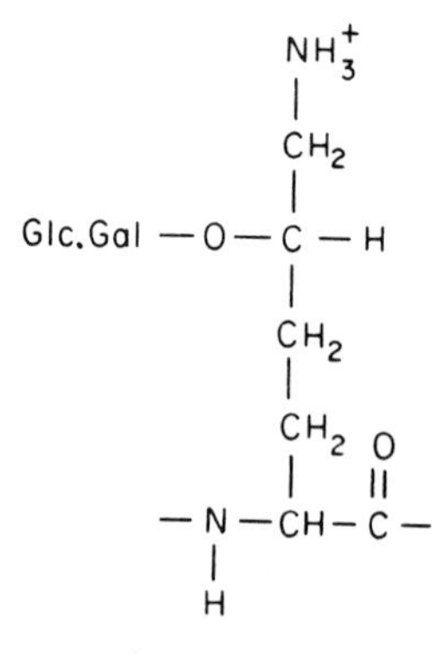

All of these processes take place on bound polyribosomes (Fessler & Fessler 1978) and after folding into helices, the pro-$\alpha 1$ and pro-$\alpha 2$ chains are secreted, via the Golgi apparatus, into the extracellular spaces of connective tissues. The secreted product is procollagen.

(b) Cleavage of procollagen. The three chains of procollagen contain sequences not present in mature collagen. For many years, these were believed to be located exclusively at the N-terminal regions of pro-$\alpha 1$ and pro-$\alpha 2$. However, separate procollagen peptidases have now been identified which act specifically at the N- or C-terminal regions of procollagen. Once the terminal sequences are removed, forming mature $\alpha 1$ chains, M_r 95 000, and $\alpha 2$ chains, M_r 90 000, tropocollagen may be formed.

(c) Assembly of tropocollagen fibres. The mature chains assemble spontaneously into tropocollagen fibres, M_r 285 000. These are extremely long rods, 300 nm in length, in the form of left-handed, type II triple helices. This structure is possible because of the repeated-Hyp.Gly.-sequence in tropocollagen. The hydroxyproline residues repel each other, thus effecting stabilization of the triple helix by keeping the polypeptide partially apart. The glycine is essential because its small size enables it to fit into the internal positions of the tropocollagen helix. Overall, the complex helix is maintained by hydrogen bonding.

(d) The final structure of collagen. The tropocollagen fibres then assemble spontaneously into collagen fibres, which fill the extracellular spaces of connective tissues. Collagen has a fundamental periodicity, resulting from the arrangement of tropocollagen in a quarter-staggered array, thus:

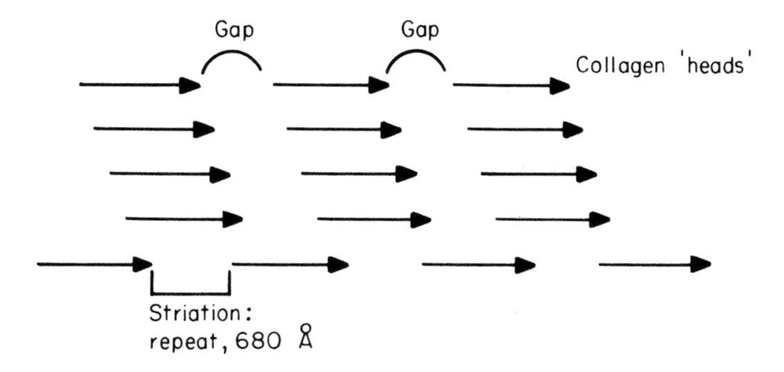

The gaps give collagen a very regular striation when examined under the electron microscope. The gaps are important nucleation sites during the mineralization phase of bone formation, when they are filled with hydroxylapatite, $Ca_{10}(PO_4)_6(OH)_2$. The last stage of collagen formation is extensive cross-linking between the tropocollagen fibres. These are mainly formed between the ϵ-NH_2 side-chains of lysine, either as aldol or lysinonorleucine cross-links.

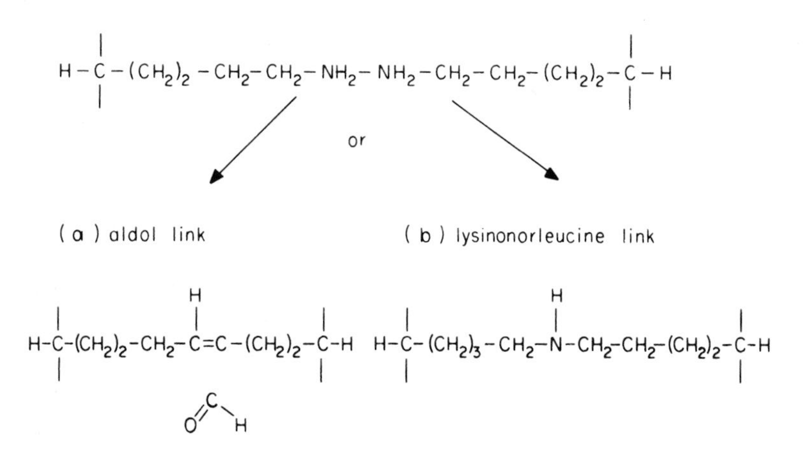

Elastin is a connective protein with remarkable two-way stretch, ideally suiting it as a structural component of the walls of blood vessels. Unlike collagen, elastin contains little if any hydroxyproline, hydroxylysine or polar amino acids. It is very rich in proline, glycine and nonpolar amino acids, such as alanine, valine and leucine. Repeated sequences, such as -Lys.Ala.Ala.Lys- and -Pro.Gly.Val.Gly.Pro-, are abundant. The remarkable elasticity of elastin still remains to be fully explained, but during post-translational modifications, component chains are linked in fours round a molecule of desmosine. This novel cross-linking agent is a condensation product of four lysine residues, each one of which is a component of a separate polypeptide chain.

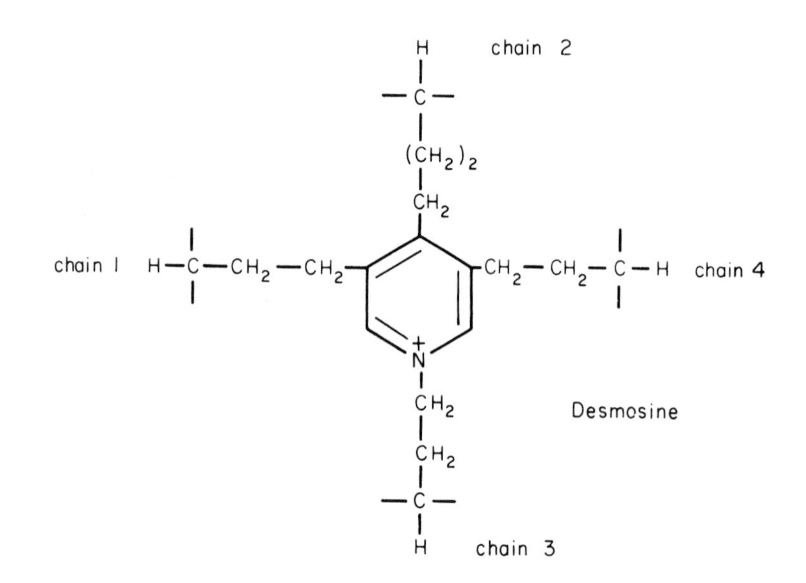

Additional cross-linking is provided by lysinonorleucine linkages, as in collagen.

9.4.3 Other cleavage reactions

Selective cleavage reactions are commonplace in the post-translational modification of eukaryotic proteins. Most digestive enzymes and polypeptide hormones are secreted as inactive zymogens and activated by limited proteolysis. This topic is discussed in more detail later in this chapter.

However, there are many examples of cascade reactions in which a series of proteins are sequentially activated by proteolysis, resulting in the initiation of important biological responses.

A relatively simple cascade reaction is provided by the renin–angiotensin system, necessary for regulation of the synthesis of aldosterone, the principal mineralocorticoid (for a review see Reid *et al.* 1978). The juxtamedullary cells in the macula densa of the kidney are sensitive to blood volume, i.e. tonicity with respect fo Na^+ ions. A high blood volume stimulates the release of a kidney enzyme, renin, which cleaves a decapeptide from the N-terminal end of an α_2 globulin, angiotensinogen, secreted into the plasma by the liver. This cleavage product is angiotensin I, which is further cleaved to an active octapeptide, angiotensin II, by a converting enzyme in plasma. Angiotensin II regulates blood volume in two ways. First, it is a powerful vasopressor and controls muscular tone in arteries, especially in the kidney. Second, it is bound to specific receptors in the zona glomerulosa of the adrenal cortex and stimulates the synthesis and secretion of aldosterone. The entire process is kept under control by specific proteases, angiotensinases, in plasma and liver, which rapidly degrade both renin and angiotensin II. The cleavage sites in angiotensinogen are as follows:

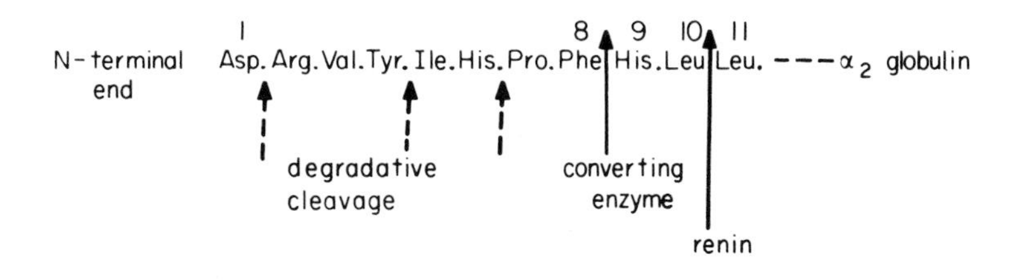

The most complex cascade reaction is the clotting process in blood, the end-product of which is the conversion of soluble fibrinogen into insoluble fibres of fibrin, the essential matrix of the blood clot (for a review see Esnuof 1977). By international agreement, most components in the cascade are designated a distinctive Roman numeral; in this nomenclature, Ca^{2+} ions are factor IV, prothrombin is factor II and fibrinogen is factor I, etc. Many factors are named after individuals who were the first to be found with inherited defects in the synthesis of specific components of the cascade, e.g. Christmas factor. In clinical terms, the antihaemophilic factor VIII is of paramount importance. Probably most of the factors are trypsin-like proteases, with serine in their active centre. The synthesis of many factors requires vitamin K. The cascade can be triggered in two ways, by an intrinsic pathway, initiated directly in plasma, or by an extrinsic pathway, initiated by a factor released from damaged cells. The latter

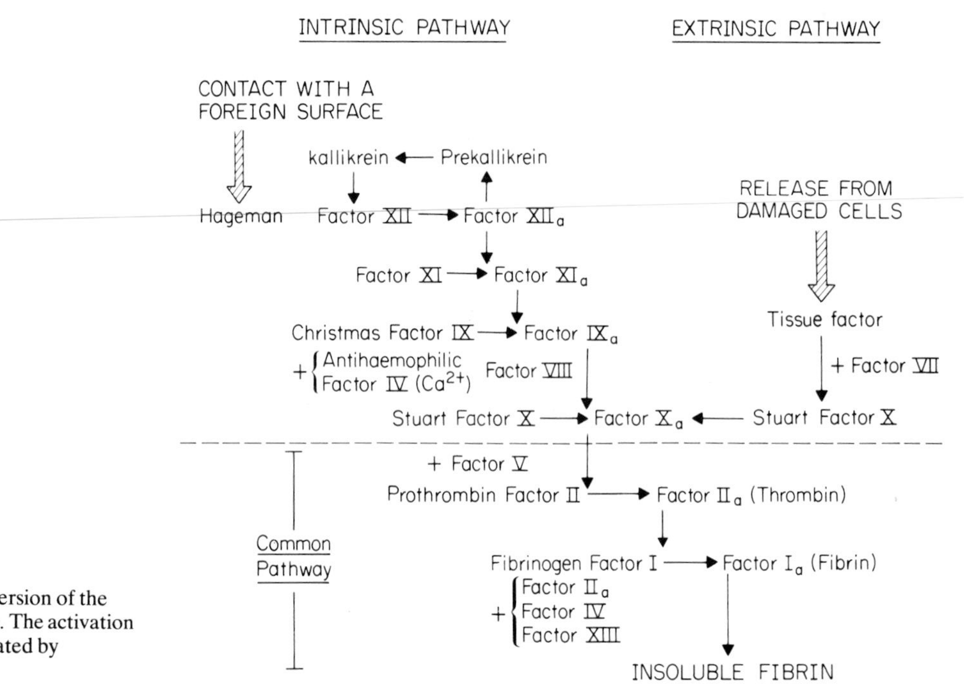

Fig. 9.16. A simplified version of the process of blood clotting. The activation of a given factor is indicated by subscript a.

part of the cascade is common, irrespective of the triggering mechanism (Fig. 9.16). The intrinsic pathway is initiated by contact of factor XII with a foreign surface, such as the modified or damaged lining of an artery. On meeting such a foreign surface, factor XII undergoes a conformational change and acquires enzymic activity. Activated factor XII then cleaves plasma prekallikrein to the enzyme, kallikrein, which activates even more factor XII by a different mechanism, involving selective bond cleavage. Subsequent reactions in the intrinsic pathway lead to the activation of factor X to factor X_a, a key enzyme in the latter stages of the clotting process. The extrinsic pathway is initiated by direct activation of factor X by tissue factor and factor VII. Tissue factor is a lipoprotein complex, released by damaged cells into the plasma; brain, lung, placenta and testis are particularly rich in tissue factor.

In the concluding stages of the cascade, prothrombin (factor II) is converted into thrombin (factor II_a) by factor X_a. Magnusson *et al.* (1975) have determined the amino acid sequence of prothrombin and identified the cleavage sites (Fig. 9.17). This is a major achievement, because the presence of no less than 12 disulphide bridges in prothrombin raises enormous problems in sequence analysis. Factor X_a cleaves -Arg.Thr- and -Arg. Ile- sequences in prothrombin and

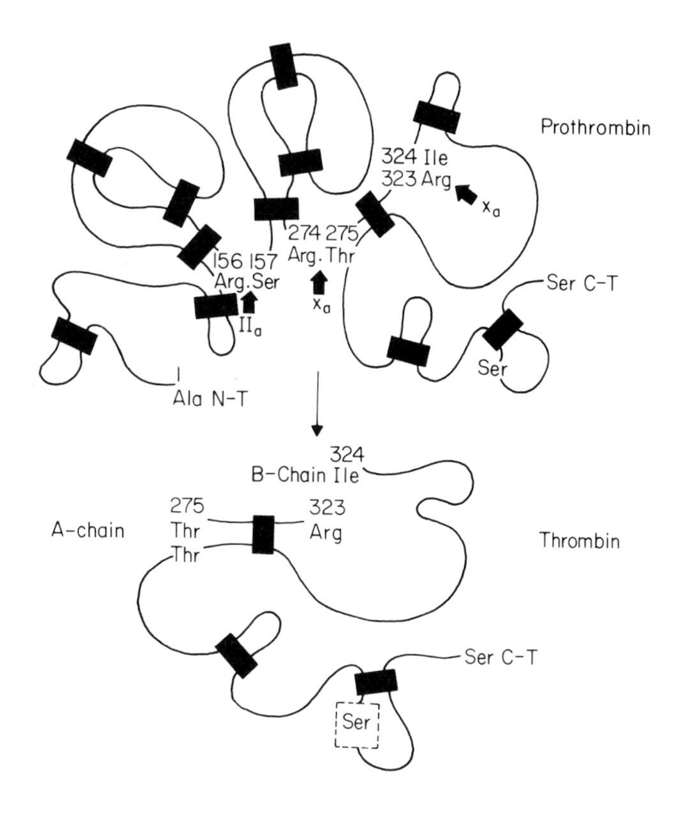

Fig. 9.17. The activation of prothrombin to thrombin. For simplicity, the numerous disulphide bridges are indicated by solid boxes. The amino acids are numbered from the N-terminal residue (N-T) as 1; C-T is C-terminal residue. The cleavage sites by Factor X_a and thrombin itself, II_a, are indicated by arrows. The enzyme active site, containing serine, is enclosed in the dashed box. (The figure is based on the work of Magnusson *et al.* 1975.)

aided by autocatalytic cleavage of an -Arg.Ser- sequence by thrombin itself, the active enzyme is released from the C-terminal region of prothrombin. Thrombin has a short A chain, linked by a single disulphide bond to a longer B chain, which contains the serine residue of the active centre of the enzyme.

Fibrinogen (factor I) is a soluble, rod-shaped protein, M_r 340 000, composed of six peptide chains, in three pairs, called Aα, Bλ and γ, interconnected by disulphide bonds. Thrombin cleaves off an A peptide of 18 residues from each α chain and a B peptide of 20 residues from each β chain. With these short sequences removed, fibrinogen has been converted into a fibrin monomer containing six polypeptide chains of similar length. The monomers associate spontaneously into a relatively insoluble product, fibrin. Under the electron microscope, fibrin has a periodic structure, suggesting that the fibrin monomers are laid down in a half-staggered array:

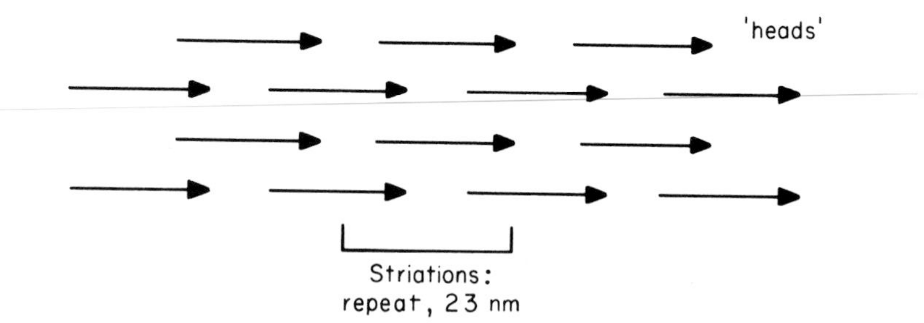

The last stage of the cascade is to stabilize the clot of fibrin. Thrombin activates factor XIII by selective cleavage in the presence of Ca^{2+} ions, forming factor XIII_a, a transamidase enzyme. This enzyme promotes the formation of peptide bonds between glutamine and lysine residues present in different chains of the fibrin fibre:

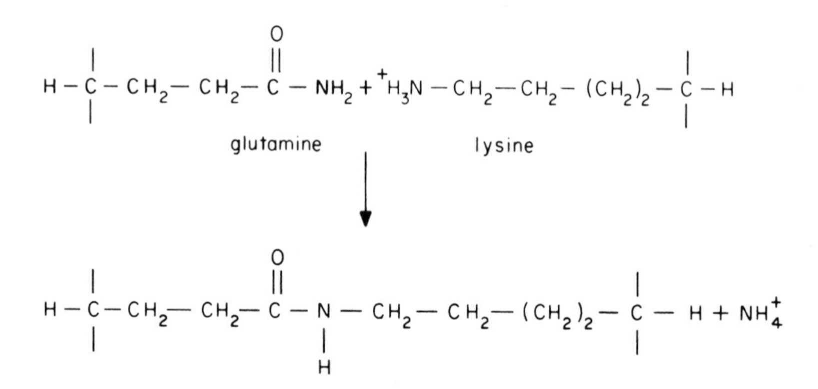

This lateral cross-linking stabilizes the fibrin clot and the remarkable cascade of reactions is completed.

9.4.4 Secretion and the signal hypothesis

Based on the classical work of Palade on the secretion of enzyme zymogens (or inactive precursors) of digestive enzymes by the acinar cells of the pancreas, the secretion of proteins is well understood (Fig. 9.18). In essence, proteins for secretion are synthesized exclusively on membrane-bound polyribosomes. The proteins enter the cisternae or channels within the membranes and are directed towards the Golgi apparatus. In this structure, the

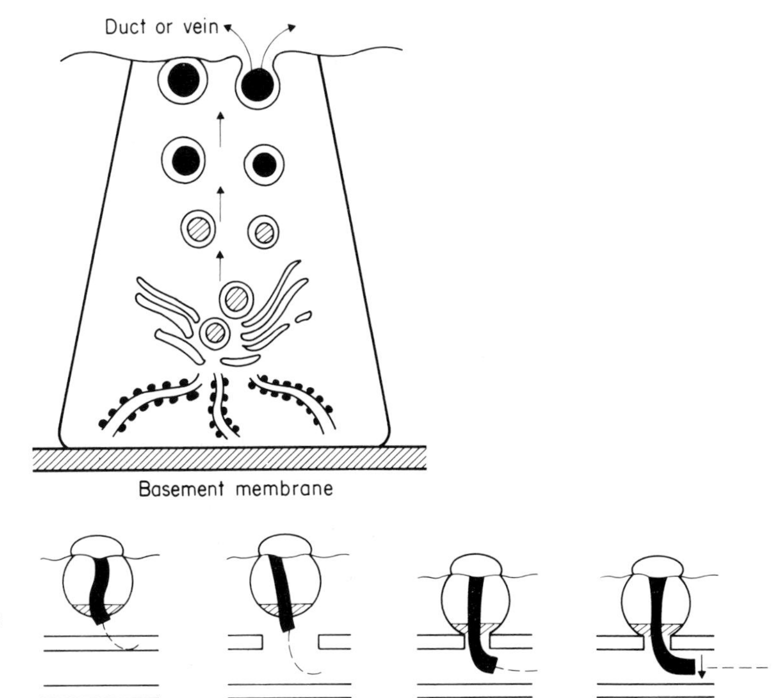

Fig. 9.18. The essential features of the secretory process. The upper diagram indicates the structures involved and the migration of secretory zymogens towards the apex of the cell, where membrane fusion releases the zymogens. The lower diagram illustrates the essential features of the signal hypothesis of Blobel and Dobberstein (1975). Hydrophobic N-terminal sequence of protein, - - -; remaining protein sequence, ▬ ; attachment sites on the large ribosomal subunit, ▨. The signal sequence enables the protein to enter the channels or cisternae in the rough endoplasmic reticulum.

secretory proteins are enclosed within membranes, forming condensing vacuoles. As these proceed towards the apical region of the cell, they become denser and more granular, forming zymogen granules. Fusion of the plasma membrane and the surrounding membrane of the zymogen granules ultimately releases the zymogens to the exterior of the secretory cell.

The synthesis and secretion of proteins as inactive precursors or zymogens prevents the secretory cell from self destruction or self stimulation, in the cases of digestive enzymes or polypeptide hormones, respectively. Most secretory proteins are emptied into ducts, but polypeptide hormones are secreted directly into the bloodstream. In addition, the site of zymogen activation by limited proteolysis can vary. Digestive enzymes, e.g. chymotrypsin, are invariably activated outside their cell of origin, but polypeptide hormones, e.g. proinsulin, are activated by cleavage within the zymogen granules.

It remained a puzzle for many years how secretory proteins enter the cisternal spaces directly after synthesis. An imaginative explanation was presented by Blobel and Dobberstein (1975) in the signal hypothesis. They suggested that the initial N-terminal or signal sequence of secretory proteins was very hydrophobic and could thus freely penetrate the membranes enclosing the cisternae. The signal sequence then opened up channels in the membrane and helped to bring the large ribosomal subunit into contact with the membrane. The large subunit furthered this attachment by means of recognition or binding sites on its lower surface. As protein synthesis proceeded, the remaining portion of secretory protein was pulled into the cisternae behind the signal sequence. Finally, as the synthesis of the secretory protein neared its conclusion, the signal sequence was cleaved off by enzymes associated with the membranes. All available evidence supports the complete acceptance of the signal hypothesis. Many different types of eukaryotic proteins, from enzymes, polypeptide hormones, immunoglobulin chains to plasma constituents are synthesized as precursors with longer N-terminal or signal sequences. There are examples of 'pre-proteins', such as prelysozyme and pre-immunoglobulin G light chain where the N-terminal of the pre-sequence is the initiating methionine residue and cleavage of a single peptide bond releases the mature and active proteins. However, there are rather more complex precursors, the 'pre-proproteins', including pre-proparathormone (parathormone previously known as parathyroid hormone) and pre-proalbumin in which there are essentially two precursor sequences. The pre-sequence contains the initiating methionine residue and probably represents the signal sequence; this sequence is cleaved very early in biosynthesis, and certainly well before the second peptide bond cleavage of the pro-sequence, which releases the active protein. The pro-sequences are quite stable in many hormone precursors, so that proinsulin and proparathormone are readily isolated. The amino acid sequences of some of the precursors of secretory proteins are given in Fig. 9.19. There is no homology in these signal sequences with respect to either length or amino acid sequence; the only invariant feature of these sequences is their hydrophobicity. Furthermore, there is no evidence that cleaved signal sequences are reutilized.

Fig. 9.19. The hydrophobic or signal sequences of certain secretory proteins. The cleavage point of pre-sequences, probably the actual signal itself, is shown as a hatched arrow; other arrows indicate the cleavage points of pro-sequences, necessary for stabilization and inactivation of the proteins during the secretory process. The initiating methionine residue is enclosed in a dashed box and the N-terminal residue of the mature, active protein is enclosed in an undashed box.

Pre-lysozyme: Met Arg.Ser.Leu.Leu.Ile.Leu.Val.Leu.Cys.Phe.Leu.Pro.Leu.Ala.Ala.Leu.Gly ⇓ Lys

Pre-IgG light chain: Met Asp.Met.Arg.Ala.Pro.Ala.Gln.Ile.Phe.Gly.Phe.Leu.Leu.Leu.Leu.Phe.Pro.Gly.Thr.Arg.Cys ⇓ Asp

Pre-proalbumin: Met Lys.Trp.Val.Thr.Phe.Leu.Leu.Leu.Leu.Phe.Ile.Ser.Gly.Ser.Ala.Phe.Ser ⇓ Arg.Gly.Val.Phe.Arg.Arg ↓ Glu

Pre-proparathormone: Met Met.Ser.Ala.Lys.Asp.Met.Val.Lys.Val.Met.Ile.Val.Met.Leu.Ala.Ile.Val.Gly.Leu.Ala.Arg.Ser.Asp.Gly ⇓ Lys.Ser.Val.Lys.Lys.Arg ↓ Ala

The enzymes necessary for these selective cleavages have yet to be identified; certainly non-specific proteases, such as cathepsins, cannot fulfil these cleavage functions. Indirect support for the signal hypothesis comes from the fact that the α- and β-globin chains, which are not secreted, do not contain signal sequences. The only inconsistent result with respect to the signal hypothesis was the apparent absence of an N-terminal signal sequence in ovalbumin, despite the fact that this secretory protein is synthesized in most birds on membrane-bound polyribosomes. An interesting explanation of this paradox has recently been presented (Lingappa *et al.* 1979); ovalbumin does contain a hydrophobic signal sequence but it is positioned internally rather than at the N-terminal end. This internal signal remains uncleaved and this must draw ovalbumin through the membranes of the endoplasmic reticulum in a hairpin-like manner during the secretion process. Lingappa *et al.* (1979) also made the interesting observation that the internal signal sequence in ovalbumin has considerable sequence homology with other secreted egg proteins, pre-ovomucoid and pre-lysozyme; the latter two precursors have conventional signal sequences at their N-terminal ends.

There may well be even more types of signal sequences because many proteins destined for chloroplasts, mitochondria and peroxisomes are made on free polyribosomes in the cytoplasm, yet they reach their appropriate locations within eukaryotic cells very precisely (for a review see Leader 1979).

Despite the widespread acceptance of the signal hypothesis, Davis and Tai (1980) have rightly drawn attention to aspects of protein secretion which are still not fully understood. In particular, the basic molecular mechanism that ensures a unidirectional transfer of the growing polypeptide chain has yet to be elucidated. In addition, it is still not known whether the signal sequence is first inserted into membrane lipids linearly or as a loop. Further, the channels enabling the polypeptide to gain access to the cisternal spaces may be permanently present or alternatively, may be induced to open up by the presence of the hydrophobic signal sequence. Last, the transport of proteins across membranes is generally believed to be an active transport mechanism, but neither the machinery itself nor the source of its energy are currently understood.

9.4.5 Glycosylation

Since glycoproteins are so common among eukaryotic proteins, e.g. blood proteins, immunoglobulins and most polypeptide hormones, considerable effort has been devoted to the elucidation of glycosylation mechanisms. A major advance (Behrens & Leloir 1970) was the identification of long-chain, unsaturated isoprenoid alcohols or dolichols; they contain 16–20 isoprenoid units and have the general structure:

$$CH_3-\overset{\overset{CH_3}{|}}{C}=CH-CH_2\left[CH_2-\overset{\overset{CH_3}{|}}{C}=CH.CH_2\right]CH_2-\overset{\overset{CH_3}{|}}{C}H-CH_2-CH_2-OH$$

isoprenoid units, 14–18

I I

Dolichols are particularly concentrated in the Golgi apparatus and the membranes of the endoplasmic reticulum, but are absent from the mitochondrial and plasma membranes. This limited intracellular distribution is in keeping with their role in the process of secretion.

The mechanisms for glycosylation are complex but many common features are now emerging (for a review see Waechter & Lennarz 1976). Glycosylation needs activated or energized sugars, formed by glycosyl-1-phosphate nucleotidyltransferases, according to the general equation:

$$XTP + sugar-1-phosphate \rightarrow XDP-sugar + PP_i.$$

The most common forms of activated sugars for glycosylation reactions are UDP-glucose, UDP-galactosamine, UDP-N-acetylglucosamine and GDP-mannose.

A large number of soluble glycoproteins share a common oligosaccharide chain, namely $\alpha(\text{Man})_n-\beta-\text{Man}-(1\rightarrow4)-\beta-\text{GlcNAc}-(1\rightarrow4)-\text{GlcNAc}$, linked to an asparagine residue. It is useful to use the biosynthesis of this mannose- and N-acetylglucosamine-containing oligosaccharide as a model for glycosylation reactions. Several of the enzymes involved still need to be fully characterized, but they are an integral feature of membranes of the Golgi and endoplasmic reticulum.

(1) UDP-N-acetylglucosamine (α-linkage) is attached to dolichol phosphate via a high-energy pyrophosphate bond.

(2) A second molecule of UDP-N-acetylglucosamine (α-linkage) is then attached forming a disaccharide, $\beta-\text{GlcNAc}-(1\rightarrow4)-\text{GlcNAc}$, attached to the dolichol moiety. Two essential points should be made here. First, the growing oligosaccharide remains attached to dolichol until the last step of the glycosylation process.

Second, that at each step involving the transfer of an activated glycosyl residue to the growing oligosaccharide chain, there is an *inversion* of configuration, i.e. an incoming UDP-N-acetylglucosamine with an α-linkage will assume a β-linkage in the growing oligosaccharide chain.

(3) GDP-mannose is then added repeatedly to complete a long-chain structure of oligosaccharide linked to dolichol.

(4) Finally, the oligosaccharide is attached to the protein via an asparagine residue, with the displacement of dolichol pyrophosphate. This sequence of events has been found in most eukaryotic cells and is best visualized by examining Fig. 9.20.

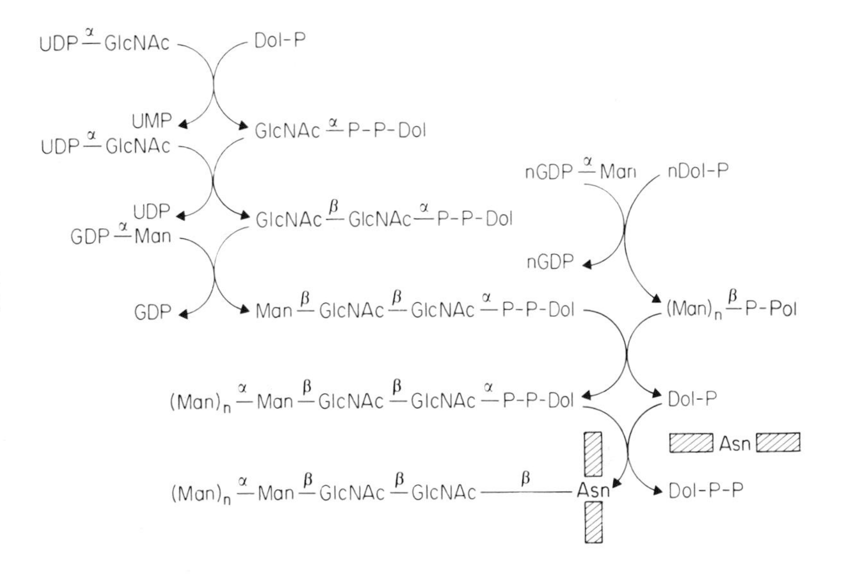

Fig. 9.20. The synthesis and attachment of an oligosaccharide commonly formed during glycoprotein biosynthesis. Note that, with the exception of the very first step, the steric configuration of a sugar residue is inverted during its transfer from an activated precursor. Dol-P is dolichol phosphate; -P-P-Dol is a pyrophosphate bond between a sugar residue and dolichol. The hatched box represents a protein with an asparagine residue.

Sialic acids (or N-acetylneuraminic acids) are often present as the terminal residues of glycoproteins in cell membranes. These terminal residues are added in the activated form as CMP-N-acetylneuraminic acid.

The classical work of Wald established the biochemical role of vitamin A (retinol) in the visual process, by the participation of its oxidised (aldehyde) form, all-*trans*-retinal, in the visual cycle. Particularly in the young, however, retinol may have a wider significance, as a deficiency results in retarded growth. It is now clear that retinol plays a part in the process of glycosylation in many tissues, including the epidermis, liver, intestinal mucosa, tracheal epithelium and cornea. It has proved possible to maintain retinol-responsive keratinocytes in culture and recent work has proved conclusively that retinol can mimic the role of dolichol in certain cells. In these cultured cells, the synthesis of mannosylretinyl phosphate has been detected; furthermore, the mannose residue can be transferred into covalent linkage with membrane proteins.

9.4.6 Arrangement of subunits and addition of prosthetic groups

The addition of prosthetic groups is an important aspect of post-translational modifications. Carboxypeptidase A is a single polypeptide chain of 307 amino acids residues, with a tightly bound zinc ion which is essential for enzymic activity. The zinc ion is located in a groove near the surface of the molecule, where it is coordinated to a tetrahedral array of two histidine side-chains, a glutamic acid side-chain and a water molecule. X-ray diffraction studies indicate that the coordinated residues are brought together from distant regions of the primary sequence. Clearly, addition of the zinc prosthetic group has a profound influence on the structure and function of the enzyme. In myoglobin, the iron atom in the prosthetic haem group has six coordination positions; four are occupied by the nitrogen atoms of the four pyrrole rings of protoporphyrin, the fifth by oxygen. Also nearby is a second or distal histidine residue, essential for oxygen binding. Elegant X-ray crystallographic studies by the research group of Kendrew (e.g. Kendrew 1961) showed that distant regions of the myoglobin were brought into a close and functional association by the addition of the haem prosthetic group. The presence of haem enables myoglobin to refold into a biologically active molecule with an extremely high content of α-helix.

Many eukaryotic proteins have a complex quaternary or subunit structure and subunit interactions have a remarkable influence on the final activity of the protein. This was forcibly emphasized for the first time by the remarkable work of Perutz and his co-workers in their determination of the three-dimensional structure of haemoglobin (e.g. Perutz 1964). Although myoglobin and haemoglobin are both oxygen transporters, they differ fundamentally in their interaction with oxygen. In particular, the oxygen dissociation curve of myoglobin is hyperbolic but that of haemoglobin is sigmoidal. Furthermore, haemoglobin can also bind CO_2 and its oxygen affinity is reduced by organic phosphates, such as 2,3-diphosphoglyceric acid. These differences were difficult to explain because early X-ray work showed that the conformations of the single chain of myoglobin and the α or β chains of haemoglobin were remarkably similar, despite many differences in their amino acid sequences. The explanation was discovered by Perutz (1970), who, in his brilliant paper, cogently argued that the distinctive properties of haemoglobin were attributable solely to subunit interactions between the two α-globin and the two β-globin chains. In essence, haemoglobin was an allosteric protein whereas myoglobin was not. Subtle intramolecular interactions changed the myoglobin-like conformations of the individual subunits of haemoglobin, resulting in profound effects on the biological properties of the protein as a whole.

Lebherz and Shackleford (1979) have examined the synthesis and assembly of the tetrameric enzyme, aldolase, in cell-free systems. These authors come to the interesting conclusion that the distinctive subunits of a protein are made concomitantly and assembled immediately into the final quaternary structure. In other words, different subunits cannot be translated asynchronously and assembled by exchange with other appropriate, but preformed, subunits. Earlier studies on the translation of the mRNAs for α- and β-globin suggest that structurally dissimilar subunits are always synthesized precisely together (Lodish 1976).

9.4.7 Additional covalent modifications

The covalent modification of eukaryotic proteins after their translation can be achieved by innumerable mechanisms (for a review see Krebs & Beavo 1979). However, phosphorylation and dephosphorylation reactions are the most important and these will be briefly discussed.

Two classical pieces of work laid the foundations for deeper insights into the mechanisms of hormone action, metabolic regulation and neurotransmission. First, Sutherland discovered the cyclic nucleotide, cyclic AMP, and suggested that it is an obligatory mediator or 'second messenger' in many hormonal responses. Second, Krebs established that cyclic AMP activated a protein kinase and set in train a cascade of reactions involving phosphorylation and dephosphorylation of proteins. As reviewed by Rubin and Rosen (1975), the inactive holoenzyme of protein kinase, M_r 174000, is composed of a regulatory (R) dimer, M_r 98000 and two, identical catalytic (C) subunits, M_r 38000. A general scheme for the hormonal stimulation of certain enzymes has been proposed by Krebs and the prime example is the stimulation of glycogenolysis in liver by adrenalin (Fig. 9.21). This cascade illustrates an important tenet of metabolic regulation by protein kinases, namely that proteins can be activated or deactivated by phosphorylation. In the example cited, there is coordinated activation of phosphorylase and inactivation of glycogen synthetase.

The simple view presented in Fig. 9.21 of the interaction of adrenalin with its membrane-bound receptor, and the subsequent activation of adenyl cyclase, needs extension in the light of recent developments. As reviewed by Ross and Gilman (1980) it is now becoming abundantly clear that the hormonal stimulation of membrane-bound adenyl cyclase also requires the presence of guanine nucleotides, in addition to the substrate, ATP. This is a general requirement for hormonal responses mediated by membrane phenomena and promoted by the synthesis of second messengers, such as cAMP. Indeed, analogues of GTP, such as Gpp(NH)p (guanyl-5′-yl-imidodiphosphate) and CTP-γ-S (guanosine-5′-0-

Fig. 9.21. The phosphorylation cascade involved in the stimulation of glycogenolysis in liver by adrenalin. The current model envisages a close association between a membrane receptor for adrenalin [hatched box] and a membrane-bound adenyl cyclase, [black box]. Phosphorylation is indicated by [P] and active enzymes have the subscript a.

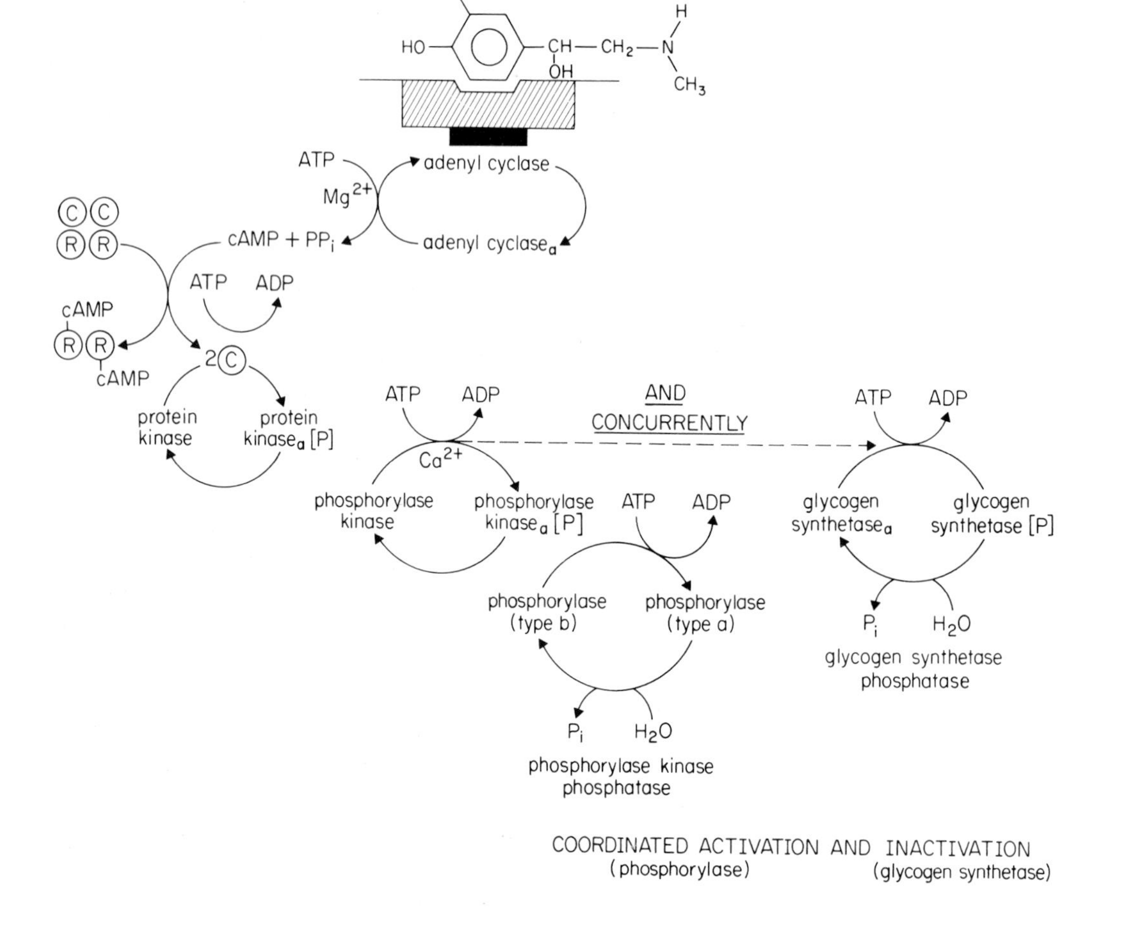

(3-thiophosphate)), and GTP itself, in certain circumstances, can stimulate adenyl cyclase directly, even in the absence of hormones. Interestingly, the cation, F^-, is another general stimulator of adenyl cyclase, as indeed are pathogenic toxins; of these, choleragen from *Vibrio cholerae* has been intensively studied.

The general involvement of both adenyl cyclase and GTP in the mode of action of hormones and neurotransmitters via membrane-bound receptors has been carefully appraised by Rodbell (1980). He envisages that membrane receptors form oligomeric complexes with GTP-regulatory proteins and prevent GTP binding to these proteins. Hormones and neurotransmitters act by releasing the inhibitory constraints imposed by the receptors, thus permitting the GTP-regulatory proteins to interact with, and ultimately to control, the activities of membrane-associated enzymes, including adenyl cyclase. In his parlance, his concept may apply generally to 'membrane signal transduction' involving surface (membrane) receptors. In this sense, transduction means the modulation (either stimulation or inhibition) of adenyl cyclase as the consequence of changes in the configuration and activity of the adjacent receptor sites by hormones or other regulatory molecules. Transduction is thus the transfer of regulatory signals through membranes and is possible only when the components in the regulatory complex are tightly integrated and juxtaposed. Rodbell (1980) proposes three classes of components within the lipid framework of the cell membrane. First, the receptor (R) component on the outer membrane, with a specific affinity and selectivity for one type of hormone or neurotransmitter. Second, the catalytic (C) units of adenyl cyclase. Third, nucleotide (N) regulatory components of adenyl cyclase, for binding GTP and responsible for mediating the effects of GTP and the various hormones on the activity of the C units. Together, C and N units constitute the holoenzyme of adenyl cyclase. Two types of N component are distinguishable in terms of function and response; type N_s evoke stimulation of adenyl cyclase by GTP, whereas type N_i evoke cyclase inhibition by GTP. As a corollary to the two N_s and N_i GTP-binding components, two classes of receptors (R) are proposed; R_a, binding activating hormones for adenyl cyclase and linked to N_s components, and R_i, binding inhibitory hormones for adenyl cyclase and linked to N_i components. The complexes of R_i and N_i need Na^+ ions to collectively inhibit adenyl cyclase; by contrast, R_a and N_a, which together activate adenyl cyclase, have no such ion requirement. The R_aN_s and R_iN_i complexes, side by side in the membrane are presented in Fig.

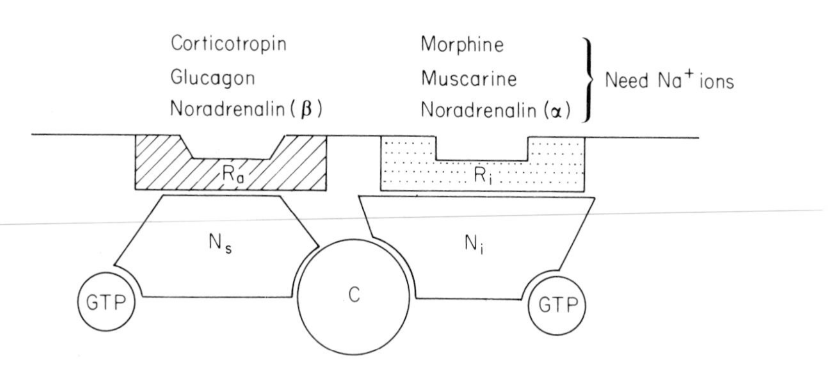

Fig. 9.22. Two adjacent receptor sites in a membrane, with either subsequent activation or inhibition of associated adenyl cyclase. The left-hand receptor, R_a, is associated with the activating form of GTP-binding components, N_s, and leads to a stimulation of adenyl cyclase. The right-hand receptor, R_i, is associated with an inhibitory form of a GTP-binding component, N_i, and leads to an inhibition of adenyl cyclase. (Redrawn, with permission, from Rodbell 1980.) The catalytic subunit of adenyl cyclase is depicted as C. Examples of hormones working by R_a or R_i receptors are given. For the significance of α- and β-noradrenalin receptors, see Table 9.6.

9.22. This is a most plausible concept and satisfactorily explains most aspects of the regulation of adenyl cyclase by a variety of hormones and pharmacological agents.

All cAMP-dependent protein kinase systems are kept under tight control by three safety devices (Krebs & Beavo 1979). First, most cells contain a phosphodiesterase, capable of inactivating cAMP by its hydrolysis to AMP. Second, there is a ubiquitous protein kinase inhibitor, a heat-stable protein, M_r 11 000, which can associate with the catalytic (C) subunit and inactivate it; the mechanism of this inhibition remains to be clarified. Third, protein phosphatases abound in eukaryotic cells. One protein phosphatase, M_r 35 000, has been purified to homogeneity from several tissues and this dephosphorylates a wide range of substrates. It is also most likely that specific protein phosphatases exist, such as glycogen synthetase phosphatase (Fig. 9.21).

The tetrameric form of cAMP-dependent protein kinase discussed earlier is present in all eukaryotic cells and is responsible for the initiation of many cascade reactions involving phosphorylation. All proteins studied thus far are phosphorylated at a serine residue, in the sequences -Lys.Arg.X.X.Ser- or -Arg.Arg.X.Ser-.

Later work has established the presence of four more categories of protein kinases.

(1) *cGMP-dependent protein kinases.* Cyclic GMP is another biologically active cyclic nucleotide and it activates a special class of

protein kinases. Two such enzymes have been purified to homogeneity from the lung and the silk gland of *Bombyx mori*. They are structurally different from cAMP-dependent protein kinases. The cGMP-stimulated enzymes, M_r 150000, are composed of two identical subunits which are not dissociated by the nucleotide activator; furthermore, these enzymes are insensitive to the heat-stable inhibitor of cAMP-dependent protein kinases.

(2) *Ca^{2+} ion-dependent protein kinases.* The best example is phosphorylase kinase (Fig. 9.21); the only other known example is myosin (light chain) kinase.

(3) *Double-stranded* RNA-dependent protein kinase. The best example of such an enzyme has already been described in section 9.3.8 with respect to interferon.

(4) *Protein kinases with no known activator.* The only one studied in detail so far is pyruvate dehydrogenase kinase (Barrera *et al.* 1972). The phosphorylation sites, three in all, are in distinctive sequences, namely -His.Gly.His.Ser.Met-, -Pro.Gly.Val.Ser.Tyr- and -Met.Gly.Thr.Ser.Val.-.

The protein kinases have been discussed until now only with respect to enzymes involved in intermediary metabolism. However, it is now clear from very exciting and new research that cyclic

Table 9.6. A survey of biologically important protein kinases. Histamine, H_1 and H_2, and noradrenalin, α and β, indicate different types of receptors in the central nervous system with different responses.

Type of protein in kinase	Typical activators	Typical substrates
cAMP-dependent	Adrenalin Glucagon Vasopressin, oxytocin Neurotransmitters, e.g. dopamine, histamine (H_2) noradrenalin (β)	Glycogen phosphorylase Glycogen synthetase Fructose-1,6-biphosphatase Pyruvate kinase Phosphofructokinase Hormone-sensitive lipase Histones Synaptic membrane proteins
cGMP-dependent	Nuerotransmitters, e.g. acetylcholine, histamine (H_1) noradrenalin (α)	Synaptic membrane proteins
Double-stranded RNA-dependent	With interferon	Synthetase for the translational inhibitor, pppApApA
Ca^{2+} ion-dependent	Exclusively Ca^{2+} ions	Phosphorylase kinase Myosin (light chain) kinase
Independent (i.e. no known activator)	None	Pyruvate dehydrogenase kinase

nucleotides and nucleotide-dependent protein kinases also play a vital part in neurotransmission (for a review see Greengard 1979). This work particularly broadens the physiological and biochemical functions of cGMP-dependent kinases.

An abbreviated summary of current research on protein kinases, their activators and their substrates is presented in Table 9.6. From this it is evident that protein kinases are critically important regulators. Some years ago the yin-yang hypothesis had a rather fashionable following (for a review see Goldberg & Maddox 1977). This hypothesis suggested a reciprocal relationship between cAMP and cGMP; as one increased, the other concomitantly decreased. There is no sound evidence to support this idea. It is also remarkable that it is fifty years since insulin was discovered by Banting and Best and yet we still do not fully understand how it works. The present impasse is ably reviewed by Czech (1977), but it may turn out that insulin responses are largely mediated by the concerted action of Ca^{2+} ion-dependent and cGMP-dependent protein kinases.

Calmodulin

The importance of Ca^{2+} ions in stimulating certain protein kinases provides a suitable introduction for a ubiquitous Ca^{2+}-binding protein, calmodulin (for a review see Means & Dedman 1980). Calmodulin is present in all eukaryotic cells and contains four calcium ion-binding sites per polypeptide chain (M_r 17 000). The most distinctive features of the protein are the post-translational addition of three methyl groups to the lysine at position 115 of the 148 amino acid sequence and its close structural homology with the troponin-C component of skeletal muscle. Trimethylation of cytochrome c is necessary for its interaction with cytochrome oxidase and the similar modification of calmodulin is believed to be vital for the performance of its diverse biological functions. In most cells, the synthesis of calmodulin is constitutive, and hence its influence on biochemical reactions is mediated by changes in the flux or distribution of Ca^{2+} ions. Using monospecific antibodies, calmodulin is widely distributed within eukaryotic cells, being especially concentrated in the actin-like filaments of the cytoskeleton. However, the nucleus contains little, if any, calmodulin. The diverse functions of Ca^{2+} ions and calmodulin are only just beginning to be understood, but the range of processes reflected by this calcium ion-binding protein is impressive. It is now abundantly clear that calmodulin has a profound influence on the assembly of microtubules from α- and β-tubulin monomers, and conversely, on the breakdown of microtubules. Not surprisingly, calmodulin also plays a vital part in the uptake and distribution of Ca^{2+} ions. Many secretory processes (e.g. release of neurotransmitters and the secretion of ions by the stomach and duodenum) are regulated by calmodulin. The influence of calmodulin on intermediary metabolism and intracellular regulation extends far beyond its stimulation of certain protein kinases (Table 9.6); for example, the activities of myosin (light chain) protein kinase, phosphorylase kinase, NAD^+ kinase and tryptophan oxygenase, to mention but a few, are all modified by calmodulin. Undoubtedly, calmodulin has a profound influence on the function and regulation of all eukaryotic cells.

Chapter 10 Organelles

10.1 THE EXTRANUCLEAR GENES OF EUKARYOTES

The study of the molecular biology of organelles has arisen from the fusion of two subjects. One is the non-Mendelian inheritance of certain genetic traits in eukaryotes and the other is the biochemistry of nucleic acids in mitochondria and chloroplasts. The fusion process has led to one of the most exciting areas of current eukaryotic molecular biology. The literature takes a little getting used to because, as with any union of geneticists and biochemists, conflicting nomenclatures have arisen. Fortunately, there is one superb book (Gillham 1978) and two detailed review articles (Birkby 1978, Tzagoloff & Macino 1979) to guide the newcomer through the literature; wherever possible, we have limited references to one of these three sources. By way of introduction, we consider the general question of extranuclear genes and survey briefly the nature of their replicons.

Non-Mendelian or cytoplasmic inheritance is evidenced by a genetic trait which (a) is not linked to any known nuclear gene, and (b) segregates in an anomalous fashion following meiosis. This is sometimes referred to as maternal or uniparental inheritance. Certainly most non-Mendelian inheritance is uniparental and maternal. However, there are exceptions in the genetics of higher plants. Examples of Mendelian and non-Mendelian inheritance in baker's yeast (*Saccharomyces cerevisiae*; referred to from now on simply as yeast) are shown in Fig. 10.1.

The phenomenon of non-Mendelian inheritance raises a number of intriguing questions: one concerns the nature of the genetic element that contains the genes; another regards the explanation of the non-Mendelian patterns. We shall return to the former question in the following paragraph, here we merely note that the genomes of mitochondria and chloroplasts account for many, but by no means all, of these genes. There is an obvious answer to the second question. Take, for example, the fertilization of a female gamete (egg) by a male gamete (sperm). The sperm, we might imagine, only contributes nuclear material to the zygote. All the cytoplasm, including the mitochondria and their genes, comes from the egg. The most casual consideration of the problem immediately suggests that this view is far too simplistic to account for non-Mendelian inheritance in general. For one thing, in the case of yeast, the two gametes are indistinguishable except for the state of the *a*/α gene; for another, the inheritance of chloroplast (plastid) genes is frequently biparen-

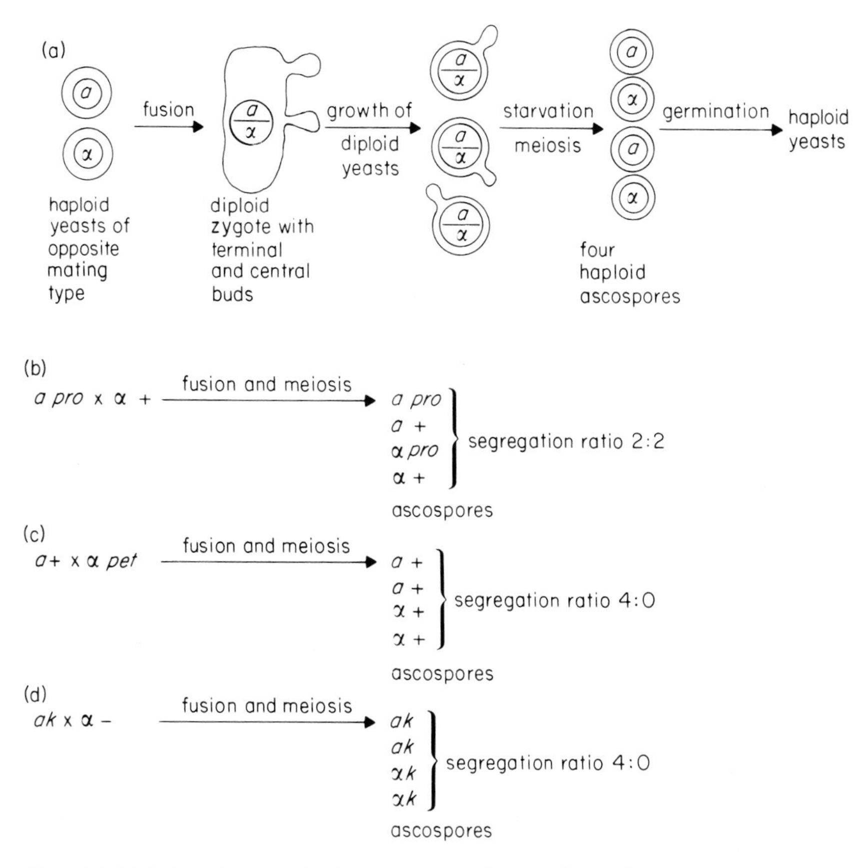

Fig. 10.1. Meiosis and segregation in yeast. (a) Summary of the formation of a zygote and consequence of meiosis. The circle represents the nucleus; *a* and *α* are the two mating types. (b) Segregation of alleles for a nuclear gene; *pro* represents a requirement for proline and + prototrophy (lack of that requirement). (c) Segregation of a mitochondrial gene. Here a respiratory deficient petite (*pet*) strain is crossed with a wild type (+); the + phenotype (derived from the *a* parent) is present in all the products of meiosis. (d) Another example of segregation of a uniparentally inherited phenotype, killer (*k*). The lack of killer factor in one parent is represented by −.

tal. Probably the most familiar example is in the garden varieties of the 'geranium' (*Pelargonium* x *zonale*) but it is true of plants representative of all the major vascular types. In one extreme case, the Japanese cedar (*Cryptomeria japonica*), although the inheritance is biparental, the commonest pattern is paternal. This may not be quite so unusual as it might appear; *C. japonica* is the only gymnosperm (conifer) in which plastid inheritance has been studied and paternal inheritance could be common in this group. Of the many possible mechanisms for non-Mendelian inheritance, the one that is most generally applicable, and for which there is some experimental evidence (Birkby 1978), is the selective degradation of the extranuclear DNA inherited from one parent. This could be readily achieved by a eukaryotic counterpart of restriction and modification (see section 3.4.2).

Some examples of extranuclear genetic elements are listed in Table 10.1. With the exception of the organelle genomes, this is the only opportunity in this book for referring to some of these extremely fascinating systems and, for this reason, we present here a short summary of the genetic systems summarized in the table. *Paramoecium,* a large single-celled protozoan and a member of the Ciliophora, is genetically complicated. The vegetative cells contain two nuclei, one of which, the macronucleus, contains amplified DNA sequences that are the sites of transcription. The smaller micronucleus defines the genotype and directs the production of a macronucleus in the zygote. The sexual system is further complicated by the existence of a large number of mating types. However, from the present point of view, the importance of *Paramoecium* lies in the

Table 10.1. Summary of some examples of extranuclear genetic elements in eukaryotic cells (see text for a discussion).

Organism or group and reference	Name of gene(s) and/or functions	Biochemical nature of genetic element(s)
Paramoecium aurelia (a)	Killer factors, κ, π, μ, λ, σ, γ, η	Symbiotic bacteria
	R bodies in killers	Bacteriophages
All eukaryotes (b)	Many (not all) mitochondrial genes	mtDNA
Eukaryotic algae and plants (b)	Many (not all) chloroplast genes	cpDNA
Trypanosomes (c)	Amplified mitochondrial genes	Kinetoplast DNA
Yeast (b, d, e)	ρ factor	mtDNA
	Killer	dsRNA
	URE-3; impaired regulation by NH_3 of pyrimidine biosynthesis	Unknown
	ψ; reinforces certain suppressors	Unknown
	π; resistance to oligomycin and venturicidin	Unknown (? episome)
	SUP1-16; a super-suppressor	Unknown (? episome)
	Plasmid with no established function	2 μm DNA
Neurospora crassa (d)	*poky (mi-1)* and other mutations suppressible by *f*	mtDNA
	mi-3; suppressible by *suI* but not *f*	Unknown
	Stopper; suppressible by neither *f* nor *suI*	Unknown
Other fungi (d)	Many examples	Unknown

mtDNA, mitochondrial DNA; cpDNA, chloroplast DNA; dsRNA, double-stranded RNA. References are to books and reviews: (a) Preer *et al.* (1974); (b) Birkby (1978); (c) Borst *et al.* (1976); (d) Fincham *et al.* (1979); (e) Wickner (1976).

large variety of 'killer strains' which are capable of killing paramoecia lacking the killer factors. The killer factors are small intracellular bacteria, some of which can be grown axenically (just). The killer systems are special cases of mutualism or symbiosis. The DNA-containing R bodies are probably the phages for the killer symbionts although some of these have an extraordinary morphology quite unlike that of any conventional group of phages. Mitochondria and chloroplasts resemble the killer factors of paramoecia in that they contain DNA, tRNA and ribosomes and can, therefore, direct their own gene expression. They differ fundamentally in one respect. Essential components of these organelles are the products of nuclear genes. Although it is possible that these organelles evolved from what were originally symbiotic prokaryotes of some sort, they cannot be thought of as such in a functioning cell of a present day organism. In contrast to the genuinely symbiotic bacteria (including the *Paramoecium* killer factors), which show a state of physiological mutualism and interdependence with their host, chloroplasts and mitochondria are in a state of genetic interdependence with their host. The kinetoplast is discussed further in section 10.2.1.

Yeast contains several extranuclear genetic systems. In addition to its mitochondrial genes, which are encoded by mitochondrial DNA (mtDNA, sometimes referred to as ρ in the case of yeast), there are several other genetic determinants. The killer strains of yeast are designated K^+R^+. They contain genes K (for a factor which is lethal to other yeasts) and R (resistance to the factor). The other three possibilities are K^-R^- (sensitive), K^-R^+ (neutral) and K^+R^- (self-killer; this seemingly suicidal genotype can be studied in a suitable nuclear background that suppresses K expression). The killer factor consists of a covalently closed double-stranded RNA molecule, i.e. an RNA plasmid. It is probably best thought of as an example of a temperate RNA virus. *URE*-3 and the ψ factor are two examples of cytoplasmic genes, at present in search of a coding macromolecule. The π factor is an interesting example of a genetic element which can move from place to place in the chromosome. It produces resistance to the two respiratory inhibitors oligomycin and venturicidin; in other words, it is involved with the function of the mitochondrion. However, these two resistances map in one of the nuclear chromosomes. Loss of π is associated with sensitivity to the antibiotics; however, venturicidin-resistant strains can be obtained from such organisms. The resistance now maps at a different locus and oligomycin resistance is not regained. The suggestion is that π is an episome-like element, or possibly a transposon, which can be moved from one site to another and which requires a specific site for expression of an otherwise cryptic oligomycin resistance gene. The suppressor, *SUP1-16,* which confers an exceptional degree of nonsense suppression of a type characteristic of yeast, is an example of another gene that can be mapped at two different loci. The mechanism involved in these events is certainly going to prove of great interest and will have implications far beyond the confines of yeast genetics. The many well-established examples of genetic instability in higher plants could be due to comparable processes. Finally, for yeast, there is the well-described plasmid invariably referred to as the '2 micron' (2μ) plasmid/DNA. A genetic role for this remains to be proved although it might be involved in mobilization of π. The plasmid has an unusual structure consisting of two unique sequences separated by two copies of an inverted repeat. Recombination occurs between the inverted repeats so that a preparation of 2μ DNA consists of an equilibrium mixture of two species in which the unique sequences are in either of their possible relative orientations.

Neurospora crassa is another fungus that competes with yeast for the volume of its genetic literature. Several morphological mutations in *Neurospora* map on extranuclear genes. Many of these are described as *mi* (maternally inherited) and can be grouped (Table 10.1) according to their suppression or otherwise by one of the nuclear mutations *f* and *suI*. The *poky* mutants are classical, respiratory-defective mitochondrial mutations. The other groups are apparently unlinked to *poky*. Stopper is an interesting phenotype: it limits the extent of vegetative mycelial growth and the corresponding gene is, therefore, concerned with senescence. The table indicates that many fungi exemplify non-Mendelian inheritance. The question of how widespread such genes may be cannot be answered. There are no other eukaryotes for which the molecular genetics are as well understood as yeast and *Neurospora* and the likelihood of the detection of the likes of Stopper and *URE*-3 in a genetically poorly defined organism is slight.

The genomes of mitochondria and chloroplasts

The rest of this chapter is concerned with the genetic material and gene expression in organelles. Two such organelles have been studied by the techniques of transmission genetics. Transmission genetics is a convenient phrase to describe the subject of analysing a genome by studying recombinants following transfer or reassortment. People studying the progeny of Hfr crosses in *E. coli,* those mapping *Neurospora* chromosomes and breeders of fancy goldfish are all examples of 'transmission geneticists'. One might think that some phrase such as 'conventional genetics' might suffice. However, the methods of organelle genetics are far from conventional. The two organelles that have been extensively analysed by transmission genetics are the yeast mitochondrion and the chloroplast of the single-celled alga, *Chlamydomonas reinhardtii.* The reason for these two choices is simple; both organisms can grow in the total absence of expression of any organelle genes. Thus a non-mitochondrial yeast will grow fermentatively and an aplastic *Chlamydomonas* can grow heterotrophically. Thus not only can a wide variety of mutants be obtained but changes in gene expression in wild types following change from fermentative to respiratory growth (in yeast) or from heterotrophic to photosynthetic growth (in *Chlamydomonas*) can be studied. These then are the two organelles that have given genetic maps. However, the absence of transmission genetics does not preclude deduction of gene organization. The advent of nucleic acid hybridization, sequencing and recombinant DNA analysis have made possible, in principle, the chemical analysis of any genome. Earlier we gave an example of a detailed genetic map for systems for which no genetic exchange or recombination is possible. Perhaps the most striking example is phage Qβ (see section 6.5.6). In particular, DNA sequencing is making an enormous impact on the understanding of the genetics of the human mitochondrion, with dramatic consequences (see section 10.4.1).

10.2 STRUCTURE AND ORGANIZATION OF ORGANELLE GENOMES

10.2.1 Mitochondrial and kinetoplast DNA

Size and shape of mitochondrial DNAs

The smallest mitochondrial DNA molecules are found in higher animals and the largest ones in plants (Table 10.2). With the exception of two protozoa, *Tetrahymena* and *Paramoecium*, which have linear mtDNA, all the mtDNAs that have been studied are covalently closed supercoiled molecules which resemble plasmids (see section 3.1.4).

Table 10.2. Approximate sizes of mitochondrial DNA molecules (summarized from Gillham 1978).

Group of organisms	Length of mtDNA (μm) $M_r \times 10^{-6}$	mtDNA
Higher animals, i.e. worms, insects and vertebrates	5–6	*c*. 10
Protozoa	13–15	*c*. 50
Fungi	6–25	12–50
Chlamydomonas	4.6	10
Higher plants	30	*c*. 74

The yeast mitochondrial genome

Technical details of yeast mitochondrial genetic analysis are outside the scope of this book. Most of the methods rely upon the use of certain antibiotics, resistance (or sensitivity) to which is determined by mitochondrial genes. Recombinants are obtained by cloning diploids or by direct analysis of zygotes (Fig. 10.1). The genetic system is unusual in several respects. Of the non-respiratory mutants, some are designated ρ^-. This nomenclature refers to the absence of a minor component in total yeast DNA called the ρ factor. The usage is historical: ρ and mtDNA are synonymous. Certain ρ^+ respiratory mutants have suffered massive deletions in the mtDNA molecule but the length of the molecule is unchanged (about 22 μm). In other words, yeast mtDNA adjusts itself by gene duplication to maintain a fixed size. Another unusual feature of the genome is the 'polarity determinant', ω. The effects of different allelic states for ω determine genetic polarity, that is the probability that alleles are derived from one parent rather than the other; finally, recombination frequencies are exceptionally high. An outline genetic map of yeast mtDNA is shown in Fig. 10.2; the legend includes references to reviews of mitochondrial genetics. The most striking feature of the map is the presence of large inserts or introns (see section 8.1). The role of the inserts is puzzling. Many of them are missing from the map of mtDNA from the very closely related yeast, *S. carlsbergensis*. Quite apart from its inserts, the gene *oxi*-3 is vastly larger than the coding requirements for its fairly modest sized protein and there may be some sort of duplication in this region. Other unusual features are the splitting of the two large rRNA cistrons by a large gap and the curious properties of *var*-1; *var* means variable and this gene encodes a ribosome-associated protein that varies enormously in molecular weight in different, otherwise closely related, strains. Finally *oli*-1 raises a question regarding the degree of genetic independence in

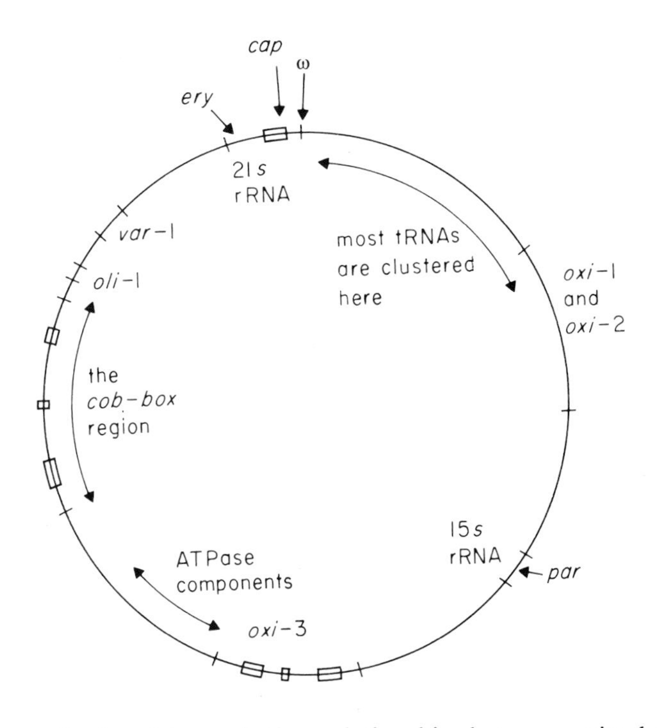

Fig. 10.2. Outline mitochondrial genetic map of yeast (*Saccharomyces cerevisiae*). The total length of the molecule corresponds to 78 kb pairs. Genes (other than ATPase, rRNAs and tRNAs) are as follows: *oxi-1,2,3* subunits of cytochrome oxidase, the *cob-box* region apo-cytochrome b (a component of cytochrome c reductase), *oli*-1 a mitochondrial protein that renders the ATPase complex resistant to oligomycin, *var*-1 a ribosome-associated protein. The boxed regions are introns. Outside the circle, *ery*, *cap* and *par* show the positions of mutations that determine resistance to erythromycin, chloramphenicol and paroromycin respectively; ω is discussed in the text. (See Borst & Grivell 1978 and Linnane & Nagley 1978 for details of the map).

mitochondria and the relationship between mitochondrial and nuclear genes; the protein corresponding to Oli-1 in *Neurospora* is a nuclear gene product. We shall return to the yeast mitochondrial genome in section 10.4.1 in which we shall see that the strange base composition of yeast mtDNA (18% GC) results in an unusual reading of the genetic code.

Kinetoplast DNA

Kinetoplasts are found in protozoa of the family *Trypanosomatidae*. These organisms are all parasitic and include the members of the genus *Trypanosoma* which are dixenous parasites of invertebrates and vertebrates and include the causative agent of sleeping sickness. The life cycle of trypanosomes is complicated. There are three flagellated forms which differ in the arrangement of the undulating outer membrane and the flagellum (Fig. 10.3(a)). The volume of the cell is mostly occupied by a single huge mitochondrion. The nucleus and a small piece of surrounding cytoplasm are located in a hole in the structure. The kinetoplast is a prominent structure, visible by light microscopy, about 1 μm in diameter. It derives its name from the fact that its position depends upon the positioning of the flagellar apparatus (as shown in the figure) although this associa-

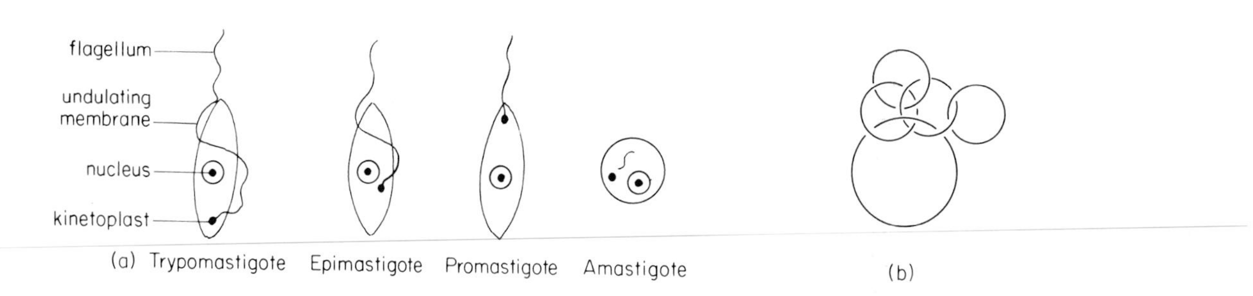

Fig. 10.3. (a) Gross structure of some of the forms of a trypanosome (redrawn after Hoare & Wallace 1966). (b) A diagram of a catenate of one large circle and four small ones. Kinetoplast DNA consists of an enormous net of small and large DNA circles linked together in this way.

tion of the structures is not reflected in any direct functional relationship between the two. The kinetoplast contains the mtDNA. The kinetoplast DNA consists of large and small circles linked together in a vast network by catenation (Fig. 10.3(b)). Catenation is the name given to the linking of closed annular structures by the type of topology typified by the links in a chain or the emblem of the Olympic Games. In the case of the most closely studied kinetoplast DNA, that of the monoxenous invertebrate parasite *Crithidia fasciculata* (Borst *et al.* 1976), the large circles (M_r 22×10^6) constitute the informational mtDNA. The main role of the small circles, which are heterogeneous, is believed to be structural.

10.2.2 Chloroplast DNA

The chloroplast DNA (cpDNA) from higher plants has a contour length of around 40–45 μm (of the order of M_r 100×10^6); in algae it is larger. The cpDNA of *Chlamydomonas* is 62 μm long (M_r 134×10^6); that of the giant unicellular alga, *Acetabularia*, may be as long as 200 μm and is comparable to a bacterial genome. The true coding capacity and the physical nature of this molecule are unknown; otherwise all cpDNAs are covalently closed.

The Chlamydomonas *chloroplast genome*

Chlamydomonas can grow photosynthetically on CO_2 as sole carbon source. However, it can dispense with the photosynthetic system if supplied with acetate. Growth on acetate in the light is termed mixotrophic growth; growth in the dark is heterotrophic. Mutants lacking an essential component of the photosynthetic apparatus have an acetate-requiring phenotype and are designated *ac*. Chloroplast mutations consist of certain *ac* types and also those resistant to certain antibiotics. A *Chlamydomonas* cell contains a single large chloroplast which is cup shaped and occupies 40% of the cell's volume. Details of the genetic system are to be found in Gillham (1978). Vegetative cells are haploid. The cells aggregate and then pair off to form the zygote (the two gametes in such a pair must be of opposite mating type; the two mating types are called mt^+ and mt^-). Most zygotes mature and produce haploid meiotic products, however, some invariably produce a vegetative diploid generation. These diploids can be used for studying recombination and segregation. Alternatively it is possible to select, among the haploids, for the rare biparentally inherited cpDNAs and to use these for genetic analysis. The methods have yielded a detailed circular linkage map for the *Chlamydomonas* chloroplast.

10.3 REPLICATION AND TRANSCRIPTION IN CHLOROPLASTS AND MITOCHONDRIA

10.3.1 DNA replication

Most mitochondria contain about 5 copies of mtDNA. Most chloroplasts of higher plants contain about 20–30copies of cpDNA, however, *Chlamydomonas* contains 80 copies and *Euglena* between 100 and 270.

The DNA polymerases of the organelles are presumed to be nuclear gene products. The mechanism of DNA replication is poorly understood. The overall mode is semi-conservative, even in the case of the linear molecule in *Paramoecium* mitochondria (see section 10.2.1; Cummings 1977). In general, the circular genomes of mitochondria and chloroplasts are believed to be replicated in the same general way as the *E. coli* chromosome (see section 4.2), however, there are exceptions. *Drosophila* mtDNA is replicated in a highly asymmetric fashion in which synthesis on one strand is almost complete before that on the other is initiated (Goddard & Wolstenholme 1978). *Chlamydomonas* cpDNA has, apparently, an alternative mode of replication by a rolling circle mechanism (see section 4.2.5, and Kung 1977 for a review) when amplification is required following transfer from heterotrophic to photosynthetic growth.

Organelle DNA is present in large numbers of copies and the copy number is not, in general, under nuclear control. Lyman and Srinivas (1978) present evidence for a role of nucleases in determining the cpDNA copy number in *Euglena*.

10.3.2 Transcription

The somewhat conflicting literature on the subunit composition and antibiotic sensitivity of organelle RNA polymerases is reviewed by Gillham (1978). In general it is probably not true to state that these enzymes are of a 'prokaryotic type'. Little is known about the regulation of transcription, but in general, it would appear that, at least in mitochondria, there are few promoters and extremely long transcripts are processed to yield tRNA, rRNA (or their precursors) and mRNA (Borst & Grivell 1978). Moreover, in HeLa cells, both DNA strands are transcribed (see the following section).

10.3.3 Transcription products

Transfer RNA

All mitochondria and chloroplasts contain genes for tRNA. In the cases which have been studied, given the special reading of the genetic code in mitochondria (see section 10.4.1) mtDNA appears to contain a complete set of tRNA cistrons and it is not necessary to invoke the importation of cytoplasmic tRNA. Likewise chloroplast tRNA, in general, is probably transcribed largely from cpDNA and does not need to be imported. Indeed in *Euglena* the chloroplasts actual export certain tRNA species to the cytoplasm (McCrea & Hershberger 1978).

Ribosomal RNA

All organellar DNAs studied so far have cistrons for the two high molecular weight rRNA species. These molecules are much smaller than the corresponding cytoplasmic rRNAs and, indeed, are in general smaller than the 16*s* and 23*s* rRNAs of bacteria; for example, the yeast rRNAs are 15*s* and 21*s* and the animal ones are 12*s* and 16*s*. Chloroplast ribosomes contain 5*s* rRNA, a gene product of cpDNA. Mitochondrial ribosomes lack 5*s* rRNA (Borst & Grivell 1978).

Messenger RNA

The knotty problem of whether it is necessary to invoke importation of mRNA into organelles is discussed by Gillham (1978). It is probably not necessary in most cases. Organelle mRNAs have poly(A) tails similar to cytoplasmic mRNAs. Chloroplast mRNA can be translated *in vitro* and the proteins can be characterized. For this reason cp mRNA was better understood for some time than mt mRNA. However, the advent of rapid DNA sequencing methods, coupled, in the case of yeast, with the development of fine structure mitochondrial genetics has led to a very detailed knowledge of gene expression in the mitochondria of yeast and man (see section 10.3.3). Of particular importance has been the understanding of the processing of the mRNA for the yeast mitochondrial *box* mRNA (Lazowska *et al.* 1980). The gene contains introns; analysis of the processing has been facilitated partly by comparison of different versions ('long and short') of the *box* region in different laboratory strains of yeasts and partly by the fact that a defect in certain *box*$^-$ mutants can be complemented by a diffusible factor with properties to be expected of a splicing enzyme (or component of such an enzyme) from different *box* mutants. The layout of the 'long' *box* region is summarized in Fig. 10.4(a). Certain intron sequences have a coding capacity. By this

Fig. 10.4. The *cob-box* region of yeast mitochondria (compare Fig. 10.2 for the position in the map). The data here refer to the 'long' form of the gene (see text; data from Lazowska *et al.* 1980).
(a) Arrangement of introns (I) and exons (E) in the gene. The exons are all approximately to scale except for E2 which is very short (see below). Exons are represented by the line and introns by boxes. The whole region corresponds to about 1.18 kb pairs. (b) Junctions between E1/I1, the I1/E2/I2 junction region and I2/E3. The non-transcribing strand is shown. Numbers in the discontinuous line are residues omitted from this figure. Underlined triplets are codons (double underlining for termination codons). The three residues marked * form a methionine codon (ATG) when I1 has been spliced out. The RNA maturase extends from this codon, through E2 into the bulk of I2 (the leucine being *C*-terminal). An internal part of the cytochrome b is represented by the codon assignments in E1, E2 (including the A*T*G*) and E3 (see text for a discussion).

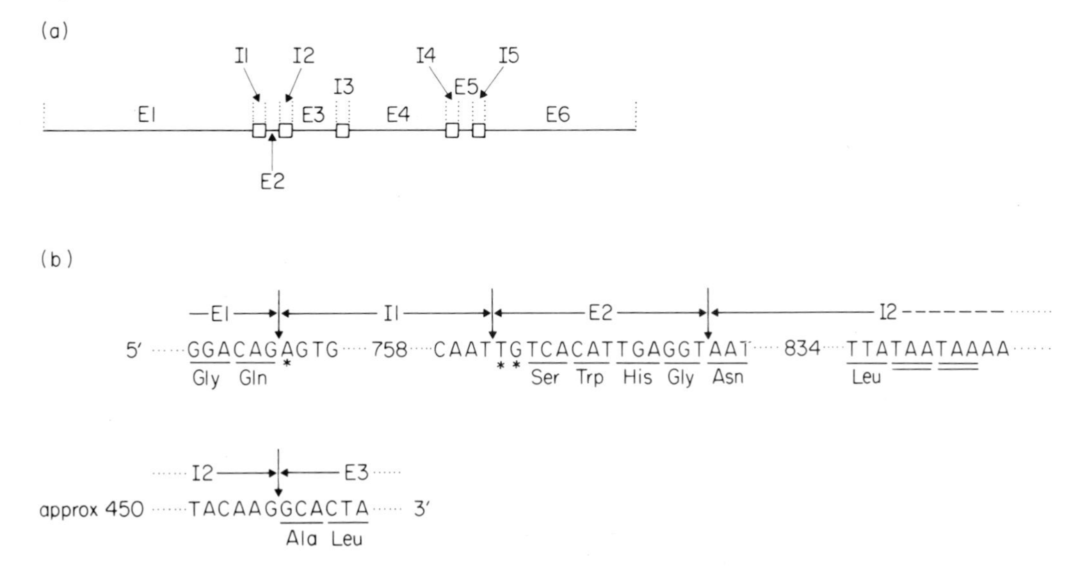

we do not mean that there is necessarily an AUG in phase with a termination codon and that in between the two there is a reasonably long and 'sensible-looking' piece of mRNA. All we mean is that in at least one of the three reading frames there is a reasonable stretch of 'sensible' sequence uninterrupted by nonsense (chain terminating) codons. The only such codons we have considered are UAA and UAG (UGA is a sense codon in mitochondria; see section 10.4.1). To follow the terminology of Lazowska *et al.* (1980) such reading frames are said to be 'open'; ones with intervening nonsense codons are 'blocked'. The bulk of intron I1 (Fig. 10.4) is blocked in all three reading frames; an open reading frame can be found at the extreme righthand end of the intron. The left-hand 2/3 of I2 has one open-reading frame in phase with the cytochrome b (exon) reading frame. Other reading frames are blocked. The situation in I4 is similar. Otherwise I5 is completely blocked and the sequence of I3 is not yet determined.

The sequences surrounding the junctions E1/I1, I1/E2/I2 and I2/E3 are shown in Fig. 10.4(b). Analysis of mutants established that the I1 region includes the coding sequence for a splicing enzyme ('maturase'). The early stages in the maturation of *cox* mRNA, therefore, involve the following steps:

(1) the whole transcript is produced;

(2) the I1sequence with its blocked reading frames is excised by a different maturase (independent studies show this to be a cytoplasmic translation product);

(3) an initiation codon (AUG) is now generated within E2 and this allows expression of the maturase in the open-reading frame of I2;

(4) I2 is now excised.

Later stages are believed to involve sequential removal of the remaining introns (possibly involving a situation in I4 similar to that in I2). Note that the very short cytochrome b coding sequence E2 plays two distinct roles: following splicing out of I1 it generates the initiating AUG and the following four codons for the *N*-terminal sequence for the maturase and also (following final maturation) it forms a part of the cytochrome when the AUG in question encodes an *internal* methionine residue.

10.3.4 The human mitochondrial genome and its transcription

Presently, the subject of organelle molecular biology is dominated by the tremendous achievement of the group of Sanger in determining the complete sequence of all 16 569 base pairs of human mtDNA (Anderson *et al.* 1981) and the complementary studies on its transcription products (Montoya *et al.* 1981, Ojala *et al.* 1981, reviewed by Attardi 1981). The findings probably apply in general to all mammalian mtDNAs. The bovine sequence is determined (but unpublished at the time of writing this book) and has many similarities and a considerable degree of homology (Anderson *et al.* 1981 discuss certain points of comparison). The two strands of mtDNA are of significantly different overall base composition and are referred to as the H (heavy) and L (light) strands. The two strands have different origins of replication. The H strand is initiated first to produce a displacement loop (D loop) of about 680 base pairs. This region contains no apparent genes; otherwise the molecule is crowded with coding regions, most of them transcribed from the H strand (Fig. 10.5). The genes are very closely crowded together with tRNA genes apparently punctuating the other structural genes.

H strand transcripts

The whole H strand is transcribed and the transcript is subsequently processed to form tRNA, rRNA and mRNA. The tRNA sequences are believed to act as signals whereby the processing RNAases recognize the clover-leaf secondary structures. The mRNA transcripts are unusual in several respects. Typically the initiation codon constitutes the extreme 5′-terminal triplet of nucleotides; also in many cases the termination codon is incomplete and relies upon the first one or two residues of the poly(A) tail for completion (see section 10.4.1 for details). There are a few cases of small overlap (one or three residues) between genes for proteins (or the putative URF proteins, Fig. 10.5) and tRNA or between tRNA genes. There is one extensive overlap of 46 residues between the putative URF A6L protein gene and the ATPase 6 gene involving two different reading frames. There is a similar 7-residue overlap between URF 4L and URF 4. Unlike yeast mtDNA, there are no introns in the sequence.

L strand transcripts

Seven of the tRNAs are L strand transcripts; there is also a putative URF gene. Otherwise the strand has no apparent coding capacity. However, three long transcripts (1, 2 and 3) and one short one (18) of the L strand are found in mitochondria (Fig. 10.4) so that more than half of the genome is transcribed in both directions.

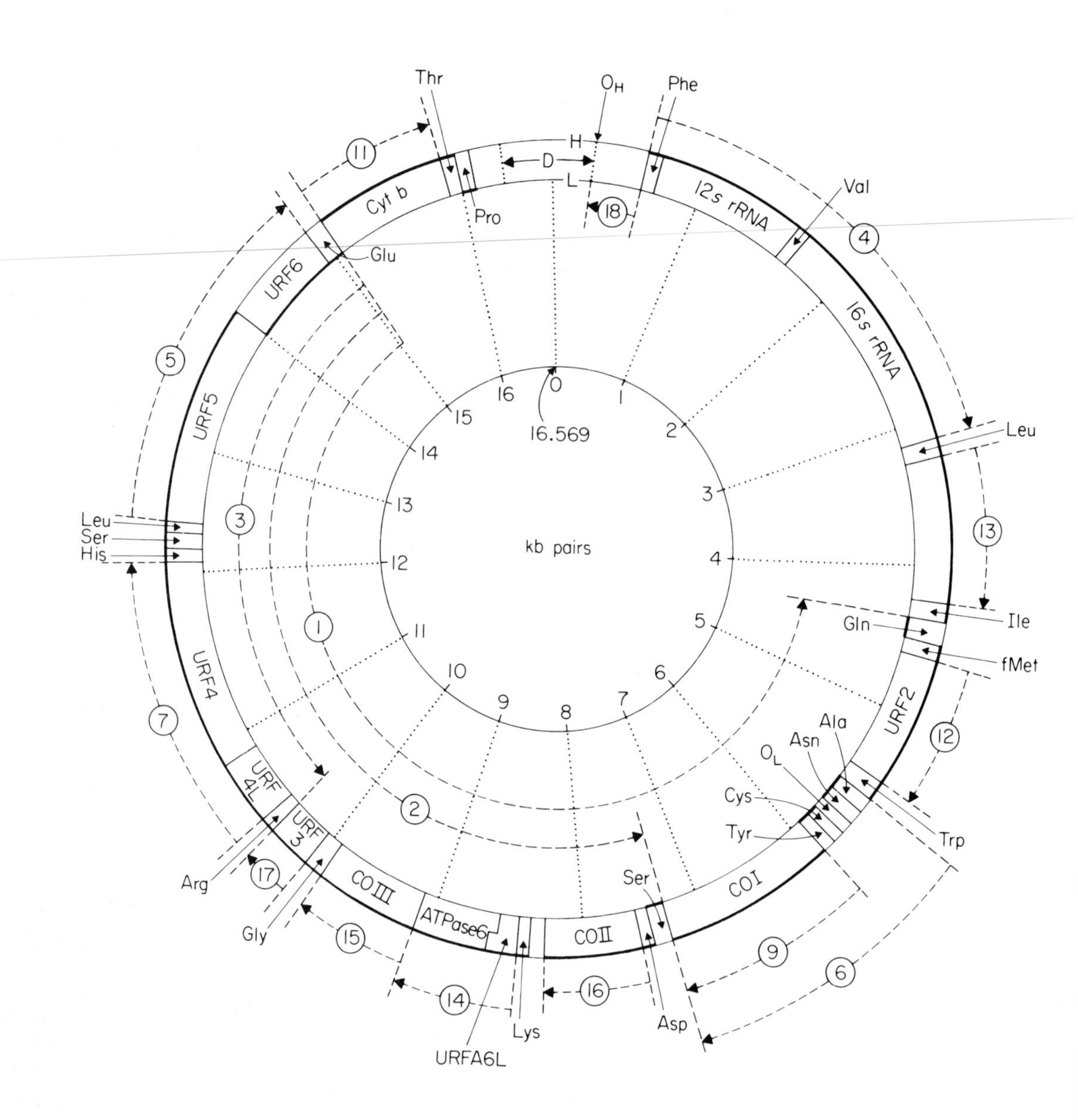

Fig. 10.5. Arrangement of genes and transcripts of human mitochondrial DNA from the complete sequence (Anderson *et al.* 1981) and analysis of the transcripts (reviewed by Attardi 1981). The molecule is drawn as two circles with the H strand on the outside and the L strand on the inside. The innermost circle is an approximate scale calibrated in kb pairs. The transcribing strand for each gene is drawn as a thick line; H strand transcription is clockwise; L strand transcription is anticlockwise. The genes are as follows. COI–COIII, subunits I–III of cytochrome oxidase; Cyt b, cytochrome b; ATPase 6, subunit 6 of that enzyme; URFs (unidentified reading frames) are coding regions for putative mitochondrial proteins; amino acid abbreviations are genes for the corresponding tRNAs; genes for rRNAs are shown as such. O_H, O_L and D are the origins for H strand and L strand replication and the D loop respectively. Numbers in circles are transcripts (other than tRNAs and rRNAs) found in mitochondria; H strand transcripts are on the outside of the circle; L strand transcripts are on the inside. See Fig. 10.6 for detailed examples of the sequence.

Gene products

The gene products consist of rRNA, the three subunits of cytochrome c oxidase, cytochrome b, subunit 6 of the ATPase, eight regions in which initiation and termination codons are in phase (the unidentified reading frames, URFs) but whose protein products are yet to be discovered and 22 tRNAs. There is no gene identified so far for $tRNA^{Met}$. Otherwise this is a complete set of tRNAs for decoding the version of the genetic code that operates in mitochondria (see section 10.4.1). One possibility is that the $tRNA^{fMet}$ precursor can be modified in two different ways to generate two different mature tRNA species. Alternatively the molecule may have such an unusual structure that it has not yet been recognized as such in the DNA sequence (see section 10.4.2).

10.4 TRANSLATION IN CHLOROPLASTS AND MITOCHONDRIA

10.4.1 The genetic code

There is no evidence that the genetic code in chloroplasts is different from the 'universal code' (see section 5.2). However, in the case of fungal and animal mitochondria there are very significant differences.

The first suggestion that the code might be read in a different way came from the observation, first made in yeast (Borst & Grivell 1978), that there is an apparent shortage of different tRNA species in mitochondria. We will return to this point shortly. There is also evidence that certain codon assignments are unusual. Examples from the human mtDNA sequence (see section 10.3.3 and Fig. 10.5) are shown in Fig. 10.6. The sequences illustrate AUA as a Met codon (Ile in the 'universal' code; see section 6.2), UGA as Trp codon (otherwise chain terminating) and AGA as a chain terminating codon (otherwise Arg). The sequences of Fig. 10.6 also illustrate the reliance upon poly(A) tailing to generate other chain terminating codons and the nature of junctions between the genes with one example of the use of a tRNA gene in punctuating the L strand transcript (see section 10.3.3).

The reading of the genetic code in the mitochondria of yeast has been deduced mainly from comparison of protein sequences and DNA sequences (Bonitz *et al.* 1980), and the reading of the code in *Neurospora* mitochondria has been deduced mainly from tRNA

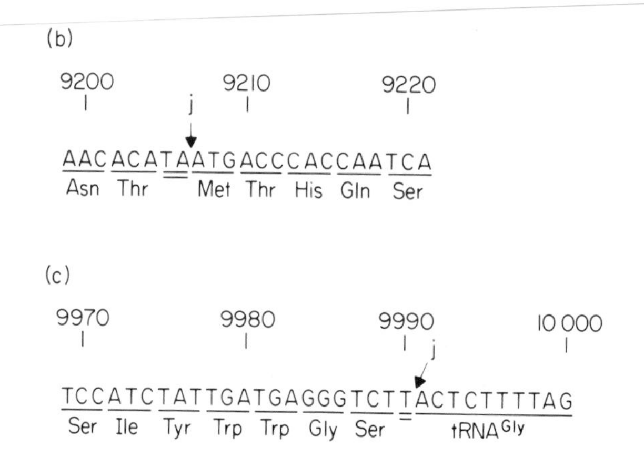

Fig. 10.6. Examples of unusual features of the genetic code taken from the sequence of the L strand of human mtDNA. Compare this with Fig. 10.5 for details of the genes involved. All the protein gene products in this figure are produced from H strand transcripts so the sequence given here is the same as that for mRNA (except that T is substituted for U). Numbers are base pairs using the same scale as Fig. 10.4. (a) The end of gene CO1 illustrating the use of AUA as a codon for Met and AGA (double underlining) as a chain terminating codon. A $tRNA^{Ser}$ gene (an L strand transcript) overlaps the COI gene by one base pair. (b) The junction between RNA 14 (the mRNA for ATPase 6) and RNA 15 (the mRNA for COIII); the junction is marked j. The chain termination codon for ATPase 6 is incomplete (double underlining). However the addition of a poly(A) tail generates the terminating codon UAA. (c) The junction between the end of COIII and $tRNA^{Gly}$; junction itself shown by j. UGA appears here as a codon for Trp. In this case the poly(A) tail is required to generate a UAA chain terminating codon from a single residue (double underlined). The j here represents the 3′-end of RNA 15; the 5′-end of this molecule is at j in (b).

sequences (Heckman *et al.* 1980). The conclusions are summarized in Table 10.3. In this table the codon families are compared with those of section 6.2. Numbers in the 'Ec' columns in the table specify the minimum number of tRNAs required according to the wobble hypothesis (see section 6.2.1 and Table 6.3). In the case of a codon family such as the AC family, at least two tRNAs (for Thr in this case) are required even though all the four ACN codons have the same meaning, the most flexible possible $tRNA^{Thr}$ would have an I in the wobble position of the anticodon and although this would recognize ACU, ACC and ACA, a second $tRNA^{Thr}$ with either C or U in the

Table 10.3. Different versions of the genetic code. The code is laid out as in Table 6.1. The codon families are headed by Ec (the 'universal' genetic code first deduced for *E. coli*), NM (*Neurospora* mitochondria), YM (yeast mitochondria) and HM (human mitochondria). Single letter abbreviations are used for amino acids; i, initiation, *CT*, chain terminating. Numbers are the minimum number of tRNA species required (see text for discussion and for references). Amino acid abbreviations: A, ala; C, cys; D, asp; E, gln; F, phe; G, gly; H, his; I, ile; K, lys; L, leu; M, met; N, asn; P, pro; Q, glu; R, arg; S, ser; T, thr; V, val; W, trp.

1st base ↓ / 2nd base →	U				C				A				G				3rd base ↓
	Ec	YM	NM	HM	Ec	YM	NM	HM	Ec	YM	NM	HM	Ec	YM	NM	HM	
U	F1 ······ L1	F1 ······ L1	F1 ······ L1	F1 ······ L1	S2	S1	S1	S1	T1 ······ *CT*	T1 ······ *CT*	T1 ······ *CT*	T1 ······ *CT*	C1 ······ *CT* ······ W1	C1 ······ W1	C1 ······ W1	C1 ······ W1	U C A G
C	L2	T1	L1	L1	P2	P1	P1	P1	H1 ······ E1	H1 ······ E1	H1 ······ E1	H1 ······ E1	R2	R1	R1	R1	U C A G
A	I1 ······ M,i2	I1 ······ M,i2	I1 ······ M,i2	I,i ······ I1 ······ M,i2	T2	T1	T1	T1	N1 ······ K1	N1 ······ K1	N1 ······ K1	N1 ······ K1	S1 ······ R1	S1 ······ R1	S1 ······ R1	S1 ······ CT	U C A G
G	V2 ······ V,i	V1	V1	V1	A2	A1	A1	A1	D1 ······ Q1	D1 ······ Q1	D1 ······ Q1	D1 ······ Q1	G2	G1	G1	G1	U C A G

wobble position of the anticodon would be required to recognize ACG. In the case of the AAN codons, for example, the situation is simpler. One molecule, $tRNA^{Gln}$, with, for example, a G in the wobble position of the anticodon could recognize the two AAPy codons, and another, $tRNA^{Lys}$, with a U in this position could recognize the AAPu codons. Two tRNAs are entered as being 'required' for AUG; one of these is for initiation and also uses GUG as an alternative initiation codon. The codes for the mitochondria include all the known differences from the Ec code and incorporate smaller numbers for the minimum tRNA requirements. In order to accommodate these smaller numbers we have to further 'slacken' the wobble rules and postulate that a U residue in the wobble position of the anticodon has no specificity whatever and will pair with anything. This explains why we can write down 1 as the number of tRNAs required in the AC family; a tRNA with the anticodon sequence 5′UGU3′ will recognize all of them. If this 'new rule' applies, however, we are left to explain how the system could distinguish between our other example, AAPy and AAPu. It was pointed out by Heckman *et al.* (1980) that a U in the wobble position in bacterial and eukaryotic cytoplasmic tRNA is invariably modified. The rule, therefore, becomes that, in the wobble position of an anticodon, a modified U (U*) will pair with either A or G and that an unmodified U has no specificity. It follows, therefore, that, in the case of the AA codons, the anticodon of $tRNA^{Gln}$ will be 3′UUG5′ and the anticodon of $tRNA^{Lys}$ will be 3′UUU*5′. Direct support comes from the sequencing data of Heckman *et al.* (1980) that shows that in the case of tRNAs recognizing unspecified codon families (such as ACN) the wobble base is indeed U whereas in cases where it is essential that codon third-residue pyrimidines are not recognized (e.g. in the case of the AAPu codons) the wobble residue is indeed U*.

With regard to codon assignments, all three mitochondrial systems employ UGA as a Trp codon (Table 10.3). There is no evidence so far that AGPu signals chain termination except in the human case; nor is there evidence for AUA as a Met codon in the fungal systems. Note that AUU is entered as an alternative initiation codon in human mtDNA; this comes from a single example—the start of URF 2 (Fig. 10.5). Other than this instance AUU is an Ile codon. The other difference is that the UC family are probably Thr (not Leu) codons in yeast mitochondria. The 'loss' of AGPu as Arg codons in human mtDNA and of the CU family as Leu codons in yeast mtDNA are among the very few changes of this type that could occur without a radical change in the code: Leu, Arg and Ser are the only amino acids that appear in more than one codon family in Table 6.1.

10.4.2 The apparatus of protein synthesis

Activation of amino acids

The aminoacyl tRNA synthetases of organelles are specific to the organelles but are the products of nuclear genes. The differences between the specificities of the organelle and cytoplasmic enzymes vary considerably from amino acid to amino acid. For comprehensive references to the aminoacyl tRNA synthetases of chloroplasts see Kung (1977).

Ribosomes

Organelle ribosomes are smaller than cytoplasmic ribosomes and have been compared with prokaryotic ribosomes. The resemblance is not great: mitochondrial ribosomes are considerably smaller than prokaryotic ribosomes. However, there is a striking correlation with prokaryotic ribosomes with respect to antibiotic sensitivity (see Gillham 1978 for a detailed summary). In general, both mitochondrial and chloroplast protein synthesis is sensitive to inhibitors of bacterial protein synthesis but is insensitive to inhibitors of cytoplasmic protein synthesis. These generalizations apply to inhibitors that bind to both the large and small subunits. The correlation is convenient because cytoplasmic or organelle protein synthesis can be inhibited selectively. It also produces a useful group of organelle mutations—resistant to antibacterial antibiotics. With the exception of specific single proteins, the ribosomal proteins of chloroplasts and mitochondria are nuclear gene products and are synthesized on cytoplasmic ribosomes and imported by the organelles.

Transfer RNA

The sequences of mitochondrial tRNAs are in general short and the generalizations about tRNA sequences (see section 1.2.9 and Fig. 1.9) do not hold. The TψC loop varies in length between three and nine bases and the bases shown as A_{14}, G_{15}, G_{18} and G_{19} in Fig. 3.8, constant in other tRNAs, are variable. Interactions involved in stabilizing the tertiary structure of Fig. 3.8(c) are not constant. For example the interaction G_{15}–U_{48} (shown as a short discontinuous line in Fig. 3.8(b)) cannot form in certain cases. All this adds up to a far looser tertiary structure. The extreme case is human mitochondrial $tRNA^{Ser}_{AGPy}$ (the gene for this molecule can be located between about 12.2 and 12.3 kb pairs in Fig. 10.4) which completely lacks the D arm altogether.

The anticodons of these molecules have already been discussed (see section 10.4.1).

Mechanism of protein synthesis

Organelle protein synthesis is initiated by fMet-$tRNA^{fMet}$ and resembles bacterial protein synthesis in this respect. The factors for protein synthesis are nuclear gene products but, in organisms that have been studied, the cytoplasm, chloroplasts and mitochondria all have their own factors (Kung 1977). Termination factors in mitochondria should form a rich field for biochemists to cultivate, in the light of the readthrough of the UGA codon (see section 10.4.1).

10.4.3 Translation products

The translation products of mitochondria are reviewed by Tzagoloff and Macino (1979); those of chloroplasts are reviewed by Gillham (1978) and Ellis and Barraclough (1978). Which of the organelle proteins are synthesized on organelle ribosomes? In the case of yeast mitochondria, some subunits of cytochrome oxidase, some components of the oligomycin-sensitive ATPase including the proteolipid that is involved in locating the enzyme complex on the membrane, one subunit of cytochrome c reductase and a few more unidentified proteins. In the case of chloroplasts, the answer is the large subunit of ribulose biphosphate carboxylase; (the smaller subunit is synthesized in the cytoplasm) and many unidentified structural proteins (see section 8.6.3). What have these proteins got in common? They are all difficult to envisage being transported through membranes either because they are very large or very hydrophobic. The problem with a hydrophobic protein is that it would tend to end up as a trans-membrane protein and would not get into the organelle compartment. If one views organelles as the remnants of symbiotic prokaryotes that have donated as many genes as possible to the nucleus, the cell has retained as organelle genes those whose products are uniquely suitable for transcription and translation *in situ*. If, on the other hand, the organelle is regarded as a cellular compartment in its evolution as well as in its present ultrastructural status, genes with the same properties have become detached from the nucleus and relocated inside the organelle. In this book, we do not come down on one side or the other in this controversy, which has achieved more in stimulating original experiment than it has in promoting disinterested logical thought.

10.4.4 Acquisition of cytoplasmic translation products

The mechanism of transport of cytoplasmic proteins into mitochondria has been reviewed by Schatz (1979) and Neupert and Schatz (1981). The proteins studied (three subunits of F_1 ATPase, cytochrome c, subunit V of cytochrome c reductase and cytochrome c peroxidase) are all present in yeast cytoplasm as precursors with a hydrophobic pre-sequence. However, the mechanism differs from the signal hypothesis (see section 9.4.4) in that the transport is not coupled to translation: the precursors are made on soluble ribosomes and make their way to the mitochondria as polypeptides, not as nascent polypeptidyl tRNA. The pre-sequence binds to a receptor on the outer membrane of the mitochondrion and this binding triggers an active transport of the polypeptides. The transport is associated with proteolysis of the pre-sequence. The details of the transport system are clearly very sophisticated because the proteins mentioned above are destined for different locations in the matrix, the inner membrane and inter-membrane space of the mitochondria. The subunits IV–VII of cytochrome c oxidase are present in the cytoplasm as a giant polyprotein precursor which is split to produce the individual polypeptides during the transport process (Poyton & McKemmie 1979).

In summary, Fig. 10.7 shows the interrelationship of macromolecular synthesis in the compartments of a eukaryotic cell. Although the involvement of DNA elements as vectors in the organization of information between the compartments is specu-

Fig. 10.7. Macromolecular exchange between the compartments of a eukaryotic cell. Processes marked (A) are found in all nucleus–cytoplasm–mitochondrion/chloroplast systems. That marked (B) is only found in some; there is no present evidence for (B?). Broken lines represent DNA transpositions. Process (1) is the migration of an unstable genetic element in a fungus (e.g. transposition of π or *SUP1–16* in yeast. Alternatively this could involve a cytoplasmic plasmid (2). Process (3) is the extension of transposition to allow the acquisition of a nuclear gene by the organelle, and process (4) is the converse (acquisition of an organelle gene by the nucleus). Processes (3) and (4) are speculative but may have occurred during evolution.

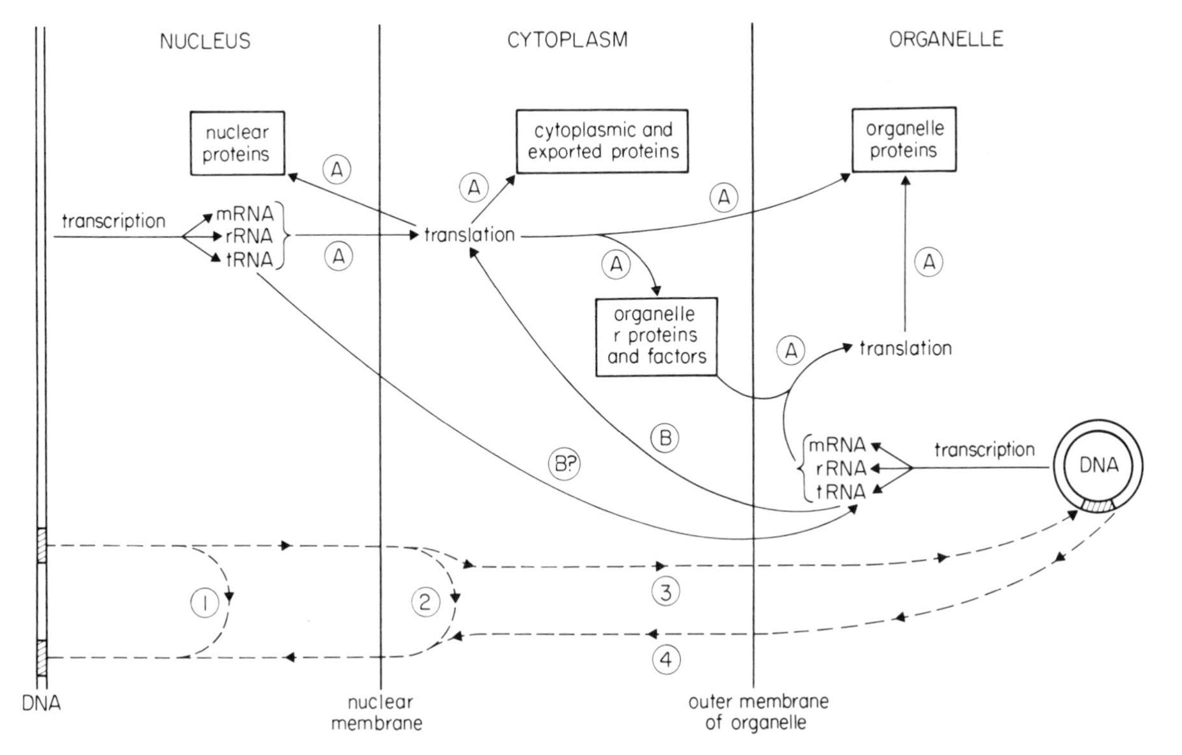

lative, the other processes are established components of the system whereby gene expression contributes to metabolic compartmentation.

10.5 THE EVOLUTIONARY ORIGINS OF CHLOROPLASTS AND MITOCHONDRIA

There are essentially two kinds of hypothesis to account for the origins of the nucleic-acid-containing organelles (chloroplasts and mitochondria). One view is that they are the result of an intracellular differentiation. The benefits of metabolic compartmentation are so great (the argument goes) that it became desirable to partition the machinery for transcription and translation in such a way that the organelles themselves acquired from the nucleus and cytoplasm the ability to make certain of their own proteins. The demerits of the hypothesis derive from the difficulty in explaining how and why the nature of genomic organization and the mechanism of translation came to differ so radically.

The alternative views are all variations on the endosymbiont hypothesis. In its simplest form this supposes that at some time a protofungal or protoanimal cell was invaded by a respiring bacterium-like organism which, through a process of increasing mutualism and interdependence, evolved into a mitochondrion. Similarly protoplant cells are supposed to have been invaded not only by such mitochondrial ancestors but also by a photosynthesizing cyanobacterium-like ('blue-green alga') organism that similarly evolved into the chloroplast. The developing symbiosis could, in principle, have involved a loss of symbiont genes with their complementation by nuclear genes. It could also have involved exchange of genes between the two (Fig. 10.7). We are not concerned here with the question of whether the endosymbiont hypothesis stands up to the most rigorous examination *per se*. Nor are we concerned with exotic variations on it (such as the possibility that the nucleus has a colonizing microbe as its remote ancestor). Rather let us accept that the endosymbiont view is an attractive hypothesis and briefly inquire what the ancestral microbes might have been like.

Major taxonomic divisions

The apparatus of gene expression and protein synthesis seems to evolve very slowly. This is perhaps to be expected as the majority of chance significant mutations in the components of the system would probably be lethal because of the essential nature of the system and the interdependence of the component parts. Moreover, experimental evidence supports the statement. If one compares the sequence (or in many cases the functional interchangeability) of, for example, rRNAs and tRNAs, one can accommodate all eukaryotes on the one hand and most prokaryotes on the other into huge taxa, for which the names 'urkingdom' or 'primary kingdom' have been proposed (Woese & Fox 1977). How many urkingdoms are there? The traditional view that there are two (prokaryotes and eukaryotes) is probably wrong. Woese and Fox (1977) and Fox *et al.* (1977) present evidence for three urkingdoms based on the sequence similarities of 16*s*/18*s* rRNA and tRNAs. The three groups are the vast majority of prokaryotes, including the cyanobacteria, termed the 'eubacteria'; a second prokaryotic group comprising methanogenic bacteria, halobacteria and certain extreme acidophiles, the 'archaebacteria'; and the eukaryotic nuclear/cytoplasmic system supposedly derived from an ancestral type of eukaryote. These organisms are termed 'urkaryotes'. The urkaryotes can be regarded as either an ancient lineage or the parts of present-day eukaryotes excluding the mitochondria and chloroplasts. Association coefficients between rRNA sequences differentiate the three urkingdoms clearly. With respect to the tRNAs, the eubacteria and urkaryotes share the common $T\psi CG$ sequence. In contrast, archaebacteria have instead either $\psi\psi C^*G^*$ or $U^*\psi C^*G$ (C^*, G^* and U^* are unidentified modified residues; the last is not T). Although not relevant to the present discussion, the archaebacteria also have plasma membranes, chemically quite different from those of eubacteria and eukaryotes.

The organelles

With respect to organelles, the chloroplasts are recognizable in this scenario. Their supposed microbial ancestors seem to have been cyanobacteria. Woese and Fox (1977) showed that duckweed chloroplasts fall clearly into the eubacteria on the basis of rRNA association coefficients. There is also independent biochemical evidence to relate chloroplasts to cyanobacteria (Stanier 1974). Chloroplasts from different groups of plants certainly differ. On the other hand the cyanobacteria are themselves a very diverse group (Carr & Waterbury 1979) and there is no reason to suppose that there was only a single cyanobacterial progenitor of all chloroplasts.

With respect to mitochondria the situation is more complicated. Their peculiar version of the genetic code seems to set them apart.

Also their rRNA sequences seem unrelated to those of the three urkingdoms (see Anderson *et al.* 1981 for a discussion and references). Moreover, the mitochondria themselves are a very diverse group of genetic systems. The genetic maps of yeast and human mitochondria (Figs 10.2 and 10.5) are not only very different but, unlike human (and other mammalian) mitochondria, the genes of yeast mtDNA are spaced out and contain introns.

It seems likely that the three urkingdoms and the supposed microbial ancestors of mitochondria diverged at a very early time in cellular evolution and that these last were a group (or groups) for which no free-living counterparts have survived.

Chapter 11 Recombinant DNA

11.1 INTRODUCTION

Recombination, *in vivo,* between homologous DNA molecules plays a central role in the generation of genetic diversity. However, since the early 1970s it has become possible to produce recombinant DNA molecules *in vitro,* and to manipulate them in an increasingly precise fashion. The central theme of recombinant DNA technology is the process of gene cloning, which consists of the production of a defined fragment of DNA and its propagation and amplification in a suitable host cell. Although the immediate objective is the construction of hybrid DNA molecules, the ultimate aims can be reduced to three basic areas. First there is the amplification of a particular DNA sequence, usually in order to facilitate analysis of its physical organization. Second it may be desired to examine the expression of a particular gene in a 'foreign' or heterologous cell or even within its normal cell, but in a more accessible form. Third the aim may be the amplification of the polypeptide encoded by a particular gene. These techniques, although only recently developed, provide a versatile and immensely powerful tool for the analysis of the organization of the genetic material and the control of its expression. In addition, recombinant DNA technology is beginning to have important applications in the applied sciences, particularly in the construction of *E. coli* strains that are capable of producing polypeptides of clinical importance (e.g. insulin, somatostatin and interferon).

A gene cloning experiment follows a well-established sequence of events. First it is necessary to produce a discrete DNA fragment in a suitable form for cloning, usually from a large and complex genomic DNA molecule. This fragment must then be introduced into a host cell to ensure its propagation and selective amplification. Obviously, even if a successful introduction could be made, the fragment would not be stably inherited by daughter cells unless it possessed a functional origin of replication. Consequently, for a foreign DNA fragment to be propagated in a cell population its replication must be coupled to that of a DNA molecule that is normally capable of promoting its own replication in the host cells (a replicon). Such a replicon is known as a vector. Three candidates present themselves as potential vectors: the chromosomal DNA of the host cell; a viral DNA; or a plasmid. Since manipulations of the intact genomic DNA *in vitro* are precluded by virtue of its size, plasmid or viral DNAs are the vectors commonly used. It is worth mentioning in this context that plasmid primes and specialized transducing phages, produced by recombination *in vivo,* have long been of value in analysing the structure and function of prokaryotic genes and, consequently, represent a predecessor of gene cloning *in vitro*. Under some circumstances, however, it is necessary to integrate a cloned DNA fragment into the genomic DNA of the host cell to ensure its propagation. The later stages of the cloning of a DNA fragment, following its joining to the vector and introduction of the hybrid molecule into the host cells are the identification of the particular cell line that carries the cloned fragment and then the structural and functional characterization of the cloned DNA. The sequence of events in the cloning of a DNA fragment are reflected in the following sections of this chapter.

This discussion of the techniques of recombinant DNA technology is largely confined to their application in *E. coli* as the host cell, since it was in this organism that the techniques were developed and are still almost exclusively being applied. However, host–vector systems have been, or are being, developed for a variety of other prokaryotic and eukaryotic cells and some mention must be given to them as they are likely to be of rapidly increasing importance.

11.2 THE GENERATION OF DNA FRAGMENTS FOR CLONING

The principal objective of any gene cloning experiment is the propagation of a defined DNA sequence as part of a vector molecule. There are a variety of methods by which this fragment may be initially generated and the choice of method is a crucial part of the strategy of the experiment. To a large extent the choice will be dictated by the size of the genome from which the fragment is to be obtained and by the ease with which clones carrying the fragment can be identified. The influence of genome size is illustrated by considering the construction of gene 'banks' or 'libraries'. The aim of such banks is to have a collection of cloned DNA sequences representative of the entire genome of an organism. Clarke and Carbon (1976) proposed that it was possible to calculate the number of clones that was necessary to constitute such a bank using the formula

$$N=\frac{\ln(1-P)}{\ln(1-f)}$$

where N is the number of clones constituting the bank, P is the probability of any given sequence being present in the bank, and f is the average size of the cloned fragments as a proportion of the size of the whole genome.

In Table 11.1 the relationship between genome size and bank size is shown. It is clear that when dealing with a comparatively small prokarytic genome only a small number of clones need be obtained to have a high probability of any particular fragment being present and if this fragment confers an easily detectable phenotype upon the host cell, then it is realistic to clone fragments which collectively represent the entire genome; an approach usually referred to as a shot-gun experiment. When dealing with a much larger eukaryotic genome, unless the object is to create a gene bank, it is normally the practice to use a population of DNA fragments enriched for the one of interest. The various methods for producing DNA fragments suitable for cloning and their enrichment are outlined below.

Table 11.1 The relationship between genome size and gene bank size. The figures are calculated using the equation of Clarke and Carbon (1976) assuming that the average size of the cloned fragments is 10 kb and that the probability of any given sequence being present in the bank is 0.99.

Organism	Approximate genome size (base pairs)	Gene bank size (no. of clones)
E. coli	3×10^6	1.4×10^3
S. cerevisiae	1×10^7	4.6×10^3
D. melanogaster	2×10^8	9.2×10^4
H. sapiens	4×10^9	1.8×10^6

11.2.1 Restriction enzymes

Type II restriction enzymes provide one of the most convenient methods for the production of DNA fragments for cloning and are invariably used for the treatment of vector DNA. The DNA fragments produced by restriction enzymes (see section 3.4.1, Fig. 3.26) may have termini with 5′ single-stranded extensions (e.g. *Eco* RI), 3′ single-stranded extensions (e.g. *Pst* I) or blunt ends (e.g. *Pvu* II) and the nature of the termini produced will influence the way in which the fragments can be joined to the vector DNA. Those restriction enzymes that recognize a hexanucleotide target sequence would be expected to cleave a DNA molecule, with a GC content of 50%, into fragments with an average size of 4096 base pairs. The most important characteristic to consider about the use of restriction enzymes is that the pattern of cleavage is specific and reproducible for any given DNA molecule; consequently, restriction enzymes generate a non-random population of fragments. There is a finite possibility that any given restriction enzyme will cleave within the DNA sequence of interest. However, if this is the case, then other restriction enzymes with different recognition sequences can be used. Alternatively, the DNA can be partially digested, in which case some fragments will be produced which still contain the intact sequence, but still have termini characteristic of the particular restriction enzyme.

An alternative strategy which produces a near-random population of DNA fragments is to use a restriction enzyme with a tetranucleotide recognition sequence (e.g. *Hae* III) which should cut a DNA molecule, on average, every 256 base pairs. A complete digest with such an enzyme would almost certainly cut within the sequence of interest, however, a partial digest would yield an almost random population of fragments. This approach can be used in the construction of a gene bank or where more than one gene is to be cloned.

11.2.2 Random shear

An alternative to partial digestion with restriction enzymes which does produce a random population of DNA fragments is mechanical shearing. DNA molecules are sufficiently rigid to be susceptible to these forces. Commonly used methods to produce populations of randomly sheared DNA are sonication, repeated extrusion through the narrow orifice of a hypodermic needle and high-speed stirring in a blender. Whichever method is used the degree of shear must be monitored by agarose gel electrophoresis so that fragments of the desired size are produced. The termini of the fragments produced by these methods are either blunt ended or have short, single-strand extensions, but these are not mutually cohesive and are consequently not so readily ligatable as those produced by restriction enzymes.

11.2.3 Synthesis of complementary DNA (cDNA)

One strategy that has proved useful in the cloning of DNA sequences that encode specific polypeptides is the technique of synthesizing complementary DNA (cDNA) to the mRNA coding for that polypeptide. This process utilizes the enzyme reverse transcriptase specified by Avian Myeloblastosis Virus (AMV); this enzyme is able to transcribe an RNA template to produce a complementary single-stranded DNA (see section 3.4.1). The technique relies on the fact that the particular mRNA can be isolated in relatively pure form. In eukaryotes this is facilitated by the fact that many mRNAs are polyadenylated and can, therefore, be readily purified by oligo (dT) cellulose chromatography. Other techniques including size fractionation on sucrose gradients, immunoprecipitation of polysomes and the use of cells hyperproducing the protein of interest can be used to produce mRNA enriched for a particular species. The consequence of this method of producing DNA fragments for cloning is that the population is already enriched for the sequence of interest and, therefore, reduces the labour involved in identifying the correct clone.

The complete process by which double-stranded cDNA is produced ready for cloning is summarized in Fig. 11.1. Reverse transcriptase is only able to transcribe the RNA template if there is a short double-stranded region to act as a primer and in this fashion resembles the DNA-dependent DNA polymerases. The requirement

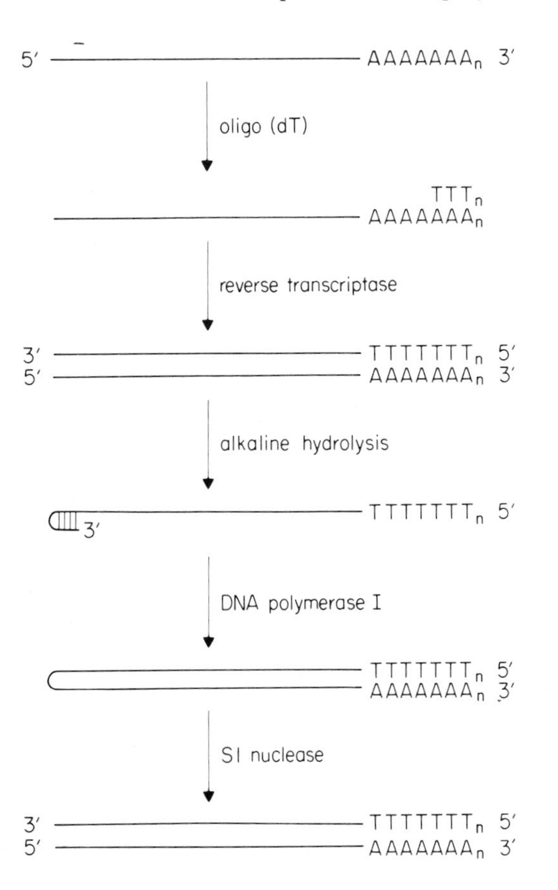

Fig. 11.1. The synthesis of double-stranded cDNA from a polyadenylated mRNA using the reverse transcriptase of avian myeloblastosis virus (AMV).

for a primer is overcome with most eukaryotic mRNAs by the fact that most of them are polyadenylated at the 3′-end and, therefore, a primer can be provided by hybridizing a short molecule of synthetic oligo (dT) to the poly (A) sequence. The oligo (dT) primer will then be extended in the 5′–3′ direction to form the single-stranded cDNA. When transcription is complete the RNA template can be destroyed by alkaline hydrolysis or separated from the cDNA on an alkaline sucrose gradient. The single-stranded cDNA is then converted to the double-stranded form by the action of DNA polymerase I. This reaction also needs to be primed by a short, double-stranded region which is in fact usually provided by the fact that the 3′-end of the single-stranded cDNA folds back to give a short hairpin loop. The double-stranded cDNA will still have this hairpin loop at one end of the molecule and this is removed by the action of S1 endonuclease which is single-strand specific.

11.2.4 Synthetic DNA

Recent advances in the synthesis of oligodeoxyribonucleotides of defined sequence have now made it possible to construct a synthetic DNA fragment that will encode a particular polypeptide, assuming of course, that the amino acid sequence of the polypeptide is known. At present this technique has been limited to the synthesis of genes encoding comparatively short polypeptides, such as somatostatin (Itakura *et al.* 1977) and the human insulin A and B chains (Crea *et al.* 1978). When designing a synthetic gene attention must be paid to the pattern of codon usage within the prospective host cell to ensure efficient translation. In addition there must be the normal signals for the initiation and termination of translation.

11.2.5 Enrichment of DNA sequences

When DNA fragments have been produced by restriction enzyme digestion it is possible to enrich for the fragment of interest, prior to cloning, thereby saving effort in the screening of the recombinant clones. Two methods are of particular value for this purpose, but they both rely on having some way of detecting the particular fragment, such as DNA/RNA hybridization or an in-vitro coupled transcription–translation system. Both preparative agarose gel electrophoresis and chromatography on RPC-5, a quaternary ammonium compound, will fractionate a restriction enzyme digest, though on different bases, so that they can be used in conjunction.

This approach was used by Tilghman *et al.* (1977) who obtained an approximately one thousandfold purification of the mouse β-globin gene prior to cloning.

11.3 CLONING VECTORS

To ensure propagation of the DNA fragment it must be joined to a vector, such as a viral or plasmid DNA molecule, which is capable of replication in the host cell. Ideally the vector should be a small molecule since this will render it more amenable to manipulations *in vitro* and will make it less likely to carry multiple cut sites for any particular restriction enzyme. The presence of single cut sites is desirable since this simplifies the process by which the DNA to be cloned may be inserted into the vector. Obviously insertion of foreign DNA into the vector should not inactivate any function essential for replication and for this reason it is useful if the vector is well characterized, both genetically and physically, so that potential sites for cloning can be identified. Since the processes by which the recombinant DNA is introduced into the host cells are inefficient it is useful if the vector carries some easily detectable marker, such as drug resistance or plaque formation. Finally, it can save effort in detecting the desired clone if host cells carrying the recombinant vector can be easily distinguished from those carrying the vector alone.

11.3.1 Plasmids

Although certain naturally occurring plasmids have been used as vectors in *E. coli,* they exhibit few of the desirable characteristics of a vector and consequently virtually all plasmid vectors have themselves been constructed by recombinant DNA techniques. The most commonly used general purpose plasmid vector which illustrates many useful characteristics is pBR322 (Bolivar *et al.* 1977). This plasmid (Fig. 11.2) has a weight of about M_r 2.8×10^6, which corresponds to a size of 4.362 kilobase pairs. The physical organization of pBR322 is well characterized, indeed its complete base sequence has been determined (Sutcliffe 1979). Cells carrying the plasmid are easily distinguished since it confers resistance to both ampicillin and tetracycline. The fact that the single restriction sites for several restriction enzymes lie within the genes encoding one or other of these resistances can be made use of to distinguish between recombinant and non-recombinant plasmids. Insertion of foreign DNA into one of these restriction sites disrupts the structural integrity of the resistance gene (insertional inactivation) and, consequently, host cells receiving the recombinant plasmid will exhibit resistance to only one antibiotic (Fig. 11.2). It is worth noting that the single *Hin*d III site of pBR322 lies within the promoter for the tetracycline resistance genes and though insertion of DNA at this site can cause insertional inactivation, residual expression of resistance may occur if the inserted DNA carries an efficient promoter. Plasmids like pBR322, which have the replication elements of the naturally occurring plasmid ColE1, are able to continue replication for a considerable time after the cells have been treated with chloramphenicol, although chromosome replication is inhibited by this treatment. Chloramphenicol amplification can lead to plasmid DNA constituting more than 50% of the total DNA of the cell and is of obvious advantage for the preparation of large amounts of plasmid DNA. Finally, pBR322 has two characteristics which facilitate the expression of cloned DNA. It is a multicopy plasmid, about 40 copies per cell, and, therefore, expression of cloned DNA may be enhanced by a gene dosage effect and also the strong promoter for ampicillin can be used to obtain transcription of cloned DNA (see section 11.7.3).

Insertional inactivation allows distinction between cells carrying recombinant and non-recombinant plasmids, but two extensions of this technique permit positive selection where it is the tetracycline resistance genes that have been inactivated. The inhibitory effect of tetracycline is reversible and so if tetracycline is added to cells growing in liquid medium, followed by cycloserine, which interferes with cell wall biosynthesis in actively growing cells, those cells which carry non-recombinant plasmids will be selectively killed. The survivors, carrying recombinant plasmids, can be detected by plating out on a medium lacking tetracycline. Alternatively, after transformation, the host cells can be plated out on a medium containing heat-treated tetracycline and a lipophilic chelating agent such as fusaric acid. The heat-treated tetracycline no longer has any antibiotic activity, but is still capable of inducing expression of tetracycline resistance and such induced cells are hypersensitive to lipophilic chelating agents and will, therefore, be selectively inhibited. In addition to these manipulative methods, several plasmid vectors have been developed which themselves permit positive selection for recombinants. The general approach is to construct a plasmid which encodes a lethal function for the host cell. Insertion of DNA into the plasmid inactivates the lethal function and, therefore, only those host cells receiving recombinant plasmids

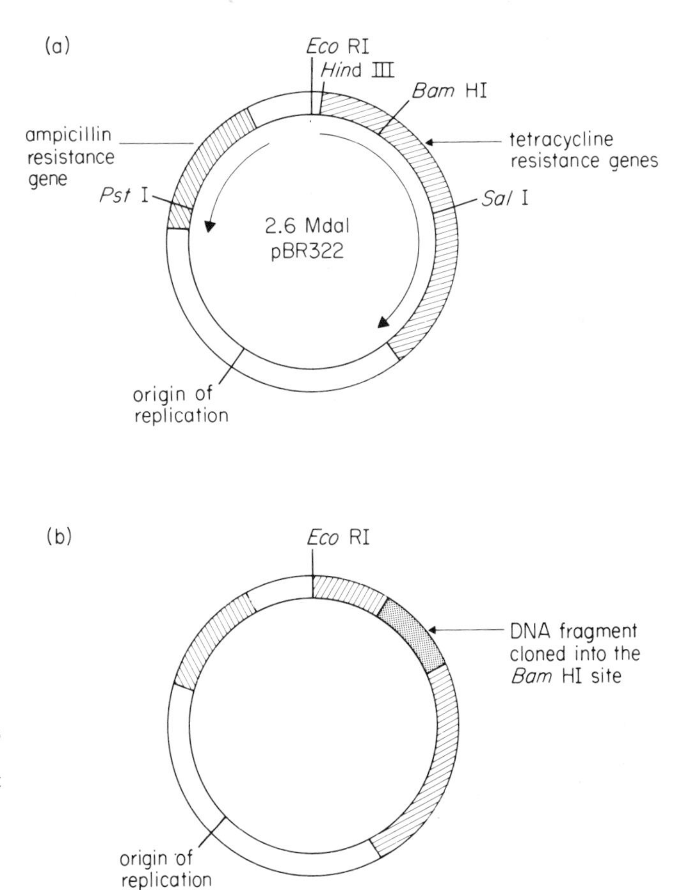

Fig. 11.2. (a) The structure of the *E. coli* plasmid vector pBR322, indicating sites for restriction enzymes with single cut sites. (b) The structure of a recombinant plasmid derived from pBR322 in which the cloned DNA has caused insertional inactivation of the tetracycline resistance genes.

will be able to grow. Plasmids which permit positive selection are typified by pKN80 (Schumann 1979), which carries an *Eco* RI fragment of phage Mu. This fragment encodes a killing function which is expressed when the plasmid is transformed into Mu-sensitive cells. Cloning of DNA into a site within the Mu fragment inactivates the killing function and thus there is a strong selection for cells being transformed with recombinant plasmids. Cells which are Mu-lysogens can be used to propagate the non-recombinant plasmid. In addition to general purpose cloning vectors there are several plasmids which have been constructed particularly with expression of the cloned DNA in mind and these expression vectors are discussed in section 11.7.3.

11.3.2 Phage vectors

λ

The most commonly used phage vectors are those derived from λ, but vectors based on M13 are gaining importance for their contribution to DNA sequencing. In spite of the size of its genome, λ is well characterized both physically and genetically and this has largely contributed to its extensive development as a vector. Some of the properties of λ are described in sections 4.1.4, 4.3.6 and 5.2.4 and the organization of the genome is summarized in Fig. 4.4. The following important characteristics relate to its use as a vector. The λ genome is a double-stranded DNA molecule with a weight of about M_r 3.2×10^7 and, consequently, it is large enough to have multiple cut sites for most commonly used restriction enzymes. The positions of the sites for *Eco* RI and *Hind* III are shown in Fig. 11.3. Before λ can be used as a vector the number of restriction sites must be reduced to simplify the subsequent in-vitro ligation reactions. There are a variety of methods by which this can be done: the replacement of segments carrying a restriction site by homologous segments from a related phage; the isolation of mutants which lack a particular restriction site; and the deletion of dispensable parts of the phage genome which include restriction sites. A second important consideration is

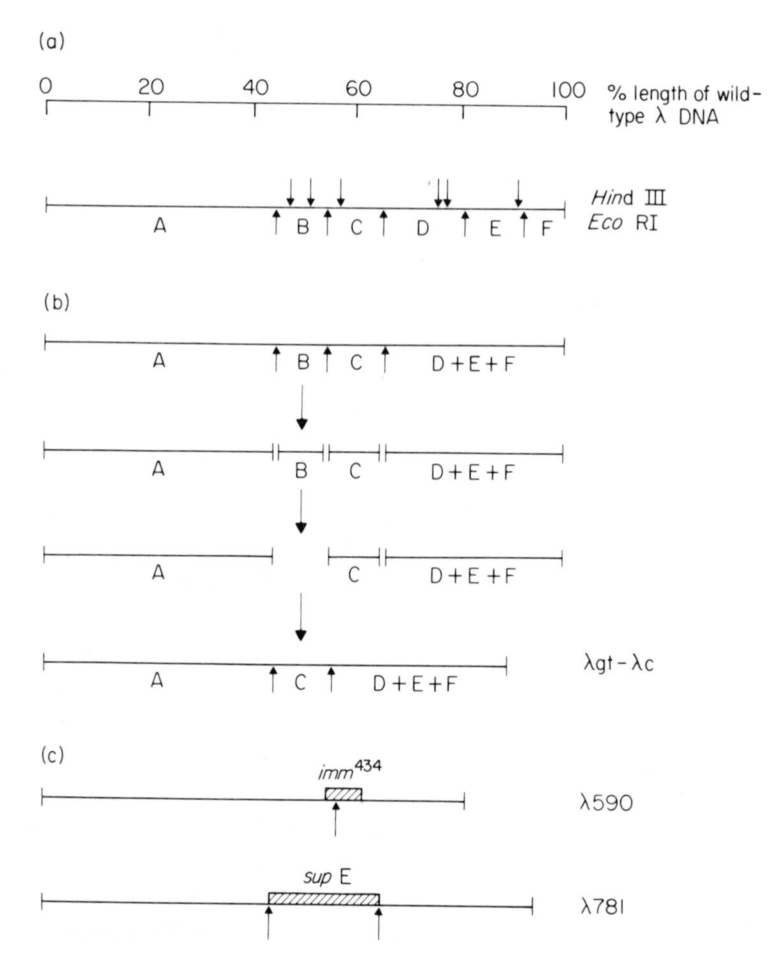

Fig. 11.3. (a) The positions of *Eco* RI and *Hind* III sites on the DNA of wild-type λ. (b) The steps in the construction of the replacement vector λgt-λc. (c) The simplified structures of the *Hind* III insertion vector λ590 and the *Eco* RI replacement vector λ781.

the relationship between the size of λ DNA and the efficiency with which it is packaged into the phage particles *in vivo*. Derivatives of λ that are smaller than 75% or larger than 105% of the size of wild-type λ are not packaged. As a result of this size-dependent packaging two basic types of λ vector have been developed, known, respectively, as insertion and replacement vectors. Insertion vectors contain only a single site for any particular restriction enzyme and generally can be used with only one restriction enzyme. The phage genome, as well as being altered to reduce the number of restriction sites to one, is drastically reduced in size to the point where it is still just capable of being packaged, in order to allow the insertion of the maximum amount of passenger DNA without exceeding the packaging limit. An insertion vector 80% of the size of wild-type λ would permit the insertion of about 9.5 kilobase pairs of DNA. A variety of insertion vectors have been constructed which permit distinction between recombinants and non-recombinants in a manner analogous to the insertional inactivation of plasmids. Murray *et al.* (1977) constructed λ 590 in which the normal immunity region of λ was replaced by the immunity region from phage 434. The immunity region carries the *c*I gene which encodes the phage repressor, (see section 5.2.4), and whilst imm^{λ} carries two *Hin*d III restriction sites, imm^{434} carries only one and this constitutes the only *Hin*d III site in λ 590. Insertion of passenger DNA into this *Eco* RI site interrupts the phage repressor gene and a consequence of this is that the recombinant phage is unable to lyse host cells and will, therefore, give rise to clear plaques. In contrast, non-recombinant λ 590 is able to lyse and gives rise to turbid plaques.

Replacement vectors contain two restriction sites for a particular restriction enzyme and these define a segment of the phage genome which is replaceable by passenger DNA. The size of the replaceable fragment is such that if the phage is reconstructed without either this fragment or a fragment of passenger DNA it will be too small to be packaged. In general, replacement vectors are capable of carrying larger fragments of passenger DNA than insertion vectors. The construction of a simple replacement vector is shown in Fig. 11.3. Wild-type λ yields six fragments, A,B,C,D,E and F, on digestion with *Eco* RI. Thomas *et al.* (1974), by a series of manipulations *in vivo* were able to obtain a derivative of λ that lacked the two right-hand restriction sites and, consequently, on restriction would yield fragments A,B,C and a composite fragment D+E+F. The fragments were then purified. Fragment B, which is not essential for lytic development, was discarded and the remaining fragments religated and transfected into *E. coli.* The resultant phage carried two *Eco* RI sites either side of fragment C, which is also non-essential, and may, therefore, be replaced by an *Eco* RI fragment of passenger DNA. Many replacement vectors have been constructed in which the replaceable fragment carries an easily detectable marker. Lambda 781 (Murray *et al.* 1977) carries an *Eco* RI fragment that encodes a mutant tRNA gene, *sup*E, which is capable of suppressing amber mutations. When this phage is used to infect cells which have an amber mutation in the structural gene for β-galactosidase the mutation is suppressed. The active enzyme is able to hydrolyse the colourless compound X-gal (5-bromo-3-chloro-2-indolyl-β-D-galactoside) to a blue derivative and consequently plaques arising from non-recombinant phage will be blue. In contrast, recombinant phage where the *sup*E fragment has been replaced by passenger DNA will give rise to colourless plaques. An additional feature of replacement vectors which simplifies the recovery of recombinant phage is the purification of the λ 'arms' away from the replaceable fragment prior to ligation. Since the two λ arms ligated together are too small to package there is a strong selection for phage carrying a segment of passenger DNA.

Recently, there have been two developments in λ vectors: production of, first, vectors which may be used with several restriction enzymes, and, second, vectors which permit positive selection for recombinant phage. Loenen and Brammar (1980) constructed a λ replacement vector which could be used with *Eco* RI, *Hind* III, *Bam* HI, *Mbo* I and *Bgl* II. The normal vector is unable to grow on a strain of *E. coli* which is lysogenic for phage P2, however, recombinant phage are able to grow on a P2 lysogen and can, therefore, be positively selected. The number of vectors derived from λ is considerable and a more detailed account of them is given by Williams and Blattner (1980).

M13

M13 is a filamentous coliphage which has a single-stranded DNA genome of about M_r 2×10^6. Upon infection of *E. coli* the single-stranded DNA is converted to a double-stranded replicative form and amplified to about 300 copies per cell. Cells infected with M13 are not lysed but continue to grow slowly, releasing phage into the medium and, therefore, M13 gives rise to turbid plaques. Messing *et al.* (1977) showed that there was a small non-essential region of the genome into which foreign DNA could be inserted and because of the filamentous nature of the phage particle there was little limitation on the size of DNA that could be packaged. The problem with M13 as a vector is that it has no known single restriction sites within the non-essential region. A derivative of M13, with improved characteristics as a vector, has been constructed (Barnes 1979). M13Hol carries the promoter proximal part of the *his* operon of *Salmonella typhimurium* inserted into the non-essential region thereby providing a selectable marker since it will complement a *hisD* mutation in *E. coli.* In addition the *his* insert contains a single *Eco* RI site into which DNA can be cloned. The value of M13 as a vector lies in the fact that the DNA packaged into the phage particles is single-stranded and, therefore, is a suitable substrate for the rapid dideoxy methods of DNA sequencing (see section 3.5.4).

11.3.3 Cosmids

Collins and Hohn (1978) showed that plasmids carrying a fragment of λ DNA, which included the *cos* site, could be used as cloning vectors in conjunction with the λ in-vitro packaging system (see section 11.5.1). *Cos* containing plasmids of less than 23 Mdal are not efficiently packaged because of their circular structure and small size. However, if the plasmids are joined to passenger DNA to increase their size and are in a concatameric form, as a proportion of them would be after ligation *in vitro*, then they represent a suitable substrate for packaging. Since only plasmids of considerably increased size are packaged there is a very strong selection for recombinant plasmids carrying very large DNA inserts and consequently there is a very low background of cells acquiring a reconstituted vector. A cosmid of 8 Mdal would require to be joined to passenger DNA of at least 15 Mdal for efficient packaging. The process also takes advantage of the inherently greater efficiency of infection, compared to transformation, for the introduction of recombinant DNA into the host cells, particularly when the plasmids are in this size range. The cosmid is maintained in the cell as a plasmid and, if it is derived from ColE1, is capable of chloramphenicol amplification to aid the purification of recombinant DNA. The capacity of cosmids for very large DNA fragments makes them extremely useful as vectors to construct gene libraries for large eukaryotic genomes.

11.3.4 Vectors for other prokaryotic hosts

Although *E. coli* host–vector systems constitute a versatile and sophisticated cloning technology there are frequently good reasons for wishing to clone in an alternative prokaryotic host. At present the most effort is going into developing vectors for use with *Bacillus subtilis* and *Streptomyces* spp.

Bacillus subtilis

There are several reasons for using *B. subtilis* as a host cell for cloning. *Bacillus* spp. are an industrially important group of microorganisms. In addition, *B. subtilis* is non-pathogenic and, therefore, represents a safe host and is Gram-positive, so providing a different intracellular environment in which to examine the expression of cloned DNA. So far no naturally occurring plasmids of *B. subtilis* have been found which have the necessary characteristics of a cloning vector. However, several plasmids of low molecular weight which also carry drug resistance markers have been isolated from *Staphylococcus aureus* and *B. cereus,* and are capable of replication and expression of their drug resistances in *B. subtilis*. Keggins *et al.* (1978) used the plasmid pUB110, originally in *Staph. aureus,* as a vector in *B. subtilis*. This plasmid has a weight of about M_r 2×10^6, specifies resistance to neomycin and contains a single *Eco* RI site. pUB110 was used to clone *Eco* RI-restricted chromosomal DNA from *B. pumilis* and *B. licheniformis* and recombinant plasmids were obtained that were able to complement tryptophan auxotrophs of *B. subtilis.* However, because of the low efficiency with which plasmid DNA produced by ligation *in vitro* transforms *B. subtilis* such shot-gun experiments are difficult to carry out. If DNA could first be cloned in *E. coli* and then transferred to *B. subtilis,* the difficulties would be largely overcome. The *Staph. aureus* plasmid pC194 confers resistance to chloramphenicol, contains a single site for *Hin*d III and will replicate in *B. subtilis.* Ehrlich (1978) digested pC194 with *Hin*d III and ligated it into the *Hin*d III site of the *E. coli* vector pBR322. The hybrid DNA was used to transform *E. coli* to ampicillin and chloramphenicol resistance. The hybrid plasmid, pHV14, also replicated in *B. subtilis,* although the ampicillin resistance specified by the pBR322 part of the molecule was not expressed. This hybrid replicon, and similarly constructed ones, can be used to clone DNA in *E. coli* and subsequently transfer it to *B. subtilis* thereby overcoming the problems associated with shot-gun experiments and also providing a vector in which the expression of genes in two very different cellular environments can be examined. Rapoport *et al.* (1979) used a hybrid replicon, derived from pHV14, to construct a gene library of *B. subtilis* DNA in *E. coli* and showed that plasmids from the library could complement many auxotrophic mutations in *B. subtilis*.

Streptomyces spp.

Over 60% of currently used antibiotics are produced by species of *Streptomyces* and recombinant DNA methods have an important role in strain improvement aimed at increased antibiotic yields and the generation of novel antibiotics by incorporating parts of different natural antibiotic biosynthetic pathways in the same organism (Hopwood 1981). Plasmid vectors possessing selectable phenotypes and single cut sites for a variety of restriction enzymes (Bibb *et al.* 1980) have been developed for use in *Streptomyces*.

Broad host range plasmids

Plasmids of the incompatibility group P, originally isolated from *Pseudomonas* spp., exhibit the ability to promote their own transfer, by conjugation, to a wide range of Gram-negative organisms and are, therefore, of interest as potential cloning vectors. The Inc P plasmid RP4 is of M_r 3.6×10^7 and has a single cut site for *Eco* RI. Jacob *et al.* (1976) used RP4 to clone DNA from *Proteus mirabilis* and *Rhizobium leguminosarum* in *E. coli* and showed that such recombinant plasmids retained the wide host range of the parental plasmid and were efficiently transmitted and maintained. The main disadvantage of using RP4 as a vector is its size. Ditta *et al.* (1980) have constructed a derivative of the related plasmid RK2 of considerably reduced molecular weight and which contain single cut sites for a variety of restriction enzymes, however, they are not self-transmissable and the extent of their host range is not yet known.

11.4 JOINING DNA MOLECULES *IN VITRO*

11.4.1 Ligation

The enzyme DNA ligase is capable of forming a covalent bond between two DNA molecules providing one has a 5′-phosphate terminus and the other a 3′-OH terminus (see sections 3.4.1 and

4.2.2). Many of the type II restriction enzymes generate DNA fragments with short, single-stranded termini which are mutually cohesive and consequently are able to form hydrogen-bonded structures with what amount to two slightly displaced, single-strand breaks with a 5′-phosphate and a 3′-OH terminus. These structures constitute a suitable substrate for DNA ligase which is then able to repair the single-strand breaks to form a single, covalently joined molecule (Fig. 11.4). In this way a vector molecule can be covalently joined to a passenger DNA molecule, provided that both have termini produced by the same restriction enzyme (or by two restriction enzymes which produce the same single-strand extensions). DNA ligase from two sources has been used for ligation *in vitro*: *E. coli* DNA ligase requires NAD as a cofactor, whilst that isolated from *E. coli* cells infected with phage T4 requires ATP. Since the length of the hydrogen-bonded regions between vector and passenger DNA is only four base pairs, the melting temperature is correspondingly low; the T_m of base-paired termini produced by *Eco* RI is about 5–6 °C (Mertz & Davis 1972). Consequently in this sort of ligation reaction a compromise must be reached between the temperature-optimum of the enzyme reaction and the tendency of the base-paired termini to dissociate. Ferretti and Sgarmella (1981) have shown that the extent of ligation is maximal at 4 °C for termini produced by *Hae* III, *Pst* I and *Eco* RI and decreases with increasing temperature. The temperature at which 50% of the maximal reaction occurs depends on the base composition of the hydrogen-bonded regions (GC pairs giving more stable structures than AT pairs).

During the ligation reaction the cohesive termini will have a finite probability of being ligated to the termini of other vector molecules, or indeed the termini of a single vector molecule may be ligated together to reconstitute the vector. Similarly, passenger DNA molecules may be ligated together or circularized. The desired reaction where a plasmid vector is being used is for the vector to be ligated to a fragment of passenger DNA and then for the hybrid molecule to be circularized. The extent to which this will happen depends on both the relative and absolute concentrations of vector and passenger DNA and has been analysed in detail by Dugaiczyk *et al.* (1975). A technique which ensures that no circularization of the vector molecule occurs and, therefore, enriches the yield of recombinant molecules is treatment of the vector with alkaline phosphatase prior to ligation. Alkaline phosphatase removes the 5′-phosphate groups from the termini of the vector, the passenger DNA is untreated. Vector molecules, therefore, cannot be ligated together

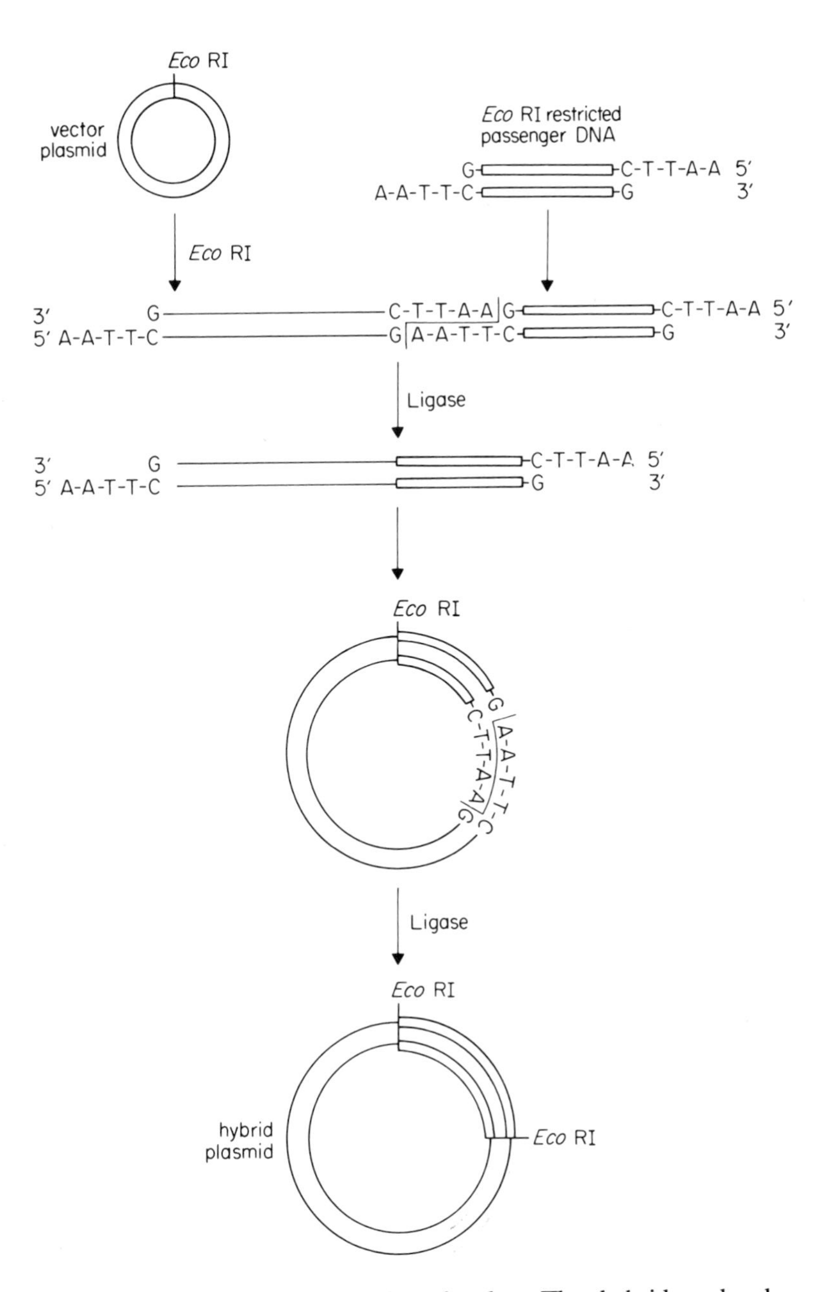

Fig. 11.4. The construction of a recombinant plasmid using restriction enzyme *Eco* RI to produce DNA molecules with cohesive ends suitable for the ligation reaction.

but only to passenger DNA molecules. The hybrid molecules produced still have one single-strand break but this can be repaired *in vivo*.

11.4.2 Blunt-end ligation

In high concentration, the ligase from T4 (see section 3.4.1) will catalyse the formation of covalent bonds between blunt-ended molecules, such as those produced by random shear, restriction enzymes like *Hae* II or the synthesis of cDNA. The enzyme still exhibits the same requirement for a 5′-phosphate and 3′-OH group. A useful extension of this technique is the attachment of synthetic

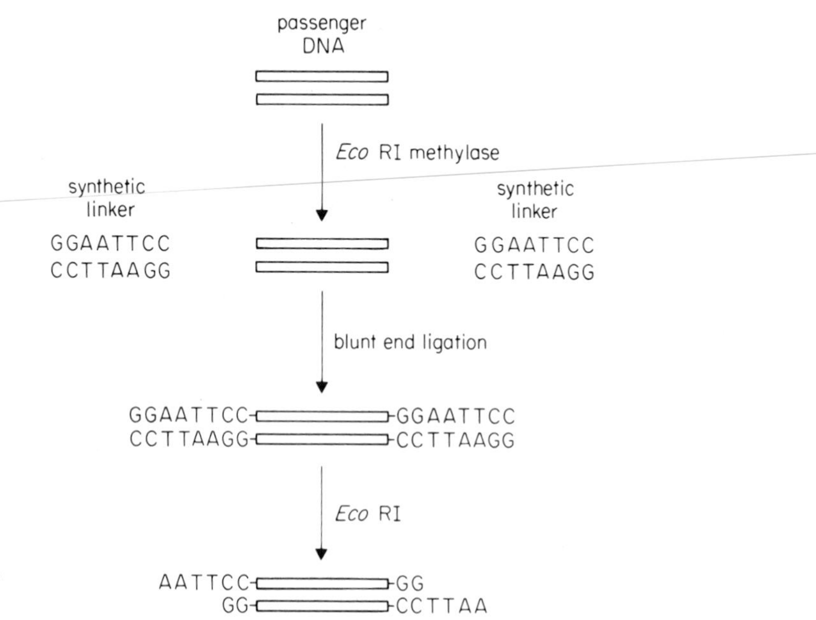

Fig. 11.5. The use of blunt-end ligation to add synthetic *Eco* RI linkers to a DNA molecule prior to its insertion into the *Eco* RI site of a vector.

linker molecules to blunt-ended fragments (Fig. 11.5). The synthetic linker is a blunt-ended oligodeoxyribonucleotide which contains the recognition site for a particular restriction enzyme. After the synthetic linkers have been attached to the passenger DNA by blunt-end ligation, it may be treated with the appropriate restriction enzyme, which will cut within the synthetic linkers thereby producing cohesive ends suitable for ligation to a similarly restricted vector molecule. Since the passenger DNA may well contain a cut site for the particular restriction enzyme, it can be protected by modification with the appropriate methylase prior to ligation to the synthetic linkers. The advantage of this approach is that it generates a chimaeric plasmid in which the passenger DNA, although produced by random shear, is now bounded by two restriction sites.

11.4.3 Homopolymer tailing

Another technique which enables the joining of blunt-ended molecules to the vector utilizes the enzyme terminal deoxynucleotidyl transferase (Chang & Bollum 1971). This enzyme will add nucleotides to the 3′-OH termini of DNA (see section 3.4.1). The preferred substrate is a single-stranded 3′-OH terminus such as is generated by *Pst* I or limited treatment with λ exonuclease, but conditions have been established (Roychoudhury *et al.* 1976) where it will add nucleotides to a 3′-OH even where the terminus has a 5′-phosphate single-strand extension or is blunt ended. The principle underlying this technique is to add a sufficiently long homopolymer extension to the passenger DNA and a complementary extension to the vector DNA that they are able to form a stable, hydrogen-bonded structure. In this fashion the passenger DNA might be given a poly(dG) extension and the vector a poly(dC) extension. Usually, the annealed structures are used directly to transform the host cells and repair of the single-strand gaps occurs *in vivo*. Judicious choice of the nature of the polydeoxyribonucleotide extensions can permit the generation of restriction sites either side of the passenger DNA in the hybrid. The way in which *Pst* I sites can be produced from poly(dC)-tailed passenger DNA annealed to *Pst* I restricted poly(dG)-tailed vector is illustrated in Fig. 11.6.

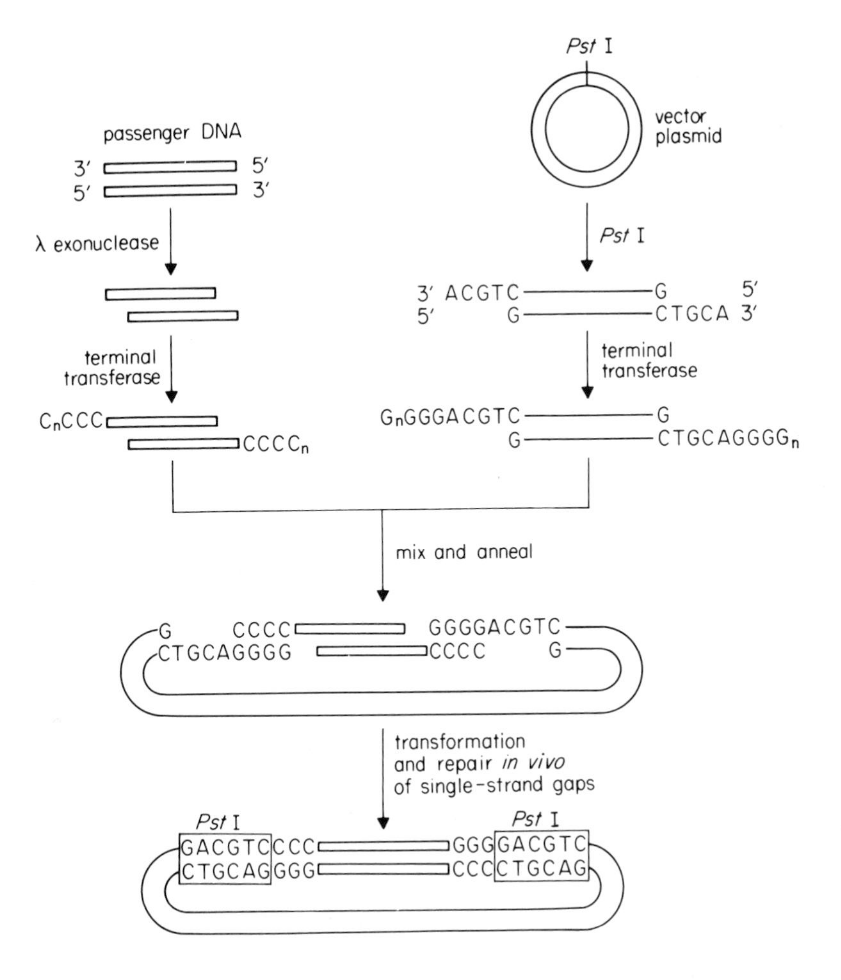

Fig. 11.6. The construction of a hybrid plasmid by the homopolymer tailing method in such a way as to generate sites for the restriction enzyme *Pst* I either side of the cloned DNA.

11.5 INTRODUCTION OF RECOMBINANT DNA INTO HOST CELLS

Since the formation of recombinant DNA molecules is carried out *in vitro*, they must subsequently be introduced into the host cells. One consideration, which applies whether the host is *E. coli* or any other bacterium, is that the incoming DNA may be subject to host-specified restriction. The recombinant molecules will not have the methylation pattern specified by the host modification system and the consequent restriction will drastically reduce the efficiency with which the recombinant vector can become established in the host cells. To overcome this problem restrictionless (r^-) mutants should be used as the host.

11.5.1 *E. coli*

Mandel and Higa (1970) showed that *E. coli* cells treated with cold calcium chloride could take up λ DNA and it was subsequently shown that similarly treated cells could also take up plasmid and chromosomal DNA. This process is referred to as transfection when phage DNA is used and transformation when plasmid or chromosomal DNA is used. The process, however, is inefficient and at best only about 1×10^7 transformants will be obtained per μg of plasmid DNA. This corresponds to about 1% of the cells being transformed and the figure is worse for λ DNA. Transformation with linear chromosomal DNA only occurs at a detectable frequency if the cells lack exonuclease V and exonuclease I and, therefore, strains with the *recB recC sbcB* genotype are normally used. Presumably the linear DNA is degraded by these exonucleases before recombination with the chromosomal DNA can take place. Transfection with λ is only slightly reduced in cells that possess exonucleases I and V, probably due to circularization by virtue of its *cos* ends, thereby protecting it from exonuclease digestion. Plasmid DNA, although existing in the cell as the supercoiled form, is usually obtained *in vitro* as a mixture of three forms: supercoiled; open-circular, and linear. The recombinant plasmid DNA produced by ligation *in vitro* will be in the open-circular and linear forms. Transformation of *E. coli* with supercoiled or open-circular plasmid DNA is not affected by exonucleases I and V, and transformation with linear plasmid DNA, although considerably reduced, is detectable. The residual transforming activity of linear plasmid DNA is presumably due to its recircularization *in vivo*. Plasmid size does have an effect on transformation efficiency, with increasing size reducing the efficiency. This can lead to a distortion of the frequency of fragments appearing in a gene bank, with large fragments being present at a disproportionately low frequency, and will also make the cloning of large fragments in a plasmid more difficult.

The ligation reaction produces a heterogeneous collection of DNA molecules, only some of which will be composed of passenger DNA linked to a vector molecule. Any passenger DNA which is not joined to a vector may be taken up by the host cell during transformation, but will be incapable of replication, and if it is linear it will be subject to exonuclease degradation. The process of transformation, therefore, selects out of the heterogeneous population those molecules which are either reconstituted vector or recombinants composed of passenger plus vector DNA. After transformation with plasmid DNA the cells are normally incubated in a rich medium for a short while to allow expression of plasmid-borne genes before being plated out on selective media.

An alternative approach to the transfection of λ DNA is the process of packaging *in vitro* (Hohn & Muray 1977). During the normal lytic cycle of lambda, replication of phage DNA gives rise to a concatameric molecule, which is cleaved at the *cos* sites into monomeric molecules during the process of packaging into the phage heads. The recombinant λ DNA produced by ligation *in vitro* will contain a proportion of concatameric molecules which can act as a substrate for the packaging reaction, provided that the *cos* sites are separated by DNA which is in the size packaging range of λ (23–32 Mdal). In practice, the packaging system *in vitro* is composed of a mixed lysate from two induced λ lysogens, both of which are blocked in the process of producing mature phage particles by amber mutations. Each lysogen is blocked at a different point in the process so that in a mixture of the two, genetic complementation will occur and any exogenous recombinant λ DNA can be packaged into phage heads. The advantage of packaging *in vitro* and subsequent infection is that it is an inherently more efficient method for introducing recombinant DNA into the cells. Plasmid vectors, known as cosmids, which contain the λ *cos* site have been developed to make use of the efficiency of the packaging system *in vitro*.

11.5.2 Other prokaryotic hosts

Bacillus subtilis and *Streptomyces* spp. are among a considerable number of microorganisms for which well-established transformation systems exist. Although transformation of *B. subtilis* with chromosomal DNA is efficient there are problems with plasmid DNA. In fact only concatameric plasmid DNA has any significant transforming activity (Canosi *et al.* 1978) and since the predominant species from a ligation *in vitro* will be recombinant plasmid and recircularized plasmid in the open-circular form, the conventional transformation system represents an extremely inefficient method for the introduction of recombinant DNA into the host cells. Several approaches have been adopted to overcome this problem, including the use of hybrid replicons. Chang and Cohen (1979) found that protoplasts of *B. subtilis* could be transformed at high efficiency with plasmid DNA. Alternatively, the plasmid markers can be rescued by recombination with homologous resident plasmids or by integration into the chromosomal DNA (Young 1980).

A different approach to the introduction of transforming DNA into Gram-positive bacteria (and also into fungi; see section 11.8) was suggested by the phenomenon of protoplast fusion. Protoplasts, produced by treatment of cells with a cell wall lytic enzyme in isotonic medium, can be fused if they are treated with a chemical 'fusogen' of which the most widely use is polyethylene glycol in high concentration (40–50%). This technique is used for obtaining very high recombination frequencies for formal genetic analysis and strain construction. Essentially the same techniques will allow fusion of protoplasts with liposomes (artificially produced, unilamellar, phospholipid vesicles) containing trapped transforming DNA. In principle, the use of liposomes can be used for very high efficiency transformation of any organism provided it is possible to regenerate viable cells by recovery of the protoplasts (Fraley & Papahadjopoulos 1981). Among the bacteria the technique is likely to have a major impact on the use of recombinant DNA methods for strain improvement and construction in the genus *Streptomyces*. These filamentous Gram-positive bacteria are already the source of the majority of clinically used antibiotics. Recent developments in *Streptomyces* genetics are reviewed by Hopwood (1981).

11.6 DETECTION AND CHARACTERIZATION OF RECOMBINANTS

Following the construction of recombinant plasmid or phage and their subsequent introduction into the host cells, there still remains the problem of identifying the particular clone which carries the DNA fragment of interest. The extent of the problem depends largely upon the strategy used to generate the recombinants. Treatment of the vector with alkaline phosphatase, the use of positive selection vectors and similar techniques described previously ensure that a very high proportion of the clones will actually carry recombinant DNA rather than the reconstituted vector. Similarly, enrichment of the DNA fragments prior to cloning or the use of cDNA produced from a specific mRNA will ensure that a large proportion of the recombinants will carry the desired fragment. Even then some method must be available for detecting the actual clone of interest, although the problem is considerably less than if a shot-gun cloning has been carried out or a gene bank is being screened for a particular fragment.

11.6.1 Identification of recombinants

Obviously, if the cloned DNA confers some easily detectable phenotype upon the host cell, such as drug resistance, prototrophy or immunity to a phage, selection of the clone by conventional techniques represents no problem. However, this is rarely the case and other methods for identifying the clones must be used.

Nucleic acid hybridization

A variety of methods based on nucleic acid hybridization are available to detect whether a particular DNA sequence is present in a clone. This approach does not require that the cloned DNA is expressed, only that the particular DNA sequence is present. Such methods demand that a 'probe', either DNA or RNA, which has extensive sequence homology with the cloned fragment, is available. Such situations would arise where a purified mRNA (the same used to make cDNA) can be isolated, or where a clone has been obtained which carries a portion of the fragment of interest. Additionally, if part of the amino acid sequence of a polypeptide encoded by the cloned fragment is known, all the possible oligonucleotides encoding the oligopeptide can be synthesized to be used as probes. This approach was one of those used by Goeddel *et al.* (1980) to detect clones carrying DNA sequences coding for human fibroblast interferon. Isolation of plasmid or phage DNA from a large number of clones for screening by nucleic acid hybridization would be a laborious task. Grunstein and Hogness (1975) developed a technique by which colonies resulting from transformation with recombinant plasmid DNA can be screened *in situ*. Following transformation and growth of the colonies on a master plate, the colonies are replica plated on to a second plate, the surface of which is covered with a cellulose nitrate filter. The replica colonies grow on the surface of the filter, which is then removed from the plate and the colonies are lysed *in situ* by alkali treatment, which also serves to denature the DNA. Protein is removed by treatment with proteinase K and the denatured DNA is irreversibly bound to the filter by heating it at 80 °C. The filter is then incubated with the RNA or DNA probe, which has been radioactively labelled. After hydridization the filter is subjected to radioautography to detect areas of hybridization which will of course correlate with the position of colonies on the master plate. A similar technique has been developed for use with phage plaques (Benton & Davis 1977).

On many occasions a suitably pure RNA probe is not available. The procedure known as hybrid-arrested translation (Paterson *et al.* 1977), although technically more demanding, enables a comparatively impure mRNA preparation to be used for the detection of a particular clone. DNA from the clones to be screened is denatured and unpurified mRNA, which will contain RNA homologous to the cloned fragment, is added. The mixture is then incubated *in vitro* in a translation system containing radioactively labelled amino acids. If the clone contains DNA which will hybridize to any of the mRNA, this species will no longer be available for translation. Consequently, the specific protein is not made in the presence of DNA from the desired clone. The presence, or absence, of the particular protein is detected by immunoprecipitation or polyacrylamide gel electrophoresis and radioautography.

Immunochemical methods

Detection of a clone by immunochemical methods obviously requires transcription of the cloned DNA and translation to produce a polypeptide with recognizable antigenic determinants, and depends on the availability of the appropriate antibody. One of the most widely used immunochemical techniques is that developed by Broome and Gilbert (1978). After transformation and growth of the clones on a master plate they are replica plated. The replica colonies are lysed *in situ* so that any proteins whose synthesis is directed by a cloned fragment will be released from the cells. A polyvinyl sheet, coated with immune serum containing antibodies to the protein of interest, is layed on the surface of the replica plate and any of the specific antigen present will absorb to the antibody on the polyvinyl sheet. The sheet is then immersed in a solution of radioactively labelled (^{125}I) antibody which will bind to any antigen bound to the polyvinyl sheet, since most antigens have more than one antigenic site. After radioautography the antigen-producing colonies on the master plate can be identified.

Structural methods

When none of the immunochemical or nucleic acid hybridization techniques can be used, the only remaining approach is to determine the size of the cloned DNA and also, possibly, to determine its restriction pattern with a variety of restriction enzymes. Detection of the desired recombinant by this method depends on previous knowledge of the size of the DNA fragment and its restriction pattern. Fortunately, the task of isolating plasmid DNA from a large number of clones for sizing of the inserted fragment has been simplified. Sufficient plasmid DNA can be isolated from a single colony to estimate the size of the plasmid by agarose gel electrophoresis. Similarly, methods have been developed for the extraction of plasmid DNA from very small volumes (1–2 ml) of culture for restriction analysis to be carried out.

11.6.2 Structural analysis

Once the desired clone has been obtained, the recombinant phage or plasmid can be isolated in large amounts so the structure of the recombinants can be determined. Obviously it will be necessary to determine the size of the cloned fragment and to establish a restriction map for it. If the recombinant was produced by restriction digestion and subsequent ligation then it is an easy matter to cut the fragment out of the recombinant molecule with the same restriction enzyme that was originally used to generate the fragment. Where the recombinant was produced from cDNA or randomly sheared DNA, estimation of the size of the cloned fragment and its restriction pattern will rely on the fact that the vector is already well characterized. Ultimately it may be desired to determine the base sequence of the cloned fragment.

11.6.3 Functional analysis

In many cases it will be of interest to examine the expression of the cloned DNA by determining which RNA species are transcribed from it and which polypeptides are synthesized. Under most conditions the RNA transcripts and polypeptides whose synthesis is directed by the cloned DNA will be obscured by the normal RNAs and proteins of the host cell. This problem is overcome by the use of mutants of *E. coli* which produce minicells. These mutants are defective in cell division and continuously produce small cells which lack any chromosomal DNA but contain copies of any multicopy plasmid that is present in the parental cells. Though the minicells are non-viable they are capable of carrying on transcription and translation for a while and, consequently, will predominantly synthesize RNAs and polypeptides specified by the plasmid DNA. If the minicells are incubated with suitable radioactive RNA precursors or amino acids, the plasmid-specified RNAs and polypeptides can be detected by gel electrophoresis and radioautography. The expression of genes cloned in a phage vector can be examined in cells that have been exposed to extremely high doses of u.v. The normal background of chromosomal gene expression is drastically reduced by the extensive DNA damage caused by the u.v. irradiation.

11.7 EXPRESSION OF CLONED DNA SEQUENCES

Expression of cloned DNA requires that the sequence encoding a polypeptide is accurately and efficiently transcribed, that the RNA product is faithfully translated and that the polypeptide produced is able to adopt its correct 2°, 3° and even 4° structures. Obviously there are numerous points in this process where problems may arise in the expression of a cloned DNA sequence in a heterologous cellular environment. Efficient and accurate transciption requires that the host cell RNA polymerase should recognize a promoter and transcriptional start-signal upstream and in the correct reading frame from the structural gene and also a transcription termination site downstream. Translation will require not only an initiation codon, but also the untranslated region of the RNA involved in ribosome binding and a termination codon. Even if the polypeptide is accurately synthesized there may be several steps before the mature protein is obtained; these might include proteolytic cleavage of a segment of the polypeptide, formation of disulphide bonds and assembly of monomers into a functional polymeric protein. Finally, the cloned DNA sequence itself must be stably maintained in the host cell and not suffer any structural alterations. There are, in fact, an increasing number of observations that some cloned DNA sequences, particularly in plasmids, are intrinsically unstable in a heterologous cellular environment and are subject to deletions and other structural alterations. In a few cases this can be explained by the fact that the cloned DNA sequence encodes a product that is toxic to the host cell, in which case it is not surprising that deletion derivatives of the plasmid are selected which are no longer capable of expressing this function. Additionally, with a homologous piece of DNA or one carrying reiterated sequences, the structural integrity of the recombinant may be lost by inter- or intra-molecular recombination, although this may be prevented by using a recombination deficient host. However, there are cases of recombinant plasmid instability where no obvious explanation is available.

11.7.1 Expression of cloned prokaryotic DNA

Since Chang and Cohen (1974) demonstrated that a cloned gene from *Staph. aureus,* encoding resistance to ampicillin, could be expressed in *E. coli*, there have been numerous examples of other genes from both Gram-negative and Gram-positive organisms being successfully cloned and expressed. This suggests that there is no fundamental barrier to heterospecific gene expression in *E. coli.* However, attempts to achieve the expression of genes from Gram-negative organisms in *B. subtilis* have not met with the same success, though genes from other Gram-positive organisms have been expressed. There is some evidence to suggest that although the transcriptional and translational processes in *E. coli* and *B. subtilis* are essentially similar, there may be slight differences in promoter recognition (Haldenwang & Losick 1980) and translation (Stallcup *et al.* 1974) between the two species. Recently, however, Rubin *et al.* (1980) have demonstrated that the gene from *E. coli* encoding thymidylate synthetase, cloned in pBR322, could transform a thymine-requiring strain of *B. subtilis* to prototrophy. Since pBR322 cannot replicate in *B. subtilis* it was concluded that the cloned DNA was integrated into the chromosome.

11.7.2 Expression of cloned eukaryotic DNA

In 1976 Struhl *et al.* demonstrated that a cloned fragment of yeast DNA was able to complement an *E. coli his*B mutation, which causes a lack of the enzyme glycerol phosphate dehydrogenase. Initially, however, it was not clear whether this apparent complementation was due to suppression of the *his*B mutation or whether the yeast enzyme was actually being synthesized. Clear evidence for the functional expression of a cloned eukaryotic gene was obtained in 1978 by Alton *et al.*, who showed that the gene from *Neurospora crassa*, encoding catabolic dehydroquinase, was efficiently transcribed in *E. coli* from a promoter on the *N. crassa* DNA and that the monomeric polypeptide was assembled into an active multimeric enzyme. Of course, it cannot be relied upon that a DNA fragment generated by restriction digestion will carry the required structural gene and also an endogenous promoter which will be recognized by the host polymerase, or the necessary downstream sequences that ensure efficient interaction between the mRNA and the ribosomes. Indeed, the very method of construction of cDNA ensures that this will not be the case. The fact that eukaryotic DNA sequences without endogenous promoters might be expressed in *E. coli* was shown by Chang *et al.* (1978), who demonstrated the phenotypic expression of a DNA sequence coding for mouse dihydrofolate reductase. Since the cloned fragment had been produced by the synthesis of cDNA it obviously did not carry its own promoter. In fact the cDNA had been cloned into the *Pst* I site of pBR322 in the B-lactamase gene and was probably being transcribed from the powerful β-lactamase promoter. Surprisingly, however, it was found that the cloned dihydrofolate reductase gene was not in the correct reading frame for the β-lactamase promoter.

Many eukaryotic genes do not have a continuous coding sequence, but in fact are divided into coding regions (exons) and non-coding regions (introns) and the initial RNA transcript must be processed to give the mature mRNA (see section 8.3.4). Consequently, many eukaryotic genes may not be cloned in *E. coli* in a functional form when restriction enzyme digestion is used to generate the DNA fragments for cloning. In turn, this problem is overcome by using cDNA for cloning since this will carry the continuous coding sequence uninterrupted by introns.

11.7.3 Expression vectors

A common objective of gene cloning experiments is the expression of a gene in a heterologous cellular environment and, in most cases, to achieve as high a level of expression as possible. This is obviously the case where eukaryotic genes encoding polypeptides of pharmaceutical interest are concerned. A variety of strategies have been arrived at to enhance expression of cloned DNA and most rely on coupling the cloned gene to a powerful prokaryotic promoter to achieve efficient transcription.

Copy-number mutants

One way of increasing the expression of a cloned gene which is already expressed to some extent is to increase the gene dosage. A plasmid vector like pBR322 is present at about 30 copies per host cell, but plasmid vectors have been developed which have a temperature-sensitive mutation in control of replication. Uhlin *et al.* (1979) described two vectors derived from the naturally occurring plasmid R1, which are temperature-sensitive, runaway-replication mutants. At 30 °C they exist at about 10–25 copies per cell, but if the temperature is increased above 35 °C, within four generations the copy number may have risen to over 2000, a process which is eventually lethal to the cell. This increase in copy number was also shown to amplify the expression of a cloned β-lactamase gene.

Cloning into the Pst *I site of pBR322*

Although pBR322 was constructed as a general purpose vector, there are certain features of the *Pst* I site on the plasmid which make it useful for obtaining the expression of cloned DNA. The single *Pst* I site lies in the ampicillin-resistance gene at a position corresponding to amino acid residues 181–182 of the β-lactamase. The idea behind cloning into this site is that the cloned sequence will be transcribed from the promoter of the β-lactamase gene and that a fused protein will be synthesized bearing the first 181 amino acids of β-lactamase joined at the carboxyl terminus to the polypeptide specified by the cloned DNA sequence. If a cDNA fragment is inserted into the *Pst* I site by the homopolymer tailing technique, in independently isolated clones the number of base pairs between the β-lactamase and the cloned cDNA will very probably be different and there will be a good chance that one of the clones will have the cDNA in the correct reading frame. Finally, the β-lactamase is a periplasmic protein; it is initially synthesized as pre-β-lactamase with a 23 amino acid amino-terminal leader sequence which directs transport to the periplasm and is cleaved off as the polypeptide is transported across the membrane. It is likely that any polypeptide fused to the pre-β-lactamase would also be transported to the periplasm, thereby facilitating its purification. Villa-Komaroff *et al.*

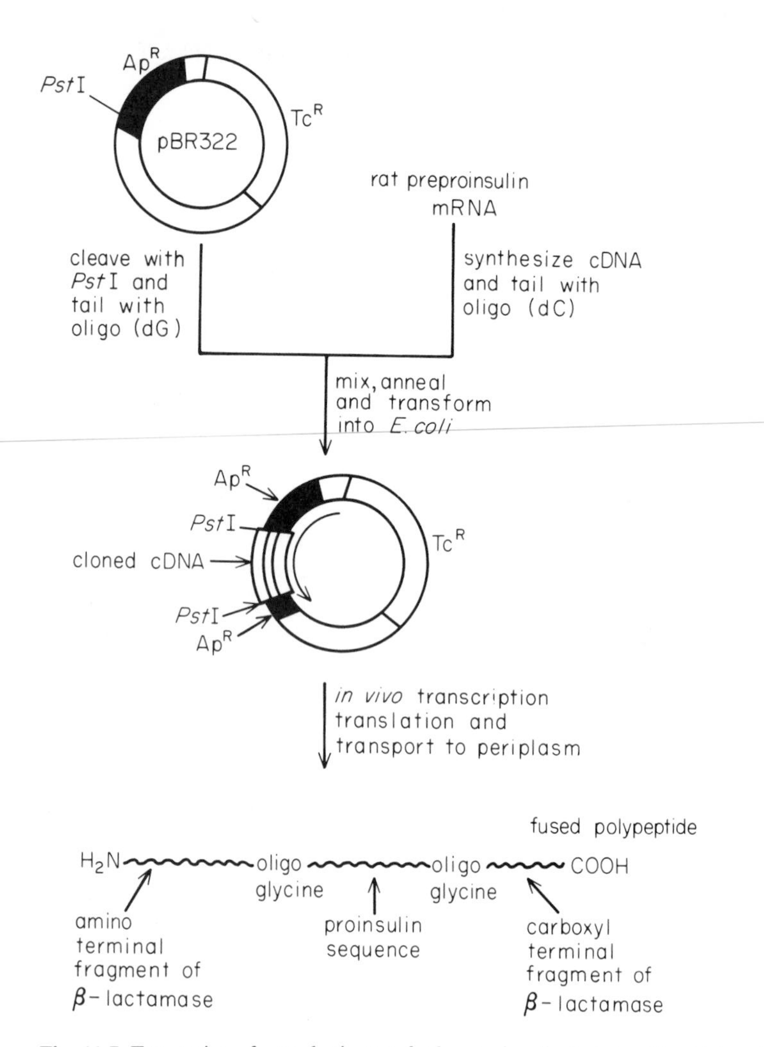

Fig. 11.7. Expression of a synthetic gene for human insulin A chain (Goeddel *et al.* (1979b) in *E. coli.*

(1978) used this approach to achieve expression of proinsulin in *E. coli* (Fig. 11.7). The mRNA for preproinsulin was isolated and cDNA prepared to it, this was cloned into the *Pst* I site of pBR322 by the homopolymer tailing technique. A clone was isolated that produced a fused protein which had antigenic determinants for both β-lactamase and proinsulin and which was transported to the periplasm. Talmadge and Gilbert (1980) have constructed a series of vectors in which the cloned DNA may be coupled directly to the β-lactamase leader sequence in any reading frame so that the polypeptide product is not fused to a large fragment of β-lactamase. Talmadge *et al.* (1980) have shown that when cDNA for preproinsulin is inserted into one of these vectors so that it is coupled to the DNA sequence encoding only a part of the β-lactamase leader sequence, *E. coli* is able to recognize the eukaryotic leader sequence and correctly cleave it to produce proinsulin from preproinsulin.

Transcription of cloned genes from prokaryotic promoters

In addition to the β-lactamase promoter of pBR322 a number of other prokaryotic promoters have been used to obtain expression of cloned genes, these include the *lac* and *trp* promoters of *E. coli* and the left promoter of λ (see section 5.2.4). Goeddel *et al.* (1979b) constructed plasmids in which the chemically synthesized genes for the A and B chains of human insulin were fused to a part of the *E. coli* Lac operon, which comprised the *lac* promoter and operator together with a large part of the structural gene for β-galactosidase (Fig. 11.8). Clones were obtained which produced large amounts (20% of the total cellular soluble protein) of a hybrid protein which was composed of a large part of the β-galactosidase polypeptide fused to either the insulin A or B chain. The synthetic genes had been so designed that the DNA sequences coding for the A and B chains were preceded by the codon for methionine. The fused proteins could, therefore, be treated with cyanogen bromide which cleaves poly-

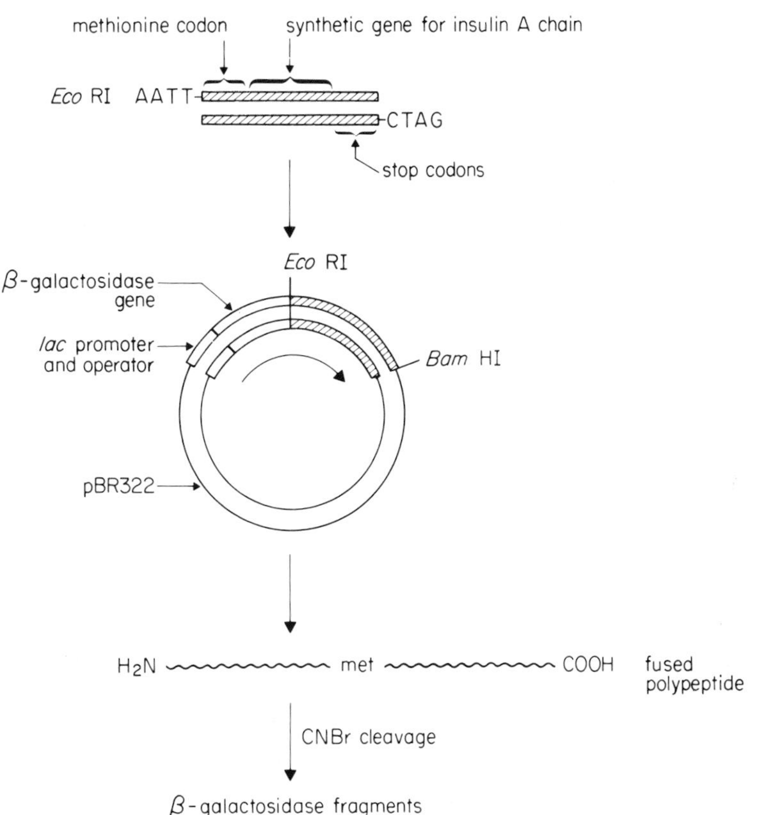

Fig. 11.8. Expression of cloned cDNA encoding rat proinsulin in *E. coli* (Villa-Komaroff *et al.* 1978).

peptides at methionine residues, to yield the mature A and B chains (neither of which contains a methionine residue).

Obviously, it is desirable to achieve efficient expression of a cloned gene in such a way that the particular polypeptide is produced in a mature form, not fused to some other protein. Roberts *et al.* (1979) described a method by which, in principle, the *lac* promoter can be placed at any distance in front of a cloned gene. The 'portable promoter' fragment which they used carries the *lac* promoter and the untranslated region prior to the translational start signal for the β-galactosidase structural gene, so that gene-promoter fusions should result in the production of a native protein. One result of their experiments was to emphasize the importance of the untranslated region, downstream from the promoter, which specifies the Shine–Dalgarno sequences in the mRNA (see section 6.5.1): these sequences are involved in the binding of the ribosomes to the mRNA and are crucial for the efficient translation of the message.

Goeddel *et al.* (1980) described the isolation of a clone which carried the DNA sequence encoding the precursor form of human fibroblast interferon, in which the mature fibroblast interferon (166 amino acids) is preceded by a leader sequence of 21 amino acids. To express high levels of the mature form of interferon in *E. coli,* initiation of translation must start at the methionine codon (amino acid 1) of the mature protein, rather than at the initiation codon of the leader sequence. By a series of manipulations *in vitro* a fragment of DNA was constructed from the original cloned gene, in which the first three base pairs constituted the methionine codon of the mature protein and this fragment was inserted into plasmid vectors, such that it was under the control of the *E. coli lac* or *trp* promoters. In both cases the promoter fragment contained the regions specifying the ribosome binding site, but lacked an initiation codon. The *trp* promoter leads to measurably higher levels of expression of the mature fibroblast interferon gene than the *lac* promoter. In an attempt to further increase expression, a plasmid was constructed with three successive *trp* promoter fragments; expression was increased a further 4- to 5-fold (about 20 200 molecules of interferon per *E. coli* cell). This increase in expression was probably due to the fact that the *trp* promoter was being derepressed because the *trp* repressor was being titrated out by the multiple copies of the *trp* operator.

11.8 CLONING IN EUKARYOTES

11.8.1 Cloning in yeast

Cloning in yeast is attractive not only because of the possibility of the manipulation of commercially important strains, but also because it is a eukaryotic host which is susceptible to many of the classical techniques of bacterial genetics. A basic prerequisite of such a system is that there should be some effective method of introducing recombinant DNA into the yeast cell and ensuring its maintenance. Hinnen *et al.* (1978) showed that sphaeroplasts of *Saccharomyces cerevisiae* could be transformed with a chimaeric plasmid pYeLeu10. This plasmid was composed of the yeast *leu*2 gene cloned in the *E. coli* plasmid ColE1 and on transformation into a *leu*2^- strain of *S. cerevisiae,* prototrophic colonies were obtained. In all cases examined the plasmid was usually, though not always, integrated into the chromosomal DNA at the *leu*2 locus and, in addition, was stably maintained.

Subsequently, a variety of yeast vectors have been developed. Beggs (1978) constructed chimaeric plasmids in *E. coli* composed of pMB9 (an *E. coli* plasmid vector), a yeast plasmid and fragments of yeast nuclear DNA capable of complementing a *leu* mutation in *E. coli.* The yeast 2μ plasmid (see section 10.1) is present in many strains of *S. cerevisiae* but has no known function. There are about 50–100 copies per cell and it replicates under nuclear control. These chimaeric plasmids could be used to transform a Leu$^-$ strain of *S. cerevisiae* to prototrophy and at a considerably greater frequency than that obtained for pYeLeu10, presumably because there was no need for integration, since these chimaeras contain the replication origin of the 2μ plasmid. They also demonstrated that the plasmid was maintained in the cytoplasm.

Struhl *et al.* (1979) constructed a recombinant plasmid containing a 1.4 kb fragment of yeast nuclear DNA which carries the *trp*1 gene from the centromere region of chromosome IV, inserted into the *Eco* RI site of pBR322. This plasmid would transform a Trp$^-$ strain of *S. cerevisiae* at high frequency and was maintained in the cytoplasm. Presumably, since pBR322 cannot replicate in yeast (Beggs 1978), the yeast centromere region must carry an origin of replication. Beggs *et al.* (1980) determined whether yeast cells were able to process an RNA transcript from a foreign DNA fragment carrying a gene with introns. They introduced the complete rabbit β-globin gene which contains two introns into the plasmid composed of pMB9, the yeast 2μ plasmid and the *leu2* gene. The cloned β-globin gene was transcribed in the yeast cells but was about 20–40 nucleotides shorter at the 5′-end and extended to about halfway through the second intron. There was no evidence of processing, however, the processing of heterologous RNA does occur in other systems (see the following section).

11.8.2 Cloning in mammalian cells

Current methods of cloning in mammalian cells largely rely on the integration of the foreign DNA into the genome; however, attempts are being made to develop vectors which are capable of being maintained in an extra-chromosomal state.

Microinjection

DNA fragments can be microinjected into mammalian cells grown in tissue culture, but the number of cells that can be treated in this way is obviously limited and any expression of the DNA is transient. However, a few cells receiving the injected DNA will integrate it into the chromosomal DNA where it can be stably maintained and expressed.

Viral vectors

Simian virus 40 (SV40) is a papovavirus which has been used extensively as a vector. The virus particles contain a double-stranded DNA molecule of about M_r 3×10^6, however, because packaging of the DNA into the viral particles will not occur if the size of the molecule is significantly increased, deletion mutants of SV40 must be used as vectors. The advantage that SV40 offers as a vector is that foreign DNA can be placed under the control of viral regulatory elements and will either become integrated in permissive cells or replicate to cause lysis in non-permissive cells. The deletion derivatives of SV40 are defective and are unable to lyse host cells after infection unless a 'helper' virus is simultaneously used during infection. The presence of the 'helper' virus permits the replication and packaging of the recombinant virus which can then be isolated in sufficient amounts to infect a larger population of cells so that expression of the foreign DNA can be investigated.

Cotransformation

The principle of cotransformation is to transform cells simultaneously with two species of DNA, one of which confers a selectable phenotype. Since within the cell population only a few cells are capable of taking up exogenous DNA and these cells can take up more than one molecule, selection for cells carrying the marker will also select for many cells which carry the second species of DNA. In practice, the selectable DNA sequence is usually the thymidine kinase (tk) gene from the virus *Herpes simplex*. Wigler *et al.* (1979) showed that purified tk gene could be used to transform mouse cells in the presence of a variety of other DNAs, i.e. ϕX174, pBR322 and rabbit β-globin sequences. Transformants could be detected which carried multiple integrated copies of the foreign DNA and it was stably maintained through several generations. The basis of the cotransformation system rests on HAT medium (Szybalski *et al.* 1966). HAT medium contains hypoxanthine, amethopterin (also known as methotrexate) and thymidine. Amethopterin (see section 2.2.5) suppresses the endogenous synthesis of purines and pyrimidines. The missing purines are supplied by hypoxanthine. Cells which possess tk are able to phosphorylate thymidine and can, therefore, grow on HAT medium, but tk^- cells cannot.

Expression of heterologous DNA in mammalian cells

The first report of the production of a discrete, functional mRNA transcribed from a gene cloned into SV40 in a mammalian cell concerned the synthesis of rabbit β-globin in monkey kidney cells (Mulligan *et al.* 1979). Rabbit β-globin cDNA (i.e. lacking introns) was inserted into SV40 in the region of the genome coding for the major viral capsid protein. The recombinant virus multiplied efficiently in monkey kidney cell cultures in the presence of a helper virus and was transcribed to yield cytoplasmic, polyadenylated mRNAs which contained the β-globin coding sequence and substantial quantities of β-globin polypeptide were synthesized.

The ability of host cells to recognize and correctly process RNA transcripts from foreign genes containing introns was established for a chimaera of the mouse β-globin sequence carried in SV40 (Chu & Sharp 1981). The chimaeric molecule was created in such a fashion that the first intron of the gene encoding the SV40 major capsid protein was fused to part of the second intron of the mouse β-globin gene, thereby creating a hybrid intron. The hybrid gene was transcribed in monkey kidney cells, the RNA transcript was correctly processed and the mRNA was polyadenylated.

11.8.3 Cloning in plant cells

At present, interest in cloning in plant cells is focused on two possible vector systems: the caulimoviruses, of which the best-studied representative is Cauliflower Mosaic Virus and the Ti plasmid of the bacterium *Agrobacterium tumefaciens*. Development of vectors using caulimoviruses will depend upon the further functional characterization of the viral genome, but the Ti plasmid has already been used to transfer a piece of foreign DNA into the genome of plant cells.

The bacterium *A. tumefaciens* causes Crown Gall tumour in a wide range of dicotyledenous plants; tumorigenicity depends on the presence in the bacterium of the Ti plasmid. Infection of a plant with *A. tumefaciens* leads to the formation of a tumour in which the plant cells produce one of two unusual amino acids, nopaline or octopine, which the bacterium is able to utilize as a nitrogen source. The molecular basis of tumour induction is the transfer of a well-defined segment of the Ti plasmid to the plant cells, where it is integrated and stably maintained in the plant cell DNA. Hervalsteens *et al.* (1980) investigated the possibility of using the Ti plasmid as a vector by inserting the bacterial transposon Tn7 into the Ti plasmid within the segment that is incorporated into the plant cell DNA, in particular into the region that codes for nopaline production. They were able to show that nopaline synthesis could not be detected in tumours caused by infection with *A. tumefaciens* carrying this plasmid. The cells of a bacterium-free tissue culture from such a tumour were found to have Tn7 present in their genomic DNA.

11.9 CONCLUDING REMARKS

The past decade has witnessed a revolutionary advance in molecular biology, most spectacularly in the field of gene cloning, which has far-reaching implications, not only for basic research but also for the applied sciences. The legal constraints that were initially applied in many countries to recombinant DNA technology, which at the time represented a wise move to contain a conjectural hazard, are now being relaxed, as conjecture is being replaced by information. This will, in part, contribute to the expansion, further sophistication and application of the techniques over the next few years. Although, at present, the main contribution of this newly acquired ability to manipulate genes has been to our understanding of the structure and function of the genetic material, it must shortly make its impact on medicine and industry. The contribution to medicine will include the production of clinically important compounds (e.g. vaccines, antibiotics, interferon, hormones) and hopefully extend to the diagnosis and treatment of hereditary disorders, whilst in industry recombinant DNA technology is likely to assist in the development of new processes for the production of foods, fuels and plastics that are not so demanding of non-renewable resources.

NOTE ADDED IN PROOF

To keep abreast of the fast-moving field of recombinant DNA technology see the series of volumes *Genetic Engineering: Principles and Methods* (Eds J.K. Setlow & A. Hollaender, Plenum Press, New York), and, in particular, vol. 2 (1980) and vol. 3 (1981).

Glossary

Genetic and Molecular Biological Terms

The terms listed here are either taken as read in the rest of the text or are the subject of some slight confusion in the literature. For definitions of ideas introduced in the book (for example operon) refer to the text via the index.

allele one of a set of possible forms of a given gene.

ascospore fungal meiospore (*see* meiosis).

ascus a sac containing ascospores.

bacteriophage bacterial virus.

chromatid one of two daughter chromosomes found in a structure that arises after chromosome duplication. The chromatids are joined at a single structure (the centromere) which itself subsequently divides with the result that the two sister chromatids become separate chromosomes.

chromatin the DNA-protein material of eukaryotic chromosomes.

chromoneme the major DNA molecule of bacteria or phages. This word has never become popular and this book follows current practice in referring to the bacterial chromoneme as a 'chromosome'.

chromosome strictly, one of several structures visible when the nuclei of eukaryotic cells undergoing meiosis or (in certain organisms) mitosis are examined microscopically after treatment with certain histological stains. The word is commonly used to refer also to chromonemes (*q.v.*) and viral genomes.

cis and trans genetic terms used in the description of pseudo-allelism. For the purposes of this book, the following should suffice. In the cis configuration two alleles are on the same homologue; in the trans configuration they are on different ones. 'Homologue' refers to one of the two chromosomes that pair during meiosis. However, cis and trans can refer also to viral alleles (present on different viral chromosomes during multiple infection) and to bacterial alleles (for example in a partial diploid).

cistron strictly a gene with the following property. Of any two alleles, the same one (often the wild type) is dominant irrespective of whether the arangement is cis or trans. In common with the rest of the molecular literature, this book uses cistron as a synonym for structural gene.

diploid containing two of each type of chromosome (except for sex chromosomes) with the possibility of allelic differences between the two chromosomes of a pair. Thus, although a bacterium may contain two or more *copies* of its chromosome, it is, nevertheless, haploid.

episome a genetic element that can exist either as an autonomous replicon (plasmid) or integrated within another replicon (usually a bacterial chromosome).

F factor (fertility factor) a particular plasmid that enables a bacterium to act as donor during conjugation.

F′ (F prime) an F factor that contains chromosomal genes.

gamete haploid germ cell.

gene a hereditary unit (defined chemically by a DNA sequence) that occupies a defined locus and which can have a specific effect on phenotype and can mutate to allelic variants.

genome all the genes of a single gamete or haploid organism (including a bacterium) or virus.

Gram stain a microscopist's stain for bacteria. For the purposes of this book, Gram-negative bacteria have a cell envelope consisting of plasma membrane, a relatively thin cell wall (peptidoglycan) and outer membrane. Gram-positive bacteria have a thicker cell wall and no outer membrane.

haploid containing only one set of alleles.

heterozygote individual that has inherited different alleles at one or more loci and therefore does not breed true.

homozygote individual that has inherited identical alleles in corresponding loci and, hence, breeds true.

insertion sequence (IS) a transposable genetic element lacking any structural gene other than one required for transposition (and possibly not even such genes).

locus position occupied by a gene in a chromosome.

meiosis the process whereby a single chromosome duplication, followed by two nuclear divisions results in a halving of the zygotic chromosome number (usually producing haploid products of meiosis from diploids). Examples of meiosis are gametogenesis in animals and the formation of sexual spores ('meiospores') in fungi.

merodiploid partial diploid (*q.v.*)

mitosis the process of nuclear division in a eukaryotic cell.

nucleoid DNA-rich region of bacterial cell, the counterpart of the eukaryotic nucleus but (unlike a nucleus) lacking a membrane.

partial diploid a cell (usually a bacterium) in which a part of the genome is present in different allelic versions, e.g. by harbouring a plasmid prime (*q.v.*).

phage bacteriophage (*q.v.*).

phenotype the properties of an organism that can be observed and are due to the expression of its inherited characters (genotype) in its environment.

plasmid extrachromosomal hereditary element.

plasmid prime a plasmid that includes chromosomal genes.

replicon strictly, a structural gene that controls initiation and replication of DNA but, by association, the whole DNA sequence (such as the bacterial chromosome) that is replicated under the influence of a single replicon system.

R plasmid (R factor) a plasmid that carries genes for resistance to antibiotic drugs.

sedimentation coefficient (S-value) the constant (symbol S) in the differential equation that describes the sedimentation of a molecule or particle in a centrifuge; it has the dimensions of time. Certain particles and molecules are conventionally referred to by their S-values (units *s*) and these are sedimentation coefficients under standard conditions (infinite dilution in water at 20 °C).

somatic cells cells other than germ cells.

spore a resting (non-vegetative) cell, usually more resistant than vegetative cells to heat and designed for dispersal. They may be sexual (*see* meiosis) or asexual (e.g. fungal conidia and bacterial spores).

structural gene a sequence of DNA that is expressed to form a single polypeptide chain or, in the case of structural genes for tRNA and rRNA, an RNA molecule.

trans *see* cis and trans.

transposable genetic element a genetic element capable of being transposed from one replicon to another or from one site to another within a replicon.

transposon a transposable genetic element that includes structural genes in addition to those required for transposition.

virus an obligate intracellular parasite which contains RNA or DNA (not both) and has an absolute dependence on its host for gene expression.

zygote diploid cell (or sometimes also the organism derived from the cell) that arises from the union of gametes of opposite mating type.

Abbreviations and units

We have used, wherever possible, standard biochemical abbreviations and symbols. Abbreviations for bases, nucleosides and nucleotides are tabulated on p. 49.

bp	base pair
kbp	1000 bp
M_r	molecular weight (relative molecular mass)
s	Svedberg unit, 10^{-13} s (*see* sedimentation coefficient)
S-value	sedimentation coefficient

References

Abelson J. (1979) RNA processing and the intervening sequence problem. *A. Rev. Biochem.* **48,** 1035–69.

Adhya S. & Gottesman M. (1978) Control of transcription termination. *A. Rev. Biochem.* **47,** 967–96.

Allan J., Hartman P.G., Crane-Robinson C. & Aviles F.X. (1980) The structure of histone H1 and its location in chromatin. *Nature, Lond.* **288,** 675–9.

Altman S. (1981) Transfer RNA processing enzymes. *Cell* **23,** 3–4.

Alton N.K., Hantala J.A., Giles N.H., Kushner S.R. & Vapnek D. (1978) Transcription and translation in *E. coli* of hybrid plasmids containing the catabolic dehydroquinase gene from *Neurospora crassa. Gene* **4,** 241–59.

Alton T.H. & Lodish H.F. (1977) Translational control of protein synthesis during the early stages of differentiation of the slime mold *Dictyostelium discoideum. Cell* **12,** 301–10.

Alwine J.C., Kemp D.J., Parker B.A., Reiser J., Renart J., Stark G.R. & Wahl G.M. (1980) Detection of specific RNAs or specific fragments of DNA by fractionation in gels and transfer to diazobenzyloxymethyl paper. *Meth. Enzymol.* **68,** 220–42.

Ames B.N. (1978) Environmental chemicals causing cancer and genetic birth defects. In: *DNA Repair Mechanisms,* ICN-UCLA Symposia on Molecular and Cellular Biology, vol. IX (P.C. Hanawalt, E.C. Friedberg & C.F. Fox eds), pp. 691–8. Academic Press, New York.

Anderson K.V. & Lengyel J.H. (1980) Changing rates of histone mRNA synthesis and turnover in *Drosophila* embryos. *Cell* **21,** 717–27.

Anderson S., Bankier A.T., Barrell B.G., de Bruijn M.H.L., Coulson A.R., Drouin J., Eperon I.C., Nierlich D.P., Roe B.A., Sanger F., Schreier P.H., Smith A.J.H., Staden R. & Young I.G. (1981) Sequence and organization of the human mitochondrial genome. *Nature, Lond.* **290,** 457–65.

Anderson W.F., Bosch L., Cohn W.E., Lodish H., Merrick W.C., Weissbach H. & Wittmann I.G. (1977). Meeting report: Fogarty Center—NIH Workshop, Bethesda Md, Oct. 1976. *FEBS Lett.* **76,** 1–10.

Anfinsen C.B. (1964) On the possibility of predicting tertiary structure from primary sequence. In: *New Perspectives in Biology* (M. Sela ed.), pp. 42–50. American Elsevier, New York.

Arnott S. (1970) The geometry of nucleic acids. *Prog. Biophys. molec. Biol.* **21,** 267–319.

Attardi G. (1981) Organization and expression of the mammalian mitochondrial genome: a lesson in economy. *Trends biochem. Sci.* **6,** 86–9, 100–3.

Backendorf C., Overbeek G.P., van Boom J.H., van der Marel G., Veejeman G. & van Duin J. (1980) Role of 16-S RNA in ribosome messenger recognition. *Eur. J. Biochem.* **110,** 599–604.

Baralle F.E. & Brownlee G.G. (1978) AUG is the only recognisable signal sequence in the 5′ non-coding regions of eukaryotic mRNA. *Nature, Lond.* **274,** 84–7.

Barnes W.M. (1979) Construction of a M13 histidine transducing phage: a single stranded cloning vehicle with one EcoRI site. *Gene* **5,** 127–39.

Barr M.L. & Bertram E.G. (1949) A morphological difference between neurones of the male and female and the behaviour of the nuclear satellite during accelerated nucleoprotein synthesis. *Nature, Lond.* **163,** 676–8.

Barrera C.R., Namihara G., Hamilton L., Munk P., Eley M.H., Linn T.C. & Reed L.J. (1972) α-Keto acid dehydrogenase complexes. *Arch. Biochem. Biophys.* **148,** 343–58.

Baulieu E.E., Godeau F., Schorderet M. & Schorderet-Slatkine S. (1978) Steroid-induced meiotic division in *Xenopus laevis* oocytes: surface and calcium. *Nature, Lond.* **275,** 593–8.

Beach D., Piper M. & Shall S. (1980) Isolation of chromosomal origins of replication in yeast. *Nature, Lond.* **284,** 185–7.

Beale D. & Lehmann H. (1965) Abnormal haemoglobins and the genetic code. *Nature, Lond.* **207,** 259–61.

Beck B.D. (1979) Polymerization of the bacterial elongation factor for protein synthesis, EF-Tu. *Eur. J. Biochem.* **97,** 495–502.

Beggs J.D. (1978) Transformation of yeast by a replicating hybrid plasmid. *Nature, Lond.* **275,** 104–9.

Beggs J.D., van den Berg J., van Ooyen A. & Weissmann C. (1980) Abnormal expression of chromosomal β-globin gene in *Saccharomyces cerevisiae. Nature, Lond.* **283,** 835–40.

Behe M. & Felsenfeld G. (1981) Effects of methylation on a synthetic polynucleotide: the B-Z transition in poly(dG-m^{5}dC).poly(dG-m^{5}dC). *Proc. natn. Acad. Sci. U.S.A.* **78,** 1619–23.

Behrens N.H. & Leloir L.F. (1970) Dolichol monophosphate glucose: an intermediate of glucose transfer in liver. *Proc. natn. Acad. Sci. U.S.A.* **66,** 153–9.

Bell G.I., Pictet R.L., Rutter W.J., Cordell B., Tischer R.E. & Goodman H.M. (1980) Sequence of the human insulin gene. *Nature, Lond.* **284,** 26–32.

Bellard M., Gannon F. & Chambon P. (1977) Nucleosome structure III. The structure and transcriptional activity of the chromatin containing ovalbumin and globin genes in the chick oviduct. *Cold Spring Harbor Symp. quant. Biol.* **42,** 779–91.

Benbow R.M., Creaux C.B., Joenje H., Krauss H.R., Lennox R.W., Nelson E.M., Wang N.S. & White S.H. (1978) Eukaryotic DNA replication: the first steps towards a multi-enzyme system from *Xenopus laevis. Cold Spring Harb. Symp. quant. Biol.* **43,** 597–602.

Benne R. & Hershey J.W.B. (1978) The mechanism of action of protein synthesis initiation factors from rabbit reticulocytes. *J. biol. Chem.* **253,** 3078–87.

Bennett G.N., Schweingruber M.E., Brown K.D., Squires C.

& Yanofsky C. (1976) Nucleotide sequence of region preceding *trp* mRNA initiation site and its role in promoter and operator function. *Proc. natn. Acad. Sci. U.S.A.* **73,** 2351–5.

Benoist C., O'Hare K., Breathnath R. & Chambon P. (1980) The ovalbumin gene—sequence of putative control regions. *Nucleic Acids Res.* **8,** 127–42.

Benton W.D. & Davis R.W. (1977) Screening recombinant clones by hybridization to single plaques *in situ. Science, N.Y.* **196,** 180–2.

Berezney R. & Coffey D.S. (1974) Identification of a nuclear protein matrix. *Biochem. biophys. Res. Commun.* **60,** 1410–17.

Berezney R. & Coffey D.S. (1977) The nuclear protein matrix: isolation, structure and functions. *Adv. Enz. Regulation* **14,** 63–100.

Bergmann J.E. & Lodish H.E. (1979) Translation of capped and uncapped vesicular stomatis virus and reovirus mRNAs. *J. biol. Chem.* **254,** 459–68.

Berk A.J. & Sharp P.A. (1977) Sizing and mapping of early adenovirus mRNAs by gel electrophoresis of S1 endonuclease-digested hybrids. *Cell* **12,** 721–32.

Beyer A.L., Miller O.L. & McKnight S.L. (1980) Ribonucleoprotein structure in nascent hnRNA is non-random and sequence-dependent. *Cell* **20,** 75–84.

Bhalla R.V., Gerach D., Modak M.J., Prensky W. & Marcus S. (1976) Preparation of [^{125}I]-dCTP and its uses as a substrate for RNA- and DNA-directed DNA synthesis. *Biochem. biophys. Res. Commun.* **72,** 513–21.

Bibb M., Schottel J.L. & Cohen S.N. (1980) A DNA cloning system for interspecies gene transfer in antibiotic-producing *Streptomyces. Nature, Lond.* **284,** 526–31.

Bielka H. (1980) The eukaryotic ribosome. *Trends biochem. Sci.* **3,** 156–8.

Birkby C.W. jr. (1978) Transmission genetics of mitochondria and chloroplasts and mitochondria. *A. Rev. Genet.* **12,** 471–512.

Birnie G.D. (Ed.) (1972) *Subcellular Components—Preparation and Fractionation,* 2e. Butterworths, London.

Bishop J.O., Morton J.G., Rosbach M. & Richardson M. (1974) Three abundance classes of HeLa cell messenger RNA. *Nature, Lond.* **250,** 199–204.

Blobel G. (1973) A protein of molecular weight 78,000 bound to the polyadenylate region of eukaryotic messenger RNAs. *Proc. natn. Acad. Sci. U.S.A.* **70,** 924–8.

Blobel G. & Dobberstein B. (1975) Transfer of proteins across membranes I: presence of protoelytically processed and unprocessed nascent immunoglobulin light chains on membrane-bound polyribosomes of murine myeloma. *J. Cell Biol.* **67,** 835–51.

Blumenthal T. & Carmichael G.G (1979) RNA replication. Function and structure of Qβ replicase. *A. Rev. Biochem.* **48,** 525–48.

Blundell T.L. & Humbel R.E. (1980) Hormone families: pancreatic hormones and homologous growth factors. *Nature, Lond.* **287,** 781–7.

Boedtker H. (1968) Molecular weight and conformation of RNA. *Meth. Enzymol.* **XII,** 429–58.

Bolden A., Noy G.P. & Weissbach A. (1977) DNA polymerase of mitochondria is a γ-polymerase. *J. biol. Chem.* **252.** 3351–56.

Bolivar F., Rodriguez R.L., Greene P.J., Betlach M.C., Heyneker H.L. & Boyer H.W. (1977) Construction and characterization of new cloning vehicles II. A new multipurpose cloning system. *Gene* **2,** 95–113.

Bonitz S., Berlani R., Coruzzi G., Li M., Macino G., Nobrega F.G., Nobrega M.P., Thalenfeld B.E. & Tzagaloff A. (1980) Codon recognition rules in yeast mitochondria. *Proc. natn. Acad. Sci. U.S.A.* **77,** 3167–70.

Borst P., Fairlamb A.H., Fase-Fowler F., Hoeijmakers J.H.J. & Weislogel P.O. (1976) The structure of kinetoplast DNA. In: *The Genetic Function of Mitochondrial DNA* (C. Saccone & A.M. Kroon eds), pp. 59–81. Elsevier North-Holland, Amsterdam.

Borst P. & Grivell L.A. (1978) The mitochondrial genome of yeast. *Cell* **15,** 705–23.

Borst P. & Grivell L.A. (1981) One gene's intron is another gene's exon. *Nature, Lond.* **289,** 439–40.

Bosch L. & van der Hofstad G.A.J.M. (1979) Initiation of protein synthesis in prokaryotes. *Meth. Enzymol.* **LX,** 11–15.

Boseley P.G., Moss T. & Birnstiel M.L. (1980) 5′ Labeling and poly(dA) tailing. *Meth. Enzymol.* **65,** 478–99.

Bostock C. (1980) A function for satellite DNA. *Trends biochem. Sci.* **5,** 117–19.

Bradshaw R.A. (1978) Nerve growth factor. *A. Rev. Biochem.* **47,** 191–216.

Breathnath R., Benoist C., O'Hare K., Gannon F. & Chambon P. (1978) Ovalbumin gene: evidence for a leader sequence in mRNA and DNA sequences at the exon–intron boundaries. *Proc. natn. Acad. Sci. U.S.A.* **75,** 4853–7.

Brestlow E. (1979) Chemistry and biology of the neurophysins. *A. Rev. Biochem.* **48,** 251–74.

Brimacombe R., Stöffler G. & Wittmann H.G. (1978) Ribosome structure. *A. Rev. Biochem.* **47,** 217–49.

Britten R.J. & Davidson E.H. (1969) Gene regulation for higher cells: a theory. *Science, N.Y.* **165,** 349–57.

Britten R.J., Graham D.E. & Neufeld B.R. (1974) Analysis of repeating DNA sequences by reassociation. *Mech. Enzymol.* **XXIX,** 363–418.

Britten R.J. & Kohne D.E. (1968) Repeated sequences in DNA. *Science, N.Y.* **161,** 529–40.

Brooks R.F., Bennett D.C. & Smith J.A. (1980) Mammalian cell cycles need two random transitions. *Cell* **19,** 493–504.

Broome S. & Gilbert W. (1978) Immunological screening method to detect specific translation products. *Proc. natn. Acad. Sci. U.S.A.* **75,** 2746–9.

Brown E.L., Belagaje R., Ryan M.J. & Khorana H.G. (1980) Chemical synthesis and cloning of a tyrosine tRNA gene. *Meth. Enzymol.* **68,** 109–51.

Brown N.L. (1979) Nucleic acid sequencing. In: *Companion to Biochemistry* (A.T. Bull, J.R. Lagnado, J.O. Thomas & K.F. Tipton eds), vol. 2, pp. 1–48. Longman, London.

Brown N.L. & Smith M. (1977) Cleavage specificity of the restriction endonuclease isolated from *Haemophilus gallinarum (Hga* I). *Proc. natn. Acad. Sci. U.S.A.* **74,** 3213–16.

Brun G. & Weissbach A. (1978) Initiation of HeLa cell DNA synthesis in a subnuclear system. *Proc. natn. Acad. Sci. U.S.A.* **75,** 5913–35.

Buckler-White A.J., Humphrey G.W. & Pigiet V. (1980) Association of polyoma T antigen and DNA with the nuclear matrix from lytically infected 3T6 cells. *Cell* **22,** 37–46.

Bukhari A.I. (1981) Models of DNA transposition. *Trends biochem. Sci.* **6,** 56–60.

Bukhari A.I., Shapiro J.A. & Adhya S.C. (Eds) (1977) *DNA insertion elements, plasmids and episomes.* Cold Spring Harbor Laboratory, Cold Spring Harbor NY.

Burdon R.H. & Adams R.L.P. (1980) Eukaryotic DNA methylation. *Trends biochem. Sci.* **5,** 294–7.

Burke W. & Gangman W.L. (1975) Temporal order in yeast chromosome replication. *Cell* **5,** 263–9.

Calos M.P. & Miller J.H. (1980) Transposable elements. *Cell* **20,** 279–95.

Campbell A. (1961) Conditions for the existence of bacteriophage. *Evolution* **15,** 153–65.

Campbell A. (1979) Structure of complex operons. In: *Biological Regulation and Development. Vol. 1. Gene Expression* (R.F. Goldberger ed.), pp. 19–56. Plenum Press, New York.

Canosi V., Morelli G. & Trautner T.A. (1978) The relationship between molecular structure and transformation efficiency of some *S. aureus* plasmids isolated from *B. subtilis. Molec. gen. Genet.* **166,** 259–67.

Carmichael G.C. & McMaster G.K. (1980) The analysis of nucleic acids in gels using glyoxal and acridine orange. *Meth. Enzymol.* **65,** 380–91.

Carpenter G. & Cohen S. (1979) Epidermal growth factor. *A. Rev. Biochem.* **48,** 193–216.

Carr N.G. & Waterbury J.B. (1979) Differentiation in filamentous cyanobacteria *and* Developmental patterns of pleurocapsalean cyanobacteria (consecutive chapters). In: *Developmental Biology of Prokaryotes* (J.H. Parish ed.), pp. 167–226. Blackwell Scientific Publications, Oxford.

Casadaban M.J. & Cohen S.N. (1979) Lactose genes fused to exogenous promoters in one step using a Mu-*lac* bacteriophage: *in vivo* probe for transcriptional control sequences. *Proc. natn. Acad. Sci. U.S.A.* **76,** 4530–3.

Caskey C.T. (1977) Peptide chain termination. In: *Molecular Mechanisms of Protein Biosynthesis* (H. Weissbach & S. Pestka eds), pp. 443–65. Academic Press, New York.

Caskey C.T. (1980) Peptide chain termination. *Trends biochem. Sci.* **5,** 234–7.

Cech T.R., Potter D. & Pardue M.L. (1977) Chromatin structure in living cells. *Cold Spring Harb. Symp. quant. Biol.* **42,** 191–8.

Chaleff D.T. & Fink G.R. (1980) Genetic events associated with an insertion mutation in yeast. *Cell* **21,** 227–37.

Challberg M.D., Desidero S.V. & Kelly T.J. (1980) Adenovirus DNA replication *in vitro. Proc. natn. Acad. Sci. U.S.A.* **77,** 5105–9.

Chamberlin M.J. (1974) The selectivity of transcription. *A. Rev. Biochem.* **43,** 721–75.

Chambliss G.H. (1979) The molecular biology of sporulation of *Bacillus subtilis.* In: *Developmental Biology of Prokaryotes* (J.H. Parish ed.), pp. 57–71. Blackwell Scientific Publications, Oxford.

Chambliss G., Craven G.R., Davies J., Davis K., Kahan L. & Nomura M. (Eds) (1980) *Ribosomes: Structure, Function and Genetics.* University Park Press, Baltimore.

Chambon P. (1975) Eukaryotic RNA polymerases. *A. Rev. Biochem.* **44,** 613–38.

Chambon P. (1977) The molecular biology of the eukaryotic genome is coming of age. *Cold Spring Harb. Symp. quant. Biol.* **42,** 1209–33.

Champney W.S. & Kushner S.R. (1976) A proposal for a uniform nomenclature for the genetics of bacterial protein synthesis. *Molec. gen. Genet.* **147,** 145–51.

Champoux J.J. (1978) Proteins that affect DNA conformation. *A. Rev. Biochem.* **47,** 449–79.

Chang A.C.Y. & Cohen S.N. (1974) Genome construction between bacterial species *in vitro. Proc. natn. Acad. Sci. U.S.A.* **71,** 1030–4.

Chang A.C.Y., Nunberg J.H., Kaufman R.J., Ehrlich H.A., Schimke R.T. & Cohen S.N. (1978) Phenotypic expression in *E. coli* of a DNA sequence coding for mouse dihydrofolate reductase. *Nature, Lond.* **275,** 617–24.

Chang L.M.S. & Bollum F.J. (1971) Enzymatic synthesis of oligodeoxynucleotides. *Biochemistry* **10,** 536–42.

Chang L.M.S., Lurie K. & Plevani P. (1978) A stimulatory factor for yeast DNA polymerase. *Cold Spring Harb. Symp. quant. Biol.* **43,** 587–95.

Chang S. & Cohen S.N. (1979) High frequency transformation of *Bacillus subtilis* protoplasts by plasmid DNA. *Molec. gen. Genet.* **168,** 111–15.

Chantrenne H. (1977) Oocyte injection. *Nature, Lond.* **269,** 202.

Chatterjee N.K., Dickerman H.W. & Beach T.A. (1977) Isolation of a distinct pool of polyribosomes from nucleus of adenovirus-infected HeLa cells. *Arch. Biochem. Biophys.* **183,** 228–37.

Chrétien M., Benjannet S., Fossard G., Gianoulakis C., Crine P., Lis M. & Seidah N.G. (1979) From β-lipoprotein to β-endorphin and 'pro-opio-melanocortin'. *Can. J. Biochem.* **57**, 1111–21.

Chu G. & Sharp P.A. (1981) A gene chimaera of SV40 and mouse β-globin is transcribed and properly spliced. *Nature, Lond.* **289**, 378–82.

Church G.M., Slonimski P.P. & Gilbert W. (1979) Pleiotypic mutations within two mitochondrial cytochrome genes block mRNA processing. *Cell* **18**, 1209–15.

Clark A.J. & Volkert M.R. (1978) A new classification of pathways repairing pyrimidine dimer damage in DNA. In: *DNA Repair Mechanisms* ICN-UCLA Symposia on Molecular and Cellular Biology vol. IX. (P.C. Hanawalt, E.C. Friedberg & C.F. Fox eds), pp. 57–72. Academic Press, New York.

Clark B. (1980) The elongation step of protein synthesis. *Trends biochem. Sci.* **5**, 207–9.

Clark L. & Carbon J. (1976) A colony bank containing synthetic ColE1 hybrid plasmids representative of the entire *E. coli* genome. *Cell* **9**, 91–9.

Cobianchi F., Riva S., Mastroemi G., Spadari S., Pedralli-Noy G. & Falaschi A. (1978) Enhancement of the rate of DNA polymerase α activity on duplex DNA by a DNA-binding protein and a DNA-dependent ATPase in mammalian cells. *Cold Spring Harb. Symp. quant. Biol.* **43**, 639–47.

Cochet M., Gannon F., Hen R., Marateaux L., Perrin F. & Chambon P. (1979) Organization and sequence studies of the 17-piece chicken ovalbumin gene. *Nature, Lond.* **282**, 567–84.

Collins J. & Hohn B. (1978) Cosmids: a type of plasmid gene-cloning vector that is packagable *in vitro* in bacteriophage λ heads. *Proc. natn. Acad. Sci. U.S.A.* **75**, 4242–6.

Cortese R., Harland R. & Melton D. (1980) Transcription of tRNA genes *in vivo*: single-stranded compared to double-stranded templates. *Proc. natn. Acad. Sci. U.S.A.* **77**, 4147–51.

Cozzarelli N.R. (1980) DNA topoisomerases. *Cell* **22**, 327–8.

Craig N.L. & Roberts J.W. (1980) *E. coli* recA protein-directed cleavage of phage λ repressor requires polynucleotide. *Nature, Lond.* **283**, 26–30.

Crawford I.P. & Stauffer G.V. (1980) Regulation of tryptophan biosynthesis. *A. Rev. Biochem.* **49**, 163–95.

Crawford L.V. (1980) The T-antigens of simian virus 40 and polyoma virus: their role in transformation. *Trends biochem. Sci.* **5**, 39–42.

Crea R., Kraszewski A., Hirose T. & Itakura K. (1978) Chemical synthesis of genes for human insulin. *Proc. natn. Acad. Sci. U.S.A.* **75**, 5765–9.

Crews S., Ojala D., Posakony J., Nishigushi J. & Attardi G. (1979) Nucleotide sequence of a region of human mitochondrial DNA containing the precisely identified origin of replication. *Nature, Lond.* **277**, 192–8.

Crick F.H.C. (1966) Codon–anticodon pairing: the wobble hypothesis. *J. molec. Biol.* **19**, 548–55.

Cummings D.J. (1977) Evidence for semi-conservative replication of mitochondrial DNA from *Paramecium aurelia. J. molec. Biol.* **117**, 273–7.

Cunningham P.P., Shibita T., DasGupta C. & Radding C.M. (1979) Single strands induce recA protein to unwind duplex DNA for homologous pairing. *Nature, Lond.* **281**, 191–5.

Curtis P.J., Mantei N. & Weissmann C. (1977) Characterization and kinetics of synthesis of 15S β-globin RNA, a putative precursor of β-globin mRNA. *Cold Spring Harb. Symp. quant. Biol.* **42**, 971–84.

Czech M.P. (1977) Molecular basis of insulin action. *A. Rev. Biochem.* **46**, 359–84.

Dale R.M.K., Livingstone D.C. & Ward D.C. (1974) The synthesis and enzymatic polymerization of nucleotides containing mercury: potential tool for nucleic acid sequencing and structural analysis. *Proc. natn. Acad. Sci. U.S.A* **70**, 2238–42.

Danchin A. & Ullmann A. (1980) The coordinate expression of polycistronic operons in bacteria. *Trends biochem. Sci.* **5**, 51–2.

Darnell J.E. (1975) The origin of mRNA and the structure of the mammalian chromosome. *Harvey Lect.* **69**, 1–40.

Darnell J.E. (1978) Implications of RNA.RNA splicing in evolution of eukaryotic cells. *Science* **202**, 1257–60.

Darnell J.E., Evans R.M., Fraser N., Goldberg S., Nevins J., Saldit-Georgieff M., Schwartz H., Weber J. & Ziff E. (1977) The definition of the transcription units for mRNA. *Cold Spring Harb. Symp. quant. Biol.* **42**, 515–22.

Davidson E.H. (1976) *Gene Activity in Early Development.* Academic Press, New York.

Davidson E.H. & Britten R.J. (1973) Organization, transcription and regulation in the animal genome. *Quart. Rev. Biol.* **48**, 565–613.

Davidson E.H., Hough B.R., Amenson C.S. & Britten R.J. (1973) General interspersion of repetitive and non-repetitive sequence elements in the DNA of *Xenopus. J. molec. Biol.* **77**, 1–23.

Davies R.J.H. (1976) Nucleic Acids. *A. Rep. Chem. Soc.* (B)**73**, 375–96.

Davis B.D. & Tai P-C. (1980) The mechanism of protein secretion across membranes. *Nature, Lond.* **283**, 433–8.

Davis M.M., Kim S.K. & Hood L. (1980) Immunoglobulin class switching: developmentally regulated DNA rearrangements during differentiation. *Cell* **22**, 1–2.

Davis R.W., Schreier P.W. & Büchel D.E. (1977) Nucleotide sequence of the attachment site of coliphage lambda. *Nature, Lond.* **270**, 757–60.

Davis R.W., Simon M. & Davidson N. (1974) Electron microscope methods for mapping regions of base sequence homology in nucleic acids. *Meth. Enzymol.* **21**, 413–30.

de Pamphilis M.L. & Wasserman P.M. (1980) Replication of eukaryotic chromosomes; a close-up of the replication fork. *A. Rev. Biochem.* **49**, 627–66.

de Robertis E.M. & Olson M.V. (1979) Transcription and processing of cloned yeast tyrosine tRNA genes microinjected into frog oocytes. *Nature, Lond.* **278,** 137–43.

Deisseroth A. & Hendrick D. (1979) Activation of phenotypic expression of human globin genes from nonerythroid cells by chromosome-dependent transfer to tetraploid mouse erythroleukaemia cells. *Proc. natn. Acad. Sci. U.S.A.* **76,** 2185–9.

Dell'Orco R.T., Crissman H.A., Steinkamp J.A. & Kraemer P.M. (1975) Population analysis of arrested human diploid fibroblasts by flow microfluorometry. *Exp. Cell Res.* **92,** 271–4.

Derynck R., Content J., de Clerq E., Volckaert G., Taverner R.J., Devos R. & Fiers W. (1980) Isolation and structure of a human fibroblast interferon gene. *Nature, Lond.* **285,** 542–7.

Dickson R.C., Abelson J., Barnes W.M. & Reznikoff W.S. (1975) Genetic regulation: the Lac control region. *Science, N.Y.* **187,** 27–35.

Ditta G., Stanfield S., Corbin D. & Helinski D.R. (1980) Broad host range cloning system for Gram-negative bacteria: construction of a gene bank of *Rhizobium melitoti. Proc. natn. Acad. Sci. U.S.A.* **77,** 7347–51.

Dohme F. & Nierhaus K.H. (1976) Role of 5S RNA in assembly and function of the 50S subunit from *Escherichia coli. Proc. natn. Acad. Sci. U.S.A.* **73,** 2221–5.

Donachie W.D. (1979) The cell cycle of *Escherichia coli.* In: *Developmental Biology of Prokaryotes* (J.H. Parish ed.), pp. 11–35. Blackwell Scientific Publications, Oxford.

Doolittle W.F. (1978) Gene pieces: were they ever together? *Nature, Lond.* **272,** 581–2.

Doolittle W.F. (1980) Revolutionary concepts in evolutionary cell biology. *Trends biochem. Sci.* **5,** 146–9.

Doolittle W.F. & Sapienza C. (1980) Selfish genes, the phenotype paradigm and genome evolution. *Nature, Lond.* **284,** 601–3.

Dugaiczyk A., Boyer H.W. & Goodman H.M. (1975) Ligation of EcoRI endonuclease-generated DNA fragments into linear and circular structures. *J. molec. Biol.* **96,** 171–84.

Dugaiczyk A., Woo S.L.C., Colbert D.A., Lai E.C., Mace M.L. & O'Malley B.W. (1979) The ovalbumin gene: cloning and molecular organization of the entire gene. *Proc. natn. Acad. Sci. U.S.A.* **76,** 2253–7.

Dunnick W., Rabbitts T.H. & Milstein C. (1980) An immunoglobulin deletion mutant with implications for the heavy-chain switch and RNA splicing. *Nature, Lond.* **286,** 669–72.

Efstradiatis A., Vournakis J.N., Donnis-Keller H., Chaconas G., Dougall D.K. & Kafatos F.C. (1977) End labeling of enzymatically decapped mRNA. *Nucl. Acids Res.* **4,** 4165–74.

Ehrlich D. (1978) DNA cloning in *Bacillus subtilis. Proc. natn. Acad. Sci. U.S.A.* **75,** 1433–6.

Eigner J. (1968) Molecular weight and conformation of DNA. *Meth. Enzymol.* **XII,** 386–429.

Elgin S.C.R., Serunian L.A. & Silver L.M. (1977) Distribution patterns of *Drosophila* non-histone chromosomal proteins. *Cold Spring Harb. Symp. quant. Biol.* **42,** 839–42.

Ellis R.J. (1979) The most abundant protein in the world. *Trends Biochem. Sci.* **4,** 241–4.

Ellis R.J. & Barraclough R. (1978) Synthesis and transport of chloroplast proteins inside and outside the cell. In: *Chloroplast Development* (G. Akoyunoglu & J.H. Argyroudi-Akoyunoglu eds), pp. 185–94. Elsevier North-Holland, Amsterdam.

Emmerson P.T. & West S.C. (1978) In: *DNA Repair Mechanisms* ICN-UCLA Symposia on Molecular and Cellular Biology vol. IX. (P.C. Hanawalt, E.C. Friedberg & C.F. Fox eds), pp. 367–70. Academic Press, New York.

Erdmann V.A. (1976) Structure and function of 5S and 5.8S RNA. *Progr. Nucl. Acid Res. molec. Biol.* **18,** 45–90.

Esnuof M.P. (1977) Biochemistry of blood coagulation. *Br. med. Bull.* **33,** 213–18.

Evans R.M., Fraser N., Ziff E., Weber J., Wilson M. & Darnell J.E. (1977) The initiation sites for RNA transcription in Ad2 DNA. *Cell* **12,** 733–9.

Evans R.M., Weber J., Ziff E. & Darnell J.E. (1979) Premature termination during adenovirus transcription. *Nature, Lond.* **278,** 367–70.

Ewig R.A.G. & Kohn K.W. (1977) DNA damage and repair in mouse leukaemia L1210 cells treated with nitrogen mustard, 1,3-bis(2-chloroethyl)-1-nitrosourea, and other nitrosoureas. *Cancer Res.* **37,** 2114–22.

Falaschi A., Cobianchi F. & Riva S. (1980) DNA-binding proteins and DNA-unwinding enzymes in eukaryotes. *Trends biochem. Sci.* **5,** 154–7.

Fareed G.C. & Davoli D. (1977) Molecular biology of papova viruses. *A. Rev. Biochem.* **46,** 471–522.

Fareed G.C. & Kasamatsu H. (1980) Electron microscopic methods for locating the origin and termination points for DNA replication. *Meth. Enzymol.* **65,** 709–17.

Fawcett D.W. (1966) *The Cell, its Organelles and Inclusions: an Atlas of Fine Structure.* W.B. Saunders, Philadelphia, PA.

Federoff N.V. (1979a) Deletion mutants of *Xenopus laevis* 5S ribosomal RNA. *Cell* **16,** 551–63.

Federoff N.V. (1979b) On spacers. *Cell* **16,** 697–710.

Ferretti L. & Sgaramella V. (1981) Temperature dependence of the joining by T4 DNA ligase of termini produced by the type II restriction endonucleases. *Nucleic Acids Res.* **9,** 85–93.

Fersht A.R. (1980) Enzymic editing mechanisms in protein synthesis and DNA replication. *Trends biochem. Sci.* **5,** 262–5.

Fessler J.H. & Fessler L.I. (1978) Biosynthesis of procollagen. *A. Rev. Biochem.* **47,** 129–62.

Fiil N.P., Willumsen B.M., Friesen J.D. & von Myenberg K. (1977) Interaction of alleles of the *relA, relC* and *spoT* genes in *Escherichia coli*: analysis of the interconversion of GTP, ppGpp and pppGpp. *Molec. gen. Genet.* **150,** 87–101.

Files J.G., Weber K., Coulondre C. & Miller J.H. (1975) Identification of the UUG codon as a translational initiation codon *in vivo*. *J. molec. Biol.* **95,** 327–30.

Filipowicz W. & Haenni A-L. (1979) Binding of ribosomes to 5′-terminal leader sequences of eukaryotic messenger RNAs. *Proc. natn. Acad. Sci. U.S.A.* **76,** 3111–15.

Finch J.T. & Klug A. (1976) Solenoidal model for superstructure in chromatin. *Proc. natn. Acad. Sci. U.S.A.* **73,** 1897–901.

Fincham J.R.S., Day P.R. & Radford A. (1979) *Fungal Genetics,* 4e. Blackwell Scientific Publications, Oxford.

Foe V.E. (1977) Modulation of ribosomal RNA synthesis in *Oncopeltus fasciatus*: an electron microscope study of the relationship between changes in chromatin structure and transcriptional activity. *Cold Spring Harb. Symp. quant. Biol.* **42,** 723–39.

Fox G.E., Magrum L.J., Balch W.E., Wolfe R.S. & Woese C.R. (1977) Classification of methanogenic bacteria by 16*s* ribosomal RNA characterization. *Proc. natn. Acad. Sci. U.S.A.* **74,** 4537–41.

Fraley R. & Papahadjopoulos D. (1981) New generation liposomes: the engineering of an efficient vehicle for intracellular delivery of nucleic acids. *Trends biochem. Sci.* **6,** 77–80.

Friedberg E.C., Bonura T., Cone R., Simmonds R. & Anderson C. (1978) Base excision repair of DNA. In: *DNA Repair Mechanisms,* ICN-UCLA Symposia on Molecular and Cellular Biology, vol. IX (P.C. Hanawalt, E.C. Friedberg & C.F. Fox eds), pp. 163–73. Academic Press, New York.

Friedman R.M. (1977) Antiviral activity of interferons. *Bact. Rev.* **41,** 543–67.

Friend C., Scher W., Holland J.G. & Sato T. (1971) Hemoglobin synthesis in murine virus-induced leukaemia cells *in vitro*: stimulation of erythroid differentiation by dimethyl sulphoxide. *Proc. natn. Acad. Sci. U.S.A.* **68,** 378–82.

Friendman J. & Huberman E. (1981) Postreplication repair and the susceptibility of Chinese hamster cells to cytotoxic and mutagenic effects of alkylating agents. *Proc. natn. Acad. Sci. U.S.A.* **77,** 6072–6.

Gadski R.A. & Chae C-B. (1976) Mode of reconstitution of chicken erythrocyte and reticulocyte chromatin. *Biochemistry* **15,** 3812–17.

Gadski R.A. & Chae C-B. (1978) Mode of chromatin reconstitution. Elements controlling globin gene transcription. *Biochemistry* **17,** 869–74.

Gall J.G., Cohen E.H. & Atherton D.D. (1973) The satellite DNAs of *Drosophila virilis*. *Cold Spring Harb. Symp. quant. Biol.* **38,** 417–21.

Gambino R., Metafora S., Felicetti L. & Raisman J. (1973) Properties of the ribosomal salt wash from unfertilized and fertilized sea urchin eggs and its effect on natural mRNA translation. *Biochim. biophys. Acta* **313,** 377–91.

Gannon F., O'Hare K., Perrin F., le Pennec J.P., Benoist C., Cochet M., Breathnath R., Royal A., Garapin A., Cami B. & Chambon P. (1979) Organization and sequences at the 5′ end of a cloned complete ovalbumin gene. *Nature, Lond.* **278,** 428–34.

Garcia-Bellido A., Lawrence P.A. & Morata G. (1979) Compartments in animal development. *Scient. Amer.* **241,** 90–100.

Garcia Ruiz J.F., Ingram R. & Hanson R.W. (1978) Changes in hepatic messenger RNA for phosphoenolpyruvate carboxykinase (GTP) during development. *Proc. natn, Acad. Sci. U.S.A.* **75,** 4189–93.

Gardner J.F. (1979) Regulation of the threonine operon: tandem threonine and isoleucine codons in the control region and translational control of transcriptional termination. *Proc. natn. Acad. Sci. U.S.A.* **76,** 1706–10.

Garel A. & Axel R. (1976) Selective digestion of transcriptionally active ovalbumin genes from oviduct nuclei. *Proc. natn. Acad. Sci. U.S.A.* **73,** 3966–71.

Garel A. & Axel R. (1977) The structure of transcriptionally active ovalbumin genes in chromatin. *Cold Spring Harb. Symp. quant. Biol.* **42,** 701–8.

Gazit B., Panet A. & Cedar H. (1980) Reconstruction of a deoxyribonuclease I-sensitive structure on an active gene. *Proc. natn. Acad. Sci. U.S.A.* **77,** 1787–90.

Geller A.I. & Rich A. (1980) A UGA termination suppression $tRNA^{Trp}$ active in rabbit reticulocytes. *Nature, Lond.* **283,** 41–6.

Generoso W.M., Shelby M.D. & de Serres F.J. (Eds) (1980) *DNA repair and mutagenesis in eukaryotes.* Plenum Press, New York and London.

Ghosh P.K., Reddy V.B., Piatak M., Lebowitz P. & Weissman S.M. (1980) Determination of RNA sequences by primer directed synthesis and sequencing of their cDNA transcripts. *Meth. Enzymol.* **65,** 580–95.

Gilbert W. (1978) Why genes in pieces? *Nature, Lond.* **271,** 501–2.

Gillham N.W. (1978) *Organelle Heredity*. Raven Press, New York.

Gilmour R.S. & Paul J. (1971) Tissue-specific transcription of the globin gene in isolated chromatin. *Proc. natn. Acad. Sci. U.S.A.* **70,** 3440–2.

Glover D.M. & Hogness D.S. (1977) A novel arrangement of the 18S and 28S sequences in a repeating unit of *Drosophila melanogaster* rDNA. *Cell* **10,** 167–76.

Goddard J.M. & Wolstenholme D.R. (1978) Origin and direction of replication in mitochondrial DNA molecules from *Drosophila melanogaster. Proc. natn. Acad. Sci. U.S.A.* **75,** 3886–90.

Goeddel D.V., Heyneker H.L., Hozumi T., Arentzen R., Itakura K., Yansura D.G., Ross M.J., Mozarri G., Crea R. & Seeburg P.H. (1979a). Direct expression in *Escherichia coli* of a DNA sequence coding for human growth hormone. *Nature, Lond.* **281,** 544–8.

Goeddel D.V., Kleid D.G., Bolivar F., Heyneker H., Yansura D.G., Crea R., Hirose T., Kraszewski A., Hakura K. & Riggs A.D. (1979b) Expression in *Escherichia coli* of chemically synthesized genes for human insulin. *Proc. natn. Acad. Sci. U.S.A.* **76,** 106–10.

Goeddel D.V., Sheperd H.M., Yelverton E., Leung D., Crea R., Sloma A. & Pestka S. (1980) Synthesis of human fibroblast interferon by *E. coli. Nucl. Acids Res.* **8,** 4057–74.

Goldberg N.D. & Maddox M.K. (1977) Cyclic GMP metabolism and involvement in biological regulation. *A. Rev. Biochem.* **46,** 823–96.

Gorini L. (1971) Ribosomal discrimination of tRNAs. *Nature New Biol., Lond.* **234,** 261–4.

Gorini L. (1974) Streptomycin and misreading of the genetic code. In: *Ribosomes* (M. Nomura, A. Tissières & P. Lengyel eds), pp. 791–803. Cold Spring Harbor Publications, Cold Spring Harbor, N.Y.

Gottesfield J.M. & Bloomer L.S. (1980) Nonrandom alignment of nucleosomes on 5S RNA genes of *X. laevis. Cell* **21,** 751–60.

Gottesman S. (1981) Genetic control of the SOS system in *E. coli. Cell* **23,** 1–2.

Gozes I., Walker M.D., Kaye A.M. & Littauer U.Z. (1977) Synthesis of tubulin and actin by neuronal and glial nuclear preparations from developing rat brain. *J. biol. Chem.* **252,** 1819–26.

Green M.M. (1977) The case for DNA insertion mutations in *Drosophila.* In: *DNA insertion elements, plasmids and episomes* (A.I. Bukhari, J.A. Shapiro & S.C. Adhya eds), pp. 437–45. Cold Spring Harbor Laboratory, Cold Spring Harbor, N.Y.

Green M.M. (1980) Transposable elements in *Drosophila* and other diptera. *A. Rev. Genet.* **14,** 109–20.

Greengard P. (1979) Some chemical aspects of neurotransmitter action. *Trends pharmacol. Sci.* **1,** 27–9.

Gronenmeyer H. & Pongs O. (1980) Localization of ecdysone on polytene chromosomes of *Drosophila melanogaster. Proc. natn. Acad. Sci. U.S.A.* **77,** 2102–8.

Gross H.J., Domdey H., Lossow C., Jank P., Raba M., Alberty H. & Sänger H.L. (1978) Nucleotide sequence and secondary structure of potato spindle tuber viroid. *Nature, Lond.* **273,** 203–8.

Gross-Bellard M., Oudet P. & Chambon P. (1973) Isolation of high molecular-weight DNA from mammalian cells. *Eur. J. Biochem.* **36,** 32–8.

Grosscheld R. & Birnstiel M.L. (1980) Spacer DNA sequences upstream of the T-A-T-A-A-T-A sequence are essential for promotion of H2A histone gene transcription *in vivo. Proc. natn. Acad. Sci. U.S.A.* **77,** 7102–6.

Grossman L.& Moldave K. (1971) Section VI. Gene localization techniques. *Meth. Enzymol.* **XXI,** 238–71.

Grossman L. & Moldave K. (1980) A.M. Makam and W. Gilbert—sequencing end-labeled DNA with base-specific chemical cleavages. *Meth. Enzymol.* **65,** 499.

Grunberg-Manago M., Buckingham R.H., Cooperman B.S. & Hershey J.W.B. (1978) Structure and function of the translation machinery. *Symp. Soc. gen. Microbiol.* **28,** 25–110.

Grunstein M. & Hogness D.S. (1975) Colony hybridization: a method for the isolation of cloned DNAs that contain a specific gene. *Proc. natn. Acad. Sci. U.S.A.* **72,** 3961–5.

Gupta R.C., Roe B.A. & Randerath K. (1979) The nucleotide sequence of human $tRNA^{Gly}$ (anticodon GCC). *Nucl. Acids Res.* **7,** 959–70.

Gurdon J.B., Lane C.D., Woodland H.R. & Marbaix G. (1971) Use of frog eggs and oocytes for the study of messenger RNA and its translation. *Nature, Lond.* **233,** 177–82.

Gurdon J.B., Woodland H.R. & Lingrel J.B. (1974). The translation of mammalian globin mRNA injected into fertilized eggs of *Xenopus laevis. Devl. Biol.* **39,** 125–33.

Guyette W.A., Matusik R.J. & Rosen J.M. (1979) Prolactin-mediated transcriptional and post-transcriptional control of casein gene expression. *Cell* **17,** 1013–23.

Haber J.E., Rogers D.T. & McCusher J.H. (1980) Homothallic conversions of yeast mating-type genes occur by intrachromosomal recombination. *Cell* **22,** 277–89.

Haldenwang W.G. & Losick R. (1980) Novel RNA polymerase σ factor from *Bacillus subtilis. Proc. natn. Acad. Sci. U.S.A.* **77,** 7000–4.

Hamada H., Muramatsu M., Urano Y., Onishi T. & Kominami R. (1979) In vitro synthesis of 5S RNA precursor by isolated nuclei of rat liver and HeLa cells. *Biochemistry* **17,** 163–73.

Hamer D.H. & Leder P. (1979) Expression of the chromosomal mouse β^{maj} globin gene in SV 40. *Nature, Lond.* **281,** 35–40.

Hanawalt P.C., Cooper P.K., Genesan A.K. & Smith C.A. (1979) DNA repair in bacteria and mammalian cells. *A. Rev. Biochem.* **48,** 783–836.

Hardy S.J.S. (1979) The structure and function of ribosomes. In: *Companion to Biochemistry* (A.T. Bull, J.R. Lagnado, J.O. Thomas & K.F. Tipton eds), vol. 2, pp. 109–36. Longman, London.

Harris R.J. & Pestka S. (1977) Peptide bond formation. In: *Molecular Mechanisms of Protein Biosynthesis* (H. Weissbach & S. Pestka eds), pp. 513–442.

Haselkorn R. & Rouvière-Yaniv J. (1976) Cyanobacterial DNA-binding protein related to *Escherichia coli* HU. *Proc. natn. Acad. Sci. U.S.A.* **73,** 1917–20.

Hastie N.D. & Bishop J.O. (1976) The expression of three abundance classes of messenger in mouse tissues. *Cell* **9,** 761–74.

Hayaishio O. & Ueda K. (1977) Poly(ADP)ribose and ADP-ribosylation of proteins. *A. Rev. Biochem.* **46,** 95–116.

Heckman J.E., Sarnoff J., Alzner-deWeerd B., Yin S. & BajBhandary U.L. (1980) Novel features of the genetic code and codon reading patterns in *Neurospora crassa* mitochondria based on sequences of six mitochondrial tRNAs. *Proc. natn. Acad. Sci. U.S.A.* **77,** 3159–63.

Held W.A. & Nomura M. (1973) Rate-determining step in the

reconstitution of *Escherichia coli* 30S ribosomal subunits. *Biochemistry, N.Y.* **12,** 3273–81.

Henderson J.F. & Paterson A.R.P. (1973) *Nucleotide Metabolism—An Introduction.* Academic Press, New York.

Hendrickson S.L., Wu J-S. R. & Johnson L.F. (1980) Cell cycle regulation of dihydrofolate reductase mRNA metabolism in mouse fibroblasts. *Proc. natn. Acad. Sci. U.S.A.* **77,** 5140–4.

Herskowitz I. & Hagen D. (1980) The lysis–lysogeny decision of phage λ: explicit programming and responsiveness. *A. Rev. Genet.* **14,** 399–445.

Hervalsteens J-P., Van Vliet F., De Beuckeler M., Depicker A., Engler G., Lemmers M., Hosters M., Van Montagu M. & Schell J. (1980) The *Agrobacterium tumefaciens* Ti plasmid as a host vector system for introducing foreign DNA in plant cells. *Nature, Lond.* **287,** 654–6.

Hicks J.B., Strathern J.N. & Hershowitz I. (1977) The cassette model for mating type interconversion. In: *DNA insertion elements, plasmids and episomes* (A.I. Bukhari, J.A. Shapiro & S.C. Adhya eds), pp. 457–62. Cold Spring Harbor Laboratory, Cold Spring Harbor, N.Y.

Hinnen A., Hicks J.B. & Fink G.R. (1978) Transformation of yeast. *Proc. natn. Acad. Sci. U.S.A.* **75,** 1929–33.

Hinton D.M., Baez J.A. & Gumport R.I. (1978) T4 RNA ligase joins 2′-deoxyribonucleoside 3′,5′-biphosphates to oligodeoxyribonucleotides. *Biochemistry, N.Y.* **17,** 5091–7.

Hoare C.A. & Wallace F.G. (1966) Developmental stages of trypanosomatid flagellates: a new terminology. *Nature, Lond.* **212,** 1385–6.

Hohn B. & Murray K. (1977) Packaging recombinant DNA molecules into bacteriophage particles *in vitro. Proc. natn. Acad. Sci. U.S.A.* **74,** 3259–63.

Holliday R. (1964) A mechanism for gene conversion in fungi. *Genet. Res.* **5,** 282–304.

Holmgren A. (1981) Thioredoxin: structure and functions. *Trends biochem. Sci.* **6,** 26–9.

Honda B.M. & Roeder R.G. (1980) Association of 5S gene transcription factor with 5S RNA and altered levels of the factor during cell differentiation. *Cell* **22,** 119–26.

Honjo T. & Kataoka T. (1978) Organization of immunoglobulin heavy chain genes and allelic deletion model. *Proc. natn. Acad. Sci. U.S.A.* **75,** 2140–4.

Hopkins C.R. (1980) Epidermal growth factor and mitogenesis. *Nature, Lond.* **286,** 205–6.

Hopkins J.D., Clements M.B., Liang T-Y., Isberg R.R. & Syvanen M. (1980) Recombination genes on the *Escherichia coli* sex factor specific for transposable elements. *Proc. natn. Acad. Sci. U.S.A.* **77,** 2814–18.

Hopper A.K., Schultz L.D. & Shapiro R.A. (1980) Processing of intervening sequences: a new yeast mutant which fails to excise intervening sequences from precursor tRNAs. *Cell* **19,** 741–51.

Hopwood D.A. (1981) Genetic studies of antibiotics and other secondary metabolites. *Symp. Soc. gen. Microbiol.* **31,** 187–218.

Hosbach H.A., Silberklane M. & McCarthy B.J. (1980) Evolution of a *D. melanogaster* glutamate tRNA cluster. *Cell* **21,** 169–78.

Howe M. (1980) The invertible G segment of phage Mu. *Cell* **21,** 605–6.

Hozumi N. & Tonegawa S. (1976) Evidence for somatic rearrangement of immunoglobin genes coding for variable and constant regions. *Proc. natn. Acad. Sci. U.S.A.* **73,** 3628–32.

Hunt T. (1980) The initiation of protein synthesis. *Trends biochem. Sci.* **5,** 178–81.

Inman R.B. (1974) Denaturation mapping of DNA. *Meth. Enzymol.* **XXIX,** 451–8.

Isenberg I. (1979) Histones. *A. Rev. Biochem.* **48,** 159–91.

Isono S. & Isono K. (1975) Role of ribosomal protein S1 in protein synthesis: effects of its addition to *Bacillus stearothermophilus* cell-free system. *Eur. J. Biochem.* **56,** 15–22.

Itakura K., Hirose T., Crea R., Riggs A.D., Heyneker H., Bolivar F. & Boyer H.W. (1977) Expression in *Escherichia coli* of a chemically synthesized gene for the hormone somatostatin. *Science, N.Y.* **198,** 1056–63.

Ivarie R.D., Fan W.J-W. & Tomkins G.M. (1975) Analysis of the induction and deinduction of tyrosine aminotransferase in enucleated HTC cells. *J. cell. Physiol.* **85,** 357–64.

Jacob A.E., Creswell J.M., Hedges R.W., Coetzee J.N. & Beringer J.E. (1976) Properties of plasmids constructed by *in vitro* insertion of DNA from *Rhizobium leguminosarum* or *Proteus mirabilis* into RP4. *Molec. gen. Genet.* **147,** 315–23.

Jacobson M.F. & Baltimore D. (1968) Polypeptide cleavages in the formation of poliovirus proteins. *Proc. natn. Acad. Sci. U.S.A.* **61,** 77–84.

Jazowska J., Jacq C. & Slonimski P.P. (1980) Sequence of introns and flanking exons in wild-type and *box 3* mutants of cytochrome b reveals an interlaced splicing protein coded by an intron. *Cell* **22,** 333–48.

Jeffreys A.J. & Flavell R.A. (1977) The rabbit β-globin gene contains a large insert in the coding sequence. *Cell* **12,** 1097–108.

Jensen E.V. & Jacobsen H.I. (1962) Basic guides to the mechanism of estrogen action. *Recent Progr. Horm. Res.* **18,** 387–414.

Johnson P.A. & Grossman L.I. (1977) Electrophoresis in agarose gels. Optimizing separations of conformational isomers of double- and single-stranded DNAs. *Biochemistry* **16,** 4217–25.

Jolly S.O. & Bogorad L. (1980) Preferential transcription of cloned maize chloroplast DNA sequences by maize chloroplast RNA polymerase. *Cell* **77,** 822–6.

Jukes T.H. (1977) The amino acid code. In: *Comprehensive Biochemistry* vol. 24 (M. Florkin, A. Neuberger & L.L.M. van Deenen eds), pp. 235–93. Elsevier/North-Holland, Amsterdam.

Jungman R.A. & Kranias E.G. (1977) Nuclear phosphoprotein kinases and the regulation of gene transcription. *Int. J. Biochem.* **8,** 819–30.

Kahman R. & Kamp D. (1979) Nucleotide sequence of the attachment sites of bacteriophage Mu DNA. *Nature, Lond.* **280,** 247–50.

Kamp D., Kahman R., Zipser D., Broker T.R. & Chow L.T. (1978) Inversion of the G segment of phage Mu controls infectivity. *Nature, Lond.* **271,** 577–80.

Kantor J.A., Turner P.H. & Neinhuis A. (1980) Beta thalassaemia: Mutations which affect processing of the β-globin mRNA. *Cell* **21,** 149–57.

Kari C., Török I. & Travers A. (1977) ppGpp cycle in *Escherichia coli. Molec. gen. Genet.* **150,** 249–55.

Karlson P. & Clever U. (1960) Induktion von Puff-Veränderungen in den Speicheldrüssenchromosomen von *Chironomus Tentans* durch Ecdyson. *Expt. Cell Res.* **20,** 623–6.

Kedes L.H. (1979) Histone genes and histone messengers. *A. Rev. Biochem.* **48,** 837–70.

Keggins K.M., Lovett P.S. & Duvall E.J. (1978) Molecular cloning of genetically active fragments of *Bacillus* DNA in *Bacillus subtilis* and properties of the vector plasmid pUB110. *Proc. natn. Acad. Sci. U.S.A.* **75,** 1423–7.

Kendrew J.C. (1961) Three dimensional structure of a protein. *Sci. Amer.* **205,** 96–110.

Kenyon C.J. & Walker G.C. (1980) DNA-damaging reagents stimulate gene expression at specific loci in *Escherichia coli. Proc. natn. Acad. Sci. U.S.A.* **77,** 2819–23.

Kenyon C.J. & Walker G.C. (1981) Expression of the *E. coli uvrA* gene is inducible. *Nature, Lond.* **289,** 808–10.

Kersten H., Sandig L. & Arnold H.H. (1975) Tetrahydrofolate-dependent 5-methyluracil-tRNA transferase activity in *B. subtilis. FEBS Lett.* **55,** 57–60.

Kimball R.F. (1980) Relationship between repair processes and mutation induction in bacteria. In: *DNA Repair and Mutagenesis in Eukaryotes* (W.M. Generoso, M.D. Shelby & F.J. de Serres eds), pp. 1–23. Plenum Press, New York.

Kimura A., Lewis R.V., Stern A.V., Rossier J., Stein S. & Udenfried S. (1980) Probable precursors of Leu enkephalin and Met enkephalin in adrenal medulla: peptides of 3–5 kilodaltons. *Proc. natn. Acad. Sci. U.S.A.* **78,** 1681–5.

King R.C. (1975) Chromosome linkages of human genes. In: *Handbook of Genetics* (R.C. King ed.), pp. 537–44. Plenum Press, New York.

King R.J.B. & Mainwaring W.I.P. (1974) *Steroid–Cell Interactions.* Butterworths, London.

Kitamura N., Semler B.L., Rothberg P.G., Larsen G.R., Adler C.L., Dorner A.J., Emini E.A., Hanecak R., Lee J.J., van der Werf S., Anderson C.W. & Wimmer E. (1981) Primary structure, gene organization and polypeptide expression of poliovirus RNA. *Nature, Lond.* **291,** 547–53.

Klar A.J.S., McIndoo J., Strathern J.N. & Hicks J.B. (1980) Evidence for a physical interaction between the transposed and substituted sequences during mating type gene transposition in yeast. *Cell* **22,** 291–8.

Kleckner N. (1977) Translocatable elements in prokaryotes. *Cell* **11,** 11–23.

Klein A. & Bonhoeffer F. (1972) DNA Replication. *A. Rev. Biochem.* **41,** 301–32.

Kleinschmidt A.K. (1968) Monolayer techniques in electron microscopy of nucleic acids. *Meth. Enzymol.* **12B,** 361–79.

Konkel D.A., Tilghman S.M. & Leder P. (1978) The sequence of the chromosomal mouse β-globin major gene; homologies in capping, splicing and poly(A) sites. *Cell* **15,** 1125–32.

Korn L.J. & Brown D.D. (1978) Nucleotide sequence of *Xenopus borealis* oocyte 5S RNA: comparison of sequences that flank several related eukaryotic genes. *Cell* **15,** 1145–56.

Kornberg A. (1979) *DNA Replication.* W.H. Freeman & Co., San Francisco.

Kornberg R.D. (1974) Chromatin structure: a repeating unit of histones and DNA. *Science, N.Y.* **184,** 868–71.

Kornberg R.D. (1977) Structure of chromatin. *A. Rev. Biochem.* **46,** 931–54.

Kozak M. (1980a) Influence of mRNA secondary structure on binding and migration of 40S ribosomal subunits. *Cell* **19,** 79–90.

Kozak M. (1980b) Evaluation for the 'scanning model' for initiation of protein synthesis in eucaryotes. *Cell* **22,** 7–8.

Kozak M. & Shatkin A.J. (1978a) Migration of 40*s* ribosomal subunits on messenger RNA in the presence of edeine. *J. biol. Chem.* **253,** 6568–77.

Kozak M. & Shatkin A.J. (1978b) Identification of features in 5′ terminal fragments from reovirus mRNA which are important for ribosome binding. *Cell* **13,** 201–12.

Krebs E.G. & Beavo J.A. (1979) Phosphorylation and dephosphorylation of enzymes. *A. Rev. Biochem.* **48,** 923–59.

Kreigstein H.J. & Hogness D.S. (1974) Mechanism of replication in *Drosophila* chromosomes: structure of replication forks and evidence for bidirectionality. *Proc. natn. Acad. Sci. U.S.A.* **71,** 135–9.

Kressman A., Clarkson S.G., Pirotta V. & Birnstiel M.L. (1978) Transcription of cloned tRNA gene fragments and subfragments injected into the oocyte nucleus of *Xenopus laevis. Proc. natn. Acad. Sci. U.S.A.* **75,** 1176–80.

Krupp G. & Gross H.J. (1979) Rapid RNA sequencing: nucleases from *Staphylococcus aureus* and *Neurospora crassa* discriminate between uridine and cytidine. *Nucl. Acids Res.* **6,** 3481–503.

Kuehn G., Affolter H-U., Atmar V.J., Seebeck T., Gubler U.

& Braun R. (1979) Polyamine-mediated phosphorylation of a nucleolar protein from *Physarum polycephalum* that stimulates rRNA synthesis. *Proc. natn. Acad. Sci. U.S.A.* **76,** 2541–5.

Kung S-d. (1977) Expression of chloroplast genomes in higher plants. *A. Rev. Plant Physiol.* **28,** 401–37.

Kurland C.G. (1977) Structure and function of the bacterial ribosome. *A. Rev. Biochem.* **46,** 173–200.

Kutherlapti T.S., Creagan R.P. & Ruddle F.H. (1974) Progress in human gene mapping by somatic cell hybridization. In: *The Cell Nucleus* (H. Busch ed.), vol. 2, pp. 209–22. Academic Press, New York.

Lagerkvist V. (1978) 'Two out of three': an alternative method for codon reading. *Proc. natn. Acad. Sci. U.S.A.* **75,** 1759–62.

Lake J.A. (1977) Aminoacyl-tRNA binding at the recognition site in the first step of one elongation cycle of protein synthesis. *Proc. natn. Acad. Sci. U.S.A.* **74,** 1903–7.

Lamb M.M. & Daneholt B. (1979) Characterization of active transcription units in Balbiani rings of *Chironomus tentans*. *Cell* **17,** 835–48.

Lampen J.O. (1978) Phospholipoproteins in enzyme excretion by bacteria. *Symp. Soc. gen. Microbiol.* **28,** 231–47.

Landy A. & Ross W. (1977) Viral integration and excision: structure of the lambda *att* sites. *Science, N.Y.* **197,** 1147–60.

Laskey R.A. & Earnshaw W.C. (1980) Nucleosome assembly. *Nature, Lond.* **286,** 763–7.

Laskey R.A., Mills A.D. & Morris N. (1977) Assembly of SV 40 chromatin in a cell-free system from *Xenopus* eggs. *Cell* **10,** 237–43.

Lawley P.D. (1976a) Carcinogenesis by alkylating agents. In: *Chemical Carcinogens,* ACS monograph 173 (C.E. Searle ed.), pp. 83–244. American Chemical Society, Washington D.C.

Lawley P.D. (1976b) Methylation of DNA by carcinogens: some applications of chemical analytical methods. In: *Screening Tests in Chemical Carcinogenesis,* IARC Scientific Publication 12 (R. Montesano, H. Bartsch & L. Tomatis eds), pp. 181–208. Internat. Agency for Res. in Cancer, Lyons, France.

Lawley P.D. & Brookes P. (1965) Molecular mechanism of the cytotoxic action of difunctional alkylating agents and of resistance to this action. *Nature, Lond.* **206,** 480–3.

Lazier C.B. (1978) Ontogeny of the vitellogenic response to oestradiol and of the soluble nuclear oestrogen receptor in embryonic chick liver. *Biochem. J.* **174,** 143–52.

Lazowska J., Jacq C. & Slonimski P.P. (1980) Sequence of introns and flanking exons in wild-type and *box3* mutants of cytochrome b reveals an interlaced splicing protein coded by an intron. *Cell* **22,** 333–48.

Leader D.P. (1979) Protein synthesis on membrane-bound polyribosomes. *Trends biochem. Sci.* **4,** 205–8.

Lebherz H.G. & Shackleford J.E. (1979) Mechanisms for the genesis of aldolase tetramers in cell-free protein synthesizing systems and *in vivo*. *J. biol. Chem.* **254,** 4227–32.

Lee F. & Yanofsky C. (1977) Transcription termination at the *trp* operon attenuators of *Escherichia coli* and *Salmonella typhimurium*: RNA secondary structure and regulation of termination. *Proc. natn. Acad. Sci. U.S.A.* **74,** 5365–9.

Lee-Huang S., Sierra J.M., Naranjo R., Filipowicz W. & Ochoa S. (1977) Eukaryotic oligonucleotides affecting mRNA translation. *Arch. Biochem. Biophys.* **180,** 276–87.

Lerner M.R., Boyle J.A., Mount S.M., Wolin S.L. & Steitz J.A. (1980) Are snRNAs involved in splicing? *Nature, Lond.* **283,** 220–4.

Levy-Wilson B., Conner C. & Dixon G.H. (1979) A subset of trout testis ribosomes enriched in transcribed DNA sequences contains high mobility group proteins as major structural components. *J. biol. Chem.* **254,** 609–20.

Lewin B. (1977) *Gene Expression 3. Plasmids and Phages*. John Wiley & Sons, New York.

Lewin B. (1980) Alternative for splicing: an intron-coded protein. *Cell* **22,** 645–6.

Lilley D.M.J. (1980) The inverted repeat is a recognizable structural feature in supercoiled DNA molecules. *Proc. natn. Acad. Sci. U.S.A.* **77,** 6486–72.

Lim L. & Canellakis E.S. (1970) Adenine-rich polymer associated with rabbit reticulocyte messenger RNA. *Nature, Lond.* **227,** 710–12.

Lin J.J.C., Kanazawa H., Ozozs J. & Wu H.C. (1978) an *Escherichia coli* mutant with an amino acid alteration within the signal sequence of outer membrane prolipoprotein. *Proc. natn. Acad. Sci. U.S.A.* **75,** 4891–5.

Lin Y. (1979) Environmental regulation of gene expression: *in vitro* translation of winter flounder antifreeze messenger RNA. *J. biol. Chem.* **254,** 1422–6.

Lingappa V.R., Lingappa J.R. & Blobel G. (1979) Chicken ovalbumin contains an internal signal sequence. *Nature, Lond.* **281,** 117–21.

Linn S. (1978) Workshop summary: enzymology of base excision repair. In: *DNA Repair Mechanisms,* ICN-UCLA Symposia on Molecular and Cellular Biology, vol. IX (P.C. Hanawalt, E.C. Friedberg & C.F. Fox eds), pp. 175–8. Academic Press, New York.

Linnane A.W. & Nagley P. (1978) Mitochondrial genetics in perspective: the derivation of a genetic and physical map of the yeast mitochondrial genome. *Plasmid* **1,** 324–45.

Livingstone D.M. (1977) Inheritance of the 2μm plasmid from *Saccharomyces*. *Genetics* **86,** 73–84.

Ljunquist E. & Bukhari A.I. (1977) State of prophage Mu DNA upon induction. *Proc. natn. Acad. Sci. U.S.A.* **74,** 3143–7.

Lobban P.E. & Kaiser A.D. (1973) Enzymatic end-to-end joining of DNA molecules. *J. molec. Biol.* **78,** 453–71.

Lodish H.F. (1976) Translational control of protein synthesis. *A. Rev. Biochem.* **45,** 40–72.

Loenen W.A.M. & Brammar W.J. (1980) A bacteriophage lambda vector for cloning large DNA fragments made with several restriction enzymes. *Gene* **20**, 249–59.

Low R.L., Arai K-I. & Kornberg A. (1981) Conservation of the primosome in successive stages of ϕX174 DNA replication. *Proc. natn. Acad. Sci. U.S.A.* **78**, 1436–40.

Lührmann R., Eckhardt H. & Stöffler G. (1979) Codon–anticodon interaction at the ribosomal peptide-site. *Nature, Lond.* **280**, 423–5.

Lyman H. & Srinivas U.K. (1978) Regulation of chloroplast DNA synthesis: possible role of chloroplast nucleases in *Euglena*. In: *Chloroplast Development* (G. Akoyunoglou & J.H. Agryoudi-Akoyunoglou eds), pp. 593–607. Elsevier North-Holland, Amsterdam.

Lyon M.F. (1961) Sex chromatin in the X-chromosome of the mouse (*Mus musculus* L.). *Nature, Lond.* **190**, 372–6.

McCann J. & Ames B.N. (1976) Detection of carcinogens as mutagens in the *Salmonella*/microsome test: assay of 300 chemicals. Discussion. *Proc. natn. Acad. Sci. U.S.A.* **72**, 5135–9.

McClintock B. (1958) The suppressor-mutator system of control of gene action in maize. *Carnegie Inst. Wash. Yrb.* **57**, 415–29.

McCrea J.M. & Hershberger C.L. (1978) Chloroplast DNA codes for tRNA from cytoplasmic polyribosomes. *Nature, Lond.* **274**, 717–19.

McDonell M.W., Simon M.N. & Studier F.W. (1977) Analysis of restriction fragments of T7 DNA by electrophoresis in neutral and alkaline gels. *J. molec. Biol.* **110**, 119–46.

McEntee K. (1977) Protein X is the product of the *recA* gene of *Escherichia coli. Proc. natn. Acad. Sci. U.S.A.* **74**, 5275–9.

McKay R. (1980) Movable genes. *Nature, Lond.* **287**, 188–9.

McKnight G.S. & Palmiter R.D. (1979) Transcriptional regulation of the ovalbumin and conalbumin genes by steroid hormones in chick oviduct. *J. biol. Chem.* **254**, 9050–8.

McKune K. & Holmes A.M. (1979) Studies on the processivity of highly purified calf thymus 8S and 7.3S DNA polymerase α. *Biochem. biophys. Res. Commun.* **90**, 864–70.

Macino G., Coruzzi G., Nobrega F.G., Li M. & Tzagaloff A. (1979) Use of the UGA terminator as a tryptophan codon in yeast mitochondria. *Proc. natn. Acad. Sci. U.S.A.* **76**, 3784–5.

MacNaughton M., Freeman K.B. & Bishop J.O. (1974) A precursor of haemoglobin mRNA in nuclei of immature duck red blood cells. *Cell* **1**, 117–25.

Magnusson S., Petersen T.E., Sottru-Jensen L. & Claeys H. (1975) The secondary structure of prothrombin. In: *Proteases and Biological Control* (E. Reich, D.B. Rifkin & E. Shaw eds), pp. 139–49. Cold Spring Harbor Press, Cold Spring Harbor, N.Y.

Mainwaring W.I.P. (1979) The androgens. In: *Reproduction in Mammals* (C.R. Austin & R.V. Short eds), vol. 7, pp. 117–56. Cambridge University Press, Cambridge.

Mainwaring W.I.P. & Irving R.A. (1979) Different means of hormonal control in cells responsive to different metabolites of testosterone. In: *Pharmacological Actions of Steroid Hormones* (F. di Carlo, E. Mellazotti & W.I.P. Mainwaring eds), pp. 154–63. Raven Press, New York.

Malcolm A.D.B (1977) Binding the *lac* repressor. *Nature, Lond.* **268**, 196–7.

Mandel J-L. & Chambon P. (1974) Animal DNA-dependent RNA polymerase I. Studies on the reaction parameters of transcription *in vitro* of simian virus 40 DNA by mammalian RNA polymerases AI and B. *Eur. J. Biochem.* **41**, 367–78.

Mandel M. & Higa A. (1970) Calcium dependent bacteriophage infection. *J. molec. Biol.* **53**, 159–62.

Maniatis T. & Efstratiadis A. (1980) Fractionation of low molecular weight DNA or RNA in polyacrylamide gels containing 98% formamide or 7M Urea. *Meth. Enzymol.* **65**, 299–305.

Maniatis T., Hardison R.C., Lacy E., Lauer J., O'Connell C., Quon D., Sim G.K & Efstradiatis A. (1978) The isolation of structural genes from libraries of eukaryotic DNA. *Cell* **15**, 687–701.

Maniatis T., Jeffrey A. & Kleid D.G. (1975) Nucleotide sequence of the rightward operator of phage λ. *Proc. natn. Acad. Sci. U.S.A.* **72**, 1184–8.

Maniatis T., Sim G.K., Efstratiadis A. & Kafatos F.C. (1976) Amplification and characterization of a β-globin gene synthesized *in vitro*. *Cell* **8**, 163–82.

Manley J.L., Fire A., Cano A., Sharp P.A. & Gefter M.L. (1980) DNA-dependent transcription of adenovirus genes in a soluble whole-cell extract. *Proc. natn. Acad. Sci. U.S.A.* **77**, 3855–9.

Manley J.L., Sharp P.A. & Gefter M.L. (1979) RNA synthesis in isolated nuclei: *in vitro* initiation of adenovirus 2 major late mRNA precursor. *Proc. natn. Acad. Sci. U.S.A.* **76**, 160–4.

Matsumoto Y-I., Yasuda H., Mita S., Maranouchi T. & Yamada M-A. (1980) Evidence for the involvement of H1 histone phosphorylation in chromosome condensation. *Nature, Lond.* **284**, 181–3.

Matusik R.J. & Rosen J.M. (1978) Prolactin induction of casein mRNA in organ cultures: a model system for studying hormone regulation of gene expression. *J. biol. Chem.* **253**, 2343–7.

Maxam A.M. & Gilbert W. (1980) Sequencing end-labelled DNA with base-specific chemical cleavages. *Meth. Enzymol.* **65**, 499–560.

Means A.R. & Dedman J.R. (1980) Calmodulin—an intracellular calcium receptor. *Nature, Lond.* **285**, 73–7.

Meilhac M., Kedinger C., Chambon P., Govidan V., Faulstick H. & Weiland T. (1970) Amanitin binding to calf thymus RNA polymerase. *FEBS Lett.* **9**, 258–60.

Melli M., Spinelli G. & Arnold E. (1977) Synthesis of histone messenger RNA of HeLa cells during the cell cycle. *Cell* **12**, 167–74.

Melton D.A., di Robertis E.M. & Cortese R. (1980) Order and intracellular location of events involved in the maturation of a spliced tRNA. *Nature, Lond.* **284**, 143–8.

Mertz J.E. & Davis R.W. (1972) Cleavage of DNA by R1 restriction endonuclease generates cohesive ends. *Proc. natn. Acad. Sci. U.S.A.* **69**, 3370–4.

Mertz J.E. & Gurdon J.B. (1977) purified DNA's are transcribed after microinjection into *Xenopus* oocytes. *Proc. natn. Acad. Sci. U.S.A.* **74**, 1502–6.

Meselson M. & Stahl F.W. (1958) The semi-conservative nature of DNA replication. *Proc. natn. Acad. Sci. U.S.A.* **44**, 671–7.

Messing J., Crea R. & Seeburg P.H. (1981) A system for shotgun DNA sequencing. *Nucl. Acids Res.* **9**, 309–21.

Messing J., Gronenborn B., Müller-Hill B. & Hofschneider P.H. (1977) Filamentous coliphage M13 as a cloning vehicle: insertion of a HindIII fragment in the *lac* regulatory region in M13 replicative form *in vitro*. *Proc. natn. Acad. Sci. U.S.A.* **74**, 3642–6.

Meyer M., Beck E., Hansen F.G., Bergmans H.E.N., Messer W., von Meyenberg K. & Schaller H. (1979) Nucleotide sequence of the origin of replication of the *Escherichia coli* K-12 chromosome. *Proc. natn. Acad. Sci. U.S.A.* **76**, 580–4.

Meyer M., de Jong M.A., Woldringh C.L. & Nanninga N. (1976) Factors affecting the release of folded chromosomes from *Escherichia coli*. *Eur. J. Biochem.* **63**, 469–75.

Meynhas O. & Perry R.P. (1979) Relationship between size, stability and abundance of messenger RNA of mouse L cells. *Cell* **16**, 139–48.

Miller O.L. jr. & Beatty B.R. (1969) Portrait of a gene. *J. cell. Physiol.* **74** (suppl. 1), 225–36.

Miller O.L. jr., Hamkalo B.A. & Thomas C.A. (1975) Visualization of bacterial genes in action. *Science, N.Y.* **169**, 392–5.

Milstein C., Brownlee G.G., Cartwright E.M., Jarvis J.M. & Proudfoot M.J. (1974) Sequence analysis of immunoglobulin light chain messenger RNA. *Nature, Lond.* **252**, 354–9.

Minkley E.G. & Pribnow D. (1973) Transcription of the early region of bacteriophage T7: selective initiation with ninucleotides. *J. molec. Biol.* **77**, 255–77.

Mintz B. & Illmensee K. (1975) Normal genetically mosaic mice produced from malignant teratocarcinoma cells. *Proc. natn. Acad. Sci. U.S.A.* **72**, 3585–9.

Molgaard H.V. (1980) Assembly of immunoglobulin heavy chain genes. *Nature, Lond.* **286**, 657–9.

Mong S., Huang C.H., Prestayko A.W. & Crooke S.T. (1980) Interaction of *cis*-diamminedichloroplatinum(II) with PM-2 DNA. *Cancer Res.* **40**, 3313–17.

Montoya J., Ojala D. & Attardi G. (1981) Distinctive features Of the 5′-terminal sequences of the human mitochondrial mRNAs. *Nature, Lond.* **290**, 465–70.

Morata G. & Lawrence P.A. (1977) Homeotic genes, compartments and cell determination in *Drosophila*. *Nature, Lond.* **265**, 211–16.

Morgan E.A., Ikemura T. & Nomura M. (1977) Identification of spacer tRNA genes in individual rRNA transcription units of *E. coli*. *Proc. natn. Acad. Sci. U.S.A.* **74**, 2710–14.

Mulligan R.C., Howard B.H. & Berg P. (1979) Synthesis of rabbit β-globin in cultured monkey kidney cells following infection with a SV40 β-globin recombinant genome. *Nature, Lond.* **277**, 108–14.

Murray N.E., Brammar W.J. & Murray K. (1977) Lambdoid phages that simplify the recovery of *in vitro* recombinants. *Molec. gen. Genet.* **150**, 53–61.

Musso R., di Lauro R., Rosenberg M. & de Crombrugghe B. (1977) Nucleotide sequence of the operator–promoter region of the galactose operon of *Escherichia coli*. *Proc. natn. Acad. Sci. U.S.A.* **74**, 106–12.

Nakanishi S., Inoue A., Kita T., Nakamura M., Chang A.C.Y., Cohen S.N. & Numa S. (1979) Nucleotide sequence of cloned cDNA for bovine corticotropin-lipoprotein precursor. *Nature, Lond.* **278**, 423–7.

Nakanishi S., Inoue A., Kita T., Numa S., Chang A.C.Y., Cohen S.N., Nunberg J. & Schimke R.T. (1978) Construction of bacterial plasmids that contain the nucleotide sequence for bovine corticotropin-β-lipoprotein precursor. *Proc. natn. Acad. Sci. U.S.A.* **75**, 6021–5.

Nakashima K., Darzynkiewicz E. & Shatkin A.J. (1980) Proximity of mRNA 5′-region and 18S rRNA in eukaryotic initiation complexes. *Nature, Lond.* **286**, 226–8.

Nakhasi H.L. & Qasba P.K. (1979) Quantitation of milk proteins and their mRNAs in rat mammary gland at various stages of gestation and lactation. *J. biol. Chem.* **254**, 6016–25.

Narang S.A., Hsiung H.M. & Brousseau R. (1980) Improved phosphotriester method for the synthesis of gene fragments. *Meth. Enzymol.* **68**, 90–109.

Neupert W. & Schatz G. (1981) How proteins are transported into mitochondria. *Trends. biochem. Sci.* **6**, 1–4.

Nevins J.R. & Winkler J.J. (1980) Regulation of early adenovirus transcription: a protein product of early region 2 specifically represses region 4 transcription. *Proc. natn. Acad. Sci. U.S.A.* **77**, 1893–7.

Ng S.Y., Parker C.S. & Roeder R.G. (1979) Transcription of cloned *Xenopus* 5S RNA genes by *X. laevis* RNA polymerase III in reconstituted systems. *Proc. natn. Acad. Sci. U.S.A.* **76**, 136–40.

Nierlich D.P. (1978) The regulation of bacterial growth, RNA and protein synthesis. *A. Rev. Microbiol.* **32**, 393–432.

Nurse P. (1980) Cell cycle control—both deterministic and probabilistic. *Nature, Lond.* **286**, 9–10.

Ochoa S. & de Haro C. (1979) Regulation of protein synthesis in eukaryotes. *A. Rev. Biochem.* **48**, 549–80.

Ogawa T. & Okazaki T. (1980) Discontinuous DNA replication. *Ann. Rev. Biochem.* **49**, 421–57.

Ohtsubo H. & Ohtsubo E. (1978) Nucleotide sequence of an insertion element, IS1. *Proc. natn. Acad. Sci. U.S.A.* **75**, 615–19.

Ojala D., Montoya J. & Attardi G. (1981) tRNA punctuation model of RNA processing in human mitochondria. *Nature, Lond.* **290**, 470–4.

Okada N., Noguchi S., Kasai H., Shindo-Okada N., Ohgi T., Goto T. & Nishimura S. (1978) Novel mechanism of post-transcriptional modification of tRNA: insertion of base Q precursor into tRNA by tRNA transglycosidase reaction. *Nucl. Acids Res.* Special publication **5**, s447–s448.

Olins A.L. & Olins D.E. (1974) Spheroid chromatin units (ν bodies). *Science, N.Y.* **183**, 330–2.

Olsson M. & Lindahl T. (1980) Repair of alkylated DNA in *Escherichia coli*. *J. biol. Chem.* **255**, 10569–71.

O'Malley B.W. & McGuire W.L. (1968) Progesterone-induced synthesis of a new species of nuclear RNA. *Endocrinology* **84**, 63–8.

O'Malley B.W., Tsai M-J., Tsai S.Y. & Towle H.C. (1977) Regulation of gene expression in chick oviduct. *Cold Spring Harb. Symp. quant. Biol.* **42**, 605–15.

Orgel L.E. & Crick F.H.C. (1980) Selfish DNA: the ultimate parasite. *Nature, Lond.* **284**, 604–7.

Orkin S.H. (1977) *In vitro* synthesis of a DNA probe for anti-sense globin sequences. *J. biol. Chem.* **252**, 5606–8.

Orkin S.H. (1978) Fidelity of globin ribonucleic acid synthesis in isolated nuclei; asymmetric gene expression. *Biochemistry* **17**, 487–92.

Oudet P., Gross-Bellard M. & Chambon P. (1975) Electron microscopic and biochemical evidence that chromatin structure is a repeating unit. *Cell* **4**, 281–300.

Paigen K., Swan R.T., Tomino S. & Ganschow R.E. (1975). The molecular genetics of mammalian glucuronidase. *J. cell. Physiol.* **85**, 379–95.

Palmiter R.D., Moore P.B., Mulvihill E.R. & Emtage S. (1976) A significant lag in the induction of ovalbumin messenger RNA by steroid hormones: a receptor translocation hypothesis. *Cell* **8**, 557–72.

Panayotatos N. & Wells R.D. (1981) Cruciform structures in supercoiled DNA. *Nature, Lond.* **289**, 466–70.

Pardee A.B. (1974) A restriction point for control of normal animal cell proliferation. *Proc. natn. Acad. Sci. U.S.A.* **71**, 1286–90.

Pardee A.B., Dubrow R., Hamlin J.L. & Kretzien R.F. (1978) The cell cycle. *A. Rev. Biochem.* **47**, 715–50.

Pardoll D.M., Vogelstein B. & Coffey D.S. (1980) A fixed site of DNA replication in eukaryotic cells. *Cell* **19**, 527–36.

Pardue M.L. & Gall J.G. (1970) Chromosomal localization of mouse satellite DNA. *Science, N.Y.* **168**, 1351–8.

Parish J.H. (1972) *Theory and Practice of Experiments with Nucleic Acids*. Longman, London.

Parish J.H. (1979) Introduction *and* Myxobacteria. In: *Developmental Biology of Prokaryotes* (J.H. Parish ed.), pp. 1–10, 227–54. Blackwell Scientific Publications, Oxford.

Parker C.S. & Roeder R.G. (1977) Selective and accurate transcription of the *Xenopus laevis* 5S RNA genes in isolated chromatin by purified RNA polymerase III. *Proc. natn. Acad. Sci. U.S.A.* **74**, 44–8.

Parker M.G. & Mainwaring W.I.P. (1977) Effects of androgens on the complexity of poly(A) RNA in rat prostate. *Cell* **12**, 401–7.

Paterson B.M., Roberts B.E. & Kuff E.L. (1977) Structural gene identification and mapping by DNA.mRNA hybrid arrested cell-free translation. *Proc. natn. Acad. Sci. U.S.A.* **74**, 4370–4.

Pederson T. & Bhorjee J.S. (1979) Evidence for a role of RNA in eukaryotic chromatin structure; metabolically stable, small nuclear RNA species are covalently linked to chromosomal DNA in HeLa cells. *J. molec. Biol.* **128**, 451–80.

Pellegrini M. & Cantor C.R. (1977) Affinity labeling of ribosomes. In: *Molecular Mechanisms of Protein Biosynthesis* (H. Weissbach & S. Pestka eds), pp. 203–44.

Perry R.P. (1976) Processing of RNA. *A. Rev. Biochem.* **45**, 605–29.

Perutz M.F. (1964) The hemoglobin molecule. *Sci. Amer.* **211**, 64–76.

Perutz M.F. (1970) Stereochemistry of cooperative effects in haemoglobin. *Nature, Lond.* **228**, 726–39.

Peterson J.L. & McConkey E.H. (1976) Non-histone chromosomal proteins from HeLa cells: a survey by high resolution, two-dimensional electrophoresis. *J. biol. Chem.* **251**, 548–54.

Petterson U., Tibbetts C. & Phillipson L. (1976) Hybridization maps of early and late mRNA sequences on the adenovirus type 2 genome. *J. molec. Biol.* **101**, 479–502.

Pollack J.K. & Sutton R. (1980) The differentiation of animal mitochondria during development. *Trends biochem. Sci.* **5**, 23–7.

Pollock T.J., Tessman I. & Tessman E.S. (1978) Potential for variability through multiple gene products of bacteriophage ϕX174. *Nature, Lond.* **274**, 34–7.

Potter H. & Dressler D. (1977) On the mechanism of genetic recombination: the naturation of recombination intermediates. *Proc. natn. Acad. Sci. U.S.A.* **74**, 4168–72.

Potter H. & Dressler D. (1978) *In vitro* system from *Escherichia coli* that catalyzes generalized genetic recombination. *Proc. natn. Acad. Sci. U.S.A.* **75**, 3698–702.

Poyton R.O. & McKemmie E. (1979) A polyprotein precursor to all four cytoplasmically translated subunits of cytochrome c oxidase from *Saccharomyces cerevisiae*. *J. biol. Chem.* **254**, 6763–71.

Preer J.R. jr., Preer L.B. & Jurand A. (1974) Kappa and other symbionts in *Paramecium aurelia*. *Bact. Rev.* **38**, 113–63.

Prescott D.M. (1976) *Reproduction of Eukaryotic Cells*. Academic Press, New York.

Pribnow D. (1979) Genetic control signals in DNA. In: *Biological Regulation and Development. Vol. 1. Gene Expression* (R.F. Goldberger ed.), pp. 219–78. Plenum Press, New York and London.

Primrose S.B. (1974) *Introduction to Modern Virology*, 1e. Blackwell Scientific Publications, Oxford.

Proudfoot N. (1980) Pseudogenes. *Nature, Lond.* **286**, 840–1.

Ptashne M., Bachman K., Humayun M.Z., Jeffrey A., Mauer R., Meyer B. & Sauer R.T. (1976) Autoregulaton and function of a repressor in bacteriophage lambda. *Science, N.Y.* **194**, 156–61.

Rabbits T.H. & Forster A. (1978) Evidence for noncontiguous variable and constant region genes in both germ line and myeloma DNA. *Cell* **13**, 319–27.

Rabbits T.H., Forster A., Dunnick W. & Bentley D.L. (1980) The role of gene deletion in the immunoglobulin heavy chain switch. *Nature, Lond.* **283**, 351–6.

Radding C.M. (1978) Genetic recombination: strand transfer and mismatch repair. *A. Rev. Biochem.* **47**, 847–82.

Rapoport G., Klier A., Billaut A., Fargette F. & Dedouder R. (1979) Construction of a colony bank of *E. coli* containing hybrid plasmids representative of the *Bacillus subtilis* 168 genome. *Molec. gen. Genet.* **176**, 239–45.

Reeder R.H. (1973) Transcription of chromatin by bacterial RNA polymerase. *J. molec. Biol.* **80**, 229–41.

Reeder R.H., Wahn H.L., Botchan P., Hipskind R. & Sollner-Webb B. (1977) Ribosomal genes and their proteins from *Xenopus. Cold Spring Harb. Symp. quant. Biol.* **42**, 1167–77.

Reese C.B. (1978) The chemical synthesis of oligo- and polynucleotides by the phosphotriester approach. *Tetrahedron* **34**, 3143–79.

Reeves R. (1977) Structure of *Xenopus* ribosomal gene chromatin during changes in genomic transcription rates. *Cold Spring Harb. Symp. quant. Biol.* **42**, 709–22.

Reid I.A., Morris B.J. & Ganong W.F. (1978) The renin–angiotensin system. *A. Rev. Physiol.* **40**, 377–410.

Rennie P.S., Symes E.K. & Mainwaring W.I.P. (1975) The androgenic regulation of the activities of enzymes engaged in the synthesis of deoxyribonucleic acid in rat ventral prostate. *Biochem. J.* **152**, 1–16.

Revel M. (1977) Initiation of messenger RNA translation into protein and some aspects of its regulation. In: *Molecular Mechanisms of Protein Biosynthesis* (H. Weissbach & S. Pestka eds), pp. 243–321. Academic Press, New York.

Revel M. & Groner Y. (1978) Post-transcriptional and translational controls of gene expression in eukaryotes. *A. Rev. Biochem.* **47**, 1079–126.

Rigby P.W.J., Dieckmann M., Rhodes C. & Berg P. (1977) Labelling deoxyribonucleic acid to high specific activity *in vitro* by nick translation with DNA polymerase I. *J. molec. Biol.* **113**, 237–51.

Riggs A.D., Bourgeouis S. & Cohn M. (1970) The *lac* repressor-operator interaction. III Kinetic studies. *J. molec. Biol.* **53**, 401–17.

Roberts J.D., Ladner J.E., Finch J.T., Rhodes D., Brown R.S., Clark B.F.C & Klug A. (1974) Structure of yeast phenylalanine tRNA at 3 Å resolution. *Nature, Lond.* **250**, 546–51.

Roberts R.J. (1980a) Directory of restriction endonucleases. *Meth. Enzymol.* **68**, 27–41.

Roberts R.J. (1980b) Small RNAs and splicing. *Nature, Lond.* **283**, 132–3.

Roberts T.M., Kacich R. & Ptashne M. (1979) A general method for maximizing the expression of a cloned gene. *Proc. natn. Acad. Sci. U.S.A.* **76**, 760–4.

Rodbell M. (1980) The role of hormone-receptors and GTP-regulatory proteins in membrane transduction. *Nature, Lond.* **284**, 17–22.

Rodney S.A. (1979) Biosynthesis of lung surfactant during fetal and early postnatal development. *Trends biochem. Sci.* **4**, 189–91.

Roeder G.S. & Fink G.R. (1980) DNA rearrangements associated with a transposable element in yeast. *Cell* **21**, 239–49.

Rogers S.G. & Weiss B. (1980) Exonuclease III of *Escherichia coli* K-12, an AP endonuclease. *Meth. Enzymol.* **65**, 201–16.

Rosen J., Ohtsubo H. & Ohtsubo E. (1979) The nucleotide sequence of the region surrounding the replication origin of an R100 resistance factor derivative. *Molec. gen. Genet.* **171**, 287–93.

Rosenberg R., Court D., Shimatake H., Brady C. & Wulff D.L. (1978) The relationship between function and DNA sequence in an intercistronic regulatory region in phage λ. *Nature, Lond.* **272**, 414–22.

Ross E.M. & Gilman A.G. (1980) Biochemical properties of hormone-sensitive adenylate cyclase. *A. Rev. Biochem.* **49**, 533–64.

Roychoudhury R., Jay E. & Wu R. (1976) Terminal labelling and addition of homopolymer tracts to duplex DNA fragments by terminal deoxynucleotidyl transferase. *Nucl. Acids Res.* **3**, 863–77.

Rubin C.S. & Rosen O.M. (1975) Protein phosphorylation. *A. Rev. Biochem.* **44**, 831–7.

Rubin E.M., Wilson G.A. & Young F.E. (1980) Expression of thymidylate synthetase activity in *Bacillus subtilis* upon integration of a cloned gene from *Escherichia coli. Gene* **10**, 227–35.

Russell D.H. (1971) Putrescine and spermidine biosynthesis in the development of normal and anucleolate mutants of *Xenopus laevis. Proc. natn. Acad. Sci. U.S.A.* **68**, 523–7.

Rutter W.J., Pictet R.L. & Morris P.W. (1973) Towards molecular mechanisms of developmental processes. *A. Rev. Biochem.* **42**, 601–46.

Sakano H., Hüppi K., Heinrich G. & Tonegawa S. (1979a) Sequences at the somatic recombination sites of immunoglobin light-chain genes. *Nature, Lond.* **280**, 288–94.

Sakano H., Rogers J.H., Hüppi K., Brack C., Tranneker A., Maki R., Wall R. & Tonegawa A.S. (1979b) Domains and the hinge region of an immunoglobulin heavy chain are encoded in separate DNA segments. *Nature, Lond.* **277,** 627–33.

Sakonju S., Bogenhagen D.F. & Brown D.D. (1980) A control region in the centre of the 5S RNA gene directs specific initiation of transcription. I. The 5′ border of the region. *Cell* **19,** 13–25.

Sanger F., Air G.M., Barrell B.G., Brown N.L., Coulson A.R., Fiddes J.C., Hutchinson C.A. III, Slocombe P.M. & Smith M. (1977a) Nucleotide sequence of bacteriophage φX174 DNA. *Nature, Lond.* **265,** 687–95.

Sanger F., Coulson A.R., Friedmann T., Air G.M., Barrell B.G., Brown N.L., Fiddes J.C., Hutchinson C.A. III, Slocombe P.M. & Smith M. (1978) The nucleotide sequence of bacteriophage φX174. *J. molec. Biol.* **125,** 225–46.

Sanger F., Nicklen S. & Coulson A.R. (1977b) DNA sequencing with chain-terminating inhibitors. *Proc. natn. Acad. Sci. U.S.A.* **74,** 5463–7.

Savage C.R., Hash J.H. & Cohen S. (1973) Epidermal growth factor: location of disulphide bonds. *J. biol. Chem.* **248,** 7669–72.

Savageau M.A. (1976) *Biochemical Systems Analysis.* Addison-Wesley, Reading, Mass.

Savageau M.A. (1979) Autogenous and classical regulation of gene expression: a general theory and experimental evidence. In: *Biological Regulation and Development. Vol. I. Gene Expression* (R.F. Goldberger ed.), pp. 57–108. Plenum Press, New York and London.

Schally A.V., Coy D.H. & Meyers C.A. (1978) Hypothalamic regulatory hormones. *A. Rev. Biochem.* **47,** 89–128.

Schatz G. (1979) How mitochondria import proteins from the cytoplasm. *FEBS Lett.* **103,** 203–11.

Schimke R.T., Alt F.W., Kellems R.E., Kaufman R.J. & Bertino J.R. (1977) Amplification of dihydrofolate reductase genes in methotrexate-resistant cultured mouse cells. *Cold Spring Harb. Symp. quant. Biol.* **42,** 649–57.

Schimmel P.R. (1979) Understanding the recognition of transfer RNAs by aminoacyl transfer RNA synthetases. *Adv. Enzymol.* **49,** 187–222.

Schimmel P.R. & Söll D. (1979) Aminoacyl-tRNA synthetases: general features and recognition of transfer RNAs. *A. Rev. Biochem.* **48,** 601–48.

Schmidt O., Mao J-I., Silverman S., Hoverman B. & Söll D. (1978) Specific transcription of eukaryotic tRNA genes in *Xenopus* germinal vesicle extracts. *Proc. natn. Acad. Sci. U.S.A.* **75,** 4819–23.

Schumann W. (1979) Construction of an HPaI and HindII plasmid vector allowing direct selection of transformants harbouring recombinant plasmids. *Molec. gen. Genet.* **174,** 221–4.

Schwartz Z. & Kössel H. (1980) The primary structure of 16*s* rDNA from *Zea mays* chloroplast is homologous to *E. coli* 16*s* rRNA. *Nature, Lond.* **283,** 739–41.

Schwartz R.J., Kuhn R.W., Buller R.E., Schrader W.T. & O'Malley B.W. (1976) Progesterone-binding components of chick oviduct. *In vitro* effects of purified hormone-receptor complexes on the initiation of RNA synthesis in chromatin. *J. biol. Chem.* **251,** 5166–77.

Schwartz R.J., Tsai M-J., Tsai S.Y. & O'Malley B.W. (1975) Effect of estrogen on gene expression in chick oviduct. V. Changes in the number of RNA polymerase binding and initiation sites in chromatin. *J. biol. Chem.* **250,** 5175–82.

Scovassi A.I., Plevani P. & Bertazzoni U. (1980) Eukaryotic DNA polymerases. *Trends biochem. Sci.* **5,** 335–7.

Seeberg E. (1978) Reconstitution of an *Escherichia coli* repair endonuclease activity from the separated $uvrA^+$ and $uvrB^+/uvrC^+$ gene products. *Proc. natn. Acad. Sci. U.S.A.* **75,** 2569–73.

Seidman J.G. & Leder P. (1980) A mutant immunoglobulin light chain is formed by aberrand DNA- and RNA-splicing events. *Nature, Lond.* **286,** 779–83.

Seidman J.G. & McClain W.H. (1975) Three steps in conversion of large precursor RNA into serine and proline transfer RNAs. *Proc. natn. Acad. Sci. U.S.A.* **72,** 1491–5.

Sekeris C. & Lang N. (1964) Induction of DOPA-decarboxylase activity by insect messenger RNA in an *in vitro* amino acid incorporating system from rat liver. *Life Sci.* **3,** 625–32.

Shafritz D.A., Weinstein J.A., Safer B., Merrick W.C., Weber L.A., Hickey E.D. & Baglioni C. (1976) Evidence for role of m^7G^5phosphate group in recognition of eukaryotic mRNA by initiation factor IF-M_3. *Nature, Lond.* **261,** 291–4.

Shapiro S. (1977) DNA insertion elements and the evolution of chromosome primary structure. *Trends biochem. Sci.* **2,** 176–180.

Shapiro J.A. (1979) Molecular model for the transposition and replication of bacteriophage Mu and other transposable elements. *Proc. natn. Acad. Sci. U.S.A.* **76,** 1933–7.

Sharp P.A. (1981) Speculations on RNA splicing. *Cell* **23,** 643–6.

Sharp P.A., Berk A.J. & Berget S.M. (1980) Transcriptional maps of adenovirus. *Meth. Enzymol.* **65,** 750–68.

Shatkin A.J. (1976) Capping of eucaryotic mRNAs. *Cell* **9,** 645–53.

Shaw D.L., Walker J.E., Northrop F.D., Barrell B.G., Godson G.N. & Fiddes J.C. (1978) Gene K, a new overlapping gene in bacteriophage G4. *Nature, Lond.* **272,** 510–15.

Sheimin R., Humbert J. & Pearlman R.E. (1978) Some aspects of eukaryotic DNA replication. *A. Rev. Biochem.* **47,** 277–316.

Shibita T., DasGupta C., Cunningham R.P. & Radding C.M. (1979a) Purified *Escherichia coli recA* protein catalyzes homologous pairing of superhelical DNA and single-stranded fragments. *Proc. natn. Acad. Sci. U.S.A.* **76,** 1638–42.

Shibita T., Cunningham R.P., DasGupta C. & Radding C.M. (1979b) Homologous pairing in genetic recombination: complexes of recA protein and DNA. *Proc. natn. Acad. Sci. U.S.A.* **76,** 5100–4.

Shine J. & Dalgarno L. (1975) Determinant of cistron specificity in bacterial ribosomes. *Nature, Lond.* **254,** 34–8.

Siebenlist U., Simpson R. & Gilbert W. (1980) *E. coli* RNA polymerase interacts homologously with two different promoters. *Cell* **20,** 269–81.

Simon M., Zieg J., Silverman M., Mandel G. & Doolittle R. (1980) Phase variation: evolution of a controlling element. *Science, N.Y.* **209,** 1370–4.

Sklar R. & Strauss B. (1981) Removal of O^6-methylguanine from DNA of normal and xeroderma pigmentosum-derived cell lines. *Nature, Lond.* **289,** 417–20.

Slack J. (1978) Chemical waves in *Drosophila. Nature, Lond.* **271,** 403.

Smith A.J.H. (1980) DNA sequence analysis by primed synthesis. *Meth. Enzymol.* **65,** 560–80.

Smith I. (1977) Genetics of the translation process. In: *Molecular Mechanisms of Protein Biosynthesis* (H. Weissbach & S. Pestka eds), pp. 629–700. Academic Press, New York.

Smith J.A. & Martin L. (1973) Do cells cycle? *Proc. natn. Acad. Sci. U.S.A.* **70,** 1263–7.

Sobell H. (1972) Molecular mechanism for genetic recombination. *Proc. natn. Acad. Sci. U.S.A.* **69,** 2483–7.

Sobell H.M., Tsai C-C., Jain S.C. & Gilbert S.G. (1977) Visualization of drug–nucleic acid interactions at atomic resolution. III. Unifying structural concepts in understanding drug–DNA interactions and their broader implications in understanding protein–DNA interactions. *J. molec. Biol.* **144,** 333–65.

Soeda E., Arrand J.E., Smolar N., Walsh J.E. & Griffen B.E. (1980) Coding potential and regulatory signals of the polyoma virus genome. *Nature, Lond.* **283,** 445–53.

Sonenberg N., Rupprecht K.M., Hecht S.M. & Shatkin A.J. (1979) Eukaryotic mRNA cap binding protein: purification by affinity chromatography on Sepharose-complex m^7 GDP. *Proc. natn. Acad. Sci. U.S.A.* **76,** 4345–9.

Southern E.M. (1975) Detection of specific sequences among DNA fragments separated by gel electrophoresis. *J. molec. Biol.* **98,** 503–17.

Southern E.M. (1980) Gel electrophoresis of restriction fragments. *Meth. Enzymol.* **68,** 152–82.

Spadari S. & Weissbach A. (1975) RNA-primed DNA synthesis: specific catalysis by HeLa cell DNA polymerase α. *Proc. natn. Acad. Sci. U.S.A.* **72,** 503–7.

Sprague K.U., Larson D. & Morton D. (1980) 5′ Flanking sequence signals are required for activity of silkworm alanine tRNA genes in homologous *in vitro* transcription systems. *Cell* **22,** 171–8.

Stahl F.W. (1978) Summary. *Cold Spring Harb. Symp. quant. Biol.* **43,** 1353–6.

Stalder J., Groudine M., Dodgson J.B., Engel J.B. & Weintraub H. (1980) Hb switching in chickens. *Cell* **19,** 973–80.

Stallcup M.R., Sharrock W.J. & Babinowitz J.C. (1974) Ribosome and messenger specificity in protein synthesis by bacteria. *Biochem. biophys. Res. Commun.* **58,** 92–8.

Stanfield S. & Helinski D.R. (1976) Small circular DNA in *Drosophila melanogaster. Cell* **9,** 333–45.

Stanier R.Y. (1974) The origins of photosynthesis in eukaryotes. *Symp. Soc. gen. Microbiol.* **24,** 219–40.

Steege D.A. (1977) The 5′-terminal nucleotide sequence of the *Escherichia coli* lactose repressor messenger RNA: features of translational initiation and reinitiation sites. *Proc. natn. Acad. Sci. U.S.A.* **74,** 4163–7.

Stein G.S., Spelsberg T.C. & Kleinsmith L.J. (1974) Nonhistone chromosomal proteins and gene regulation. *Science, N.Y.* **183,** 817–24.

Stein G.S., Stein J.L., Park W.D., Detke S., Lichter A.C., Shephard E.A., Jansing R.L. & Phillips I.R. (1977) Regulation of histone gene expression in HeLa S_3 cells. *Cold Spring Harb. Symp. quant. Biol.* **42,** 1107–19.

Stein J.P., Catterall J.F., Kristo P., Means A.R. & O'Malley B.W. (1980) Ovomucoid intervening sequences specify functional domains and generate protein polymorphism. *Cell* **21,** 681–7.

Steiner D.F., Cunningham D., Spiegelman L. & Aten B. (1967) *Science, N.Y.* **157,** 697–700.

Steitz J.A. & Jakes K. (1975) How ribosomes select initiator regions in mRNA:base pair formation between the 3′ terminus of 16S rRNA and the mRNA during initiation of protein synthesis in *Escherichia coli. Proc. natn. Acad. Sci. U.S.A.* **72,** 4734–8.

Steitz J.A., Sprague K.U., Steege D.A., Yuan R.C., Laughrea M., Moore P.B. & Wahba A.J. (1977) RNA.RNA and RNA.protein interactions during the initiation of protein synthesis. In: *Nucleic-acid protein recognition* (H.J. Vogel ed.), pp. 493–508. Academic Press, New York.

Steitz J.A. & Steege D.A. (1977) Characterization of two mRNA:rRNA complexes implicated in the initiation of protein biosynthesis. *J. molec. Biol.* **114,** 545–58.

Stinchcomb D.T., Thomas M., Kelly J., Selker E. & Davis R.W. (1980) Eukaryotic DNA segments capable of autonomous replication in yeast. *Proc. natn. Acad. Sci. U.S.A.* **77,** 4559–63.

Stöffler G. & Wittmann H.G. (1977) Primary structure and three-dimensional arrangement of proteins within the *Escherichia coli* ribosome. In: *Molecular Mechanisms of Protein Biosynthesis* (H. Weissbach & S. Pestka eds), pp. 117–202. Academic Press, New York.

Struhl K., Cameron J.R. & Davis R.W. (1976) Functional genetic expression of eukaryotic DNA in *Escherichia coli. Proc. natn. Acad. Sci. U.S.A.* **73,** 1471–5.

Struhl K., Stinchcomb D.T., Scerer S. & Davis R.W. (1979) High frequency transformation of yeast: autonomous replication of hybrid DNA molecules. *Proc. natn. Acad. Sci. U.S.A.* **76,** 1035–9.

Stüber D. & Bujard H. (1981) Organization of transcriptional signals in plasmids pBR322 and pACYC184. *Proc. natn. Acad. Sci. U.S.A.* **78,** 167–71.

Sugimoto K., Oka A., Sugisaki H., Takanami M., Nishinura A., Yasuda Y. & Hirota Y. (1979) Nucleotide sequence of *Escherichia coli* K-12 replication origin. *Proc. natn. Acad. Sci. U.S.A.* **76,** 575–9.

Susuki Y. & Onshima Y. (1977) Isolation and characterization of the silk fibroin gene with its flanking sequences. *Cold Spring Harb. Symp. quant. Biol.* **42,** 947–57.

Sutcliffe J.G. (1979) Complete nucleotide sequence of the *Escherichia coli* plasmid pBR322. *Cold Spring Harb. Symp. quant. Biol.* **43,** 77–90.

Sutherland B.M. (1978) Enzymatic photoreactivation of DNA. In: *DNA Repair Mechanisms,* ICN-UCLA Symposia on Molecular and Cellular Biology, vol. IX (P.C. Hanawalt, E.C. Friedberg & C.F. Fox eds), pp. 113–28. Academic Press, New York.

Sykes J., Metcalf E. & Pickering J.D. (1977) The nature of the proteins in 'chloramphenicol particles' from *Escherichia coli* A19 (Hfr *rel met rns*), *and,* The nature of the proteins present in the 'relaxed particles' from methionine-starved *Escherichia coli* A19 (Hfr *rel met rns*) (consecutive papers). *J. gen. Microbiol.* **98,** 1–28.

Szybalski W., Szybalski E.H. & Ragni G. (1966) Genetic studies with human cell lines. *Nat. Cancer Inst. Monograph* **7,** 75–88.

Tai P-C., Wallace B.J. & Davis B.D. (1978) Streptomycin causes misreading of natural messenger RNA by interaction with ribosomes after initiation. *Proc. natn. Acad. Sci. U.S.A.* **75,** 275–9.

Talkington C.A., Nishioka Y. & Leder P. (1980) *In vitro* transcription of normal, mutant and truncated mouse α-globin genes. *Proc. natn. Acad. Sci. U.S.A.* **77,** 7132–6.

Talmadge K. & Gilbert W. (1980) Construction of plasmid vectors with unique PstI cloning sites in a signal sequence coding region. *Gene* **12,** 235–41.

Talmadge K., Kaufman J. & Gilbert W. (1980) Bacteria mature preproinsulin to proinsulin. *Proc. natn. Acad. Sci. U.S.A.* **77,** 3988–92.

Tanemura S. & Bauerle R. (1979) Suppression of a deletion mutation in the glutamine aminotransferase region of the *Salmonella typhimurium trpD* gene by mutations in *pheA* and *tyrA*. *J. Bact.* **139,** 573–82.

Tartof K.D. (1979) Evolution of transcribed and spacer sequences in the ribosomal RNA genes of *Drosophila. Cell* **17,** 607–14.

Tata J.R. (1976) The expression of the vitellogenin gene. *Cell* **9,** 1–4.

Tekamp P.A., Valenzuela P., Maynard T., Bell G.I. & Rutter W.J. (1979) Specific gene transcription in yeast nuclei and chromatin by added homologous RNA polymerases I and III. *J. biol. Chem.* **254,** 953–63.

Telford J.L., Kressmann A., Koski R.A., Grosschedl R., Müller F., Clarkson S.G. & Birnstiel M.L. (1979) Delineation of a promoter for RNA polymerase III by means of a functional test. *Proc. natn. Acad. Sci. U.S.A.* **76,** 2590–4.

Thelander L. & Reichard P. (1979) Reduction of ribonucleotides. *A. Rev. Biochem.* **48,** 133–58.

Thomas M., Cameron J.R. & Davis R.W. (1974) Viable molecular hybrids of bacteriophage lambda and eukaryotic DNA. *Proc. natn. Acad. Sci. U.S.A.* **71,** 4579–83.

Thomas P.S. (1980) Hybridization of RNA and small DNA fragments transferred to nitrocellulose. *Proc. natn. Acad. Sci. U.S.A.* **77,** 5201–5.

Tilghman S.M., Curtis P.J., Tiemeier D.C., Leder P. & Weissmann C. (1978) The intervening sequence of a mouse β-globin gene is transcribed within the 15S-globin mRNA precursor. *Proc. natn. Acad. Sci. U.S.A.* **75,** 1309–13.

Tilghman S.M., Tiemeier D.C., Edgell M.H., Seidman J.G., Leder A., Enquist L.W., Norman B. & Leder P. (1977) Cloning of specific segments of the mammalian genome: bacteriophage λ containing mouse globin and surrounding gene sequences. *Proc. natn. Acad. Sci. U.S.A.* **74,** 4406–10.

Timasheff S.N. (1979) The *in vitro* assembly of microtubules from purified brain tubulin. *Trends biochem. Sci.* **4,** 61–5.

Tonegawa S., Maxam A.M., Tizard R., Bernard O. & Gilbert W. (1978) Sequence of a mouse germ-line gene for a variable region of an immunoglobulin light chain. *Proc. natn. Acad. Sci. U.S.A.* **75,** 1485–9.

Towle H.C., Tsai M-J., Tsai M.Y. & O'Malley B.W. (1977) Effect of estrogen on gene expression in the chick oviduct: preferential initiation and asymmetrical transcription of specific chromatin genes. *J. biol. Chem.* **252,** 2396–404.

Traub P. & Nomura M. (1969) Structure and function of *Escherichia coli* ribosomes. IV Mechanism of assembly of 30S ribosomes studied in vitro. *J. molec. Biol.* **40,** 391–413.

Travers A.A., Kari C. & Mace A.F. (1981) Transcriptional regulation of bacterial RNA polymerase. *Symp. Soc. gen. Microbiol.* **31,** 111–30.

Trendelenburg M.F., Mathis D. & Chambon P. (1980) Transcription units of chicken ovalbumin genes observed after injection of cloned complete genes into *Xenopus* oocyte nuclei. *Proc. natn. Acad. Sci. U.S.A.* **77,** 5984–8.

Tsai S.Y., Tsai M-J., Schwartz R.J., Kalimi M., Clark J. & O'Malley B.W. (1975) Effect of estrogen on gene expression in the chick oviduct: nuclear receptor levels and the initiation of transcription. *Proc. natn. Acad. Sci. U.S.A.* **72,** 4228–32.

Tyler B.M., Deleo A.B. & Magasanik B. (1974) Activation of transcription of *hut* DNA by glutamine synthetase. *Proc. natn. Acad. Sci. U.S.A.* **71,** 225–9.

Tzagoloff A. & Macino G. (1979) Mitochondrial genes and translation products. *A. Rev. Biochem.* **48,** 419–41.

Uhlin B.E., Molin S., Gustaffson P. & Nordström K. (1979) Plasmids with temperature-dependent copy number for amplification of cloned genes and their products. *Gene* **6**, 91–106.

van de Putte P., Cramer S. & Giphart–Gassler M. (1980) Invertible DNA determines host specificity of bacteriophage Mu. *Nature, Lond.* **286**, 218–22.

Villa-Komaroff L., Efstratiadis A., Broome S., Lomedico P., Tizard R., Naber S.P., Chick W.L. & Gilbert W. (1978) A bacterial clone synthesizing proinsulin. *Proc. natn. Acad. Sci. U.S.A.* **75**, 3727–31.

Vinograd J. & Hearst J.E. (1962) Equilibrium sedimentation of macromolecules and viruses in a density gradient. *Fortschr. Chem. org. NatStoffe* **20**, 372–422.

Vodkin M.H. & Fink G.R. (1973) A nucleic acid associated with a killer strain of yeast. *Proc. natn. Acad. Sci. U.S.A.* **70**, 1069–72.

Vogelstein B., Pardoll D.M. & Coffey D.S. (1980) Supercoiled loops and eukaryotic DNA replication. *Cell* **22**, 79–85.

Vogt V.M. (1980) Purification and properties of S_1 nuclease from *Aspergillus. Meth. Enzymol.* **65**, 248–55.

von Hippel P.H. (1979) On the molecular bases of the specificity of interaction of transcriptional proteins with DNA. In: *Biological Regulation and Development. Vol. 1.* (R.F. Goldberger ed.), pp. 279–348. Plenum Press, New York.

Waechter C.J. & Lennarz W.J. (1976) The role of polyprenol-linked sugars in glycoprotein synthesis. *A. Rev. Biochem.* **45**, 95–112.

Wahli W., Dawid I.B., Wyler T., Jaggi R.J., Weber R. & Ryffel G.U. (1979) Vitellogenin in *Xenopus laevis* is encoded in a small family of genes. *Cell* **16**, 535–49.

Wallis J.W., Hereford L. & Grunstein M. (1980) Histone 2B genes of yeast encode two different proteins. *Cell* **22**, 799–805.

Wang H-J., Quigley G.J., Kolpak F.J., Crawford J.L., van Boom J.H., van der Marcel G. & Rich A. (1979) Molecular structure of a left-handed DNA fragment at atomic resolution. *Nature, Lond.* **282**, 680–6.

Waring M.J. (1970) Variation of the supercoils in closed circular DNA by binding of antibiotics and drugs: evidence for models involving intercalation. *J. molec. Biol.* **54**, 247–79.

Wasylyk B., Kedinger C., Corden J., Brison O. & Chambon P. (1980) Specific *in vitro* initiation of transcription on conalbumin and ovalbumin genes and comparison with adenovirus-2 early and late genes. *Nature, Lond.* **285**, 367–73.

Weatherall D.J. & Clegg J.B. (1979) Recent developments in the molecular genetics of human haemoglobin. *Cell* **16**, 467–79.

Weil R.A., Luse D.S., Segall J. & Roeder R.G. (1979) Selective and accurate initiation of transcription at the Ad2 major late promoter in a soluble system dependent on purified RNA polymerase II and DNA. *Cell* **18**, 469–84.

Weissbach A. (1977) Eukaryotic DNA polymerases. *A. Rev. Biochem.* **46**, 25–47.

Welfle H., Stahl J. & Bielka H. (1972) Studies on proteins of animal ribosomes. XIII. Enumeration of ribosomal proteins of rat liver. *FEBS Lett.* **26**, 228–36.

West S.C., Cassuto E. & Howard-Flanders P. (1981) Homologous pairing can occur before DNA strand separation in general genetic recombination. *Nature, Lond.* **290**, 29–33.

White R.L. & Hogness D.S. (1977) R loop mapping of the 18S and 28S sequences in the long and short repeating units of *Drosophila melanogaster* rDNA. *Cell* **10**, 177–92.

Wickens M.P. & Laskey R.A. (1981) Expression of cloned genes in cell-free systems and in microinjected *Xenopus* oocytes. In: *Genetic Engineering,* vol. 1 (R. Williamson ed.), pp. 103–67. Academic Press, London and New York.

Wickner R.B. (1976) Killer of *Saccharomyces cerevisiae*: a double-stranded ribonucleic acid plasmid. *Bact. Rev.* **40**, 757–73.

Wickner S.H. (1978) DNA replication proteins of *Escherichia coli. A Rev. Biochem.* **47**, 1163–91.

Wigler M., Sweet R., Sim G., Wold B., Pellicer A., Lacy E., Maniatis T., Silverstein S. & Axel R. (1979) Transformation of mammalian cells with genes from prokaryotes and eukaryotes. *Cell* **16**, 777–85.

Willetts N. & Skurray R. (1980) The conjugation system of F-like plasmids. *A. Rev. Genet.* **14**, 41–76.

Williams B.G. & Blattner R.F. (1980) Bacteriophage lambda vectors for DNA cloning. In: *Genetic Engineering,* vol. 2 (J.K. Setlow & A. Hollander eds), pp. 201–81. Plenum Press, New York.

Williams B.R.G. & Kerr I.M. (1980) The 2–5A(pppA$^{2'}$p$^{5'}$A$^{2'}$p$^{5'}$A) system in interferon-treated and control cells. *Trends biochem. Sci.* **5**, 138–40.

Williams P.A. (1981) Catabolic plasmids. *Trends biochem. Sci.* **6**, 23–6.

Williamson A.R. (1976) The biological origin of antibody diversity. *A. Rev. Biochem.* **45**, 467–500.

Williamson V.M., Young E.T. & Ciriacy M. (1981) Transposable elements associated with constitutive expression of yeast alcohol dehydrogenase II. *Cell* **23**, 605–14.

Wilson D.A. & Thomas C.A. (1974) Macromolecular rDNA in *Tetrahymina pyriformis* is a palindrome. *J. molec. Biol.* **104**, 421–53.

Witkin E.M. (1976) ultraviolet mutagenesis and inducible DNA repair in *Escherichia coli. Bact. Rev.* **40**, 869–907.

Woese C.R. & Fox G.E. (1977) Phylogenetic structure of the prokaryotic domain: the primary kingdoms. *Proc. natn. Acad. Sci. U.S.A.* **74**, 5088–90.

Woodland H.R., Flynn J.M. & Wyllie A.J. (1979) Utilization of stored mRNA in *Xenopus* embryos and its replacement by newly synthesized transcripts: histone H1 synthesis using interspecies hybrids. *Cell* **18**, 165–71.

Woodland H.R., Gurdon J.B. & Lingrel J.B. (1974) The translation of mammalian globin mRNA injected into fertilized eggs of *Xenopus laevis*: II: the distribution of globin synthesis in different tissues. *Devl. Biol.* **39**, 134–40.

Wool I.G. (1979) The structure and function of eukaryotic ribosomes. *A. Rev. Biochem.* **48**, 719–54.

Wrede P., Wood N.H. & Rich A. (1979) Initiator tRNAs have a unique anticodon loop configuration. *Proc. natn. Acad. Sci. U.S.A.* **76**, 3289–93.

Wu C., Bingham P.M., Livak K.J., Holmgren R. & Elgin S.C.R. (1979a) The chromatin structure of specific genes: I. evidence for higher order domains of defined DNA sequence. *Cell* **16**, 797–806.

Wu C., Wong Y-C. & Elgin S.C.R. (1979b) The chromatin structure of specific genes: II. disruption of chromatin structure during gene activity. *Cell* **16**, 807–14.

Wurmbach P. & Nierhaus K.H. (1979) Isolation of the protein synthesis initiation factors EF-Tu, EF-Ts and EF-G from *Escherichia coli*. *Meth. Enzymol.* **LX**, 593–606.

Yamamoto K.R. & Alberts B.M. (1976) Steroid receptors: elements for modulation of eukaryotic transcription. *A. Rev. Biochem.* **45**, 721–46.

Yamane T. & Hopfield J.J. (1977) Experimental evidence for kinetic proofreading in the aminoacylation of tRNA by synthetase. *Proc. natn. Acad. Sci. U.S.A.* **74**, 2246–50.

Yanofsky C. (1981) Attenuation in the control of expression of bacterial genomes. *Nature, Lond.* **289**, 751–8.

Yanofsky C., Carlton B.C., Guest J.R., Helinsky D.R. & Henning U. (1964) On the colinearity of gene structure and protein structure. *Proc. natn. Acad. Sci. U.S.A.* **51**, 266–72.

Yanofsky C. & Ito J. (1966) Nonsense codons and polarity in the tryptophan operon. *J. molec. Biol.* **21**, 313–34.

Yates J.L. (1979) Role of ribosomal protein S12 in discrimination of aminoacyl-tRNA. *J. biol. Chem.* **254**, 11550–4.

Yates J.L., Arfsten A.E. & Nomura M. (1980) *In vitro* expression of *Escherichia coli* ribosomal protein genes: autogenous inhibition of translation. *Proc. natn. Acad. Sci. U.S.A.* **77**, 1837–41.

Yates J.L. & Nomura M. (1980) *E. coli* protein L4 is a feedback regulatory protein. *Cell* **21**, 517–22.

Young F.E. (1980) Impact of cloning in *Bacillus subtilis* on fundamental and industrial microbiology. *J. gen. Microbiol.* **119**, 1–15.

Young N.S., Benz E.J., Kantor J.A., Kretschmer P. & Neinhuis A.W. (1978) Hemoglobin switching in sheep; only the γ gene is in the active conformation in fetal liver but all the β and γ genes are in the active conformation in bone marrow. *Proc. natn. Acad. Sci. U.S.A.* **75**, 5884–8.

Zasloff M. & Felsenfeld G. (1977) Use of mercury-substituted ribonucleoside triphosphates can lead to artefacts in the analysis of *in vitro* chromatin transcripts. *Biochem. biophys. Res. Commun.* **75**, 598–603.

Zengel J.M., Mueckl D. & Lindahl L. (1980) Protein L4 of the *E. coli* ribosome regulates an eleven gene r protein operon. *Cell* **21**, 523–5.

Zieg J., Silverman M., Hilmen M. & Simon M. (1977) Recombinational switch for gene expression. *Science, N.Y.* **196**, 170–2.

Ziff E.B. (1980) Transcription and RNA processing by the DNA tumour viruses. *Nature, Lond.* **287**, 491–9.

Ziff E.B. & Evans R.M. (1978) Coincidence of the promoter and capped 5′ terminus from adenovirus 2 major late transcription unit. *Cell* **15**, 1463–75.

Zipser D. (1970) Polarity and translational punctuation. In: *The Lactose Operon* (J.R. Beckwith & D. Zipser eds), pp. 221–32. Cold Spring Harbor Publications, Cold Spring Harbor, N.Y.

Zorbach W.W. & Tipson R.S. (1968) *Synthetic procedures in nucleic acid chemistry*, vol. 1. Interscience, New York.

Zorbach W.W. & Tipson R.S. (1973) *Synthetic Procedures in Nucleic Acid Chemistry*, vol. 2. Interscience, New York.

Zurawski G., Brown K., Killingly D. & Yanofsky C. (1978) Nucleotide sequence of the leader region of the phenylalanine operon of *Escherichia coli*. *Proc. natn. Acad. Sci. U.S.A.* **75**, 4271–5.

Index